P9-CMZ-061

HOLOGRAPHIC VISIONS

HOLOGRAPHIC VISIONS

A History of New Science

SEAN F. JOHNSTON
University of Glasgow

Great Clarendon Street, Oxford OX2 6DP

Oxford University Press is a department of the University of Oxford.
It furthers the University's objective of excellence in research, scholarship,
and education by publishing worldwide in

Oxford New York

Auckland Cape Town Dar es Salaam Hong Kong Karachi
Kuala Lumpur Madrid Melbourne Mexico City Nairobi
New Delhi Shanghai Taipei Toronto

With offices in

Argentina Austria Brazil Chile Czech Republic France Greece
Guatemala Hungary Italy Japan Poland Portugal Singapore
South Korea Switzerland Thailand Turkey Ukraine Vietnam

Published in the United States
by Oxford University Press Inc., New York

First published 2006

British Library Cataloguing in Publication Data
Data available

Library of Congress Cataloging in Publication Data

Johnston, Sean, 1956–
Holographic visions : a history of new science / Sean F. Johnston.
p. cm.
ISBN-13: 978–0–19–857122–3 (alk. paper)
ISBN-10: 0–19–857122–4 (alk. paper)
1. Holography—History. I. Title.
QC449.J66 2006
621.36′75—dc22

2005036594

Typeset by Newgen Imaging Systems (P) Ltd., Chennai, India
Printed in Great Britain
on acid-free paper by
Biddles Ltd, King's Lynn, Norfolk

ISBN 0–19–857122–4 978–0–19–857122–3

1 3 5 7 9 10 8 6 4 2

Dedicated to my family: Libby, Daniel, and Samuel

PREFACE

Holography hit the world with a bang in 1964. This book tracks that explosion, from the near-silent burning of the slow fuse that lit it to the repercussions and fading embers that followed. The history of holography can be visualized as a sublime fireworks show in which there have been impressive bursts intermingled with unexpected fizzles and more than a few duds.

Yet holography has simultaneously been an iceberg-like subject, hidden and mysterious. Much of its early activity gestated in laboratories engaged in classified research, both in America and the Soviet Union; much, too, in guarded processing techniques and the secretive business practices of its major application, anti-counterfeiting; and, more recently, in Asia, where differences in language and business culture limited information flow to rumour or scientific meetings. And some aspects, such as holographic erotica, remain ever obscure. Politics, culture, commercial secrecy, and even propriety have concealed the subject, making it peculiarly vulnerable to myths and misunderstandings.

Both metaphors—of fireworks and icebergs—are inadequate, and capture only restricted perspectives of the subject. Over its first sixty years, holography has run hot and cold, a subject rich in episodes of parallel discovery, priority disputes, intellectual accomplishment, suspicion, local victories, and lost opportunities.

Why attempt a history of what some still see as an immature subject? One reason is because notions of maturity carry questionable assumptions. Holography is a young science that illustrates how new subjects come to be. It provides answers to questions such as, How does a scientific subject materialize? How does its content stabilize? How do those who practice it come to recognize themselves as a distinct group? And how do its definitions and products depend on their environments?

This is a study of how new science develops. Holography is an unusual and important example of post-war science constructed from divergent visions. It grew beyond its founding communities to generate conflicting interpretations of success. This visionary subject exemplifies how science, technology and wider culture are woven inextricably in the modern world.

There is a rich thread of information to unravel, and the project is timely. As a subject that has not enjoyed uncontentious success, archives of holography are few and historical information is ephemeral. And most of the early practitioners—known as *holographers* since about 1966—were attracted to the field early in their careers. Many are still alive, with memories, documents, and holograms. As an African writer has said of his continent's oral history, 'quand un vieillard meurt, c'est une bibliothèque qui brûle' (when an old person dies, a library burns).[1] The same can be said of entire technical communities. This project has sought to capture the spirit, diversity and insights of some of those libraries.

And yet this book inevitably chases shadows. I focus on the subject and its creators as much as on their remarkable products. What any text, film, or computer representation of the subject cannot do adequately is demonstrate the visual wonder of viewing a good hologram. Like the enjoyment of a stage play, concert, or surprising magic trick, the appreciation of a hologram is a product of its time, context, and audience. It is both a permanent product and a fleeting perceptual experience. The few illustrations of holograms in this book are therefore intended as mere hints—unsatisfactory evocations—of that intense visual experience.

Holograms represent the culmination of optical physics. The creation of those unsettling images is inherently a private activity hidden in dark, equipment-filled rooms but, in recent years, also a high-volume product manufactured increasingly by assembly-line processes. This juxtaposition of the private and the public, the dimly perceived and the brightly-lit, the painstakingly created and cheaply produced, is an important feature of this modern technology.

The communities and individuals that created holography—producing holograms, developing markets and becoming its consumers—each occupied separate spaces, invoked different understandings, and attached different meanings to their subject. For that reason, readers, too, may find themselves attracted to particular parts of that history. The four sections of the book explore complementary aspects of creation. Their overall aim is to explain the emergence and evolution of the ideas, products, communities, and markets that collectively defined the new field of holography.

Sean Johnston
Dumfries, Scotland
December 2005

[1] Voiced by the Malian writer Amadou Hampâté Bâ (1901–91) at a 1962 UNESCO meeting.

ACKNOWLEDGEMENTS

One of the most satisfying aspects of writing *Holographic Visions* was the opportunity to get to know researchers, artists, and entrepreneurs who had played crucial roles in the formation of a new subject. Attempting to write a history of a field in ferment could not be envisaged without the contribution of its practitioners. I am grateful for the invaluable information and perspectives provided by the following persons:

Dr Carl Aleksoff
Dr Brian Athey
Dr Norm Barnett
Eric Begleiter
Prof. Stephen Benton
Dr Margaret Benyon
Prof. Hans Bjelkhagen
Jeff Blyth
Dr Roger Brentnall
Dr Nigel Briggs
Don Broadbent
Dr Douglas Brumm
Harriet Casdin-Silver
Prof. H. John Caulfield
Clark Charnetsky
Dr Gary Cochran
Betsy Connors
Melissa Crenshaw
Lloyd G. Cross
Duncan Croucher
Gary Cullen
Ed Dietrich
Prof. Yuri Denisyuk
Frank Denton
Mary Dentschuk
Vincent DiBiase
Dr William Fagan
Dr Arthur Funkhouser
Dr Devi Garibashvili
John W. C. Gates
Donald Gillespie
Prof. Joseph Goodman
Dr Kenneth Haines
Prof. Parameswaran Hariharan
Dr Alan Hodgson
Randy James
Prof. Tung Jeong
Michael Kan
Lewis Kontnik
Dr Adam Kozma
Yasumasa Kamata
Mathias Lauk
Prof. Roger Lessard
Prof. Emmett Leith
Dr Carl Leonard
Prof. Dr Adolf Lohmann
Ana MacArthur
Dr Vladimir Markov
Steve McGrew
Rob Munday
Dr Ambjörn Naeve
Ana Maria Nicholson
Harry Owen
Prof. Nicholas Phillips
Dr David Pizzanelli
Richard Rallison
Amanda Ranalli
Dr Martin Richardson
Andrea Robertson
Jonathan Ross
Jason Sapan
Graham Saxby
Prof. Dr Johannes Schwider
Larry Siebert
Dr Chris Slinger
Prof. Marat Soskin
Prof. Dmitry Staselko
Dr Karl Stetson
Dr Jim Trolinger
Prof. Jumpei Tsujiuchi
Fred Unterseher
Juris Upatnieks
Dr Charles Vest
Prof. Peter Waddell
Michael Waller-Bridge
Dr C. Roy Worthington
Prof. Leonid Yaroslavsky
Dr Richard Zech
Gary Zellerbach

Some fifty of them were interviewed in person. I am also grateful to the many other scientists, artists, engineers, artisans, educators, administrators, entrepreneurs, collectors, and enthusiasts who responded to more specific queries, including:

Dr Edmund Akopov
Dr Albert Baez
Dr Lawrence Bartell
Dr Kaveh Bazargan
Sunny Baines
Hugh Brady
Phillippe Boissonet
Hugh Brady
John Brown
William H. Carter
Prof. Pierre Chavel
Jay S. Chivian
Dr Georges Dyens
Andrew Ellard
John Fairstein
Dr Albert Friesem
Nancy Gorglione
Prof. Pál Greguss
Jeff Hecht
Dr Leonid Akopov
John Howard
Stephanie Hunt
Prof. Guy Indebetouw
Pearl John
Prof. Dieter Jung
Dr Sergey Kostyukevych
Prof. Antoine Labeyrie
Ian Lancaster
Dr Jean-Louis Le Gouët
Larry Lieberman
Mònica López
Mark Lucente
Prof. Manoranjan De
Jim McIntyre
Mike Medora
Odile Meulien
Rod Murray
Craig Newswanger
Jeff Allen
Ron Olson
Dr Andrew Pepper
Jerry Pethick
Al Razutis
Daryl Sharp
Bernd Simson
Steven Smith
Dr Noel Stephens
Prof. Dr George Stroke
Prof. H. Henry Stroke
Vivi Tornari
Prof. Douglas Tyler
Doris Vila
Dr Weston Vivian
Jon Vogel
Peter Woodd

Tracing the history of holography has been helped and hampered by the rapid evolution of the field. On the one hand, access to company records and the reminiscences of former employees has been helped by the decline in some commercial domains. Thus the bankruptcy of the New York Museum of Holography in 1992, and the subsequent purchase of its records through auction by the MIT Museum, has made available a rich collection of business, artistic, and practitioner records. Similarly, members of firms no longer active in the business of holography have provided candid and informative assessments of their experiences in the field. Individual holographers, too, having been involved in a series of start-up companies or short-lived research projects, have been a fruitful source of information. On the other hand, holography is still seen by some observers as a young subject, too youthful to be thought of historically or (perhaps) even to deserve a history. As a result, only a handful of long-term and representative archives have been established, and few firms or individual holographers have kept more than isolated mementos of their personal involvement.

Archives of holography are still mainly in private hands and accessible document collections are scarce. I thank Joan Parks Whitlow (Registrar and Collections Manager), Mary Leen (Acting Director), and Deborah Douglas (Curator of Science and

Technology) of the MIT Museum for generous access to the records of the New York Museum of Holography, Arthur Daemmrich of the Chemical Heritage Foundation and Maril Cappelli of GRC for data on the Gordon Research Conferences, and the archivists and librarians at Imperial College, London; the Science Museum library, London; Birkbeck College, London; and the Bentley Historical Library, University of Michigan. Selections from private collections of documents or images were provided by Lawrence Bartell, Jeff Blyth, Don Broadbent, Clark Charnetski, Gary Cullen, Yuri Denisyuk, Art Funkhouser, Kenneth Haines, Emmett Leith, Carl Leonard, Adolf Lohmann, Ana MacArthur, Ambjörn Naeve, David Pizzanelli, Larry Siebert, Juris Upatnieks, and Michael Waller-Bridge. I am very grateful to Hans Bjelkhagen, Gary Cochran, Jonathan Ross, and Graham Saxby for the opportunity to examine and borrow from their collections, and especially to Emmett Leith, Juris Upatnieks, and Adolf Lohmann who, over a period of many months, patiently answered my numerous questions, provided interpretations, helped to locate further contacts, and commented on my published papers and individual chapters of the draft manuscript. Margaret Benyon, Yuri Denisyuk, Steve McGrew, Rob Munday, Karl Stetson, and Fred Unterseher also read and commented helpfully on portions of the draft manuscript, although the content, perspectives, and conclusions are my own. At Oxford University Press, Sönke Adlung (Senior Commissioning Editor, Physics), Anita Petrie (Production Editor, Science) and their colleagues shaped and shepherded the final product. But most of all I thank my wife, Elisabeth McNair-Johnston, for her encouragement, ideas, and research assistance.

Some of the material in this book has been presented in preliminary form in seminars, conference proceedings, and journal articles, where participants, editors, referees, and readers have helped to refine it. Those sources include: 'Reconstructing the history of holography', *Proceedings of the SPIE 5005*, 455–64 (2003), and presented at *Practical Holography XVII*, Santa Clara, CA, January 2003; 'Scientific instruments and their cultures', at the *Gesellschaft Deutscher Chemiker Annual Meeting*, Munich, 6 October 2003; 'Commercial holography: midlife crisis or wise maturity?', at the *Holopack-Holoprint 2003* Meeting, Vancouver, Canada, 18–20 November 2003; 'Telling tales: George Stroke and the historiography of holography', *History and Technology* 20 (2004): 29–51; 'Holoscopes, wave photographs and lensless photography: changing visions of holography', at the *International Conference on Optical Holography*, Kiev, Ukraine, 24–26 May 2004; 'From classified research to public access holography: equipment as social liberator', at the *British Society for the History of Science Annual Meeting*, Liverpool, 27 June 2004; 'Holography: from science to subcultures', *Optics and Photonics News* 15 (2004): 36–41; 'Stephen Benton on holography, Polaroid and MIT', *Optics and Photonics News* 15 (2004): 32–35; 'Shifting perspectives: holography and the emergence of technical communities', *Technology and Culture* 46 (2005): 77–103; 'Scientists, artists and artisans: the making of holographic communities', at the meeting of the Royal Photographic Society Holography Group, 30 April 2005; 'Explosion with a slow-burning fuse: the origins of holography in Ann Arbor, Michigan', at the *International Conference on Holography, Optical Recording and Processing of Information*, Varna, Bulgaria, May 2005; 'From white elephant to Nobel Prize: Dennis

Gabor's wavefront reconstruction', *Historical Studies in the Physical and Biological Sciences* 36 (2005): 35–70; 'Attributions of scientific and technical progress: the case of holography', *History and Technology* 21 (2005): 367–92; and, 'Absorbing new subjects: holography as an analog of photography', *Physics in Perspective* 8 (2006, forthcoming).

I am grateful for research grants provided generously by the British Academy, the Shearwater Foundation, the Carnegie Trust, the Royal Society, and the American Institute of Physics' Friends of the Center for the History of Physics, which made possible the archival work, interviews, and document collation that form the basis of this work. Permissions to reproduce figures are individually cited, but I also acknowledge the permission to quote from archival material granted by Hilary McEwan, Archivist, Imperial College Library Archives and Special Collections (London, UK) for the Dennis Gabor archives; the Trustees of the Science Museum (London, UK) for the Gordon Rogers archives; Deborah Douglas, Curator of the Science and Technology Collection, MIT Museum (Cambridge, MA, USA) for the Museum of Holography Archives; Karen L. Jania, Reference Archivist, Bentley Historical Library (Ann Arbor, MI, USA) for assorted archives of the Willow Run Laboratories, Electro-Optical Sciences Lab and Institute of Science and Technology of the University of Michigan; Arthur Daemmrich, Chemical Heritage Foundation (Philadelphia, PA, USA) for the Gordon Research Conference archive; and the Director of the IEEE History Center at Rutgers University for permission to quote from their oral histories.

CONTENTS

List of Figures xvii
Acronyms xx

1. **Introduction: Seeking Coherence** 1

PART I: CREATING A SUBJECT 13

2. **Wavefront Reconstruction in England and Beyond** 15
   - 2.1 Introduction 15
   - 2.2 Holoscopy 22
   - 2.3 'A new microscopic principle' 30
   - 2.4 Microscopy by reconstructed wavefronts 33
   - 2.5 The Diffraction Microscope at Imperial College 34
   - 2.6 Gordon Rogers and 'D.M.' 37
   - 2.7 The Californian connection 43
   - 2.8 Adolf Lohmann in Germany 48
   - 2.9 The decline of diffraction microscopy 51
3. **Wave Photography in the Soviet Union** 60
   - 3.1 The Vavilov State Optical Institute and the backdrop of Soviet science 60
   - 3.2 Yuri Denisyuk and his *Kandidat* research 63
   - 3.3 Wave photographs 71
   - 3.4 Pause and reception 74
4. **Lensless Photography in America** 78
   - 4.1 The Willow Run Laboratories and optical processing 78
   - 4.2 From optical processing to wavefront reconstruction 95
   - 4.3 Lensless photography 99
   - 4.4 Three-dimensional wavefront reconstruction 102

5. Constructing Holography **120**

5.1 Introduction 120

5.2 George Stroke and the packaging of holography 123

5.3 The Nobel Prize and historiographical validation 142

5.4 Patents, priority, and profits 145

5.5 Finding coherence 147

PART II: CREATING A MEDIUM 149

6. Early Exploitation **151**

6.1 Satisfying sponsors at the University of Michigan 151

6.2 Seducing investors at Conductron 159

6.3 Ann Arbor, 'holography capital of the world' 168

6.4 Seeking applications further afield 175

6.5 Expansion in the East 180

7. Technology of the Sublime: The Versatile Hologram **189**

7.1 Prospects and problems 189

7.2 Inventing and reinventing holographic interferometry 191

7.3 The Americanization of reflection holography 200

7.4 Image-plane holograms 202

7.5 360° holograms 204

7.6 Pulsed holograms 206

7.7 Rainbow holograms 209

7.8 Holographic stereograms 212

7.9 Computer-generated holograms 216

7.10 Photosensitive materials 220

7.11 The medium and its message 227

PART III: CREATING AN IDENTITY 229

8. Defining the Scientific Holographer **231**

8.1 Reshaping optical engineering for holographers 231

8.2 Carving a niche with journals 235

8.3 Meetings as social nuclei 242

8.4 Defining the holographer 250

9. Culture and Counterculture: The Artisan Holographer 253

9.1 Introduction 253

9.2 Challenging the orthodox optical laboratory: material culture and community identity 262

9.3 Training artisans: the birth of schools 271

9.4 Transmitting the counterculture: practical publications 280

9.5 Shaping and reshaping an identity 283

10. Aesthetic Holographers and Their Art 287

10.1 Introduction 287

10.2 Artist–scientist collaborations 290

10.3 Artists and artisans 303

10.4 Formalizing the art: accredited schools 305

10.5 Distinguishing subcultures 308

10.6 Enchanting audiences through exhibitions 312

10.7 Critiques from mainstream art 317

11. Building Holographic Communities 325

11.1 Uncertain identities 325

11.2 Strengthening networks 327

11.3 Special congregations: symposia 330

11.4 Special places: museums 335

PART IV: CREATING A MARKET 347

12. Commercialization and Ubiquity 349

12.1 Making holography pay 349

12.2 Entrepreneurs and cottage industry 356

12.3 Optimistic investment: the Ilford story 367

12.4 Embossed holograms and profitability 372
12.5 Patents and commercial holography 378
12.6 Security and its influences 386

13. The Hologram and Popular Culture **393**

13.1 Shock and awe 393
13.2 From the sublime to kitsch 398
13.3 Persistent ideas 402
13.4 Holograms as analogy and paradigm 409
13.5 Conflicting values and competing meanings 413

14. Conclusion: Creative Visions **415**

14.1 Through a glass darkly: visions of scientific genesis 415
14.2 Assessing progress: visions of successful technology 418
14.3 Preserving and predicting: visions of the past and future 433
14.4 Careers in light: visions of technical identity 438
14.5 Imagined and invented futures 441

Bibliography 447

Interviews 447
Archives consulted 448
Books 449
Papers and articles 454
Unpublished communications 479

Appendix 489

Publication statistics 489

Index 491

LIST OF FIGURES

2.1	Dennis Gabor at Siemens & Halske AG, Berlin, late 1920s (Imperial College archives Gabor collection)	17
2.2	BTH Research Laboratories Building 52 (background)	19
2.3	Gabor patent illustrations	28
2.4	Gabor 1948 conference poster	30
2.5	Gabor in middle age	35
2.6	Zone plate	39
2.7	Pheasants' eyes created by diffraction from dust particles and lens imperfections	40
2.8	Albert Baez, 1956	45
2.9	Hussein El-Sum, 1968	47
2.10	Gabor's two-hologram interference microscope of 1951	55
3.1	Denisyuk as a junior scientist, *c.*1966–7	75
3.2	Denisyuk as renowned scientist, 1970	75
4.1	Willow Run Laboratories, 1962	80
4.2	Synthetic Aperture Radar (SAR) image of Willow Run airport, one of thousands produced during development of SAR at WRL and its successor, ERIM, *c.*1973. The Willow Run Labs are at the bottom (east) of the airport runways	82
4.3	Optical processor at Willow Run, *c.*1960	88
4.4	Emmett Leith, 1958	89
4.5	Willow Run 'blockhouse'	93
4.6	(a) SAR geometry and (b) signal record	94
4.7	Portion of chirped SAR photographic record	95
4.8	Early methods used by Leith and Upatnieks to introduce an off-axis reference beam	98
4.9	First newspaper images of hologram (left) and reconstruction of a two-dimensional greyscale transparency	102
4.10	Page from Juris Upatnieks' laboratory notebook	105
4.11	Juris Upatnieks and Emmett Leith at optical table, Dec 1963	110
4.12	*Toy train*	114
4.13	Spectra-Physics advertisement using one of the iconic train holograms	117
5.1	Organization of University of Michigan Electro-Optical Laboratory *c.*1965–7	126

5.2 Organization of the University of Michigan Optics Group of the WRL Radar & Optics Laboratory *c.*1965 126
5.3 U-M Electro-Optical Sciences Laboratory members 129
5.4 Emmett Leith, 1966 132
5.5 George Stroke, 1966 132
5.6 Dennis Gabor next to his pulsed portrait hologram 141
5.7 Concepts contributing to holography, 1947–66 148
6.1 Optical organizations and companies in the Ann Arbor region, 1946–2005 161
6.2 Conductron optical processor, 1965 163
6.3 Gary Cochran setting up to record an airplane model at Conductron Corporation, 1967 164
6.4 Dave Wender preparing to record a short animated holographic movie at Conductron Corporation, 1966 167
6.5 Brochure advertising pulsed holography, McDonnell Douglas, 1972 169
6.6 GC Optronics advertisement *c.*1969 175
7.1 Hammer blow, illustrating time-averaged holographic interferometry 194
7.2 Pulsed hologram of bullet in flight 197
7.3 First pulsed hologram of a human subject: Larry Siebert self-portrait, Conductron, 31 Oct. 1967 208
7.4 Soviet pulsed hologram: Dmitri Staselko and A. Smirnov, 1970 208
7.5 'Multiplex' hologram recording geometry 215
7.6 Illustration of (1) Gabor, (2) Leith-Upatnieks, and (3) Denisyuk hologram recording geometries 216
7.7 Adolf Lohmann (center) with Byron Brown and Ronald Kay of IBM 217
7.8 Computer-generated binary hologram and its reconstruction, Lohmann et al., 1967 218
7.9 Victor Komar, Deputy Director of NIKFI, with holographic movie film, 1986 222
8.1 Attendance at the thirteen Gordon Research Conferences 245
8.2 Scientific holographers: attendees at the first Gordon Research Conference 247
8.3 Hans Bjelkhagen recording a reflection hologram of the Swedish crown 247
8.4 Juris Upatnieks at ERIM, Ann Arbor, MI, 1979 248
8.5 Karl Stetson at United Technologies Research Center, East Hartford, CT, 1984 248
8.6 Experimental laser for hologram recording, NIKFI, Moscow, *c.*1985 248
8.7 Svetlana Soboleva in Moscow laboratory, 1989 249
9.1 The peace symbol as the subject of a computer-generated Fourier-transform hologram 253
9.2 Lloyd Cross 260
9.3 Pethick sandbox table 265
9.4 Artisans and proselytizers 268

9.5 Entrepreneurs—members of the Multiplex Company 1973 275
9.6 Genealogy of San Francisco holography 279
9.7 New York School of Holography advertisement *c.*1976 279
9.8 'Public access holography' via the sand table, 1982 282
10.1 Fritz Goro, with Juris Upatnieks and Emmett Leith, hologram for *Life* magazine, 1966 291
10.2 Bruce Nauman, *Making Faces*, 1968 292
10.3 Margaret Benyon, 1970 294
10.4 Benyon exhibition, Nottingham University Art Gallery, 1971 296
10.5 Harriet Casdin-Silver with *Equivocal Forks*, *c.*1977 302
10.6 Sam Moree at the New York School of Holography 304
10.7 Rudie Berkhout, 1979 305
10.8 Exhibition poster, Museum of Holography, 1976 314
10.9 Mobile hologram display, Ukraine, early 1980s 314
10.10 Exhibition poster melding art with technological wonder, based on *The Kiss II* multiplex hologram by Lloyd Cross 316
10.11 bookmark for the Hologram Gallery, Amsterdam 316
11.1 Nick Phillips in Bulgaria with Russian holographic plates, 1990 326
11.2 Tung Jeong at first Summer School on Holography, Lake Forest, Illinois, 1972 331
11.3 Attendees at the first *International Symposium on Display Holography*, Lake Forest College, 1982 333
11.4 Stephen Benton with *Crystal Beginnings*, *c.*1980 334
11.5 Rosemary (Posy) Jackson at the New York Museum of Holography, 1976 336
11.6 Preparing for the opening of the New York Museum of Holography, 1976 337
12.1 An 18 × 24 in. pulsed ruby laser hologram for Cartier's, New York 353
12.2 John Perry at the Living and Teaching Center, Vermont, late 1980s 366
12.3 Steve McGrew at Holex Corporation, 1978 374
12.4 Migration of some of the primary holography patents and firms 377
12.5 Advertisement for the Atari Cosmos game system, 1981 382
12.6 *New Scientist* response to news of methods of counterfeiting security holograms 388
13.1 Holographers François Mazzero and Pascal Barr viewing hologram 394
13.2 Erotic hologram: *Pam and Helen*, Multiplex Company, 1974 397
13.3 Marketing erotica: Peter Claudius and the 'Holoblo', 1976 397
13.4 Iconic representations: a hologram in *Star Wars* 407
13.5 A mark of establishment: hologram of Queen Elizabeth II 414
14.1 Emmett Leith with 1966 Fritz Goro model, 2003 439
14.2 Dennis and Marjorie Gabor preparing for his portrait 445
A.1 Publications per year, 1945–2001 inclusive 490

ACRONYMS

ABNH	American Bank Note Holographics
AEI	Associated Electrical Industries
AIP	American Institute of Physics
ARO	Army Research Office
ARPA	Advanced Research Projects Agency
AWRE	Atomic Weapons Research Establishment
BAAS	British Association for the Advancement of Science
BTH	British Thomson-Houston
BRH	Bureau of Radiological Health
BUORL	Boston University Optical Research Laboratory
CAVS	Center for Advanced Visual Studies
CGH	Computer-Generated Hologram
CSIRO	Commonwealth Scientific and Industrial Research Organisation
DARPA	Defense Advanced Research Projects Agency
DCG	dichromated gelatin
DoD	Department of Defense
DOE	Diffractive Optical Element
DOVID	Diffractive Optically Variable Image Device
DSIR	Department of Scientific and Industrial Research
EE	Electrical Engineering
EOSL	Electro-Optical Sciences Laboratory
ERIM	Environmental Research Institute of Michigan
GRC	Gordon Research Conferences
HNDT	holographic non-destructive testing
HOE	holographic optical element
HUD	head-up display
IEEE	Institute of Electrical and Electronics Engineers
IHMA	International Hologram Manufacturers' Association
IST	Institute of Science and Technology
JOSA	*Journal of the Optical Society of America*
JPL	Jet Propulsion Laboratory
L.A.S.E.R.	Laser Arts Society for Education and Research
MIT	Massachusetts Institute of Technology
MoH	Museum of Holography, New York

NASA	National Aeronautics and Space Administration
NPL	National Physical Laboratory
NSF	National Science Foundation
ONR	Office of Naval Research
OSA	Optical Society of America
OSRD	Office of Scientific Research and Development
RCA	Royal College of Art
RPS	Royal Photographic Society
SAR	Synthetic Aperture Radar
SDS	Students for a Democratic Society
SPIE	Society of Photo-Optical Instrumentation Engineers
SUNY	State University of New York
U-M	University of Michigan
USBC	US Bank Note Corporation
WRL	Willow Run Laboratory

1

Introduction: Seeking Coherence

> In a Boston hotel suite recently, a few dozen normally sedate scientists and engineers were playing with a toy locomotive, a toy train-conductor and other such items.
>
> The train wasn't really there at all. But if you stood in exactly the right place and looked into a piece of equipment you would have seen it, real as life. The toys had been 'reconstructed' by a technique that looks simple, yet is one of the most sophisticated developments in modern science.
>
> The 'reconstruction' was done with a gas laser made by Perkin-Elmer Corp., and a 'hologram', a special photographic plate made by researchers at the University of Michigan.[1]

In a few sentences, the introduction to this 1964 news item framed a young subject in terms that were to endure for a decade: as an activity filled with childlike wonder, as an outgrowth of photography, as an expression of modern science, and as an illustration of industrial and scientific collaboration.

And yet the field was not entirely new, nor did it consistently attract positive assessments. Sir Lawrence Bragg of Cambridge University had written in 1948 that it seemed 'something of a miracle to me that it should work'.[2] Eight years later, Dennis Gabor, who had conceived the basic idea as an improvement for microscopic imaging, was deeply discouraged and worried about being seen as 'the bright boy who produces brilliant dud ideas'.[3] Holography, as this new subject came to be known, attracted new audiences and definitions in the decades that followed. By 1966, George Stroke touted his own work in the field as 'a hair-thin balance between sophisticated mathematics and extremely refined experimentation [. . .], new miracles comparable to a successful Apollo moon shot in this field'.[4] And a decade later, for George Lucas, writing the script of *Star Wars*, the hologram had become a science-fictional device.

[1] Novotny, George V., 'The little train that wasn't', *Electronics*, 37 (30), 30 Nov. 1964: 86–9.

[2] Bragg, W. L. to D. Gabor, letter, 5 Jul. 1948, IC GABOR EL/1.

[3] Gabor, Dennis to T. E. Allibone, letter, 22 Mar. 1954, IC GABOR MA/6/1.

[4] Stroke, George W., 'Press Release: Breakthrough in "lensless photography" sets stage for 3-dimensional home color television and a possible multi-million industrial explosion in electro-optics', 12 Mar. 1966, Leith collection.

> He sits up and sees a twelve-inch three-dimensional hologram of Leia Organa, the Rebel senator, being projected from the face of little Artoo. The image is a rainbow of colors as it flickers and jiggles in the dimly lit garage. Luke's mouth hangs open in awe.[5]

Emmett Leith, responsible for the 1964 explosion of interest that was a by-product of classified radar research, described it during the 1980s as 'a somewhat narrow field, but [. . .] big enough to spend a lifetime in'.[6] And for art critic Edward Lucie-Smith, during the 1990s it was 'a pariah art form' for which 'triumph is probably certain, though not always in the forms that present-day artist-holographers suppose'.[7]

Group after group was seduced by holography. Engineers became enthused by the potential of the hologram to reveal subtle properties of the human-made world, using it to detect mechanical distortions and short-lived events. Corporations invested in what they confidently presumed would be a growth industry. And everyone who saw a hologram was awed by its ability to reconstruct a stunningly realistic image. Over a half-century, holography engendered wonder in new viewers, provoked repeated waves of commercial enthusiasm from entrepreneurs, and generated enduring satisfaction and frustration among holographers—the self-identified scientists, engineers, artists, artisans, and hobbyists who made it their own.

But like the multiple perspectives offered by the hologram, the history of holography cannot be captured from any one of these viewpoints. Each community had a different vision of its problems and potential, and each shaped a history and forecasts to fit. Personal reminiscences jostle with dry scientific papers and expansive company forecasts; throngs briefly captivated by large exhibitions contrast with the solitary workers who spent long hours in optical laboratories. Each successive wave and community reinterpreted the subject and reshaped its goals, meanings, and successes. How has this creation of history from individual experiences occurred? More generally, how do communities construct their past and imagine their future?

Hundreds—perhaps thousands—of capsule histories of holography have been written since the 1950s, appearing in newspapers, magazines, conference proceedings, scientific papers, introductions to books, and practitioners' folklore. Tracing those accounts, and understanding their underlying assumptions and ideals, is one of the purposes of this book. All have been short summaries or local in perspective. They have been written at distinct times for diverse audiences and often came to dramatically different conclusions about which ideas, events, players, and products were important. Physicist Paul Kirkpatrick, for example, who had explored the field with his Stanford University students during the early 1950s, downplayed the work of the subsequent generation by vaunting a 200 year genealogy for holography. He was nevertheless acutely aware of the subjectivity of categorizing the field and its history, arguing that definitions of holography 'had better be clearly stabilised before its malleable period passes'.[8] Indeed,

[5] Lucas, George, 'Star Wars script (revised fourth draft)', 15 Jan. 1976.
[6] Hecht, Jeff, 'Applications pioneer interview: Emmett Leith', *Lasers & Applications*, 5 Apr. 1986: 56–8.
[7] Lucie-Smith, Edward, 'A 3D triumph for the future', *The Spectator*, 277 (8777), 5 Oct. 1996: 57–8.
[8] Kirkpatrick, Paul, 'History of holography', *Proceedings of the SPIE*, 15 (1968): 9–12.

physicist George Stroke of the University of Michigan energetically advanced his own version of what he called 'the now historical account' to support his priority claims and to influence the awarding of the Nobel Prize in 1971.[9] By contrast, physicist Yuri Denisyuk, at the Vavilov Institute in Leningrad, struggled to communicate the meaning and significance of his own work at home and abroad until it was rehabilitated by developments in the West. And by the early 1970s, yet another perspective was being experienced and told through the eyes of a small counterculture community in San Francisco, an account that has since become part of the oral folklore of the subject.[10] Over the following twenty years, these accounts were swamped by a wave of tales about entrepreneurs, proprietary processes, and business rivalries. And at technical conferences, it has been the longest-lived and best-funded participants whose stories have held sway almost as mythic tales.[11]

Historians construct stories from such sources that they hope are more revealing, critical, or representative than those told by practitioners themselves. What divides and unites these accounts? How are they embedded in their cultural context? And what can they tell us about the nature of new knowledge, new techniques, new products, and new social groups? How, indeed, can we capture the collection of perspectives that show how a new science becomes established?

There are particular difficulties in seeking answers about holography, relating to human recollections, documentary sources, and the scale of explanation.

Human memories preserve and interpret past events selectively. But these individual and community visions, even when objectively inaccurate, can be revealing about ambitions and interpretations, being framed in ways that vaunt a sense of personal or group identity. Individual recollections, triggered perhaps by contemporary written accounts, can also be used alongside archival sources to help locate or fill in documentary evidence.

A second problem concerning oral histories is that the sampling tends to be biased towards accounts of success. The survivors available for interview or correspondence are frequently those individuals and firms that had a long period of activity in holography or who feel that their contribution was significant. Most such accounts emphasize achievements, but it is necessary to collate perceived failures as well as successes—and there have been many commercial failures, and more than a few technical dead-ends, in the field of holography. By contrast, the members of research groups or companies that failed to achieve their goals may have retained few records or wish not to recall the past. Attributions of success and failure made by participants may surreptitiously influence subsequent analyses.

This asymmetry of historical accounts, an entirely typical but under-represented situation in histories of science and technology, has been increasingly criticized by historians

[9] Stroke, George W. to D. Gabor, telegram, Nov. 1971, Paris, IC GABOR MS/15.

[10] Cross, Lloyd, 'The Story of Multiplex', transcription from audio recording, Naeve collection, Spring 1976.

[11] Gabor, Dennis, 'Holography, 1948–71', in: Nobel Prize Committee (ed.), *Les Prix Nobel en 1971* (Stockholm, 1971), pp. 169–201; Denisyuk, Yu N. and V. Gurikov, 'Advancement of holography, investigations by Soviet scientists', *History and Technology* 8 (1992): 127–32; and Leith, E. N., 'Overview of the development of holography', *The Journal of Imaging Science and Technology* 41 (1997): 201–4.

and sociologists.[12] This bias can be compensated to a degree by actively tracing practitioners rather than relying on a self-selecting sample to present themselves for interview and by consciously studying 'successful' and 'unsuccessful' endeavours on equal terms, but the task becomes more difficult with time: achievements are well documented, but unproductive activities are not and may eventually be forgotten. If scholars were to attempt to write of these events decades after the fact the documentary evidence would be uneven at best and seriously misleading at worst. Subjects having contentious success, such as holography, are particularly vulnerable. For this reason, a contemporary history of holography can capture aspects of the subject that subsequent attempts may be unable to.

A related reason for attempting to write a history within the lifetime of its participants is that holography has been shared among marginal technical communities. The field has not become an established profession or generated a consensual history like the three big engineering professions in America, namely civil, mechanical, and electrical engineering.[13] Instead, some subcultures of holography (e.g. university scientists) have documented their successful work (but less often their unsuccessful forays) in papers, while others (e.g. the artisans, teachers of short courses, and commercial holographers) have left their mark through works of art or commercial products or by founding companies or institutions. But much is ephemeral. A sea of information circulates orally at scientific and engineering conferences, each of which emphasizes a different flavour of holographic activity and is captured only inadequately by published conference proceedings.

A separate problem concerns the scale of historical research and explanation. Despite the importance of timely data gathering, many histories are written many years after the fact because they can provide a sense of the large-scale influences that were not obvious at the time. To cite one example, the rapid growth in scientific research through the 1950s and 1960s was widely interpreted by contemporary commentators as inevitable characteristics of a modern technological society, but we recognize more readily today that the cold war era introduced an unusual scale of science funding by military sponsors, with significant knock-on effects. This context was important for the history of holography and also flavours alternative accounts of the history of the laser[14] and of satellite

[12] More contentiously, the Strong Programme of the Edinburgh School of historians, sociologists, and philosophers of science argued that explanations of the truth or error of scientific statements must be analysed by similar methods and generally invoke a combination of rational and sociological explanations. For influential early discussions, see Bloor, David, *Knowledge and Social Imagery* (London: Routledge and Kegan Paul, 1976) and Barnes, Barry, *Interests and the Growth of Knowledge* (London: Routledge and Kegan Paul, 1977).

[13] On the establishment of American engineering professions, see Noble, David F., *America by Design: Science, Technology, and the Rise of Corporate Capitalism* (New York: Knopf, 1979).

[14] Contrast, for example, a professional historian's account (Bromberg, Joan Lisa, *The Laser in America, 1950–70* (Cambridge, MA: MIT Press, 1991)), a popular science writer's account of the first three years (Hecht, Jeff, *Beam: The Race to Make the Laser* (Oxford: Oxford University Press, 2005)) and personal accounts of participants (Townes, Charles H., *How the Laser Happened: Adventures of a Scientist* (Oxford: Oxford University Press, 2002); Maiman, Theodore H., *The Laser Odyssey* (Blaine, WA: Laser Press, 2000); and Taylor, Nick, *Laser: The Inventor, the Nobel Laureate, and the Thirty-Year Patent War* (New York; London: Simon & Schuster, 2000)).

surveillance systems, which involved overlapping technical communities that skirted holography.[15] Distance from the events can thus provide an alternative perspective and a distinct narrative line. On the other hand, this distance can also produce a myopic and misleading view, obscuring the complexity of detail that influenced the trajectory of development. Too often, this oversimplification has encouraged technical history to be related in the form of a moral tale: triumph over adversity, tragic failure, or inevitable progress. This style of history writing, popularized with confident Victorian accounts such as Samuel Smiles' *Lives of the Engineers*, is still battled by historians of science and technology today.[16]

Striving to develop reliable explanations based on empirical evidence, historians in recent years have pursued fine-grained studies and shied away from attempting to paint big picture narratives. Microhistories, spanning relatively short time scales to provide a convincingly detailed account of individual episodes or communities, can nevertheless downplay larger-scale social factors. For a subject like holography, which has enrolled an unusual spectrum of interest groups, this approach is inadequate. The subject has been too wide-ranging to be comprehensible by microstudies. There is a degree of commonality among its subcultures that deserves explanation. There are features of its emergence, community by community and application by application, that may provide insights for other nascent technical subjects. Moreover, wider-reaching studies of science and technology can be applied to new purposes: there is a need to understand the dynamics of new subjects, new disciplines, and new professions for policy making and fostering and funding of innovation. There are, therefore, tensions straining the history of science and technology: between technical accuracy on the one hand and faithfulness to the spirit of an age on the other; between understanding details and widespread effects; between practitioners' intellectual history and wider social and economic accounts (often still discussed crudely in terms of sociologist Robert Merton's categories of *internal* and *external* factors of science).[17]

There is a final and more intractable difficulty in writing a history of holography. For the casual observer, the main point of interest is the hologram itself, but these artefacts have been produced by an elaborate network of workers, expertise, institutions, and funding, most of which is invisible. This problem has been addressed recently by

[15] On varying accounts of the CORONA surveillance satellite project of the 1960s, see Day, Dwayne A., John M. Logsdon, and Brian Latell, *Eye in the Sky: The Story of the Corona Spy Satellites* (Washington, DC: Smithsonian Institution Press, 1998); Richelson, Jeffrey, *The Wizards of Langley: Inside the CIA's Directorate of Science and Technology* (Boulder, CO: Westview Press, 2001); Taubman, Philip, *Secret Empire: Eisenhower, the CIA, and the Hidden Story of America's Space Espionage* (New York: Simon & Schuster, 2003); and Lewis, Jonathan E., *Spy Capitalism: Itek and the CIA* (New Haven: Yale University Press, 2002). These focus on the political, administrative, and business histories of the subject rather than its technological dimensions.

[16] For excellent introductions to the historiography of science and technology, respectively, see Kragh, Helge, *An Introduction to the Historiography of Science* (Cambridge: Cambridge University Press, 1987) and Staudenmaier, John M., *Technology's Storytellers: Reweaving the Human Fabric* (Cambridge, MA: MIT Press, 1985).

[17] See, for example, Merton, Robert K. and Norman William Storer, *The Sociology of Science: Theoretical and Empirical Investigations* (Chicago; London: University of Chicago Press, 1973) and the earlier Merton, Robert K., *Social Theory and Social Structure* Glencoe, Ill. (1957).

increasing attention to the histories of scientific instruments, notably the work of Peter Galison.[18] But holography covers more ground and has generated products beyond science itself. The inadequate melding of material culture with its social and intellectual context is a common problem in the history of science and technology, and it is illustrated by the sometimes distinct perspectives and subcultures of practitioners, historians, and museum curators.[19] It is peculiarly awkward for an intensely visual medium like holography. Holograms, particularly of the first generation, must be viewed using special optical arrangements sometimes involving expensive equipment. The need for display maintenance means that they are less frequently exhibited than, say, early computers or steam engines. Exhibitions have declined since the 1980s because museums and galleries cannot demonstrate holograms as a straightforward illustration of progress or industrial relevance or even as convincing precursors of newer technologies and art forms. This uncomfortable categorization has sometimes caused the history of holography to be judged as a detour and an irrelevance, a false step.

Writing about the unusual case of holography is therefore fraught with difficulty. This book correspondingly attempts to build a sense of a grand narrative from a number of intermeshed small narratives. And it adopts a comprehensive approach to combining oral and documentary histories that has been illustrated successfully by other historians of the recent past, for example, Gabriel Hecht in writing about French nuclear power and, much earlier, Martin Jay in writing a history of the Frankfurt School of social research.[20] A combination of techniques—including correspondence and interviews with well over one hundred current and past practitioners, fieldwork at a variety of conferences and laboratories, study of published documents, and research in public and private archival collections—has explored multiple sources and perspectives.[21]

This book seeks coherence: an accurate account of an emerging subject from multiple perspectives. A Picasso-like portrait is obligatory to capture its essence. These differing visions of holography come not only from its participants and audiences but also from different disciplinary gazes. Aspects of this account will appeal to distinct audiences and so readers may prefer to dip into specific chapters.

For general readers, this is a history of a captivating subject that emerged from hidden laboratories to influence wider scientific thought. Appropriated by artists as an aesthetic

[18] Galison, Peter, *Image and Logic: A Material Culture of Microphysics* (Chicago: University of Chicago Press, 1997).

[19] For example, the text and photographs in this book cannot adequately provide a full appreciation of the primary product, the hologram. On the other hand, collections of holograms illustrate the impact and range of uses of the medium but under-represent its intellectual and social context. Analytical studies and collections of material culture can be allied, but the merging of 'textuality' and 'materiality' has been constrained, in part, by the distinct backgrounds and goals of scholars in the humanities and museum curators, as mentioned in note 24 of this chapter.

[20] Hecht, Gabrielle, *The Radiance of France: Nuclear Power and National Identity After World War II* (Cambridge, MA; London: The MIT Press, 1998); Jay, Martin, *The Dialectical Imagination: A History of the Frankfurt School and the Institute of Social Research 1923–50* (London: Heinemann, 1973).

[21] Research databases record some 35,000 scientific and engineering papers on holography internationally. See the appendix for a bibliometric analysis, the bibliography for a list of interviewees, and the acknowledgements for a list of correspondents.

medium and by laypeople through science fiction film and television, its ideas and practices were translated and mutated in their passage between subcultures. Those subcultures ranged from the positivist scientists and industrial technologists of the 1940s, to 1950s engineers engaged in classified research, to the late 1960s counterculture, to public-access holographers of the 1970s, to entrepreneurs of the 1980s, and thence to popular culture.[22] Holography illustrates the post-war enthusiasm for science and technology and how that excitement was generated, sustained, and channelled. But it also shows, unusually, how this momentum carried into the domains of fine art, cottage industry, and big business over the succeeding decades. The subject became enmeshed with the domains of science, commerce, the military, and popular culture during the late twentieth century. General readers will find a wide-ranging story anchored in contemporary history and exploring a variety of cultural domains.

This comprehensive history of holography also challenges existing accounts. Can we attribute holography to a handful of geniuses, for example, or to the inevitable march of progress? These contrary positions are caricatures of some of the brief histories that have been written of the subject. Practitioners, would-be scientists, and the general public still crave heroes and evidence of the triumph of the intellect, but this detailed account supports neither stance.

Holographers will find a detailed account here of the history of their emerging science, technology, and art and the individuals and groups that created them. The book evaluates information from multiple sources about seminal workers, the dissemination of ideas, patent disputes, and business activities, and illustrates how successive communities made holography their own. Nevertheless, in a field that has attracted several thousand participants, this history can merely sample the research threads, companies, and events that comprise the subject. Whether workers have been involved for months or decades in the field, they will discover ideas, episodes, and applications that have been unknown to them in detail. The cases also illustrate the creation and reshaping of folklore and provide consequent insights into the generation of historical myths. This recasting of personal recollections into a wider historical context can be a powerful, if unsettling, experience.

This book is also relevant to current scholarship in the humanities and social sciences. The history of holography is a fertile case study in the domain of *science studies*, an interdisciplinary melding of history and philosophy of science with more recent perspectives on social studies of science and technology, combined with insights from economic, political, and cultural studies. The history, philosophy, and sociology of science were recognized increasingly by academic posts from the 1950s, although too often as merely a

[22] Positivism, first enunciated by Auguste Comte in the 1830s and remaining the most popular philosophy of science for over a century, sees scientific knowledge as naturally progressive, with physics and astronomy representing the most advanced and quantitative of the sciences. Buttressed by Victorian technological innovation, these views led to the related claims of technological determinism, which holds that technology advances inexorably in intellectual and economic respects and impels an irresistible social adaptation. These themes are explored in Chapter 14.

bridge between what C. P. Snow described as the two cultures of the Arts and Sciences.[23] This collection of subjects still tends to be pursued along distinct disciplinary lines, generating diverse approaches, separate research problems, and sometimes competing interpretations. To date, this grouping of science studies cannot itself be described as a discipline, although it is taught in differing flavours within a variety of university contexts.[24]

The field of holography has, to date, attracted limited study along these lines. Sarah Maline, for example, has explored the aesthetics of holography from the perspective of art history and written a PhD dissertation on the topic.[25] PhD dissertations by Andrew Pepper, Margaret Benyon, Paula Dawson and others have also explored fine art holography from a contemporary art perspective.[26] Ivan Tchalakov has presented an anthropological study of a Bulgarian holography laboratory along the lines developed by sociologist of science Bruno Latour,[27] and Susan Gamble, a fine-art holographer, has

[23] The views of prominent researchers and their discipline-building activities *c.*1960 are illustrated by the discussion following Crombie, A. C. and M. A. Hoskin, 'A note on history of science as an academic discipline', in: A. C. Crombie (ed.), *Scientific Change: Historical Studies in the Intellectual, Social and Technical Conditions for Scientific Discovery and Technical Invention, from Antiquity to the Present* (London: Heinemann 1963), pp. 757–94. On the cultural division between different modes of thought, see Snow, C. P., *The Two Cultures; and, A Second Look—An Expanded Version of the Two Cultures and the Scientific Revolution* (Cambridge: Cambridge University Press, 1964).

[24] For example, under the name *Science Studies*, it is taught at the University of Glasgow as part of the Liberal Arts programme, at Edinburgh through the Sociology Department, and at Bath through the Science Studies Centre; as *History and Philosophy of Science* (HPS) in the Philosophy School at University of Leeds, as an HPS department at the University of Pittsburgh, and at the University of California, Davis, as an HPS programme through separate History and Philosophy departments; as *Science and Technology Studies* (STS) in the School of Humanities and Sciences at Stanford University; and until recently as *Science and Technology Dynamics*, a policy-oriented subject in the Natural Science Faculty at the University of Amsterdam but later recreated as an STS programme in the Social Science Faculty. Most such departments were established or significantly reshaped from the 1970s. A pro-science perspective, arising from the mid-1980s with support from British scientists, is the *Public Understanding of Science*, supported, for example, by chairs at Imperial College, London, and Oxford University. Yet another and much longer-established tradition emerging from antiquarian interest in the material culture of science has been supported by a distinct community of historians of scientific instruments, particularly at major museums such as the Science Museum in London, Smithsonian Museum in Washington, and Deutsches Museum in Munich. The interactions between these professional communities and their distinct perspectives are subjects worthy of study in their own right, and fascinating recent work in this direction has been presented in Mayer, Anna-K., 'Setting up a discipline, II: British history of science and "the end of ideology", 1931–48', *Studies in History and Philosophy of Science* 35 (2004): 41–72.

[25] Maline, Sarah Radley, *Art Holography, 1968–93: A Theatre of the Absurd*, PhD thesis, Art History, University of Texas at Austin (1995).

[26] Pepper, Andrew, *Drawing in Space: A Holographic System to Simultaneously Display Drawn Images on a Flat Surface and in Three Dimensional Space*, PhD thesis, Fine Arts, University of Reading (1988); Benyon, Margaret, *How is Holography Art?*, PhD thesis, Royal College of Art (1994); Dawson, Paula, *The Concrete Holographic Image: An Examination of Spatial and Temporal Properties and Their Application in a Religious Art Work*, PhD thesis, Fine Arts, University of New South Wales (2000).

[27] See Tchalakov, Ivan, 'The object and the other in holographic research—approaching passivity and responsibility of human actors', *Science, Technology and Human Values* 29 (2002): 64–87. On the actor-network theory (ANT) that underlies his analysis, see, for example, Latour, Bruno and Steve Woolgar, *Laboratory Life: The Construction of Scientific Facts* (Princeton, NJ; Chichester: Princeton University Press, 1986); Latour, Bruno, *Science in Action: How to Follow Scientists and Engineers through Society* (Milton Keynes: Open University Press, 1987); and, the edited volume Bijker, Wiebe E. and John Law, *Shaping Technology/Building Society: Studies in Sociotechnical Change* (Cambridge, MA: MIT Press, 1992).

written a dissertation on the historical antecedents of the hologram from the viewpoint of history of science.[28] Holography, and indeed the flourishing field of modern optics, has only begun to attract the attention it deserves as a subject of science studies.

The broader study of the relationship between new sciences and their practitioners has gained increasing interest, however. Timothy Lenoir, for example, has written on the cultural dimensions behind the production of scientific disciplines, especially in late nineteenth-century Germany.[29] He concludes from his case studies and drawing on the work of Michel Foucault and Pierre Bourdieu, that disciplines are not monolithic or created by individuals. Instead, they are shaped by separate and sometimes conflicting disciplinary programmes 'generated simultaneously within political discourse and ideological discourse, and are therefore best understood as discourses of power as well as instruments of knowledge production'.[30] While twentieth-century disciplines, too, commonly involved mixtures of academic, industrial, and governmental influences, it is also clear that much science and technology never attains disciplinary status, nor is all (or even most) science created as part of a disciplinary programme. Foucault drew an explicit link between intellectual disciplines as organized forms of knowledge on the one hand and the standardization of behaviours by technical training or education on the other, showing how social power accrues to members of an organized technical community.[31] The muted effects of power relations are evident in the would-be discipline of holography.

But not all technological and scientific subjects become disciplines. In a more relevant sociological vein, Terry Shinn and others have argued that many technical activities can be interpreted as *research technologies* that fail to develop the attributes of a discipline such as academic stability or the status and visibility of a technical profession.[32] Klaus Hentschel, for example, in arguing that spectroscopy was not a discipline before the mid-twentieth century, suggests that it amounted to a bundle of research technologies.[33] In my own previous work, I have made similar points about the histories of photometry, radiometry, and colorimetry.[34] This research-technology perspective is also consistent with the detailed account of holography and holographers provided here. More

[28] Gamble, Susan, *The Hologram and its Antecedents 1891–1965: The Illusory History of a Three-dimensional Illusion*, PhD thesis, History of Science, Cambridge University (2002).

[29] Lenoir, Timothy, *Instituting Science: The Cultural Production of Scientific Disciplines* (Stanford: Stanford University Press, 1997).

[30] Lenoir, Timothy, *Instituting Science: The Cultural Production of Scientific Disciplines* (Stanford: Stanford University Press, 1997), pp. 45–74, quotation p. 62.

[31] Foucault, Michel and Alan Sheridan, *Discipline and Punish: The Birth of the Prison* (London: Allen Lane, 1977).

[32] On research technologies, several of which concern branches of optics, see Joerges, Bernward and Terry Shinn (eds), *Instrumentation: Between Science, State and Industry* (Dordrecht: Kluwer Academic Press, 2001).

[33] Hentschel, Klaus, *Mapping the Spectrum: Techniques of Visual Representation in Research and Teaching* (Oxford: Oxford University Press, 2002), pp. 420–5.

[34] For an extended case study that frames the special career attributes in the slightly wider category of *peripheral science*, see Johnston, Sean F., *A History of Light and Colour Measurement: Science in the Shadows* (Bristol: Institute of Physics Publishing, 2001).

important, the case of holography extends the research-technology analysis by examining how ideas and techniques crossed an unusually broad range of scientific and cultural boundaries.

This book is a contribution to that ongoing dialogue, providing an extensive and detailed historical account pursued and interpreted along distinct lines. The empirical history within these pages transcends the intellectual, social, and economic, seeking a coherent treatment based on multiple perspectives. The format of the book mirrors the expansion of the subject itself: initially localized in particular problems and understandings; later splitting to explore historical episodes not only intellectually, but socially as well; and still later converging around commercial applications until defocusing into multiple and inconsistent interpretations. Physical optics, with its attention to coherence, interference, focusing, defocusing, and real and virtual images, provides a metaphor for the subject itself. The presentation is roughly chronological to emphasize that the history of this subject was highly contingent on the context, the local working environments, fortuitous accidents, and cascading clusters of experiences, skills, and resources. The narrative highlights these multiple causes and effects and illustrates the nature and direction of historical forces that shaped holography.

The book is divided into four parts, envisaging the development of holography as (I) an intellectual subject, (II) a visual medium and scientific technology, (III) a social practice, and (IV) an economic and cultural activity.

Part I, 'Creating a Subject', explores how an intellectual basis for the field developed from three distinct origins in Britain, the Soviet Union, and America, respectively. The detailed narrative examines how concepts and intellectual judgements developed in disparate working contexts. Why did the subject develop where and when it did? Emmett Leith, one of its key innovators, has suggested modestly that the underlying concepts could have been derived or discovered from any number of ideas circulating after the Second World War.[35] Others, by contrast, depict an inexorable evolution of ideas, with some genealogies of holography tracking its essential ideas back to the emerging science of optical interference from the mid-nineteenth century.[36] The evidence marshalled in this book suggests, however, that the historical actors were primed both by their environments and by then-current intellectual problems. Many of them unquestionably were highly creative and insightful individuals as well and were superbly equipped to explore the intellectual territory that opened for them. However, I argue that it was the particular post-war context of science and industry, combined with a spirit of military exploration in the context of the cold war, that explains the nurturing and flourishing of those ideas. While delving into the evolution of theoretical concepts, the chapters of Section I illustrate how intellectual concerns were mixed inextricably with research sponsorship,

[35] Leith, Emmett N., 'Reflections on the origin and subsequent course of holography', *Proceedings of the SPIE* 5005 (2003): 431–8.

[36] Scanlon, M. J. B., 'Holography: a simple physical account', *GEC Review* 8 (1992): 47–57.

career building, and institutional politics. It draws upon perspectives from the philosophy of science (particularly epistemology) and on the sociology of scientific knowledge (SSK).[37]

Part II, 'Creating a Medium', focuses on the decade-long (1964–73) frenetic research to extend the technical capabilities of holography. It concentrates on how experimental skills diffused between government, commercial, and self-financed laboratories and explores how the craft of holography yielded early applications and optimistic forecasts. It applies insights from the sociology of scientific and technological practice and is informed by understandings common to the Social Construction of Technology (SCOT)[38] and research-technology.

Part III, 'Creating an Identity', adopts perspectives from the sociology of professions to explore the growth of communities that came to call themselves holographers.[39] These technical groups were unusually disparate, and individual chapters chart their separate rise: scientist–engineer holographers, who colonized an existing but mundane specialty (optical engineering) to transform it into a high-tech would-be discipline; fine-art or aesthetic holographers, who appropriated the subject, extending its capabilities and purposes in the process; and artisanal holographers, who grew from a distinct counterculture community to transform the subject during the 1970s with radical methods and goals. The section documents the efforts by these groups to stabilize the occupation, profession, and discipline of holography, and their strategies of cooperation during the 1980s and beyond.

Finally, Section IV, 'Creating a Market', focuses on the commercial sprouting of holography alongside its evolving popular understanding. As with the other aspects of emergence, there was a co-evolution of both the cultural meaning of holograms and viable commercial products. The chapters document the broadening perceptions of the hologram in popular culture, ranging from erotica and children's toys to metaphors of brain organization and new-age understandings of holism. This concluding section of the book also examines the judgements of success and progress that have surrounded holography by revisiting the separate visions of its interest groups.

Together, these varied disciplinary perspectives, combined with exploration of the divergent visions of its practitioners, seek to provide a rounded picture of how the new science of holography came to be.

[37] A broad range of philosophers and sociologists have considered the social influences on scientific belief, notably Thomas Kuhn (Kuhn, Thomas S., *The Structure of Scientific Revolutions* (Chicago: University of Chicago Press, 1962)), David Bloor (note 12), Harry Collins (Collins, H. M., *Changing Order: Replication and Induction in Scientific Practice* (London; Beverly Hills: Sage Publications, (1985)) and Bruno Latour (note 27).

[38] On SCOT, see MacKenzie, Donald A. and Judy Wajcman (eds), *The Social Shaping of Technology: How the Refrigerator Got Its Hum* (Milton Keynes, England; Philadelphia: Open University Press 1985) and Pinch, Trevor J. and Wiebe E. Bijker, 'The social construction of facts and artifacts: or how the sociology of science and the sociology of technology might benefit each other', in: W. E. Bijker, T. P. Hughes, and T. J. Pinch (eds), *The Social Construction of Technological Systems* (Cambridge, MA: MIT Press, 1987), pp. 17–50.

[39] See Abbott, Andrew Delano, *The System of Professions: An Essay on the Division of Expert Labor* (Chicago: University of Chicago Press, 1988) and MacDonald, Keith M., *The Sociology of the Professions* (London: Sage Publications, 1995). A previous study along these lines is Divall, Colin and Sean F. Johnston, *Scaling Up: The Institution of Chemical Engineers and the Rise of a New Profession* (Dordrecht: Kluwer Academic, 2000).

PART I
Creating a Subject

Tracing the origins of a subject invites controversy. The genesis of new ideas can be more tentative and groping than we care to remember and novel perspectives may mutate into more familiar guises that suggest erroneous origins. Furthermore, conceptual pigeon-holes may be constructed anew by each investigator, making any similarity or generality of early ideas difficult to discern. Similarities that we identify may be a construction of culture as well as of intellectual reality. So, exploring how intellectual conceptions and cultural contexts shape a young subject can challenge our assumptions about knowledge and its acquisition. For the individual researchers and technical communities described in this book, establishing this new pattern of knowledge took time; seeing the various insights as coherent required the merging of experiences from different backgrounds and disciplines. And, as later chapters relate, the pattern of knowledge continued to change form over the subsequent decades.

The first part of this book attempts to answer the question, 'Where did the ideas of holography come from?' It deals with far-flung clusters of researchers who came to see themselves as builders of a new subject. These groups revitalized a mundane branch of physics, exploring aspects of optics and imaging in new and profoundly different ways. Each began with specific goals and constructed different vocabularies and physical explanations.

Three locations seeded this research, which was to coalesce into a recognized subject by the mid-1960s: an industrial electrical laboratory in Rugby, England (discussed in Chapter 2); a State Scientific Institute in what was then Leningrad, USSR (Chapter 3); and, a classified research laboratory in Willow Run, Michigan, USA (Chapter 4). From these influences, a combined subject began to crystallize along with a contested history (Chapter 5). The context is crucial to the story: the separate working cultures generated distinct visions of the field. Those perspectives played a role in its convincing synthesis during the mid-1960s, illustrating how conservative cultural forces encouraged a sense of continuity. Over two decades, insights and individuals intersected and interfered to reconstruct a new, more general subject: holography. These chapters stress two themes: first, that during the formative stages of the new subject that would become holography, conceptions were each independent, tentative, and dissimilar; and second, obtaining useful experimental results was perceived as discouragingly problematic in each case.

For two decades, therefore (1947–66), investigators were on a quest for coherence: intellectual coherence based on theoretical concepts; social coherence based on a uniformity of method and purpose; and optical coherence, an elusive quality of light sources that was to prove crucial for the nascent subject.

2

Wavefront Reconstruction in England and Beyond

2.1 INTRODUCTION

In September 1948, the *New York Times* carried the first-ever news story introducing the hologram.

> **NEW MICROSCOPE LIMNS MOLECULE**
>
> **Britons impressed by paper combining optical principle with electron method**
>
> BRIGHTON, England: Sir Lawrence Bragg, Nobel prize winner in physics, said today that he had been moved from incredulity to admiration of the ingeniousness of the new super-microscope that was described here to the physics section of the British Association for the Advancement of Science.
>
> Dr. D. Gabor, Hungarian scientist, now working for the British–Thomson–Houston electrical engineering concern, explained how by combining the electron microscope with a new optical principle it could be made to resolve the pattern of atoms in the molecule or the details of a virus.
>
> Introducing his new principle of 'wave-front reconstruction', Dr Gabor said the electron microscope had reached its technical limit [. . .]. His new device, which he called a 'diffraction microscope', gets around this difficulty by a two-step process. In the first an electron photograph, which he calls a 'hologram' is taken. This has no visual resemblance to the object under examination. In the second step, the likeness is restored by a reconstruction, or synthesis, carried out with light waves.
>
> [. . .] The second source is a pinhole of monochromatic light, which illumines the hologram and resolves the diffracted picture (in the demonstration it looked like a futuristic tapestry) into an image at least as clear as the earlier radio pictures.
>
> The electron part of the diffraction microscope is still only in the paper stage. This is an engineering problem to produce a pinpoint beam. Dr. Gabor,

> however, demonstrated the optical part and convinced physicists of the validity of the wave-front reconstruction principle.[1]

But the hologram—the futuristic tapestry that mysteriously recreated images out of thin air—had only the briefest flash of publicity. The *Times* article captured what was to be the peak of popular attention about holograms for the next sixteen years. The moment of visibility was flanked by an energetic run-up and a slower decline. When Dennis Gabor had devised his new concept for optical imaging the previous year, it did not spring into existence fully formed. It was a tentative and uncertain scheme to improve the limitations of one of his company's products. Over the next decade, the evolving method went by a variety of names: holoscopy, wavefront reconstruction, interference microscopy, diffraction microscopy, and even Gaboroscopy, reflecting changing evaluations of its content and purpose. And Gabor, a well-connected and creative research engineer, worked actively to publicize and exploit his concept, marshalling influential mentors, mounting public demonstrations for peers, publishing in scholarly journals, obtaining research grants, and collaborating in experimental verifications.

Yet, the scheme failed to capture the attention of many researchers or to generate any commercial interest. Despite his strong and varied efforts to promote it, Gabor's theory was repeatedly deemed unintuitive and baffling; the technique was appraised by those who investigated it to be of dubious practicality and, at best, constrained to a narrow branch of science. By the late 1950s, Gabor's subject was judged by its handful of practitioners to be a white elephant. Nevertheless, the concept was later rehabilitated and Gabor was to win the Nobel Prize in Physics for the work in 1971. Part of Gabor's concept was reused, generalized, and repackaged for a new generation of practitioners.

Creating a new science is seldom a one-man job, nor do clever ideas necessarily succeed. This chapter recounts Gabor's fostering of wavefront reconstruction not to vaunt the achievements of a neglected pioneer—which is what his Nobel Prize, in effect, accomplished for a different audience—but rather to probe how new subjects and their histories become established. The clarity of hindsight is a familiar aphorism, but identifies a psychological and analytical mirage: like many episodes in the history of science and technology, Gabor's work appears less orderly and decisive on close examination. Practitioners renewed Gabor's reputation retrospectively during the late 1960s because his work could be represented as the foundation on which to construct their new subject of holography. Had that subject been differently configured, Gabor's historical relevance might have been assessed less favourably. This first narrative chapter is not intended to judge objective success or failure; instead, it focuses on the difficulties experienced in constructing a meaningful subject, a practical application, and a viable technical community from Gabor's ideas through the late 1940s and early 1950s.

***

[1] 'New Microscope Limns Molecule: Britons impressed by paper combining optical principle with electron method', *New York Times*, 15 Sep. 1948: 35.

Fig. 2.1. Dennis Gabor at Siemens & Halske AG, Berlin, late 1920s (Imperial College archives Gabor collection).

Like many of his generation, Dennis Gabor had had a fertile but unsettled early career. Gábor Dénes (to use his national style of address) was born on 5 June 1900 in Budapest, Hungary. He served briefly with the Hungarian artillery in northern Italy during the First World War, and started his higher education at the Engineering University just as the First World War ended in late 1918.[2]

When he received call-up papers for further military service, Gábor left Hungary and continued studies after second-year exams at the Technische Hochschule Berlin-Charlottenburg, obtaining a Diploma in Electrical Engineering at the end of 1923. The following year he started work there for a Doctor of Engineering (Dr-Ing.) under the engineering Professor Matthias. His thesis research consisted of the construction of one of the first high-voltage, high-speed cathode ray oscillographs, which he used to study electrical transients in power transmission lines. Upon receiving his degree in early 1927, Gábor joined Siemens & Halske (S&H) AG, Berlin, the pre-eminent German electrotechnology company, working in one of the physical laboratories to develop high-pressure mercury and cadmium vapour lamps (see Figure 2.1).

Gábor was of Jewish origin although brought up, as he was later to write to a magazine editor, in an

> anti-religious atmosphere [. . .] at a time when the young middle class of that country fell passionately in love with western thought. 'Kultur', technical civilisation and political progress were worshipped with equal undiscriminating fervour, quite unimaginable for those who were born in the old countries in the West, with hundreds of years of creative tradition behind them. It was like adolescent nationalism, during which Hungary produced that 'unpardonably brilliant Hungarian emigration' (to quote John Strachey), all born between 1879 and 1908.[3]

[2] Allibone, T. E., 'Dennis Gabor: A biographical memorial lecture', *holosphere* 10 (1981): 1, 4–6, and Gabor, Dennis to Appointment Committee for Chair in Electron Physics, typewritten CV, 19 Feb. 1958, IC GABOR GB/1.

[3] Gabor, Dennis to T. Raison, letter, 15 Jul. 1961, IC GABOR MN/5-6; Edson, Lee, 'A Gabor named Dennis seeks Utopia', *Think*, Jan.–Feb. 1970: 23–7; Gabor, Dennis to T. Raison, letter, 15 Jul. 1961, IC GABOR MN/5-6.

Among the notable Hungarians of that generation were the physicist/physiologist Georg von Békésy (1899–1972), aerodynamicist Theodore von Kármán (1881–1963), author Arthur Koestler (1905–83), the computer pioneer John von Neumann (1903–57), and nuclear physicists Leo Szilard (1898–1964), Edward Teller (1908–2003), and Eugene Wigner (1902–95). Gábor (1900–79) was also part of that restless wave. Indeed, Gábor, von Kármán, von Neumann, Szilard, Teller, and Wigner were born in the same quarter of Budapest. Gabor later credited his father Bertalan Gunsberg (changing his surname to Gábor in 1899), the head of the Hungarian General Coal Mining Company, as instilling in him a love of the romance of science and invention.

With the political rise of Adolf Hitler, Gábor began looking for other employment in 1932. Late that year, he wrote to an acquaintance at Metropolitan-Vickers in England, a major electrical manufacturer, about their mutual work in cathode ray oscillographs and then to a manager there, T. Edward Allibone, the following March:

> If I would choose, I could continue my work in the Siemens laboratorys. / An other interesting and materially very satisfactory position was offered to me by an other big firm in Berlin. / But for some reasons, I would like best to get to England [sic].[4]

Allibone remembered Gábor and his work from a visit that he had made to Siemensstadt in 1930 but, observing that 'it is almost impossible for foreigners to obtain a post in England owing to the present condition of unemployment', was unable to offer him a post.[5]

Gábor left S&H in 1933 soon after Hitler came to power, returning to Budapest for a year to work on his own invention, a new design of gas discharge lamp. He offered the rights to the lamp to electrical manufacturers around the world via distribution agreements through a Swiss firm, Universag Technische AG. In 1934 he approached two other British companies, Associated Electrical Industries (AEI) and British Thomson-Houston (BTH). Both BTH and Metropolitan Vickers were, in fact, sister companies under the AEI umbrella, and were to be important to him in his subsequent career.

Despite their connections, the three firms had distinct company cultures and origins. The BTH Company had appeared at the end of the nineteenth century during the ferment of electrical manufacturing firms and electrification of towns around the world. BTH was formed in 1896, with most of its capital held by American Thomson-Houston (then part of General Electric) and administered by American directors. The passing of the Power Act in Britain in 1900 allowed BTH to win new contracts to supply power to large areas. In its early years, it produced steam turbines, components for trolley buses, and incandescent lamps. Unlike many British firms, it was not highly dependent on military contracts during

[4] Gabor, Dennis to T. E. Allibone, letter, 9 Mar. 1933, IC GABOR MS/6/1.

[5] Thomas Edward Allibone (1903–2003, born in Sheffield), who had become a respected scientist at Metropolitan Vickers, was later seconded to work on the Manhattan project in America during the Second World War, subsequently returning to head the AEI research station at Aldermaston. A lifelong friend, he was to write Gabor's necrology a half-century later (Allibone, T. E., 'Dennis Gabor', *Biographical Memoirs of the Fellows of the Royal Society* 26 (1980): 107–47).

Fig. 2.2. BTH Research Laboratories Building 52 (background), constructed in 1924 and where Gabor and Williams worked (S. Johnston photo).

the First World War. As a consequence, the company was able to expand quickly, while conforming to the post-war business trend of consolidation and amalgamation. Metropolitan Vickers, a rival of BTH, had itself been created by amalgamation in 1919, and merged with BTH in 1928 to consolidate their similar markets. Both were amalgamated into AEI Ltd the following year but retained their names, sites, and personnel.

During the early 1930s, the Metrovick portion of AEI was concerned mainly with heavy engineering, so upon Gábor's second contact Allibone recommended him to the research director at BTH, Hugh Warren. Gábor reached an inventor's agreement with the company and anglicized his name upon moving to Rugby, where BTH was based. In his first year there he was paid £550, and two years later became a full-time employee of the firm. He began researching a variety of projects often related to electric lighting, which, during the Depression when Gabor joined the firm, were the company's principal hope for new products.[6]

Thus, technical workers such as Dennis Gabor at BTH (see Figure 2.2) were immersed in a rapidly fluctuating business environment, competing with sister companies as much as with other British and foreign firms. This immersion in, and awareness of, the international electrical market was well appreciated by the directors of the company; the electrical

[6] BTH survived the depression with the aid of income from its electric lamp products, but also contributed to the development of jet engines and radar equipment. Continuing rivalry with Metrovick after the war weakened the AEI group of companies. Both names were subsumed in AEI in 1960. AEI itself eventually was absorbed into GEC (General Electric Company) in 1967, another of its old rivals in the electric lamp industry which had first proposed amalgamation forty years earlier (Jones, Robert and Oliver Marriott, *Anatomy of a Merger: A History of G.E.C, A.E.I. and English Electric* (London: Cape, 1970)).

industry had had this flavour since its growth in the late nineteenth century.[7] A multilingual research engineer with a strong patent record, Gabor embodied the international perspective that the firm had espoused since its creation.[8]

Dennis Gabor's background and environment shaped his intellectual horizons. BTH was unusual in the British context in being part of a commercial group having research labs. BTH and Metrovick each had laboratories, and its consolidating parent, AEI, establishing a unique British industrial laboratory after the Second World War dedicated to long-term research. Consequently, Gabor was fortunate to interact with a diverse collection of researchers while maintaining his contacts with émigré scientists from continental Europe.

During the Second World War he was classified as an Enemy Alien and excluded from war work at BTH, notably development of the magnetron tube used for radar. He worked in a hut outside the security fence at the firm. Nevertheless, 'a continuous stream of colleagues poured into the hut to derive inspiration from his fertile mind' recalled Allibone years later.[9] Because he was deprived of security clearance, Gabor was privileged in escaping research and development on short-term piecemeal military work and to apply himself to projects with longer lead times to be polished after the war.[10] Because such projects were never brought to commercial fabrication, Gabor also escaped the relative drudgery of production engineering.

As the company shifted back to civilian products at the end of the war, Gabor found his work gaining more attention. He later recalled that his three post-war years at BTH were a period of 'intense and happy activity' during which time he worked with his 'friend and collaborator' Ivor Williams on three lines simultaneously. The projects had foundations in electrical and optical theory and in practical invention, suggesting his range and depth.

> The first was frequency compression, a first, still incomplete but promising realisation of the principle which I found in my theoretical work.
>
> The second was an outlet for the inventor, three-dimensional cinematography. Just before the war, that great enthusiast and successful businessman, Oscar Deutsch, gave an address at a BTH dinner, giving us a vision of the cinema of the future, which had to be of course 3-dimensional. The Director of Research, the late Sir Hugh Warren, encouraged us to 'exercise our minds' on the problem. With I. Williams I produced by 1948 what I think was quite an

[7] For an excellent study of the emergence of the electrical power industry, see Hughes, Thomas Parke, *Networks of Power: Electrification in Western Society, 1880–1930* (Baltimore: Johns Hopkins University Press, 1983).

[8] His stilted written English became flawless upon moving to Britain, as were his German, and, of course, Hungarian. Gabor was also relatively at ease in French, and spoke English with a precise Hungarian accent (Gabor, Dennis, 'Gabor interviewed by Rex Keating', *c.*1963, IC GABOR P/1).

[9] Allibone, T. E., 'Dennis Gabor 1900–79', Memorial service address, Holy Trinity Church, 15 Mar. 1979.

[10] Gabor worked, for example, on a method of detecting aircraft by the heat emitted by their engines. His invention was a 'relay screen' consisting of a thin plastic film coated on one side with a layer that absorbed infrared radiation and on the other with a fluorescent dye. Infrared light focused on the film caused it to expand and, via intervening optics, revealed a fluorescent image in an eyepiece; see Allibone, T. E., 'Dennis Gabor: A biographical memorial lecture', *holosphere* 10 (1981): 1, 4–6. Such infrared research and development was also ongoing in Germany and America during the war, and was later generously funded in America, particularly at the Infrared Laboratory of the Willow Run Laboratories, discussed in Chapter 4.

impressive show. But Oscar Deutsch had died in the meantime, and it was decided not to pursue the subject further. This was perhaps just as well, because [. . .] the scheme was possible with small screens only, and the advent of the large screen would have killed it in all probability.

My third and last scheme in the BTH was 'microscopy by reconstructed wavefronts'. In 1927 I had missed a wonderful opportunity for starting electron microscopy on the ground floor. Twenty years later I wanted to come back to it, and make the instrument realise its full potentiality; the resolution of atomic lattices. Electron lenses could not be corrected. My idea was to break off the imaging process with electrons at a point where the picture was unintelligible, but contained the full information, and finish it by light optics, correcting the aberrations of the electron lenses by optical lenses. The results were spectacular, but it turned out that the ultimate aim could not be achieved.[11]

This triplet of projects stood out from other mundane post-war work at BTH, but they each had problems, being either incomplete, limited, or unachievable. So, from the vantage point of 1961, his most creative undertakings during the post-war years at BTH appeared to Gabor to have been dead ends. But even then, Gabor recalled one as *spectacular*: it was rich in theoretical promise and miraculous in demonstration. That latter project, in broader form, was to bring him fame late in life and forms the historical starting point for this book.[12]

This last of Gabor's post-war BTH projects grew from an idea that interested only a handful of specialists: a method for improving electron microscopes. His PhD project, the cathode-ray oscillograph with the magnetic lens system at its heart, had in fact been cannibalized by others to construct the first electron microscope a few years after he had left Germany. Ernst Ruska and Max Knoll had produced the first electron microscope there during the early 1930s.[13]

Such microscopes were modelled on their optical analogues. In a transmission electron microscope, electrons are accelerated through a vacuum and focused on the microscope sample, which absorbs or scatters them and, when focused by a second magnetic lens, creates an image on a photographic plate or fluorescent screen. The advantage of such devices was that—as first established in the 1920s—such fast-moving electrons can be

[11] Gabor, Dennis to T. Raison, letter, 15 Jul. 1961, IC GABOR MN/5–6. Oscar Deutsch (1893–1941), creator of the Odeon cinema chain from 1931, had produced a chain of 248 cinemas within a decade.

[12] As we shall see, by the late 1960s Gabor and others had identified all three projects as components of the new science of holography. His ideas for frequency compression could be recast as a practical implementation of information theory, a subject that he and others were to explore more intimately during the 1950s as a prelude to a more general theory of imaging; his patents on stereoscopic cinema sensitized him to the properties and limitations of three-dimensional imaging; and, as discussed later, his scheme for improving the electron microscope contained insights and mathematical formalism that donated intellectual components to the subject that was to become 'holography'. For Gabor at this time, however, these interests were separate and framed specifically in terms of practical implementations.

[13] Ernst August Friedrich Ruska (1908–88) studied at the Technische Hochschule Berlin-Charlottenburg. In 1931 he investigated the electromagnetic electron lens, and used several in series to produce the first electron microscope two years later. Siemens-Reiniger-Werke AG, which Ruska joined in 1937, brought out the first commercial electron microscope in 1939. Ruska won the Nobel Prize for improvements to microscopy in 1986. Max Knoll first described the concept of the *scanning* electron microscope in 1935. By that time, independent research groups were pursuing electron microscope development in Britain, America, Canada, France, Sweden, and Belgium.

considered as waves having a wavelength some one-hundred-thousand times shorter than that of visible light. In principle, electron microscopes can, therefore, yield images having correspondingly higher resolution. Where optical microscopes are limited to a spatial resolution of some 200 nanometres (nm), transmission electron microscopes of the early 1940s could resolve features of about 5 nm or, in the measuring units of the day, 50 angstroms (Å).[14] Despite this clear advantage, the developers realized that electron microscopes could be much better. The ratio of wavelengths is some 10^5:1, and resolution would be well below 0.01 nm if electron lenses could be designed that were as effective as glass lenses.

Stinging at his lost opportunity to have invented the first electron microscope, Gabor was interested in contributing to its further development in England. From 1943, he consequently was in close communication with Michael E. Haine of Metropolitan Vickers, BTH's sister firm. Haine was working on the development of the company's own commercial electron microscope and Gabor proffered various ideas on how to improve their design. Gabor wrote *The Electron Microscope* on the still-new instruments in 1944, concluding it with a chapter entitled 'The ultimate limit of electron microscopy'.[15]

Over a period of four years, the two developed a closer working relationship. When Haine moved to the new AEI laboratory in Aldermaston, their correspondence continued. The research station at Aldermaston had been set up in a converted country house in Berkshire in 1947—a matter of some pride for the chairman of AEI as a rare example of a fundamental research department divorced from applied research.[16] There Haine and his colleagues carried out experiments and development with occasional suggestions from Gabor. In early 1947, however, Gabor devised the dramatically novel approach that he believed could transform the imaging of electron microscopes.

2.2 HOLOSCOPY

Gabor conceived the scheme, according to his own account, while waiting for a tennis court at Easter 1947.[17] However, he much later pushed back the origins of his ideas, recounting them as a lifelong enquiry:

> I first became interested in the problems which ultimately led to holography a long time ago, at the age of seventeen, when I had learned about Huygens' Principle, and read Abbe's

[14] For an insightful analysis of the early standardization and application of electron microscopes, see Rasmussen, Nicolas, *Picture Control: the Electron Microscope and the Transformation of Biology in America, 1940–60* (Stanford, CA: Stanford University Press, 1997).

[15] Gabor, Dennis, *The Electron Microscope: Its Development, Present Performance, and Future Possibilities* (London: Electronic Engineering, 1948).

[16] Jones, Robert and Oliver Marriott, *Anatomy of a Merger: A History of G.E.C, A.E.I. and English Electric* (London: Cape, 1970), p. 252.

[17] The accounts of the genesis of his ideas were nearly all recorded many years after the event. See, for example, Gabor, Dennis to T. Raison, letter, 15 Jul. 1961, IC GABOR MN/5-6; Gabor, Dennis to I. Williams, letters, IC B/GABOR MA/13/1; Gabor, Dennis 'Holography, 1948–71', in: Nobel Prize Committee (ed.), *Les Prix Nobel En 1971* (Stockholm, 1971), pp. 169–201; Gabor, Dennis 'Holography, past, present and future', *Proceedings of the SPIE-The International Society for Optical Engineering* 25 (1971): 129–34.

> Theory of the Microscope. I asked myself a question:- 'When we take a photograph, the image appears in the plane of the plate. But by Huygens' Principle the information which goes into the image must be there in *every* plane before the plate, also in the plane before the lens. How can it be there in that uniform whiteness? Why can we not extract it?' Of course I said, white light is a complete jumble. But could we not do it at least with monochromatic light?[18]

In 1947, Gabor was familiar with contemporary thinking about such questions, which conceived of separating the imaging process into two stages: one to record, and the other to reconstitute, the image. Others, notably the Polish physicist Miecislav Wolfke (1883–1947) in 1920 and Sir Lawrence Bragg from 1939, had imagined a two-step imaging technique, using x-rays as the first step and visible light as the second. Indeed, the idea of dividing the imaging process into two parts—an optical transformation, followed by a second transformation to form the image—had been used as a conceptual convenience by the German microscope designer Ernst Abbe (1840–1905) during the late nineteenth century. Wolfke, who had been one of Abbe's doctoral students, realized that this process could pay dividends if implemented physically. If x-ray and visible wavelengths were used in the two stages, there would be an inbuilt magnification of about 10,000 owing to the ratio of their wavelengths. However, he also realized that detectors of visible light, such as photographic plates, recorded only the intensity of light, not its phase. And without information about the phase and amplitude of the light waves, the information in the wavefront of light could not, in general, be deduced.

As hinted in his 1961 recollection above, Gabor also envisaged a two-step process, but using *electron* waves and light waves. In his scheme, an electron microscope would be used merely to cast the shadow of the object. That shadow, according to physical optics, would be surrounded by fringes of optical interference, because the electrons themselves, considered as waves having a very short wavelength, would interfere constructively and destructively to yield light and dark regions. The wavefront passing very close to the object would be diffracted, or deviated, towards portions of the wave that are undeviated further away. In this way, interference fringes would ring the shadow of an opaque object. The first step of the imaging process was to record this interference pattern on photographic film. In effect, the interference pattern would encode information about both the phase and magnitude of the wavefront.

The second step was to mount and illuminate that film in a special optical apparatus. The optical arrangement would cause a beam of visible light to be diffracted by the recorded interference pattern. According to Gabor, this should reconstruct a wavefront precisely like the one that originally had recorded the fringe pattern. In the process, this reconstructed wavefront would recreate an image of the original object. This circuitous and seemingly pointless encoding and decoding of the image would be worthwhile, he

[18] Gabor, Dennis to N. Calder, letters, IC GABOR EM/7. This retelling of the origins evokes Albert Einstein's descriptions of how he developed the theory of special relativity from schoolboy speculations of travelling alongside a beam of light. Huygens' Principle, discussed further in §3.2, states that every point on a wave acts as a source of spherical waves and that the wave pattern at a later time is merely the sum of these individual wavelets.

argued, because optical lenses could correct for aberrations (image imperfections) created by electron microscope lenses. Electron microscope lenses, unlike optical lenses, could not readily be designed to overcome spherical aberration, which causes an unavoidable smearing of focused images and limits the ultimate resolution attainable. With his two-step wavefront reconstruction microscopy, however, Gabor suggested that this limitation could be overcome by designing the second half of the system—the optical reconstruction portion—to correct for such defects. He hoped that it might be possible to achieve a resolution of 1 Å (0.1 nm), to resolve images of atoms themselves.

Gabor's tennis-court revelation took time to refine in terms that others could appreciate. His labelling of the new principle, process, or technique took time to settle, too. Although he conceived the idea in early May 1947, it took two months for the new Research Director at BTH, L. J. Davies, to approve time to investigate Gabor's idea. Gabor's concept of wavefront reconstruction did not appeal immediately to his superiors at BTH. The work appeared to be primarily optical design, while the company specialized in electrical engineering design and electrical products.

Gabor and Williams began optical experiments that July and continued into 1948. Gabor's intention from the beginning was to perform a feasibility experiment on the optical reconstruction concept and then, if successful, to collaborate with personnel at the AEI Research Laboratory in Aldermaston to work on the electron microscope portion of the two-step process. Gabor intended to liaise with the Director, T. E. Allibone, engineers Jim Dyson and Tom Mulvey, and have the project overseen by his friend Michael Haine.[19]

From the outset Gabor's scheme raised questions. A fundamental concern about the concept was whether the available sources of waves—either the electron beam or visible light—would be sufficiently coherent. Two wave trains are defined to be coherent if they have the same frequency and phase. In practice, coherence is measured by the ability of two wavefronts to interfere. As physicist Adolf Lohmann puts it, 'coherence is something that describes how well two partners are willing and able to collaborate. Here [it means that] two light beams, coming from two particular directions, are willing and able to create interference fringes. The better the coherence, the higher the contrast of the fringes'.[20]

If two such wavefronts are perfectly coherent (i.e. have precisely the same frequency and well-defined phase) they will generate interference fringes when combined, ranging in intensity from zero to some maximum. Two less coherent wavefronts will generate a washed-out fringe pattern having lower contrast and non-coherent or incoherent light will not generate any visible interference fringes at all.[21]

[19] Gabor, Dennis to T. E. Allibone, letters, 1932–54, IC GABOR MA/6/1.

[20] Lohmann, Adolf W. to SFJ, fax, 19 Feb. 2004, SFJ collection.

[21] Strictly speaking, the intensity will fall to zero only if the two waves have equal amplitudes; if not, the intensity fluctuations, or contrast, will be smaller. Concepts of optical coherence were explored by Emil Wolf during the 1950s and further studied during the early 1960s after the advent of the laser. See, for example, Glauber, R. J., 'Coherent and incoherent states of the radiation field', *Physical Review* 131 (1963): 2766; Beran, M. J. and G. B. Parrent Jr., *Theory of Partial Coherence* (Eaglewood Cliffs, NJ: Prentice-Hall, 1964); and Mandel, L. and E. Wolf, 'Coherence properties of optical fields', *Reviews of Modern Physics* 37 (1965): 231.

The coherence properties of the electron beam were still not well characterized. And even obtaining such a coherent source of *visible* light was difficult enough. To describe a light source as having a well-defined frequency—sometimes referred to as *temporal coherence*—is to say that it is monochromatic. If such light is dispersed by a prism, its spectrum will be seen to consist of a single narrow spectral line. A light source having a well-defined phase (a condition described as *spatial coherence*) is also unusual. It requires that the wave be obtained from nearly a point source, where the wave train is unmixed with others. Spatial coherence measures, in effect, how the light amplitudes at two locations are related to each other. An incandescent lamp, for example, is highly incoherent. Its light consists of a wide range of frequencies, all generated by independent atoms within the hot filament. Those waves have no constant phase relationship to each other and the intensity distribution of the frequencies it produces depends on the filament temperature. The only light sources suitable for Gabor's concept were akin to neon tubes: gas lamps that generate a series of discrete frequencies or pure colours. If such a lamp were filtered to remove all but one frequency of light, and passed through a pinhole to select one clean wave train, then it could be made reasonably coherent. That is to say, if the wavefront of light were to be divided into two portions, those parts would generate interference fringes when brought together again, as long as the difference in the optical paths was not too large—certainly no more than a fraction of a millimetre. This degree of coherence (a property later dubbed the *coherence length*) was adequate to test Gabor's concept, because the two interfering portions of the wave consisted of light diffracted around the edge of an object and the other light proceeding directly to the photographic plate without diffraction. For flat or very shallow microscopic objects, Gabor reasoned that wavefront reconstruction microscopy should generate enough interference fringes to work.

The drawback of creating a coherent light source with optical filters and pinholes was that this arrangement reduces the intensity dramatically. Gabor and Williams consequently found their feasibility experiments difficult. They discovered that the best compromise between coherence and intensity could be obtained from a high-pressure mercury lamp. If the light was filtered to use only a single emission line of mercury, they could obtain a coherence length of only 0.1 mm, enough to produce about 200 fringes. To achieve the spatial coherence that they needed, they illuminated a pinhole of merely 3 microns (μm) diameter. Gabor observed that this left them with 'enough light to make holograms of about 1 cm diameter of objects, which were microphotographs of about 1 mm diameter, with exposures of a few minutes, on the most sensitive emulsions'.[22] The photographic record of the diffraction pattern, which Gabor dubbed a *hologram* from Greek roots to denote *whole picture*, was tiny and yielded a reconstructed image that required a microscope for viewing.

[22] Gabor, Dennis, 'Holography, 1948–71', in: Nobel Prize Committee (ed.), *Les Prix Nobel en 1971* (Stockholm, Imprimerie Royale PA Norstedt & Soner: 1971), pp. 169–201, p. 16.

Gabor recalled eighteen years later, 'When I started holography, neither I nor my assistant, Ivor Williams, had any experience in optics, yet we got it going in half a year'.[23] They had decided to test the principle by creating a hologram with visible light instead of the electron beam. Their first image was a chart consisting of black letters on a white background, which they reduced photographically to create a transparent microphotograph. This was then used to create a diffraction pattern, or hologram, which looked like a fuzzy, ringed version of the original—the futuristic tapestry mentioned in the *New York Times*. When the hologram transparency was itself put in place and illuminated by the optical system, they found that the nearly illegible pattern miraculously generated an image that could be viewed through a microscope eyepiece. The quality of the image was not as good as the original microphotograph, but was discernibly more legible than the hologram. But the quality of the reconstruction was marred by spurious patterns and also seemed to depend on how the hologram had been chemically processed to yield a transparency and how the final photograph was generated (a higher contrast print made it more legible). Moreover, the first half of the proposed instrument—the electron beam that was to generate the hologram—was, as yet, entirely unexplored. Nevertheless, by the end of 1947 Gabor was encouraged enough by their first results to spread the news to prominent researchers.

In December Gabor visited Cambridge to see Sir Lawrence Bragg, who had made his own reputation from the studies he and his father made of x-ray diffraction before the First World War.[24] The Braggs' work had generated a method for inferring the structure of crystals, showing that regular lattices of atoms would diffract short-wavelength light to form a regular series of spots that can be decoded to determine the atomic planes making up the crystal. During the early 1940s, Bragg had been extending his work to x-ray imaging. He recently had developed the concept of an x-ray microscope, and, with associate Charles Bunn, explored a practical implementation.[25]

The basic idea of Bragg's x-ray microscope was a double-diffraction process. A crystal lattice would diffract x-rays to form a regular pattern of spots that could be recorded on photographic film; that pattern could then be used to diffract light to form an image of the original crystal structure. However, there were severe limitations to this scheme: the x-ray diffraction pattern did not record all the information about the diffracted waves; it

[23] Gabor, Dennis to P. Goldmark, letter, 3 Jun. 1967, IC GABOR LA/9. No notebooks or equipment survive from this work, despite Gabor's own attempts to locate them for the Smithsonian museum during 1965–6. Allibone had borrowed the equipment for a time years earlier, but the keeper of demonstration equipment there could not recall the apparatus. The best holograms would have been kept with the equipment, but Ivor Williams, by then Manager of the Central Research Laboratory at AEI, kept back some for sentimental reasons. He was nevertheless unable to locate any of them for the museum (Williams, Ivor to D. Gabor, letter, 20 Dec. 1966, IC B/GABOR MA/13/1). The Gabor archives hold copies of the enlarged photographic prints made of the holograms and reconstructions (IC GABOR D/39).

[24] Sir William Lawrence Bragg FRS (1890–1971) had worked with his father William Henry Bragg (1862–1942) on x-ray analysis of crystal structure, winning them the Nobel Prize in 1915. For the next four years, the younger Bragg (the youngest ever Laureate) was Technical Advisor in sound ranging for the British army, subsequently taking up a Professorship in Physics at the University of Manchester for the succeeding eighteen years, Directorship of the National Physical Laboratory 1937–9 and then the Cavendish chair in Physics until 1953.

[25] Bragg, W. L., 'A new type of X-ray microscope', *Nature* 143 (1939): 678; Bragg, W. L., 'The X-ray microscope', *Nature* 149 (1942): 470.

saved the *intensity* but not the *phase*. Phase information, which provided spatial information about the relative position of points on the object, had to be inferred by limiting observation to a narrow class of regular objects such as crystal lattices, which altered the phase in regular steps. Bragg's x-ray microscope combined elegance with guesswork: it achieved results only for objects that diffract waves in simple, or calculable, ways.

Gabor's own concept owed some similarities to Bragg's ideas in that both were two-step processes and sought to extend the resolution of spectroscopy—Bragg by employing short-wavelength light rays (x-rays) and Gabor via electron waves. However, Gabor's concept provided more information than did Bragg's, because the hologram encoded information about not only intensity but also phase. The fringe pattern of the hologram recorded the comparison between the wave diffracted by the object and the original regular wavefront. Unlike Bragg's scheme, Gabor's holoscope could study a more general class of irregular objects.

Gabor had hinted of his ideas to others even before he contacted Bragg. Electron microscopist J. B. Le Poole wrote to him that January on behalf of the International Commission on Optics, asking for a short article on the optics of electron microscopy for a forthcoming ICO publication and enquiring about 'obtaining images from the diffraction pattern according to the method you spoke of in September'. Gabor replied two weeks later, providing the requested description of phase contrast in electron microscopy, and added, 'As regards the second idea, this is for the time being a very confidential matter, as I have not yet published anything on it. But I have filed a patent, and I am circulating a report on the theory among some British experts' (see Figure 2.3).[26]

In mid-January 1948, Gabor sent a report on the two-step method of 'electron-microscopical ideas' to Bragg and V. E. Cosslett, a researcher pursuing electron microscopy at the Cavendish Laboratory at Cambridge, and for internal circulation at BTH.[27] And for Le Poole, he composed the first description of his concept intended for a scientific audience:

> **Reconstruction of optical images from diffraction patterns obtained with diverging beams ('Holoscopy')**
>
> D. Gabor in 1947 started work on a new line which might lead to a method of electron microscopy with superior resolution. It is a two-step method in which first a diffraction diagram of the object is obtained by means of electrons, from which the original object is reconstructed in an optical apparatus. The general foundation of the new method is, that a wavefront can be reconstructed from any plane which it traverses, if in that plane systems of lines or small patches are marked out in which the phase of the wave coincides with a standard wave, transmission being restricted to those regions of phase coincidence, where it is made proportional to the intensity of the original by a photographic process. It is only necessary that these lines or patches shall extend over an area corresponding to the required aperture. If now this

[26] Le Poole, J. B. to D. Gabor, letter, 21 Jan. 1948, IC GABOR EL/1 and Gabor, Dennis to J. B. Le Poole, letter, 5 Feb. 1948, IC GABOR EL/1.

[27] Gabor, Dennis to Sir W. L. Bragg, letter, 19 Jan. 1948, Rugby, IC GABOR EL/1. Physicist (Vernon) Ellis Cosslett FRS, president of the Royal Microscopical Society, founded the Electron Microscopy section of the Cavendish Laboratory, Cambridge.

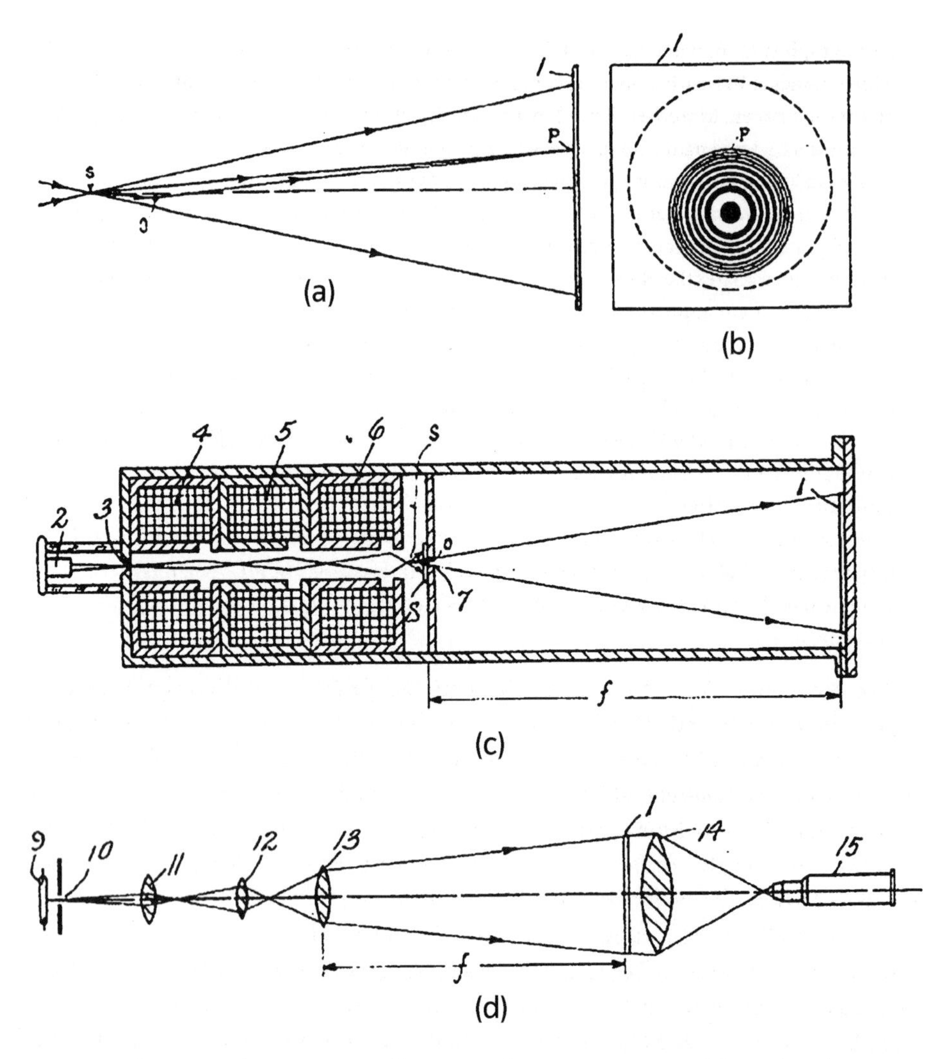

Fig. 2.3. Gabor patent illustrations. (a) Illustration of source point S, rays diffracted at object point O, and resulting interference with undeviated rays at point P. The hologram is the pattern of such interference points at surface *l*. (b) Appearance of interference pattern produced by the diverging beam and 'punctiform object' (i.e. opaque spot O). (c) Diagram of the electron-beam portion (i.e. first half) of the holoscope, which employs three electromagnetic lenses to produce a divergent beam diffracted by the object at O. (d) Diagram of the optical reconstruction portion (i.e. second half) of the holoscope, which Gabor dubbed a 'synthetizer' [sic]. The hologram *l* yields a reconstructed image visible in microscope eyepiece *15*. (Gabor, 'Improvements in and relating to Microscopy' (British Thomson-Houston: Great Britain, 1947), 9).

> photograph is illuminated by the standard wave, only those parts of it will be transmitted which coincide in relative phase and in intensity with the original wave, and the original object can be seen. Moreover three-dimensional objects may be recorded in one photograph, hence the suggested name 'holoscope' which means 'entire' or 'whole' vision.
>
> In practical applications it is suggested to make the illuminating electron wave almost stigmatic, with a Gaussian diameter of minimum confusion smaller than the resolution required. (The physical diameter can be much larger). The beam diverging from or converging into this small spot illuminates the object, which must be mounted on a thin, even membrane, which is a sufficiently good approximation of a plane parallel glass plate. The 'holoscopic' images obtained on a photographic plate at a suitable distance from the object will be in first approximation shadows of the object, but surrounded by a fine system of interference fringes, which potentially contain the whole wavefront, and its modifications by the object. The photograph is reversed or printed with certain special precautions, and introduced into an apparatus which is a sort of optical imitation of the electronic apparatus, with monochromatic light substituted for electrons.
>
> Practical work so far has been restricted to an optical model of the method, with satisfactory results. D. Gabor has succeeded in unscrambling in the second step 'holograms' which were extremely unlike the original. Preparations for the combined electronic-optical process are being made.

Gabor noted in his letter to Le Poole, 'I should be glad if you could keep this second report for yourself until March, as in the meantime I may have made more progress towards publication'.[28] This seems to be the only written reference Gabor made to *holoscopy* and the *holoscope*, although his collaborators were later to use the term from time to time. The provisional, but so far unpublished, patent specification, drafted a few months earlier in the autumn of 1947, had coined the term *hologram*, which Gabor continued to use repeatedly over the subsequent few years.[29]

Gabor and Williams continued their optical experiments during the spring of 1948, producing a new series of holograms and image reconstructions in May.[30] Gabor maintained a close correspondence with Bragg that year as the most interested and influential outside authority. On 24 March, Gabor sent him 'a short report on the first experiments, with some of the results', requesting that Bragg, as an FRS, read a paper to the Royal Society on his method. Bragg replied ambivalently the next day, 'The results are interesting. I must confess I have not thrashed out the theory of them perfectly yet, but I mean to do this when I have a little more time'.[31] Gabor demonstrated his experimental apparatus at the London Conference of the Electron Microscope Group on 7–8 April 1948 and reported to Bragg that it had had 'a very satisfactory reception' (see Figure 2.4).[32]

[28] Gabor, Dennis to J. B. le Poole, letter, 5 Feb. 1948, Rugby, IC GABOR EL/1.

[29] His contemporaries did not favour the term. Gordon Rogers (discussed later) pondered the merits of the terms 'physical shadow', 'diffraction image', or 'Fresnel diffraction pattern' with dissatisfaction.

[30] Gabor, Dennis to file, 1948, IC GABOR D/39.

[31] Bragg, W. L. to D. Gabor, letter, 27 Jul. 1948, IC GABOR EL/1.

[32] Gabor, Dennis to W. L. Bragg, letter, 19 Jan. 1948, Rugby, IC GABOR EL/1.

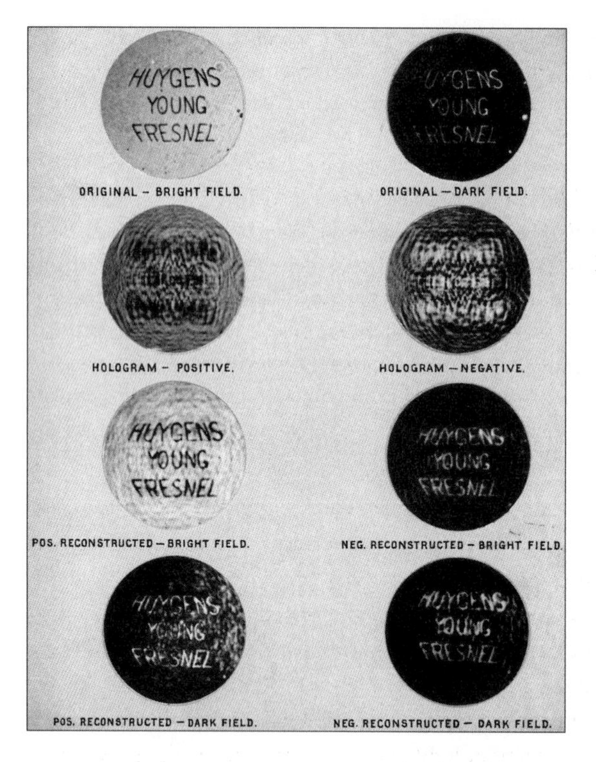

Fig. 2.4. Gabor 1948 conference poster (MIT Museum MoH collection).

2.3 'A NEW MICROSCOPIC PRINCIPLE'

Gabor had a short paper published in *Nature* in mid-May 1948 that incorporated the work he and Williams had carried out up to February. The paper was aimed at electron microscopists and framed in language and orientation that made it unlikely to be noticed by other scientists.

The principle that Gabor described was the basis for a visual optical method intended to supplement what were, by then, becoming the conventional practices of electron microscopy. Such a hybrid system was not likely to appeal either to traditional optical microscopists or to the small but growing community of electron microscopists. While electron microscopy in the late 1930s had recorded images directly on photographic plates, the latest generation of microscopes incorporated fluorescent screens so that the image could be viewed immediately and directly. A two-step process, and particularly one that required photographic processing both in the encoding and reconstruction stages, bucked the developing norms of practice.

The proposed technique did offer some intriguing possibilities, though. Like the earlier patent application and description for the ICO, the paper alluded to the exotic three-dimensional nature of wavefront reconstruction.

> It is a striking property of these diagrams that they constitute records of three-dimensional as well as of plane objects. One plane after another of extended objects can be observed in the microscope, just as if the object were really in position.[33]

Gabor highlighted what he saw as a new opportunity for microscopy: a much-improved usable depth of field, in which the image could be observed at leisure after recording. This would be a significant advantage over conventional electron microscopy, which could destroy a sample after extended exposure to electrons.

As his unpublished report and patent application had stressed, Gabor's technique was an unfamiliar and revolutionary form of imaging. It encoded and decoded the image in a two-step process in a way that made it amenable to optical correction or treatment. And it also recorded more information than did a conventional photograph: it was *three-dimensional*. But this was distinctly unlike the three-dimensional imaging that Gabor had already been investigating for television and cinema.[34] It was the apparent recreation of the object's optical properties seen by one eye through a microscope eyepiece. As the patent put it, 'viewing the source through the plate, one sees not only the source, but also the object, enlarged in the same ratio. This reconstructed virtual object may be observed or photographed by suitable viewing devices, as known in the art'.[35] But Gabor's concept was both literally and figuratively myopic: there is no hint from surviving documents that he ever linked wavefront reconstruction with stereoscopic imaging before the early 1960s.

Nor had the putative three-dimensional properties of the technique been confirmed owing to the inadequacy of Gabor's coherent light source. Furthermore, the paper intimated that the new microscopic principle had been scarcely half-tested: the optical second stage had been verified crudely, but Gabor had not yet done any work on the design of the electron microscope first-stage portion to yield a complete microscopic apparatus. Thus, the brief two-page paper failed to excite much interest amongst readers of *Nature*: the technical description was imprecise and yet overly focused on a particular community; it was tentative in its experimental verifications but optimistic in its claims; the eventual method seemed to doom microscopists to a complex, hybrid working culture that grafted optics and photography onto electronics; and the principle itself appeared counter-intuitive, suffering from a lack of easily grasped physical insights. It was a principle in limbo, promising much but, as yet, delivering little.

Hoping that a better explanation would generate more interest from microscopists and his employers, Gabor wrote to Bragg again in mid-June, excited by the intellectual and professional possibilities, to say that he had improved the 'previous rather patchy treatment' with 'a straightforward, and I think fairly complete, theory'.[36] But Gabor realized that his explanations of his concept were difficult to digest and decided that direct

[33] Gabor, Dennis, 'A new microscopic principle', *Nature* 161 (1948): 777–8, quotation p. 778.

[34] Between 1940 and 1944, Gabor had patented a series of inventions for the filming and projection of stereoscopic pictures and movies. See Tanner, P.G. and T. E. Allibone, 'The patent literature of Nobel laureate Dennis Gabor (1900–79)', *Notes and Records of the Royal Society of London* 51 (1997): 105–20, for a list and brief discussion.

[35] Gabor patent, op. cit., quotation p. 2.

[36] Gabor, Dennis to W. L. Bragg, letter, 7 Jul. 1948, IC GABOR EL/1.

demonstrations impressed his colleagues more. He arranged further shows for influential scientists,[37] including his acquaintance, the nuclear physicist Rudolph Peierls, to whom he sent a copy of the *Nature* paper along with a note:

> The short paper is the first description of a lucky find, which has made me very happy. You need not take the explanations too seriously, this tentative theory was good enough to lead me to the experiment, now I have replaced it by a more complete and coherent one. I have also perfected the experimental arrangement considerably, and now I can produce really pretty reproductions of the original from apparently hopelessly muddled diffraction diagrams. Do you still often pass by car through Rugby on your way to Cambridge? If so, I should be very glad if you could spare half an hour some time to visit me in the BTH and have a look at it [. . .]. I hope I shall be able to rouse sufficient enthusiasm in competent circles to get the funds necessary for this ambitious program.[38]

The trick of pulling a clear picture from a muddled hologram grabbed the audience. On Gabor's urging, Bragg watched and was, indeed, seduced by the experimental demonstration. He encouraged further demonstrations, writing in July, 'I think I am beginning to understand the principle, though it is still rather a miracle to me that it should work. I shall look forward very much to hearing your account'.[39]

Nevertheless, the parlour trick was hard to follow up. By late July 1948, having read Gabor's draft Royal Society paper more carefully, even Bragg was cautioning that the theory was difficult to absorb and that Gabor needed to make a stronger impact.

> I have been trying hard to understand a bit better your very exciting principle, and I think I have made some progress. My difficulty is due to my not having enough time and energy to get down to it properly, not to any lack of clarity in the exposition [. . .] The *R.S. Proceedings* should be the vehicle for really new and exciting pieces of work which appeal to a wider body of scientists than those interested in a highly specialized field. They should therefore be written in such a way that they are readily absorbed by scientists in general.

Bragg suggested the *Philosophical Magazine* or *Proceedings of the Physical Society* for the full mathematical treatment, because 'To be quite realistic, few people will read it thoroughly, but the work must be on record somewhere'.[40] Gabor grudgingly made some cuts, but in a letter a few weeks later Bragg suggested that he not waste time on further revisions until referees had had an opportunity to see it.

Even so, Gabor was eager to win supporters and asked Bragg whether he might be permitted to give preprints of the manuscript paper for the Royal Society, in the form of a BTH report, to six interested parties researching microscopy: (1) the Cavendish

[37] At this time Gabor knew, and corresponded with, Max Born, Rudolph Peierls, P.A.M. Dirac, Louis de Broglie, and Sir Charles Galton Darwin (grandson of the evolutionist); even considering the unusually rich post-war flux of émigré scientists, he was a well-connected industrial scientist indeed.

[38] Gabor, Dennis to R. Peierls, letters, 15 Jun. 1948, IC GABOR MP/2. Peierls (1907–95), a German physicist who had studied under Werner Heisenberg and emigrated to England in 1933, taught at Birmingham University before the war but was part of the British Mission for the Manhattan Project from 1943. After the war he returned to a chair at Birmingham, and subsequently to Oxford University.

[39] Bragg, Sir Lawrence to D. Gabor, letter, 5 Jul. 1948, IC GABOR EL/1.

[40] Bragg, Sir Lawrence to D. Gabor, letter, 27 Jul. 1948, IC GABOR EL/1.

Laboratory; (2) National Physical Laboratory; (3) Dr C. R. Burch, Bristol; (4) Sir Edward Appleton, Department of Scientific and Industrial Research; (5) Prof. G. D. Preston, Dundee; and (6) AEI Research Lab, Aldermaston.

2.4 MICROSCOPY BY RECONSTRUCTED WAVEFRONTS

Through such networking, and especially with the shepherding of Sir Lawrence Bragg, Gabor's concept came to the attention of a wider circle of scientists at the end of the summer of 1948. His redrafted 33-page paper—presented to the Royal Society by Bragg on Gabor's behalf in late August 1948 and published after further revisions six months later—explored possibilities at much greater length. Here the term *hologram* again made an appearance, although *holoscopy* had been retired in favour of the ungainly phrase *microscopy by reconstructed wavefronts*.

> As the photograph of a diffraction pattern taken in divergent, coherent illumination will be often used in this paper, it will be useful to introduce a special name for it, to distinguish it from the diffraction pattern itself, which will be considered as a complex function. The name 'hologram' is not unjustified, as the photograph contains the total information required for reconstructing the object, which can be two-dimensional or three-dimensional.[41]

As the abstract and text now emphasized even more emphatically, this was a *three-dimensional* imaging process.

> While the application to electron microscopy promises the direct resolution of structures which are outside the range of ordinary electron microscopes, probably the most interesting feature of the new method for light-optical applications is the possibility of recording in one photograph the data of three-dimensional objects. In the reconstruction one plane after the other can be focused, as if the object were in position, though the disturbing effect of the parts of the object outside the sharply focused plane is stronger in coherent light than in incoherent illumination. But it is very likely that in light optics, where beam splitters are available, methods can be found for providing the coherent background which will allow better separation of object planes, and more effective elimination of the effects of the 'twin wave' than the simple arrangements which have been investigated.[42]

Bragg had also been able to find a late slot for Gabor's demonstration at the September annual meeting of the British Association for the Advancement of Science (BAAS) in Brighton.[43] This brought Gabor's technique to its widest audience and culminated in the *New York Times* article that opened this chapter.

[41] Gabor, Dennis, 'Microscopy by reconstructed wavefronts', *Proceedings of the Royal Society of London, Series A* 197 (1949): 454–87, quotation p. 456.

[42] Ibid., quotation p. 486.

[43] Gabor later recalled that for his informal talk and demonstration at the Brighton meeting, BTH had 'generously paid my 3d class fare Rugby-Brighton, but the hotel for only one night, because employees were not supposed to attend meetings without giving lectures' (Gabor, Dennis to N. Calder, letters, IC GABOR EM/7).

That autumn, with the British Association meeting and the extended paper under his belt, Gabor returned to the larger project of developing the *system* of holoscopy. This project, as he had originally planned, divided neatly into two parts: the electron microscope experiments to produce a good diffraction pattern would be performed by Michael Haine and his colleagues at AEI, working under T. E. Allibone; and, the optical reconstruction apparatus, which he dubbed an 'optical synthetizer' [sic], would be designed at BTH by Williams and himself, probably with assistance from Thomas Smith, a former optician at the National Physical Laboratory recommended by Bragg.[44] Allibone helped arrange generous funding for the AEI portion from the Department of Scientific and Industrial Research (DSIR)—an unprecedented development, because the DSIR had never previously awarded such a contract to an industrial company. The grant was presumably motivated by the government's wish to help promising British technological exports during the difficult post-war period—and the electron microscope was just such a promising technology. The DSIR grant was the first government grant for what would, during the late 1960s, become a gravy train for researchers in holography.

Thus, in the space of fifteen months, Gabor had conceived a new imaging concept, had convinced his company administrators to support brief experimental tests, and had then promoted his results to an influential mentor (Sir Lawrence Bragg), and a widening collection of microscopists. With his energetic promotion and the influential endorsements of Bragg and Allibone, Gabor had published two papers, demonstrated the optical principle at the BAAS, and gained newspaper coverage and DSIR support. Together, these developments added up to new opportunities both for wavefront reconstruction and for Dennis Gabor himself (see Figure 2.5).

2.5 THE DIFFRACTION MICROSCOPE AT IMPERIAL COLLEGE

Wavefront reconstruction became a key factor in improving Gabor's career prospects. As early as 1947, he had been looking for fresh pastures. The war over, many colleagues were again on the move: back to continental Europe for some émigrés and to America for others. Post-war Britain was a country of shortages—of labour, money, raw and finished materials, and, not least, of paper for publications. Gabor had spent twelve years in the BTH Research Laboratory on a variety of projects. To his regret, and limiting his career aspirations, none of the projects had been commercialized.[45] Now in his late forties and with a record of fertile but uncommercialized ideas, Gabor was seeking new opportunities.

[44] Gabor, Dennis, 'Optical synthetizer for electron microscope', Report, British Thomson-Houston, Sep. 1948.

[45] The earliest of them, the gas discharge lamp invention that he brought to BTH in 1934, had not been pursued after 1940: it could not compete with the new fluorescent lamps then being introduced, and GEC plc, the main potential customer, eventually decided not to take up an option. An option was, however, taken up by Philips and General Electric in 1939 and 1940, respectively. See Gabor, Dennis to Appointment Committee for Chair in Electron Physics, typewritten CV, 19 Feb. 1958, IC GABOR GB/1.

Fig. 2.5. Gabor in middle age (Godfrey Argent photo; Imperial College archive)

Gabor sought the advice of his friend T. Edward Allibone, intimating as early as May 1947—within days, in fact, of conceiving wavefront reconstruction—that he was thinking of leaving BTH. Until then, Gabor had believed the company to be the best place to develop his ideas of a stereoscopic cinema system, but he now realized that the available resources there were inadequate for such a big project. More positively, he noted that the Post Office Laboratory had shown interest in his work on communication theory.[46]

So over a year later, in the midst of his first successes with wavefront reconstruction, Gabor was primed to exploit the concept. In July, he wrote to Bragg, 'I want to do my best to carry out the work myself to its conclusion. It appears that the matter is at the moment *sub judice* at the highest levels of our Company. At any rate I am determined that wherever this work will be going, I want to go with it'.[47]

Just a week later, as he was finalizing his Royal Society paper with Bragg, Imperial College in London published an advertisement for a new position, the Mullard Readership in Electronics. Gabor sounded out his friends, who unanimously advised him to apply for the post. Within days of Bragg's presentation of his paper, Gabor drafted an application for the Readership. His latest work, which he was now thinking of as the *Diffraction Microscope*, played a large part in his planned move. His initial confidantes about his career plans were people to whom he had first confided his concept. Bragg,

[46] Gabor, Dennis to T. E. Allibone, letter, 4 May 1947, IC GABOR MS/6/1.

[47] Gabor, Dennis to W. L. Bragg, letter, 7 Jul. 1948, IC GABOR EL/1.

Allibone, and Prof. S. R. Milner—all Fellows of the Royal Society—provided references. The University also requested one from his superior, L. J. Davies, whom Gabor reported as being incensed at not having been informed directly.

Gabor approached the Head of Electrical Engineering at Imperial College and member of the selection panel, Prof. Willis Jackson, about his aptitude for the post. The letter is worth quoting at length, because it reveals candidly Gabor's unusual professional strengths, his growing array of supporters, and his ambitions for wavefront reconstruction as a concept and the diffraction microscope as its implementation.

> What is Electronics, what is the Mullard Reader in Electronics supposed to teach? If it is the sort of thing that goes month after month between the covers of the American journal of this name, then I must declare that there are thousands in this country who are much better qualified to teach it than I. But a professor of Dynamics need not be necessarily a good watch-mender, and the fact that if anything went wrong with my radio, my laboratory assistant would probably mend it quicker and better than I need not necessarily disqualify me. If it is desired that the Imperial College should turn out engineers who, for instance, can be put straightaway on the development of time circuits, without having Puckle's book at their elbow, I am definitely not the man for the job. I have designed every screw in my sound compressor myself, and I made quite a bit of it with my own hands, but I left the whole amplifier to my laboratory assistant, because I knew that he could copy it quite well out of the Wireless World and Radio Handbooks.
>
> What I could teach is first principles, electromagnetism, electron dynamics, circuit dynamics, gas discharges, and what I do not know, such as feedback theory, servo systems, radar, I could learn. But I am afraid I am no longer young enough to learn the practical details. Ever since I left the University I have crammed more and more fundamental stuff in my head, more mathematics, more physics, and when it came to practical research, I could get from first principles very quickly down to the details. Of course there were always people around me who knew the routine stuff which I did not know [. . .]
>
> And most important of all, there is the new project of the electron diffraction microscope. I think I ought to inform you in confidence on the plans which are now beginning to take shape. At the Brighton meeting Sir Charles Darwin approached me with the most agreeable suggestion, that the best lens optician in the country, Mr. T. Smith FRS now retired from the NPL should do the lens design for us, paid by the NPL. Another move is, that Dr. Allibone has approached Sir Edward Appleton for money, and though this sort of thing is unprecedented, there is a chance that we shall get State assistance for scientific work in an industrial laboratory. Thus there is good hope that a sizeable team will soon work at Aldermaston on this job, with whom I want to cooperate as a paid consultant. This is a point which the Appointment Board ought to know, though they need not know the above, confidential details.[48]

From the viewpoint of the University of London and his referees, Gabor was ideally suited to his new post: He had worked in three countries, patented widely, and published in a wide range of electrical and optical domains. Gabor was duly offered the post and, with it, as he confided to Sir Charles Darwin, the opportunity to act as 'Consultant of the

[48] Gabor, Dennis to I. C. Prof. Willis Jackson, letter, 28 Sep. 1948, Rugby, IC Gabor GB/1.

A.E.I Laboratory at Aldermaston, to cooperate with them in the development of the electronic Diffraction Microscope'.[49]

Besides creating a new Electronics laboratory at Imperial College, Gabor also intended to continue his speech compression project with BTH, especially since it reflected the electronic theory aspects of his new post. At age 48, he was running at full steam.

Thus the story of wavefront reconstruction through its conception in 1947 and promotion through 1948 was that of a single protagonist, Dennis Gabor, moving between industrial and academic science and supported by a powerful ally, Sir Lawrence Bragg. But over the next decade, a handful of other investigators took up the subject and extended it in new directions. By the autumn of 1950, when Gabor had been in his Imperial College post for nearly two years, he wrote to his friend C. R. Burch at Bristol University that, despite having most of his time occupied with the writing of his lecture courses, and 'getting a laboratory going with totally inexperienced young people', a flurry of publications had demonstrated that others were taking an interest.[50] Scientific papers by Bragg, his AEI co-workers, and another British physicist, Gordon L. Rogers, had appeared next to his own in a recent issue of *Nature*. Gabor had also heard of interest at Stanford University in California. The beginning of the 1950s looked promising indeed for wavefront reconstruction as a theoretical domain and practical application.

2.6 GORDON ROGERS AND 'D.M.'

Apart from his associates at AEI, Gabor began to interact on wavefront reconstruction with Gordon Rogers. During the spring of 1948, during his first efforts to publicize his new work, Gabor had given a colloquium on wavefront reconstruction at University College Dundee attended by Gordon Rogers, who was a lecturer there.[51] Gabor also sent a copy of his BTH report on the concept to Rogers' superior at Dundee, Prof. G. D. Preston.[52] As a former PhD student of Sir Lawrence Bragg, Rogers was familiar with

[49] Gabor, Dennis to C. Darwin, letter, 29 Nov. 1948, Rugby, IC GABOR EL/2. T. Smith, a senior optician at the NPL, assisted Gabor with the optical design of his 'synthetizer' for reconstructing holograms.

[50] Gabor, Dennis to C. R. Burch, letter, 11 Sep. 1950, IC GABOR EL/1; Bragg, W. L., 'Microscopy by reconstructed wavefronts', *Nature* 166 (1950): 399–400; Haine, M. E. and J. Dyson, 'A modification to Gabor's diffraction microscope', *Nature* 166 (1950): 315–6. In their paper, Haine and Dyson proposed an improved arrangement to increase the usable field, which would reduce the requirements for electrical stability. By placing (magnetic) lenses between the object and the hologram, the diffraction pattern would be magnified and the effective resolution of the photographic film would be increased.

[51] Gordon Leonard Rogers (b. 1916) had a peripatetic early career, with lectureships at the Carnegie Laboratory of Physics at University of Dundee (1946–51); Victoria University College in Wellington, New Zealand (1952–5); in 1957 seeking jobs at Aldermaston, Nottingham, and Sheffield; a Principal Lectureship in Wave Optics at the College of Advanced Technology, Birmingham; and subsequently Professorship of Physical Optics at the rebadged University of Aston in Birmingham. The Rogers papers are archived at the Science Museum library, Imperial College.

[52] George Dawson Preston (1896–1972) specialized in metallurgy early in his career, but took up x-ray diffraction during the 1920s to study crystal structures. He gained the Chair of the Physics Department in 1943, bringing one of the first electron microscopes in the country with him from the National Physical Laboratory (NPL), and overseeing development of the technique at Dundee.

current research on improvements to microscopy, particularly Bragg's concept of an x-ray microscope. Rogers had done experimental work on Bragg's microscope and also had tried to develop stereoscopic x-ray imaging, based on his background as an ardent amateur photographer.[53] He was unusually positioned to provide useful comments on Gabor's internal report. He also gave Gabor advice on using cine camera objectives instead of microscope objective lenses and recommended improvements to Gabor's photographic technique, the choice of optical filters and photographic emulsion, and chemical processing to improve the photographic contrast.[54] Rogers was again in direct contact with Gabor about wavefront reconstruction (or specifically diffraction microscopy, which he habitually abbreviated 'D.M.') from November 1949, writing that he had been 'working hard' since reading Gabor's *Proceedings of the Royal Society* paper two months earlier.[55]

Rogers made three important contributions to the new subject of wavefront reconstruction: first, he provided a simpler and more general description of wavefront reconstruction than had Gabor; second, he published a detailed account of his experiments, which made it easier for others to repeat and extend them;[56] and third, he collaborated directly and enthusiastically with other researchers interested in extending the technique.

While Rogers became excited by the possibilities, he was critical of Gabor's general understanding and formulation of wavefront reconstruction. In particular, Rogers gradually conceived a link with physical optics that Gabor had not noticed: the hologram could be conceived as a generalized zone plate. A zone plate (see Figure 2.6) is a bull's-eye-shaped pattern of alternately opaque and transparent concentric circles designed so that the edges of the black circles diffract light towards the optical axis where they combine constructively to yield a bright spot. Unintuitively, this pattern acts much like a glass lens, producing a focused image. Unlike a lens, though, a zone plate works only for monochromatic light: its focal distance depends on wavelength.

Such optical patterns had been observed in many circumstances before, although the physical requirements were uncommon and the explanations were not always generalized.[57] Bragg, for example, wrote to Gabor in 1948,

> We were much troubled with the diffraction rings in my x-ray microscope experiments, which my assistant called 'pheasants' eyes' and you called 'peacocks' eyes'. We finally gave up photographing the diffraction image through a microscope, and made the picture direct on a

[53] Rogers, Gordon L. to files, 1942–4, Sci Mus ROGRS 4.

[54] Rogers, Gordon L., 'Comments by Dr. G. L. Rogers on Dr D. Gabor's Report B. T. H. No. L. 3696', report, Sci Mus ROGRS 6, 18 Mar. 1948.

[55] Rogers, Gordon L. to D. Gabor, letter, 12 Nov. 1949, Sci Mus ROGRS 6.

[56] Rogers, Gordon L., 'Experiments in diffraction microscopy', *Proceedings of the Royal Society (Edinburgh)* A63 (1952): 193.

[57] For example, an optical alternative to drawing and laboriously inking in such patterns was to photograph interference between monochromatic light reflections from a flat surface and a spherical lens, an arrangement noted by Isaac Newton in the seventeenth century and known as 'Newton's rings' (Wood, R. W., *Physical Optics* (New York: MacMillan, 1929), pp. 36–41). The Fresnel zone plate had been first described and coined by Lord Rayleigh in 1871.

Fig. 2.6. Zone plate (from Wood, R. W., *Physical Optics* (New York: MacMillan, 1929), p. 39).

> fine grained plate, afterwards magnifying it up. I wish I understood the theory of them. I have a vague idea that they are due to minute faults in the microscope lenses, which show up very much more when one is dealing with light from a point source (see Figure 2.7).[58]

Gabor, enthusiastic at sharing the subtleties of his technical difficulties, had quickly replied:

> The remaining pheasant's eyes, /to put the ornithology right,/ are due to dust particles and glass defects in the holograms. We have ordered 2 mm polished glass plates of more or less optical quality from Pilkington's and Ilford's have declared themselves ready to coat these with particularly dust-free emulsions, but this will take some time.[59]

These spurious patterns, claimed Gabor, were due mainly to optical imperfections in the optical components used to record the holograms. Bragg and Gabor had observed pheasant's eyes in different optical contexts and half-understood that they were caused by diffraction around point imperfections and, yet, had failed to associate them in a fundamental way with the concept of wavefront reconstruction.

While zone plates were discussed in most optical textbooks of the period, they usually served as merely a demonstration of the phase and diffractive properties of light that are important in physical optics. Most texts described a single readily observed experimental fact: that the zone plate produced a bright focus from a collimated light source because the diameters of the circles of the zone plate cause the diffracted light arriving at one particular point on the optical axis to be in phase, producing a bright focus. Less frequently reported in optics texts was the fact that such a pattern acts not only as a positive lens (yielding the so-called real focused image) but also as a negative lens (yielding a divergent virtual image seeming to emanate from a point behind the zone plate) for a

[58] Bragg, Sir Lawrence to D. Gabor, letter, 5 Jul. 1948, IC GABOR EL/1.
[59] Gabor, Dennis to W. L. Bragg, letter, 7 Jul. 1948, IC GABOR EL/1.

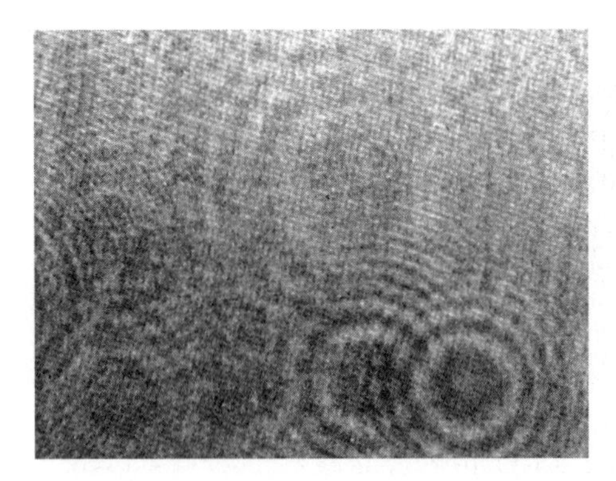

Fig. 2.7. Pheasants' eyes created by diffraction from dust particles and lens imperfections (S. Johnston photo).

monochromatic beam of light. These two images correspond to a similar troubling effect in Gabor's hologram: a primary reconstructed image and also an undesired conjugate image that overlapped it. This twin image problem, now recognized as a fundamental property of zone plates, had no obvious solution.

Rogers understood that the hologram itself is merely the superposition of many such zone plates that reconstruct the individual focused points comprising the image. His first publication on the subject was a brief note in *Nature*.[60] Nevertheless, this was an insight that developed gradually. When Rogers began his first hologram experiments in 1949, he had not yet made the connection between holograms, zone plates, and wavelength: he attempted, but failed, to discover a relationship between the focal distances of the hologram image when illuminated with two different wavelengths of light. Not until mid-April 1950 did he record his realization that a hologram obeys the lens formula.[61] Rogers also struggled to extend the description of the zone plate itself, especially the fact that conventional zone plates described in optics textbooks produced not just two foci, but a whole series. He wrote:

> There is one place where the analogy with a zone plate appears to break down. The zone plate not only has a primary focus, but a number of higher order foci corresponding to powers which are odd integral multiples of the primary power. This, it seems, arises from the practice of constructing zone-plates with abrupt changes from black to white at a sharp boundary, analogous to a square-topped wave. As is well known, a square-topped wave has a larger number of Fourier coefficients, and thus the sharper the boundaries in a zone plate, the greater the number of orders obtained. The hologram, on the other hand, produced by a continuous tone photographic process, does not have sharp alternations, but ones more closely conforming to sinusoidal variations, and thus we expect the greater part of the diffracted energy to be concentrated in the first order patterns.[62]

Later still, he and others convinced themselves that the zone plate was not merely an analogy, but the simplest case of a general relationship. Albert Baez, the following year, and Hussein El-Sum, in his PhD thesis of 1952, painstakingly verified such optical

[60] Rogers, Gordon L., 'Gabor diffraction microscopy: The hologram as a generalized zone-plate', *Nature* 166 (1950): 237–8.

[61] Rogers, Gordon L., lab notebook, 21 Jan. 1949–50, Sci Mus ROGRS 2/10 and 2/11.

[62] Rogers, Gordon L., 'Gabor diffraction microscopy: the hologram as a generalized zone-plate', *Nature* 166 (1950): 237–8.

relationships by experiment. This generalization required the mapping of concepts of information theory, or at least Fourier analysis, onto physical optics. While such concepts had a long history of interconnection going back to the late nineteenth century—and the work of Albert Michelson in America, Lord Rayleigh in England, based on earlier French work in wave optics by scientists such as François Arago, Jean-Bernard-Leon Foucault, and Armand-Hippolyte-Louis Fizeau—they were not familiar to many practising optical scientists in the mid-twentieth century.

Gabor commented on Rogers' first paper:

> I much enjoyed reading your note in NATURE. It so happened that in the meantime Jim Dyson and I, independently, also discovered the general lens property of the hologram, but you were doubtlessly the first [. . .] It is rather queer that it did not occur to me earlier that the hologram acts like an optical system, seeing that we have made use of this property in our experiments, when we illuminated it in the reconstruction from a distance different from that used in the taking. But now we are making conscious use of it.[63]

Indeed, Gabor had shown precisely this pattern in his patent application (see Figure 2.3, patent figure (a).[64]

Rogers continued to communicate closely with Gabor through the early 1950s, undoubtedly providing the intellectual incentive and critiques to refine Gabor's own ideas. During 1950, stymied by lack of progress at AEI in creating electron holograms, they considered applying wavefront reconstruction as a means of improving optical microscopy; Gabor eventually won grant money from the Royal Society Paul Instrument Fund to pursue it.[65] They also began to have misgivings about whether a single hologram could truly record the complete image, or whether two were required.[66] Over the next couple of years Gabor, with student W. P. Goss at Imperial, and Rogers alone at Dundee, pondered a variety of optical schemes to reconstruct a complete image untainted by an unfocused twin image.[67] To promote this open collaboration further, Gabor tried, but failed, to obtain a post for Rogers at AEI.[68]

[63] Gabor, Dennis to G. L. Rogers, letter, 18 Aug. 1950, Sci Mus ROGRS 6.

[64] This concept-blindness is discussed as *contextual screening* later.

[65] Gabor, Dennis to G. L. Rogers, letter, 6 Nov. 1950, Sci Mus ROGRS 6; Gabor, Dennis, 'A New Interference Microscope', mimeographed report, Imperial College, 4 Dec. 1950; Gabor, Dennis to L. J. Davis, letter, 18 May 1951, IC GABOR B/36.

[66] Gabor's ideas changed about the relationship between the conjugate image and three dimensionality: for example, he wrote, 'though, as is well known, a single hologram does not contain the full information on a general object, it entirely describes a *plane object*, hence it should be possible to obtain a *full* reconstruction, eliminating the "conjugate object", if we use the correct procedure' (Gabor, Dennis to unspecified recipient, report, 7 Jul. 1951, IC GABOR EL/2).

[67] E.g. Gabor, Dennis, 'A new interference microscope', mimeographed report, Imperial College, 4 Dec. 1950; Gabor, Dennis, ' "Diffraction microscopy. Full reconstruction by interpolation" ', report, Imperial College, 7 Jul. 1951; Gabor, Dennis, 'Microscopy by reconstructed wavefronts: II', *Proceedings of the Physical Society (London)* B64 (1951): 449–69. Goss (b. 1927) had an MSc when he was working with Gabor.

[68] In early 1951, Rogers sought Gabor's recommendation for a permanent post. Gabor suggested joining AEI, an idea Rogers supported if only for one or two years to pick up experience. Haine was unable, though, to offer Rogers a position. Gabor was not particularly surprised, but chastened that 'unfortunately the poor chap believed so much in my recommendation that he resigned his lectureship in Dundee'. He noted that Rogers would be able to find a Service job (i.e. military-related research) if he did not have moral objections

Rogers also kept up a close correspondence with Bragg about progress in implementing the diffraction microscope and collaborated on a paper that proposed an alternate solution to the twin image problem. As its introduction explained, they hoped to eliminate the effects of the unwanted image by taking a subsidiary hologram and subtracting it from the reconstruction.

> The confusion due to loss of information about *phase* in one plane, the main hologram plane, may be cleared up by measuring the *amplitude* in another plane, which we call the auxiliary hologram. In other words, two suitably separated amplitude patterns will do service for a single pattern giving both amplitude and phase. While one hologram gives two images, two holograms can be made to give one image.[69]

His idea was to make a second hologram of the conjugate image and to subtract it optically from the original by placing the two holograms in contact. However, the great precision required to register them proved problematic. Gabor's own, more complicated, schemes involving the optical subtraction or nulling of one hologram reconstruction by another (also known as optical quadrature) attracted little enthusiasm. When Gabor presented one such experimentally delicate scheme at a conference on electron microscopy at the National Bureau of Standards in Washington, DC in 1951, he attracted only comments concerning its dubious feasibility.[70] Bragg, too, remained diffident, suggesting that Gabor's solution was rather coarse for optical work.[71]

A new opportunity to promote diffraction microscopy developed in 1952, when Gabor began a collaboration with Max Born, then Professor of Mathematical Physics at the University of Edinburgh, and Emil Wolf, a young PhD, on an English-language version of Born's respected German text *Optik*.[72] The book was to be an excellent opportunity to

(Gabor, Dennis to M. E. Haine, letter, early 1951, IC B/GABOR MH 2/2). But Rogers, a devout Quaker who had planned to turn from science to 'relief work' in the closing years of the War, seems not to have favoured a Service job; his 'Radioactive dosage calculations' notebook contained an inserted 'PEACE—for war resistance and the world community' newspaper from February 1950. On the other hand, by 1965 Rogers was seeking a summer consultancy contract with the Radar Research and Development Establishment (RRDE) in Malvern, England, at £6 per day. (See Sci Mus ROGRS 4.) Thus, an opportunity for close collaboration between optical and electron microscopy researchers never developed. Rogers took up a new university post in New Zealand in 1952.

[69] Bragg, W. L. and Gordon L. Rogers, 'Elimination of the unwanted image in diffraction microscopy', *Nature* 167 (1951): 190–3. See also Rogers, Gordon L., 'Two hologram methods in diffraction microscopy', *Journal of the Optical Society of America* 56 (1956): 849–58.

[70] Gabor, Dennis, 'Progress in microscopy by reconstructed wavefronts', presented at *Conference on Electron Microscopy*, Washington, 1951.

[71] Bragg, Sir Lawrence to G. L. Rogers, letter, 25 Mar. 1952, Sci Mus ROGRS 6.

[72] Wolf (b. 1922, Prague, Czechoslovakia) had British nationality, and lived in England 1940–59, obtaining his BSc and PhD at Bristol in 1945 and 1948, respectively. He was assistant to E. H. Linfoot at Bristol from 1948 and, after an introduction from Gabor, assistant to Born in Edinburgh from 1951. In 1959 Wolf joined the University of Rochester, NY, and was promoted to Professor in 1961. The relationship between Born and Gabor, which had begun in Germany, was a close and respectful one, given the different backgrounds and status of the two men. Gabor commiserated at their over-work, suggesting that 'both you and I have chosen the wrong country to settle in after Germany' and counselling Born to seek a post in America, or at least consulting work, as Gabor himself had done that year (Gabor, Dennis to M. Born, letter, 5 Dec. 1951, IC GABOR MB/10/3). See also Wolf, Emil, 'Recollections of Max Born', *Optics News* 9 (1983): 10–14.

describe the principles of wavefront reconstruction, and Gabor eventually produced a chapter on his concept and limited experimental results.[73]

Nevertheless, as Bragg had noted earlier, Gabor's exposition did not promote his ideas with clarity. Rogers was a critic of what he saw as Gabor's excessively optimistic claims. Commenting on Gabor's draft chapter on wavefront reconstruction, he wrote, 'Fortunately, by now several people (I fancy Hopkins is one of them) have taken the snags out of the original draft, and his claims are now much more modest and reasonable. I still feel, however, that they are not very helpful, though they are no longer unsound'.[74] Born himself commented to Gabor, 'I have read more of your MS. And I think that your considerations are most ingenious. But I can at the same time not conceal that they always seem to me a little weird, and prickle my physical sensitivities'.[75]

2.7 THE CALIFORNIAN CONNECTION

Despite this unpromising reception, limited interest was developing elsewhere. First in Dundee and then in New Zealand, Rogers became a crucial conduit for information between Gabor at Imperial College and other researchers intrigued by wavefront reconstruction.[76]

Rogers struck up a correspondence with Paul Kirkpatrick of Stanford University from September 1950, after reading Kirkpatrick's proposal to apply Gabor's method to x-ray microscopy.[77] Rogers sent a box of 'assorted diffraction supplies' to get him started that December.

Kirkpatrick confided that he had no intention of pursuing wavefront reconstruction directly, preferring instead to investigate a more readily achieved goal: creating an x-ray telescope by focusing x-rays from curved cylindrical mirrors.[78] Rogers, perhaps relieved

[73] The first edition (Born, Max and Emil Wolf, *Principles of Optics: Electromagnetic Theory of Propagation, Interference and Diffraction of Light* (London; New York: Pergamon Press, 1959)), the most rigorous and canonical optics text of its generation, appeared in 1959 with Born and Wolf as principal authors. Following the irritation that both felt on the dissolution of the partnership of their publisher, Gabor decided to withdraw as co-author, providing instead chapters on diffraction and interference microscopy and on electron optics for the book. Born returned to Germany on his retirement in 1953 and a year later was awarded the Nobel Prize in Physics. The only other books of the period to discuss wavefront reconstruction were Cosslett, V. E., *Practical Electron Microscopy* (London: Butterworths, 1951) and Longhurst, Richard S., *Geometrical and Physical Optics* (London: Longmans, Green, 1957).

[74] Rogers, Gordon L. to G. D. Preston, letter, 2 Sep. 1952, Sci Mus ROGRS 6. Harold Horace Hopkins (1918–94) worked in commercial optical design during the Second World War and earned a doctorate at Imperial College, developing the first high-quality zoom lens and exploring fibre-optics imaging during the 1950s. Hopkins was a long-time, if sometimes uncomfortable, associate of Gabor at Imperial College and had collaborated initially with Born and Gabor on the book.

[75] Born, Max to D. Gabor, letter, 21 Feb. 1951, IC GABOR MB/10/3.

[76] Among the handful of researchers who made contact with him were Albéric Boivin at Laval University in Quebec and Emil Wolf in Scotland.

[77] Kirkpatrick, Paul, 'An approach to X-ray microscopy', *Nature* 166 (1950): 251; Rogers, Gordon L. to P. Kirkpatrick, letter, 13 Sep. 1950, Sci Mus ROGRS 6.

[78] Kirkpatrick, Paul to G. L. Rogers, letter, 3 Dec. 1950, Sci Mus ROGRS 6.

by the withdrawal of an experienced experimentalist and potential rival, continued to channel information between himself, Kirkpatrick, and Gabor over the next year and thereafter developed a close friendship with Albert Baez, Kirkpatrick's former student. The other investigator at Stanford was Hussein M. A. El-Sum, Kirkpatrick's PhD student, with whom Rogers appears to have had no direct correspondence until 1952.

Rogers also maintained close contact with Gabor, coaching him on collaborating with the Americans in 1951.

> If possible P. Kirkpatrick at Stanford would be a good point of call [. . .] Let me know in due course how your tour of the States is going, and how they are getting down to the technical work of making diffraction microscopy work. I bet it is a lot more slick and polished than yours truly on a shelf up in Dundee! I wonder, however, whether they are getting results any better.[79]

The experimental context was different in California, but not much better. During the early 1940s, Albert Baez had been teaching mathematics and physics at Wagner College of Staten Island, NY, as part of the Army Specialized Training Program (ASTP), rising from instructor to tenured professor in the space of four years.[80] On the advice of a friend, he moved to Stanford University in 1944, again to teach on the ASTP program with Paul Kirkpatrick, who was acting head of the Physics Department while other academics were engaged in war work. When the war ended, Baez began a PhD with Paul Kirkpatrick.[81]

Kirkpatrick himself had come to Stanford University in 1931. Having engaged in x-ray research before the war, he had been considering an idea for building an x-ray microscope. The principle was quite unlike Bragg's concept of an x-ray microscope or Gabor's scheme for an improved electron microscope. Unlike electrons, which can be focused fairly well by an electromagnetic lens (except for unavoidable spherical aberration, the fault that Gabor sought to bypass), x-rays could not be focused at all. Kirkpatrick envisaged that x-rays might be focused if reflected from surfaces at glancing incidence, which allows them to continue without absorption. In a university department impoverished by the war, Kirkpatrick and Baez pursued the project using scrap equipment. Baez (see Figure 2.8) was able to demonstrate the feasibility of constructing a focusing mirror from a slightly curved glass plate used at grazing incidence, which would produce a line image. Two such cylindrical mirrors in succession, used with their curvatures at right angles to each other, were able to focus radiation from an x-ray source to a point.

This scheme had no connection with Gabor's wavefront reconstruction concept. As they pursued their research on difficult microscopies, however, the Californian researchers

[79] Rogers, Gordon L. to D. Gabor, letter, 1 Nov. 1951, Sci Mus ROGRS 6.

[80] Baez, Albert V., 'Anecdotes about the early days of x-ray optics', *Journal of X-Ray Science and Technology* 7 (1997): 90–7. In 1951 UNESCO offered Baez a one-year position in Iraq as a physics teacher and organizer. He returned with his family to California in 1952 to resume teaching at Redlands until 1958, when he took up a position at MIT (Baez, Albert V. to SFJ, email, 13 Mar. 2003, SFJ collection).

[81] Paul Kirkpatrick (1894–1992), completed a BS in physics (1916) from Occidental College and a PhD in 1923, taught Physics and English in China 1916–18, was professor of Physics at the University of Hawaii 1923–31, and was thereafter at Stanford until 1959.

Fig. 2.8. Albert Baez, 1956 (American Institute of Physics Emilio Segrè Visual Archives).

learned of Gabor's concept and its possibilities. Baez imagined using a point source of x-rays to record an x-ray hologram. Kirkpatrick assigned another student, Hussein El-Sum, to explore this possibility as the task for his PhD dissertation. Baez took up a teaching post at a small university in southern California, the University of Redlands, with no track record in research.[82]

El-Sum studied and extended Gabor's microscopy by reconstructed wavefronts from late 1948, that is, from the time Gabor began to present his work publicly. He completed his PhD on the subject in 1952, and with Kirkpatrick published their first account of this research in an American journal the same year.[83]

Working in Paul Kirkpatrick's laboratory, El-Sum's ultimate aim was to apply Gabor's techniques to x-ray microscopy. He worked alongside members of the x-ray laboratory and Albert Baez, corresponding with Gabor and Michael Haine by post. His dissertation, like the thinking of his contemporaries, situated wavefront reconstruction firmly in the context of microscopy. However, he set his sights more widely than did the British investigators. The first 124 pages of his thesis considered the theoretical possibilities of the technique

[82] Hussein Mohammed Amin El-Sum (*c.*1925–*c.*1978) was born in Cairo, Egypt and obtained his BS degree in Physics from Cal Tech and his PhD in physics from Stanford in 1952. During the 1960s and 1970s he was active in acoustical holography through his firm, El-Sum Consultants.

[83] El-Sum, H. M. A., *Reconstructed Wave-Front Microscopy*, PhD thesis, Physics, Stanford University (1952); El-Sum, H. M. A. and Paul Kirkpatrick, 'Microscopy by reconstructed wave-fronts', *Physical Review* 85 (1952): 763.

for other types of microscopy and did so with mathematics that derived the principles in a more straightforward manner than Gabor had done. Some of his clarifications are refreshingly down-to-earth:

> The principle of diffraction microscopy can be summarized in the following way. Let the diffracted radiation be interfered with by a strong biasing uniform (background) radiation. Take a picture of the diffraction pattern. Use this picture to diffract another uniform wave to get the image of the original object.[84]

and (drawing on Rogers' insight),

> A general plane object is a collection of individual scattering points; each point yields a diffraction pattern of a zone plate, which reconstructs a single image and conjugate image. The diffraction pattern of a general object is a generalized zone plate. Its record (the hologram) will act as a positive and a negative lens to yield real and virtual images of the original object.[85]

The 75-page experimental section of El-Sum's thesis verified some characteristics of holograms that were predicted by, and yet counter-intuitive to, scientists of his generation:[86] that the hologram reconstructed two images, acting simultaneously as a converging and diverging lens; that the negative of a hologram behaved the same as the original positive hologram; and that, as Gabor had emphasized during 1948,

> it is possible to reconstruct two objects in two different planes from the same diffraction picture [. . .]. This argument can be extended to any number of objects. This means that each plane in a thick object will be reconstructed independently.[87]

In other words, as Gabor had repeatedly stressed, a hologram should reconstruct an image having *depth*. But, as Gabor had envisaged, the three dimensionality was not a matter of directly perceiving depth by two-eyed stereoscopy: El-Sum, like other contemporary microscopists, still envisaged viewing the reconstructed image either through an eyepiece or as a projection onto a surface. Thus, wavefront reconstruction continued to be shaped and marginalized according to the conception of a particular scientific community, the microscopists.

Experimentally, El-Sum (see Figure 2.9) employed different apparatus than had Gabor (e.g. a filtered incandescent lamp rather than a high-pressure mercury lamp). He also devised methods of reducing (or at least smearing) the noise from imperfections on the

[84] El-Sum, H. M. A., *Reconstructed Wave-Front Microscopy*, PhD thesis, Physics, Stanford University (1952), quotation p. 23.

[85] Ibid., quotation pp. 28–9.

[86] Gabor and El-Sum had founded their theoretical understanding of wavefront reconstruction on the Abbe theory of the microscope (1873) and Huygens' principle (1678), which explain image formation from optical interference of light waves. Most microscopists and even physicists were much more familiar with classical geometrical optics, in which light is considered in the approximation of geometrical imaging. The Abbe theory, however, had at its heart ideas linking imaging with frequency analysis and information theory, connections that Gabor and others were to elaborate from the 1950s, as discussed later.

[87] El-Sum dissertation, op. cit., quotation p. 57.

Fig. 2.9. Hussein El-Sum, 1968 (from *Proceedings of the SPIE* 15 (1968), p. 145).

optical elements by rotating the optical filter, for example, and by selecting more homogenous film emulsions.[88]

At Redlands University, Albert Baez followed the developments in Kirkpatrick's laboratory with interest. Baez was encouraged by his university administrators to introduce research by undergraduates at Redlands. The idea appealed to the Research Corporation, which provided some financial support to get started. Kirkpatrick suggested to Baez that wavefront reconstruction using visible light—the kind of experiments that Gabor had performed at BTH and Imperial since 1948—would be an ideal topic for undergraduate research, because of its low cost and ready availability of the necessary equipment. Baez hoped to piggyback on this visible optical research to extend the results to x-ray optics if and when a suitable source became available. He wrote to Rogers in early 1951,

> I have made a hologram and lensless reconstruction [. . .] It uses a Cenco 2 watt arc lamp, 24 mm microscope objective, green filter, and then object in expanding beam, shadowed onto the film. For reconstruction, it uses the same arc lamp and filter and exposed film, and then a film further away to receive the 'focused' image. I am looking ahead to 2 wavelength microscopy—using x-rays to make the hologram and light to reconstruct. I am in the process of studying the matter to see what obstacles must be overcome first. I believe that I can learn a

[88] News about the innovation spread from El-Sum to Baez to Gabor and to Rogers, who wrote: 'Before leaving London, I had a good talk with Gabor, and was delighted to hear of your very ingenious scheme for removing dust particle effects &c by rotating the lenses during exposure. The degree of accuracy required must be considerable, especially if it is desired to reach down to the theoretical limits of resolution of the process' (Rogers, Gordon L. to A. V. Baez, letter, 8 Mar. 1952, Sci Mus ROGRS 6).

lot by practicing with visible light first, but hope to jump into x-rays as soon as I understand the situation a little better.[89]

Combined with optical benches and other components already available in the undergraduate physics laboratory, the only additional item required was fine grain spectroscopic plates of the type El-Sum was using.[90] Such photographic plates were the highest-resolution film available, and allowed fine interference fringes to be recorded faithfully. As early as in the 1950 teaching term, Baez and his students were recording holograms and he was able to use the Research Corporation grant to employ five undergraduates over the summer, organizing the research much like the graduate student teams supervised by Kirkpatrick at Stanford. Baez had thus begun the first teaching of wavefront reconstruction anywhere and his activities comprised the largest research group in the subject through the 1950s.[91]

Just as the possibilities of Gabor's diffraction microscope had attracted a research grant for AEI from the DSIR a few years earlier, Baez was able to obtain further research funding for Gaboroscopy.[92] Unlike the AEI workers who were a mere 100 miles away, Baez visited V. E. Cosslett and William C. Nixon at Cambridge University, who had developed an x-ray tube having an extremely small aperture and suitable as the needed 'point source' for x-ray wavefront reconstruction experiments. Baez was able to obtain a grant from the National Research Council in Washington, DC, to purchase one of the $1000 Cosslett-Nixon tubes and to bring Nixon to Redlands for a semester to set it up, subsequently attracting grants from the American Cancer Society and the Office of Naval Research.

Baez and Bill Nixon found, however, that the x-ray source was not small enough to be deemed a 'point source'. When Nixon produced a projection radiograph of a silver grid, it showed only a single diffraction fringe. Such a hologram was too sparse to reconstruct anything like an image, and helped clarify their understanding of coherence. By comparison, the ongoing work of Dyson, Mulvey, and Haine at AEI represented a superb success; yet one which was far from adequate to maintain their enthusiasm for long.[93]

2.8 ADOLF LOHMANN IN GERMANY

Thus, Dennis Gabor's conception of a diffraction microscope attracted a handful of prepared researchers in widely separated locales during the early 1950s. The last to develop

[89] Baez, Albert V. to G. L. Rogers, letter, 16 Jan. 1951, Sci Mus ROGRS 6.

[90] Baez, Albert V., 'Anecdotes about the early days of x-ray optics', *Journal of X-Ray Science and Technology* 7 (1997): 90–7; Baez, Albert v. to SFJ, email, 13 Mar. 2003, SFJ collection.

[91] Significant publications during this period were Baez, Albert V., *Journal of the Optical Society of America* 42 (1952): 756; Baez, Albert v. 'Resolving power in diffraction microscopy', *Nature* 169 (1952): 963–4.

[92] Gordon Rogers wrote, 'I have had some correspondence with Baez who is certainly going to try D.M. [diffraction microscopy] with x-rays (he calls it "Gaboroscopy" by the way; a slight advance on "Holoscopy")' (Rogers, Gordon L. to M. E. Haine, letter, 20 Oct. 1952, Sci Mus ROGRS 6).

[93] Haine, M. E. and T. Mulvey, *Journal of the Optical Society of America* 42 (1952): 756.

an interest during that decade was a young German interested in the relationship between information and optics, a subject that Gabor himself had been studying from the mid-1940s.[94]

Adolf Lohmann (b. 1926) had been directed towards these interests almost accidentally as a teen at the end of the Second World War.[95] He learned the elements of signal processing theory from a friend, Horst Wegener, was had been selected for training in the manufacture, maintenance, and operation of newly developed Telefunken radar systems. When the war ended, Lohmann and Wegener studied physics at the nearby University of Hamburg. Both later studied for their graduate degrees there under Rudolf Fleischmann, who had been a professor at the Rëichs-Universität Strasburg, and returned to Hamburg when Strasbourg reverted to French administration. As nuclear physics, Fleischmann's specialty, was not a permitted subject until 1953, Fleischmann took up research on the optics of thin films. This subject created research topics for his students, with Wegener studying phase contrast as a means of measuring film thickness and Lohmann taking up diffraction by gratings.[96] They both gained their doctorates from the University of Hamburg in 1953, publishing a paper together that employed a signal processing approach to optics.[97]

Out of this wartime technical knowledge and post-war study in optics came a distinct analytical viewpoint. During the mid-1950s Lohmann focused on the Abbe theory of the microscope and its understanding of the image in terms of spatial frequencies.[98] According to Abbe, a microscope could be understood as a kind of optical filter. When light passes through a restricted opening the image that it produces will be unavoidably altered. The aperture of the microscope optics limits its resolution. The degradation of the image could be represented as the removal of certain spatial frequencies, because the

[94] Gabor, Dennis, 'Theory of communication', *Proceedings of the IEEE* 93 (1946): 429–57.

[95] Born in Salzwedel, Lohmann had spent his teenage years in Stade, Germany. During the last two years of the Second World War, following the general mobilization of German youth, Lohmann served as a 'FLAKhelfer' (Flieger-Abwehr-Kanone helper), or auxiliary anti-aircraft operator at night, and a high-school student during the day in Bremen. The Telefunken Doppler 39 radar taught him the principles of the Doppler effect. Lohmann's friend Wegener was a fellow flakhelfer and both were trained as anti-aircraft gunners after receiving their high school degrees. After surrendering to Allied forces in May 1945, Lohmann and Wegener were interned for a couple of months and then rejoined their families; Lohmann worked for a time at gas and electricity works near Hamburg. Following his eventual degrees, he was peripatetic, pursuing optics and holography at the Technical University of Brunswick until 1963, at the Royal Institute of Technology in Stockholm 1958–9, at the IBM Research and Development Laboratories in San Jose, CA, 1961–7, at the University of California at San Diego in La Jolla 1967–73, and thereafter at the University of Erlangen, Germany (Lohmann, Adolf W. to SFJ, fax, 11 Apr. 2005, SFJ collection; Lohmann, Adolf W. to SFJ, fax, 20 May 2003, SFJ collection; Lohmann, Adolf W. to SFJ, fax, 13 May 2003, SFJ collection; Lohmann, Adolf W. to patent attorneys, patent declaration, Jun. 1987 Haines collection).

[96] The phase contrast microscope, a means of visualizing differences in refractive index or thickness, was to yield a Nobel Prize for Frits Zernike in 1953. Diffraction gratings, consisting of fine scribed or opaque lines on glass, had been used since the 1880s to disperse light into a spectrum.

[97] Lohmann, Adolf W. and Horst Wegener, 'Theory of optical image formation using a plane waves expansion', *Zeitschrift für Physik* 143 (1955): 431–4.

[98] Lohmann, Adolf W., 'A new duality principle in optics, applied to interference microscopy, phase contrast etc.' *Optik* 11 (1954): 478 88; Lohmann Adolf W., 'The Abbe experiments as methods for measuring the absolute light phase', *Zeitschrift für Physik* 143 (1956): 533–7.

image itself can be decomposed into sinusoidal components of different frequencies. This so-called Fourier decomposition provides an elegant mathematical understanding of image formation and alteration. Abbe's theory of microscopes could be generalized mathematically to describe any kind of optical process. His generalization could also be understood as the application of familiar concepts from the field of communication theory to physical optics.[99]

After Lohmann moved to Braunschweig, where he was a junior lecturer at the Technical University, he met Heinrich Fack, an employee of the Physicalische-Technische Bundesanstalt (PTB), the central government research laboratories. He and Fack had heard of Gabor's wavefront reconstruction from an Assistant Professor's seminar report in 1953, but understood little. When Dennis Gabor had been invited to Göttingen for a seminar in 1954, the two attended for the day.[100] Lohmann recalls Gabor's talk—'filled with integrals such that the concept was drowned'—as baffling and intimidating to the audience; no questions were asked. Lohmann overheard two of the Göttingen professors referring snidely to Gabor's theory as comparable to 'onion radiation', a derided scientific claim with which Gabor had had an early association.[101]

Sensitive from his own experience to such out-of-hand rejection by arrogant academics, Lohmann remembers resolving to recuperate Gabor's work if possible, and tried analysing wavefront reconstruction in terms of signal theory.[102]Because Lohmann had done work on optical transfer functions (OTFs), he had contact with Gabor's colleague Harold H. Hopkins at Imperial College, who was well known in the field. Lohmann obtained a short travel grant and visited Hopkins, whose office was next to Gabor's, for a week in 1955. Owing to Lohmann's poor English, he spent more time with Gabor than with Hopkins.[103]

[99] Ernst Abbe (1840–1905) obtained a PhD in physics and mathematics from the University of Göttingen, and became an assistant professor of physics at the University of Jena while working part-time for Carl Zeiss, a microscope craftsman. His development of a theory of spatial filtering led to improvements in the Zeiss microscopes; Abbe later became owner of the firm and reorganized it as a foundation, instituting enlightened employment practices for its workers.

[100] Lohmann, Adolf W. to patent attorneys, patent declaration, Jun.1987, Haines collection.

[101] In 1923, Alexander G. Gurwitsch (or Gurvich), on experiments with onion roots, had hypothesized that they emitted radiation, probably at ultraviolet wavelengths, that promotes cell division in other nearby roots (see Gurwitsch, A. G. and L. D. Gurwitsch, 'Twenty years of mitogenetic radiation', *Uspechi Biol. Nauk. V.* 16 (1943): 305–34). Tiberios Reiter, a medical acquaintance of Gabor, had read such reports and inferred a more general connection with health. Gabor consequently developed a lamp that he expected would reproduce these reputed mitogenetic rays and hence this beneficial effect. An arc lamp, filtered by a coloured solution, yielded ultraviolet radiation described as useful for agriculture or for the 'destruction of pernicious tumours'. Gabor made it the subject of his first patent (German patent 578,709, 'Einrichtung zur Behandlung von lebenden Zellen mittells Lichstrahlen', filed on 7 Aug. 1927; Gabor also had it filed in France, Britain, Switzerland, and America over the following two years, for example, as US #1,856,969: 'Treating living cells with light rays'). See also Tanner, P.G. and T. E. Allibone, 'The patent literature of Nobel laureate Dennis Gabor (1900–1979)', *Notes and Records of the Royal Society of London* 51 (1997): 105–20, quotation p. 106.

[102] Lohmann, Adolf W., 'Optical single side band transmission applied to the Gabor microscope', *Optica Acta* 3 (1956): 97–9.

[103] Lohmann's paper was translated into English by Manoranjan De, Hopkins' PhD student (De, Manoranjan to SFJ, email, 31 May 2003, SFJ collection).

Lohmann suggested his own method of avoiding the twin image problem, which Gabor praised as a simple demonstration by comparison to his own complicated proposals. Gabor was enthused enough by Lohmann's work to send it to Haine and Dyson, presumably to indicate, if nothing else, that the research was not dead and buried.[104]

2.9 THE DECLINE OF DIFFRACTION MICROSCOPY

Nevertheless, by the time of Lohmann's visit in 1957, Gabor's concept had been abandoned by everyone—including, for all practical purposes, Gabor himself. For three years, 1949–52, most of the work on diffraction microscopy had been carried out at AEI by Michael Haine, by Gabor with his student Goss, and independently by Gordon Rogers in Dundee and El-Sum and Baez in California. Gabor had contact with his AEI associates through their joint research grant and urged them on at every opportunity. Writing to Haine in 1949, for example, he cajoled, 'what a pity that it was not possible to produce some sort of electron-hologram for the Delft Conference, even a little smudge would have done [. . .] I am looking forward to having a little time to think about the outstanding problems of diffraction microscopy. It is my favourite baby, and I must not neglect it'.[105] His Delft paper that year consequently rehearsed his optical results so far, looking rather hopefully at the requirements for the coherence of the light source and perfection of the optical elements.

> The objectives were full of local imperfections, and these produced a very uneven background on the photographic plate full of false detail. In order to get at the true detail it was necessary to subtract the background photographically [. . .] Even so, some 'noise' persists in the background. It is comforting to think that we shall not have to contend with this difficulty in the electronic scheme. However bad an electron lens is, it is always perfectly polished, it has neither scratches, dust nor cementing specks.

But, employing the opposite argument in his conclusion, Gabor forecast optimistically:

> Thus, it can be hoped that the diffraction principle will allow us to shift the difficulties of phase contrast, like others, from electron optics to light optics, on to the shoulders of the optician, who will be well able to cope with them.[106]

During early 1950, the prognosis from AEI had looked promising, with Haine reporting, 'Since your visit we have been concentrating very hard on the holoscope [. . .] P.S. the first

[104] Gabor, Dennis to A. Lohmann, letter, 2 Jan. 1957, London, Lohmann collection; Gabor, Dennis to A. Lohmann, letter, 5 May 1957, London, Lohmann collection.

[105] Gabor, Dennis to M. E. Haine, letter, 18 Jun. 1949, Rugby, IC GABOR MA/6/1.

[106] Gabor, Dennis, 'Problems and prospects of electron diffraction microscopy', presented at *Conference on Electron Microscopy*, Delft, 1949, pp. 4–5.

picture taken on the new system (i.e. object before objective) shows a nice 7–8 fringes over the whole of the picture with no distortion. Preliminary attempts at reconstruction of the negative show something (!)'.[107]

Those seven hard-won fringes from the electron microscope were about the best hologram achieved by the Aldermaston group. Gabor, immersed in teaching but at the peak of his enthusiasm, replied a month later, 'This *is* a reconstruction, there is no doubt about it [. . .] I have no doubt that the same hologram with somewhat better photography would give an even better result'.[108]

Haine and his colleagues had not yet given up. In fact, they had convinced themselves that they could get by with rather few visible fringes from their electron beam, as Haine reported to Rogers:

> The question of the visibility of the last fringe which you mention is one which worried us for some time but has now been cleared up. The question is better put in terms of communication theory. The random contrast (graininess) of the photographic plate can be considered as noise (it turns out to be almost completely random (white). The question is, how much of the fringe system can be of such contrast that it is lost in the noise and yet still lead to reconstruction? The conclusion we arrived at (Gabor also) was that the whole fringe system could be below noise level. You will appreciate that this means 'lost' as far as any local measurement over a small area of the system is concerned or better, 'lost' as far as could be determined with incoherent illumination.[109]

So the seven visible fringes might be accompanied by others submerged in the noisy background. But despite this theoretical breathing space, no practical improvement in results could be coaxed from the AEI apparatus.

By summer, Haine reported that further work was making slow progress owing to vibration, mechanical drift, and temperature variation of the electron microscope. Looking forward, Gabor predicted that to surpass the resolution of conventional electron microscopy by a factor of ten would require either that the electron beam be made some 100 times more monochromatic (thereby starving the image of energy) or by using an intermediary optical system of considerable complexity.[110]

Leaving the problems of electron microscope stability to the Aldermaston group, Gabor worked on optical schemes for improving the quality of reconstruction from his microscopic holograms. But, following an extended visit to America the following year,

[107] Haine, Michael E. to D. Gabor, letter, 14 Mar. 1950, Aldermaston, B/GABOR MH 2/2.

[108] Gabor, Dennis to M. E. Haine, letter, 14 Apr. 1950, Aldermaston, IC B/GABOR MH 2/2.

[109] Letter M. E. Haine to Rogers 6 Oct. 1952. Rogers replied, 'No doubt, you will by now have studied a cyclostyled M.S. by Gabor on the whole subject, intended for Born-Wolf's optics, which I have worked through with profit. The analysis there satisfactorily clears up the question of fringe visibility. The analogy with communication theory is particularly clear: coherent observation being analogous to using a tuned receiver which can pick up a periodic signal much below the noise level as "seen" by an aperiodic amplifier.' (Rogers, Gordon L. to M. E. Haine, letter, 20 Oct. 1952, Sci Mus ROGRS 6). This tantalizing connection between information theory and wavefront reconstruction was to be developed more profitably by others, notably Lohmann in Germany and Emmett Leith in America, about four years later.

[110] Gabor, Dennis, 'Problems and prospects of electron diffraction microscopy', presented at *Conference on electron microscopy*, Delft, 1949.

he returned to dispirited co-workers, and wrote to Haine:

Dear Michael,

I was very distressed by my visit yesterday. I understand that for the time being you want to drop diffraction microscopy. My feeling is that this would mean, in all probability, dropping it for good. I do not know whether the decision corresponds or not with the lapse of the A.E.I.-D.S.I.R. contract, and anyway, I have no influence on that policy. All I can do, and want to do, is to impress on you that it would be unwise to terminate all work on it at the present stage, which would be interpreted by everybody outside as the admission of complete failure.[111]

Correspondence between Gabor and Haine became more sporadic and by 1954 Gabor was frustrated to the point of desperation and wrote in confidence to Allibone at AEI.

Dear Edward,

I am writing this to you as to a friend. I feel that I have to do something about Diffraction Microscopy, because my scientific reputation is at stake, on the other hand I am in a rather worried state of mind, and there is the danger that I might damage myself by a too rash or harsh action. So if you think that the attached letter can do any good, please take official notice of it, and pass the copy on to Michael Haine, otherwise tear up both copies and let me know what you think of the situation.

The attached letter read:

I am considerably worried about the state and outlook of the work on Diffraction Microscopy. The impression has got abroad that it has led to disappointing results, and has been or is about to be abandoned. As I do not like being the bright boy who produces brilliant dud ideas, I should like to know a little more about the present position, and to give my help if possible.

Mr Dyson told me that he gave his opinion on the optical side of the Diffraction Microscopy method in a report, but I have not seen this. As regards the electronic side, I am almost equally in the dark. I have not yet seen any photographs taken with coherent electrons with half-tone objects; only black and white. I have made suggestions how to make use of the weak fringes in the shadow zone, but I do not know what has come out of these. Moreover I have not seen any photographs taken with coherent illumination for a long time; I think that the last I have seen were taken before the Washington Conference, in Nov 1951. I understand that the last year was spent mainly improving the electron microscope on orthodox lines, but I have not yet seen any diffraction photographs taken with the improved instrument.

Are there any further difficulties, apart from those which Mr. Haine has published, and described *in extenso* in his last report on this subject, from which I had the impression that a resolution of 2.5 A still appears obtainable?

You will understand that I am very anxious about these questions. I am in constant danger of overwork, yet if it has to be, I am willing to undertake the optical work myself, if this is the bottleneck, that is to say unless there are difficulties on the electronic side of which I am not sufficiently aware.

With kind regards,
Sincerely Yours,
D. Gabor[112]

[111] Gabor, Dennis to M. E. Haine, letter, 26 Nov. 1952, IC GABOR MH/2/2.
[112] Gabor, Dennis to T. E. Allibone, letter, 22 Mar. 1954, IC GABOR MA/6/1.

Allibone, down with a cold, peevishly suggested a visit, because 'Your letter reads as though you are almost ignorant of our present position,—which you should not be as a consultant on this work'.[113] The subsequent visit seemed to resolve nothing, however. Haine was not interested in further seemingly fruitless diversions from their incremental improvements to conventional transmission electron microscopy.

Gabor kept trying to reinvigorate the cooperation at AEI. In an information-packed post-Christmas 1955 letter to Haine, Gabor enthused that his own work at Imperial was going well:

> In the optical microscope we are now quite near to reconstruction [. . .] we have the two part-beams exactly parallel and capable of interference. We have also an almost-perfect quadrature prism, and most of the ghosts are eliminated. It took unearthly patience, but my chap Goss has got it![114]

Gabor continued work at Imperial College with assistant W. P. Goss between 1951 and 1956 on the more complex optical reconstruction apparatus. Gabor's implementation was a form of Mach-Zehnder interferometer (see Figure 2.10), an optical arrangement invented by Ludwig Zehnder in 1890 and reinvented a year later by Ernst Mach, who used it to study supersonic sound waves.[115] While capable of displaying exquisitely subtle variations in optical density, the interferometer was painfully difficult to adjust. Adolf Lohmann, who was offered a position with Zeiss in 1952, was to have begun by developing methods and advice for customers who bought their Mach-Zehnder interferometers. He studied the theory of the device, and recalled curtly:

> The two lenses, two reflectors and two beam splitters had highly interrelated kinematic adjustment rules. And at those times the coherence length of the mercury or sodium light were fairly short. I preferred to remain in academia. [. . .] The experimental implication is a nightmare.[116]

Beyond the instrumental complications, Eric Ash, working on the design of electron lenses for correcting the aberrations in electron microscopes as one of Gabor's graduate students from 1949 until 1952, has suggested that experimental engineering was not Gabor's forte:

> He was one of only two or three people I have met that I would describe as a genius. He was warm, but he was an awful supervisor. He had difficulties communicating on the same wavelength as ordinary mortals. Secondly, he was clearly a physicist, although he had been brought up as an engineer. He had the illusion that he was an engineer, which is the inverse of what

[113] Allibone, T. E. to D. Gabor, letter, 26 Mar. 1954, Aldermaston, IC Gabor MA/6/1

[114] Gabor, Dennis to M. E. Haine, letter, 28 Dec. 1955, IC GABOR MS/6/3; Gabor, Dennis, Pat. No. 2,770,166 'Improvements in and relating to optical apparatus for producing multiple interference patterns' (1956), assigned to National Research Development Corporation, UK.

[115] Zehnder, Ludwig, 'Ein neuer Interfernz-Refraktor', *Zeitschrift für Instrumentenkunde* 11 (1891): 275–85.; Mach, Ernst, 'Modifikation und Ausfuhrung des Jamin-Interferenz-Refraktometers', *Ann. Akad. Wiss (Wien)* 28 (1891): 223–4.

[116] Lohmann, Adolf to SFJ, fax, 3 Apr. 2005, SFJ collection.

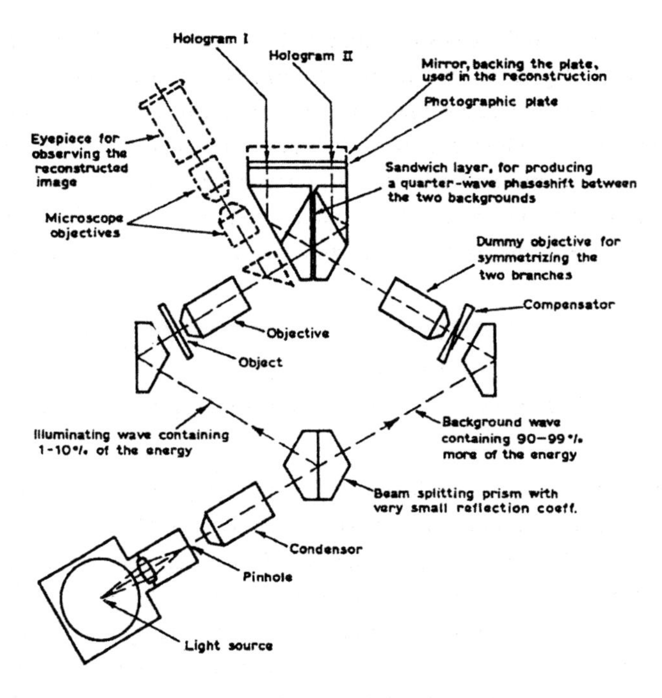

Fig. 2.10. Gabor's two-hologram interference microscope of 1951 (from Gabor, 'Interference microscope with total wavefront reconstruction', *Journal of the Optical Society of America* 56 (1966), 849–58, p. 852)

one expects [. . .] He was absolutely hopeless on experiments [. . .] For many years I enjoyed the reputation of being the only chap who got a Ph.D. out of him.[117]

Through this period, Gabor, Rogers, and Bragg hoped that a practical implementation of the diffraction microscope could be achieved and even commercialized. Writing to Gabor in 1951, Rogers encouraged further promotion, asking whether anyone had 'tried to "sell" the thing, and if so have you got patents on reconstructors!!'.[118]

But Gabor's commercial hopes for the quadrature method could not be sustained. Although he had filed British and American patents on the method in 1951, his experiments with Goss eventually convinced him that the method was impracticable. Gabor noted a decade later that they had performed few experiments with the apparatus because, by the time they had learned how to make good quadrature prisms, Goss's scholarship had nearly run out.[119] Gabor confided to Gordon Rogers, 'There is just no

[117] Nebeker, Frederik, 'Eric Ash, Electrical Engineer, an oral history conducted in 1994 by Frederik Nebeker, IEEE History Center, Rutgers University, New Brunswick, NJ, USA.' www.ieee.org/organizations/history_center/oral_histories/transcripts/ash.html, accessed 10 Nov. 2004.

[118] Rogers, Gordon L. to D. Gabor, letter, 1 Nov. 1951, Sci Mus ROGRS 6.

[119] As a consequence, their paper served as a historical priority claim. Gabor observed that 'we started this work too early—ten years before the invention of the laser' and that the most important innovation in the interim was the use of diffuse radiation by Leith and Upatnieks. Gabor, Dennis and W. P. Goss, 'Interference microscope with total wavefront reconstruction', *Journal of the Optical Society of America* 56 (1966): 849–58, p. 561.

business in this sort of thing, I am sorry to say. I wish I had not wasted so much time on patent applications myself!'[120]

In the same way, the complementary electron interference project, intended to generate electron wave holograms, was effectively dead at AEI. By 1958, Allibone publicly narrated the work in historical terms, dismissing it as one of many white elephants that the company had produced.

> We spent a great amount of time investigating this idea, solving very many different problems in sequence, such as keeping the specimen free from contamination for half an hour and free from vibration to the order of 1 Å and holding the voltage constant to 0.1 V in 100 000 V for half an hour, but the best holograms we could produce failed to give us a reconstructed image as good as the image we could then achieve by direct microscopy and we were obliged to drop the work. To that extent, therefore, it can be regarded as having been unsuccessful, but out of it we have learned so much about microscopy that the E.M. 6 has been produced capable of a resolution of 5 Å.[121]

For eight years from 1954, the bulk of Gabor's consultancy work for AEI focused on the theory of nuclear fusion, while diffraction microscopy was shelved quietly by both sides. Indeed, acrimony over its brief history flared only once in the interim, following a 1961 article by Gabor in *New Scientist*. Allibone wrote to Gabor to complain that he had misrepresented the failure of wavefront reconstruction.

> It said that 'the results were spectacular, but unfortunately trivial disturbances, such as vibrations and stray magnetic fields, have proved so far an insuperable drawback'. Surely this is absolutely wrong; if I have remembered correctly, the failure of the whole principle of electron microscopy by reconstructed wavefronts was the confusion caused by the second image [. . .] In other words, if we overcame those trivial disturbances we could still not make a success of electron microscopy by reconstructive wavefronts. I do think it is important that this misconcept be adjusted . . .[122]

Gabor, in a detailed letter to his friend five days later, replied that he was in complete disagreement with Allibone, Haine, and Dyson at AEI. Any remaining opportunities for collaborative research dissolved when the Aldermaston research station was closed in 1963, transferring its work to Gabor's old laboratory at BTH in Rugby and to Metrovick in Manchester.[123]

[120] Gabor, Dennis to G. L. Rogers, letter, 23 May 1958, Sci Mus ROGRS MS 1014/11. Gabor later publicly claimed prejudice on the part of microscope manufacturers, describing his microscope as 'an interesting device which could not only take three-dimensional photographs, but could also observe objects with 1/10 or even 1/50 of the light which is necessary in ordinary instruments. This was offered in 1956 to British, German, and American firms but nobody wanted to make it. Before the invention of the laser there was no interest in holography' (Gabor, Dennis to *Sunday Times*, letter, 14 Apr. 1968, IC B/GABOR/11/111).

[121] Allibone, T. E., 'White and black elephants at Aldermaston', *Journal of Electronics and Control* 4 (1958): 179–92.

[122] Allibone, T. E. to D. Gabor, letter, 13 Sep. 1961, IC GABOR MS/6/3.

[123] Allibone, then aged 60, was reappointed as scientific advisor to AEI, but he soon left for a position as chief scientist at the Central Electricity Generating Board. Jim Dyson became Superintendent of the Light Division at NPL, Tom Mulvey (b. 1922) eventually rose to the position of Professor of Electronic Engineering at Aston University, and Ivor Williams became Manager of the AEI Central Research Lab in Rugby and Haine its Director.

The other researchers were following a similar trajectory of pessimism and dismissal. Gordon Rogers, who had been the most enthusiastic and consistent investigator of diffraction microscopy in Britain, also had found his faith in Gabor's concept waning. Like Gabor, he had had suspicions as early as 1950 that the technique was fundamentally limited and that a single hologram could not reconstruct an unambiguous complete image, suggesting that the limitation probably followed from the second law of thermodynamics.[124] Their separate attempts to find a theoretical and practical solution through pairs of holograms were inconclusive even by 1952. Though continuing to explore the technique himself sporadically over five years Rogers had given up on optical applications. He confided his pessimism to Gabor in 1954, particularly about the AEI experimental problems, the lack of interest from other scientists, and the strong constraints on the kind of microscopic object that could be examined by a diffraction microscope.[125] Gabor remained optimistic at that point, replying,

> You write that 'Diffraction Microscopy seems quite dead as a method for aiding the electron microscope'. If it is temporarily dead, it is so by neglect and for no other reason. Dyson just could not take any interest in it, and Haine was discouraged by his attitude. I wish they had taken you at that time! Now I have undertaken to do the reconstructions myself. It may have been a lightheaded promise, because I am up to the neck in work, but I cannot leave this job in the state it is.[126]

Yet everyone else had had enough. Discussing the problem of the seemingly unavoidable twin image and the serious degradation caused by dust and imperfections in the optics, Rogers reiterated to Albert Baez in 1956, 'As far I am concerned, I am quite happy to let Diffraction Microscopy die a natural death. I see relatively little future for it, and am looking forward to doing something else'.[127] His final attempt to salvage something from the technique was a proposal during 1955–7 for analysing ionospheric measurements according to the ideas of wavefront reconstruction.[128] Kirkpatrick and El-Sum published their own work in the mid 1950s,[129] but Baez ceased his own research and teaching on Gaboroscopy when he left Redlands University in 1958 to take up a post at MIT. While Hussein El-Sum continued to publish, only a handful of others showed brief interest, such as a scheme for

[124] Rogers, Gordon L. to S. L. Bragg, letter, 5 Sep. 1950, Sci Mus ROGRS 6 and Rogers, Gordon L. to A. Boivin, letter, 1951, Sci Mus ROGRS 6.

[125] 'The object must be *very small* compared with the "gaps". [. . .] Letters don't do so badly, ditto microscope scales, but more exacting objects would be tricky'. (Rogers, Gordon L. to D. Gabor, letter, 8 Nov. 1954, Sci Mus ROGRS 6). Rogers' point was that Gabor's concept required objects occupying only a tiny portion of the optical field so that an adequately intense undiffracted wave could mix with the diffracted portion to yield interference.

[126] Gabor, Dennis to G. L. Rogers, letter, 25 Oct. 1954, Sci Mus ROGRS 6.

[127] Rogers, Gordon L. to A. V. Baez, letter, 19 Jul. 1956, Sci Mus ROGRS 6.

[128] Rogers suggested that the ionosphere could be mapped by substituting radio waves, rather than electron waves, as the first stage of wavefront reconstruction: 'The principal advantage of the method lies in the ability to change wavelength half-way, and in its applicability to radiations not easily focused [. . .] Use is here made of a third advantage, ability to correct for the aberrations in the electron process.' (Rogers, Gordon L., 'Diffraction microscopy and the ionosphere', draft paper, Jan. 1957, Sci Mus ROGRS 4).

[129] Kirkpatrick, P. and H. M. A. El-Sum, 'Image formation by reconstructed wavefronts I. Physical principles and methods of refinement', *Journal of the Optical Society of America* 46 (1956): 825–31.

holographic optical information storage devised that year by an engineer at IBM in California.[130] Tellingly, even Gabor left his work with Goss unpublished until 1966.[131]

Thus, wavefront reconstruction had had a shaky ten-year run and a decisive termination. The new subject appeared to outsiders to have been evaluated fairly, but found wanting on intellectual grounds. To most microscopists, it seemed arcane, complex, and unpromising, in part because of Gabor's expository style. Yet Gabor had worked hard to develop interest in his concept and used it effectively to advance his career. He had gained crucial support during the early months from Sir Lawrence Bragg, Gordon Rogers, and T. E. Allibone. Gabor's colleagues at AEI and his student W. P. Goss had struggled to obtain experimental results, and the practice and theory had been extended further by Hussein El-Sum, Albert Baez, and Adolf Lohmann. Collectively, these researchers had competently defined the nature and the problems of wavefront reconstruction.

What were the boundaries of their new subject? All of them constructed it as a new form of *microscopy*: an imaging technique for microscopic samples. There are at least four explanations for this perceptual pigeon-holing: first, Gabor's conception had begun with the problems he perceived for electron microscopy; second, his holoscope had formal similarities to preceding instrument concepts (the Abbe theory of spatial frequencies and the Bragg x-ray microscope); third, he had promoted wavefront reconstruction specifically to microscopists via demonstrations and papers; and fourth, those who took an interest in Gabor's work were themselves seeking to improve microscopy. This constraining view of the subject thus followed from its disciplinary origins and perceptions of its similarity to earlier research.

There was a second constraint on how they conceived the boundaries of their subject, namely their implicit assumptions. As the writings of Gabor and El-Sum indicate, both were well aware that wavefront reconstruction would record three dimensions of a sample and, yet, neither ever mused about stereoscopic imaging. This was natural, considering their labelling of the technique as a microscopy. Microscopes had associated traits that may have seemed inescapable: they were traditionally optical devices centred on a single axis and used an eyepiece for viewing. Such physical limitations could well have hindered ideas about more creative geometries. Less obviously, they may have sensed subliminally that the limited coherence length of their light sources confined the sample size to microscopic dimensions. So, in practice, stereoscopy was not even conceivable for them. Historically speaking, such speculations are unproductive; we know only of those ideas that were evaluated and committed to print, rather than those that never were imagined.

The most important conclusions drawn by this band of investigators concerned the drawbacks of their new subject in clear terms: most saw it as fatally flawed by the

[130] El-Sum, Hussein M. A., Pat. No. 3,083,6155 'Optical apparatus for making and reconstructing holograms' (1960), assigned to Lockheed Aircraft Corp; Kay, Ronald H., Pat. No. 2,982,176 'Information storage and retrieval system' (1961), assigned to IBM Corp.

[131] Gabor, Dennis and W. P. Goss, 'Interference microscope with total wavefront reconstruction', *Journal of the Optical Society of America* 56 (1966): 849–58. In 1964, after the explosion of interest in America, Gabor had obtained American funding to wrap up the project (Gabor, Dennis and W. P. Goss to report, 1964, US Army Contract (European Research Office) DA-91-591-EUC-3886 OI 652-1251, at IC GABOR GD/2/5).

twin-image problem, in which the fuzzy second image seemed doomed to overlap the desired image, rendering the technique unsatisfactory for any practical use. Yet, at least three other technical reasons were to circulate for the demise of wavefront reconstruction: limitations of the electron source (Allibone and Gabor); complexity and inadequacy of optical solutions for removing the conjugate image (Haine, Rogers, Bragg, Lohmann, Gabor, and Goss); and, later, the inadequate coherence of the available light sources (Gabor).

In their various ways, these evaluations were constrained by their investigators' histories, backgrounds, and working contexts. Their disciplines and intellectual starting points—their working context—screened them, creating perceptual barriers that restricted conceptions of the new subject.[132] The power of this *contextual screening* is clear when we see that Dennis Gabor, a highly creative inventor with direct recent experience in both stereoscopic imaging and information theory, failed to make connections between these subjects and his work on wavefront reconstruction. From the standpoint of Gabor and the others, the intellectual environment of wavefront reconstruction had been thoroughly explored; they could scarcely recognize the barriers imposed by their working cultures or the intellectual territory that might offer other routes.[133]

There the matter rested in Britain while, in the Soviet Union and America, little-known work—one a diversion to pursue a research project for an advanced degree and the other a by-product of military research—was getting under way. In the years during which the British and Californian momentum had dissipated in pessimistic evaluations and mutual recrimination, others were reinventing similar ideas from separate starting points and constructing the subject along crucially different lines.

[132] Another contemporary example of this effect in modern optics was the discovery of phase contrast about which Frits Zernike wrote, 'On looking back to this event, I am impressed by the great limitations of the human mind. How quick are we to learn, that is, to imitate what others have done or thought before. And how slow to understand, that is, to see the deeper connections. Slowest of all, however, are we in inventing new connections or even in applying old ideas in a new field' (Zernike, Frits, 'How I discovered phase contrast', *Nobel Lecture, Nobel Prize for Physics* (1953)).

[133] Two other expressions of this effect are, first, Ludwig Fleck's notion of a 'thought collective' and, second, Terry Shinn's description of narrow-niche technologies as being 'embedded' in a specific application. Fleck imagines a style of thought to be created through education and training by a narrow social group (Harwood, Jonathan, 'Ludwig Fleck and the sociology of knowledge', *Social Studies of Science* 16 (1986): 173–87). Shinn describes the repeated dis-embedding and re-embedding of a technology in a new context as a characteristic feature of a research-technology (Shinn, Terry, 'Strange cooperations: the U.S. Research-Technology perspective, 1900–1955', in: B. Joerges and T. Shinn (eds), *Instrumentation: Between Science, State and Industry* (Dordrecht, 2001), pp. 69–95). I argue here that, after seeming dead ends as a contextually screened (or 'embedded') subject, holography began to cross scientific and cultural boundaries to become a research-technology.

3

Wave Photography in the Soviet Union

By the late 1950s, the original workers in wavefront reconstruction had moved on to other research. Dennis Gabor was promoted to professor at Imperial College, studying problems of nuclear fusion and writing about the interactions between science and society. The AEI team focused on improving their commercial electron microscope design. Paul Kirkpatrick was pursuing glancing-incidence reflectors as the basis of an x-ray telescope and Albert Baez and Gordon Rogers were newly ensconced in academic posts that gave them little time, or immediate interest, for the problems of holograms.

At about this time, though, something like the subject of wavefront reconstruction was being reinvented in a different context. A lone researcher was beginning vaguely similar explorations at the principal Soviet optical research centre in Leningrad.

3.1 THE VAVILOV STATE OPTICAL INSTITUTE AND THE BACKDROP OF SOVIET SCIENCE

To an extent unmatched in other countries, scientific research was highly centralized in the Soviet Union. During the 1920s and 1930s, Russian science and technology had been transformed by *Sovietization*. St Petersburg, Russia (renamed Petrograd in 1914, Leningrad in 1924, and reverting to St Petersburg in 1991) had been a centre for research in physics and astronomy from the end of the nineteenth century and indeed had become the centre for Russian scientific culture. Peter the Great had founded the Russian Academy of Sciences there in 1724–5, based on the French model established sixty-eight years earlier. From its origin, the Academy was a state institution, with salaries paid to forty full members as in the Académie Française des Sciences. Over the next 200 years, the original building on the banks of the Neva River was joined by museums and other institutes.

After the 1917 Russian Revolution, the abandoned estates of the Russian aristocracy were also transformed into space for the growing laboratories, research units, and institutes. The Academy of Sciences was expanded to include technologists as well as scientists, and its membership was increased to ninety academicians, along with some seventy honorary and 300 corresponding members. Between 1929 and 1937, the number of students at universities and technical schools increased by a factor of five. The number of higher educational institutions rose at a similar pace, from about 300 to over 2000. Soviet historians referred to the growth in Russian science and technology between 1922 and 1928 as the Scientific-Cultural Revolution.[1] As historian Loren Graham notes, the Soviet government (and subsequent Soviet historians of science) saw science as playing a key role in society and interpreted science and politics as interwoven.[2]

During this reorganization, the locations of administration and research altered. When the national capital was moved to Moscow in the early post-revolution years, Petrograd/Leningrad remained important for scientific activities because of its university, Polytechnic Institute, the Physico-Technical Institute and the State Optical Institute (having the Russian acronym GOI).[3] Historically, most of the scientific activities in the Union of Soviet Socialist Republics (The Soviet Union) were concentrated in Moscow, Leningrad, and Novosibirsk. From 1934, when the Academy of Sciences was moved formally from Leningrad to Moscow to make it directly subordinate to the government, the original concentration of scientific activities at Leningrad were extended and diversified at a number of other centres, including Kiev, Kharkov, and Dniepropetrovsk in the Ukraine and Sverdlovsk in the Urals.[4]

Leningrad became the major centre for optics, but GOI, the State Optical Institute, was founded at an unpropitious time. The Russian Revolution initially exacerbated the economic and social chaos that had been begun by the First World War and the country's partial occupation by Germany. On the other hand, the Revolution created opportunities for enthusiastic scientists. As early as December 1918, immediately after the end of the First World War and in the early months following the Revolution, the new Commissariat of Education established the GOI institute as a model for institutions in the new Soviet Union. In 1922, the now consolidated Soviet state allocated a large fund for the purchase of equipment abroad, and GOI became the largest optical research centre in Russia and one of the best-equipped institutes of the country.

Intended to serve the various needs of the State, the Institute was organized to study six branches of optics: (1) spectroscopy; (2) opto-technology; (3) optical chemistry (especially the chemistry of optical glass); (4) photometry; (5) photography; and

[1] Medvedev, Zhores A., *Soviet Science* (Oxford: Oxford University Press, 1979), pp. 10, 14, 30. By contrast, Soviet historians referred to the 1960s as the 'Scientific-Technical revolution' (*Nauchno-Tekhnicheskaia Revoliutsiia*, or NTR), describing its key characteristics as the technological application of science as a direct productive force, a worldwide phenomenon and one expected to hasten the development of socialism (Roberg, Jeffrey L., *Soviet Science Under Control: The Struggle for Influence* (Basingstoke: MacMillan, 1998)). This period has been recognized more commonly in the west as a continuation of post–Second World War interactions between government, industry, and science characterized by the establishment of 'big science' and what President Eisenhower labelled the Military–Industrial Complex. See, for example, de Solla Price, Derek, *Little Science, Big Science* (Washington: Columbia University Press, 1963); Galison, Peter and Bruce Hevley (eds), *Big Science: The Growth of Large-Scale Research* (Stanford: Stanford University Press, 1992); Weinberg, A. M., *Reflections on Big Science* (Boston: MIT Press, 1967). Chapters 4 and 6 discuss the consequences of these large-scale trends for American optics.

[2] Graham, Loren R., 'Russian & Soviet Science and Technology', *History of Science Society Newsletter* 18 (1989): 1.

[3] Boyarchuk, A. A. and L. V. Keldysh, 'From a physics laboratory to the Division of General Physics and Astronomy', *Physics Uspekhi* 42 (1999): 1183–91.

[4] Crowther, J. G., *Soviet Science* (London: Kegan Paul, Trench, Trubner & Co., 1936).

(6) pyrometry. In 1931 it employed 240 workers and played a major part in developing a domestic optical industry in Stalin's Soviet Union. By 1936, applied physical optics was also becoming an important line of study as a separate department of the Institute. Its activities mirrored and extended those taking place in other centres, notably standards laboratories in France, Germany, America, and England.[5]

In the department of physical optics the director, Vladimir P. Linnik, invented new types of interferometer and sought to improve optical microscope design. In the early 1930s, Linnik had done basic research on X-ray diffraction and interferometry. His research interests consequently had intellectual links with the earlier work of Ernst Abbe and with the ongoing studies of Bragg and (later) Gabor in England. The work of Linnik's colleagues ranged from the theoretical to the practical: Alexander Lebedev (director of applied physical optics) worked on polarization interferometers for chemical applications, J. Ossipov fabricated interferometer plates and M. Romanova measured the metre in terms of the wavelength of light.[6] During the Second World War and the siege of Leningrad, the Institute focused on the supply of optical instruments to the Soviet army and navy, and on camouflage design. After the war, the GOI was named after one of its most prominent workers, S. I. Vavilov. Its Director from 1932 to 1945, Vavilov had focused his own research on luminescence and non-linear optics.[7] During this period, researchers at GOI had produced internationally recognized work ranging from the discovery of the Cerenkov radiation from subatomic particles (Pavel A. Cerenkov) to innovative telescope designs (Dmitri D. Maksutov).

From the end of the First World War, then, the Vavilov Institute had grown to become a highly coherent optical research and production centre without parallel in either Britain or America. Such centres had received a strong international impetus by the First World War, during which the weaknesses in national optical industries were identified. Yet, the intensive concentration on all aspects of optics—particularly the combination of physical optics, spectroscopy, and emulsion chemistry—made the GOI quite unlike state organizations in

[5] The Bureau des Poids et Mesures, Physikalisch-Technische Reichsanstalt, National Bureau of Standards and National Physical Laboratory were among the inter-war research centres most concerned with applying physical optics to practical applications. From the 1890s, the wavelength of monochromatic light had been identified as the basis for metrological measurements; see, for example, Michelson, Albert A., *Studies in Optics* (Chicago: University of Chicago Press, 1927).

[6] Crowther, op. cit.

[7] Sergei Ivanovitch Vavilov (1891–1951) was a prominent optical physicist, administrator, and popularizer of science. Like a number of his contemporaries, Vavilov divided his activities between Leningrad and Moscow, some 400 miles away. An Academician from 1932, he administered the State Optical Institute and the Biological Institute in Moscow and was a member of numerous committees. The recipient of several Soviet prizes, Vavilov was President of the USSR Academy of Sciences 1945–51, and a deputy of the Supreme Soviet of the Russian Federation from 1938 to 1947. See Kojevnikov, Alexei, 'President of Stalin's Academy: the mask and responsibility of Sergei Vavilov', *Isis* 87 (1996): 18–50. Like many scientists of his generation, Vavilov was immersed in a fluctuating political environment. His elder brother, Nikolai (1887–1943), was also an Academician and biologist but became a foe of Trofim Lysenko, the autodidact agriculturalist who rose to power in Stalin's Soviet Union. Nikolai Vavilov consequently was arrested in 1942 and accused of undermining socialist agricultural reforms and died of pneumonia in prison a year later. During this period (September 1941–January 1944), GOI personnel were largely evacuated further back from enemy lines, while the citizens of Leningrad, including the adolescent Yuri Denisyuk, were under siege by Nazi forces.

other countries. For example, the National Physical Laboratory (NPL) in Britain, founded in 1899, was tasked with standardization and applied research for the benefit of British industry; its optical work was limited to photometry, optical instrument design, and some spectroscopy. It seldom strayed into theoretical studies or modern physics.[8] Similarly, the Bureau of Standards in Washington, DC (later known as the National Bureau of Standards (NBS) and renamed the National Institute of Standards and Technology (NIST) in 1988), founded in 1901, also focused on economically useful applied technology with relatively little work in physical optics. By contrast, activities at the Vavilov more consistently ranged from basic scientific research to applied optics and to fabrication and testing to meet changing demands of the Soviet, and eventually post-Soviet, economy.

After the Second World War, most branches of science and technology that had military orientation or applications received high state priorities. The Academy of Sciences established new research institutes and, as in the west, the direct government funding for science increased dramatically. The Academy grew to an enormous and complex organization that dominated not only the pure sciences, but the applied sciences and technology as well. Over the following three decades, the Institute expanded to become the largest optical institution in the world. By the late 1970s the S. I. Vavilov State Optical Institute had affiliates and laboratories in other Soviet cities and employed some 12,000 employees, about 800 of whom had, or were seeking, advanced degrees. Unlike their American and British counterparts, most practising Soviet scientists were associated with a research institute and senior research workers were also affiliated with universities for teaching and supervision of research students. This influence of researchers on advanced teaching increased after the Second World War: the Moscow Physical-Technical Institute, for example, was founded in 1951 with research scientists in charge of education.[9] Zhores Medvedev characterizes scientists and technologists during this time as the new privileged elite. The number of students nearly doubled compared to the pre-war levels and 'almost all demobilized soldiers who had a secondary education were absorbed by the enlarged network of higher technical schools and universities'.[10]

3.2 YURI DENISYUK AND HIS *KANDIDAT* RESEARCH

One such worker was Yuri Nicholaevitch Denisyuk, who joined the Vavilov Institute in 1954. Born in Sotchi, on the Black Sea, in 1927, he grew up in Leningrad and obtained his

[8] On optics at the NPL and NBS, see Johnston, Sean F., *A History of Light and Colour Measurement: Science in the Shadows* (Bristol: Institute of Physics Publishing, 2001), Chap. 5.

[9] Graham, Loren R., 'The place of the Academy of Sciences system in the overall organization of Soviet science', in: J. R. Thomas and U. M. Kruse-Vaucienne (eds), *Soviet Science and Technology: Domestic and Foreign Perspectives* (Washington DC: National Academy of Sciences, 1977)

[10] Medvedev, op. cit., quotation p. 44. The new institutes often were established within the Academy system because of the higher status it conferred, providing higher staff salaries and locations in attractive major cities such as Leningrad, Moscow, or Kiev.

first degree from the Department of Physical Engineering at the Leningrad Institute of Fine Mechanics and Optics in May 1954. Rather than fulfilling his boyhood hopes of pursuing fundamental physics, however, he began work at the Vavilov Institute that year, where he was to stay for the following thirty-four years. During his first seven years at the Vavilov, until December 1961, he worked under Alexander E. Elkin in the field of optical instrumentation for the Soviet Navy, 'occupied', as he later recalled, 'with very dull work relating to the development of conventional optical devices consisting of lenses and prisms'.[11]

Like many of his contemporaries among the large wave of post-war technical workers, Denisyuk decided to pursue an advanced degree. From 1958, Elkin provided time for Denisyuk to do research for a *kandidat* thesis and recommended another colleague in the laboratory, Dr Eugenii Iudin, as supervisor.[12] Although Iudin died not long afterwards, Denisyuk was able to continue his studies without formal supervision, although overseen and supported by Elkin, over the following two and a half years (December 1958–June 1961). He was supported by a small stipend and well supplied with material resources purchased for the ongoing submarine research at the Vavilov Institute. The title of his eventual thesis (in translation) was *On Reflection of Optical Properties of an Object in the Wavefield of the Radiation Scattered By It.*[13]

This research dealt with some of the questions that also had interested Dennis Gabor, although Denisyuk, working and thinking in a very different context, was initially unaware of any connection. The theme of his dissertation was the complete reproduction of all the optical properties that constituted an image. By recording and recreating the wavefield of light, he hoped to develop a more complete form of photography.

The inspiration for Denisyuk's ideas is unclear. As he worked independently, there were no colleagues to corroborate his early explorations and his motivations cannot be independently reconstructed. Just as Gabor recounted his own moment of inspiration on a tennis court or alternately recalled childhood musings, Denisyuk subsequently proffered his own origin tale. He recalls having been inspired for his thesis work by the writings of the Russian science fiction author Ivan Antonovich Yefremov. Indeed, two

[11] Denisyuk, Yu N., 'My way in holography', *Leonardo* 25 (1992): 425–30, p. 425.

[12] Denisyuk was confirmed as a *Kandidat* in 1964, and 'Doctor of Physics and Mathematics' in 1971. The 'Candidate of Science' degree, a more rigorously assessed qualification than the western PhD, is its rough intellectual equivalent. Those studying for *kandidats* were commonly salaried research institute workers rather than graduate students on a scholarship and the degree is ordinarily obtained after several years of practical experience. The 'Doctor' degree normally required a second and lengthier thesis and was even more widely and openly evaluated by peers. At least until the 1970s, the Doctorate normally was achieved only in late middle age and by only some one-sixth to one-tenth of *kandidats*. The system of qualification had been made more complex and multi-staged during the Khrushchev period, when Denisyuk was pursuing his first wave of research. As in the west, the award of an advanced degree greatly facilitated promotion to a senior administrative or scientific post and had an even more exaggerated effect on salary. See Medvedev, op. cit., pp. 78–83 and Rabkin, Yakov M., 'Science studies as an area of scientific exchange', in: J. R. Thomas and U. M. Kruse-Vaucienne (eds), *Soviet Science and Technology: Domestic and Foreign Perspectives* (Washington DC: National Academy of Sciences 1977).

[13] Denisyuk, Yu N. to SFJ, email, 16 Apr. 2003, SFJ collection; Denisyuk, Yu N. to SFJ, email, 17 Apr. 2003, SFJ collection.

of Yefremov's stories are plausible triggers. Published after the war, they provided a potent mixture of scientific idealism, advice to would-be Soviet scientists, and—most curious of all—a remarkably detailed discussion of unusual optical devices that are central to the plots.

In 'Shadow of the Past', for example, Yefremov recounts how palaeontologists working in Central Asia find dark slabs of petrified rosin in which pictures from prehistory—ghostly images of dinosaurs and cave dwellers—are mysteriously recorded. To understand the effect, the chief palaeontologist studies a text on the theory of colour photography and discovers a nineteenth-century letter from early photographer Joseph Niepce to Louis Daguerre that seemed to provide the key.

> Nikitin let out a muffled cry and pressed his clenched fists to his temples, as if to lock in his head the sudden thought that had struck him. 'When the image obtained was examined under a definite angle of falling light, a very beautiful and interesting sight was to be seen. This phenomenon should be placed in connexion with the Newton effect of coloured rings; it is possible that some definite part of the spectrum acts upon rosin, creating minute variations in the thickness of layers . . .'. Thus the precious thread explaining the ghost of the dinosaur continued to unravel itself. Thin and fragile at first, it gradually became more and more firm and trustworthy. Reading on, Nikitin discovered that the structure of the smooth surface of photographic plates changes under the action of stationary light waves, and that these waves produce definite colour images, as distinct from the usual black and white images produced by the chemical action of light on photographic plates covered with silver bromide. These imprints of complex reflections of light waves, which remain invisible even when strongly magnified, have one peculiarity—they reflect light of only a definite colour and when the prototype is lighted under a strictly definite angle. The sum total of these imprints produces a marvellous image in natural colours.[14]

Another story, 'Stellar Ships', also set in contemporary Central Asia, described the discovery of an ancient artefact left by alien visitors. The story's culmination was the palaeontologist's unravelling of its optical secret:

> He placed the mysterious instrument, or part of an instrument, beneath the glare of a special microscopic bulb and turned it about, trying to detect some detail of construction that had so far escaped detection. Suddenly, in the circle of the reverse side of the disc, he caught a glimpse of something that showed faintly under the opaque film. With bated breath, he strained his eyes to see what that something was, changing the angle of the disc. And then, through the cloudy film which time had imprinted on the transparent substance of the circle, he saw, or imagined he saw, a pair of eyes that looked straight up at him.

[14] Yefremov, Ivan Antonovich, 'Shadow of the past', in: I. A. Yefremov, *Stories* (Moscow: Foreign Languages Publishing House, 1954), pp. 34–5. Yefremov (1907–72) grew up in St Petersburg. Orphaned early, he went through the Russian Civil War as the protégé of a Red Army Regiment and, later, of Academician P. P. Sushkin, a palaeontologist. Yefremov, too, became a palaeontologist and, from 1944, a writer of science fiction. Both 'Shadows of the Past' and 'Stellar Ships' are semi-autobiographical, with the characters Nikitin and Davidov, respectively, having careers very like his own and set in the contemporary Soviet Union. The stories thus provided Denisyuk with an inspirational and highly relevant message directly from a prominent scientist in his own city. Interestingly, Dennis Gabor, like Denisyuk, was an admirer of Yefremov's fiction; see, for example, Gabor, Dennis, *Inventing the Future* (London: Secker & Warburg, 1963), pp. 208–11.

After polishing the surface of the heavy 5 in. diameter disc, the two scientists decided that it was a special kind of optical lens that produced a three-dimensional picture from one side.

> From the deep bottom of the absolutely transparent layer, magnified by some mysterious optical device to its natural proportions, there looked out at them a strange but an undoubtedly human face. Its dominating feature was its huge prominent eyes, which looked straight ahead. They were like pools reflecting the eternal mystery of creation, glowing with intelligence and an intense will, like two powerful rays directed through the brittle wall of glass into the endless vistas of universal space.[15]

Denisyuk describes this passage, seemingly a prescient description of three-dimensional imagery from a hologram, as a strong motivation for his research to create such an image by modern optics. Whether they were an incentive to explore or a subsequently identified similarity, Yefremov's stories unquestionably capture the essential character of Denisyuk's later developments.

From the outset, then, Denisyuk was concerned with the problem of general imaging. In 1958, he 'carried out some investigations to develop image display devices which could reproduce an absolute illusion of the presence of the objects displayed'.[16] Realistic three dimensionality was a feature that conventional photography could not produce. Initially he took up the ideas of Gabriel Lippmann, a turn-of-the-century French physicist who had proposed a form of three-dimensional photography based on a special aperture system consisting of small lenses.[17] The fly's eye arrangement of these 'integral photographs' had theoretical shortcomings, however: in order to obtain high-quality near-continuous shifts of parallax as the observer moved, the lenses would have to be made as small as possible; however, as lenses were made smaller, their individual resolution diminished, thereby constraining image quality. The fly's eye arrangement could only ever yield a compromise three-dimensional imaging system, trading off three dimensionality for blurriness.

Through 1958 and into 1959, Denisyuk studied the theoretical problem at a more fundamental level. His initial ideas were guided by Huygens' Principle. Writing in the seventeenth century, Huygens had considered how waves of light propagated and reasoned that, at any point, a wavefront could be thought of as an infinite collection of point sources that radiated light spherically. The combination of those spherical waves, adding constructively where the crests met and destructively where a crest met a trough, would yield a new wavefront. By this mental construction the propagation of light could be followed moment by moment and imagined as a stepwise process.

[15] Yefremov, Ivan Antonovich, 'Stellar Ships', in: I. A. Yefremov, *Stories* (Moscow: Foreign Languages Publishing House, 1954), pp. 258–9.

[16] Denisyuk, Yu N., *Fundamentals of Holography* (Moscow: Mir, 1984), p. 64.

[17] Lippmann, Gabriel, 'Epreuves réversibles photographies intégrales', *Comptes Rendus de l'Académie Français* 146 (1908): 446–51. Besides investigating such schemes for 'integral photographs', Lippmann (1845–1921) won the Nobel Prize for Physics in 1908 for his method of reproducing colours photographically based on the phenomenon of optical interference, discussed later.

Denisyuk imagined this as a key for recreating any image. Put another way, any arbitrary wavefront of light could be generated by positioning a series of spherical sources at the appropriate positions in space or, equivalently, generating the appropriate phase. In very general terms, Denisyuk recalls realizing that the wave field of light (i.e. the interacting waves extending through space) emanating from an object could be reproduced in two fundamentally different ways: either by a system of 'phase controllable coherent emitters' or as a 'passive structure'.[18] The former system imagined creating a wavefront from individual sources having controllable intensity, phase, and frequency, rather like a radio transmitter and a network of antennas. Such a scheme appeared to Denisyuk to be clearly unfeasible. A faithful recreation of a wavefront would require thousands or millions of such emitters. Any practical combination of such emitters would, even more than the fly's eye lenses, be limited in number and so would constrain the observer to extremely poor definition and to a limited number of discrete perspectives. And coherent sources of light (having controllable phase) were limited to filtered emission lamps. On the other hand, the passive method—some kind of optical device that could be constructed to shape or control a wavefront of light—was worth investigating further.

Denisyuk began to reason that the full optical information about a complex object could be recorded by combining its light with that from a reference wave. From the outset, Denisyuk focused his thinking on standing waves in space: a situation in which the phase and amplitude of the wave field was constant with time and position. Such a situation could be produced via two counter-propagating waves. Where the two waves overlapped, the result would be a standing wave—an unchanging field having fixed intensity and phase at every point—which he could record as a two-dimensional slice on photographic emulsion. Just like Huygens' explanation of an arbitrary wavefront, a photographic slice of the wavefield should capture information that could then be used to generate the wave anew at a later time. Gabor had conceived something similar, albeit in another geometry and with very different goals in mind.

Initially Denisyuk envisaged trying, like Gabor before him, to record a cross-section of this wavefield of light as an interference pattern in a thin emulsion of photographic film. But unlike Gabor, he arranged the reference wave to be transmitted toward the object wave, instead of coming from the same direction. This imposed a serious limitation, because it meant that the photographic plate would be nearly parallel with the wavefront. In this configuration, instead of the relatively large interference fringes that Gabor had recorded, Denisyuk's fringes would form surfaces nearly parallel to the emulsion. As a result, the interference peaks would be separated by the crests and troughs of the wave, that is, half a wavelength of light, or about 0.25 μm, apart. Realizing that the wave crests

[18] Denisyuk, Yu N., 'The work of the State Optical Institute on holography', *Soviet Journal of Optical Technology* 34 (1967): 706–10, quotation p. 706. This article provides an historical account of Denisyuk's early researches and serves as a priority claim for (a) formulating the technical requirements for general imaging; (b) the simple and elegant reflection technique, which subsequently was explored by various American researchers (see §7.3), and (c) the combination of the Lippmann, Gabor, and Leith-Upatnieks techniques (discussed in Chapter 4) of imaging into a more general and practical form.

would be about half a wavelength apart, he sought a photographic emulsion considerably *thinner*: about 0.1 μm thick in order to avoid recording merely a muddy smear. This was considerably thinner than any available emulsion.[19] Denisyuk was unaware of Gabor's work at this time, and did not consider other arrangements of the two interfering beams of light that would have produced a more practicable spacing of the interference fringes. Another practical worry was whether variations in the emulsion would disturb the phase and amplitude of the wave travelling through it and prevent recreation of the wave field.[20]

Denisyuk had other concerns, too. He realized that when the reference wave propagates in a direction nearly opposite the wave reflected from the object, the result would be an interference pattern filling a volume of space. He was unsure, though, whether the interference pattern recorded in the slice of space occupied by the film emulsion could reconstruct the entire wavefield. Even if such an interference pattern could be recorded on the flat surface, it was unclear how it could act as a source of spherical waves to recreate the wavefront. Huygens' principle seemed to provide no guidance as to the appropriate approach to follow. He was therefore at an impasse. Uncertain theoretical foundations matched his experimental difficulties.

Denisyuk discovered, however, that Gabriel Lippmann previously had considered some of the theoretical and experimental problems that interested him. Besides his fly's eye system for recording three-dimensional images, Lippmann had devised a method of colour photography in the 1890s based on the interference of light waves, a concept that appeared to have much in common with the problem that Denisyuk envisaged.[21] Indeed, Lippmann's method was essentially that described by Yefremov in 'Shadow of the Past'. It involved placing a thick photographic emulsion in contact with a mirror consisting of liquid mercury. Light focused by a camera lens would travel through the emulsion and then reflect from the mercury surface back into the emulsion from the other side, setting up a standing wave for each of the wavelengths in the light source. This standing wave was recorded as an interference pattern through the entire thickness of the emulsion.

[19] Denisyuk, Yu N., *Fundamentals of Holography* (Moscow: Mir, 1984).

[20] Denisyuk, Yu N. to SFJ, email, 3 May 2003, SFJ collection.

[21] Lippmann, Gabriel, 'La photographie des couleurs', *Comptes Rendus de L'Academie des Sciences* 112 (1891): 274; Lippmann, Gabriel, 'Photographie des couleurs', *Journal de Physique* 3 (1894): 97–106; Lippmann Gabriel, 'Sur la théorie de la photographie des couleurs simples et composés', *Journal de Physique* 3-e serie 3 (1894): 97–107; Lippmann Gabriel, 'Colour photography', 1908 Nobel lecture, in: *Nobel Lectures, Physics 1901–1921* (Amsterdam: Elsevier, 1967); Lippmann, Gabriel, 'Epreuves réversibles photographies intégrales', *Comptes Rendus de l'Académie Français* 146 (1908): 446–51. While Lippmann developed a reasonably practical method of colour photography based on the technique, and indeed was awarded the 1908 Nobel Prize in Physics for it, some of the essential findings were published a half-century earlier. John Herschel and Edmond Becquerel had been the first to state that a 'volume photograph' (made with a thick emulsion) could reproduce the colour of images even without dyes, although they did not explain the effect (Becquerel, E., 'De l'image photographique colorée du spectre solaire', *Annales de Chimie et de Physique* 22 (1848): 451; Herschel, J., 'On the chemical action of the rays of the solar spectrum on preparations of silver and other substances', *Philosophical Transactions* 130 (1840): 1–59). Lord Rayleigh later suggested that the phenomenon might be explained as the recording of standing waves in the emulsion, which would selectively reinforce the original wavelength after the plate was developed (Rayleigh, Lord, *Philosophical Magazine* 7 (1887): 149). For a good treatment of the evolution of ideas relating to Lippmann's colour process and its explanations, see Connes, Pierre, 'Silver salts and standing waves: the history of interference colour photography', *Journal of Optics* 18 (1987): 147–66.

The recording of these multiple wave crests was not a hindrance, but rather an advantage: after suitable chemical development of the monochrome film, the emulsion would act simultaneously as a photographic image—being dark where the light had been intense and focused—and also as a colour-selective filter, because each standing wave in the emulsion would selectively reinforce a single wavelength. The result was a full-colour rendition of the original image with the full spectrum of wavelengths in the original source (at least so far as the emulsion was sensitive to those wavelengths).[22]

Thus, Lippmann photography was superior in colour rendition to experimental three-colour techniques of that period, but suffered from two serious disadvantages. First, the thick, fine-grained emulsion was very slow or insensitive making long exposures necessary even in bright light and, second, the colour reproduction varied when viewed at angles not exactly perpendicular to the film, because the effective thickness of the standing-wave pattern changed with viewing angle.

Denisyuk recalls being encouraged by reading Lippmann's published confirmation that the method reproduced the full spectrum of the original light source when his special photograph was illuminated with white light (i.e. light incorporating the full visual spectrum); it suggested that this spectral property of the Lippmann photograph was only one of the general properties of volume interference photographs. He suspected that a generalization of the technique could record other properties of wave fields, namely the phase and amplitude distributions. Lippmann's scheme of using a thick emulsion appeared to be the key to recording the shape of a wavefront.

Just as importantly, Denisyuk was able to adapt the practical details of Lippmann photography to his new generalization. He conceived his first experimental objects as a generalized version of Lippmann's liquid mercury mirror: he would use spherical convex mirrors, producing a spherical interference pattern in space. As he initially conceived his new method, Denisyuk concluded that it would create a structure in the emulsion that was identical to the optical properties of the original object. Thus he had identified a candidate for the passive structure he was seeking: it would be an unusually thick photographic emulsion that would record interference layers that modelled the surface of a reflecting object.

Denisyuk began to prepare optical equipment and the necessary processing chemistry from mid-1958. He also studied French to be able to read Lippmann's original papers. By December 1958, he had begun his first experiments to test the ideas. He recalls that, although the suppliers of Navy funds for submarine research 'did not count money', he was undemanding: 'my experimental set-up was so simple that I could do it using my stipend and friendly connections in the Institute'.[23] He employed a filtered mercury lamp

[22] The layers corresponding to each colour do not overlap in quite the simple way described above. As Lippmann showed more generally, the structure of these layers in the emulsion corresponded to the Fourier transform of the original spectrum. Lippmann colour photography therefore provides a faithful and objective reproduction of the optical frequencies in the original source; every other form of colour photography is based on three dyes that combine to obtain an approximately correct subjective colour sensation for the average human eye.

[23] Denisyuk, Yu N. to SFJ, email, 16 Apr. 2003, SFJ collection. He added that, several years later, financial support from the Navy and other sources made possible very large reflection holograms, but 'it was not a

to produce monochromatic light (as Gabor had done a decade earlier, and as, in fact, researchers in interference optics had conventionally been doing since the 1880s). His subjects were distinctly unlike those of Gabor: instead of microscopic black and white transparencies, Denisyuk selected shallow spherical convex mirrors, which he positioned behind a photographic plate. This was a close analogue of the Lippmann colour process, but highly simplified: he was using monochromatic light instead of white light, a curved mirror instead of a flat mirror, and a parallel (collimated) beam instead of a focused image. Light from the mercury lamp passed through the emulsion (thus serving as what was dubbed the 'reference' wave) and then reflected from the mirror back to the emulsion (becoming the 'object' wave). The resulting standing-wave interference pattern was recorded through the depth of the thick emulsion.

Despite a seemingly straightforward experimental extension of Lippmann's method, Denisyuk's initial work until early 1959 was discouraging. He had looked to Lippmann's writings to recreate or further develop the chemistry to develop these special plates, but found that the high-resolution plates needed to record the very closely spaced fringe patterns of the standing waves were far too insensitive to record the fringe pattern. Where Lippmann had been able to make his exposures in bright sunlight, Denisyuk was restricted to a dim, highly filtered lamp. Denisyuk's sources of information—early twentieth-century papers by Lippmann and, later, those of the German investigator E. Valenta and American optical scientist Harold E. Ives—indicated that they must have acquired tacit knowledge about photographic emulsions that they did not clearly communicate in their published works.[24]

Denisyuk collaborated with Dr Rebekka R. Protas, a specialist in silver-halide emulsions at the Vavilov Institute, to improve the sensitivity of the plates a thousandfold. Together, Denisyuk and Protas experimented on some 200 emulsion formulas, adjusting the chemistry and processing to alter silver content, size of the silver-halide grains, and emulsion thickness to further optimize and hypersensitize the emulsion. There were subtle requirements of the emulsion far beyond those of regular photography: it had to maintain a nearly unchanged thickness after it was developed so as to avoid shifting the colour of the reproduced image and the size of photosensitive grains was crucial to the recording of the standing waves and to the resulting fringe quality. They further discovered that the additional chemical steps of gold sensitization and triethanolamine hypersensitization made the emulsion faster while leaving its resolution unaffected.[25] The photochemical expertise at the Vavilov Institute was crucial to the project.

From 1959, Denisyuk began using the improved emulsion to record standing-wave patterns from his shallow mirrors and found that his scheme did, in fact, work. Once the

science. Science developed in a very small room with help from [an unskilled assistant], Vera Rongonen' (Denisyuk, Yu N. to SFJ, email, 16 Apr. 2003, SFJ collection).

[24] Ives, Harold E., 'An experimental study of the Lippmann color photograph', *Astrophysical Journal* 27 (1908): 323–32; Lippmann, Gabriel, 'Photographie des couleurs', *Journal de Physique* 3 (1894): 97–106; Valenta, E., *Die Photographie in natüralichen Farben* (Germany: Halle, 1912).

[25] Denisyuk, Yu N. and R. R. Protas, 'Improved Lippmann photographic plates for the record of standing light waves (in Russian)', *Optika i Spektroskopija* 14 (1963): 721–5.

photographic plate was processed, it reflected collimated mercury light just as the original spherical mirror had done, producing divergent light. His small flat plates acted just like the original convex mirrors, showing that 'a spatial photograph of the standing wave pattern does in fact provide a surprisingly complete reproduction of the recorded wave field'.[26] And just as Rogers and Baez had carefully checked the properties of their Gabor holograms, Denisyuk verified the characteristics of his own type. Denisyuk's plates did indeed act as mirrors, but a peculiar kind of mirror that changed its focal length with wavelength. This colour-dependent property was similar to Gordon Rogers' studies of zone plate holograms made nearly a decade earlier.

Denisyuk heard of this prior work tardily, as he was completing his dissertation. He recalls learning of Gabor's work around 1960, through a paper by H. M. A. El-Sum, which had been brought back from a conference in Stockholm by a colleague.[27] His first publications correspondingly tried to relate the earlier work to his own.

3.3 WAVE PHOTOGRAPHS

Denisyuk's first paper was published in 1962. Two other papers (bearing the same title, but expanded content, 'On the reflection of optical properties of an object in the wave field of light scattered by it') were published in 1962 and 1965. In his first three-page paper, submitted in February 1962, Denisyuk announced 'a phenomenon discovered by the author, wherein the reflecting properties of an object are manifested with extraordinary fidelity'.[28]

Citing the work of Gabor and Lippmann, Denisyuk described how objects scatter light falling on them back towards the original source. The result is interference between the original wave and the scattered wave, yielding a standing-wave pattern that can be photographed.[29] Denisyuk described his 'wave photograph' as 'a unique kind of optical equivalent of the object':

> If radiation from the same source that illuminated the object during exposure is allowed to impinge on this structure, it will reflect this radiation in such fashion that the wave field of the reflected radiation will be identical to the wave field of the radiation reflected by the object.[30]

[26] Denisyuk, 'The work of the State Optical Institute on holography', op. cit., quotation p. 707.

[27] Denisyuk, Yu N. and V. Gurikov, 'Advancement of Holography, Investigations by Soviet Scientists', *History and Technology* 8 (1992): 127–32. Denisyuk never had personal contact with El-Sum or Gabor, although he received a telegram from Gabor after his award of the Nobel Prize, for which Denisyuk had provided testimony for the Nobel committee judges.

[28] Denisyuk, Yu N., 'On the reflection of optical properties of an object in the wave field of light scattered by it', *Doklady Akademii Nauk SSSR* 144 (1962): 1275–8, p. 1275. This is an English translation of the Russian original.

[29] Strictly speaking, the scatter in Denisyuk's implementation is more properly described as *specular* or mirror-like reflection. He had not attempted to record diffusely reflecting objects. Optical scattering from objects was studied in America a couple of years later, as discussed in Chapter 4.

[30] Denisyuk, Yu N., 'On the reflection of optical properties of an object in the wave field of light scattered by it', *Doklady Akademii Nauk SSSR* 144 (1962): 1275–8, quotation p. 1275.

Indeed, he portrayed the wave photograph as a model of the original reflecting object. The first images publicized by Denisyuk were of reflections from convex mirrors and from a reflective micrometer scale.[31] The paper summarized his dissertation experiment in which he had used the collimated light from a mercury lamp to illuminate the very fine-grained Lippmann emulsion and reflecting objects to record the standing-wave pattern. He had subsequently photographically processed and then bleached the emulsion to make the reconstructed image as bright as possible.[32] The result was a wave photograph that acted like the original spherical mirror: thus the photographic plate was measured to reflect light and to focus it at the expected focal length.

This visualization of the photographic plate as an optical reproduction of the object surfaces was deeply ingrained. Eight years later, Denisyuk summarized the effect as follows:

> The antinodal surfaces, recorded by the wave photograph, are transformed, after development and bleaching, into curved mirrors of a special kind. In accordance with the standing wave formation mechanism, the shape of these mirrors is such that the phase of the source S on their surface is equal to the phase of the wave scattered by the object O. The reflection coefficient of these mirrors, which depends on the intensity of the standing wave, is proportional to the amplitude of the wave reflected from the object.[33]

Just as Gabor had generalized his understanding gradually, Denisyuk explained his scheme according to different mathematical formalisms. In his first publications, he developed the imaging theory in terms of the scattering from individual points and as a decomposition in terms of three-dimensional harmonics.[34] This also suggested to him that the proposed method might be limited to recording shallow portions of reflecting surfaces.

According to this line of thinking, Denisyuk moved conceptually from something resembling Lippmann's concept to ideas that focused on the recording of *volumes* of standing light waves. Instead of considering Huygens' situation of the progress of a wavefront, he was led to thinking about the properties of an entire wavefield in a region of space. His ideas were simultaneously constrained and daring: constrained by

[31] 'Holographic Filament': a rust-coloured, coin-shaped hologram onto which 'a convex mirror is recorded [. . .] the existence of the hologram becomes evident when you observe the image of a filament of an incandescent lamp in the light reflected by the hologram. The image of the filament formed by the hologram is enlarged and colored' (from letter by Denisyuk to the Museum of Holography, 1991: item MOH-1993.47.180).

[32] Bleaching the emulsion reduces the absorption of light, but retains information in the hologram: depending on the chemical bleaching processes involved, the originally dark portions of the emulsion remain either of altered optical density or produce surface variations. Either or both physical effects change the phase of a wavefront of light passing through the emulsion and so the resulting *phase hologram* modulates the wavefront to reconstruct an image.

[33] Denisyuk, 'The work of the State Optical Institute on holography', op.cit., quotation p. 707.

[34] Denisyuk, Yu N., 'On the reflection of optical properties of an object in the wave field of light scattered by it', *Doklady Akademii Nauk SSSR* 144 (1962): 1275–8; Denisyuk, Yu N., 'On reflection of the optical properties of an object in wavefield of radiation scattered by it', *Optika i Spektroskopija* 15 (1963): 522–32.

presumptions about the suitable optical geometry (being contextually screened by the idea that the reference and object waves should combine at the photographic plate from opposite directions), but daring in subsequently envisaging the Lippmann method as a particular case of a more general solution.[35] Denisyuk's improvements in imaging seemingly had little in common with Gabor's. His implementation avoided Gabor's problems of a troublesome conjugate image and the need for objects that caused little blockage of the optical beam, but revealed a separate constellation of experimental problems.

Two months later Denisyuk submitted a much longer paper with the same title to *Optika i Spectroskopija (Optics & Spectroscopy)*.[36] In it, Denisyuk noted that diffraction gratings, too, could be recorded in this way.[37] Importantly, he observed that the technique was limited by the brightness of the mercury source and by the monochromaticity of its light, just as Gabor had hinted over a decade earlier. Thus, Denisyuk emphasized two attributes that were crucially important to wavefront reconstruction and wave photography, the characteristics of the recording medium and the spectral coherence of the light source.

Denisyuk's technique was in several ways more general than what Gabor had conceived. Both had begun with Huygens' principle and had considered interference phenomena. But Denisyuk conceived his studies in terms of recording a spatial relationship—a standing wave in a thick photographic emulsion. He described this *volume* effect as a phenomenon as yet inadequately named or described: 'the astonishing capacity of the wave fields to depict material objects with a degree of accuracy never attained before'.[38] Gabor, by contrast, had conceived a more stepwise *process*, in which information was transformed successively from an image to a two-dimensional pattern and back to an image. Denisyuk's method was not encumbered with a conjugate image, as was Gabor's. And although Denisyuk did not emphasize it, his emulsion also could reconstruct a three-dimensional image in white light. However, the light source available to him (the high-pressure mercury arc lamp) had a coherence length of a few tenths of a millimetre and so only very shallow objects such as curved mirrors or coins could be recorded. Thus, both Gabor and Denisyuk recognized that in principle their techniques could provide much more complete information than a photograph, but neither publicly suggested that deep three-dimensional imaging might be feasible even when they later learned of the much greater coherence of lasers.

[35] Denisyuk's envisaging of counter-propagating waves appears entirely logical and inevitable when his goals are considered: he sought to capture the minor deviations from a plane wavefront; his emulsion was meant to sample that surface and so was parallel to it. For the same reason he employed collimated light to obtain a plane wavefront and shallow objects that would modulate the wavefront by not much more than the emulsion thickness.

[36] Denisyuk, Yu N., 'On reflection of the optical properties of an object in wavefield of radiation scattered by it', *Optika i Spektroskopija* 15 (1963): 522–32.

[37] While other labs (notably the NPL in England) were then investigating interference fringes as a source of the fine parallel lines needed to record a photographic diffraction grating, Denisyuk's implementation yielded a *reflective* grating directly.

[38] Denisyuk, *Fundamentals of Holography*, op. cit., quotation p. 54.

3.4 PAUSE AND RECEPTION

Unlike Gabor, who had been able to publicize his concepts of wavefront reconstruction and encourage others to investigate the technique, Denisyuk had little influence. After completing his thesis studies in 1961, he returned to Navy-related work on infrared receivers and radar processing. As chief of a new laboratory set up in June 1961, his opportunities to pursue wave photography were limited. He recalls, 'We were unable to photograph more complex objects since we had no lasers at that time. Although the advantage of using this new source of radiation was apparent to us, work on the hologram method was almost terminated in 1961 in view of the lack of definite plans for its use in motion pictures and television'.[39] Indeed, it is unclear whether Denisyuk, in fact, conceived any such potential for wave photography. But despite its novelty, Denisyuk was at a loss to suggest applications of his method other than using it to correct for chromatic aberrations in optical systems. His early papers suggest that wave photographs could serve either as efficient diffraction gratings or as wavelength-compensating optical elements. And although sophisticated and novel, neither application excited the attention of optical scientists. Like Gabor's own limited gaze, Denisyuk's (see Figures 3.1 and 3.2) technical view was contextually screened.

Denisyuk set aside wave photography for reasons other than a job change and uninspiring prospects. Compounding the decision to abandon his research in wave photography was a delay in the publication of Denisyuk's early articles in Soviet journals owing to personal enemies that he had encountered a decade earlier.

> Once in University its Party leaders (Mr. Pogarev and Mr.Sheremet) tried to denounce me, a simple student, as an American spy ([I was also reminded of] the fact that my father in law is a Jew).[42]

Anti-Semitism had been officially sanctioned from 1949 by a campaign against 'cosmopolitans', a euphemism for Jews, which, historian Jeffrey Roberg argues, 'fostered a contempt for all things not Soviet'.[43] Anti-Soviet tendencies were a convenient criticism on which to found personal attacks. On the other hand, a positive criterion was Party membership, which could provide influential connections useful for career advancement and tended to obviate suspicions of reliability. Unfortunately, Denisyuk was not a member of the Communist Party, although this was not particularly unusual.[44]

[39] Denisyuk, 'The work of the State Optical Institute on holography', op. cit., quotation p. 708.

[40] Denisyuk, Yu N. to SFJ, email, 23 Aug. 2003, SFJ collection.

[41] Denisyuk, Yu N., *Soviet Union* 9 (1970): 12–14.

[42] Denisyuk, Yu N. to SFJ, email, 13 Apr. 2003, SFJ collection.

[43] Roberg, op. cit., quotation p. 51.

[44] During the mid-1930s, for example, when Stalin's first purge of intellectuals was underway and by which time scientific institutions had been increasingly 'Sovietized', there had been only fifteen Communists among the 136 staff at the nearby Leningrad Physico-Technical Institute (later renamed the A. F. Ioffe Institute). Moreover, following the death of Stalin in 1953, Communist Party membership had become less critical to career advancement than hitherto. Indeed, even two decades earlier the number of 'red directors', or administrative secretaries of institutes who owed their position more to Party membership than to technical qualifications, had been falling. From 1956 to 1976, members of the Communist party made up a nearly constant 20% of the staff working in scientific institutes, although the fraction was

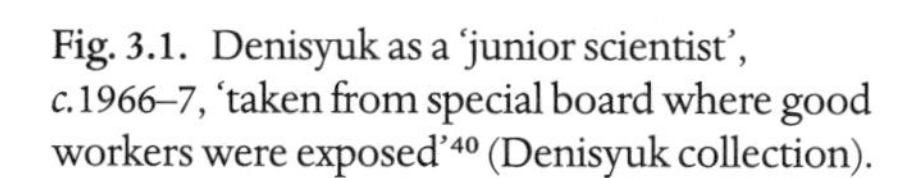

Fig. 3.1. Denisyuk as a 'junior scientist', *c.*1966–7, 'taken from special board where good workers were exposed'[40] (Denisyuk collection).

Голография. Новое в

Название работы, за которую молодой советский физик Ю. Денисюк удостоен в этом году высшей государственной награды — Ленинской премии, — для неспециалистов звучит непонятно: «Голография с записью в трехмерной среде». Недавно Комитет по делам изобретений и открытий при Совете Министров СССР, рассмотрев эту работу, признал ее крупным открытием и принял решение о внесении ее в Государственный реестр открытий СССР. По просьбе редакции ученый рассказал о том, что такое голография, что она дает и что может дать в будущем.

Fig. 3.2. Denisyuk as renowned scientist, 1970[41] (Denisyuk collection)

Such personal and political contretemps continued to hound him in his post at the Vavilov Institute. In 1961, Denisyuk asked a senior colleague of his acquaintance ('Academician L.'—presumably Alexander Lebedev) to recommend his papers for publication and was told that this would require letters of reference from four well-known scientists.[45] Denisyuk, deciding that such pre-reviews were unreasonable, bypassed him and instead approached Academician Vladimir Linnik at the Vavilov Institute for support. Linnik was able to have the first paper published in *Reports of the USSR Academy of Sciences* some two years after its completion, but Academician L., now an enemy, used his own high status to characterize the results as unworthy of attention.[46]

In a working culture steeped in the opposing forces of patronage and persecution, Denisyuk discovered that his work gained some interest among corresponding members of the Academy but was then snubbed when its support from Linnik became known. Further papers based on his thesis work appeared tardily.[47] A more powerful patron, but one who boosted Denisyuk's career somewhat later, was Petr Kapitsa.[48] Denisyuk also recalls L. A. Artsimovich and M. V. Keldysh as being important later supporters.[49]

considerably higher for members of the Academy and had risen after the Second World War. In 1970, it was estimated that about half the senior scientists at Leningrad research institutes were Party members but the number was higher among social scientists than physical scientists. This distribution of Party membership by disciplinary field, professional status, and historical period was typical of the Soviet Union as a whole.

[45] A requirement of the Candidate of Sciences degree in the Soviet Union was that the research be published in a scientific journal before the thesis was defended. This made the work much more widely available and open to criticism than its Western equivalent.

[46] Denisyuk, Yu N. to SFJ, email, 3 May 2003, SFJ collection. Vladimir Pavlovich Linnik (1889–1984), a 1914 graduate of Kiev University, joined GOI in 1926, where he researched interferometry, microscopy, and adaptive optics. He was an Academician from 1939 and was awarded numerous Soviet prizes between 1945 and 1973. Alexander Alexeyevich Lebedev (1893–1969), after graduating from Petrograd University in 1916, spent his career at GOI and at the University of Leningrad. He studied optical glass, anti-reflection coatings, electron optics, optical detection and ranging of objects, infrared physics and technology, atmospheric optics and hydro-optics, and opto-electronic instrument engineering. Lebedev was elected an Academician in 1943 and was the recipient of numerous State prizes.

[47] Denisyuk, Yu N., 'On reflection of the optical properties of an object in wavefield of radiation scattered by it', *Optika i Spektroskopija* 15 (1963): 522–32; Denisyuk, Yu N., 'On reflection of the optical properties of an object in wavefield of radiation scattered by it. II', *Optika i Spektroskopija* 18 (1965): 276–83; Denisyuk, Yu N., 'On the problem of a photograph reproducing the full illusion of the reality of the object depicted (in Russian)', *Zhurnal Nauchnoi i Prikladnoi Fotografi i Kinematografi* 11 (1966): 46–56.

[48] Denisyuk, Yu N. to SFJ, email, 3 May 2003, SFJ collection. Petr L. Kapitsa began his career at the Leningrad Physico-Technical Institute. From 1921, Kapitsa worked at Cambridge University, acquiring an international reputation for research on magnetic fields and low-temperature physics. From 1926, he was courted by Soviet authorities to return to Leningrad and, during one visit in 1934, his exit privileges were revoked. Kapitsa subsequently rebuilt his career at the Leningrad Physico-Technical Institute, while also lobbying to improve the relationship between Soviet scientists and the political elite. In the Acknowledgements of his paper, Denisyuk publicly thanked Kapitsa, Linnik, and I. V. Obreimov for 'their attention and support' (Denisyuk, Yu N., 'On reflection of the optical properties of an object in wavefield of radiation scattered by it', *Optika i Spektroskopija* 15 (1963): 522–32, p. 532). Academician Ivan Vasilievich Obreimov (1894–1981), from Leningrad, had been an early member of GOI and the first Director of the Ukrainian Technical Physics Institute from 1928, where he researched optics and spectroscopy. Incarcerated for three years by the Narodnyi Komissariat Vnutrennikh Del (NKVD, or secret police), he subsequently was Deputy Head of scientific institutes in Moscow.

[49] Denisyuk, 'My way in holography', op. cit., quotation p. 430. Lev Andreevich Artsimovich (1909–73), an outspoken plasma physicist and Academician, headed Soviet research in power by nuclear fusion in

Such institutional supporters were particularly important since Denisyuk had lost his supervisor early during his degree studies.

Thus, Denisyuk's original wave photography work seemed to attract personal and political criticism or indifference, rather than intellectual attention, at home. For several years, it gained little more attention abroad.[50] To his contemporaries, the connection with Gabor's wavefront reconstruction appeared marginal and, in any event, both schemes appeared sterile and unpromising for applications. The two techniques shared an important characteristic: at the beginning of the 1960s both were easily dismissed.

Moscow, a subject that had also been studied by Dennis Gabor during the late 1950s. On Mstislav Keldysh, chief theoretician of Soviet cosmonautics during the 1960s and President of the Soviet Academy of Sciences, see Chapter 6, note 101.

[50] See Chapters 4 and 6 for further discussion.

4

Lensless Photography in America

In his Nobel Prize speech of 1971, Dennis Gabor observed that, 'around 1955, holography went into a long hibernation'.[1] But despite Gabor's own departure from the subject, others took it up for new purposes in new locales. Each of these environments shaped a distinct variant of the subject. As the central character of this chapter, Emmett Leith, later reflected, 'holography really did not shrink in the 1955–62 period as one would erroneously conclude from the open literature. In a manner of speaking, it went underground'.[2] As we have seen, one such hidden environment all but invisible in the West was the Vavilov Institute, which hosted Denisyuk's individualistic investigations. The other important veiled site was a busy classified laboratory at the University of Michigan.

4.1 THE WILLOW RUN LABORATORIES AND OPTICAL PROCESSING

On a scale unmatched by British Thomson-Houston and the Vavilov Institute, The University of Michigan was awash with targeted research funding by the late 1940s. At the end of the Second World War, the University had already been generating income from sponsored research for a generation, having established a Department of Engineering Research in 1920 to promote industrially sponsored work. In the post–Second World War environment, though, direct federal sponsorship appeared considerably more attractive and reliable. Fresh from a record of successful applied research during the war and during the early months of the cold war conflict, the American War Office decided to continue funding research and development projects at universities during peacetime.[3] Faculty members at the University of Michigan were well positioned to exploit this new government funding. In 1946 two professors at its College of

[1] Gabor, Dennis, 'Holography, 1948–71', in: Nobel Prize Committee (ed.), *Les Prix Nobel En 1971* (Stockholm, 1971), pp. 169–201, quotation p. 176.

[2] Leith, Emmett N., 'The legacy of Dennis Gabor', *Optical Engineering* 19 (1980): 633–5.

[3] The support of American physical science by agencies devoted to military needs has been estimated as 90% of all funding for the post-war decade [Kevles, Daniel J., *The Physicists: The History of a Scientific Community in Modern America* (Cambridge, MA.: Harvard University Press, 1995)]. On the arguments behind

Engineering, William Gould Dow and Emerson W. Conlon, proposed a large-scale research project to develop an antiballistic missile system.[4] Dow later recalled,

> We were both vividly aware, as a result of our within-the-system wartime experiences, of the oncoming 'Cold War', and the threats it carried. We were equally aware that Uncle Sam's Dept of Defense fully realized the urgent need to keep in the front rank of applied science, lest we indeed lose the next war. We knew also that the higher echelons of government had a high regard for the research capabilities of our universities. This meant the DoD [Department of Defense] would very likely be willing to finance military-oriented research at the University of Michigan fairly handsomely.[5]

Rockets, even more than the new jet aircraft, appeared to be the weapons of the future. Wright-Paterson Air Force Base scientists accepted the professors' proposal for Project WIZARD, allowing the university to establish the Michigan Aeronautical Research Center (MARC) at the Willow Run Airport near Ypsilanti, Michigan the following year.[6]

Ford had built a major automobile plant at Willow Run during the 1930s, which was converted to the production of B-24 bombers during the Second World War. The factory subsequently reverted to car production, but the university inherited the airport for one dollar in 1946 leasing the landing field and larger hangars to airlines and using the drafty smaller hangars and buildings on the east side for research laboratories. From the late 1940s, other University of Michigan (U-M) laboratories were founded there and supported principally by Air Force contracts. The location suited its function. The requirements of classified research contrasted with traditional academic openness. Located some 15 miles from the Ann Arbor campus, Willow Run intellectually and physically isolated its workers. No academic teaching took place on the site, although some of the staff held dual appointments as academics in the Department of Electrical Engineering. Through the 1950s, increasing numbers of graduate students worked and undertook thesis projects there. As groups and funding mushroomed, the 150 acre site was renamed the Willow Run Research Center and, later still, the Willow Run Laboratories (WRL) (see Figure 4.1).[7]

the government funding of research, see also Wang, Jessica, *American Science in an Age of Anxiety: Scientists, Anticommunism, and the Cold War* (Chapel Hill; London: University of North Carolina Press, 1999) and Zachary, G. Pascal, *Endless Frontier: Vannevar Bush, Engineer of the American Century* (New York; London: Free Press, 1997).

[4] Bill Dow (1895–1999) had obtained his BS and EE degrees at the University of Minnesota (U-M) in 1916 and 1917, served in the First World War and worked in electrical sales and testing before joining the University of Michigan in 1926 as an instructor. During the Second World War he developed radar countermeasures at Harvard University. He became a Professor when he returned to U-M in 1945. During 38 years with the College of Engineering, he oversaw the foundation of thirteen research units and laboratories. Emerson Conlon was Chair of the Dept of Aeronautical Engineering in 1946, and had worked at the Naval Bureau of Aeronautics during the war.

[5] Morgan, Mary, 'Research firm has colourful past: U-M faculty started today's ERIM International for government projects', *Ann Arbor News*, 2 Sep. 1999, 19.

[6] The Boeing-Michigan Aeronautical Research Center (BOMARC) surface-to-air missile began testing in 1952 and was operational between the late 1950s and early 1970s.

[7] The Aeronautical Research Center was renamed the Willow Run Research Center of the Engineering Research Institute in 1950, and the Willow Run Laboratories in 1955. WRL became part of the university's new Institute of Science and Technology (IST) in November 1960 and some of its laboratories were moved to the new IST building on the U-M North Campus in the mid 1960s.

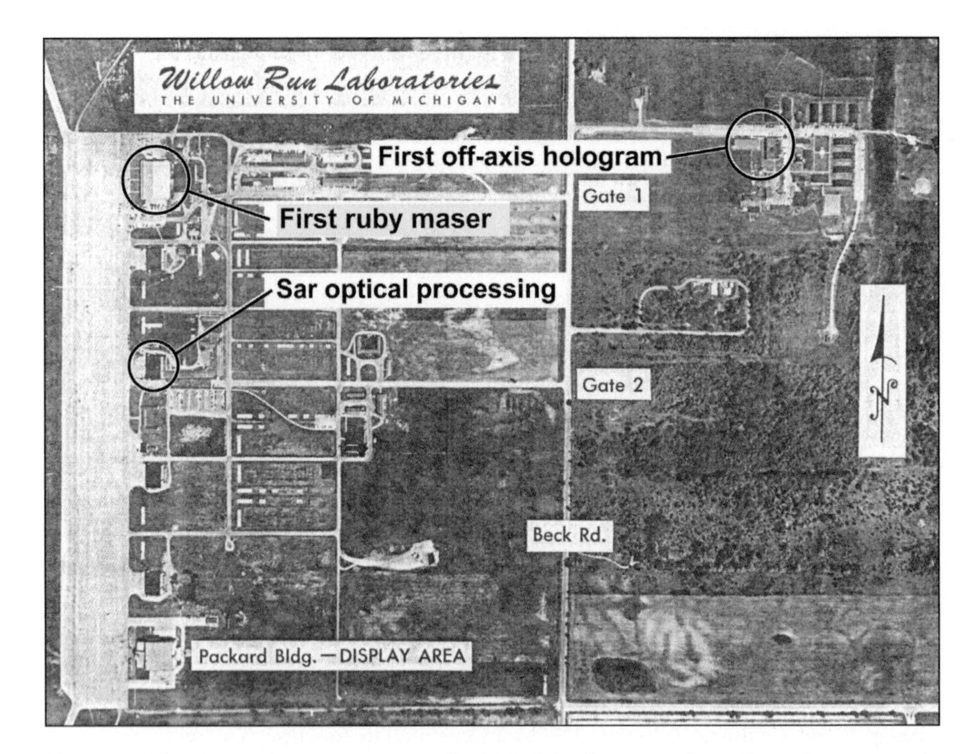

Fig. 4.1. Willow Run Laboratories, 1962 and technical developments there: first ruby maser, 1958 (the converted hangar burned down in October 2004); first successful synthetic aperture radar optical processing, 1959; first off-axis holograms of binary transparencies, 1960; first off-axis 3-D object holograms, December 1963 (Bentley Historical Library, annotations added).

The subjects investigated by the Willow Run workers covered a wide array of new technologies conceived during the war. They included radar, infrared, acoustics, optics, guidance, and data processing. An early digital computer design, the Michigan Digital Automatic Computer (MIDAC), was developed there in the early 1950s. So, too, was the first ruby maser, developed in a former aircraft hangar at Willow Run Laboratories by Chihiru Kikuchi and his group in December 1957. The Laboratories also presented summer schools for the growing community of military-contract researchers around America, and so became a locus for the developing expertise in these classified fields.

One of the areas under investigation at Willow Run was development of a variant of imaging radar that became known as Synthetic Aperture Radar (SAR). In 1951, Carl Wiley of Goodyear Aircraft Corporation had reported the principle of sharpening radar images by performing a frequency analysis of the signals acquired by an airborne coherent radar system. In such a system, the radar waves would be emitted sideways from the aircraft, rather than straight ahead or downwards as in conventional radar systems developed during the war. The signal would be pulsed, as in conventional radar, but both the time of arrival and the *frequency* of the return echoes would be measured.

The time of arrival, as with the earlier wartime radar systems, would determine the range, or straight-line distance, of the reflecting object from the aircraft. The frequency of the returned echo, however, would provide information about the distance of the object behind and ahead of the aircraft: the plane's forward motion causes a Doppler shift, raising the frequency of the echo return for objects progressively ahead of the plane and lowering the frequency for objects further behind it. Thus, a two-dimensional map of the radar reflection could be produced by plotting the arrival time of the radar echoes versus the frequency shift.

This was not a real-time operation. In practice, the return time of the signal would be analysed to infer range, or cross-track information about reflecting objects, and then the process would be repeated for the various shifted frequencies in the signal to build up a picture of range information for successive slices in the along-track (azimuth) direction behind and ahead of the flight line. The analysis required that radar signals be recorded and then successively filtered to yield an image of cross-track strips corresponding to the various Doppler shifts.

Engineers at the University of Illinois tested the idea experimentally in 1952. Flying over the Florida Keys in a C-46 aircraft supplied by the US Air Force, the researchers recorded the raw radar signal on magnetic tape and then laboriously analysed it by repeatedly playing back the tape through a tuneable electrical filter (see Figure 4.2). Chalmers Sherwin, the principal investigator, recounted:

> We put the beam out and ran along the coast of Key West, Florida, recording on our tape all the signals from that strip. We took that section back, put in the loop, went round and round and measured each frequency and adjusted it, and there was a map of Key West, Florida, all taken within the beam width of the radar.[8]

They had discovered that such frequency analysis could indeed usefully identify man-made objects in radar signals. The investigators were interested in detecting just this kind of object: submarine snorkels poking out of a background of sea waves.[9] The following year, the Illinois group communicated their encouraging results to the Willow Run laboratory during a summer study for Project WOLVERINE, an army contract covering battlefield surveillance projects. The Michigan group was immediately interested and had their own ideas about how to improve the system. Their subsequent research led eventually to a successful radar surveillance system first described in the open literature nearly a decade later.[10]

As these first investigators began to appreciate, their clever radar scheme could be improved further. Higher-resolution images could be recreated by detecting, and then coherently adding, the pulsed return signals one by one and then decomposing this

[8] Sherwin, Chalmers W. to J. Bryant, interview, 12 Jun. 1991, MIT Radiation lab, IEEE History Center.

[9] Sherwin, Chalmers W., J. P. Ruina, and R. D. Rawcliffe, 'Some early developments in synthetic aperture radar systems', *IRE Transactions on Military Electronics* MIL-6 (1962): 111–5 and Sherwin to Bryant, op. cit.

[10] Cutrona, L. J., W. E. Vivian, E. N. Leith, and G. O. Hall, 'A high-resolution radar combat-surveillance system', *IRE Transactions on Military Electronics* MIL-5 (1961): 127–31.

Fig. 4.2. Synthetic Aperture Radar (SAR) image of Willow Run airport, one of thousands produced during development of SAR at WRL and its successor, ERIM, *c.*1973. The Willow Run Laboratories are at the bottom (east) of the airport runways (Leith collection).

combined signal into its separate frequencies. In this way an optical image could be recreated from a data set recorded along the entire aircraft flight line. The ability to resolve detail in the along-track, or azimuth, direction would be determined by the length of the flight line: in effect, by recording signals as the antenna moved, the system would behave like a single very long (and hence highly directional) antenna or 'synthetic aperture'.[11] Just as a large-diameter lens is capable of focusing images having higher spatial resolution (a fact appreciated since Abbe's theory of imaging), the longer synthetic antenna allowed a smaller radar spot to be imaged. The technique therefore offered the promise of an imaging radar system that could compete with the resolution of optical cameras and one that could operate in all weather conditions. Even from these early proof-of-concept demonstrations, it is apparent that increasingly powerful generalizations were being made about the technology. SAR could be understood in terms of Doppler analysis or more potently as a synthetic aperture produced by signal summation. Over the following five years, the repeated reconceptualization of SAR was to yield new insights and generalizations.

[11] SAR is also known as 'side-looking radar' because it emits radar waves from one side of the aircraft but yields an image that appears to be recorded from directly overhead. This was a distinct advantage for obtaining imagery across national borders, in addition to radar's other military advantage of performing well in adverse weather conditions. See Jenson, Homer, L.C. Graham, Leonard J. Porcello, and Emmett N. Leith, 'Side-looking airborne radar', *Scientific American* 237 (1977): 84–95.

To pursue these ideas, the Radar Laboratory was formed in 1953 by Louis J. Cutrona and Weston E. Vivian of Willow Run,[12] funded initially by Project WOLVERINE, and thereafter principally by Project MICHIGAN, the university's largest defence department contract. Project MICHIGAN, established in 1954 as the second phase of a Navy, Air Force, and Army Research Program administered by the Army Signal Corps and which had already spent $500,000 in an exploratory phase was funded until the late 1960s. The approval of Project MICHIGAN by the University led to a reorganization of facilities to create a 'center for classified research' on the east end of Willow Run Airport.[13] Of the hundreds of technical workers who found employment at Willow Run, one of the first to become involved with the Radar Laboratory investigations was Emmett Leith. Raised and educated in Michigan, Leith joined as a Research Assistant in February 1952.[14] Having taken four standard undergraduate optics courses as a Physics major—in physical optics, two in spectroscopy, and in x-rays and crystal structure—he found himself well placed in an environment dominated by electrical engineers to undertake his first major task at Willow Run: a preliminary study of the optical processing of radar data.[15]

The elementary concepts of synthetic aperture radar outlined above had no evident connection with the work of either Gabor or Denisyuk. One reason for this was the barrier introduced by different disciplinary languages. The jargon and concepts of the three subjects could be translated only with difficulty. To use a term coined by historian of science Thomas Kuhn in a somewhat different context, the concepts underlying holoscopy, wave photography, and SAR were *incommensurable*. Kuhn used this term to describe the inability of participants to comprehend or even discuss scientific theories

[12] Louis J. (Lou) Cutrona (b. 1915) joined Willow Run in 1949, having worked previously at Bell Telephone Laboratories, Federal Telecommunications Corporation, and Sperry Gyroscope Company. Arriving with a PhD, his post was Research Engineer until 1956, when he was appointed Professor of Electrical Engineering. Weston E. (Wes) Vivian (b. 1924) joined Willow Run in 1951, gaining a PhD at the University in 1959. Both Cutrona and Vivian joined Conductron Corporation in 1962, which is discussed in §6.2. Vivian served a term as Democratic member of Congress 1965–7 and subsequently was an independent consultant.

[13] Board of Regents Proceedings May 1953, Feb 1954. The US Department of Defense contracts, with specific Army and Air Force sponsors, continue to investigate applications of coherent optics at the University of Michigan and the successor to Willow Run Laboratories, Environmental Research Institute of Michigan (ERIM), to the present day. ERIM spun-off a for-profit arm, ERIM International, in 1997. This was purchased by Veridian in 1999, and ERIM itself was renamed Altarum in the interim. General Dynamics subsequently acquired a portion of the company in 2003.

[14] Leith was born in Detroit in 1927 and obtained BS (1949) and MS (1952) degrees in physics from Wayne State University, joining WRL at an initial salary of $3950/year. Along with some of his colleagues, he continued to take evening courses at Wayne State and, during some periods, morning courses at U-M before work, 'a rather hectic existence' (Leith, Emmett N. to SFJ, email, 20 May 2003, SFJ collection). Leith remained at U-M through his entire career, being promoted to Graduate Research Assistant in April 1955, to Research Associate in September 1956, to Research Engineer at the Institute of Science and Technology (IST) in 1960 and Associate Professor of Electrical Engineering at the University in 1965, and to full Professor in May 1968. He was later appointed Schlumberger Professor in Engineering (1984–93) and Chief Scientist at ERIM, the not-for-profit private corporation that succeeded the WRL Leith completed a PhD at Wayne State University in 1978 under F. T. S. Yu on his work of the late 1950s and early 1960s.

[15] Several of Leith's professors at Wayne State University had been spectroscopists, a major growth field in American optics in the interwar and post-war period. Thus Leith had up-to-date training in new areas of optics (spectroscopy and infrared) as well as traditional optics.

developed before and after a revolution in thinking.[16] A more recent and nuanced anthropological analogy is that of Peter Galison, who describes different subcultures of physics based on distinct instrumentation. Galison characterizes workers trained within these subcultures as belonging to dissimilar technical tribes speaking different languages and communicating via compromise pidgin dialects in constrained working contexts or trading zones. Despite the common feature of two groups separated by their languages and concepts, the situation described in this chapter is rather different from both of these understandings: here, two disciplines (electrical engineering and physical optics) provided tools to establish a marginal community of workers.[17] This idea is developed further in Chapters 5 and 8.

Radar research had been generated and expanded principally by electrical engineers. As a result, the technologies proposed for SAR naturally were founded on that disciplinary viewpoint. The proof-of-concept experiment carried out by the University of Illinois had employed analogue electronics to perform frequency analysis. They had proposed more elaborate systems of analysing the radar data using electrical delay lines, analogue electrical filters, and magnetic tape recording or drum storage. Other groups, also dominated by electrical engineers, investigated the radar data using the first generation of digital computers to perform frequency analysis.

The Willow Run group started with similar ideas. They, too, conceived the radar signal analyser initially as a signal processing system having components familiar to electrical engineers: 'signal storage, phase shifters for focusing, and frequency-selective filters (not necessarily passive electronic filters) for weighting and segregation of the resolved signals; plus possibly a vector magnitude resolver', as one early report summarized it.[18] Like Wiley and Sherwin, they envisaged the key idea of the system as an exploitation of the Doppler effect. Nevertheless, they realized that such a radar imaging system had an inherent weakness: if a long observation time were to be employed (to yield a long synthetic antenna and hence a high spatial resolution for the images) the distance and angular position of the object would vary during the flight causing a more complex Doppler shift. An accurate solution would require a more complicated data analysis. The mathematical procedure of cross-correlation, requiring several thousand separate mathematical operations to generate the complete image, could neither be implemented by the analogue electronic circuitry designs available, nor could the early digital computers

[16] Kuhn, Thomas S., *The Structure of Scientific Revolutions* (Chicago; London: University of Chicago Press, 1964).

[17] Galison extends his metaphor of interlanguages to suggest that pidgins (crude trading vocabularies) may be replaced by creoles (more richly descriptive), amounting to 'full disciplinary languages big enough to grow up in professionally' (Galison, Peter, *Image and Logic: A Material Culture of Microphysics* (Chicago: University of Chicago Press, 1997), quotation p. 50). Holography, however, was less a trading zone, where two cultures barter established goods, than an isolated outpost, where borrowed resources establish a hybrid community.

[18] Vivian, W. E., L. J. Cutrona, and E. N. Leith, 'Report of Project MICHIGAN: A Doppler Technique for Obtaining Very Fine Angular Resolution from a Side-Looking Airborne Radar', Report 2144-5-T, University of Michigan Willow Run Laboratory, July 1954.

cope with such a task.[19] The initial understanding of SAR therefore envisaged a complex and challenging electronic system.

Perhaps because of the seemingly insurmountable problems of the scheme, Cutrona and Vivian were more open to other perspectives. In any case, following a discussion with Russell Varian, one of the inventors of the klystron radar source, they conceived the idea of *optical* processing to generate high-resolution images from the SAR data.[20] The approach seemed possible because of the formal similarity between the mathematics of the SAR analysis and the mathematics of optical systems. Lenses can map or transform a distribution of light in one plane into a very different distribution in another plane. Most interestingly for electrical engineers, a properly positioned lens will create the Fourier transformation of an image in another plane. If an electrical signal can be represented as a variation in intensity or transmission (e.g. by being recorded onto photographic film) then a Fourier transformation can be performed merely by placing that film at the focus of a lens and recording the distribution of output intensity on a surface at the same distance on the opposite side.[21] Unlike electrical methods, or the relatively simple and slow digital computers just then beginning to appear, the optical transform was instantaneous. Moreover, the lens could perform this feat for not just a one-dimensional signal (e.g. a time dependent signal $S(t)$) but for a *two*-dimensional signal $S(x, y)$, because the lens itself has two dimensions perpendicular to the optical axis. The principal advantage of optical processing would be that the output (Fourier) plane of this simple processor could be manipulated in sophisticated ways to perform mathematical operations on the signals. This manipulation was merely a matter of optically masking (or 'spatially filtering') portions of the intensity pattern there. So an optical processor could be conceived as a deceptively simple optical device consisting of a photographic input, a lens system, a

[19] For a more detailed overview of the technical challenges of SAR and holography, see Leith, Emmett N., 'A short history of the Optics Group of the Willow Run Laboratories', in: A. Consortini (ed.), *Trends in Optics* (New York, 1996), pp. 1–26.

[20] Russell Varian (1898–1959) invented the klystron tube, a high-frequency amplifier for generating microwaves, with his brother Sigurd and William Hansen at Stanford University in 1938. The concepts incorporated in the term *optical processing* became a recognized part of optics in the 1950s via several investigators, including Dennis Gabor and Adolf Lohmann as already discussed. On the cognitive and experimental content of this ill-defined subject (which integrates the communication theory concepts of filtering and linear systems theory) and its relationship with the later subject of holography, see Leith, Emmett N., 'The evolution of information optics', *IEEE Journal of Selected Topics in Quantum Electronics* 6 (2000): 1297–304.

[21] Some aspects of Fourier optics had been explored during the late nineteenth century, particularly by Ernst Abbe (Abbe, Ernst, *Archiv. Mikroskopische Anat.* 9 (1873): 413) and Lord Rayleigh (Rayleigh, Lord, 'On the theory of optical images, with special reference to the microscope', *Philosophical Magazine* 42 (Series 5) (1896): 167). The advantages of a Fourier description was developed further during the 1940s, particularly in Duffieux, P. M., *L'intégrale de Fourier et ses Applications à l'Optique* (Besançon: Faculté des Sciences, Besançon; Societé Anonyme des Imprimeries Oberthur: Rennes, 1946) and in Schade, Otto. H., 'Electro-optical characteristics of television systems', *RCA Review* 9 (1948): 5 (Part I), 245 (Part II), 490 (Part III), 653 (Part IV). Neither of these publications became widely available. The specific investigations by Gabor, Lohmann, Leith, and others during the 1950s added to a growing theoretical generalization that spread widely by the early 1960s, as illustrated by O'Neill, E. L., *Introduction to Statistical Optics* (Reading, MA: Addison-Wesley Publishing Co., 1963); Françon, M., *Modern Applications of Physical Optics* (New York: John Wiley, 1963); Linfoot, E. H., *Fourier Methods in Optical Image Evaluation* (London: Focal Press, 1964); and Mertz, Lawrence, *Transformations in Optics* (New York: John Wiley, 1965).

mask plane (that might consist of a photographically produced mask or 'filter function'), and a means of recording the output on photographic film.[22]

Given the choice of what portion of the proposed programme to undertake, Emmett Leith chose to work on the optical processor. As a junior engineer, Leith was opting for an area of relative advantage because optical processing was an unusual and distinctly non-disciplinary approach dissociated from the context of electrical engineering. He consequently contributed to an internal report on optical processing initiated by his more senior colleagues, although he felt that the method was impracticable and likely to fail.

The three initially proposed a system based on incoherent processing, that is, employing a classical optical system rather than one in which the performance relied on the wave nature of light and coherent radiation.[23] The result was a theoretical study with an analysis of stability requirements and ideas for implementations.[24] This became a template for the development of a practical optical processing system for SAR data and consequently portions were classified as secret until the late 1960s.[25]

Upon completion of their initial study during the summer of 1954, the project expanded to develop a prototype system. As the group enlarged, others developed the radar transmitter and receiver, and the flight stabilization system to compensate for buffeting and irregularities of the aircraft's motion. Leith continued to work on the optical processor. Over the following couple of years, a chain of insights successively transformed his understanding and conception of optical processing. As he began to investigate the optical properties of the proposed system in 1954, Leith reconceived what had been seen as a complex and unintuitive mathematical processor into more familiar terms: diffraction gratings.[26] A diffraction grating is an optical device usually consisting of regular straight lines scribed on glass. Its line spacing is constant or, in terms of communication theory, its spatial frequency is fixed. The lines diffract light to form spectra of different *orders*. Each spectrum consists of an orderly progression of the frequencies in the diffracted light. Part of the light is undeviated and forms the zeroth order; the rest of the light is diffracted to either side to form first order spectra (sometimes denoted $+1$ and -1) and, depending on how the grating is constructed, higher order spectra may also be produced.

In musing about gratings, Leith discovered that his analysis seemed to disagree with the patterns observed in the prototype processor and so he began to consider the

[22] In practice, such a system is far from instantaneous, because of the need to first record the input signal onto film and to record and chemically process the output signal in the same way. Nevertheless, optical processing promised the efficient processing of large quantities of data of the kind expected from synthetic aperture radar.

[23] In an *incoherent* optical processing system, only light intensities can be combined; in a *coherent* optical processor, the light wave's amplitude and phase are available and so complex mathematical operations can be performed. This provides considerably greater scope for useful mathematical operations.

[24] Vivian, W. E., L. J. Cutrona, and E. N. Leith, 'Report of Project MICHIGAN: A Doppler Technique for Obtaining Very Fine Angular Resolution from a Side-Looking Airborne Radar', Report 2144-5-T, University of Michigan Willow Run Laboratory, July 1954.

[25] Cutrona, L. J., Emmett N. Leith, L. J. Porcello and W. E. Vivian, 'On the application of coherent optical processing techniques to synthetic-aperture radar', *Proceedings of the IEEE* 54 (1966): 1026.

[26] Leith, 'A short history of the Optics Group of the Willow Run Laboratories', op. cit., quotation p. 6.

performance of the system for an ideal simple case: monochromatic light from a point source. Such a source was precisely the coherent light that Dennis Gabor had employed seven years earlier.

In 1955 Leith was joined by recent graduate Len Porcello and the two spent a couple of months studying the properties of coherent and non-coherent radiation in the optical cross-correlator.[27] They produced a detailed and extensive memo analysing the mathematics of the process, which they discussed in terms of classical physical optics. But Lou Cutrona, who became interested in the work, made a connection to communication theory: he told them about a paper given at their annual radar symposium by Ed O'Neill, who linked optics with communication theory. 'From that time on', recalls Leith,

> all of our optical processing techniques were developed within the Fourier (i.e., communication or linear systems) framework. We extensively used Fourier transforms, linear systems theory, spatial filtering, transfer functions, point spread functions, etc to describe our systems and our techniques.[28]

Thus, by a unique and indirect route, Leith had reached a theoretical and practical position occupied and tentatively explored by Gabor, Rogers, El-Sum, and Lohmann. He had developed an understanding of optical processing from the discipline of optics and had adopted coherent light as the preferred implementation. But his additional communication theory perspective could readily be exploited in an environment replete with electrical engineers: it fit neatly with radar theory and it provided a satisfying and elegant description of the process of signal analysis.[29] Thus, the grafting of physical optics and communication theory flourished at Willow Run, although it had been transplanted in relatively unfertile ground elsewhere.

In September of that year, Leith was struck by a further extension of this merged approach: he realized that the waves diffracted by the data record were in fact optical replicas of the original radar waves.

> The most interesting thing along the way was, when I was looking at the results of the analysis, something suddenly dawned on me, which I thought was most astonishing: if you look at the

[27] Leonard J. (Len) Porcello (b. 1934) joined Willow Run as a Research Assistant, receiving an MS degree in Physics and MSEE degree in Electrical Engineering from University of Michigan in 1957 and 1959, respectively. Another Research Assistant and co-author with the Optics Group, Carmen J. Palermo (b. 1933), was with Willow Run between 1956–62 and 1964–6. He also received his MS from University of Michigan in 1958 and his PhD there in 1962.

[28] Leith, Emmett N. to SFJ, email, 8 Jan. 2004, SFJ collection. See, for example, O'Neill, E. L., 'Spatial filtering in optics', *IRE Transactions on Information Theory* IT-2 (1956): 56 and O'Neill, E. L. (ed.), *Communication and Information Theory Aspects of Modern Optics* (Syracuse, NY: General Electric, 1962). The work of a communication theorist, Peter Elias, was also seminal in promoting this general perspective; see Elias, Peter, 'Optics and communication theory', *Journal of the Optical Society of America* 43 (1953): 229.

[29] Among the electrical engineers in the group by the early 1960s were Len Porcello, Carmen Palermo, Bill Brown, Ken Haines, Karl Stetson, Adam Kozma (dual mechanical and electrical), Anthony (Bud) VanderLugt, and Bob Harger. Three in particular—Brown, Palermo, and Harger—were PhD specialists in communication theory and powerful theorists (Leith, Emmett N. to SFJ, email, 8 Jan. 2004, SFJ collection).

Fig. 4.3. Optical processor at Willow Run, *c.*1960. The large aperture optics were mounted on steel slabs fastened to granite supports and steel I-beams. The large defect-free lenses cost some $10,000 to design and a further $10,000 to fabricate, and were optimized for the wavelength of mercury green light (546 nm) (Leith collection).

> data record and considered it being illuminated with the beam of coherent light, the field that emanated from it in the first diffracted order was in fact an optical reconstruction of the original radar field which was captured by the synthetic aperture radar as it flew along its flight path.[30]

This recreated field of light that emerged from the optical processor (see Figure 4.3) was much smaller in scale than the original radar wave field, because the respective wavelengths (about 0.00005 cm (500 nm) versus 3 cm) were in the ratio of 1 : 60000.

This two-step process, with a shift of wavelength in between, had strong similarities with Gabor's wavefront reconstruction concept of 1947. The resemblance becomes more obvious if concepts familiar to electrical engineers are translated into terms common to optical physicists. For example, the information recorded by SAR systems amounted to two distinct properties of the signal: the echo return times and frequency spectrum. A complementary way of understanding this was as a time-dependent amplitude (the signal strength) and phase shift (the Doppler frequency shift), which were to be transformed into a spatially dependent amplitude (the image). In fact, the radar signal of a small radar-reflecting object, when recorded onto photographic film, appeared to be a portion of a zone plate. Leith developed the theory of SAR from this physical optics viewpoint, rather than beginning from concepts then familiar to the radar world, such as cross-correlation and Doppler filtering. His advantage lay in having a foot in two worlds: that of electrical engineers and that of optical physicists. Leith (see Figure 4.4) was a crucial intermediary between two mutually uncomprehending groups establishing an intellectual trade.[31] Most of this recasting of the subject was done between October 1955

[30] Leith, Emmett N. to SFJ, interview, 22 Jan. 2003, Santa Clara, CA, SFJ collection.

[31] Here Galison's perspective is again apt; see Galison, op. cit., pp. 803–44.

Fig. 4.4. Emmett Leith, 1958 (Leith collection).

and April the following year. By early 1956 he had drafted a 36-page quarterly report to the sponsor that summarized his new perspective.[32]

Leith's report made little impact at the time. The cultures of physical optics and electrical engineering shared few common features in their intellectual bases, concepts or practice. As he later described in a belated PhD thesis, 'for two or three years the idea languished, having met nearly complete indifference by the synthetic aperture radar community'. He blamed three reasons: (1) the abstruse tone of the memo; (2) its limited circulation (which constituted 'a respectable burial'); and (3) its physical optics viewpoint, 'which was strange to radar engineers [. . .] The new viewpoint was a radical departure from the conventional'.[33]

In short, the new perspective was contextually screened from electronics engineers. Some five months later, Leith finally learned that his reconception of SAR had strong similarities with Dennis Gabor's work. In October 1956, Leith came across a recent paper on wavefront reconstruction by Kirkpatrick and El-Sum,[34] and from there looked up the earlier papers of Gabor and Rogers. The discovery simultaneously valourized and sidelined his work. Leith reports disappointment that his ideas were not unique, but was encouraged that the ideas had been deemed important enough for publication and derived from a distinctly different foundation.[35] In some respects, Leith's work started

[32] Leith, Emmett N., 'A data processing system viewed as an optical model of a radar system', memo to W. A. Blikken of Willow Run Laboratories, University of Michigan, 22 May 1956.

[33] Leith, Emmett N., *The Origin and Development of the Carrier Frequency and Achromatic Concepts in Holography*, PhD thesis, Electrical Engineering, Wayne State University (1978), p. 6.

[34] Kirkpatrick, P. and H. M. A. El-Sum, 'Image formation by reconstructed wavefronts I. Physical principles and methods of refinement', *Journal of the Optical Society of America* 46 (1956): 825–31.

[35] Leith, 'A short history of the Optics Group at the Willow Run Laboratories', op. cit, quotation p. 9.

from a direction opposite Gabor's: while Leith sought to reconstruct microwave radar (i.e. very long wavelength) interference patterns using visible light, Gabor had aimed at reconstructing electron (very short wavelength) interference patterns and El-Sum had focused on reconstructing x-ray (also short wavelength) interference patterns. The notion of a 'carrier frequency' was also a new and significant feature in Leith's analysis. Just as importantly, the insights that he had developed allowed the Optics Group to simplify and considerably improve their SAR optical processor design.

Despite an increasingly general conception of SAR and optical processing, the Willow Run SAR system was initially unsuccessful. The first eight data-collection flights during 1957 yielded no image when optically processed. Funding appeared unlikely to continue for what had taken some two-dozen researchers two years of development, but the ninth flight yielded impressive imagery. The team subsequently realized that aircraft buffeting by air turbulence made the data too irregular to reconstruct, and for some time they were able to obtain adequate signal records only by flying during cloudless nights around 4 a.m.[36] Detecting spurious motions of the aircraft and putting corrections into the recorded data became the major problem in getting SAR to work well and years of work and millions of dollars went into solving the problem. By 1960 the system had been not only vindicated but also adopted as the new orthodoxy within the secret community of SAR research.[37] Like Gabor before him, Leith's greater attention from contemporaries could be attributed to the gradually increasing clarity of viewpoint and to the potent effect of successful demonstration.[38]

At least part of this success was soon made public. By April 1960, the Army was eager to announce the success of the new 'AN/UPD-1 (XPM-1) High-resolution Radar Combat-Surveillance System'—the delivered version of the coherent optical processing system combined with radar electronics produced by Texas Instruments. The system was a potent demonstration of how militarily funded research could yield novel and powerful new technologies. As newspapers reported:

> **Willow Run Research Paying Big Dividends**
> Disclosure that the University has developed an airborne radar system hailed as a major improvement in combat surveillance focuses national attention on the research program

[36] Leith, 'A short history of the Optics Group of the Willow Run Laboratories', op. cit., quotation p. 12.

[37] Willow Run's optical method of processing SAR data became the standard during the 1960s. Only in the late 1970s did digital processing systems, developed first for the American SEASAT satellite, begin to usurp the place of the optical method. As late as 1980, optical processing was the almost unanimous choice in SAR systems, with proponents of digital systems arguing with relatively little success that they could overcome the apparent limitations (Ausherman, Dale A., 'Digital versus optical in synthetic aperture radar (SAR) data processing', *Optical Engineering* 19 (1980): 157–67). For example at ERIM, the organization that succeeded Willow Run Labs, optical processing was used for the high-volume SAR analysis through the 1980s and into the early 1990s because, although computers could do more sophisticated processing by then, the underlying culture of analysis took time to adapt to digital information (Aleksoff, Carl to SFJ, interview, 9 Sep. 2003, Ann Arbor, SFJ collection). See also Leith, 'The evolution of information optics', op. cit., p. 1300.

[38] Leith, *The Origin and Development of the Carrier Frequency and Achromatic Concepts in Holography*, op. cit., quotation pp. 6–7.

being carried out at the Willow Run Laboratories of the U-M [. . .] The new radar system was developed for the Army under Project Michigan, which includes about 75 percent of the many studies going on at the laboratories under contracts with government and industry.[39]

A Spy in the Sky! Army Unveils Radar Camera: It Shoots Enemy Thru Clouds of Smoke
[. . .] The revolutionary aspects of the new army system include its ability to make a 'sideways' gathering of information and its ability to make a photographic record of areas many miles distant without the distortion of distance—its pictures look like aerial photographs taken from directly overhead.[40]

Radar Photos From Afar to Pierce Iron Curtain
[. . .] After the plane returns to base, the film is taken to the processing van, where it is developed and fed into a special processor which converts the signals into a picture.[41]

The Army's New Eyes: U-M Scientists Develop Far-Sighted Radar Unit.
[. . .] The plane receives the signals, and equipment in the van processes them.[42]

Unsurprisingly, the public reports of this classified technology transmitted little more than a sense of awe about the power of modern science. But this message, extended back to the wartime achievement of the atomic bomb, was also to be the pattern for future developments—especially wavefront reconstruction—at Willow Run.

This glimpse of the products of Willow Run revealed little. Published reports remained coy about the hidden activities and technical foundations of SAR. Indeed, while the new side-looking radar system was touted to the public as being a major achievement, its connection with optical processing remained secret, and could not be pieced together from the breathless newspaper accounts. The Willow Run personnel were prevented from drawing attention to the link between radar and optical processing. The key papers concerning the method appeared only years later.[43] The newspaper coverage was a

[39] 'Willow Run Research Paying Big Dividends', *Ann Arbor News*, 20 Apr. 1960, 4.

[40] Dodd, Philip, 'A spy in the sky! Army unveils radar camera: it shoots enemy thru clouds of smoke', *Chicago Daily Tribune*, 20 Apr. 1960, 3.

[41] 'Radar Photos From Afar to Pierce Iron Curtain', *Detroit News*, 20 Apr. 1960, 1.

[42] Pearson, Jean, 'The Army's New Eyes: U-M Scientists develop far-sighted radar unit', *Detroit Free Press*, 20 Apr. 1960.

[43] The original papers were presented at symposia having restricted attendance, such as the *National Convention on Military Electronics* in 1959 (Cutrona, L. J., E. N. Leith, C. J. Palermo, and L. J. Porcello, 'Optical data processing and filtering systems', *IRE Transactions on Information Theory* IT-6 (1960): 386–400; Cutrona, L. J., Emmett N. Leith, and L. J. Porcello, 'Data processing by optical techniques', presented at *National Convention on Military Electronics*, Washington DC, 1959); and the *6th Annual Radar Symposium* in 1960 (Leith, Emmett N. and C. J. Palermo, 'Spatial filtering for ambiguity suppression, and bandwidth reduction (SECRET)', presented at *6th Annual Radar Symposium*, Ann Arbor, MI, 1960). Papers that linked optical processing with radar data were routinely rated 'classified' or 'secret'. Leith continued to present classified papers for military sponsors thereafter, but papers began to appear in the open literature from the mid-1960s (Cutrona, L. J., E. N. Leith, L. J. Porcello and W. E. Vivian, 'On the application of modern optical techniques to radar data processing', presented at *Proceedings of the 9th AGARD Symposium on Opto-Electronic Components and Devices*, Paris, 1965; Kozma, Adam, Emmett N. Leith, and Norman G. Massey, 'Tilted-plane optical processor', *Applied Optics* 11 (1972): 1766–77). The Willow Run research was not uniquely constrained, of course: such restricted symposia also heard classified or secret papers from Bell Telephone Laboratories, Convair Division of General Dynamics, US Naval Research Laboratory, and Motorola and Texas Instruments, among other organizations.

welcome recognition for their achievements but, as Leith recalls, the close association with classified research was a nuisance:

> One thing that was a real pain, a real sticker, was the association of optics and radar; that was *secret*. If you worked in optics, you couldn't mention radar, and the other way around; it didn't work. And there were some nasty anecdotes: we wrote a paper, a bunch of us—there was Lou Cutrona, myself, and Carm Palermo and Len Porcello, and they said, 'that paper will be classified unless Cutrona's name and Leith's name come off it. It was exactly the name—because I worked in radar, and therefore I couldn't write about optics, but the other guys, who didn't work in radar, could. They worked in radar too, but they just didn't write papers [. . .] You could talk about one or the other, but not both in the same breath.[44]

Nevertheless, at least to the coterie of radar researchers, the success of the U-M SAR system meant that optical processing now was recognized as an elegant and widely applicable method, and the Willow Run Optics and Radar Laboratory continued to further explore its possibilities.[45] The Optics Group worked in a surprisingly unconstrained and well-funded environment. The directions being pursued in optical processing were a sideline of the SAR work that could be accommodated because of the size of the contract, Leith's productive generation of new ideas, and the willingness of the military sponsors to seek, and to explore, new approaches.

The laboratory and its goals both become more coherent. The Optics Group moved about a half mile further east of the airport runway to the 'blockhouse' in 1961 (see Figure 4.5). The large windowless building was radiation and blast protected, so its solid construction made it ideal for the purposes of coherent optical work. By 1962, the members of the Optics Group comprised some fifteen scientists, engineers, technicians, and support staff. Leith and one associate, an optical engineer responsible for packaging their optical processor, embodied the optical expertise.[46] The remainder of the staff had been chosen for their prior knowledge of communication theory and picked up physical optics on the job. All, though, were coordinating the expertise from two disciplines: electrical engineering and physical optics. The essential tool in each of the half-dozen laboratories and the foundation of the practical knowledge developed by the Optics Group was the mercury arc lamp, the source of their coherent light. The new environment was thus coherent in terms of facilities and viewpoint.

The Group also began to see their intellectual terrain in more coherent terms. From the time of the first SAR system field tests, coherent optical processing had appeared to

[44] Leith, Emmett N., Donald Gillespie and Brian Athey to SFJ, interview, 11 Sep. 2003, Ann Arbor, SFJ collection.

[45] Cutrona, L. J., E. N. Leith, C. J. Palermo, and L. J. Porcello, 'Optical data processing and filtering systems', *IRE Transactions on Information Theory* IT-6 (1960): 386–400.

[46] One of the important but shadowy workers was Norman G. Massey, whom Leith describes as 'developing doctrine' for optical processing. Massey managed the optical technicians and junior engineers, and set up, optimized, and maintained the optical processors. Understanding intimately the equipment limitations and capabilities, he was able to design new optical processing configurations to overcome deficiencies in the data (Leith, Emmett N. to SFJ, email, 4 Mar. 2003, Ann Arbor, SFJ collection). This experimental and procedural expertise was an essential complement to the theoretical insights developed by the Optics Group.

Fig. 4.5. Willow Run 'blockhouse'. It had been built as the control centre for air defence during the late 1950s, part of a military installation of anti-aircraft missiles for the protection of Detroit from Russian bombers, and was occupied by the Optics group in 1962 (Upatnieks collection).

have numerous possibilities to this diverse group of workers and their enthusiasm had been strongly boosted by the success of the original SAR processor. A new application, for example, and one that profoundly influenced Leith's later thinking about wavefront reconstruction was the compression of chirp pulses.[47] A chirp pulse is a signal designed to have a smoothly varying frequency. At audio frequencies, such a descending tone sounds like a chirp. In SAR, a chirp was produced merely by the motion of the aircraft: as the plane approached, passed, and then moved away from a point that reflected radar waves, the return echo would decrease in frequency according to the Doppler effect. On the photographic record of the signal, this would appear as a slice of a Fresnel zone plate in the along-track direction (see Figure 4.6).

Leith proposed that the radar pulses themselves be swept in frequency, so that the radiated pulse was also chirped and that the pulses, instead of being compressed in the usual way in the radar receiver, be recorded in uncompressed form on the signal film. This longer coded pulse allowed certain limitations of the short simple pulse to be overcome: it meant that the echo would chirp even if the aircraft were not moving. In practice, this meant that the signal recorded onto a strip of photographic film (see Figure 4.7) would now show such an echo as the slice of a Fresnel zone plate in the cross-track direction.

[47] See Leith, *The Origin and Development of the Carrier Frequency and Achromatic Concepts in Holography*, op. cit., Chapter 3.

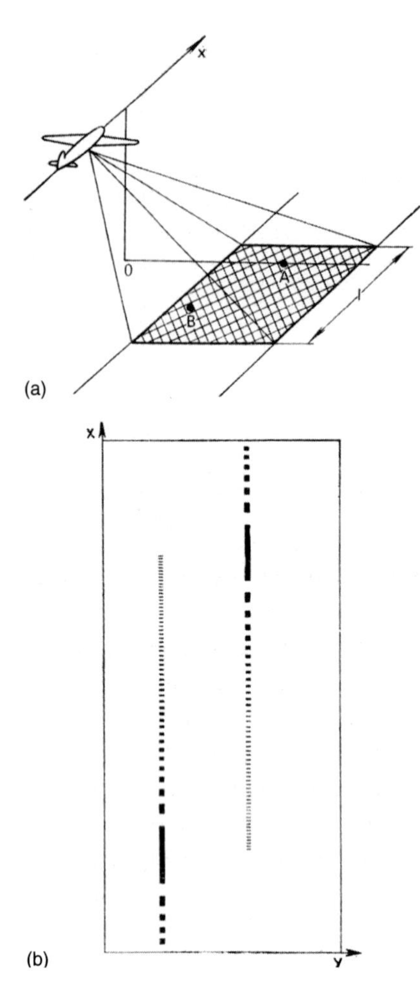

Fig. 4.6. (a) SAR geometry and (b) signal record from Ostrovsky.[48]

When recorded from a moving airplane, the result was therefore a two-dimensional zone plate, although one that was elongated in the along-track direction. As Gordon Rogers had discovered some seven years earlier in a different context, this recorded pattern acted exactly like a lens to reconstruct a point image of the original radar-reflecting object.[49] The addition of a radar chirp pulse to the transmitter allowed the optics of the processor to be fully integrated. This optical configuration, developed in 1958, proved elegant and practical removing the need for a complicated electrical filtering arrangement. The process of bringing the zone plates to a focus meant that the normally separate operations of pulse compression and synthetic aperture data processing now became two aspects of a single,

[48] Ostrovsky, Y. I., *Holography and its Application* (Moscow: Mir Publishers, 1977), Fig. 78, p. 180.

[49] Because the zone plate was elongated, however, it acted as an anamorphic lens, that is, one that has a different effective focal length along perpendicular axes. This effect, caused by the varying range of the reflected radar signals, was compensated by another sophisticated optical arrangement, the 'tilted plane processor'.

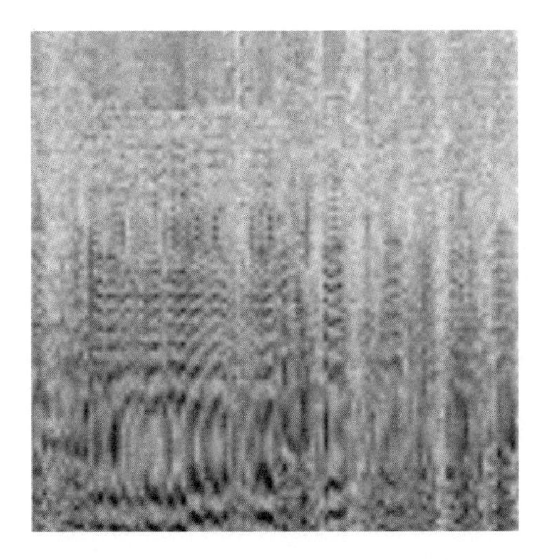

Fig. 4.7. Portion of chirped SAR photographic record (Leith collection).

two-dimensional operation. The optical processing system was no more complicated than before, so the pulse compression operation was obtained for free.[50]

Buoyed by such insights, the workers in the Optics Group began to explore a wide range of applications: other concepts in SAR data processing; optical processing of underwater acoustic signals from sonar; optical analysis of seismic data; and methods of character recognition by using sophisticated optical matched filtering, which was closely related to the problem of chirp compression. Thus numerous applications sprang from a developing body of theoretical work and experimental expertise. The coherent optical processor, just like the digital computer a decade later, came to be seen as a general-purpose machine for solving diverse real-world problems.[51]

4.2 FROM OPTICAL PROCESSING TO WAVEFRONT RECONSTRUCTION

Gabor's wavefront reconstruction seemed to be a good candidate for problem solving and applications. After having discovered the Kirkpatrick–El-Sum paper in late 1956, Leith recalls feeling that it would be interesting to pursue the subject further. He was not overly hopeful, though. His first impression was that his SAR work might provide an answer to Gabor's twin-image problem: the presence of a carrier frequency might make

[50] For further details, see Leith, 'The evolution of information optics', op. cit., p. 1298.

[51] For his own contributions in creating systems for the optical processing of radar signals at the Vavilov Institute, Yuri Denisyuk was awarded the National Prize in 1989. Denisyuk later discovered, comparing results with Leith, that both had been involved in designing optical processors for synthetic aperture radar adopting similar methods: where the Willow Run group initially had used a conical (axicon) lens, Denisyuk had opted for a set of crossed cylindrical lenses (Denisyuk, Yu N. to SFJ, email, 3 May 2003, SFJ collection).

it possible to separate the primary and conjugate images. On the other hand, he felt that SAR work could not be applied directly to optics and most importantly to the creation of optical holograms: a photographic plate records only the time average of the optical wave, rather than the instantaneous time signal with phase and amplitude as with radar detectors.

For the next four years, 1956–60, Leith mused intermittently about various approaches to get around the problem of superposed images.

> My first impression was that if you introduced an off-axis reference beam so that the hologram became like a diffraction grating with various diffractive orders; [. . .] within each order you would have all three images that would be totally inseparable, [. . .] three sets of inseparable images instead of just one. That was my thought, I thought: that won't work.[52]

It was not until 1960 that he decided to pursue the idea experimentally with new colleague Juris Upatnieks, who had joined the Optics Group that year as a Research Assistant.[53] Like all new staff members at Willow Run who initially did not have security clearance, Upatnieks was 'kept off to one side, and followed around so that [he didn't] look into anything [he] shouldn't look into'.[54] As a result, Upatnieks was sidelined to Leith's studies of wavefront reconstruction and in late 1960 they set out to duplicate Gabor's original work.[55] Just like Gabor's 1948 audiences, Leith and Upatnieks experienced the miracle of wavefront reconstruction—the excitement of reconstructing an image from nowhere.

> It was most fascinating, because here you had a piece of film that had nothing but garbage on it, or very fuzzy images at best, and then you looked downstream where they came to a focus, and there you saw a real, nice sharp image, and there was nothing producing it—there was an image but no optical elements—kind of like a grin without a cat by Lewis Carroll's analogy, you might say; [. . .] and that was most fascinating, even though it was nothing new: it was what Gabor had done.[56]

These images were in-line holograms: crude, tiny and two-dimensional reconstructions of a binary transparency—like pen-and-ink drawings rather than photographs, solid black on a clear background. As Gabor and others had found, the image was seriously marred by optical noise, the most serious of which was the out-of-focus twin or conjugate image formed when light is diffracted.

[52] Leith, Emmett N. to SFJ, interview, 22 Jan 2003, Santa Clara, CA, SFJ collection. See also Leith, 'A short history of the Optics Group of the Willow Run Laboratories', op. cit., p. 14.

[53] Upatnieks (b. 1936), who had emigrated from Latvia in 1951, was a junior engineer at Goodyear Aircraft Corp from 1957–9 and completed a Bachelor's degree in Electrical Engineering from the University of Akron in Ohio in 1960. Like Leith, he remained firmly rooted in Ann Arbor through his career. Upatnieks was promoted to Research Associate in the IST in 1965 when he gained an MS in Engineering from the University, to Associate Research Engineer in July 1967, and to lecturer in the Electrical Engineering Department from 1971. He was subsequently a Research Engineer at ERIM from its formation in Jan 1973, when the Willow Run Laboratory was rebadged 'in order to provide for the orderly total separation of the Willow Run research facilities from the University' (Board of Regents Proceedings July 1972).

[54] Leith, Emmett N. to SFJ, interview, 22 Jan. 2003, Santa Clara, CA, SFJ collection.

[55] One of their early holograms, in fact, reproduced the 'Huygens, Young, Kirchoff, Fresnel' text that Gabor had employed fifteen years earlier [MIT Museum items 1993.061.003.1-.5].

[56] Leith, Emmett N. to SFJ, interview, 22 Jan. 2003, Santa Clara, CA, SFJ collection.

Based on Leith's earlier ruminations, he and Upatnieks then began to investigate ways of separating these images to produce a clean two-dimensional image unmarred by background noise. Their first scheme, demonstrated in February and March 1961, was to use a diffraction grating to diffract the light from a mercury arc lamp into two separate beams adding, as Upatnieks put it, 'an offset frequency to the hologram'.[57] When the two beams were recombined by a second diffraction grating onto the photographic plate, the interference yielded a fine fringe pattern. In the parlance of communication theory, this was a carrier wave, and the alternating fringes could be described as having a spatial frequency of about 20 lines per millimetre. This idea, which Leith and Upatnieks initially called the carrier frequency approach in line with the jargon of their colleagues, came to be known as the off-axis technique by optical physicists (see Figure 4.8). Upatnieks recalls the importance of discovering the advantage of using diffraction gratings as beam splitters:

> We had worked for considerable time trying to demonstrate the use of offset reference beam holography without success since [the mercury lamp] light sources did not have sufficient brightness and/or coherence to perform the experiment. Then we came upon the idea of using a diffraction grating to split the light beam into several parts, and a lens to image it. Since imaging does not require coherence, the offset method could be used with light sources having low coherence. This was a breakthrough in our experimental work.[58]

They then tried placing a simple object—a wire—in one of the beams. At the plane where the beams recombined, they again placed a photographic plate to record the interference pattern. When the exposed and processed plate was subsequently positioned in the beam from the mercury arc, various images appeared: one, like Gabor's, reconstructed the twin images and noise, and others, diffracted to each side, showed images of the original wire, free of its twin at last.[59]

Thus, electrical engineering and physical optics were merged on a daily basis at WRL. Carl Aleksoff, then an undergraduate summer student there, recalls that, during 1961 or 1962, 'Emmett often had open on his desk the holography page from Born & Wolf [which had been written by Dennis Gabor]. That's what we had at the back of our minds; "that's been done". That motivated our thinking—or at least Emmett's'.[60] Leith remembers,

> It got exciting, and absorbed more and more of my attention [. . .] We got some really lousy reconstructions; they were noisy, for no good reason that we could think of, but it was just a matter of making the system a little bit better, and the optics a little cleaner and so on, and gradually it got better and better.[61]

[57] Upatnieks, Juris, 'Monthly progress report for February 1961', Institute of Science and Technology, University of Michigan, 27 Feb. 1961.

[58] Upatnieks, Juris to SFJ, email, 15 May 2003, SFJ collection. With the more coherent laser source, they were able to use prisms or mirror arrangements to combine the reference and object beams, as shown in Figure 4.8.

[59] In the parlance of physical optics, the combined image was the zeroth diffracted order, or the equivalent of Gabor's in-line hologram. The separate primary and conjugate images appeared in the $+1$ and -1 diffracted orders of this mutated diffraction grating or hologram.

[60] Aleksoff, Carl to SFJ, interview, 9 Sep. 2003, Ann Arbor, SFJ collection

[61] Leith, Emmett N. to SFJ, interview, 22 Jan. 2003, Santa Clara, CA, SFJ collection.

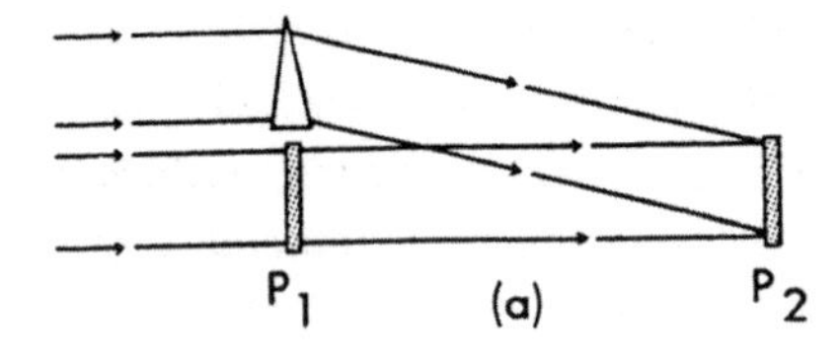

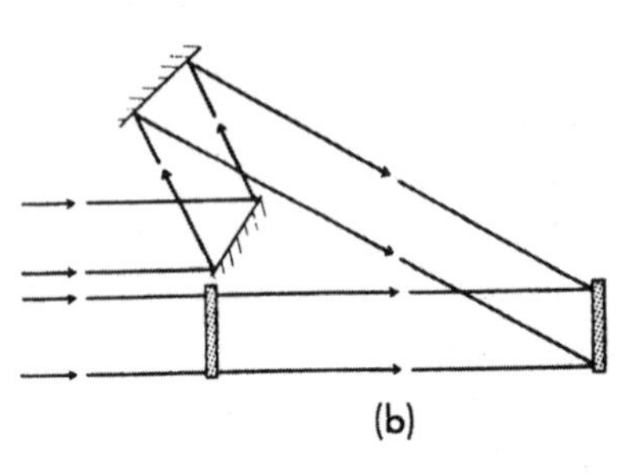

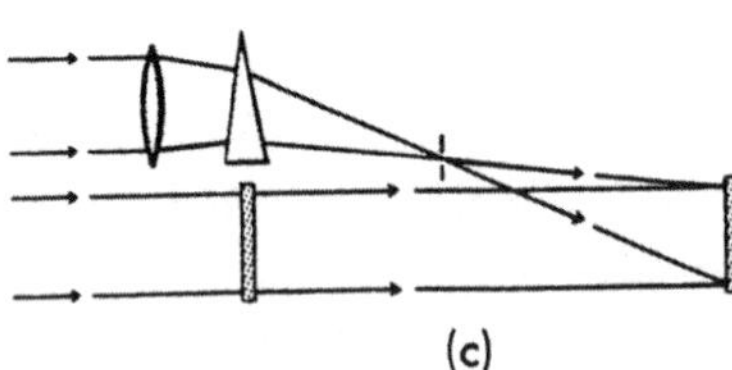

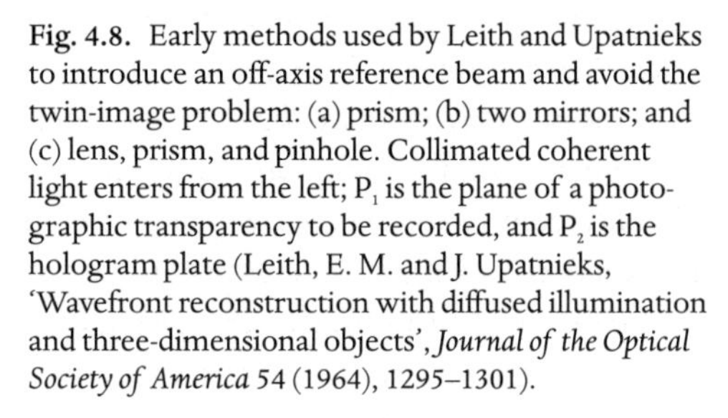
Fig. 4.8. Early methods used by Leith and Upatnieks to introduce an off-axis reference beam and avoid the twin-image problem: (a) prism; (b) two mirrors; and (c) lens, prism, and pinhole. Collimated coherent light enters from the left; P_1 is the plane of a photographic transparency to be recorded, and P_2 is the hologram plate (Leith, E. M. and J. Upatnieks, 'Wavefront reconstruction with diffused illumination and three-dimensional objects', *Journal of the Optical Society of America* 54 (1964), 1295–1301).

Nevertheless, Leith was ambivalent about the practical significance of their results. The published illustrations looked little better than Gabor's work over a decade earlier. Their optical configuration was delicate: unlike Gabor's original approach, using lenses on-axis, Leith and Upatnieks were using two separated beams of coherent light that had to be kept mechanically stable in order to recombine and interfere usefully at the photographic plate. They had created a form of interferometer, which, as optical physicists well knew, demanded mechanical stability for its component parts to within a fraction of a wavelength of light.[62] Moreover, their need to create a carrier frequency or fine fringe pattern on the hologram meant that this sensitivity was amplified: the finer the fringe pattern, the smaller the mechanical displacement that would destroy it. To add to their problems, the resolution of the photographic plate had to be higher than for a regular image. To usefully separate the twin images, the fringe pattern had to be so fine that high-resolution spectroscopic plates were required. They also recognized that the specifications of the photographic plate were further constrained by the need to record both subtle and extreme variations in intensity.[63] And finally, the various optical requirements left little light for recording the essential fringe information of the hologram; using their

[62] In fact, as Gabor realized by 1956, his two-hologram interference microscope was fatally sensitive to such instabilities of alignment. His scheme also had the severe disadvantage of requiring precisely fabricated prisms.

[63] In other words, the dynamic range of the emulsion had to be high in order to record the intense zeroeth order and the relatively weak first diffracted orders. Photographic emulsions have the highest dynamic range of any commonly used recording medium.

monochromatic mercury lamps, the photographic film was further starved of energy, leading to long exposure times. Where implementing Gabor's technique had seemed elegantly simple, the technical constraints imposed rising restrictions on the technique developed by Leith and Upatnieks. Their proof-of-concept exploration was almost immediately laden with difficult experimental questions. Perhaps unsurprisingly, these first Willow Run results in wavefront reconstruction, like Leith's 1956 report on optical processing, provoked little interest among their Willow Run colleagues: as Leith remembers, 'the interest rose as the quality got better, and that was a slow process'.[64]

The pair produced two papers on the improvements they had achieved over the previous six months, written during late 1960 and the summer of 1961. In the first, they described their off-axis geometry for wavefront reconstruction, which solved the conjugate image problem that had plagued Gabor and the others; in the second paper, they described wavefront reconstruction in terms of communication theory.[65] The theory presented in the memo and subsequent paper was essentially the theory of the chirp filter of radar applied to wavefront reconstruction.

But there the project was interrupted: Juris Upatnieks took a leave of absence for military service.[66] The pause hints at their sense of limited significance for the work. Leith recalls that their project was mostly suspended while Upatnieks was away, and that 'other related things, mostly of a classified nature and primarily in the area of optical processing', occupied his time.[67]

Having made important improvements but remaining unknown to virtually everyone, Leith's and Upatnieks' work was in limbo for eighteen months. Coincidentally, Denisyuk in Leningrad had put aside wave photography at about the same time because his Candidate of Science work had been completed and he was taking up leadership of a new laboratory. So, having accomplished major improvements but still little recognized by colleagues, international research on wavefront reconstruction ceased entirely.

4.3 LENSLESS PHOTOGRAPHY

Leith and Upatnieks had already succeeded in obtaining good holograms of stark black-and-white binary transparencies. Within a couple of days of Upatnieks' return in early November 1962, he and Leith resumed the project by making holograms of more

[64] Leith, Emmett N. to SFJ, interview, 22 Jan. 2003, Santa Clara, CA, SFJ collection.

[65] Leith, Emmett N. and Juris Upatnieks, 'New techniques in wavefront reconstruction', *Journal of the Optical Society of America* 51 (1961): 1469; Leith, Emmett N. and Juris Upatnieks, 'A communication theory of reconstructed wavefronts', memo to file, 10 Apr. 1961, Upatnieks collection; Leith, Emmett N. and Juris Upatnieks, 'Reconstructed wavefronts and communication theory', *Journal of the Optical Society of America* 52 (1962): 1123–30.

[66] Upatnieks was away from Willow Run from May 1961 to November 1962 as a Lieutenant serving as an instructor in Microwave Calibration at the US Army Ordnance School, Aberdeen proving ground, Maryland (Upatnieks, Juris, 'Artist files,' MIT Museum 31/928, 1979).

[67] Leith, Emmett N. to SFJ, email, 20 May 2003, SFJ collection.

complex objects than before, consisting of several words of opaque lettering on transparent glass.[68] They also attempted something new: the recording of holograms of greyscale (continuous tone) transparencies. Again, progress was incremental and undramatic: they struggled with the dim illumination provided by their mercury arc source and obtained their first result in late December using a helium–neon laser.[69] Over several months, and by paying close attention to experimental details, they eventually obtained excellent results. They wrote a paper during the summer of 1963, showing the first high-quality reconstructed images of two-dimensional objects.[70] In contrast to their care concerning experimental details, they chose their subjects hastily:

> We had black and white pictures of a bookmark, my daughter as a little girl, and then there was a picture of Juris's brother. We showed some excellent pictures, in fact one of the reviewers said, 'did you get these pictures mixed up?' The reconstruction was indistinguishable from the original.[71]

Reports of their work on reconstructing images of binary transparencies, delivered at the Optical Society of America (OSA) meeting of 1961 and reported in the subsequent 1962 paper had attracted mild interest, but their achievement in reconstructing greyscale images was recognized more widely.[72] They had shifted the technique to a new

[68] They used a Mach-Zehnder interferometer with the transparency in one leg, a mercury arc lamp as the source, and a cylindrical lens to focus the slide in one dimension. (Upatnieks, Juris, 'Development of holography from November 1962 to January 1965', Report for patent attorneys, Willow Run Laboratories, Feb. 1965).

[69] Upatnieks, Juris to SFJ, email, 24 Jun. 2003, SFJ collection; Upatnieks, Juris to SFJ, email, 9 Jan. 2005, SFJ collection; Upatnieks, Juris, 'Computation Notebook #768', lab notebook, beginning 16 Jan. 1963, Upatnieks collection, p. 3; Upatnieks, Juris to SFJ, email, 28 Jan. 2005, SFJ collection. They began with a 100 W GE mercury arc lamp behind a green filter, with the light imagined onto a slit with a combination of spherical and cylindrical lenses and made one-dimensional holograms with it. They also tried a much brighter high-pressure mercury arc lamp with a pinhole, but were unable to record two-dimensional holograms with it because of insufficient intensity and coherence length. Most of their wavefront reconstruction work used the mercury lamp from November 1962 to early January 1963, and mid-January to late February 1963, but favoured the laser during early January, February to May, and after October 1963. The use of lasers is discussed at greater length later.

[70] Leith, Emmett N. and Juris Upatnieks, 'Wavefront reconstruction with continuous-tone transparencies', *Journal of the Optical Society of America* 53 (1963): 522; Upatnieks, Juris and Emmett N. Leith, 'Wave front reconstruction with continuous tone objects', *Journal of the Optical Society of America* 53 (1963): 1377–1381.

[71] Leith, Emmett N. to SFJ, interview, 22 Jan. 2003, Santa Clara, CA, SFJ collection. The continuous-tone two-dimensional holograms of Kim Leith and Ojars Upatnieks that appeared in the December 1963 *JOSA* article had been preceded by earlier attempts that suffered more from coherent-light artefacts. The first such hologram recorded, however, was of a secretary at the University of Michigan. This could not be published because she left employment without signing a model release. (MIT MoH Archive collection items 1993.061.005, 1993.061.006 and 1993.061.008).

[72] There is no record of correspondence between the first generation of wavefront reconstruction researchers (Gabor and his AEI collaborators, Rogers, El-Sum, and Baez) and Leith and Upatnieks during this period. Gabor and Rogers both corresponded with Leith for the first time in 1965 (Leith, Emmett N. to D. Gabor, letter, 26 Mar. 1965, Leith collection; Rogers, Gordon L. to E. N. Leith, letter, 6 Jul. 1965, Sci Mus ROGRS 4). However, Adolf Lohmann remembers being aware of the Willow Run work from 1961, when he heard Leith present a paper at the OSA conference in Los Angeles and was an anonymous reviewer of their important 1964 *JOSA* paper (Lohmann, Adolf W. to SFJ, fax, 30 Jun 2004, SFJ collection); Leith and Upatnieks refer to Lohmann's work in their 1962 paper. Leith's awareness of Denisyuk's work at this time is unclear. He refers to Lippmann photography in their 1964 paper and cites Denisyuk's 1962 paper in a 1964 patent disclosure, but little more until late 1965; see §7.3 for further discussion.

environment and made it interesting to a new audience. For the first time, imaging by wavefront reconstruction could be related to the wider world. Wavefront reconstruction was no longer just a work-around for dissatisfied microscopists or a convenient way of decoding radar data: it was about *pictures*.

The impressive quality of these two-dimensional greyscale images suggested a news story to the OSA, which was due to publish the 1963 Leith–Upatnieks paper. The public relations department of the umbrella organization, the American Institute of Physics, came up with the theme of 'lensless photography' for a press release. Incorporating the reconstructed image of Leith's young daughter, Kim, the four-page press release was timed to coincide with the paper's publication in *JOSA* (see Figure 4.9). Newspapers across America picked up the story in early December 1963, usually with few mutations.

The press release reshaped the meaning of Leith's and Upatnieks' accomplishment through its text and illustrations. The text emphasized that their technique was a sophisticated form of photography. It translated the unfamiliar components of wavefront reconstruction into more conventional objects: a 'blurred photographic negative', an 'optical system', 'projector-like device', 'projection screen', and 'camera-like device'. The illustrations of the diffraction hologram and its reconstruction were unfortunately of identical size and format, suggesting that one mapped directly onto the other. This misidentification was consolidated by the description of the hologram as a 'blurred image' (in fact, the holograms of greyscale images were unrecognizable, unlike Gabor's in-line holograms of binary images). Even worse, the image of the Kim's head was at about the same position as a large interference ring on the hologram, suggesting some correspondence between the large-scale noise and the reconstructed image. Many editors and subsequent newspaper readers must have searched in vain for further points of comparison between the hologram and reconstructed image, failing to perceive their more complex relationship.

Two American newspapers, the *New York Times* and *Wall Street Journal*, gave the story particularly careful attention despite the premium on news stories at that time: coverage of the Kennedy assassination, which had occurred two weeks before the OSA press release, dominated newspaper stories that month.[74] The *Times* interviewed Leith by telephone for over an hour and produced a half-page story.[75] The *Wall Street Journal* article focused more on applications in microscopy, but devoted two-thirds of its space to a description of the principles of lensless photography.[76] Both articles described the

[73] American Institute of Physics, 'Press release: Lensless optics system makes clear photographs', 5 Dec. 1963, 2.

[74] For example, the story ('New camera operating without lens shows scientific promise', *New York Times International Edition*, 11 Dec. 1963, 14) appeared in the International Edition of the newspaper in the shadow of a competing story: 'Oswald Assassin, F.B.I. Concludes'.

[75] Osmundsen, John A., 'Scientists' camera has no lens', *New York Times*, 5 Dec. 1963, 55. Unfortunately, the subtitle read 'It records the phone pattern directly on photographic film', where 'phone' should have read 'phase', undoubtedly confusing even more readers.

[76] 'Lensless photography uses laser beams to enlarge negatives, microscope slides', *Wall Street Journal*, 5 Dec. 1963, 28.

Fig. 4.9. First newspaper images of hologram (left) and reconstruction of a two-dimensional greyscale transparency (right) illustrating 'lensless photography', and based on an AIP press release, December 5 1963 (Leith collection)[73].

unfamiliar process as a two-step extension of photography, with the hologram identified as a smudged, blurred or 'unclarified' negative. This identification as an analogue of photography persisted in subsequent news items. The *New York Times* devoted a half-page article in its *Sunday Photography* section to holography.[77] The concepts and their interpretations made difficult reading then, and appear equally disorienting today. Those first reports of lensless photography portrayed its novelty, its reliance on high science and its perplexing nature. Like the newspaper accounts of SAR three years earlier the stories conveyed a sense of amazement but little else.

4.4 THREE-DIMENSIONAL WAVEFRONT RECONSTRUCTION

It is ironic that this first public coverage of lensless photography occurred at precisely the time that Leith and Upatnieks were privately making a dramatic extension to the technique, based on a new choice of object. From September 1963, Leith and Upatnieks

[77] Deschin, Jacob, 'No-lens pictures: photographic technique employs light alone', *New York Times*, Dec. 15, 1963, 143.

began serious attempts to record holograms of diffusely illuminated transparencies and, from December, diffusely reflecting three-dimensional objects.

Both innovations were a conceptual and experimental leap. The use of diffuse illumination was arguably as important as the decision to record three-dimensional objects. As physicist Howard Smith has observed, until that point Leith, Upatnieks, and their predecessors had made holograms only of thin transparent objects.[78] The resulting shadowgrams could not be viewed easily without additional optical components. However, by placing a diffuser such as ground glass behind their transparencies, Leith and Upatnieks recorded a combined hologram of both. When reconstructed, the wavefront appeared as if the slide was illuminated on a light box: every part of the transparency became visible to an unaided eye even when not positioned directly on the optical axis.

There were other more subtle advantages, too. Because the diffuser scatters light to the photographic plate, the shadow of the object no longer appears. Instead, each portion of the plate receives light from every portion of the object and, as a result, any part of the hologram can reconstruct the entire image, albeit with a lower resolution. Moreover, the spurious pheasant's eyes that spoiled the quality of Gabor's early holograms were now less of a problem, because diffuse illumination smeared out such interference fringes over the entire plate. And a final important advantage was that the hologram now became much less dependent on the emulsion characteristics. In Gabor's technique the photographic plate had to have a well-controlled contrast so that it provided a faithful darkening proportional to the intensity of light. Any non-linearity would cause a serious increase in distortion and noise of the reconstructed image. Upatnieks remembers that their early 1963 efforts were directed at improving image sharpness and trying to reduce the levels of coherent noise in the images, and to understanding the effect of photographic film linearity on image quality. They mused about the effect that diffuse light would have on recording a hologram.

> Toward the end of our Spring 1963 effort, we discussed recording diffuse light. The primary concern was that diffuse light amplitude in the signal beam could be described only in terms of statistics, meaning that light at some spots would be very intense and certainly fall outside the linear range of the film. We wondered what effect this would have and expected a degraded image. Otherwise it could not be much of a problem since our model of an object consisted of an assembly of points. Diffuse illumination would only introduce irregular brightness and phase for each point.[79]

With diffuse illumination and off-axis geometry, Leith and Upatnieks knew that their images would have a dynamic range (the range of intensities from brightest to dimmest) far greater than conventional photographs. They also anticipated somewhat *poorer* spatial resolution. In addition, diffuse illumination raised further questions: for example, would the coherence of light be preserved after reflection from diffuse objects? Their colleague

[78] Smith, Howard M., *Principles of Holography* (New York: Wiley-Interscience, 1969), pp. 7–9.
[79] Upatnieks, Juris to SFJ, email, 12 Jan. 2005, SFJ collection.

Karl Stetson recalls, 'there was a lot of misunderstanding about laser light in that laboratory at that time, and about what coherence really meant'.[80]

By December 1963 they had answers, at least in principle. As Leith explained to a newspaper reporter, 'there is still a kind of coherence remaining in the light beam after this diffusion (with milk glass etc) occurs, and this is exactly the kind we need. Our light must be monochromatic, and there must be spatial coherence in the sense that the phase at one point in space is time-invariant with respect to the phase at another point in space. However the point-by-point variation of this phase can be a completely random function, as it becomes when the light is diffused'.[81] Upatnieks verified experimentally that their laser had a coherence length of about 10 in., but could be used to record a somewhat shallower depth.[82] (see Figure 4.10).

But it was the recording of solid objects that was the greatest leap. Experimentally, diffusely reflecting objects required a different optical arrangement from that used for transparencies, and one that deviated even more clearly from Gabor's original in-line approach. Instead of having components mounted on solid optical rails, they had to be independently positioned on a plane, which made them more vulnerable to vibration.[83] The angle between the beams also tended to be larger, which eventually demanded higher-resolution photographic film to record the fringes. But most crucially, the new investigations required a light source having much greater coherence than the mercury arc lamp: instead of a coherence length of fractions of a millimetre, the source had to have several centimetres of coherence to record interference across the depth and breadth of the object.

The laser had not been a part of their initial wavefront reconstruction work begun by Leith and Upatnieks nearly four years earlier, although a tenacious mythology was to arise later concerning it.[84] In fact, the use of an off-axis reference beam did not in itself demand a laser or even require a more coherent light source, as Leith and Upatnieks had demonstrated since 1961. Earlier investigators—Gabor, Rogers, El-Sum, Baez, and Lohmann—could have obtained high-quality reconstructions in the same way: simply by ensuring that the two optical paths (later denoted the object beam and reference beam) had nearly equal pathlengths and that the interferometer be an achromatic one, that is, having the beams split and recombined by means of diffraction gratings.

[80] Stetson, Karl A. to SFJ, email, 30 Oct. 2004, SFJ collection. George Stroke subsequently claimed to have told Leith about the preservation of coherence after diffuse reflection, an assertion that cannot, unfortunately, be fully assessed from extant documents (Aspray, William, 'George Wilhelm Stroke, Electrical Engineer, an oral history conducted in 1993 by William Aspray, IEEE History Center, Rutgers University, New Brunswick, NJ, USA.' (1993)). On Stroke's priority claims concerning coherence and diffuse reflection, see Chapter 5.

[81] Wixom, Charles to L. H. Dulberger, letter, 11 Dec. 1963, Leith collection.

[82] Upatnieks, Juris, 'Computation Notebook #768', lab notebook, 29 Jan. 1963, Upatnieks collection.

[83] Their first holograms of solid objects, though, employed a single expanded laser beam to illuminate the object and a single mirror beside it, which reflected their diffusely and specularly reflected light, respectively, to the photographic plate. In this geometry, only the stability of the object and mirror were critical. Splitting the unexpanded laser beam into two or more parts was a later innovation.

[84] On origin tales, see Chapter 5. The laser's coherence was, of course, essential for producing holograms of *three-dimensional* objects. But as described later, the laser did not provide a *simple* alternative to the mercury lamp.

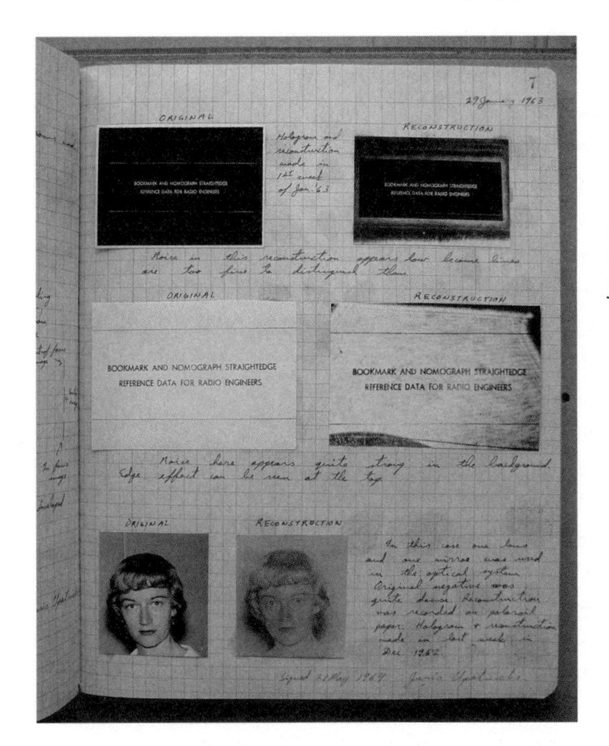

Fig. 4.10. Page from Juris Upatnieks' laboratory notebook, 29 January 1963 (Upatnieks collection)

This was hardly a novel realization in itself; equal pathlength interferometers had been the norm in investigations of physical optics through most of the nineteenth century, and the study and application of unequal pathlength interferometers had been a major research theme in the career of the American physicist Albert A. Michelson from the 1880s.[85] Instead, what had limited previous investigators was literally a narrow-minded view: the assumption (or perhaps merely comfortable oversight) that imaging inevitably required axial optical systems—lenses aligned along a single axis. Certainly microscopy researchers (including Kirkpatrick, El-Sum, Gabor and, before them, Abbe) concerned themselves exclusively with rotationally symmetric optics. As discussed in §2.9, contextual screening limited their ability to conceive developments of the method beyond their planned applications. They had a blind spot concerning an off-axis approach, even a symmetrically off-axis one as employed originally by Leith and Upatnieks.

Coherent light sources were becoming more available during the early 1960s.[86] The generation of coherent radiation by stimulated emission had been discussed in a different

[85] For a career overview of his research, see Michelson, Albert A., *Studies in Optics* (Chicago: University of Chicago Press, 1927). On Michelson as a research technologist focusing on instrument concepts, see Johnston, Sean F., 'An unconvincing transformation? Michelson's interferential spectroscopy', *Nuncius* 18, fasc. 2 (2003): 803–23.

[86] On the military connections with such sources, see Forman, Paul, 'Inventing the maser in postwar America', *Osiris* 7 (1992): 105–34 and Seidel, Robert W., 'From glow to flow: A history of military laser research and development', *Historical Studies in the Physical and Biological Sciences* 17 (1986): 111–48.

context from the early 1950s, and the first maser (an acronym for Microwave Amplification by Stimulated Emission of Radiation) was invented in 1954. This was radiation of very long wavelength and quite disconnected from the optical processing work underway at Willow Run. Maser designs proliferated during the late 1950s and a handful of widely separated investigators began to envisage a visible-light version or *laser* (for Light Amplification by Stimulated Emission of Radiation). While the first ruby *maser* was developed at WRL, Theodore Maiman at Hughes Research Laboratories produced the first pulsed ruby *laser* in May 1960. However, its brief pulse was ill adapted to the optical processing then underway at Willow Run. Optical processing work demanded a continuous source with reliable intensity that could be aligned optically and recorded straightforwardly on photographic film. A more suitable variant was the continuous wave (CW) laser invented by a group under Ali Javan at Bell Laboratories in New Jersey in December 1960, but this generated only a beam of invisible infrared radiation. It was more than a year later, in June 1961, that the helium–neon laser was adapted to generate red light. Commercialized within months, the HeNe laser rapidly became the most common and inexpensive type of laser available.[87] Yet for that first year or two, lasers proved unfamiliar, temperamental, and problematic. There was considerable uncertainty, for example, about the practical spatial coherence of these novel light sources.[88]

Nevertheless, workers in coherent optics were eager to investigate lasers for optical processing. The Optics Group had no direct experience with lasers at this point, but nearly a decade of highly relevant experience with the mercury arc lamp. Its half-dozen laboratories, each populated by a couple of researchers studying one or another aspect of the military application of optical processing, looked forward to lasers as a tool of convenience.[89]

The WRL was not a privileged recipient of the first developmental lasers, though. The Optics Group procured lasers on the open market for evaluation in late 1962, at about the time that Juris Upatnieks returned from his 18 months' military service.[90] Karl Stetson, who had been hired by Wendell Blicken, the Head of the Optics Group, in September that year, helped him choose a thermopile to measure the power of the laser. When the Model 100 HeNe Spectra-Physics/Perkin Elmer arrived, Stetson's first job at the

[87] On the early history of the laser, see Bromberg, Joan Lisa, *The Laser in America, 1950–70* (Cambridge, MA: MIT Press, 1991) and Hecht, Jeff, *Beam: The Race to Make the Laser* (Oxford: Oxford University Press, 2005).

[88] The pulsed ruby laser was a tiny device, scarcely a centimetre long; as a result, its light would not interfere if beams were displaced by more than such a distance. Moreover, measuring the optical coherence of a pulsed source was not trivial. The first HeNe lasers also had low coherence, because the laser beams were a combination of different optical modes, representing different optical resonances caused by various geometrical paths through the laser cavity, some of which were unstable owing to thermal or mechanical changes to the alignment of the laser mirrors.

[89] Stetson, Karl A., 'The problems of holographic interferometry', *Experimental Mechanics* 39 (1999): 249–55.

[90] Stetson, Karl A. and J. M. Marks, 'Use of a Helium-Neon laser as a light source in processing side-looking radar data', report DDC No AD 336 356, Willow Run Laboratory, 1963. This report was written a matter of months after the first commercial HeNe laser became available.

laboratory was its care and tuning. Upatnieks and Leith pressed their colleague Anthony VanderLugt's new 3 mW helium-neon laser into service in early December 1962.[91] Upatnieks recalls that he borrowed the laser beam by redirecting it with a mirror to a set-up near VanderLugt's experiment. Stetson remembers Leith and Upatnieks recording and reconstructing the first hologram of a continuous-tone object by laser on Christmas Eve 1962, with 'results that were astonishing: legible reconstructions were obtained immediately, although not of very high quality'. [92] By late January Upatnieks had recorded reconstructions of a variety of two-dimensional greyscale holograms.[93]

But the laser appeared to be a dubious replacement for the mercury lamp. In substituting the bright new light source, Leith and Upatnieks had anticipated clear advantages. The newly available helium–neon lasers were considerably brighter than their mercury arc sources, so exposure times would be correspondingly shorter. They also hoped that the laser's much longer coherence length would make it possible to form interference patterns between two beams of light that were of considerably different optical pathlength, so that three-dimensional objects could be recorded. Yet, they could not interpret their first experimental results as progress. They were appalled to find that the greater coherence also led to much worsened optical noise. Stray reflections off optical components and scattering from imperfections such as air bubbles in lenses or dusty surfaces caused prominent interference patterns on the recording film. When the hologram was illuminated with a laser, the reconstructed image also was bathed in dramatic, shifting fluctuations in intensity. This swamping by laser 'speckle' made the much brighter laser images virtually unusable at first and particularly difficult to photograph (speckle is caused by interference from random reflections from diffuse surfaces, which transforms the reflected beam into a sea of bright spots that depend critically on the viewer's position and the aperture of the eye or camera). It took time for them to learn how to tame the laser. Karl Stetson recalls:

> The lasers had been purchased for the optical processors used for converting the recorded radar data to pictorial images, of course. The shock they got when they used them for that purpose was profound, because the multiple reflections and scattering in their optical systems, which could be tolerated with filtered mercury arc lamps was impossible to tolerate with laser light due to its vastly greater coherence. So they did continue to use filtered mercury arc lamps with the

[91] A. B. VanderLugt (1937–2000), who completed his first two degrees at University of Michigan and later a PhD at the University of Reading, England, joined WRL in 1959 and became an authority on optical matched filtering based on the optical processing principles being developed there. Such spatial filters, developed from 1961 and declassified in 1963, offered complete phase and amplitude control and grew from the work Leith and Upatnieks were doing on the communication theory of wavefront reconstruction. VanderLugt used the holographic complex spatial filter for object recognition and character reading and Kozma for matched filtering of radar data.

[92] Upatnieks, Juris, 'Development of holography from November 1962 to January 1965', Report for patent attorneys, Willow Run Laboratories, Feb. 1965.

[93] Upatnieks, Juris to SFJ, email, 24 Jun. 2003, SFJ collection; Stetson, Karl A. to SFJ, email, 30 Oct. 2004, SFJ collection; Upatnieks, Juris to SFJ, email, 9 Jan. 2005, SFJ collection; Upatnieks, Juris, 'Computation Notebook #768', lab notebook, 29 Jan. 1963, Upatnieks collection, pp. 6–8.

> processors for some time after the lasers were purchased. There were numerous steps taken to clean up and simplify the radar processors so that they could use the laser light and thus speed up the processing. That involved, generally, removing as many lenses and optical surfaces as possible, and the joke was that someday Emmett would find a way to eliminate the film.[94]

Leith's recollection is similar:

> First we stuck the laser into our synthetic-aperture radar processor, and the results were horrible. We asked 'is this what we were waiting for?' The problem was noise. There was stray light in our optical processors, reflected from the lens barrels and other places [. . .] To get rid of that noise, we put stops here, blocks there and tape somewhere else—mostly tape—to get rid of the reflections. That gave us something that looked better, although there was residual scattering from bubbles inside the lenses and scratches, and so on. If those problems are taken care of, the laser source becomes acceptable.[95]

Lasers were equally problematic for recording holograms. Progress on wavefront reconstruction with mercury lamps had been incremental and unimpressive. The first holograms were scarcely identifiable as such, except by those adepts with faith in the process and coached in what to see. The same proved true, too, of Leith's and Upatnieks' holograms made with the laser. They alternated between mercury lamp and laser light sources over a couple of months and gradually reduced the light scattered from their components, but opted for the laser as the preferred source. Upatnieks spent the early months of 1963 improving the quality of the images, moving to high-resolution film and a liquid gate (a glass-liquid-film sandwich) to reduce the effects of emulsion irregularities.

It was their latest low-noise, high-resolution images made with the laser, submitted in April 1963 to the *Journal of the Optical Society of America* (*JOSA*) that triggered the lensless photography publicity.[96]

The greater coherence of the laser allowed Leith and Upatnieks to expand their experiments. They obtained a Spectra-Physics / Perkin Elmer laser for their own laboratory that spring and during March Upatnieks spent a day trying to record a hologram of a small three-dimensional object, consisting of several steel rods.[97] But his first attempts,

[94] Stetson, Karl A. to SFJ, email, 30 Oct. 2004, SFJ collection.

[95] Hecht, Jeff, 'Applications pioneer interview: Emmett Leith', *Lasers & Applications*, 5 Apr. 1986: 56–8.

[96] Upatnieks, Juris, 'Computation Notebook #768', lab notebook, 1 Apr. 1963, Upatnieks collection, p. 16; ibid., 25 Apr. 1963, p. 19; Leith, Emmett N. and Juris Upatnieks, 'Wavefront reconstruction with continuous-tone transparencies', *Journal of the Optical Society of America* 53 (1963): 522. Intriguingly, Leith suggests that the Willow Run laser holograms may not have been the first: he recalls that 'sometime about 1962 I attended a conference where a paper described making holograms with a laser. This was before we had any lasers, since they were not yet in the market place and the only persons to have them were the ones who worked in laser development laboratories. The conference had no published proceedings, and I have no record of any details. The holograms were extremely noisy and thus of low quality' (Leith, Emmett N. to SFJ, email, 4 Mar. 2003, Ann Arbor, SFJ collection; Leith, 'A short history of the Optics Group of the Willow Run Laboratories', op. cit., quotation p. 7). No further details or corroboration have been discovered.

[97] The Spectra-Physics and Perkin Elmer companies initially made lasers as a joint venture. Spectra-Physics lacked the high quality optics required for the lasers, which Perkin Elmer, a dominant firm for speciality optics and instrumentation, could provide. They were on good terms with the company, as suggested by the 1965 Spectra-Physics advertisement in Figure 4.13. The lab had developed close connections with Perkin Elmer from about 1958, when the company began to supply the optics for their optical processors, and via close relations with Abe Offner, their lens designer. Second, the Willow Run Labs director, Fred

between other work and with a hastily set up experiment, failed. Being involved in other research for sponsors, they did not give the attempts high priority. At the time, they were more actively pursuing optical pulse compression, a technique of considerable importance for the SAR project, and were still trying to reduce the coherent noise in their images.[98] In June, they tried superimposing several holograms on a single photographic plate, rotating it between exposures of various greyscale transparencies. After a break to return to classified research, they returned to holography in the early autumn, investigating methodically how resolution and brightness were affected by variables such as the off-axis angle, film type, and orientation.

In December 1963 Leith and Upatnieks again turned to three-dimensional objects, but this time persisted until it began to work. These behind-the-scenes experimental achievements were beginning to appear the very month that Leith and Upatnieks entered the news for lensless photography (see Figure 4.11), the story about their reconstruction of two-dimensional greyscale images. Indeed, that publicity seems to have spurred their work. When Leith was interviewed about that significant but lesser accomplishment, he intimated to the *New York Times* reporter the possibilities of holograms of solid objects. Whether the reporter misunderstood or merely mistrusted the claims, he ignored the potential story and stuck to the original conceptual frame of two-dimensional lensless photography. Upatnieks recalls that their work on three-dimensional objects began immediately after the 5 December press release, 'since reporters quizzed us about making holograms of reflecting objects and their 3-D image qualities. We had not done that yet, but these properties were obvious to us'.[99]

Upatnieks obtained the first good three-dimensional hologram less than a week later, with another hologram made on 19 December 1963, showing 'partial reconstruction of several three-dimensional objects'.[100] Despite this trigger point, the point of origin is elusive. Leith recalls, 'Everything came slowly. People ask "well, what was your first hologram, which is the first hologram?" [. . .] there wasn't a first hologram; it was an evolutionary process'.[101]

The first successful exposures of these objects made in December 1963 were on small photographic glass plates, too narrow to look through with both eyes:

> You saw almost no parallax, you saw no three-dimensionality, no stereo. We knew that that image that we were looking at was three-dimensional, of course it was, so we cured that by

Llewellyn, was a personal friend of Dick Perkin, founder of P-E [Leith, Emmett N. to SFJ, email, 20 Aug. 2003, SFJ collection].

[98] Upatnieks, Juris to SFJ, email, 12 Jan. 2005, SFJ collection. Upatnieks' notebook records that photography of three-dimensional objects was begun at the beginning of December.

[99] Upatnieks, Juris to SFJ, email, 9 Jan. 2005, SFJ collection.

[100] Upatnieks, Juris, 'Development of holography from November 1962 to January 1965', Report for patent attorneys, Willow Run Laboratories, Feb. 1965, 4; he also recalled, 'First high-quality hologram showing excellent 3-D properties and parallax was obtained about December 22' (Upatnieks, Juris, 'Computation Notebook #768', lab notebook, 20 Feb. 1964, Upatnieks collection, p.31).

[101] Leith, Emmett N. to SFJ, interview, 22 Jan. 2003, Santa Clara, CA, SFJ collection. On holograms as relics, see §14.3. Despite incremental advances, a point of origin had to be defined for legal purposes. One of

Fig. 4.11 Juris Upatnieks and Emmett Leith at optical table, December 1963. The apparatus, described in press reports as a hologram camera, was in fact a plate holder intended to allow multiple exposures and was probably never used. The actual apparatus was distributed over a large solid table.[102] (University of Michigan News Service, Leith collection).

> [. . .] going to plates that were [. . .] strip plates—Kodak sold them—and we cut them to this length so two eyes could get in, and sure enough, they were three-dimensional as prescribed. But the realization was unbelievable; just seeing it there, seeing it happen, was most astonishing.[103]

Seeing that three-dimensional reconstruction was remarkable, unsettling, and sublime. Intellectually obvious or not, viewing a three-dimensional image produced from a jumbled interference pattern on a flat plate was memorable for all who witnessed it. Unfortunately for museum curators and historians, there are few holographic icons to venerate. Like their earlier choice of subjects for greyscale transparency holograms, the first exposures of solid subjects consisted of convenient found items. 'We looked in desperation for an object that would be suitable and found scrap metal, resistors, capacitors, little piles', Leith recalls.[104] Indeed, even those early holograms produced in their lab (and retained in dusty cabinets for decades) seem perversely mundane: they show tools, models, and figurines.

During the last week of December 1963, they moved from a custom-made steel optical bench to a 4 × 5 foot granite table. Three days before Christmas, they had obtained their first high-quality reconstructions of a three-dimensional object, a toy train.

the earliest Dec 1963 holograms of an opaque three-dimensional object was lodged with patent attorney Bob Washburn in 1964 in case any interference claims were lodged [Leith, Emmett N. to SFJ, 20 May 2004, SFJ collection].

[102] Upatnieks, Juris to SFJ, email, 30 Sep. 2003, SFJ collection.

[103] Leith, Emmett N. to SFJ, interview, 22 Jan. 2003, Santa Clara, CA, SFJ collection.

[104] Ibid. More variants of the train hologram were made, many as give-aways, 'following the junk pile images. The first had just the train; the second had a hex-wrench in front to highlight the depth effect. Later, the bicyclist replaced the hex-wrench, a ceramic bird was put in back, and then a name-plaque was put in front' (Leith, Emmett N., 'Materials collected on 5/19/93 by S.A. Benton, at Ann Arbor', 19 May 1993, MIT 1993.061.004).

But such objects were mechanically unstable, and sometimes moved during the long exposures: 'we had a lot of failures', says Leith. 'Unfortunately, toy train holography is a destructive process; to make a good toy train hologram you should fill the innards of the train with epoxy to give stability and then epoxy the train to a section of track'.[105] Their early single-beam set-ups were problematic, too, because the reference and object beams were uneven in intensity, causing the reconstructions to be dim and variable. Upatnieks recalls that they made many improvements between December and the following May.

The pair gradually became aware of further technical constraints for recording holograms of solid objects: the first lasers tended to drift from single-mode to multi-mode operation causing the coherence length to vary laterally across the beam and so made the holograms of objects more difficult to record. The laser's characteristics were fickle, and could degrade disturbingly, making it little better than an optically noisy mercury lamp. With colleague Fred Rotz, they monitored the laser beam using a photomultiplier detector and spectrum analyser, adjusting the laser mirrors for best stability.

There were also problems at the other end of their apparatus. Their experimental set-up wasted much of the available light. The inefficiency was worsened because off-axis holograms demanded film having higher resolution to record the fine interference fringes. The cheap and convenient 35 mm film that had been used for the earlier work had to be replaced by glass plates. The slow Kodak 649F emulsion—a photographic formulation designed for the high-resolution work of spectroscopists—required some 10 min. of exposure from their laser to adequately darken the plate.[106] This made the rigidity of the optical arrangements all the more critical. Thus far, their experiments with off-axis holograms had arranged the two beams side-by-side on an optical rail, a set-up that was inherently stable. But the holograms of reflecting objects demanded a larger surface and separate optical components. And the stability of the apparatus over this time period—especially for the mirrors and object itself—was also more important than they had appreciated. Commercially available optical mounts proved too shaky to maintain distances to better than a wavelength of light over such long time periods. And temperature variations, mechanical creep or small vibrations could readily shift the interference fringes and smear the hologram. Karl Stetson recalls, 'they had no idea of the need for vibration isolation but were recording holograms with fixtures on optical rails on ordinary tables', with the importance of vibration isolation gradually recognized by Stetson and his colleague Robert Powell.[107]

[105] Leith, Emmett N. to SFJ, interview, 22 Jan. 2003, Santa Clara, CA, SFJ collection. The first hologram widely seen and reported, that of the model train, had been produced using a machined model borrowed from an electronics technician at Willow Run. See, for example, the hologram collection at the MIT Museum archives (most of which were originally collected by the New York Museum of Holography).

[106] Until late 1963, they used ordinary films such as Kodak high contrast copy film, Pan-X or SO243, which provided some thirty times lower resolution than the slow 649F emulsion. Upatnieks carried out extensive resolution tests that summer and began using 649F plates for their diffusely illuminated holograms. The commercial availability of argon ion lasers from 1965, which had powers of the order of watts rather than milliwatts, were cited by their manufacturers as a breakthrough allowing exposures as short as 10 sec.

[107] Stetson, Karl A. to SFJ, email, 30 Oct. 2004, SFJ collection. Stetson began recording holograms with Leith and Upatnieks from Sep. 1964.

The depth of the technique emerged gradually. There was no unique moment of discovery, no sudden perception of a breakthrough. Almost imperceptibly, and by subtle iterations, Leith and Upatnieks crossed a threshold of awareness. Indeed, its very significance may be illusory. The step from two- to three-dimensional imaging was, suggest the inventors retrospectively, intellectually trivial. Leith states dismissively, 'Gabor said "it's three-dimensional"; he had always said so. We knew it was supposed to happen'.[108] Similarly, Upatnieks recalls:

> The property of holographic images being three-dimensional was obvious to us and I am sure to Prof. Gabor. In fact, it was so obvious that initially we considered it to be a waste of time to show it experimentally. [. . . But] once we succeeded, we were fascinated by the reality of the image and spent hours looking at it, and showed it to our colleagues.[109]

Perhaps Leith's and Upatnieks' nonchalance about their road to three-dimensional imagery is a reaction to the subsequent years of celebrity triggered by the publicity concerning it. Nevertheless, lensless photography did not sweep the world overnight. Like the invention itself, publicity took time to develop.

Three-dimensional holograms were slow to make an impact beyond Willow Run. The first published descriptions of three-dimensional imagery appeared inconspicuously at Christmas 1963, barely three weeks after Leith and Upatnieks had begun work in earnest on 3-D holograms and a few days after their first high-quality results. Aimed at a technically aware audience, a one-page article in *Electronics* downplayed the association with photography. Mentioning numerous applications, the story ended with a hitherto unmentioned capability:

> Leith points out that it is possible to record opaque three-dimensional objects, using reflected light, and that this has been done. The image reconstructed from the hologram can be photographed, completing the second step. In addition it is possible to view the reconstructed image directly by placing the eye so as to intercept the light emerging from the hologram. A three-dimensional projection is formed, having the effect of stereo projection, though only one hologram is used.[110]

These were still merely hints to describe the revolutionary technology in terms familiar to readers. A longer article that month in *Science Fortnightly* also vaunted the three-dimensional capabilities of the new technique. In it, Leith emphasized microscopic applications, but again tried to place the latest findings into a wider context. Unfortunately, the properties of the off-axis hologram seemed too novel for straightforward description even for amateur photographers:

> The resulting hologram is then capable of projecting a three-dimensional image in space—and no screen is required. The projected 'image' hangs in midair. The technique is limited to

[108] Leith, Emmett N. to SFJ, interview, 22 Jan. 2003, Santa Clara, CA, SFJ collection.

[109] Upatnieks, Juris to SFJ, email, 24 Jun. 2003, SFJ collection.

[110] Dulberger, Leon H. and Charles Wixom, 'Lensless optical system uses laser: opaque 3-D objects may be imaged without lenses using reflected light', *Electronics*, 27 Dec. 1963.

> indoor photography, since it requires coherent (monochromatic, in phase) light to work. Sunlight is not coherent.[111]

Scepticism or disbelief arrested many readers. Upatnieks recalls 'numerous doubts expressed by inquiring reporters', and their consequential decision to make higher-quality holograms as demonstration pieces.[112] By March 1964, he and Leith had produced the widely seen hologram of a toy train (see Figure 4.12).[113] With impressive holograms to show off, the news about three-dimensional imaging during the winter of 1963–4 began to raise the profile of their latest research, something that had not occurred at Willow Run since the announcement of the SAR system in 1960.

> It was a type of imagery that had never before been seen. People sat up and took notice, people in the laboratory looked at it in astonishment, the management came in and looked at it, and the Director came in, people outside the university came and looked at it.[114]

Over those first three months, word-of-mouth accounts of visitors and reports in the popular press began to raise attention. Seeing was believing for the reporters, but explaining the process was another matter. The photographic analogy launched by the AIP press office proved difficult to quash. One early attempt was a Sunday article in the local Ann Arbor newspaper. Besides descriptions of side-looking radar and death-ray lasers, the article devoted a few paragraphs to lensless photography. It emphasized attributes that had been unmentioned in previous reports.

> The light bounces from the subject, into a mirror and onto film. The result is a transparency that looks to the eye like a buttermilk sky. But when laser light is played upon it, the original scene takes shape in three dimensions. An unusual property of the transparency is that the whole or any part of it contains the entire picture. Tear it up and any fragment of it will reveal the total picture under laser light.[115]

The lensless photograph, or off-axis hologram, was becoming ever more curious. It showed no trace of an image. It could be created only by special laser light. It was a full picture contained within any fragment—an even more intriguing and enduring concept, but also imprecisely disseminated.[116] In its three-dimensional guise, lensless photography began to look very unlike photography indeed.

Nevertheless, the local eyewitness accounts had little impact beyond Ann Arbor. The galvanizing event was a presentation at the final session of the spring meeting of the OSA

[111] 'Laser photographic process uses no lenses, produces 3-D images', *Science Fortnightly*, 1 (9), 25 Dec 1963: 1–2.

[112] Upatnieks, Juris to SFJ, email, 25 Jun. 2003, SFJ collection.

[113] Upatnieks, Juris, 'Computation Notebook #768', lab notebook, 18 Jun. 1964, Upatnieks collection, pp. 4–20 and 4–21. Most of the holograms were recorded in March and May 1964.

[114] Leith, Emmett N. to SFJ, interview, 22 Jan. 2003, Santa Clara, CA, SFJ collection. Despite security measures for classified research at WRL, it was not uncommon for visitors to have relatively easy access to the site.

[115] Lutz, William W., 'New discoveries at Michigan universities', *The Detroit News—The Passing Show*, Sunday Feb. 23, 1964, 1.

[116] On the later cultural repercussions of this property of Leith-Upatnieks holograms, see Chapter 13.

Fig. 4.12 *Toy Train*, Emmett Leith and Juris Upatnieks, March 1964, one of several variants of the first widely seen hologram (Upatnieks collection).

in 1964. On Friday, 3 April, in a session on 'Information Handling by Optics' held at the Sheraton Hotel in Washington DC, Juris Upatnieks described their latest work. Having been written late the previous year, the abstract released in March again framed their research as an extension of photography, although clearly describing the novel three-dimensional features of the image:

> Photography of three-dimensional objects has been achieved by an extension of the two-beam wavefront-reconstruction technique described earlier by the authors. In taking the picture, the scene to be photographed is illuminated with coherent light. A photographic plate intercepts a portion of the light reflected from the scene and at the same time receives a portion of the direct beam, which is reflected onto the plate by means of a mirror. If desired, ground glass may be placed in the incident beam, thereby diffusing the light that illuminates the scene. The reconstructed virtual image can be observed by looking through the hologram when it is placed in a beam of coherent light; upon doing this, one sees that all the visual properties of the original scene are preserved; parallax between near and more distant objects, a requirement to refocus the eyes when viewing objects in different parts of the scene, and, finally, a stereo effect equal to that obtained by ordinary stereo photography. The real image can be recorded by placing a film at the desired plane of the three-dimensional image.[117]

This was a down-to-earth and astonishing announcement. In lucid language divorced from any specialist jargon and concepts, the abstract announced optical characteristics that surprised many practising physicists: experimental evidence that a ground glass plate would not destroy optical coherence and the creation of a reconstructed image that was dramatically superior to conventional stereoscopic images, evincing the properties of parallax, focus at different planes, and binocular depth.[118] The optical phenomena raised

[117] Upatnieks, Juris and Emmett N. Leith, 'Lensless, three-dimensional photography by wavefront reconstruction', *Journal of the Optical Society of America* 54 (1964): 579–80.

[118] By contrast, Gabor's papers had highlighted only the localization of image focus at different depth. Gabor's work in stereoscopic cinema relied on the then conventional ideas of two separate images to provide binocular depth cues for the illusion of three-dimensional imagery.

Table 4.1 Leith and Upatnieks research in wavefront reconstruction, 1960–4[120]

- August 1960–May 1961: repetition of Gabor experiments; investigation of off-axis geometry to solve the twin-image problem; study of the relationship with communication theory (about four months in total).
- November 1962–May 1963: hologram recording and reconstruction of transparencies with plane wave illumination, increasingly replacing mercury lamp with laser illumination. One abortive attempt at recording the hologram of a three-dimensional object.
- May 1963–December 1963: improving hologram image resolution, mainly by recording point or line images; pulse compression experiments; optical simulation of SAR signal film.
- September 1963–November 1963: recording holograms of diffusely reflecting objects.
- December 1963–May 1964: recording holograms of three-dimensional objects and diffusely illuminated transparencies and recording their reconstruction with high resolution.
- From May 1964: holography of microscopic objects and their magnified reconstruction; lens correction by holograms.

questions for all optical scientists, ranging from the properties of laser light, to imaging and emulsion properties, and to the nature of stereoscopic vision. The abstract described a change of perspective in more than one sense. Leith has noted retrospectively that 'the abstract, more than any other document, including the papers and the news releases, [. . .] set in motion the great explosion in holographic activity'.[119] But it was an explosion triggered by a slow-burning fuse (see Table 4.1).

It was the subsequent demonstration to optical specialists that caused the greatest impact. Juris Upatnieks presented their 15-min. paper and announced that an example of their work was on display. The laser manufacturer Spectra-Physics, which had a display suite, allowed them to display a hologram of a toy train there. When Leith and Upatnieks reached the room, they found a long queue of optical scientists out of the suite, down the hall, and around the corner. Many of these specialists were disbelieving or confused. The toy train appeared perfectly real and yet could not be touched behind the photographic plate. Several questioned where it was hidden or sought the mirrors that had produced the illusion. One feared that his eyes had been damaged by the laser light; lasers as sources of illumination were still a distinct novelty, even for optical scientists.

In the following weeks, telephone calls and letters from researchers seeking details, clarifications, and diagnoses of problems inundated Leith and Upatnieks. Many of the physicists and engineers who attempted to recreate their work encountered difficulties, as they had relatively little experience with coherent light, optical processing, or lasers. Indeed, the most closely related specialty was interferometry (providing the requisite

[119] Leith, Emmett N. to SFJ, email, 4 Mar. 2003, Ann Arbor, SFJ collection.

[120] Upatnieks, Juris to SFJ, email, 14 Jan. 2005, SFJ collection, based on notebook records and Upatnieks, Juris, 'Development of holography from November 1962 to January 1965', Report for patent attorneys, Willow Run Laboratories, Feb. 1965. Upatnieks recalls that they did not keep daily records, and began to keep regular notes only later when a summary monthly progress report was required. Notebook #768 was a formal record that he started at the suggestion of patent attorneys as a basis for any future patent applications, with later notebook pages witnessed and signed by a colleague [Upatnieks, Juris to SFJ, email, 28 Jan. 2005, SFJ collection].

skills and hardware) and spectroscopy (for the photographic emulsions and processing necessary). Both these subcultures were relatively small and specialized; acquiring the necessary skills consequently took time. Most investigators were, at best, familiar with one or the other but not both.

Besides the demands of other optical scientists, Leith and Upatnieks found their time increasingly taken up with explanations and prognostications for the popular press. The April 1964 OSA meeting and subsequent *JOSA* paper that autumn introduced the subject to optical scientists, but further conference presentations continued to spread the word to a wider range of practitioners. In November, a Symposium on Optical and Electro-Optical Information Processing Technology was held in Boston, Massachusetts, cosponsored by the Office of Naval Research (ONR) and local chapters of the Professional Technical Group on Electronic Computers of the IEEE, the OSA, and Association for Computing Machinery. Like the dramatic April OSA meeting, it was accompanied by an exhibition of Leith and Upatnieks' holograms in an upstairs suite. The pair reached successively wider technical audiences in the following months, presenting papers to optical engineers, automotive engineers, electronics engineers, and instrumentation engineers.[121] Each new body of spectators evinced the same startled reactions. At the National Electronics Conference in Chicago in 1965, for example, a reporter observed, 'long lines of scientists saw a chess board with five men, a toy railroad scene and a model of an army tank. The objects were really back at the university's laboratory in Ann Arbor'.[122] The new viewers were just as amazed as the optical scientists had been over a year earlier.

> 'It is the most startling thing I have seen', said a nuclear scientist from Argonne National laboratory. A navy captain said, 'It's fantastic. I don't see how it can be done'. Many of those who view the picture ask to touch the photographic plate to assure themselves that it is indeed flat and that the Michigan exhibitors are not trying to pull a fast one'.[123]

Laser Focus magazine dubbed the new technique lasography, but failed to win adherents.[124] Optical firms such as Spectra-Physics—eager to promote a practical application for lasers—piggybacked on the acclaim of the University of Michigan work (see Figure 4.13).

[121] Leith, Emmett N. and Juris Upatnieks, 'Holograms: their properties and uses', *SPIE Technical Symposium* 10 (1965): 3–6; Leith, Emmett N., 'Applications of holography', presented at *Automotive Engineering Congress*, Detroit, Michigan, 1968; Leith, Emmett N. and Juris Upatnieks, 'Modern holography', presented at *1968 IEEE international convention*, New York, 1968; Leith, Emmett N. and Juris Upatnieks, 'Some recent results in holography', presented at *Northeast Electronics Research and Engineering Meeting*, Boston, MA, 1968; Leith, Emmett N., 'Holography: An introduction and survey', presented at *Proceedings of the 7th Annual Biomedical Sciences Instrumentation Symposium on Imagery in Medicine*, University of Michigan, Ann Arbor, MI, US, 1969.

[122] 'National Electronics Conference, Chicago', *Pontiac Press*, 28 Oct. 1965.

[123] 'Photo shows work of new optic system: Viewers call exhibit startling, fantastic', *Chicago Tribune*, 27 Oct. 1965.

[124] '3-D lasography—the month old giant', *Laser Focus*, 1 Jan. 1965.

Fig. 4.13 Spectra-Physics advertisement using one of the iconic train holograms by Leith and Upatnieks to promote newly available argon ion lasers (*Physics Today*, May 1965, Funkhouser collection).

By a combination of demonstrations, conference communications, and journal publications, Leith and Upatnieks continued to spread the word. They submitted an expanded version of their OSA meeting paper to *JOSA* in June 1964, which was published that November, nearly a year after their first successes with 3-D imaging.[125] That publication,

[125] Leith, Emmett N. and Juris Upatnieks, 'Wavefront reconstruction with diffused illumination and three dimensional objects', *Journal of the Optical Society of America* 54 (1964): 1295–301.

too, was preceded by a press release. In it, the American Institute of Physics publicity office vaunted the new capabilities of wavefront reconstruction achieved over the previous year. It focused on the unfamiliar three dimensionality of the imaging and especially parallax. The ability to look around and over the objects in a reconstructed scene demarcated the Leith–Upatnieks technique from any previous form of stereoscopy.[126] The association with the laser accentuated the modernity and mystique surrounding the technique. A final frisson of the exotic for uninitiated readers—but representing the developers' own insights concerning their method—was provided near the end of the release.

> The process can be thought of as capturing and storing the light rays and releasing them at some later time, whereupon the imaging process is carried to completion.[127]

Gabor had emphasised the two-step nature of the process. For Willow Run researchers, two-step imaging was an essential property of SAR systems. But wavefront reconstruction originating and ending with visible light made the idea of storing a wavefront more obvious. It was also, unknown to the Willow Run workers, a theme that had been developed by Yuri Denisyuk in Leningrad a year or two earlier. But the notion of a window with a memory was particularly apt for the Leith–Upatnieks hologram.

A reporter covering the Boston conference in November gave a first-hand report of seeing a hologram that recited what was to become a familiar litany of counterintuitive properties:

> When you looked at the hologram, illuminated from behind by a gas laser, you saw the train and conductor toys right there on the table, in three dimensions. If you wanted to see what was behind the little man, or in front of the toy locomotive, you simply moved your head to see them. No need for viewing glasses, double images or squinting [. . .]
>
> Because there are no lenses, each point on the object is recorded all over the photographic plate. So you can take a hologram and cut it in half—or in a dozen pieces—and each piece will still show the entire object, from a slightly different point of view, with only a little loss in sharpness.[128]

The hologram had undergone a metamorphosis from an incomprehensible intermediate component in Gabor's holoscope into an exciting but even more baffling artefact. The off-axis hologram or lensless photograph was able to recreate disturbingly realistic three-dimensional images; it somehow encoded details of the entire scene in its every portion; and it froze time, releasing that image when prompted again by the laser beam, itself an exotic new technology. Wavefront reconstruction had been recast as a vision of the

[126] The off-axis hologram stood in sharp contrast to familiar but derided 3-D technologies: stereoscopic cinema had been a short-term fad during the early 1950s and had required viewers to wear either polarizing or two-colour glasses; the Victorian stereoscope had had periodic revivals, the biggest being during the 1940s and 1950s as the Stereo Realist camera; and in the early 1960s it was reaching a new audience in its resurgence as a children's stereo viewer marketed through toy shops or breakfast cereal promotions. All four prior technologies had the well-appreciated limitation of a fixed stereo view *without* parallax.

[127] American Institute of Physics, 'Press release: Objects behind others now visible in 3-D pictures made by new method', 25 Oct. 1964.

[128] Novotny, George V., 'The little train that wasn't', *Electronics*, 37 (30), Nov. 30, 1964: 86–9.

future, no longer constrained by the goals of microscopists or even hinting at the optical processing research that had given it birth. Divorced from its roots, it seemed guaranteed to provide continuing awe, commercial applications, and success.

And yet the rate of accommodation was slow. The members of the early expert audiences, and the wider public, struggled to fit this stunning new imagery into their understanding and expectations of optics. Their varying responses illustrate the inexorably gradual pace of intellectual adaptation and cultural mutation. To appreciate and comprehend the hologram required profound shifts: theoretical for some, cultural for others. The practitioners were challenged as much as their audiences were by lensless photography. Leith, Upatnieks, and soon their colleagues in the Optics Group, found themselves straddling a professional no-man's land. They were beginning a transition from the underground—a working environment of traditional secrecy at WRL—to public visibility. They also were straddling disciplines. The electrical engineers at WRL were merging their background in communication theory with new experimental and theoretical knowledge from physical optics, generating a new subject in the process.[129] And by representing their unintuitive technique as an extension of photography, they were making it intelligible for wider audiences.

The extraordinary new technology, germinating literally in the dark and sprouting in the hidden field of optical processing was grafting onto existing scientific cultures and sowing the seeds of new ones. The 1964 Boston meeting was a key event in that process of growth: it brought the first eyewitness reports of the new imaging technology directly to the general public; it astonished members of cognate disciplines beyond optics itself; and it simultaneously clouded the achievements of the Willow Run workers with rival claims from a colleague at the University of Michigan, George W. Stroke.

[129] This hybrid subject proved stable. According to Brian Athey, a close associate in later years, Leith consistently employed a successful technique of problem solving and idea generation through his career, a combination of linear Fourier theory and a building block approach to optical components akin 'to Feynman diagrams in particle physics' (Athey, Brian to SFJ, interview, 2 Aug. 2004, Glasgow, SFJ collection). This modular, linear systems approach applied to physical optics appears very similar to the methodology employed by electrical engineers.

5

Constructing Holography

5.1 INTRODUCTION

Over the span of two decades, 1947–1966, the subject of holography was constructed as a collage from distinct perspectives. Dennis Gabor's concept of holoscopy or wavefront reconstruction had led a handful of investigators to pursue novel forms of microscopy; Yuri Denisyuk's wave photography provided a different slant on the possibilities of general imaging influenced by Lippmann photography, but initially convinced few colleagues; and Leith and Upatnieks, alongside the separate work of Adolf Lohmann, had reconceived wavefront reconstruction according to communication theory. Following the excitement of seeing three-dimensional imaging, hundreds of researchers began to connect these three dissimilar formulations from mid-1964. They succeeded in constructing a new, more general subject by the end of 1966. Holography, a field with extraordinary intellectual and commercial potential, had been born.

Its birth was not a simple one, though. The race to explore fresh intellectual territory yielded not only new perspectives, but also controversy. The participants sought to identify their new field, and their roles in it, according to an intellectual lineage. Social and economic factors contributed to its definition. Until the mid-1960s, origin stories about these subjects, if they were told at all, had been told in terms of failures. But after that time these stories served a new purpose. They could now tell a tale of triumph over adversity; of tenacious and even heroic workers toiling in obscurity; of the success and inevitable progress of well-funded science; of a chain of insights leading to a general principle; and of moments of insight that supported patent claims. Such stories argued that the rapidly expanding subject must have had a discrete point of origin. But by reconnecting the separate dots—extrapolating backwards in time—workers in the field developed contrasting views of that origin and of their own lineage. One danger was having the tail wag the dog: differing assessments of the relevance and significance of precursor ideas and rehabilitated earlier work could lead to profoundly different claims. These differing claims are significant for two reasons. First, they clouded the historical events, imbuing them with meanings that had been absent originally. Second, they affected the very conception of the subject itself. Tracing the priority disputes and patent claims can therefore help to clarify the public history and shaping of the subject. This chapter consequently focuses on how holography was constructed or packaged.

Origin myths in science and technology can seldom be tracked closely owing to the dearth of available evidence concerning the construction, subliminal or intentional, of such accounts. A subject young enough to have many of its early participants still available for direct contact, holography illustrates how the history and explanation of a subject are woven from multiple accounts and contemporary interpretations. As a burgeoning field, it offered many opportunities for intellectual exploration, professional advancement, and corporate profit. And like many new sciences, the concepts and activities associated with holography became surrounded by myths and myth making.

Because of the implications for demonstrating early involvement in the field, even the origins of the term holography itself were the source of a remarkable amount of acrimony. George W. Stroke claimed to have invented the term in 1964, 'after several weeks of thinking about a word which would for all time associate Gabor and his hologram for scientific and public use', formally proposing it at a meeting on Optical Data Processing in Boston that November.[1] Other accounts credit Stroke's graduate student David Falconer, Gordon Rogers, or Gabor himself.[2] Falconer claimed to have coined the word while a student in George Stroke's lab.[3] Rogers separately claimed to have generalized Gabor's older noun *hologram* to yield both a noun (*holography*) and a verb (to *holograph*). In 1949 he had sent Gabor a letter in the form of a 4 × 6 mm microscope cover slide, which read,

> Dear Dr Gabor, I write you a short message by 'holograph' which is the system of conveying information by hologram. I trust you will be able to decipher this and the other messages. If you haven't already thought of this idea, I may be the first to convey information by this means. I hereby warn you that some of the holograms are of three-dimensional objects, so that the whole picture will not be in focus at once. Looking forward to your holographed reply.
>
> Yours sincerely, G. L. Rogers.
>
> P.S. added 11/10/49—Holography is a very *slow* means of communication![4]

Rogers himself was uncertain, though, that Gabor had ever been able to decipher the holographed letter and does not appear to have used the term in notes or publications. The importance of priority became significant to Rogers only after the term had seemingly been reinvented and appropriated by one or two others: association with the

[1] Rogers, Gordon L., 'The word "holography" ', *Scientific Research* 8 Jan (1968): 57–8.

[2] LoVetri, Joan, IDS thesis, MIT Museum 41/1238, (1982); Lehmann, Matt, 'Holography—the early days', unpublished account to R. Jackson, MIT Museum 29/790, 2 Jun. 1982.

[3] Cochran, Cary D. to SFJ, interview, 6 and 8 Sep. 2003, Ann Arbor, MI, SFJ collection. Falconer left U-M in 1965 to join Conductron and was later with Stanford Research Institute.

[4] Rogers, Gordon L. to D. Gabor, 'holographed' letter, 27 Sep. 1949, Dundee, Rogers, Gordon L., 'The word 'holography', *Scientific Research* 8 Jan (1968): 57–8.

coining of the word retroactively rehabilitated his hard work. Rogers' archives record the production of that hologram in September 1949 but not his use of the term holography in any extant documents. Indeed, Rogers was to write soon afterward,

> I shed no tears over the loss of the term 'hologram'; it was too like the word 'holograph', which has been in good standing with quite another meaning in Scots Law for some centuries. But it is not so easy to find a substitute [. . .] On the whole I would favour concocting a new term altogether, possibly with the phrase 'pattern' tacked onto the end.[5]

The origins and use of terminology can suggest not only seminal workers, but allegiances, too. For reasons suggested by the following discussion, the members of Willow Run Radar and Optics Lab preferred to use Gabor's term *wavefront reconstruction* until the mid-1960s. Thereafter, Soviet investigators were left as the principal users of the term as others adopted *holography*.[6]

So terminology provides important clues about the classification and research affiliations. The first historical summary of the subject, and arguably its earliest and most influential packaging, was disseminated by George Stroke while a professor of Electro-Optics at the University of Michigan. Stroke's account supported particular priority claims and the contentions embroiled several workers active in the fields of holography and information processing during the mid-1960s. They transcended mere personality conflicts, however. Stroke's assertions influenced the early historiography of holography and the awarding of the Nobel Prize in Physics to Dennis Gabor in 1971 (historiography is defined as the methodology of historical research or the written historical accounts themselves; both definitions are pursued in this chapter). Just as importantly, his claims in the public domain denied the Prize to other contenders, particularly to rivals at the Willow Run Laboratories. While they were little publicly documented at the time, some cases of conflict became widely known, and, at times, historically distorted and privately mythologized. The cases were most frequent during the first decade of holography's explosive expansion, when there were few precedents and everything to play for.

An extended discussion of these episodes reveals the importance and extraordinary allure of intellectual priority for practising scientists. Establishment of precedence can

[5] See Rogers, Gordon L. to M. E. Haine, letter, 25 Jun. 1951, Sci Mus ROGRS 6, Technical Correspondence (quoted) and Rogers, Gordon L. to M. E. Haine, letter, 20 Oct. 1952, Sci Mus ROGRS 6, Technical Correspondence. Nor did Rogers care for Gabor's choice, 'holoscopy'. Despite their aptness, the terms hologram/holography are not analogous to 'photogram/ photograph/ photography', where a photogram is a contact-printed collage but, by contrast, a 'holograph' is a handwritten or self-produced document. The tardiness of dictionaries to be updated, and the consequent potential for confusion, is illustrated by a bizarre 1968 newspaper item that announced Emmett Leith as an expert in handwritten legal documents; while correctly identifying him as a professor of electrical engineering working in the areas of radar, microwave, and communication theory, the article declared that 'he and a colleague are responsible for the renewed interest at U-M in holography, the drafting of a document such as a will totally in the handwriting of the person in whose name it appears and therefore not requiring attestation of witnesses' ['Handwritten will expert to speak', *Jackson Citizen's Patriot*, 24 Mar. 1968].

[6] Leith first employed the term 'holography' in a *Physics Today* article and a conference paper at a SPIE meeting in August 1965.

provide professional and social status, bring support in the form of a succession of research contracts and consultancies, and can mean the difference between a reputation that long outlives its subject or a subsequent career in obscurity. More generally, the rows concerning the history of holography reveal the tenacity of insiders' accounts and the difficulty of deconstructing such claims. The subject can serve as a caution to historians who frequently have fewer sources of evidence for unravelling contemporary accounts and establishing historical veracity. The episodes provide an example of the additional constraints on scientific openness during the cold war period and the incompleteness of reviewed papers in documenting priority and the development of concepts.

The creation of the new subject also entrained the formation of new technical communities, an aspect explored in later chapters. The holographers most in the public eye during the late 1960s and the key figures portrayed in early accounts of its history were Dennis Gabor, Emmett Leith, and George W. Stroke.[7]

5.2 GEORGE STROKE AND THE PACKAGING OF HOLOGRAPHY

Between 1960 and 1964, when the Optical Lab of WRL was fresh from its successes in synthetic aperture radar and beginning research in wavefront reconstruction, the University of Michigan (U-M) was reorganizing its administration. The University had recognized for some time that Willow Run, while generating considerable income, was poorly integrated with academic functions: although several of the Electrical Engineering department professors held joint appointments with WRL, it remained physically isolated from the campus fifteen miles away. Administrators arranged for a gradual merging of the WRL classified activities with undergraduate teaching.

One move in this direction was the establishment of a professorship in Electro-Optical Sciences and the appointment of George Stroke to fill it. In negotiating with Stroke, William Dow, head of the Electrical Engineering Department explained:

> Our need for competence in Applied Optical science is the basis for our interest in you. My initial interest in your capabilities arose because optical science is rapidly becoming of great importance in dealing with new electronically-related devices, systems, and observational data inputs. The microwave spectrum merges through infrared into the optical spectrum. Expanding systems uses of the infrared spectrum and of optical quantum electronics—lasers—require of electrical engineers an understanding of optical science not heretofore necessary. Quite apart from that, optical data processing principles and techniques that originated here are giving a new multi-thousand-channel processing resource that our students and our research staff must be able to use [. . .] thus I came to feel that your presence among us would

[7] Yuri Denisyuk was little mentioned in Western accounts of the mid-1960s, but was rehabilitated at home and abroad by 1970. See §6.5 for a further discussion of Soviet activities.

> substantially strengthen our research and instructional capabilities in an area common to basic and applied science that is making a powerful new impact.[8]

James Wilson, acting director of the Institute of Science and Technology, which oversaw WRL, announced that Stroke's post was intended to promote the transfer of expertise from the hidden world of contract research to academe, enabling 'graduate students to work with modern developments in optics which have been advanced in industry and in defense and the space programs conducted by the federal government'. Indeed, Wilson added,

> The University is pioneering in this field. New knowledge in optics is being developed and used in industry with dramatic results—especially in electronic information processing, light amplification and control with lasers, and communications. Universities are doing research in this field, but little in the way of formal teaching related to it.[9]

George W. Stroke (1924–) appeared to be an ideal candidate. Like Gabor, Leith, and a handful of other post-war researchers, Stroke had a career background that straddled the disciplines of physics and electro-optical engineering. Born in Yugoslavia, he spent his youth in France and the period 1943–7 as an apprentice optical worker in British Palestine, completing an engineering optics diploma at the École Supérieure d'Optique in Paris in 1949. In 1952, he immigrated to America and obtained a post as a Research Associate with George R. Harrison, Dean of Science at MIT and Head of its Spectroscopy Laboratory.[10] There his technical experience was deepened considerably: he assisted in the development of an interferometric servo-controlled ruling engine, a sophisticated device for precisely scribing a glass plate with fine, parallel lines to create a diffraction grating, itself intended as the heart of a spectrometer to disperse light into a spectrum.[11] Stroke completed work on the theory of diffraction gratings for his *Docteur ès Sciences*, received in Paris in 1960, and returned to MIT with a wider circle of academic contacts and solid mathematical training added to his practical experience in engineering optics, but had a gradual falling out with Harrison.[12]

[8] Dow, W.G. to G. W. Stroke, letter, 30 Nov. 1962, Ann Arbor, Bentley Historical Library, Institute of Science and Technology Box 21: Personnel. See Figure 5.1 for the administrative structure at the University of Michigan.

[9] 'IST establishes lab in new study field', *Ann Arbor News*, 2 Oct. 1963, 37.

[10] George R. Harrison (1898–1979) spent his career at MIT, becoming Professor of Experimental Physics there in 1930 and Dean of Science in 1942 until his retirement in 1964. He was head of the MIT Spectroscopy Laboratory, chief of the Optics Division of the National Defense Research Committee (NDRC) during the Second World War, and later headed the Office of Scientific Research and Development (OSRD) Office of Field Service. George Stroke's brother Henry had completed his MS in Physics at MIT in 1952, and completed PhD studies there in 1954.

[11] Harrison, G. R. and G. W. Stroke, 'Interferometric control of grating ruling with continuous carriage advance', *Journal of the Optical Society of America* 45 (1955): 112–21.

[12] Stroke suggested that mere jealousy was at the root of their conflict, recalling, 'I was the only person in his career who ever became a professor from among his many, many brilliant collaborators. He could not stand it' (Aspray, William, 'George Wilhelm Stroke, Electrical Engineer, an oral history conducted in 1993 by William Aspray, IEEE History Center, Rutgers University, New Brunswick, NJ, USA.' (1993)). Stroke later provided an authoritative 350-page article on diffraction gratings based on his work with Harrison (Stroke, George W., 'Diffraction Gratings', in: S. W. Flugge (ed.), *Handbuch der Physik* (Berlin: Springer-Verlag, 1967), pp. 26–754).

Stroke was advised by friends to seek a permanent post at Michigan where he could benefit from burgeoning funding to embark on the design of diffraction gratings for the space programme and astronomy. The Office of Naval Research (ONR) and NASA were identified as likely sponsors.[13] After protracted negotiations, his new post was to be funded jointly by the Electrical Engineering (EE) department and the Institute of Science and Technology (IST) at the U-M North Campus and with the expectation that he would consult at WRL.

By that time, Willow Run had some one thousand employees. WRL had been incorporated as a functional unit into the IST in 1960 and the IST building on the U-M North Campus housed various laboratories that were interdisciplinary or, in some cases, too large to fit into academic departments.[14] WRL remained physically isolated on the airport site during the early 1960s, although closely tied to the EE department with several of the EE professors holding joint appointments with WRL.

5.2.1 Leith, Stroke, and priority disputes

George Stroke fitted awkwardly into the ongoing activities. He had been recruited initially for the post in Electro-Optics (see Figure 5.1) in 1962 and visited WRL (see Figure 5.2) that summer. Arthur Funkhouser, a student who had done an undergraduate dissertation in physics under Stroke between 1960 to 1962 at MIT, decided to pursue an MS under his supervision at U-M. Stroke arranged for him to enter the EE department, joining Funkhouser there in the summer of 1963.[15] Stroke began setting up an optics laboratory in the new IST building at the U-M North Campus in November. His former technician at MIT, Frank Denton, joined him as project engineer in February 1964, completing the lab for development work on a diffraction grating ruling engine, an extension of the work that had occupied Stroke and Denton in the MIT Spectroscopy Laboratory and funded by the Office of Naval Research but also by the National Science Foundation (NSF), not one of WRL's usual sponsors.[16]

[13] By the early 1960s, the U-M had more contracts with NASA than did any other American university, largely owing to its early involvement in guided missile research (Miller, James, *'Democracy is in the Streets': From Port Huron to the Siege of Chicago* (New York; London: Simon & Schuster, 1987)).

[14] The IST, proposed to the U-M Board of Regents in late 1958, had budgets approved in 1959 and opened in 1960.

[15] U-M Board of Regents Proceedings, Sep. 1962, pp. 895–6. William Dow, Head of Electrical Engineering, had first courted Stroke in December 1961, intending that he 'accept and carry out substantial group leadership in the work of the Radar laboratory of the IST in the area of [his] specialized preparation as related to their work in optical data processing', which Stroke rejected. Instead, he stressed his desire for a separate laboratory and funding not tied to WRL (Dow, W.G. to G. W. Stroke, letter, 30 Nov. 1962, Ann Arbor, Bentley Historical Library, Institute of Science and Technology Box 21: Personnel). As a result, Stroke declined to move to Michigan after a summer 1962 visit, and finally assumed his position at U-M in August 1963.

[16] The National Science Foundation (NSF), established in 1950, was promoted by some scientists as a necessary alternative to the prevailing model of wide scale military funding of American physical science after Second World War, which, they argued, constrained the directions and openness of basic research. While the NSF bill required that research results not be secret, it required that grant recipients sign a loyalty affidavit. Moreover, the early NSF could attract significant funding only by convincing Congress of the

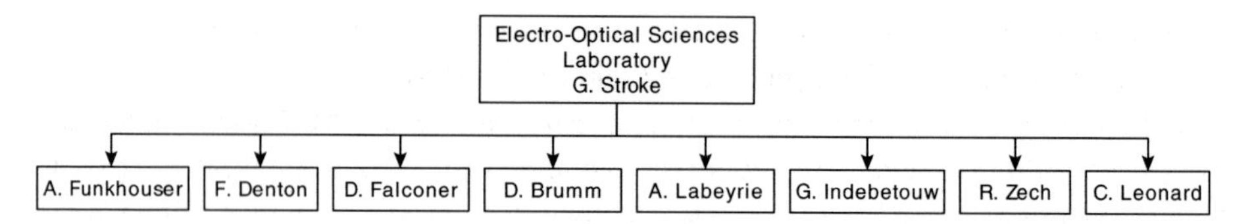

Fig. 5.1. Organization of University of Michigan Electro-Optical Laboratory *c.*1965–7 (Labeyrie, Indebetouw, Zech, and Leonard joined the laboratory in 1966). The diagram omits a secretary and graduate students of shorter duration.

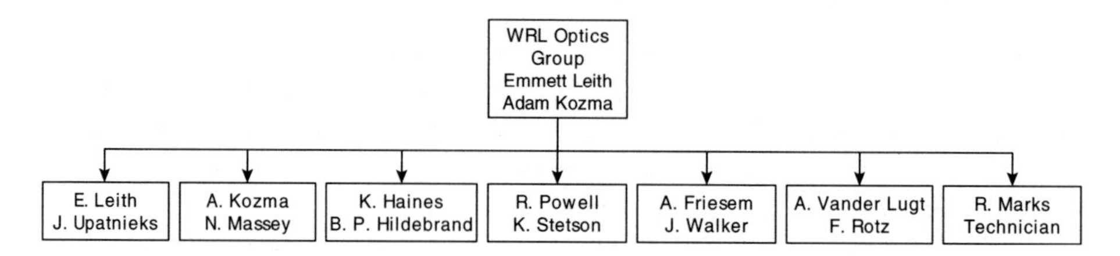

Fig. 5.2. Organization of the University of Michigan Optics Group of the WRL Radar & Optics Laboratory *c.*1965. Diagram omits other junior researchers, a draftsman, electronics technician, photographer, and two secretaries.

Denton recalls Stroke as a researcher obsessive about his work, who 'fell in love with holography', and began to use the ruling engine space for holography, too.[17] The new group expanded to include students Robert Restrick, Douglas Brumm, and David Falconer from the physics department and was allocated offices upstairs in the IST building.[18]

The separate locations of the new lab and WRL were significant and illustrate how the physical space mapped onto administrative and intellectual partitioning of the developing subject. On his arrival at U-M, Stroke apparently had foreseen little involvement with Willow Run's ongoing research on wavefront reconstruction. He planned initially to fit into the ecology of the University by pursuing the areas he knew best (gratings and optical instrumentation), taking up new areas that he perceived as cognate (relativistic optics and quantum electronics), and engaging in some shared work concerning electrical communication. From late summer 1963, however, Stroke began to visit WRL about

military utility of basic research. Thus both Willow Run's classified contracts and the NSF money flowing into Stroke's lab were entwined with the cold war concerns. Until the mid-1960s, NSF grants represented only 10% of government-funded basic research. See Wang, Jessica, *American Science in an Age of Anxiety: Scientists, Anticommunism, and the Cold War* (Chapel Hill; London: University of North Carolina Press, 1999), especially pp. 254–62.

[17] Denton, Frank to SFJ, telephone interview, 1 May 2003, SFJ collection.

[18] Denton (b 1935) emigrated from England in 1958 and rose at MIT from lab assistant to technician, and to project engineer at the University of Michigan. He left the University to form his own optical company, Photo-Technical Research (PTR) in 1967. On PTR, see §6.3.

twice a month, taking particular interest in the work of Leith and Upatnieks and the matched filtering research being pursued by Anthony VanderLugt.[19]

In these visits in the first months after taking up his University of Michigan post, Stroke suggested, 'Emmet [sic], for example, could head up the Electro-Optical Communications laboratory in a more formal way'.[20] He proposed that the Willow Run work on wavefront reconstruction be combined with, or transferred to, his planned lab, and to seek joint funding from the National Science Foundation (NSF). Stroke and Leith signed notes of agreement to this effect in early December—just days, in fact, after the influential press release on lensless photography and the beginning of Leith and Upatnieks' concentrated work on the holograms of solid objects:

> the Radar Lab and the Electro-Optical Sciences Lab have strong common interests and plan to have joint unclassified projects in areas such as optical data processing, optical and x-ray holograms, etc. It is planned that these joint projects will be conducted on North Campus, this certainly being the case if part of the Radar lab moves to North Campus under any circumstances. From an organizational viewpoint the Radar laboratory has formed a subgroup of its optics group which, as part of the Radar Lab, will take part in these joint efforts in cooperation with the Electro-Optical Sciences Lab.[21]

Leith and other members of the WRL Optics Group felt that the aim of bringing the work to an academic environment was worthwhile and most were pleased to participate. Stroke's relative prominence in optics contrasted with those at Willow Run, who for the most part were, in Leith's own words, 'rather narrow specialists in a relatively unknown area of optics'.[22] They also recognised that they were disadvantaged by the secrecy surrounding their classified research, compared to colleagues at the town campus of the University of Michigan. Nevertheless, they saw Stroke's knowledge as peripheral to the expertise developed at Willow Run.

Stroke, however, had already begun complaining privately to administrators about competition for funding and support from Willow Run and, indeed, tensions in its operating culture.

> It might be added, unfortunately, that the lack of a Ph.D. by such people as Leith, is not only a lack in adequate terminal training in areas outside narrow specializations, but it shows up in a

[19] The notion of spatial filtering for 'deblurring' of images in coherent light had been discussed a decade earlier by French investigators under whom Stroke studied [Maréchal, André and P. Croce, 'Amélioration de la perception des détails des images par filtrage optique des fréquences spatiales', in: (ed.), *Problems in Contemporary Optics* (Arcetri-Firenze, 1956), pp. 76–82], but Stroke was able to extend their solution during the late 1960s by identifying holography as a general case. Such work had also been proceeding at WRL since the late 1950s, particularly by VanderLugt.

[20] Stroke, George W. to J. T. Wilson, memo, 27 Aug. 1963, Ann Arbor, Bentley Historical Library Institute of Science and Technology Box 21: Electro-Optical Lab.

[21] Brown, William M., George W. Stroke, and Emmett N. Leith to file, memo, 10 Dec. 1963, Ann Arbor, Bentley Historical Library Institute of Science and Technology Box 21: Electro-Optical Lab.

[22] Leith, Emmett N. to WRL administrators, memo, *c.*Summer 1965, Ann Arbor, Michigan, Leith collection.

> lack of appreciation of the process of Ph.D. thesis and more generally basic educational responsibilities and basic scientific research work, such as those on which a great university must insist.[23]

The working cultures were indeed different: the WRL Optics Group consisted of some eighteen engineers experienced in the theory and application of optical processing, engaged in full-time contract research and supplemented by graduate students for summer employment. The new Electro-Optical Sciences Laboratory (EOSL) (see Figure 5.3) consisted of Professor Stroke, an engineer/technician and a band of graduate students. Stroke combined research with teaching and academic supervision; the Optics Group operated less hierarchically than did EOSL and had a narrower distribution of ages.

Overt disagreements concerning the relative status of Stroke's EOSL and the Radar and Optics Laboratory of WRL began almost immediately. The NSF grant application, closely based on Leith's unsuccessful proposal submitted the previous summer to the US Army Research Office, was found to the surprise of WRL administrators to list George Stroke as project Director and senior author.[24] Within a few months, the Director of the IST was warning the Vice President of Research of looming problems, noting that Prof. Stroke

> came to Ann Arbor either completely unprepared to appreciate the long history of optical work at Willow Run and its reputation and seniority in its special fields, or unaware of its abilities and reputation [. . .] In any case, he elected to build on the Willow Run work but has been singularly unsuccessful in carrying the Willow Run people with him. They, I feel rightly, wanted to be full partners and co-equals and he has felt that tenure and departmental connection on his part leaves them subordinates [. . .] He seems temperamentally unsuited to being the bridge and campus leader that the Willow Run people wanted.[25]

Even more significantly, Stroke began voicing his concern, first privately to Leith's superiors, and then publicly, that he was not receiving due recognition for contributions to the work at Willow Run. He claimed that he had conferred to Leith the idea of diffuse illumination of objects when making holograms of two-dimensional transparencies, although Leith and Upatnieks had initiated that research in April 1963, at a point when Stroke had only once met Leith and in a public setting. However, the Leith–Upatnieks work on diffuse illumination of three-dimensional objects got well underway in December 1963 at the very time that administrators were negotiating collaborative work between the two men. Stroke later suggested that their research on the holograms of solid objects was triggered by insights from him; Leith, on the other hand, denied any

[23] Stroke, George W. to J. T. Wilson, memo, 30 Sep. 1963, Bentley Historical Library Institute of Science and Technology Box 21: Electro-Optical Lab.

[24] Leith, Emmett N., 'Proposal for Applications of the Wavefront Reconstruction Technique', Grant proposal ORA-63-1255-PB1, Institute of Science and Technology, University of Michigan to US Army Research Office, Durham, NC, 28 May 1963.

[25] Wilson, J. T. to A. G. Norman, *c.*Mar. 1964, Ann Arbor, Bentley Historical Library Institute of Science and Technology Box 21: Electro-Optical Lab.

Fig. 5.3. U-M Electro-Optical Sciences Laboratory members, July 1965: From left to right: Paul Peters (student), Frank Denton (engineer), Douglas Brumm (student), Nancy Pruitt (secretary) George Stroke, Arthur Funkhouser (student), Rolf McClellan (student), and Robert Restrick (student) (Funkhouser collection).

correlation and Upatnieks recalls that the trigger for reviving experimental work that had begun abortively that summer was reporters' interest.[26] It may be that the brief interactions between Leith and Stroke, increasing during December 1963, made both realize the as-yet half-demonstrated potential for laser wavefront reconstruction, spurring Leith to accelerate experimental work and Stroke to pre-emptively claim conceptual priority and administrative seniority.

In any event, rising unease appears to have been established and mutually recognized as soon as the memorandum of collaboration had been signed. Stroke wrote the following April to the university news service to claim credit for originating and developing much of the Optics Group's wavefront reconstruction and spatial filtering work.[27] Willow Run staff warily continued to maintain relations, with Leith participating in a summer school on coherent optics organized by Stroke in 1964.

The most troubling event in the eyes of the Willow Run researchers occurred that autumn. Returning from summer travel in September, Stroke learned of an experiment that Upatnieks had recently performed with a student on microscopy, in which they had enlarged images holographically.[28] Stroke asked Leith for time in his laboratory to

[26] Aspray, op. cit.; Upatnieks, Juris to SFJ, email, 9 Jan. 2005, SFJ collection.

[27] Stroke, George W. to C. W. Wixom, letter, 11 Apr. 1964, Leith collection.

[28] By employing a reconstructing laser beam that was more divergent than the original reference beam the image was correspondingly magnified.

reproduce the results. When Stroke's request was rebuffed, he asked to borrow photographs of the results to use for his teaching at the University of Michigan, which Leith and Upatnieks, with reservations, provided. A week later he suggested to Leith that they present a joint paper on the work; this, too, they rejected. Leith, now mistrustful, specified by letter that the loaned photos were not to be circulated beyond Stroke's class.[29] Stroke, seemingly adopting the role of senior researcher in group collaboration, nevertheless wrote a paper himself that included Leith and Upatniek's results and presented it at the Boston meeting that November—but with scarce mention of their production of the work.[30] Towards the end of the accompanying press release he noted:

> Aside from Dr Stroke and D. G. Falconer, a graduate assistant in the Physics Department at the University of Michigan, a number of their colleagues, associates and other experimenters are engaged in the development of lensless photography in view also of other applications at the University of Michigan. Among those are E. N. Leith, J. Upatnieks, D. Brumm and A. Funkhouser.[31]

Thus, lensless photography had been appropriated and Leith and Upatnieks of the Optics Group were publicly categorized with Stroke's graduate students. Stroke published the photographs and the work in another journal that autumn.[32]

Leith, Upatnieks, and others from the Optics Group who had attended the same session of the Boston meeting were aghast and members of Stroke's own lab recall being surprised by his contentions.[33] Following a series of mutual recriminations, meetings with administrators and letters by Leith to journal editors, the Head of Electrical Engineering recommended that IST support for George Stroke be phased out.[34] Bypassing these behind-the-scenes attempts to defuse the rows, Stroke presented the affair publicly as a case of inappropriate usurpation of seniority cloaked in secrecy:

> Various rumors appear to have recently originated about some aspects of the 'lensless photography' work being carried out at the University of Michigan.
>
> [. . .] In no way should Professor Stroke's apparent silence in the face of a one-sided spreading of rumors be interpreted as other than a traditional desire of a scientist and senior faculty

[29] Leith, Emmett N. to G. W. Stroke, letter, 20 Oct. 1964, Leith collection.

[30] Stroke, George W., 'Theoretical and experimental foundations of electro-optical image formation, image modulation, and wave-front reconstruction imaging', presented at *Symposium on Optical and Electro-Optical Information Processing Technology*, Boston, MA, 1964.

[31] Stroke, George W., 'New optical principle of "lensless" x-ray microscopy', Press release for Symposium on Optical and Electro-Optical Information Processing, Boston, University of Michigan, 10 Nov. 1964.

[32] Stroke, George W. and D. G. Falconer, 'Attainment of high resolutions in wavefront-reconstruction imaging', *Physics Letters* 13 (1964): 306–9.

[33] Denton, Frank to SFJ, telephone interview, 1 May 2003, SFJ collection; Funkhouser, Arthur to SFJ, email, 9 Apr. 2003, SFJ collection.

[34] Leith, Emmett N. to G. W. Stroke, letter, 30 Nov. 1964, Leith collection; Leith, Emmett N. to WRL administrators, memo, *c.*Summer 1965, Ann Arbor, Michigan, Leith collection; Leith, Emmett N. to SFJ, email, 17 Mar. 2003, Ann Arbor, SFJ collection; Farris, H. W. to J. T. Wilson, 21 Jan. 1965, Ann Arbor, Bentley Historical Library Institute of Science and Technology Box 21: Electro-Optical Lab.

member to maintain the dignity of the faculty and the University when faced with rumors apparently originating from junior staff.

[. . .] The extent of such collaborative efforts, carried out to a large part in a framework of military secrecy, cannot be revealed without infringement of national interests. It now appears that some one-sided advantage may have been taken by some party of the collaborative effort, in the face of Professor Stroke's strict adherence and respect for the secrecy rules.[35]

As suggested by his papers and press releases, Stroke was a more practiced communicator than was Leith. His holograms, too, supported his public stance subliminally. When Leith was shown in the popular press demonstrating the hologram of a military tank, Stroke proffered himself alongside his own hologram of a statue of Abraham Lincoln (Figures 5.4 and 5.5). Similarly, Stroke's later announcement of advances in colour holography was illustrated with a hologram of soldiers in front of a red, white, and blue American flag.

Leith was not the only individual at the University of Michigan that claimed grievances: Karl Stetson wrote to the executive secretary of the Optical Society of America (OSA) to complain that Stroke had plagiarized the holographic interferometry results that he had obtained with Robert Powell.[36] And C. Roy Worthington, an assistant professor of biophysics in the U-M Physics Department also protested to his departmental chairman. Worthington used x-ray diffraction to study biology and his PhD student, John Winthrop, was investigating the diffraction imaging by faceted insect eyes. Together, the two developed a theory of Fourier transform holograms for x-ray wavelengths, which they submitted to *Physics Letters* in February 1965.[37] Shortly afterwards, Winthrop told his friend David Falconer, George Stroke's PhD student, of their work. Stroke made contact with Worthington to learn more. Worthington recalls that he and Winthrop were surprised to discover that Stroke later submitted his own paper on the subject to *Physics Letters* and subsequently claimed lensless Fourier transform holography as his own in papers and a book.[38] Nevertheless, the environment was not conducive to correcting what Worthington saw as a misattribution of credit.

The Physics Dept was quite aloof, with a past international background in modern physics [. . .] They had never heard of Dennis Gabor, Emmett Leith, George Stroke or holography.

[35] Stroke, George W., 'To whom it may concern: lensless photography work at the University of Michigan', News release, Apr. 1965.

[36] Stetson, Karl A. to M. A. Warga, 7 Mar. 1966, Ann Arbor, Bentley Historical Library Institute of Science and Technology Box 21: Electro-Optical Lab.

[37] Winthrop, John T. and C. Roy Worthington, *Physics Letters* 15 (1965): 124–6. The work was more fully described in Winthrop, John T. and C. Roy Worthington, 'Theory of Fresnel images I. Plane periodic objects in monochromatic light', *Journal of the Optical Society of America* 55 (1965): 373–81 and Winthrop, John T., *The Formation of Diffraction Images: Fresnel Images, Compound Eye, and Holographic Microscopy*, PhD thesis, University of Michigan (1966). Winthrop subsequently worked at American Optical, where he became chief lens designer.

[38] Stroke, George W., 'Lensless Fourier-transform method for optical holography', *Applied Physics Letters* 6 (1965): 201–3. Worthington further asserts that Stroke made a retroactive priority claim by altering the published content of a OSA conference abstract after its original draft and presentation in Feb. 1965 (Stroke, George W., 'Three advances in Fourier transform holography', *Journal of the Optical Society of America* 55 (1965): 1566 (conference abstract)).

Fig. 5.4. Emmett Leith, 1966 (National Aeronautics and Space Administration).

Fig. 5.5. George Stroke, 1966 (University of Michigan News Service).

> There was zero interaction with Willow Run. To be honest, Willow Run was a mystery to me [. . .] I remember discussing my version of the [Fourier Transform Holography] story with the Dept Chairman, David Dennison, who was or had been an important scientist. He said forget it. The truth will be revealed in due course.[39]

[39] Worthington, C. Roy to SFJ, letter, 21 Sep. 2004, SFJ collection; Worthington, C. Roy to SFJ, email, 29 Sep. 2004, SFJ collection. Dennison spent his entire career at U-M, gaining a name for himself by clarifying the nature of the spin of the proton in 1927.

But researchers beyond the University of Michigan also complained. An administrator reported that workers at Bell Telephone Laboratories had grumbled privately to U-M engineers that Stroke had misrepresented his work with them.[40] And somewhat later, Adolf Lohmann found illustrations from his papers on spatial filtering reproduced without attribution in a grant report by Stroke.[41] Later still, electron microscopist Albert Crewe, who had loaned Stroke an image for experiments on 'holographic deblurring', denounced him for improperly claiming that the technique rendered the DNA helix visible. The controversy, publicly reported in *Science* in November 1971, tarnished Stroke's reputation and that of electron microscopists too, while Stroke proffered privately that the microscopists should be grateful for the publicity.[42] The chemist Lawrence Bartell, citing the 'immense egos' of both Stroke and Crewe, subsequently published an account of their falling out based on his personal experience.[43]

The President of the OSA read another such complaint from a researcher at the University of Rochester that Stroke had denigrated Leith and Upatnieks in his abstract for an OSA meeting, but had then failed to support the allegations in the presentation itself, thereby insinuating unsubstantiated claims into a permanent record. He replied, 'We had other communications from people about the Stroke problem and it was discussed at the meeting of the Board of Directors of the OSA. It was felt that OSA should not take any action [. . .] it was our opinion that the truth of scientific status and priority is generally well recognized and the field recognizes the position of Leith and Upatnieks vs. Stroke'.[44] But such discussions took place behind closed doors, leading other researchers such as Hussein El-Sum to probe WRL administrators about just what was going on at the University of Michigan.[45]

Stroke had himself taken the dispute beyond the University, and found effective means of disseminating his version of events. He claimed that Leith and Upatnieks had not acknowledged his contributions to their work, pleading privately to the editor of the *Journal of the Optical Society of America* (JOSA), in which he sought to publish results, that they 'had been carrying out officially and under salary [. . .] as a joint project' work that was rightfully his own as Director of the Electro-Optical Sciences Laboratory.[46]

His claims may have had some validity in the European context: there, a full professor commonly could claim co-authorship of a publication if he had specified the goal and scrutinized the first draft of a manuscript. It may be that he interpreted the signing of the

[40] Snell, J. to R. Evaldson, letter, 4 Apr. 1966, Leith collection.

[41] Lohmann, Adolf W. to SFJ, fax, 19 Aug. 2003, SFJ collection; Stroke, George W., 'Optical image deblurring methods', *Naval Research Reviews* (1971): 14–20.

[42] Nelson, Bryce, 'DNA double helix: photo sends controversy spiraling', *Science* 173 (1971): 800–1; Stroke, George W. to D. Gabor, telegram, Nov. 1971, Paris, IC Gabor MS/15.

[43] Bartell, Lawrence S., 'A brief history of holographic hubris and hilarity', *The Chemical Intelligencer*, Oct. 1998: 53–6.

[44] Williams, Van Zandt to J. Shewell, letter, 23 Mar. 1966, Bentley Historical Library Institute of Science and Technology Box 21: Electro-Optical Lab.

[45] Snell, J. to R. Evaldson, letter, 4 Apr. 1966, Leith collection.

[46] Stroke, George W. to D. L. MacAdam, letter, 12 Dec. 1964, Ann Arbor, Michigan, Leith collection.

memorandum of collaboration in December 1963 as just such a statement of hierarchy and supervision. Moreover, authorship would be limited to those holding a master's degree or to an exceptional undergraduate or technician. According to such ranking, Leith's MS degree was both administratively and intellectually inferior to Stroke's PhD and may have been used to justify Stroke's adaptation of Leith's project proposal to gain his 1964 NSF contract. Nevertheless, such intellectual appropriation did not occur for other Europeans working in the same field and having a clearly established hierarchical relationship with senior colleagues, such as Serge Lowenthal (working in Paris under André Maréchal), F. T. Arecchi (in Florence, under G. Toraldo di Francia), and Emil Wolf (in Edinburgh, under Max Born).[47] Stroke later cited his own early career experiences as relevant to the U-M situation: recalling having worked as an apprentice optical worker in Palestine, he noted that Prof Emanuel Goldberg, 'was a typical German professor, and had lived for a few years in France. The best chance that you had was to work with a kind of father figure. Not father in the family sense, but similarly. That's how you do things'.[48] Nevertheless, the WRL workers were neither inexperienced nor young compared to Stroke. The members of his own laboratory report that Stroke developed a paternalistic relationship with them, but this never developed for his WRL colleagues who were already in place when he arrived. Nor, despite his interpretation of the memorandum of agreement, were they administratively under his direction. Nevertheless, Stroke's written claims appear both ambiguous and misleading concerning such hierarchies.

Via informal lectures, conference presentations, papers, an academic monograph, and popular articles, Stroke publicized his version of the history of holography. In a survey of holography written for *Technology Review* in 1967, the biography supplied by Stroke stated that he had 'first initiated research in 3-D holography in 1962, and has since continuously stimulated much of the research in this field by numerous scientific contributions, the first book in the field and by world-wide lectures'.[49] His article provoked a response from E. G. Loewen, Director of the Grating and Metrology Laboratory at Bausch and Lomb in Rochester, NY, who accused Stroke of 'unsubstantiated slander in having accused Leith and Upatnieks of gross plagiarism'.[50]

Through 1965 and 1966, Stroke's laboratory continued to pursue development of both the grating ruling engine and holography. Students from further afield—Antoine Labeyrie from France, and Guy Indebetouw from Belgium—joined Stroke's laboratory

[47] Lohmann, Adolf to SFJ, email, 4 Nov. 2003, SFJ collection.

[48] Aspray, op. cit. Goldberg (1881–1970), born in Moscow, had received a PhD in chemistry in 1906 from Wilhelm Ostwald's institute in Leipzig. By 1933 he was head of Zeiss Ikon, the world's largest camera firm, in Dresden. Thereafter he spent four years in France and the remainder of his career in Tel Aviv, where he operated a successful optical instrumentation company. Among his achievements was the 1927 'Statistical Machine', a system of microfilm records combined with data retrieval cards; in effect, the invention was an early type of optical memory employing a form of character recognition. See, for example, Buckland, Michael K., 'Histories, heritages and the past: the case of Emanuel Goldberg', presented at *Second Conference on the History and Heritage of Scientific and Technical Information Systems*, Philadelphia, 2002.

[49] Stroke, George W., 'Recent advances in holography', *Technology Review* 69 (1967): 16–22.

[50] Loewen, E. G. to Editor *Technology Review*, 1967, Bentley Historical Library Institute of Science and Technology Box 21: Electro-Optical Lab.

that year.[51] Stroke's contentions increasingly isolated him; some of his students switched to the WRL Optics group, which in 1966 finally relocated to the same building.[52] The IST building was dubbed 'Fort Apache' by its new residents, perhaps as much for its continuing skirmishes as for its neo-Aztec exterior.[53]

Despite the rising boundaries, relations continued to deteriorate. Stroke complained to H. W. Farris, Head of the Electrical Engineering Department, that Leith had never cited his papers in a suitably reciprocal fashion—a belated reversal of accusations that had been made by Leith for two years.[54] Stroke further claimed that 'academic and scholarly contributions' and 'academic gentility' would suffer because of what he characterized as Leith's attempts at 'industrial and commercial monopolizing'.[55]

Commercial activities, increasingly appealing for the lab personnel, provoked departmental confrontations: Stroke interpreted such activity as misappropriation of laboratory information and time and dismissed engineer Frank Denton and a graduate student from his laboratory and further petitioned for the dismissal of his secretary and Don Gillespie, who worked in another U-M laboratory, for engaging in commercial activities outside their university jobs.[56]

The most visible opportunity to disseminate his claims for a wide audience was Stroke's *An Introduction to Coherent Optics and Holography*, the first book to outline a

[51] Labeyrie, Antoine E. to SFJ, email, 22 Jul. 03, SFJ collection, Leonard, Carl to SFJ, email, 12 May 2003, SFJ collection. Labeyrie (b. 1943) spent nine months at the University of Michigan as a Research Fellow following an engineering degree and 'license de physique' in Paris. He returned to France, working for a time on holographic diffraction gratings for the Jobin-Yvon spectrometer company, and later specialized in interferometric astronomy. He become Professor of Observational Astrophysics at the Collège de France in 1991 and Director of the Observatoire de Haute-Provence in 1995. Guy Indebetouw obtained his PhD at Berne, and collaborated with Stroke on image deblurring, later working in holographic microscopy and becoming a Professor of Physics at Virginia Tech in 1978.

[52] The Radar & Optics and Analytic groups at Willow Run were moved to the University of Michigan North Campus in 1966 as part of William Brown's long-term plan to make their activities more academically visible—part of the plan, in fact, that had brought George Stroke to the university in 1963. They occupied a building in the new IST complex, setting up Leith's laboratory some 200 ft from Stroke's Electro-Optical Sciences Laboratory, the two buildings being connected by a short passage. From this period until Stroke's departure from the university in 1967, the two were situated within walking distance. The Analytic Group of Willow Run was relocated immediately opposite Stroke's office.

[53] Wolff, Michael, 'The birth of holography: a new process creates an industry', *Innovations* (1969): 4–15.

[54] The mutual disregard was mirrored in file systems: Leith's casually organized files, clippings, and ephemera reveal few mentions of Stroke, but he maintained an orderly tabulated collection of documents relating to episodes in a 6 in. thick 'Stroke file'.

[55] Stroke, George W. to H. W. Farris, memo, 24 Mar. 1966, Ann Arbor, Leith collection.

[56] Denton, Frank to SFJ, telephone, 1 May 2003, SFJ collection; Funkhouser, Arthur to SFJ, email, 9 Apr. 2003, SFJ collection; Gillespie, Donald to SFJ, interview, 29 Aug. and 4–6 Sep. 2003, Ann Arbor, MI, SFJ collection. The bulk of the papers in the Electro-Optical Sciences Lab file concern the Stroke allegations between 1964 and 1967. Stroke complained to university officials about the start-up business of Gillespie (Jodon Instruments), who worked for Chihiro Kikuchi in nuclear engineering, and that of Frank Denton, his secretary Nancy Pruitt, and math graduate student Rolf McClellan (PTR) in an attempt to have them barred from future university employment. This led to a final wave of administrative wrangles hastening Stroke's departure from the University. U-M administrators dissolved his Electro-Optics Lab in early 1967; academically isolated, he resigned in July 1967 and took a post at the State University of New York at Stony Brook.

history of the young subject.[57] Announced publicly from the summer of 1964 and written during 1965—by which time relations with Willow Run had foundered—Stroke's book portrayed a genealogy of holography that minimized the work of Leith and Upatnieks and vaunted his own.

The book led to public rebuke for Stroke. A review by Emil Wolf, who had collaborated with Dennis Gabor and Max Born on the early writing of their canonical text *Principles of Optics* some fifteen years earlier, savaged it as a hastily prepared 'first'.[58] After illustrating a number of cases of 'misstatements, false mathematical arguments and confusing notation', he turned to the 'disturbing' attribution of credits:

> Anyone familiar with the field of holography, or anyone who is not but who will take the trouble to check the literature, will know that one of the first major contributions to this field, after Gabor's pioneering publications, was the separation of the twin images by a clever arrangement first described in the published literature by E. N. Leith and J. Upatnieks. Yet this contribution is completely ignored in the first few sections of the chapter on holography and is played down elsewhere. Moreover, several figures in this book show the Leith-Upatnieks arrangement without acknowledgement. On the other hand, alleged contributions of the author are frequently stressed, even if they never reached the print [. . .] This sort of referencing, which presents unsupported claims and not a real reference at all, evokes a question of ethics. The reader should reflect on the scientific climate that would be created if other authors were to adopt such a way of referencing.[59]

Wolf concluded that the chief merit of the 'very disappointing book' was in making available reprints of Gabor's three pioneering papers on holography. The review attracted the praise of John Howard, the editor of *Applied Optics*, who intimated, 'I'm not sure I would have the courage to carry such a review, considering the uproar that would probably ensue. We chose Gabor (who was rather non committal) as a way to evade backlash'.[60]

5.2.2 Stroke, Gabor, and a historical lineage

It was through Dennis Gabor, however, that Stroke's historiography was further promoted. George Stroke had first met Gabor in the summer of 1964 and an amicable and mutually beneficial relationship developed between them. Indeed, Gabor later reflected that George Stroke had been the first researcher to contact him after Leith's and

[57] Stroke, George W., *An Introduction to Coherent Optics and Holography* (New York: Academic Press, 1966).

[58] Wolf had made a consulting visit to the U-M in December 1963, after Stroke had suggested to the IST Director that Wolf be sounded out as a possible addition to the University of Michigan faculty (Stroke, George W. to J. T. Wilson, memo, 12 Dec. 1963, Bentley Historical Library Institute of Science and Technology Box 21: Electro-Optical Lab).

[59] E. Wolf, review of *An Introduction to Coherent Optics and Holography* in: *Journal of the Optical Society of America* 57 (1967). Another reviewer observed, 'The chapter on holography lists 94 references which are noted so often as to make them useless. The continual references to the author's own papers are particularly annoying' (Morgenstern, Arthur L., 'Review of *An Introduction to Coherent Optics and Holography*', *Electronics*, 16 May 1966).

[60] Howard, John M. to E. N. Leith, letter, 5 May 1967, Leith collection. *Applied Optics* was later to become the principal organ for publications on holography.

Upatnieks' revived wavefront reconstruction became known.[61] Stroke began corresponding with Gabor regularly and significantly offered one-quarter of the royalties for his book to Gabor, presumably in exchange for permission to include three of Gabor's seminal papers as appendices. Selling 3665 copies in the first year alone—a respectable number for such a text—this proved profitable for both.[62] Stroke's activities seemed designed to limit contacts between, and to alienate, Gabor and Leith, on one occasion reporting 'a gross attempt at slander' of Gabor by another member of the Willow Run Labs, who inadvertently had had his letter displayed on a notice board near Stroke's office.[63]

The episode hints at tensions between Stroke, Leith, and Gabor. It also suggests their differing perceptions of their relative contributions to the subject and thus to its developing history. Adam Kozma, the senior WRL researcher whose letter was made public, had moved with his family to London in 1966 to study for a PhD at Imperial College under Harold H. Hopkins. Kozma (b *c.*1930–) had been second-in-command of thirty-four staff members in the WRL Optics Group; he and Leith complemented each other well, with Kozma excelling at the administrative tasks that Leith disliked. Kozma, whom Leith describes as a 'highly perceptive, outspoken' and 'highly moral person—the "ethicist" of the laboratory', had written a newsy but unguarded letter to his WRL colleagues. He had described one of Gabor's lectures as 'witty [. . .] excellent dramatic stage presence [. . .] he led his audience along his theme extremely well'. In a later section concerning an after-lecture discussion with Gabor and Hopkins, however, Kozma had suggested that Gabor's revival of work on his two-hologram interference microscope was behind the times.

> There we talked generally about holograms and spatial filtering. Gabor discussed an idea he has about increasing the resolving power of the electron microscope, using a very crude spatial filtering technique. Basically it's phase equalizing using a ½ wave phase flipping filter and an equalizing amplitude filter. At a cursory glance, I would guess that there is not much hope of success especially the way in which he wants to proceed. I got into an argument with him about diffusely illuminated holograms which we finally settled by my allowing him to gracefully back down. It is amazing how little he knows about coherent optics and holography. At least this is my impression.[64]

Back at the University of Michigan, Stroke had discovered a copy of Kozma's letter posted on a notice board opposite his office. He sent a copy to Gabor by first-class mail, describing it as 'disgraceful'. Stroke followed it up with a trans-Atlantic telephone call the

[61] Kozma, Adam to E. N. Leith, letter, 3 Jan. 1967, London, Leith collection.

[62] Stroke, George W. to D. Gabor, 23 Apr. 1967, IC GABOR MR7/5. Gabor was approaching compulsory retirement as professor, although permitted a two-year extension of his chair until 1967. Having entered academe late in life, Gabor was conscious that his pension would be small and planned to augment his income with consultancies, as he had done periodically since the 1950s.

[63] See IC Gabor ML/4 for correspondence between Stroke and Gabor, Gabor and Leith, and Leith and Kozma.

[64] Kozma, Adam to E. N. Leith, letter, 7 Oct. 1966, IC GABOR ML/4.

next morning, dictating to Gabor the imprudent paragraph from Kozma's breezy, largely complimentary two-page letter to Leith.[65]

Gabor confronted Kozma the next day and immediately wrote to Leith about the 'extremely silly and indiscreet letter', describing Kozma as 'very contrite'. He offered Kozma 'every chance of living it down' and 'every opportunity to help George Jull [a former student and then visiting Professor from a Canadian university] and me in building up a small optical laboratory', adding,

> I presume that this silly letter [which, at that point, had not been viewed by Gabor in its entirety] is an outcome of the unhealthy tension which has developed between your laboratory and George Stroke. For my part I have no desire to be a judge in this situation; I am indebted both to you and your team and to George Stroke for having so beautifully expanded and developed the field of holography. There is room in this field for more than one team, and I hope also for me, because I am trying for a come-back, as you will probably soon hear from Holotron. If a deal with Holotron comes about, I hope that there will be friendly cooperation between us.[66]

Kozma, having scarcely settled in England and with a PhD at stake, agreed, in effect, to recant, writing a second and apologetic letter to Leith to be posted publicly at the University of Michigan; Leith did his best to placate the offended Gabor; and Stroke, in his turn, continued to keep Gabor abreast of developments at the University of Michigan.[67] George Stroke thus became a vocal critic of the WRL workers and a confidante to Gabor. Gabor proved magnanimous in accepting Kozma's apology, later publishing with him and recommending him to others, perhaps aided by their intellectual interests and common national backgrounds.

During early 1967 U-M administrators took the decision to reorganize the Electro-Optical Sciences Laboratory; and that summer, Stroke made the transition from the University of Michigan to a professorship at SUNY Stony Brook.[68] Gabor was then consulting for about half the year at CBS Laboratories in Stamford, Connecticut, where he was organizing his own holography laboratory. Gabor's correspondence with its

[65] Stroke, George W. to D. Gabor, letter, telegram and telephone conversation, 1 Nov. 1966, Ann Arbor, IC Gabor ML/4.

[66] Gabor, Dennis to E. N. Leith, letter, 3 Nov. 1966, London, IC Gabor ML/4.

[67] Leith sought to improve relations with Gabor by circumventing Stroke, writing, 'I was pleased to learn that you will consult for Holotron; perhaps we can have more contact than in the past, if this is agreeable to you. Also, I hope we can get together while participating in the forthcoming symposia. At the time of our symposium last April, George Stroke, according to Prof. Farris, arranged a busy schedule which left you little time for other things. I would like to see a get-together, perhaps a dinner arrangement, involving you, myself, and some of my colleagues such as Prof. Cutrona and Worthington, who are actively engaged in holography, providing this can be arranged' (Leith, Emmett N. to D. Gabor, letter, 6 Dec. 1966, London, IC Gabor ML/4).

[68] Smith, Allen F. and Gordon Van Wylen to G. W. Stroke, Dec. 1966–Jan. 1967, Ann Arbor, Bentley Historical Library Institute of Science and Technology Box 21: Electro-Optical Science Laboratory; Dow, W.G. to G. W. Stroke, memo, 26 Jan. 1967, Ann Arbor, Bentley Historical Library Institute of Science and Technology Box 21: Electro-Optical Lab; see also note 56. Stroke's brother Henry, also a physicist, interpreted his decision to leave the University of Michigan as a matter of his resistance to military research, something that George Stroke had encountered, and disliked, at his previous post at MIT (Stroke, H. Henry to SFJ, email, 6 Dec. 2004, SFJ collection).

Director, his close friend and countryman Peter Goldmark, occasionally mention Stroke as a resource available to both Gabor and CBS Labs.[69] Gabor began to rely further on Stroke as an expert on the mushrooming literature and latest developments in laser holography, a part of the field in which Gabor himself had, as yet, no direct experience. Stroke did consult for CBS Laboratories at Gabor's request.[70]

Thus, over a period of four years, George Stroke was embroiled in rising controversy at the University of Michigan and effectively promoted a contested conception of the subject and its history. As one of the few practitioners describing the historical trajectory of holography during the 1960s, he was influential in framing and presenting this version of the subject to scientists and the wider public, favouring what he described as 'broad and historical' accounts of holography, such as a Japanese seminar in 1967 about which he subsequently reported of himself, 'Dr. Stroke, a pioneering contributor to laser holography, discussed many of the varied developments for which he was responsible'.[71]

Dennis Gabor relied on such accounts for his own lectures and papers synthesizing an overview of the subject. Even after Stroke's departure from the University of Michigan, Leith seemingly still felt vulnerable in this battle of historiographies, given the hidden nature of the early work at Willow Run. He intimated to Gabor.

> I am glad that, after so many years of secrecy, we were allowed to communicate this work to you. In 1956, when I reinterpreted the process in terms of holography (or wavefront reconstruction, as it was then commonly called), I wondered if some day we would meet you and, if so, be able to tell you of this work. By now, this wish has been fairly well consummated. There are, however, yet other things to add: as our government permits, we will publish them and thus reveal them to you.[72]

Gabor expressed his distress at being between what he called the two 'hostile camps' and sought historical compromise by bracketing Leith and Stroke from 1962, while giving Leith full credit for work before that time, a chronology not accepted by Leith and unsupported by documentary evidence. Through early 1969, Gabor attempted to conciliate, stating that, despite Stroke's obvious ambition, 'I would never have judged [him to be] a stealer of merits', and counselled generosity on Leith's part.[73]

Gabor found himself in an awkward position in such incidents, because they involved not just contentions between Leith and Stroke but about the history of his own work. George Stroke sought to rehabilitate Gabor's reputation in holography by a particular

[69] Peter Carl Goldmark (1906–77) was raised in Budapest and part of the same 'restless generation' of emigrants, working at a British electrical firm in the early 1930s and later researching early colour television systems, like Gabor himself. The two had first met in Germany when Goldmark was a first-year student and became Gabor's laboratory assistant (Edson, Lee, 'A Gabor named Dennis seeks Utopia', *Think*, (January–February 1970): 23–7). Between the 1950s and 1970s they met and corresponded frequently. Gabor's consulting work for the CBS Laboratories is among the best represented in his archive.

[70] Gabor, Dennis to P. Goldmark, letters, 1966–7, IC GABOR LA/9.

[71] Stroke, George W., 'U.S.-Japan seminar on holography, Tokyo and Kyoto, 2–6 October 1967', *Applied Optics* 7 (1968): 622.

[72] Leith, Emmett N. to D. Gabor, letter, 21 Feb. 1969, Ann Arbor, IC Gabor ML/4.

[73] Gabor, Dennis to E. N. Leith, letter, 22 Apr. 1969, Leith collection.

reading of the historical record: he wished to demonstrate that the papers of the late 1940s and early 1950s encompassed the subsequent work by Leith and others. This went so far as arguing that there was not, in fact, any technical shortcoming in Gabor's work at all, but merely inadequate interpretation.[74]

However, this claim did not sit well with others who had also worked in the field during that period and formed their own accounts of why wavefront reconstruction had failed. As we have seen, Michael Haine and his AEI colleagues judged diffraction microscopy to be unworkable because of the twin-image problem, which polluted the desired sharp image with an overlapping out-of-focus image. And for Gordon Rogers, Gabor's in-line technique was fundamentally incomplete, requiring two complementary holograms for total image reconstruction and, thus, critically hampered by the practical complexities of such an arrangement. Stroke's contentions also upset him and in 1966 Rogers proposed a paper that would attack his claims. Gabor was awkwardly placed; to support Rogers would be to reopen the disagreements of colleagues regarding the decline of diffraction microscopy, while to support Stroke created breaches not only with the new generation in Michigan and beyond, but with those earlier colleagues as well. Gabor counselled, 'I am still of the opinion that you would be best advised *not* to publish your paper. I cannot see that it clears up the issue at all, and I am sure that it would be a start for some acrimonious polemics with G. W. Stroke'. Gabor sought rather ineffectually to reconcile the Stroke and Rogers accounts, which brought him implicitly to justify Stroke's position. Seeming to recognize this impasse, Gabor ended, 'May I also say, with some regret, that your holograms are so poor that I would not publish them in your place. I am fully agreed with what you say in your first letter of June 27, that "we did not miss quite so many obvious points 15 years ago as our American friends seem to think". Agreed, but almost anybody reading your paper would get the opposite impression!'[75]

As Gabor's closing flourish hints, he, Rogers, and others, such as Paul Kirkpatrick at Stanford, were irritated by the publicity garnered by the Willow Run researchers. Competing stories had been developing since the explosion in Ann Arbor. Immediately after the Leith and Upatnieks OSA demonstration in 1964, Rogers had pounced to alert Gabor of the American work, taking the opportunity to dub Gabor 'the grandfather of coherence optics' and to portray the new work squarely in the context of diffraction microscopy.

> I have been greatly interested to see the work of Leith and his co-workers in America on diffraction microscopy, especially the little device with the Fresnel bi-prism. This is one of great interest and I am making a further study of the process. It might appear that the result so

[74] See Stroke, George W., D. B. Brumm, A. T. Funkhouser, A. Labeyrie, and R. C. Restrick, 'On the absence of phase-recording or 'twin-image' separation problems in 'Gabor' (in-line) holography', *British Journal of Applied Physics* 17 (1966): 497–500.

[75] Gabor, Dennis to G. L. Rogers, letter, 5 Aug. 1966, Sci Mus MS 1014/15. Rogers could only retort impotently, 'while I do attach some importance to history and to the correct use of words, these points could be left until a definitive treatise is written on Holography in fifty years time' (Rogers, Gordon L. to D. Gabor, letter, 31 Oct. 1966, Sci Mus MS 1014/16).

Fig. 5.6. Dennis Gabor (right) next to his pulsed portrait hologram recorded by R. Rhinehart of McDonnell Douglas Corporation, October 1971, first displayed at the German Society for Photography on 10 November 1971, immediately before the award of his Nobel Prize for Physics (Boeing Inc: London Science Museum Collection).

> produced is in fact a contradiction of the proposition that two holograms are required. This, however, is not the case as a double system of coding is in fact employed in their system and this enables both amplitude and phase information to be independently coded on the same piece of film. I also note that they were using a gas laser which makes it possible to have coherent wave fronts of considerable area without prohibitively long exposures.[76]

Even two years later, Paul Kirkpatrick managed to construct a history of the subject that failed even to mention Leith and his co-workers.[77] What is beyond dispute in these conflicting accounts is that the roles of the first generation of holographers—Gabor, Rogers, El-Sum, Baez, and Kirkpatrick—were rehabilitated by the rise in popularity of holography, even if privately discomfited by Stroke's version of the story.[78]

A more appealing explanation that glossed over the details was that wavefront reconstruction had been reborn owing to the invention of the laser. This mythology was augmented in no small part by Gabor (see Figure 5.6), who observed in his Nobel Prize speech, 'When the laser became available in 1962, Leith and Upatnieks could at once produce results far superior to mine, by a new, simple and very effective method of eliminating the second image [. . .] made possible by the great coherence length of the helium-neon laser'.[79] This mis-telling of the story, unconscious as it almost certainly was, nevertheless provided a satisfying explanation for why the directions taken up at Willow Run had not been pursued earlier by Gabor and his collaborators.

[76] Rogers, Gordon L. to D. Gabor, letter, 5 May 1964, Sci Mus MS 1014/13.

[77] Kirkpatrick, Paul, 'History of holography', *Proceedings of the SPIE, Holography* 15 (1968): 9–12.

[78] Gabor was appreciative of the fame that had come to him late in life via the U-M work and wrote to Leith, 'You will be pleased to hear that the revival of holography which was initiated by you now brings me one honour after the other. After the Albert Michelson Medal of the Franklin Institute I have now been awarded the Rumford Medal of the Royal Society [. . .] Many thanks!' (Gabor, Dennis to E. N. Leith, letter, 21 Nov. 1968, Leith collection).

[79] Gabor, Dennis, 'Holography, 1948–71', in: Nobel Prize Committee (ed.), *Les Prix Nobel En 1971* (Stockholm, 1971), pp. 169–201.

5.3 THE NOBEL PRIZE AND HISTORIOGRAPHICAL VALIDATION

The personal contentions between Stroke and Leith, driven by the common desire for professional recognition, have afflicted many episodes in science. However, the Stroke controversies were significant in shaping the early accounts of a burgeoning scientific field transformed by the achievements at the Willow Run Laboratories and the Vavilov Optics Institute and one in which a growing number of observers were predicting the award of a Nobel Prize by the late 1960s. There were, for example, hints to Gabor from friends such as H. M. A. El-Sum: 'While in Stockholm, I heard of other anticipated nice things which are kept quiet at the present, and I hope to be the first to congratulate you next year for winning the biggest of all international awards'.[80] But T. E. Allibone recalls that Gabor himself was pessimistic.

> Gabor had been on the Nobel Prize list for some time. Indeed he had always recognised that the original invention of holography might one day be so rewarded; he and I hoped in Aldermaston years that if a resolution of 1–2A were ever achieved, he and the team producing the electron hologram and the optical reconstruction might share a Prize. After the explosion of holography in the 1960s the recognition of his basic work and of the brilliant inventions which followed appeared to be well justified. However [. . .] in 1968 in a mood of depression he said that the days of awards for inventions were passed.[81]

In later years, Stroke claimed to have been instrumental in gaining the Nobel Prize for Dennis Gabor by persistent lobbying.[82] Adolf Lohmann has commented that Stroke may indeed have influenced the awarding of the Prize in subtle and more obvious ways.[83] The Nobel Prizes are susceptible to influence without breaking its own rules and procedures. In the case of Gabor's nomination, the Nobel Committee for Physics sent out several hundred enquiries to a carefully selected cross section of the physics community in the autumn of 1970. Recipients were requested to complete forms with the names of one, two, or three worthy candidates for the Prize, to be returned by mid-January 1971. Thereafter, the committee had some eight months to identify the most suitable candidate, or combination of candidates, and then to submit their proposal for acceptance by the full Physics Section of the Royal Academy of Sciences in Stockholm, Sweden. Once approved, the successful candidate was announced in early October. The opinions of the Nobel committee could be influenced in various ways.

- By submitting a strong proposal for the candidate if invited to do so by the committee members, that is, if already acknowledged as a recognized participant in the subject, as George Stroke was.

[80] El-Sum, H. M. A. to D. Gabor, letter, 26 Nov. 1969, IC GABOR EM/8.

[81] Allibone, T. E., 'Dennis Gabor 1900–1979', *Biographical Memoirs of Fellows of the Royal Society* 26 (1980): 107–40.

[82] Aspray, op. cit.

[83] Lohmann, Adolf W. to SFJ, fax, 9 Oct. 2003, SFJ collection.

- By an individual contacting those already known to be invited, namely previous Nobelists. Stroke notes that he lobbied physicists Alexandr Prokhorov and Alfred Kastler, winners of the Nobel Prize in 1964 and 1966, respectively.
- By promoting the candidate and his field, for example by review articles. In the years prior to Gabor's Nobel Prize, Stroke collaborated with him in popularizations and surveys of the field that linked their names in historical accounts.[84]

More subtle methods of influence were also possible. The candidate might be 'talked up' in a conversation with a Swedish colleague concerning the forthcoming Nobel Prize or alternatively another candidate might be denigrated. The views expressed in such coffee-break chats could modulate opinions in the Nobel Committee about the mood of the physics community. In the uncommon cases of multiple candidates for the Prize, such influence—and the Nobel Committee's need to rely more than usual on informed outside advice—could be important.[85]

In the case of the Nobel Prize for holography, Stroke publicly and privately promoted a collection of claims: first, that holography was an important, escalating field; second, that Gabor was its primary progenitor; third, that Leith had made rather obvious extensions based merely on the availability of the laser; and fourth, that Stroke himself had constructed the modern subject by introducing a mathematical treatment for a subcategory that he dubbed lensless Fourier transform holography.[86] In the Nobel Prize procedure of identifying one, two, or three plausible candidates, Stroke thus vaunted a pair (Gabor and himself) while simultaneously deprecating a triplet (Gabor, Leith, and anyone else) because if Leith were admitted to have made major contributions, then arguably so had a handful of other individuals, including Stroke. This strategy required not only the career boosting common to many scientists but also the disparagement of a formidable assortment of adversaries at the University of Michigan and beyond.

[84] Gabor, Dennis, W. E. Kock, and George W. Stroke, 'Holography', *Science* 173 (1971): 11–23. Significantly, the accounts that portrayed Stroke's version of events were his own writings, the historical narratives that he edited with Dennis Gabor, and a popular book by one of his collaborators (Kock, Winston E., *Lasers and Holography; an Introduction to Coherent Optics* (Garden City, New York: Doubleday, 1969)]. Win Kock, who had worked on early electronic music, microwave lenses, acoustics, and solid state physics in a distinguished career at firms such as Bell Telephone Laboratories, the Bendix Corporation and NASA's Electronics Research Center, had joined Stroke to take part in a US–Japanese seminar on holography in 1967 funded by the NSF. Kock, Stroke, and Dennis Gabor collaborated on a popular article in the year of Gabor's Nobel Prize, when Kock himself was near retirement (Gabor, Dennis, W. E. Kock and George W. Stroke, 'Holography', *Science* 173 (1971): 11–23). Kock's idiosyncratic account attributes Denisyuk's research to Stroke and co-workers and indeed makes no mention of non-American work.

[85] Lohmann, Adolf W. to SFJ, fax, 9 Oct. 2003, SFJ collection. An example is the awarding of the Nobel Prize for the development of the electron microscope. Eight contenders from the German companies Siemens and AEG had published on the technique. The Nobel Committee did not award the Prize until only one remained alive: E. Ruska, then in his eighties, enjoyed the prize for only a year before his own death.

[86] Here, too, priority is contested. Stroke's work on the problem of image deblurring enjoyed some success but was an extension of research on spatial filters begun by VanderLugt at Willow Run under military sponsorship (Goodman, Joseph W., *Introduction to Fourier Optics* (New York: McGraw-Hill, 1968), Chapter 8).

Beyond the seminal Emmett Leith, potential claimants to the prize could well have included Yuri Denisyuk and perhaps Gordon Rogers or Adolf Lohmann or Juris Upatnieks. Like Leith, Denisyuk independently had developed a distinct and powerful version of wavefront reconstruction from first principles. This triumvirate is uncontentious unless Leith and Denisyuk are portrayed as having been unaware of possibilities or poorly motivated. Stroke's accounts actively disparaged Leith, and largely ignored, or belittled the generality, of Denisyuk's accomplishments.

The other candidates for the third man of the Nobel Prize are admittedly arguable. Upatnieks had performed the experiments conceived by Leith, co-developed certain concepts, and co-authored most of the important papers. Lohmann had first publicized the signal processing connection with holography and resolved Gabor's twin-image problem, although imperfectly. Rogers had been the earliest proselytizer and clarifier of concepts, conceiving the hologram as a generalized zone plate. But because the prize was limited to three winners, Stroke required a clear historiography to support his own claims to the award at the expense of those others.

Whether or not Stroke's claim of overt influence with the Prize Committee is sustainable, there is no doubt that he publicized his version of historical events effectively. Following the news of the award of the Prize to Gabor alone, his correspondence with Gabor remained effusive in its support and equally restless to have Gabor correct publicly what Stroke portrayed as historical misconceptions. He edited Gabor's Nobel Prize lecture, which provided an officially recognized version of the history of the subject to 1971. Stroke's two dense pages of amendments emphasized his own contributions at the expense of Leith, Denisyuk, and others.[87]

Following the prize, their names remained linked. George Stroke arranged for a visiting professorship for Gabor at the New York Institute of Technology. The two collaborated, too, on a National Science Foundation film on holography and other research. Gabor's subsequent popular writings continued to be influenced by Stroke's historiography.[88] After Gabor's death in 1979, articles authored by Stroke continued to link their names, with Stroke portrayed as a conduit for revitalizing, generalizing, and extending Gabor's ideas.[89] Stroke's most candid and contentious public statement about his career in holography was an interview for the IEEE twenty-two years after the Prize.[90]

[87] Stroke, George W. to D. Gabor, letter, 26 Nov. 1971, IC Gabor MS/15; Stroke, George W. to D. Gabor, telegram, Nov. 1971, Paris, IC Gabor MS/15.

[88] See, for example, Gabor, Dennis, *Holographie 1973: Vortrag, gehalten an dem Mentorenabend der Carl Friedrich von Siemens Stiftung in München-Nymphenburg am 15. Juni 1973* (München: Die Stiftung, 1973).

[89] Stroke, George W., 'Optical engineering', in: (ed.), *Encyclopaedia Britannica Yearbook of Science and the Future* 1980), pp. Similarly, in a letter to the Director of the Museum of Holography, Winston Kock acted as intermediary in passing on Stroke's suggestion that their joint paper with Gabor be 'prominently displayed' in the museum (Gabor, Dennis, W. E. Kock, and George W. Stroke, 'Holography', *Science* 173 (1971): 11–23, Kock, Winston E. to P. Jackson, letter, 1 Sep. 1977, MIT Museum 30/909). On the Museum of Holography, see §11.4.

[90] Aspray, op. cit.

5.4 PATENTS, PRIORITY, AND PROFITS

Priority disputes were important in shaping the historiography of holography: they contested the significance of intellectual insights, and the cause-and-effect relationship between them. Patent disputes are closely related and were just as significant. Even more simplistically and artificially than priority disputes, they defined the sequence of discovery and application, focusing on concepts that could be embodied in inventions.

In a patent deposition, Adolf Lohmann alluded to the importance of personalities in priority disputes:

> Gabor in later years did not follow the literature in holography personally [. . .] instead he relied on advice and judgment by others, who sometimes let selfishness influence their presentations to Gabor. From Gabor's statements it became clear to me [. . .] that Gabor was not entirely immune against those influences. [In a joint paper by Gabor and Stroke[91]] there are passages which might be misinterpreted by an innocent reader not knowledgeable about the history of holography.[92]

These types of disagreements are linked by concerns of status or profit. It is significant that academic priority was not the only issue in the Stroke–Leith controversies. Just two weeks after the explosive OSA presentation in April 1964, the potential income from a successful patent for off-axis holography was being discussed by all participants. Stroke wrote to the Chairman of the U-M Research Policy Subcommittee as his disputes with Leith and Upatnieks were gestating, suggesting that a windfall of money for Willow Run researchers would strain academic relations. His concerns illustrate the high expectations that researchers were already placing on the new technology.

> A procedure has been initiated, under the direction of Dr Fred Llewellyn at Willow Run, to acquire a patent through the Battelle Institute for some wavefront reconstruction work carried out by two research staff members, Mr Emmett Leith and Mr Juris Upatnieks [. . .]
>
> According to the best information received, the patent, to be obtained through the Battelle Institute should not only benefit the university, but also involve 15% return to the two research staff members whose names figure on the patent application.
>
> In conversations, figures as high as possibly $10 million are mentioned, when describing the possible benefit involved for this work and also to justify any of the procedures used, when they are questioned in terms of ethics and implications on University freedoms.
>
> [. . .] In fact, there have already arisen elements of disagreement in conversations as to the stand to take, especially when the fact was not brought out that there has been input by other research staff and indeed faculty in the specific case noted above, an input which has gone unrecognised in the patent which has been drawn up.
>
> [. . .] Let me stress that the best personal relations exist between the research staff at Willow Run and myself in the various aspects of this work.[93]

[91] Gabor, Dennis and George W. Stroke, 'The theory of deep holograms', *Proceedings of the Royal Society of London, Series A (Mathematical and Physical Sciences)* 304 (1968): 275–89.

[92] Lohmann, Adolf W. to patent attorneys, patent declaration, Jun. 1987, Haines collection.

[93] Stroke, George W. to C. G. Brown, letter, 16 Apr. 1964, Ann Arbor, Leith collection.

The key dimension of this emerging controversy was the degree of innovation represented by the Leith–Upatnieks work. Dennis Gabor was enrolled by Battelle to judge the originality of the research, a situation that may have made him uneasy because of its potential side effect on his rehabilitated intellectual status.[94] Significantly, the Battelle patents asserted a different historiography than did Stroke: according to Battelle, Leith and Upatnieks were the sole inventors of the novel technique of off-axis holography.

Patent disputes also introduced new players, who added their own glosses to the historical accounts. The most important was Lou Cutrona, who had moved from being originator and supervisor of the WRL optical processing work to co-found a spin-off company, Conductron.[95] In 1966, when the Battelle patent for the Leith–Upatnieks work was being investigated, Cutrona wrote an internal memo to show that 'a number of techniques developed for optical radar processors were identical and served the same purpose as the use of an off-axis reference function in recording a hologram'. He argued that.

> Inasmuch as the techniques and purpose of the non-normal reference function in holography is identical with the techniques and purpose used in radar processing, it is believed that it is not a significantly new step to use the same techniques in holography which were previously developed for the radar case, or in any event, the technique of the reference function for holography is one which persons skilled in the high resolution radar art would independently arrive at because of its similarity of technique and purpose with the technique used in the radar case. For these reasons, it is recommended to the Examiner of patents that a separate patent for the use of an off axis reference function not be granted, but that this technique be considered as included in the claims made in the four referenced patents.[96]

The financial consequences of such a re-evaluation of the innovation are clear. But Cutrona was able to support his claim about the ability of a skilled worker to discover holography from SAR: Gary Cochran, the manager of Conductron's optical work, had since 1962 supervised their commercial development of an optical processing system. He had readily and enthusiastically taken up holographic research from late 1964.

> I picked up Emmett's paper on holography when he wrote it—his ground-breaking paper—and because of my background at that moment, I read the first paragraph and could have written the rest of the paper myself; he couldn't say where that came from, but we all knew. That was fascinating to me. And remember, he was not [. . .] a professor or anything at that time; he was just a guy working with Lou Cutrona (and at that time Lou was with us); he had no PhD, no way of moving forward [. . .] It really intrigued me because I thought, you know, this is what I'm doing, and I know exactly what he's saying.[97]

[94] Stroke later claimed that Gabor had been 'bribed' to sign the affidavit and that Stroke himself had counselled him not to do so (Apray, op. cit.).

[95] On Conductron, see §6.2.

[96] Cutrona, L. J., 'The relationship of off axis reference function in holography to the equivalent problem in radar processing', Conductron inter-office memo, Conductron Inc, 2 Aug. 1966. Indeed, he cited four previous patents (Pat appl. 26,916, Wendell A. Blikken et al., 'Optical plural channel signal data processor'; Pat appl. 96,052, Lous J.Cutrona et al., 'two dimensional optical data processor'; Pat appl. 116,895, Arthur L. Ingalls, 'plural channel optical data processor'; Pat appl. 116,896, Emmett N. Leith, 'Coherent optical cross correlator') as direct and obvious precursors embodying the idea of off-axis holography.

[97] Cochran, Gary D. to SFJ, interview, 6 and 8 Sep. 2003, Ann Arbor, MI, SFJ collection.

This was not a priority claim, but rather an assertion about originality. Conductron engineers—many joining the firm directly from WRL—were from much the same working culture as their former colleagues fifteen miles away.

Such were the opening salvos of commercial exploitation and patent wrangles.[98] These claims and counterclaims were not resolved until considerably later, but illustrate the contentions that almost immediately surrounded the chain of insights that led to holography.

5.5 FINDING COHERENCE

These episodes reveal the intellectual evaluations, social manoeuvres, and economic factors that consolidated an official history (see Figure 5.7) of holography. George Stroke was a key player in documenting a contentious evolutionary line for the new subject, but the strictures of the Nobel Prize and patent disputes, and an emerging community identity, also exerted pressures on interpretations of the work of Gabor, Denisyuk, Leith, and Upatnieks.[99]

There are, then, several coexisting but disconnected origin stories. One was Stroke's account, woven through survey papers and semi-popular writings. In the relationship between Gabor and Stroke, both individuals gained recognition and status. Stroke vaunted Gabor's little-remembered work; in return, Stroke gained considerable status from his association with Gabor. Beyond his influence on individual careers, Stroke had longer-enduring effects on holography and its history. His early narratives eclipsed the sparse publications of Willow Run workers and vaunted a particular reading of the historical episodes. The result was a clouded history in which the crucial claims about occupational context, professional relationships, and intellectual evolution had been divorced from documentary evidence.

Another version was the multiple-voiced counter-story of the Willow Run workers, kept alive by the oral accounts of practitioners still attending conferences forty years later. In contrast to our common expectations of history, these folk tales arguably represent a more faithful account than the publicly available documentation and accord well with archival documentary evidence. However, oral accounts may also mutate more rapidly than written reports. What some contemporary holographers dubbed the 'Stroke problem' was widely related and mythologized at conferences from which George Stroke

[98] The pursuit of commercial holography by Conductron and others is discussed further in Chapters 6 and 12.

[99] Patent evaluations (e.g. Lohmann, Adolf W. to patent attorneys, patent declaration, Jun. 1987, Haines collection and Lohmann, Adolf W. to SFJ, fax, 30 Jun. 2004, SFJ collection) illustrate the difficulties of distinguishing clear accomplishments from a spectrum of overlapping activities. Lohmann characterized the contributions of Gabor, Denisyuk, and Leith as experimentally distinct, noting that Gabor had employed a 0° angle between reference and object beams (which were not distinct); Denisyuk had employed a 180° angle between the object and reference beams and used collimated rather than diffused light; and Leith and Upatnieks had generalized the angle between the beams and used diffusely reflecting objects.

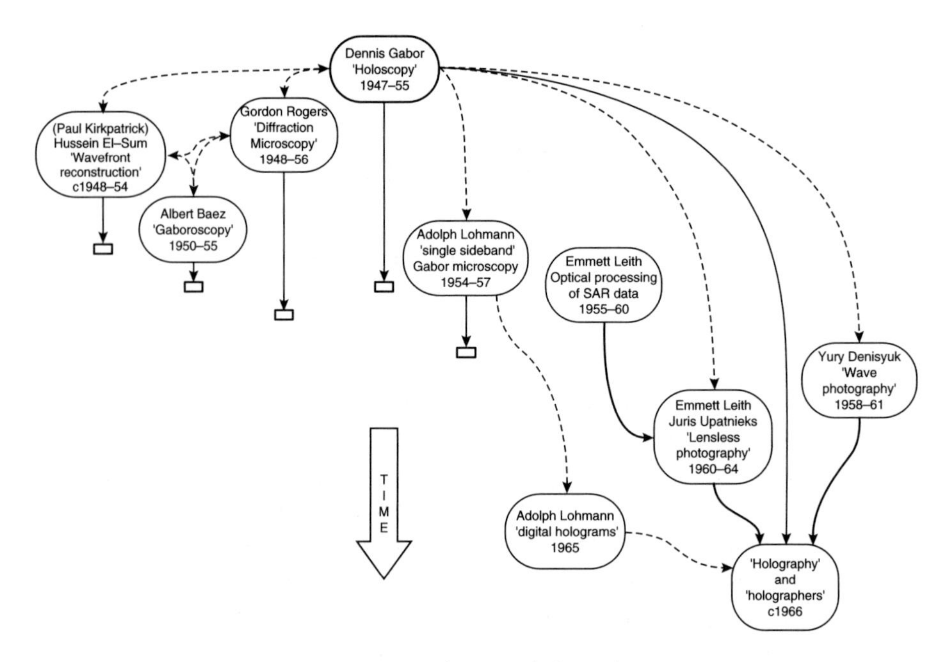

Fig. 5.7. Concepts contributing to holography, 1947–66.

was absent after the mid-1970s. While these stories served specific social purposes—privately righting perceived wrongs, serving as paradigms of bad science, and even binding together a group of practitioners—they circulated in a narrow subculture and have faded for subsequent generations.

More recently, an anodyne, Stroke-free, oral tradition tends to underplay the accomplishments of the first communities of holographers. Sustained by frequent conferences on holography attended by a coterie of several hundred participants, this later tradition promoted a simpler and enduring historiography. In it, Gabor and Leith (and latterly Yuri Denisyuk) were portrayed as the fathers of the subject and, as traditional in such tales, imbued with the qualities of genius, modesty, and generosity. While recognizably based on fact, these, too, are tales beyond the documentary evidence.

Such popular accounts de-emphasize the role of social factors in the creation of a new intellectual subject. But the priority and patent claims can be unravelled; eyewitness accounts and archival evidence can be reconciled to probe causes and to clarify the actual events. As illustrated in these past four chapters, institutional contexts and personal motivations were decisive factors in building the new subject of holography.

PART II

Creating A Medium

Holography passed from a handful of investigators in the early 1960s to a jostling assortment of university and industrial laboratories by the end of the decade. They all focused on answering two questions about the medium itself: what are holograms good for, and how can they be improved? The rate of publishing research papers on wavefront reconstruction multiplied eight-fold between 1964 and 1965 alone and trebled again by 1970.[1] This explosion of research brought the subject to new locales having new perspectives. Companies sprouted to exploit holography, riding a wave of what Jeff Hecht calls 'the sustaining forces of money, mindset and momentum'.[2] Within less than a decade, the subject was rapidly explored, applied, and newly categorized as sponsors liberally dispensed funds and most of the subsequently important types of hologram were invented. The decade from 1964 was unparalleled, then, in expanding the technological dimensions of the new subject.

This section correspondingly turns from facts to artefacts. The following chapters illustrate how expertise spread and was exploited (Chapter 6) and how the expanding set of tools defined the craft and products of holography (Chapter 7). They emphasize the shifting mix of methodologies at play: where theoretical perspectives had guided the first generation, researchers of the boom decade were impelled by experimental exploration. The new subject was shaped by how it worked and what it could do.

[1] See Appendix.

[2] Hecht, Jeff, *City of Light: The Story of Fiber Optics* (Oxford: Oxford University Press, 1999), p. 129.

6
Early Exploitation

Seeing was believing for optical scientists and that first experience of viewing holograms created a wave of enthusiasts. Many of the first wave consisted of attendees at the spring 1964 OSA meeting or those who saw holograms in the months that followed. Most of those attendees were employees of American commercial firms, universities, or government research centres. Not surprisingly, then, the initial bubble of enthusiasm engendered by the train hologram, and the consequent scramble to begin research over the next few months, was most obvious in America.

The funding climate was also propitious there. From 1964, when the National Science Foundation and Air Force accepted the first U-M research proposals for holography, financial support proved easy to obtain. American government agencies such as the National Science Foundation (NSF), National Aeronautics and Space Administration (NASA), and Department of Defense (DoD) via the Army, Navy, and Air Force were eager to back exploratory research. Large corporations, readily seduced by the potential of the technology, also set up research teams or sponsored existing organizations. Many plausible potential applications were identified during this first expansive decade; some were ultimately rejected, while others survived to struggle for economic viability. These motivations, patterns of funding, and tenuous footholds were to become a model for subsequent decades, as the seduction of holography gradually suffused downwards from large, centrally managed research and development organizations to cottage industry and to new waves of entrepreneurs further afield.[3]

6.1 SATISFYING SPONSORS AT THE UNIVERSITY OF MICHIGAN

> We haven't given much thought to [commercial applications]. We did the work solely out of personal interest, for strictly scientific purposes. However, it seems that any process that works well must find application. (Emmett Leith, December 1963)[4]

[3] The successive waves of uptake are discussed in subsequent chapters, particularly Chapter 12.
[4] Wixom, Charles to L. H. Dulberger, letter, 11 Dec. 1963, Leith collection.

At U-M, holography struggled to escape the velvet handcuffs of military sponsorship. On the one hand, research contracts were readily available; on the other, the free dissemination and wider application of the technology were not actively encouraged. This dramatic technology had had an almost imperceptible rise to prominence. The reason for this was not an overt intention to restrict access to holographic developments, but merely the unfamiliarity of the technology and its possible benefits for wider society. The conference presentations and scientific papers of Leith and Upatnieks before 1964 had made little immediate impact on colleagues back at Willow Run. One reality of military contracts was that clearance had to be sought for all publications, but the 1960–3 developments on wavefront reconstruction evoked few, if any, concerns for the administrators. Nevertheless, as Leith recalls, 'reticence was the byword': those early papers on wavefront reconstruction had to be approved by the military agency that sponsored their work, a process that introduced a delay of several months even before the papers' refereeing by scientific journals.[5] The material, which was outside their previous experience, may well have baffled the military personnel who reviewed it. The Radar Lab's director, William Brown, recalled:

> While there was some excitement about it, there perhaps wasn't as much as there should have been. While it looked like an excellent piece of coherent optics work, from a technical standpoint we couldn't be sure whether we had an important scientific tool on our hands or just a curiosity.[6]

From late 1963, though, Leith and Upatnieks were catapulted from the hidden world of military contract research to the public stage. In contrast to the lack of interest a year earlier, their employers were now excited. Fred B. Llewellyn, who had succeeded Lou Cutrona as Director of the Willow Run Laboratory in 1962, was very cooperative in accommodating the research.[7] Also the Assistant Director of the Institute of Science and Technology of which it was a part, Llewellyn became a strong supporter and sought funding and patent coverage from the Battelle Memorial Institute of Columbus, Ohio. Leith recalls the buoyant response:

> We got research funds. First of all the Air Force threw money at us, just to explore it for applications. They gave us $150,000, which was a big chunk of money at that time, just to hunt for applications, and it wasn't really a classified contract, this was a side line from the radar work we'd been doing as we went along, so this was a welcome thing, then the Battelle people gave us some money, so we had a lot of good funding.[8]

[5] Leith, Emmett N. to SFJ, email, 11 Mar. 2003, Ann Arbor, SFJ collection

[6] Wolff, Michael, 'The birth of holography: a new process creates an industry', *Innovations* (1969): 4–15, p. 7.

[7] W. M. Brown (b. 1932) joined the University in 1958, having obtained a Dr Eng. degree from Johns Hopkins University, and working previously at the Air Arm Division of Westinghouse Corp., and at the Radiation laboratory at Johns Hopkins. Before joining the University of Michigan, he had worked for a year at the Weapons Systems Evaluation Group in the Pentagon. Leith recalls him as 'a strong advocate of optics and [. . .] in large part responsible for the development of a large and strong optics group' (Leith, Emmett N., 'A short history of the Optics Group of the Willow Run Laboratories', in: A. Consortini (ed.), *Trends in Optics* (New York: Elsevier, 1996), pp. 1–26, p. 24). After twenty-four years at WRL and its successor, ERIM, most of his later career was in universities, complemented by work with industry and government-related programmes.

[8] Leith, Emmett N. to SFJ, interview, 22 Jan. 2003, Santa Clara, CA, SFJ collection.

The first grant proposal specifically intended to seek applications of what he referred to as the 'hologram technique' had been submitted a full year earlier to the US Army Research Office (ARO), when Leith sought $30,300 for a project planned to require fourteen engineering man-months and seven support man-months. In the proposal he made reference to his successful reconstruction of continuous-tone pictures, and identified five plausible applications: the improvement of electron microscopy (as Gabor had planned); x-ray microscopy (as El-Sum had envisaged); a method of encoding and decoding images 'to make the transmission process less susceptible to a atmospheric impulse noise'; a 'lensless microscope' achieved by recording and reconstructing the holographic image with a diverging beam; and a means of recording photographs with very high dynamic range.[9] That proposal was rejected by the ARO with the recommendation that it be resubmitted with less emphasis on applications, because the ARO was charged with developing basic technology rather than investigating practical uses. A similar version of the proposal was funded, though, by the NSF in 1964 under the name of George Stroke, as discussed in §5.2.1.

Subsequent applications were successful and military contracts played an important role in parcelling out research work, controlling dissemination of knowledge, seeding subsequent multiple sourcing of expertise, and promoting competition in research. For example, under the broad title 'Investigation of Hologram Techniques', the Air Force funded the Willow Run Optics Group, the Conductron Corporation (discussed later) and Hughes Research Laboratories from August 1964. In the first round of contracts, seven members of the Optics Group studied the geometry of recording holograms and the dynamic range of photographic films; Conductron studied complex spatial filters and A. D. Jacobson of Hughes investigated 'the ideal and practical characteristics of the hologram method'.[10] Extensions and sponsorship under Army contracts also allowed the Willow Run Laboratories (WRL) Optics Group to pursue more wide-ranging studies, including the measurement of mechanical deformation and the storage of more than one hologram on a single plate.[11]

Beyond military contracts, Fred Llewellyn of WRL also provided seed money and, even more importantly in an explosively expanding subject, agreed to expedite the clearance of the Group's papers for publication. The clearance procedures would otherwise have hampered the work at Willow Run seriously, despite its international expertise in coherent optics.[12]

[9] Leith, Emmett N., 'Proposal for Applications of the Wavefront Reconstruction Technique', Grant proposal ORA-63-1255-PB1, Institute of Science and Technology, University of Michigan to US Army Research Office, Durham, NC, 28 May 1963.

[10] Section I: Investigation of hologram techniques, University of Michigan Contract AF 18 (600) -2779; Section II: Holograms and complex spatial filters Conductron Corp Contract AF 33(615)-1014; Section III: Hologram Studies: Hughes Research Laboratories Contract AF 33(615) -1593.

[11] Contract AF 33(615)-3100, and DA-28-043-AMC-00013(E), AF 18(600)-2779; and AF 33(615)-1452 (Aug. 964–Dec. 1965). The work involved Optics Group members Leith (as Principal Investigator), Upatnieks, Stetson, Powell, Friesem, Haines, Hildebrand, Massey, Palermo, and Marks. Early contracts were with the Air Force Avionics Laboratory; others were for different Air Force and Army labs.

[12] As discussed in §7.3 and §12.5, Battelle (later Holotron), which was concerned with patents, instigated its own research efforts. Consequently its researchers, such as Nile Hartman, published seldom and

The Willow Run Optics Group grew to about twenty-five by 1965, all wanted to do work in the subject.[13] Leith, who succeeded Wendell Blikken as Head of the Optics Group in 1962, decided that all members of the group should receive some of the research and development money to investigate applications.

As more members of the group became interested in doing research in holography, the administrators coped with the new public exposure of both the subject and the lab:

> For our first paper (1962) on holography to get approved, it had to be sent to the military agency that sponsored our work, and it took several months, maybe 6 months, to get approval for the work to be publicly presented. I imagine that the military personnel approving the work were thoroughly baffled, since the material was well out of their previous experience. After we got more spectacular results, the Willow Run directors became cooperative, providing support for the work out of overhead funds, thus allowing us to publish almost immediately. It would have been an utter disaster for us if they had not, since papers were soon being given at a furious pace by the optics community. Then there were the patent people, who wanted to not publish or to delay publication, but these efforts were more easily thwarted.[14]

As discussed in §4.1, each of the WRL groups had sprung to life one by one in response to specific military research and development contracts. This structure yielded opportunities for collaboration and the sharing of information. The coherent profile of its personnel encouraged like-minded research. WRL was staffed principally with young electrical engineers. Their principal sponsor, the Wright Patterson Air Force base in Ohio, allowed considerable freedom for the staff to pursue interests beyond the ostensible goals of the Synthetic Aperture Radar (SAR) optical processing work.[15]
The working environment was both open and stimulating. Kenneth Haines, a member of the group between 1962 and 1968, recalled Leith as an easy-going manager and inspirational motivator of the early hologram research.

> He was the kind of guy who would rush out of his office and show it to all the guys, who were about the same age. He would share it; he would say 'What do you think of this?' and we would all gather around and go off into our labs and ponder it, and he'd say, 'Why don't you do an experiment on this? Or on that?' When you get onto something exciting, and someone stimulates you, you can do things you never thought you could do; and you're all young, and pliable.[16]

encouraged the Optical Group at Willow Run to similarly delay or avoid publication of novel findings (Leith, Emmett N. to SFJ, email, 11 Mar. 2003, Ann Arbor, SFJ collection).

[13] Although Head of the Optics Group, Leith was not an enthusiastic administrator and was supported by Adam Kozma. Leith later described himself as requiring just enough money to do research, not to lead large teams (Leith, Emmett N. to SFJ, interviews, 29 Aug.–12 Sep. 2003, Ann Arbor, MI, SFJ collection).

[14] Leith, Emmett N. to SFJ, email, 11 Mar. 2003, Ann Arbor, SFJ collection. On the pursuit of holography patents, see Chapter 12.

[15] Leith, Emmett N. to SFJ, email, 6 Jun. 2003, SFJ collection; Wolff, Michael, 'The birth of holography: a new process creates an industry', *Innovations* (1969): 4–15.

[16] Haines, Kenneth to SFJ, interview, 21 Jan. 2003., Santa Clara, CA, SFJ collection. Haines (b. *c*.1938), an electrical engineer with an MSc in physiology from the University of Toronto, joined the group in 1962, initially working with Anthony VanderLugt on the optical processing of SAR data. He left Willow Run in 1966 to join Holotron to seek commercial applications of holography. He was Reader in Physics at the University of Canterbury in New Zealand during the early 1970s and returned to a variety of positions in American holography firms, notably American Bank Note Holographics (see §12.4).

The members of the Optics Group generally had engaged in projects by organising in pairs, supported by about half a dozen summer students, both undergraduate and graduate.[17] As a consequence, the members of the group tended to team up with each other on projects as a matter of convenience: Robert Powell and Karl Stetson, B. Percy Hildebrand and Ken Haines, Adam Kozma and Norman Massey (see Figure 5.2). Leith and Upatnieks remained a strong team.

The Radar & Optics Laboratory of Willow Run was a breeding ground for a new type of technical worker in modern optics, and proved fertile in producing holographers. Interestingly, several from this group of enthusiastic, hands-on engineers went on to complete advanced degrees on holography: Upatnieks gained an MS in 1965, Haines a PhD in 1966, Hildebrand and VanderLugt a PhD in 1967, Friesem in 1968, and Kozma and Stetson PhDs from Imperial College and Institut för Optisk Forskning in Stockholm, respectively, in 1969. In 1978, a decade after gaining his professorship, Leith completed his own PhD. Several of the WRL researchers went on to careers as commercial research and development engineers and entrepreneurs. With highly valuable experience under their belts, several members of the Optics Group departed after a couple of years for industrial posts.[18]

Leith and Upatnieks were also followed up rapidly by other researchers at the University of Michigan as news of their work spread in 1964. Among the first to learn details of their work, as discussed in §5.2, was Prof George Stroke and his growing collection of graduate students, setting up the U-M Electro-Optical Sciences Laboratory from the autumn of 1963.

Other researchers at the University of Michigan independently picked up on holography merely by hearing about the Willow Run work or participating with either the Optics Group or the Electro-Optical Sciences Laboratory. For example, Charles Vest, a new faculty member in Mechanical Engineering, worked with the members of the Optics Group at WRL and later shared a laboratory with Leith when Leith had become a professor of Electrical Engineering.[19] As a new Assistant Professor, Vest recalls having heard of holographic interferometry at a seminar given by Robert Powell of Willow Run.

[17] The summer students included Carl Aleksoff, Doug Brumm, Carl Leonard, and Art Funkhouser.

[18] Haines joined Holotron in Wilmington, Delaware; Hildebrand went to Battelle's Northwest laboratory in Seattle, Washington; Massey and several others went to KMS Industries in Ann Arbor, while VanderLugt joined the Electro-Optics Center of Radiation Inc. there after his PhD in England. Powell joined GCO in Ann Arbor in 1966, moved to American Optical in Massachusetts a couple of years later and launched a holography consulting firm, Powell Associates, around 1970. He later was at Oakland University in Michigan and subsequently moved back to his home state, Texas. Stetson was equally peripatetic, joining the GCA Corporation in Bedford, Massachusetts, then pursuing a PhD at the Institutet för Optisk Forskning in Stockholm while consulting for GCO. Stetson later worked at the National Physical Laboratory (NPL) in England and joined the Ford Motor Company in Dearborn, Michigan, in 1971 and the United Aircraft Research Labs (later the United Technologies Research Center) in Connecticut. In 1994 he founded Karl Stetson Associates.

[19] After completing his PhD at the University of Michigan in 1967, C. M. Vest (b. 1941) joined the Mechanical Engineering Dept there as assistant professor, becoming full professor in 1977. Researching holographic interferometry at the Radar & Optics Laboratory, he later wrote a classic text on the subject (Vest, Charles M., *Holographic Interferometry* (New York: Wiley, 1979)). He became President of MIT in 1990.

He 'drifted over' to the Radar & Optics Lab in order to 'hang around, and do a little bit of research there in a totally different field'. As the only mechanical engineer in the laboratory, Vest found Leith and his colleagues welcoming. With some start-up money from his department, and a small lab made available by Leith in the blockhouse, Vest began researching applications of holography for mechanical engineering, specifically how stress, strain, and temperature could be measured and how to get three-dimensional information from a hologram about physical phenomena. He was also assigned the management of an Advanced Research Projects Agency (ARPA) contract, coordinated by the Watertown arsenal in Boston, on the use of holographic interferometry for non-destructive testing and 'just generally had a good time'.[20] And while contracts from the Department of Defense were the bread-and-butter of the group, Vest recalls an openly exploratory spirit pervading the Optics Group and its Head:

> [L]ike a lot of really good scientists, [Emmett] never lost a bit of his childhood enthusiasm and interests; he was always happy to have other things going on around [. . .] In terms of teaching, learning, working with people, having an insatiable curiosity about the next thing that should be done, that was *him*.[21]

Alongside the contract research, the Optics Group continued to play a limited role in popularizing holography itself. Carl Leonard, originally George Stroke's student for an MS degree and then working under Emmett Leith for PhD studies during the late 1960s, recalls being tasked with weekly demonstrations for visitors and investigation of applications suggested by Leith.[22] A handful of visitors—some merely curious local residents or colleagues from other departments—would tour the laboratory every month. Leonard also worked on some of the non-military projects, such as a multi-image hologram of the model of a building for the U-M Department of Architecture. Yet such projects were clearly a sideline and suggested a lack of market awareness within the Optics Group: their NSF proposal to explore the architectural applications of holography was rejected because the reviewers felt that such display applications would be 'too commercial', not 'a better visual aid than the original model' and would provide 'no benefit other than transport'. In fact, 'The use of holography, time lapse, etc borders on trickery and probably will communicate little that bears on reality'.[23]

Still others at the University took up holography independently. For example, Norm Barnett and Ralph Grant of the Electrical Engineering Department began using

[20] Vest, Charles M. to SFJ, interview, 13 Jul. 2003, Cambridge, MA, SFJ collection. On holographic interferometry, see §7.2.

[21] Vest, Charles M. to SFJ, interview, 13 Jul. 2003, Cambridge, MA, SFJ collection.

[22] Carl Leonard (b *c*.1944) obtained BS and MS degrees in Electrical Engineering at the University of Michigan and studied for a PhD there. He worked in the Radar and Optics Laboratory from 1967 and was transferred to ERIM when it was founded in 1973. In 1979 he took a position at Scientific Applications Inc in Ann Arbor and later worked in Space Science at the University. From the 1980s, he moved increasingly away from holography towards electrical engineering, a career trajectory evinced by a number of Ann Arbor engineers (Leonard, Carl to SFJ, interview, 4 Sep. 2003, Ann Arbor, MI, SFJ collection).

[23] Leonard, Carl, Emmett N. Leith and Juris Upatnieks, 'Investigation of architectural applications of holography', University of Michigan, Apr. 1972, Leonard collection.

holography during 1965 and 1966.[24] They explored holography as a tool to study vibration and sound. The skills were nevertheless demanding; they discovered that, 'with enough faith we could get a hologram—but you really had to believe in it when you looked at it!'[25] They consulted George Stroke, located across the street in the IST building, and also Karl Stetson and Robert Powell of the WRL Optics Group, who had then been working in holographic interferometry for over two years.

Even within the University of Michigan, it was easy to lose track of the proliferating activities of holography: Barnett first met Leith at the 1968 meeting on the engineering uses of holography at the University of Strathclyde in Scotland. Apart from the consultation with IST workers, Barnett and Grant pursued an independent course, building a holographic lab in the basement below the reverberation room in one of the Electrical Engineering buildings, where it could be decoupled from the building vibrations. Their equipment included a massive granite table purchased from Arnett's Monument Works, the local tombstone supplier, and supported on rubber mounts.[26] They applied holographic interferometry promiscuously and enthusiastically: to the motion of soil and sand, to stroboscopic studies of vibration, to microphone design, and to classified work for the Navy, which they sought to render innocuous to their academic department:

> We did studies of sonar transducers in and out of water. They were acceptable at the university in that era as long as you called them 'n-stiffened cylindrical shells' [even though to informed eyes] they looked like the section of a torpedo.[27]

Much of their work was proof-of-concept research, fishing for contracts with the Office of Naval Research (ONR) or industry. They demonstrated, for example, the feasibility of recording a hologram of an object in an aquarium to show naval referees that underwater holography could be a useful technique and located a mechanical resonance in the structure of a commercial microphone that had eluded its manufacturers for years; neither of these led to development contracts, however.

Thus, most of the U-M holography research was funded by military contracts: WRL workers benefited from a continuous stream of such exploratory contracts, as did collaborators from other departments such as Charles Vest and Norm Barnett. Interest by the American military was natural, as the WRL had been almost entirely funded by contracts from the Army (sometimes via the Army Research Office (ARO) for small contracts), Air Force, and ONR since its origin.[28] This continuity was uninterrupted even after WRL

[24] Barnett (b. 1923) had completed a physics degree at U-M during the war and subsequently worked on infrared communications and acoustics. After having worked at the university in both these areas through the 1950s, sometimes in collaboration with workers at Willow Run, he joined the Electronics Defense group of the Electrical Engineering Department in the early 1960s. On Grant, see GC Optronics, discussed later.

[25] Barnett, Norm to SFJ, interview, 11 Sep. 2003, Ann Arbor, SFJ collection.

[26] George Stroke had recommended that he sandwich balsa wood sheets between brick and granite layers, but Barnett opted for granite on rubber. Within a couple of years, most workers had moved to air pistons to 'float' the massive tables. For more on optical tables, see §9.2.

[27] Barnett, Norm to SFJ, interview, 11 Sep. 2003, Ann Arbor, SFJ collection

[28] A 1965 article local newspaper article reported that Project MICHIGAN alone had received some $50 million from the Army and was currently receiving funding of some $3 m per year ('Generals tour WR Laboratories', *Ypsilanti Press*, 8 Jul. 1965).

separated from the university to become ERIM. For instance, a later example was the holographic sight, conceived at the University of Michigan in 1971 by Juris Upatnieks on an Air Force contract to suggest practical uses for holography. A number of versions were developed, first for vehicles (owing to an initial power consumption of some 30 W), helicopters, and finally for small arms.[29] This single application occupied some 30 years of development and several ERIM contracts, and yielded three American patents.

The potential for military funding holography was apparent to other researchers beyond the University of Michigan, even if they did not benefit from prior contacts. Matt Lehmann at Stanford University recalled that, when he heard of the Upatnieks-Leith paper, he made a hologram on Polaroid film with an 11 mW laser, which luckily yielded an image at his first attempt. He then made a Fourier transform hologram of an eight-line poem on a 35 mm transparency. All the information was contained in a hologram about 1 mm diameter:

> It seemed a good way to impose a secret message in an otherwise innocent photograph so we superimposed our holographic message within a window area of a photograph of the Hoover Tower. Although we had hoped to obtain CIA support for more holographic research, funds were not forthcoming. Our project, 'Witchcraft', never made it beyond the initial proposal.[30]

Joe Goodman (b. 1936), who had come to Stanford in 1958 to pursue a PhD on radar countermeasures, became a Research Associate there under Wright Huntley in 1963 to explore applications of lasers under Air Force sponsorship. Fascinated by Leith's and Upatnieks' early papers, he, Huntley, Lehmann, and David Jackson took up holography and Goodman began teaching a graduate course on Fourier optics, which eventually became a book, *Introduction to Fourier Optics*. Over the following thirty years, thirteen of his forty-nine PhD students worked on topics related to holography.

Jim Trolinger, a researcher in the aerospace industry, recalls:

> There were so many possibilities that it was hard sometimes to maintain focus on any one of them [. . .] Holography provided a capability to record in 3-D and make all kinds of interferometry possible, where it was not practical before. The basic idea was that one could record the wavefront and decide later what kind of diagnostics to employ, because the wavefront had all of the optical information in it. This feature offered a huge payoff for short-lived events, events spread out over a large volume, and events occurring in remote places, where diagnostics equipment could not be operated.[31]

[29] Upatnieks, J. and A. M. Tai, 'Development of the holographic sight', *Proceedings of the SPIE The International Society for Optical Engineering* 2968 (1997): 272–81.

[30] Lehmann, Matt, 'Holography—the early days', unpublished account to R. Jackson, MIT Museum 29/790, 2 Jun. 1982. Lehmann (d. 1997) had graduated with an engineering degree from Stanford University in 1931, worked at Walt Disney studios in the early 1940s, and later worked as official photographer at Stanford's Systems Techniques Laboratory. When Joseph Goodman's group received an Air Force contract to study holography, Lehmann joined as a photographic research engineer, subsequently generating several patents on the technique.

[31] Trolinger, James D., 'The history of the aerospace holography industry from my perspective', unpublished manuscript, May 2004, SFJ collection. The first team to use Gabor in-line holography for particle analysis from 1963 was George B. Parrent, Jr and Brian J. Thompson, later authors of a reference guide on Fourier optics (Parrent, G. B. and B. J. Thompson, *Physical Optics Notebook* (Bellingham, WA: Society of Photo-optical Instrumentation Engineers, 1971)).

Trolinger and his colleagues began holographic research at the Arnold Engineering Development Center in 1967, imaging particles in flight to measure flow fields. They later studied sprays, fuel injectors, and a wide variety of particle fields, with some moving at hypersonic speeds.[32]

Civilian aerospace engineers were as eager to exploit holography as were their Air Force counterparts. A snapshot of the scattergun approach of this research is given by the proceedings of the Ames Research Center conference held in January 1970. Sponsored by the Instrument Division of NASA and supported by the NASA Office of Advanced Research and Technology, the conference brought together fifty-five conference attendees, forty-two of whom were from NASA and ten from NASA contractors. These research teams, all variously funded under the auspices of NASA, were reporting on what had been an unfamiliar area of research for nearly all of them five years earlier. Their various projects and programmes all aimed to find and develop techniques that took advantage of the unique characteristics provided by holography. Apparent to several of them was a reinvention of a few generic ideas for special applications:

> As a first impression, the work at the separate centers appears to be highly redundant, considered from the holographic point of view. However, the work being reported is the result of the application of holographic techniques to many disciplines. From the applications point of view, there is considerably less evidence of redundancy. Funding for these activities, at least at JPL, has come largely from the discipline area where the holographic techniques are being applied. Certainly consideration should be given to this applications orientation in any effort to consolidate a holographic program within NASA.[33]

NASA funding involved its own laboratories and large American institutions and corporations that had become key players in the American defence industry over the previous two decades: TRW (which presented the most papers at the Ames conference); the Jet Propulsion Laboratory (JPL); Goddard Space Flight Center (GSFC); the Langley Research Center; Sperry Rand; the Marshall Space Flight Center; the Ames Research Center; and the Manned Space Flight Center.

6.2 SEDUCING INVESTORS AT CONDUCTRON

Alongside the explosion of research at established institutions, enthusiasm sprouted in start-up firms. A leafy town scarcely five miles across and having cows grazing at the city

[32] Holographic particle image velocimetry was used extensively but never adequately automated.

[33] Michalak, Michael W., 'Holography in the Test and Evaluation Division at Goddard Space Flight Center', presented at *Holographic Instrumentation Applications*, Ames Research Center, Moffett Field, Calif, 1970. This applications-centred, rather than discipline-centred, grouping of specialists is common to research-technologies, as defined by Shinn (Joerges, Bernward and Terry Shinn (eds), *Instrumentation: Between Science, State and Industry* (Dordrecht, 2001)).

limits, Ann Arbor was dominated by the University of Michigan Central and North campuses. The university's diverse activities in optics and infrared research provided the skilled personnel to found some of the first companies to exploit commercial holography during the 1960s (see Figure 6.1). As a result, the early commercial take-up of holography blossomed in Ann Arbor. Unlike other centres of technical expertise, Ann Arbor's commercial holography was firmly rooted in classified research. The majority of personnel, firms, equipment choices, procedures, and outlook—in short, the technical culture—derived directly from Willow Run.

The city had begun to develop an industrial base during the 1950s owing to the heavy government investment in research and development contracts with the University of Michigan. Strand Engineering (1955) was one of the first firms to appear, followed by Parke Davis (1958), the Bendix Corporation, Federal-Mogul, and Climax Molybdenum. By 1969, fifty-eight R&D companies, employing some 3000 persons, were located there. During the same period, the research funding at the University of Michigan nearly quadrupled to $62.4 million annually.[34]

The first and most important early commercial explorations of holography's potential were made at the Conductron Corporation. The company was founded in 1960 by Keeve M. (Kip) Siegel, an engineer–entrepreneur who had come to Ann Arbor in 1948 to join the Upper Atmospheric Physics Group, then part of the Michigan Aeronautical Research Center (MARC), and headed the group between 1949 and 1953.[35] As the WRL expanded during the 1950s, Siegel's group grew, mutated, and changed location. In 1957 it became part of the Electrical Engineering department and was reorganized as the Willow Run Radiation Laboratory. By 1961, it had ninety-three employees studying a wide range of classified and open projects. Siegel, a larger-than-life figure who directed the laboratory authoritatively but charismatically, occasionally threatened administrators that he would pull the laboratory out of the university to work directly for a corporate sponsor.

Siegel's formation of Conductron Corporation in the autumn of 1960 was nevertheless a surprise to most staff. He continued for over a year as Head of the Radiation Laboratory while President of Conductron despite conflicts of interest between the organizations in competition for contracts.

Conductron's employees, drawn from both the Radiation and the Radar & Optics Labs, supported the early contracts. As at WRL most of the Conductron technical staff members were either physicists or electrical engineers, several having worked as high school teachers.[36] Conductron, and two succeeding companies founded by Siegel in Ann

[34] Anon., 'Many holographic roads lead to Ann Arbor', *Laser Focus*, February 1969: 16–7.

[35] K. M. Siegel joined Willow Run as a Research Associate with a BS degree from Rensselaer Polytechnic Institute and completed an MS there in 1950.

[36] Clark Charnetski (b. 1942), responsible for hologram production at Conductron, had a BS in Physics from the University of Michigan and had taught high school physics, as had Dave Wender; their colleague Garland Bush had taught chemistry. Gary Cochran (b. 1938), head of optics research, had also taught high school physics before pursuing an MS degree from the University of Michigan and working on infrared development at Willow Run. He then studied part-time for a PhD in Physics while employed at Conductron until 1966. After completing his PhD in 1967, he returned briefly to Conductron and then followed Kip Siegel to his new firm, KMS Industries. Larry Siebert (b. 1938) had an MS in Electrical Engineering from

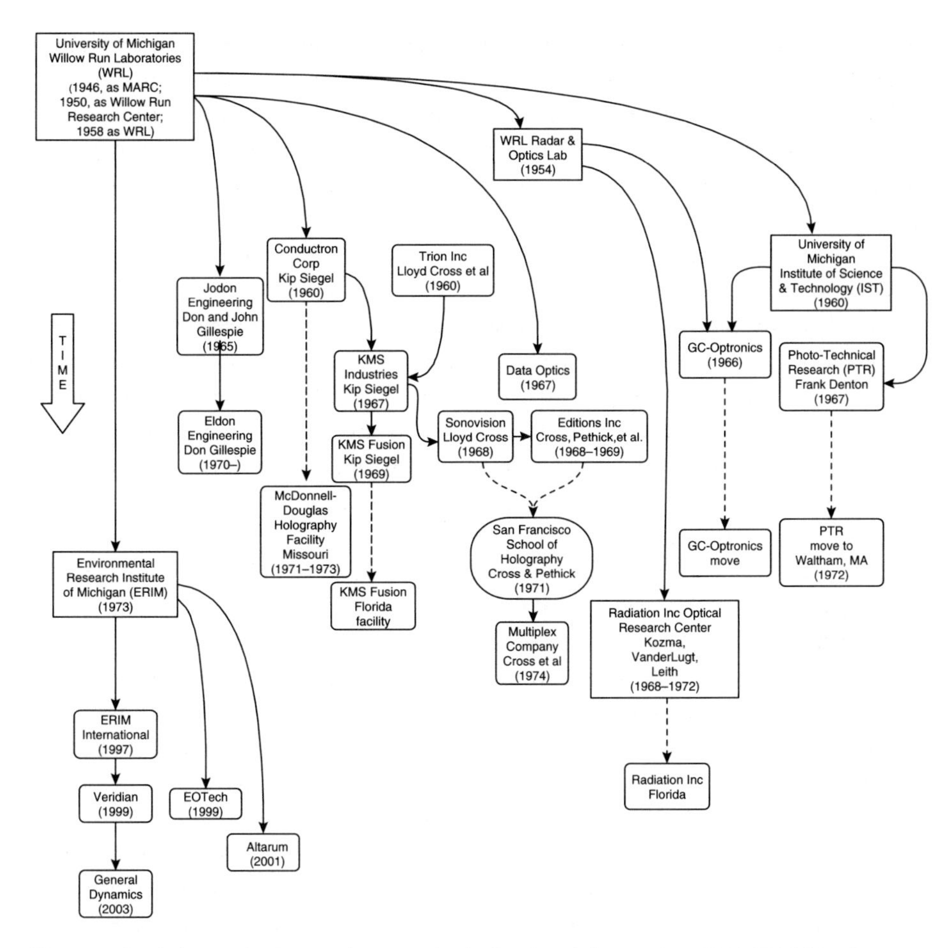

Fig. 6.1. Optical organizations and companies in the Ann Arbor region, 1946–2005. Despite some export (dotted lines indicating firms founded, or moving, beyond the region) the transfer of skills and personnel was regional.

Arbor, KMS Industries, and KMS Fusion, were cross-fertilized by employees who tagged along on his new ventures. Over the first few years of Conductron, some twenty staff of the Radiation Laboratory had left to join it, and staff numbers in the laboratory continued to dwindle through the 1960s. And when Siegel himself resigned from Conductron in 1967 after a disagreement about expansion plans with Jim McDonnell of McDonnell Douglas, some key employees (including former Willow Run researchers Wes Vivian and Lou Cutrona) formed KMS Industries with him.[37] Similarly, when Siegel later founded a

Ohio State and had been a member of the Institute of Radio Engineers (IRE), the precursor to the Institute of Electrical and Electronic Engineers (IEEE), since his student days. He had been a research engineer at North American Aviation, where he worked with laser systems, before joining Conductron in 1966.

[37] Senior, T. B. A., 'Radiation Lab History', http://www.eecs.umich.edu/RADLAB/labhistory.html, accessed 1 Sep. 2004.

new firm, KMS Fusion, to promote laser fusion as a power source, he attracted staff that had been with him at Willow Run, Conductron, or KMS Industries.[38]

Just as at the Willow Run Labs, Conductron's income derived from military research and development contracts. The company initially manufactured radar-absorbing ferrite tiles (from which the name Conductron derived) for the air intake of the F-111 bomber, work that had been initiated in the WRL Radiation Laboratory. A related contract was for the development and production of 'penetration aids'—radar decoys designed to open while travelling along their flight trajectory to simulate nuclear missiles, and thereby overload the Soviet anti-ballistic missile system.[39]

Another important activity of the new company was the manufacture of optical processors (see Figure 6.2) for synthetic aperture radar, the activity fostered and developed at the Radar and Optics Laboratory at Willow Run. The two originators of the Willow Run SAR optical processing work, Lou Cutrona and Wes Vivian, joined Conductron in 1961 to oversee such contracts.

Lou Cutrona sought income for Conductron in the same way that he had at WRL, by military development contracts for synthetic aperture radar equipment. He also built up expertise at Conductron that mirrored the expertise within the WRL Optics Group, hiring a former WRL optics production engineer, Arthur Ingalls, and a physics graduate and part-time Willow Run employee, Gary Cochran, in early 1962 to work on SAR optical processing. When Conductron won a $5 million contract for a commercial synthetic aperture radar in 1963, the optical group expanded significantly. Working independently of Willow Run, Cochran's group developed expertise in similar areas.[40]

The sponsorship, working context, and expertise proved fertile for incubating similar intellectual progeny. Supported by such lucrative military contracts during the early 1960s, the Conductron Corporation also took up holography, and by the same route that Leith had been drawn to it—via synthetic aperture radar. Cochran read the Leith–Upatnieks papers on wavefront reconstruction and connected them with his SAR work—'the whole heart and soul of holography'.[41] Because Conductron was working on the optical processing of SAR data just as Leith and others were doing at Willow Run, Cochran had available similar precision optics, stable platforms, precision mirrors, and one HeNe laser which was being investigated for use in the optical processor. He recalls, 'I just had to go down to the lab, and see if I could do this blasted thing'.[42]

[38] KMS Fusion was Siegel's last Ann Arbor venture. He died in 1975 after suffering a stroke while defending KMS Fusion's continued government funding before a Senate hearing. The company survived for the following year on Siegel's life insurance policy.

[39] Charnetski, Clark to SFJ, interview, 3 Sep. 2003, Ann Arbor, MI, SFJ collection.

[40] During the 1962–4 period, Cochran recalls mistrust from Willow Run personnel who assumed that he had stolen some of their optical processing ideas. Only later was the independent reinvention of those ideas generally accepted. Lou Cutrona, in particular, does not appear to have served as a hands-on technical intermediary between WRL and Conductron.

[41] Cochran, Gary D. to SFJ, interview, 6 and 8 Sep. 2003, Ann Arbor, MI, SFJ collection

[42] Ibid. From October 1964 he conceived and explored a wide range of ideas (Cochran, Gary D., 'New Ideas Notebook', Conductron Corporation, 1965).

Fig. 6.2. Conductron optical processor, 1965 (Cochran collection).

Cochran (see Figure 6.3) managed to make a hologram on a narrow spectroscopic plate using a very weak HeNe laser, as Leith and Upatnieks had done over a year earlier.

> Kip Siegel had a chance to see it, and he was fascinated. Kip Siegel was a promoter. He was the kind of guy, it didn't matter whether you made any money, if he could get the excitement of Wall Street bankers and get more investment capital. He saw the potential [. . .] Siegel was clearly interested in *developing the technology as a tool of investment.*[43]

Don Gillespie of Jodon Engineering Associates Inc, a small Ann Arbor business specializing in holographic equipment, recalled chatting with Siegel about business:

> I told him, 'I finally understand how to make a million—don't sell technology to scientists [as Jodon was doing]—sell technology to consumers!' Kip said 'no, you've got it almost right; what you've got to do is sell the *promise* of technology to *investors*'.[44]

Activities at Conductron focused on this goal. Cochran began to study the field and, obtaining larger plates over the next month, began to make bigger and better holograms in early 1965, often working after hours well into the evening. Kip Siegel became seduced by the potential. One late night, after having shown a visitor around the facility, he came to the laboratory and told Cochran to do whatever he wanted to do in holography. Cochran began buying materials, hiring more personnel, and exploring holography as a daytime pursuit. Among those involved with the holography projects were Carleton E. (Sandy) Thomas and Bob Buzzard (who directed much of the holography work with Cochran), Clark Charnetski (producing many of the special-order holograms), Richard Zech (colour and reflection holography),[45] Craig Dwyer (holographic emulsions and bleaching chemistry) and Larry Siebert (pulsed holography). Their investigations spanned the major problems identified for display holography during the 1960s.

[43] Cochran to SFJ, op. cit.

[44] Gillespie, Donald to SFJ, interview, 29 Aug. and 4–6 Sep., 2003, Ann Arbor, MI, SFJ collection

[45] Richard Zech obtained an MSEE degree under George Stroke in 1966, and a PhD under Emmett Leith in 1974. After leaving Conductron in 1969, he was a consultant and company officer for divisions of companies such as Harris Corporation, TRW, and McGraw-Hill.

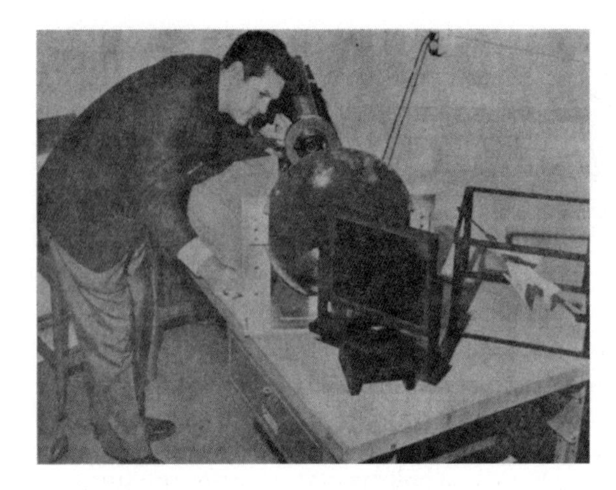

Fig. 6.3. Gary Cochran setting up to record an airplane model at Conductron Corporation, 1967 (Charnetski collection)

Facilities expanded to suit their explorations. By 1965 Conductron had a 5 × 10 ft. granite table and built a second laboratory under the front entrance steps of the building, which housed an accompanying darkroom and a 6 × 12 ft. concrete slab table on an iron framework. Because it vibrated badly during the 2 to 3 min exposures, the Conductron engineers built an array of solid concrete blocks separated by carpet padding, floating on inner tubes. Using a helium–neon and, later, an argon ion laser, they worked to produce impressive display holograms, creating 8 × 10 in. and 11 × 14 in. holograms during 1966.[46] When the two lasers were combined to make separate holographic exposures using different reference beam angles, they could create a two-colour image. While such colour holograms attracted interest and exemplified the progress being made, their display requirements (two carefully aligned lasers) made them too unwieldy to be sold. Other holograms were for commercial clients for use in trade show displays, and were illuminated with filtered mercury lamps. Yet others were made, and shown, only to visitors in the Conductron lobby.

While pursuing government contracts and moneymaking projects, Siegel directed Conductron's staff to develop demonstration pieces for holography. A consummate showman, his unrestrained predictions played a crucial role in escalating public expectations. Siegel was interested not merely in investment but in selling Conductron itself and had made overtures to the McDonnell Douglas Corporation. McDonnell Douglas bought Conductron in January 1966, but continued to give Siegel and his team free rein. The majority of Conductron stock was used to acquire McDonnell Electronics Corporation, which made aircraft simulators and spacecraft simulators for the Mercury and Gemini manned space programmes.

[46] Later notable extensions included the 1.5 × 1.0 m. hologram of the Venus de Milo recorded by Jean-Marc Fournier, Gilbert Tribillon, and Jean Charles Viénot of the Laboratoire d'Optique de Besançon, France in 1975, and later the 1 m^2 holograms produced by Nicholas Phillips at Loughborough University, England.

Holography remained an important area of both technical exploration and fund raising. By early 1967, Cochran's optics group of some forty people was responsible for SAR contracts and holography production and development, and half a dozen holographers involved intimately in extending the art. An example of Siegel's typical spiel comes from a speech given to the employees of Conductron Missouri, the newly rebadged McDonnell Electronics subsidiary that produced aircraft simulators:

> If you went to a classroom, and were taking pictures of one of your children—say, playing in kindergarten—and one of the children ran behind the other children, instead of stopping taking movies you continue taking movies, and then when you get home and you show those movies of your children, when your child has ducked behind other children, all you need to do is move your head and you can follow your child as your child is running behind other people. You'll find when you see the hologram, you'll be able to see behind other objects. There are other properties associated with holograms that you'll be able to see, demonstrated at the Hanover Fair in Germany, and the IEEE meeting in New York, and we've been demonstrating all over the world—you can take off half the picture, cut off half the picture, and you still have the ability to see the whole picture, because now you can look around corners and you have the effect of having enough intelligence every place in the picture to get the whole picture [. . .]
>
> I am hoping that by the year 1976 that the United States will have, as far as new products are concerned, only three-dimensional television and three-dimensional movies on the market. I would not expect two-dimensional processing, two-dimensional television, two-dimensional home movies to continue—that's my personal belief. I don't think people will buy things that are antiquated. If that volume comes to pass, then I think that the price will be equivalent to the price of 2-dimensional television today, assuming that the dollar does not become too inflated. In any real measure of the absolute buying power of the absolute dollar, I think you could buy a 3D set for the price you would buy a 2D set today, and I'd give similar answers on processing and similar prices on film.[47]

Siegel's use of the term *intelligence* for information hints at the close association between military and commercial research. His portrayal of holography as a future consumer industry relied on an implicit faith in scientific, technical, and economic progress that he shared with his colleagues. Yet Siegel's technical claims far exceeded the capabilities, or even expectations, of his engineers; the commercial forecasts diverged from technical extrapolations.

This impossibly optimistic outlook for holography carried Conductron through three years (1964 to 1967), until Siegel resigned from the company and left to form KMS Industries. His showmanship and promotion of the new medium flavoured subsequent forecasts for a generation.[48]

[47] Siegel, Keeve M., 'Speech to Conductron Missouri', audio recording, 14 June. 1966, C. Charnetski collection.

[48] Siegel's role in influencing forecasts of holography is discussed in §14.2.

6.2.1 Pursuing customers

On a more realistic scale, the growing variety of demonstration pieces was also used to attract commercial interest. This pursuit of commercial customers was groundbreaking, and convinced other firms that there could be a viable market for holography. For instance, Clark Charnetski credits the exhibition of a Conductron hologram made for Agfa at the Photokina trade show in Germany—but recorded on Kodak plates—with Agfa's introduction of its own high-resolution holographic emulsions in the late 1960s. Until that time, Kodak 649F had been the only suitable high-resolution emulsion and, yet, available only with difficulty in Europe.

Conductron's most important customer proved to be Kingsport Press in Tennessee, printers of *National Geographic Magazine* and *World Book Encyclopedia* and its *Science Year* annuals, produced by Field Enterprises of Chicago. In 1966, Kingsport envisaged including a hologram in the forthcoming *1967 Science Year* and approached Conductron. The project relied on copying a master hologram to make some 500,000 copies. The resulting copy would be illuminated by any small white light source and filtered through a supplied red filter. Conductron engineers and technicians had to adapt their experimental style to large-scale production. They developed methods of contact copying by adapting commercial photographic equipment and using two HeNe lasers as sources. They chose Agfa film, to the chagrin of the Kodak representative, and purchased 1000 rolls of 250 ft each. The development, production (printing, processing, and inspection), and marketing team comprised over a dozen people.[49] Recording two hologram copies at a time onto 105 mm wide film, they processed one film spool and then checked each hologram for quality while recording the next spool. The initial aim was to produce 10,000 holograms per day, but by scheduling 10 h days and gradually identifying production problems, they achieved a peak of 16,000 holograms on their best day, with the production run completed within six weeks.[50] The publication proved the best-selling edition of the *Science Year* for Field Enterprises, but no comparable orders followed.[51]

The seduction of investors, rather than sales from holographic products, remained Conductron's primary goal. These two activities nevertheless merged into one in Siegel's promotion of the firm. One of Siegel's public goals, for example, was to develop holography as a medium for recording large-scale events. An instance of his exuberant showmanship was his broadcast aim of making a holographic movie of the Olympic games. When his would-be partners at McDonnell Douglas Corporation remained sceptical of the feasibility of holographic movies, Siegel set his Conductron engineers on the problem. Bob Buzzard and Gary Cochran devised a holographic animated movie to show that such applications could be practical. Using a loop of 70 mm wide film (see Figure 6.4), they

[49] 'Conductron develops new techniques for hologram production', *Conductron-Missouri antenna* 1 (10), (Aug. 1967): 1–2.

[50] Charnetski, Clark to SFJ, interview, 3 Sep. 2003, Ann Arbor, MI, SFJ collection; Cochran, Gary D. and Robert D. Buzzard, 'The new art of holography', in: *Science Year: The World Book Science Annual 1967* (Chicago: Field Enterprises Educational Corp., 1967), pp 200–11.

[51] Cochran to SFJ, op. cit.

Fig. 6.4. Dave Wender preparing to record a short animated holographic movie at Conductron Corporation, 1966 (Charnetsky collection).

recorded several dozen individual hologram frames of an object having a cyclic motion: a model merry-go-around. By repositioning the merry-go-round for each individual exposure, they produced an animated 3-D movie. For reconstruction, they adapted a commercial projector so that the laser source was strobed with a shutter to illuminate the separate frames. The reconstructed scene was also viewed through magnifying optics to give a realistic perception of a three-dimensional changing scene. The response of viewers seems to have been subdued: the movies were shown occasionally to visitors (mainly investors or advertisers), but were denied customs clearance to be shown in foreign countries because of what was deemed sensitive technology.

The need for ever-more impressive demonstrations led the Conductron engineers to more difficult technologies. While the animated holography demonstration won a bet for Siegel from Jim McDonnell, the engineers realized that producing practical holographic movies would require a pulsed laser to record moving objects, such as people, before they could move enough to smear the interference fringes on the photosensitive plate. Larry Siebert joined the firm in 1966 to develop such a laser, and had obtained good results by the following year—but of course far short of recording cinematic holograms of athletes on a sunny day.[52]

The McDonnell Douglas marketing staff were immediately enamoured by the potential of the technique and generated considerable publicity in their search for applications. Their marketing plan targeted advertisers, liquor, and automobile companies. For example, they arranged holograms of a group of ten Conductron staff, another of a contrived scene of poker playing, and yet others of commercial products to illustrate the potential of pulsed holography. They collaborated, too, with the first wave of artists, Bruce Nauman in 1968 and Salvador Dali in 1970.[53] During 1970, the holography laboratories were relocated from Ann Arbor to St Louis, Missouri; most of the technical staff stayed

[52] Discussed in §7.6.
[53] On artists, see also §10.2.

behind. From Missouri, the new McDonnell Douglas team produced holograms for a wide variety of customers, working in some cases through free agents as middlemen. The McDonnell Douglas operation closed in 1973 when the corporation decided that a demonstrable market for pulsed holograms was too elusive to continue seeking.

Between 1965 and 1970, when Conductron closed its holography laboratories for reestablishment by new personnel at McDonnell Douglas Missouri, the Ann Arbor group created some one thousand holograms for clients ranging from Hoffman La Roche (pharmaceutical displays for trade shows) to General Motors (lobby, museum, and trade show displays) to artists (Richard Wilt and Bruce Nauman). Their subjects included plaster casts of anatomical organs (for pharmaceutical manufacturers), sharks (for an aquarium, and later purchased by Hugh Hefner for his Chicago Playboy mansion pool), aircraft models and classified images (to show off Conductron and McDonnell Douglas products)(see Figure 6.5).

What Conductron achieved was remarkable then, and since. The company developed a method to mass-produce holograms, and made a half-million for the *World Book Encyclopedia* in 1967. Its engineers created the first pulsed laser portraits of humans and interested the first artists in using holography in 1968. They pursued advertising applications creating numerous product displays. And they accomplished their primary aim of attracted investors, notably the McDonnell Douglas Corporation, which bought Conductron in 1966.[54]

6.3 ANN ARBOR, 'HOLOGRAPHY CAPITAL OF THE WORLD'

Within one small northern American city, then, holography developed several nuclei: first, the group of workers led by Emmett Leith at the U-M Willow Run Radar and Optics Laboratory; second, the academic group under George Stroke at the U-M Electro-Optical Sciences Laboratory (EOSL); and third, a team at Conductron Corporation, a mile away. It is significant that the WRL and EOSL workers afterwards were among the most peripatetic and influential in promoting holography. Remarkably, the more publicized commercial explorations of Conductron had muted consequences.

Nevertheless, other, smaller firms added to the growing commercial interest in holography. Conductron was not the only company producing holograms commercially in the mid-1960s. Frank Denton, who had been George Stroke's technician at MIT and later moved to join him at the University of Michigan, founded Photo-Technical Research

[54] Charnetski, Clark to SFJ, interview, 3 Sep. 2003, Ann Arbor, MI, SFJ collection; Cochran, Gary D. to SFJ, interview, 6 and 8 Sep. 2003, Ann Arbor, MI, SFJ collection; Goldberg, Shoshanah, 'Conductron and McDonnell Douglas', *holosphere* 15 (1987): 16–8; Siebert, L.D. to SFJ, interview, 4 Sep. 2003, Ann Arbor, MI, SFJ collection.

Fig. 6.5. Brochure advertising pulsed holography, McDonnell Douglas, 1972 (Boeing Inc; MIT Museum MoH collection).

(PTR) in 1966 to manufacture holograms and diffraction gratings commercially.[55] The company, and its staff, grew out of George Stroke's laboratory. Rolf McClellan, Stroke's PhD student, co-founded the company with Denton, and for a time included Stroke's secretary in their plans. With another of Stroke's students, Arthur Funkhouser, Denton briefly produced 'moon-glasses'—spectacles and other displays of diffractive colour formed from crossed holographic diffraction gratings. Denton's first commercial holograms were in fact produced in the university laboratory, leading Stroke to dismiss him and to pursue technician Donald Gillespie, working in another IST laboratory.[56] His firm produced the first large order of laser transmission holograms in 1966—some fourteen thousand, manufactured a few months before the half-million Conductron *Science Year* production run—for inserts in the trade magazine *Laser Focus*. Denton also produced laser transmission holograms for a subscription kit—'Things of Science' in 1968, and later supplied large quantities of holograms of chessmen to Edmund Scientific (chess figures became the iconic 1960s hologram through this product, the *Science Year* hologram and Leith's depiction of a chessboard hologram in a *Scientific American* article). These products were direct extensions of their work at the University of Michigan and employed readily available and inexpensive aerial reconnaissance film.

[55] '(Ann Arbor holography firms)', *Ann Arbor News*, 16 Feb. 1969. Gratings became their principal product.

[56] See §5.2.1.

Just as George Stroke's technician founded PTR, Don Gillespie, a former Willow Run technician working in nuclear physics at the University of Michigan, also left to form his own firm in Ann Arbor to manufacture and sell holographic equipment.[57] Gillespie took up holography from 1965, operating the part-time business out of his basement. His small company appears to have been the first to publicly advertise commercial holograms—produced with an extremely weak (0.12 mW) helium–neon laser that he had built himself—in February 1966. His table was constructed from file cabinets, a solid wooden door, bubble cell insulation, and finally a top surface of marble recovered from a demolished building. Gillespie was able to produce holograms of small objects—machine shop instruments and Hummel figurines—using exposures of some 25 min[58] and sold half a dozen for $100 each. In 1967, he and his brother John, who provided capital for expansion, moved the firm, now known as Jodon, to new premises. Gillespie built, sold, and repaired lasers, supplied holographic film, and produced optical mounts, attenuators, filters, power meters, and variable beamsplitters for the growing holographic community. In the absence of strong competition, Jodon lasers and optical mounts produced rapid company growth during the 1960s.[59] Yet another spin-off company was Data Optics formed by engineer Thomas Herman and machinist Edward Volinski in 1967, who had produced custom-made equipment for the Willow Run SAR project and subsequent optical projects in the U-M Electrical Engineering department. Their firm initially specialized in optical positioning equipment for interferometry and holography applications.

Of all of the Ann Arbor activities, arguably the most seminal were those of Lloyd G. Cross. Cross had worked at WRL from the mid-1950s in the laboratory of Chihiro Kikuchi, who first demonstrated maser action in a ruby crystal in 1957.[60] When Ted Maiman at Hughes Research Laboratories discovered laser action in this type of crystal in 1960, Cross was co-leader of a project group to design, build, install, and operate a ruby maser preamplifier for a new 85 ft diameter radio telescope located at Peach mountain, about 20 miles north of Ann Arbor. On hearing of Maiman's laser, and a second one made by Bob Collins of Bell Laboratories, Cross sought internal funding to construct his own. When funding was not immediately forthcoming, he organized an after-hours project crewed by his maser project team.[61]

[57] Donald Gillespie (b. 1934) grew up in Flint, Michigan and was educated at Flint Junior College (as were Lloyd Cross and Lee Cross of Trion Instruments) and the University of Michigan, joined the WRL's Radiation Laboratory in 1956 as a student to work under Chihiro Kikuchi and John Lamb on development of the ruby maser. He later moved with Kikuchi to manage the graduate student laboratory in the nuclear engineering department of the university and both participated in the start-up business Trion Instruments in 1961 (see below).

[58] Gillespie, Donald to SFJ, interview, 29 Aug. and 4–6 Sep. 2003, Ann Arbor, MI, SFJ collection.

[59] Following a stock coup by his brother in the early 1970s, Don Gillespie took over the European operations of Jodon and subsequently formed a separate firm, Eldon Inc., in Ann Arbor.

[60] During the late 1950s, Cross was an undergraduate physicist at Willow Run. In 1956 he was employed part-time as an 'Assistant to research', and from Aug. 1958, after obtaining his BS degree, he had a full-time post as Research Assistant and then Research Associate at Willow Run. At Willow Run he studied the measurement of magnetoresistivity, developed infrared detectors, and designed maser antennas and circuitry.

[61] According to Don Gillespie, he and Cross met Bob Collins of Bell Labs when he visited WRL for a seminar about plans for the ruby laser. They drove Collins to the airport building (at Willow Run itself) for

By early 1961, they had succeeded in constructing their own ruby laser, and the first to generate a pulse powerful enough to pierce a razor blade (a demonstration that was to become iconic by the mid-1960s). He recalls that 'the laser became the first available for viewing by the general public, and the project was immediately taken over by the laboratory administration to provide a continuous series of 15 min demonstrations to curious and sometimes disbelieving visitors from around the globe'.[62]

Together with electrical engineer Doug Linn and physicist Lee Cross (no relation), and with consultation from his former supervisor Chihiro Kikuchi and Prof. Peter Franken of the U-M Physics Department, Lloyd Cross started Trion Instruments, Inc. in 1961 to develop commercial pulsed ruby lasers, and built and sold the first one to Texas Instruments the following year.[63] The firm also researched continuous-wave lasers in ruby, calcium fluoride, and in calcium tungstate, but these proved unsuccessful. With no competition, however, sales reached some $500,000 over the next few months. In 1962, Cross sold Trion to Lear-Siegler, diversified manufacturers of executive aircraft and military and industrial systems, while remaining on staff of the new Lear-Siegler Laser Systems Center. He recalls visiting Emmett Leith's laboratory some time in 1965, and being 'completely amazed and fascinated'.[64]

All holography activities were interrelated to a degree. Cross's trajectory carried him from Willow Run to his own company and then to Kip Siegel. While Conductron had pioneered the exploration for commercial niches for holography, Kip Siegel and Jim McDonnell disagreed on future directions for Conductron after its buy-out by McDonnell Douglas. Siegel left to found KMS Industries on the opposite side of Ann Arbor with thirty others, mainly from Conductron itself. Those who left with him did so for the entrepreneurial spirit, the expectation of personal advancement, and the appeal of a charismatic leader.[65]

As he had done with Conductron, Siegel sought investors and venture capital for KMS Industries. He sought prominent researchers to join the company both for their ability to innovate and for the public relations potential. He courted Emmett Leith, who joined the firm tentatively in 1967 as a consultant but soon decided to remain at the University of Michigan. Lloyd Cross, too, joined the company that year after leaving his own firm of Trion Instruments. Gary Cochran, who also joined the new company as manager of

his return journey and at the terminal bar pumped him for further information. 'As quick as Collins was on the plane', Cross was back at Building 41 to build a laser. He borrowed a landing-lamp strobe light from the airport maintenance staff, and a few weeks later produced their first ruby laser pulse (Gillespie, Donald to SFJ, interview, 29 Aug. and 4–6 Sep., 2003, Ann Arbor, MI, SFJ collection).

[62] Cross, Lloyd to SFJ, email, 25 Oct. 2003, SFJ collection.

[63] Trion Instruments Inc / Lear-Siegler Inc, 'Trion Instruments Inc brochure', 1961.

[64] Cross, Lloyd G. and Cecil Cross, 'HoloStories: reminiscences and a prognostication on holography', *Leonardo* 25 (1992): 421–4. See also Cross, Lloyd, 'The Story of Multiplex', transcription from audio recording, Naeve collection, Spring 1976.

[65] Siegel organized KMS Industries, however, as a more hierarchical company with a tier of middle managers on the advice of his investors. As a result, the working culture was less appealing for the engineers and scientists who joined it.

optics projects, could boast for a brief period that KMS Industries had 'a fantastic group of people', including Leith, Cross, and Lou Cutrona.[66]

At KMS Industries, Cross took on the role of laser expert that Larry Siebert still fulfilled at Conductron, a few miles away. Cross developed a pulsed laser for KMS Industries, and began to explore holography both in the company laboratory and at home. His first hologram—of a champagne glass being filled from a champagne bottle, and made to celebrate the project—used a pulsed laser. A creative but unmanageable worker, Cross took holography in new and seemingly uncommercial directions. He chose to work nights rather than days, and attracted a coterie of artists to his activities. Cochran recalls:

> Lloyd became such a difficult man because he was unavailable at critical times, [and] decisions had to be made. I came in one night just to see what he was doing. And I came down into the holographic lab—he left beautiful color holograms, where you'd turn on these two or three lasers, and there would be a color image—and here were two artists, young, long haired, you know, the whole business, and they were sitting up on the table, and talking with Lloyd about what they wanted to make. And I got caught up a little bit in it, but basically he would be having anyone he wanted over, you know, and there would be these artists, planning these ventures, and nobody had any control over him whatever.[67]

At the same time, Cross was exploring the use of lasers for light shows. While trying to construct a thin beamsplitter, he discovered that it was sensitive to vibration. He consequently devised a system in which reflections from mirrors on a loudspeaker membrane reflected laser beams in Lissajous figures on a screen, oscillating in time with the music. He recalled, 'I found it was possible to make beautiful fantastic displays with extremely high contrast, and when I showed them to other people, they got excited too'.[68] With significant financial support from KMS Industries investors, Cross left to form yet another Ann Arbor company, Sonovision Inc.

The holographers at KMS Industries, like those at Conductron, began to pursue commercial niches anew. Between 1967 and 1969, they examined a widening range of applications that increasingly took them away from display holography. They studied an adaptation of the principle of SAR to oil prospecting. Carleton (Sandy) Thomas and Gary Cochran developed a fingerprint recognition system based on optical correlation, but sold only a couple to the Department of the Treasury because 'nobody wanted to put their finger into a hole.' He notes that the search for applications began to assume a more frantic tone.

> I was desperately searching for applications, and didn't find that many [. . .] It didn't matter to Kip, because he was acquiring companies based on the dream. He was talking about three-dimensional TV, and regardless of the bandwidth I said it needed, he said 'Aw, it will come

[66] Cochran, Gary D. to J. Zorn, 18 May 1967, Ann Arbor, Cochran collection; Cochran to SFJ, op. cit. Indeed, these were key individuals in Ann Arbor's development of holography, electro-optics, and optical processing.

[67] Cochran to SFJ, op. cit.

[68] Wolff, Michael, 'The birth of holography: a new process creates an industry', *Innovations* (1969): 4–15, p. 13.

along some day'. I remember listening to him give the pitch to people [. . .] It sounded wonderful, but I had no idea how to do it technically.[69]

Kip Siegel's public dreams appealed to investors even if it could no longer motivate his staff. He envisaged holograms used as street signs, going so far as to suggest that a hologram of a brick wall could serve to dissuade motorists from entering a one-way road.[70] He proposed creating holograms of tanks as a tank weapon target and gained military interest, until his engineers pointed out that a misdirected tank shell would destroy the hologram and its light source. Siegel used such visions and inventive demonstrations to attract further investors and, using the capital from the Conductron sale, acquired a wide assortment of companies. Disillusioned by this 'dog and pony show', Cochran recalls telling his associates, 'I guess my job is only to keep their attention while he picks their pockets'.[71] Cochran resisted offers to join the next start-up company, KMS Fusion, and subsequently shifted his career to automotive electronics. Conductron and KMS holographers became disillusioned and, to a man, sought new avenues in engineering.

Siegel, too, was re-evaluating his commitment to holography. By 1969 the American stock market was becoming disenchanted with conglomerates of disparate companies, which KMS Industries had become with its investments ranging from rubber dinghy companies to theodolite manufacturers. After two periods of development—1965–7 at Conductron, and 1967–9 at KMS—investor interest in holography had been piqued, but few commercially viable applications had been identified and even fewer repeat orders had been obtained. Nor had wider academic or public interest been courted: many of the Conductron holography developments were never shown beyond a sales pitch or in the restricted access of the company lobby. Thus Siegel decided that a more achievable future—and reliable sponsorship—was the development of fusion power, using powerful lasers to heat and compress deuterium pellets.

There were, nevertheless, a few well-financed firms that envisaged long-term markets in holography during the late 1960s. One that set up an Ann Arbor base was Radiation Inc. The company had been founded in 1950 by Homer Denius and George Shaw in Melbourne, Florida, as a manufacturer of space and military electronics. Located at the former Naval Air Station at the Melbourne, Florida airport, near Cape Canaveral, Radiation Inc. quickly became a premier developer of miniaturized electronic tracking and pulse-code technologies for America's new space program, and for the manned space flights themselves. Radiation's products were used on America's first communication and weather satellites and by the military for the Minuteman, Atlas, and Polaris missile systems. The company was an early entrant into the microelectronics business, and in 1963 developed its first working semiconductor for use in its digital communications equipment.

The original connection with Ann Arbor was via Joseph A. Boyd (b. 1921), who had been director of the University of Michigan IST for its first four years, 1958–62. Boyd

[69] Cochran to SFJ, op. cit.

[70] On a direct consequence of this forecasting, see §13.3.

[71] Cochran to SFJ, op. cit.

joined Radiation Inc. in 1962 and became president the following year, guiding its merger with Harris-Intertype in 1967. Carmen Palermo, who had worked on optical processors with Leith as a research assistant, also was with Radiation Inc. from 1962 to 1964 and some Conductron employees joined them during this period. In 1969, Radiation Inc. attracted senior researchers from the Willow Run Optics Laboratories, Emmett Leith (as Chief Scientist), Adam Kozma (General Manager), Anthony VanderLugt (as Optics Research Manager), and Albert Friesem and Frederick Rotz (as Principal Research Engineers), to form a branch in Ann Arbor called the Optical Research Center of Radiation Inc. and later the Harris Electro-Optics Center, to develop holographic memories.[72] The idea, which proved persistently appealing for investors, was that the hologram potentially allows extremely dense information storage. Moreover, this information can be recovered by optical matched filters, making it a form of rapid associative memory. A commercial product was not forthcoming, however, and the group disbanded in 1972.

A more successful business was founded in 1966 by two other U-M researchers, Ralph M. Grant and Joseph Crafton, and specialized in holographic non-destructive testing.[73] They had developed the idea when Karl Stetson of WRL had told Grant about the sensitivity of his holographic interferometry experiments, and Grant had envisaged using holography in an underwater sound-detector device. The concept gained Navy funding, leading to further work at GC Optronics on sonar transducers. The transducer work, in turn, led them to the idea of non-destructive testing (see Figure 6.6). Gordon Brown developed the technique into commercial products, notably a system for locating bonding faults in tires, and another for spotting unbonded regions between honeycomb sandwich panels for aircraft components.[74] These applications proved valuable for the aerospace industry, and by 1969 the firm had some sixty employees, several of whom came directly from WRL: Karl Stetson consulted, and Robert Powell and Arthur Funkhouser were project engineers there in 1967, working on projects such as pulsed holography for measuring displacements, holographic interferometry for contouring Ford car bodies, crack detection, bond analysis, and corrosion.

By the late 1960s, then, Conductron, KMS Industries, Jodon, PTR Optics, GC Optronics, Radiation Inc. and other Ann Arbor firms represented the highest international concentration of holography companies. Combined with the WRL, their original seed, Ann Arbor was justifiably promoted as the 'holography capital of the world'.[75]

[72] 'Optics center established', *Radiation Ink*, June 1969: 1.

[73] Ralph Grant had been on the faculty of Electrical Engineering at U-M and working on the Project MICHIGAN contract like Leith. He had subsequently worked with Norm Barnett in the Electrical Engineering Department.

[74] Gordon M. Brown (b. *c.*1936) obtained a BS in mechanical engineering at the General Motors Institute, 1957, and an MS in nuclear engineering at U-M in 1961. He was with GC Optronics until 1973, and then a research engineer at the Ford Motor Co. Research Laboratory until 1995, where he founded the Computer-Aided Holometry laboratory. He also founded Optical Systems Engineering in 1982.

[75] 'Holography: Ann Arbor shows how', *Detroit Free Press*, 6 Aug. 1967; Anon., 'Many holographic roads lead to Ann Arbor', *Laser Focus*, Feb. 1969: 16–7; '(Ann Arbor holography firms)', *Ann Arbor News*, 16 Feb. 1969:; Leith, Emmett N., 'Electro-optics and how it grew in Ann Arbor', *Optical Spectra*, May 1971: 25–6.

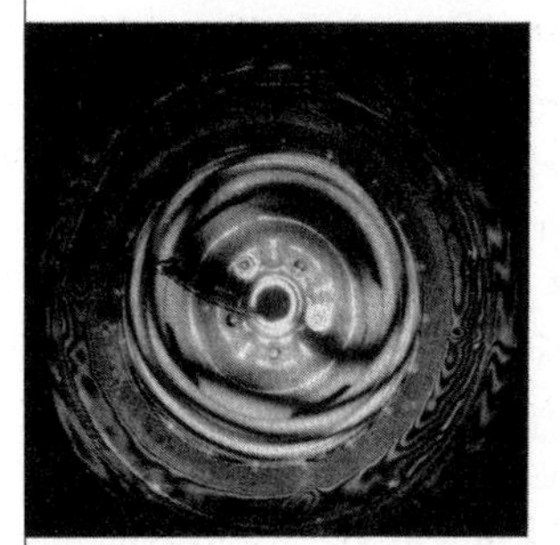

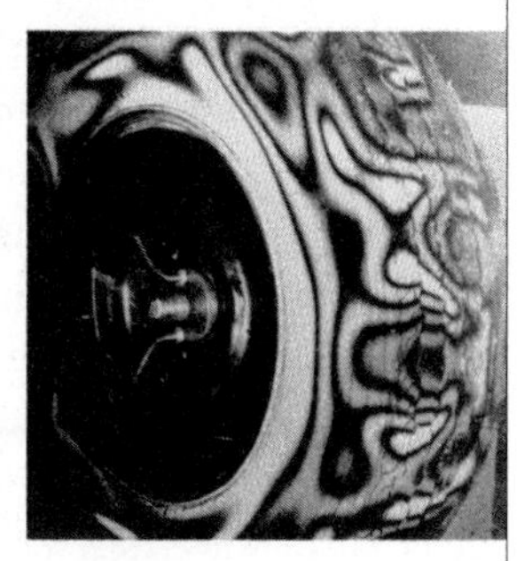

Fig. 6.6. GC Optronics advertisement *c.*1969 (Bjelkhagen collection).

6.4 SEEKING APPLICATIONS FURTHER AFIELD

The experience with SAR processing and the pool of technical labour made Ann Arbor a natural early centre for the new science and technology of holography, but other researchers and companies rapidly joined the fray. The most direct source of this transplantation was military, NASA, and NSF contracts, as mentioned earlier. And the information diffused not just through journal articles, but also by direct communication: Emmett Leith and Juris Upatnieks made the rounds of American engineering conferences from 1964, and George Stroke carried his own message to scientific seminars in Europe, sowing enthusiasm and research activity widely.

And Leith was not the only first generation researcher primed to resume research and hoping to trigger a rapid expansion of the subject. Gordon Rogers, for example, followed the published reports of the Michigan work closely, and took up the subject afresh in 1965. His career was now established as Reader in Wave Optics at the College of Advanced Technology in Birmingham, England, where he had had a post since the late 1950s.

Rogers wrote a paper on phase contrast (or bleached) holograms published in The *Journal of Optical Society of America* (*JOSA*) as early as May 1965, but found that his re-entry to the field was hampered by its more recent American roots. Kodak 649F high-resolution spectroscopic plates, the photographic plates used by Leith and Upatnieks, proved to have a three-month delivery time in Britain. The Kodak factory in Harrow attempted to duplicate the plates, but was finding that the emulsion material could vary by an order of magnitude in sensitivity from batch to batch, making repeatable experiments impossible.[76] Obtaining reprints of militarily funded research also proved difficult outside America. Rogers first contacted Emmett Leith in the summer of 1965. In his introductory letter, he suggested his competence in the subject by asking about derivations of formulas in Leith's 1962 paper, and hinted at his own insights about wavefront reconstruction. Leith appears to have been intrigued, replying, 'I infer that you have uncovered some interesting facets of wavefront reconstruction, but of course, I can only guess as to what they may be [. . .] Since our more recent work is not being done on a military contract, I have been sending copies of them out of the country'.[77] Rogers' subsequent correspondence with Leith welcomed that he was 'now working under different auspices, and have therefore greater freedom to communicate' and ranged from the technical to the social, discussing everything from beam ratios to encouraging Dennis Gabor to speak on holography at Physics Society meetings.[78]

As the Leith correspondence suggests, Rogers was resuming the activities he had undertaken in the early 1950s to promote the subject openly and actively. Unlike his earlier forays in the field, he could now link international researchers with engineers and potential customers. From early 1966, Rogers sought to interest engineering firms in the use of holography in the tool room; he advised V. E. Cosslett at Cambridge University on the appropriate materials for modern holography; and he sought a book contract to write a text on coherence optics and holography.[79] He also encouraged mechanical engineering staff at Loughborough University of Technology (subsequently an important British centre of holography research) to get involved in the subject.[80]

At about the same time, Gabor himself had found his reputation as the father of holography rising rapidly. In 1966 he noted to his friend Peter Goldmark, Director of CBS Laboratories, 'I am now a terribly popular man. I had to decline half a dozen invitations to lecture on holography in the US'.[81] Gabor began to dabble in holography again in two locations. At CBS Laboratories in New Jersey, Gabor started a small laboratory with advice from George Stroke. And from early 1967 Adam Kozma of Willow Run helped Gabor and his former student, George Jull, to set up a holography laboratory at Imperial

[76] Rogers, Gordon L. to E. N. Leith, 15 Jul. 1965, Sci Mus ROGRS 4; Rogers, Gordon L. to V. E. Coslett, 23 Jun. 1966, Sci Mus ROGRS 4.

[77] Leith, Emmett N. to G. L. Rogers, 6 Jul. 1965, Sci Mus ROGRS 4.

[78] Rogers, Gordon L. to E. N. Leith, 15 Jul. 1965, Sci Mus ROGRS 4.

[79] Rogers, Gordon L. to I. A. G. Le Bek, 15 Feb. 1966, Oliver and Boyd Ltd, Sci Mus ROGRS 4; Rogers, Gordon L. to K. E. Nicholds, 27 May 1966, Redditch, Worcs., Sci Mus ROGRS 4; Rogers, Gordon L. to V. E. Coslett, 23 Jun. 1966, Sci Mus ROGRS 4.

[80] McCann, B. and A. Denby to G. L. Rogers, Feb. 1967, Sci Mus ROGRS 4. Dr John Butters of Loughborough had visited Rogers a month earlier and after a visit and the donation of plates by Rogers the group had made their first holograms.

[81] Gabor, Dennis to P. Goldmark, letter, 2 Mar. 1966, IC GABOR LA/9.

College. However, Gabor's contributions to the subject were principally in the area of marketable ideas through his consultation for CBS laboratories. Through 1966 and 1967 he proposed ideas for a holographically generated screen for three-dimensional movies; underwater acoustic holography; holographic portraits using a scheme to give pulsed lasers adequate coherence length, and aesthetic applications of wall-hung holograms as decorations.[82]

Despite such intriguing possibilities, CBS Laboratories did not investigate holography aggressively. Skilled workers seemed difficult to obtain, and Gabor, now retired from Imperial College, was not active enough to manage them. His correspondence with Goldmark suggests his keen frustration:

> You are probably as impatient as I am with the slow progress [of] holography in your lab at a time when others, (like Holotron and KMS) are going at it hammer and tongs, and when it appears urgent to show progress beyond the present state of the art, so that we could ask for money. Of course, we have only 1-1/2 men on the job, and to direct them from a distance is more difficult than I thought. Expert holographers are few and very expensive, besides there is no guarantee that they will not be willful and un-cooperative [. . .] The ideal thing would be if you could provide two optical laboratories, similar to the one in existence, side by side, one of them with a dark room. Also, next to it, an office for the optical workers. In one of these two labs, I would like to work myself, with suitable assistance. If need be, I could do my work with one or two good technicians.[83]

It may be that Gabor imagined recreating the competition between the WRL Optics Laboratory and Electro-Optical Sciences Lab to boost output, while avoiding an atmosphere of 'wilful and un-cooperative' researchers. In any event, CBS never instituted this parallel research. Within a couple of years, only one holographic application remained active for Gabor: a scheme for panoramic holograms. In it, an artist makes a small model which viewed through a special lens makes it look as if it extended to infinity, giving 'an unlimited third dimension to the artist'. Gabor hoped to use it as 'an emotionally powerful method of teaching cultural history' but noted to an art historian that its development at CBS Laboratories had 'progressed rather slowly'.[84]

Thus, Unlike Leith, Denisyuk and a growing group of international specialists, Gabor's continually imaginative ideas had little impact on holography during the 1960s and beyond. By the early 1970s, both Rogers and Gabor (by now a Nobelist) were portrayed as the 'grand old men' of British holography, but had largely been bypassed by research in the expanding subject.[85]

One reason for Gabor's disappointment at the CBS Laboratories' lack of success was his perception of more successful exploitation elsewhere. As early as 1964, George Stroke

[82] Gabor, Dennis to P. Goldmark, letters, 1966–7, IC GABOR LA/9.
[83] Gabor, Dennis to P. Goldmark, letter, 3 Jun. 1967, IC GABOR LA/9.
[84] Gabor, Dennis to D. S. Strong, 18 Jan. 1972, IC GABOR MS/16. See also §10.2.
[85] Benyon, Margaret to SFJ, interview, 21 Jan. 2003, Santa Clara, CA, SFJ collection.

had cited predictions of a $10 million income from holography patents, but within two years, predictions had mushroomed. Gabor had confided to Goldmark in 1966:

> I think I have told you that in Ann Arbor I talked to the Battelle people, and tried to dampen their unbounded optimism. They thought that Holography may become as good a money-spinner as Xerography which has brought them so far $340 million in royalties, of which Carson has got 40%! Apparently I have not succeeded because, as the European Representative, Dr. A. McGrew told me on the telephone from Zurich, they are willing to go into it to seven figures.[86]

To Gabor, the Battelle activities appeared to be exploding and he was eager to be involved. The University of Michigan had given the institution rights to the forthcoming Leith–Upatnieks holography patents. Battelle subsequently had formed a partnership with Du Pont Inc. and Scientific Advances Inc. to found Holotron in 1966, a company intended specifically to exploit holography in numerous commercial domains. They planned to use the researchers and facilities at Du Pont and Battelle to investigate new applications and extensions of the medium.

According to Gabor, who visited to propose linkages with CBS Laboratories, the Holotron administrators were 'keen on making holography a commercial success', and were not interested in military applications. They hoped to develop two lines: three-dimensional portraits and holographic instruments.[87] From mid-1966, the company also pursued, with Coherent Radiation Laboratories, white-light lasers (lasers emitting a combination of red, green, and blue wavelengths to yield a white combination). Holotron felt that such a laser, probably based on a continuous-wave krypton ion laser, would be crucial for exploiting colour holography.[88] In fact, no firm relationship developed with Coherent or CBS, and neither Battelle/Holotron nor CBS Laboratories were to announce major developments over the remainder of the decade.

CBS Laboratories' lukewarm exploitation of holography was typical of other well-financed firms closely linked with imaging technologies. Another influential yet tentative corporate player was the Polaroid Corporation, headed by Edwin Land. A wide-ranging inventor who had made his name with the invention of inexpensive sheet polarizing material in the 1930s and instant photography in the 1950s, Land had long had an interest in photography and three-dimensional imaging. The research laboratories on Main Street in Cambridge, Massachusetts—a few streets away from the large MIT campus—were filled with chemists and engineers, but also non-traditional technical workers, many without advanced degrees.

Land adopted the same approach with holography. Stephen Benton, who worked part-time at Polaroid as an MIT undergraduate, recalls Land as an atypical director of an unusual company. He gave undergraduates opportunities to do research, with any discoveries or inventions reverting to the company. Benton had been interested in 3D since

[86] Gabor, Dennis to P. Goldmark, letter, 30 May 1966, IC GABOR LA/9.

[87] Gabor, Dennis to P. Goldmark, 27 Sep. 1966, IC GABOR LA/9. Holotron activities are further discussed in Chapter 12.

[88] Jarrett, Steven M., 'Early ion laser development', *Optics and Photonics News* 15 (2004): 24–7.

childhood, and had been introduced to Land by 'Doc' Edgerton of MIT in 1961. Benton recalls that when holography 'started going' in 1965, Land said, 'If you'd like to do this, I'd love to have a lab put together, and play with it'. And I said 'sure!'.[89]

While Land had been travelling with the photographer Ansel Adams, he encountered George Stroke giving a seminar on holography.[90] He invited Stroke to visit Polaroid in March 1965 to talk about holography. Stroke displayed a hologram made by student Douglas Brumm of a coffee cup, inspiring Benton, his wife Jeannie, and colleagues John and Mary McCann to record their first holograms at the Polaroid building that evening.

It seemed to Benton that neither Stroke, pursuing the mathematical details of Fourier holography, nor Leith and Upatnieks, 'interested in technology and instrumentation', focused on what interested both Land and Benton: three-dimensional imaging.

> There was never any pressure at Polaroid to do anything; 'here's your lab, here's your equipment', you know, as long as, whoever was involved, when you bumped into Edwin Land you had something in your pocket he hadn't seen before, *that's all it took!* Especially if you were an undergraduate, or a graduate student, someone who didn't need a lot of money, so there weren't a lot of demands. But when I did finally graduate he would start spending money on equipment and start scaling things up from little holograms to big holograms. Even though I was sort of half-time at Polaroid and half-time at Harvard after I got my PhD. I would do my research at Polaroid. As long as I had something that was fairly new (I went from monochrome to color, a little bit of animation) he was really happy.[91]

Benton did his graduate studies in optical data storage via holography, taking one summer intensive course at Ann Arbor under Emmett Leith. After obtaining his PhD in 1968, Benton was able to teach as an Assistant Professor of Physics at Harvard while continuing research at Polaroid. At least half his time at Polaroid was spent with holography, with the remainder devoted to various optical technology projects: to monitor thickness of coatings for quality control, and investigating other kinds of 3-D technologies, for example. This relatively relaxed regime led him to develop and popularize a new variant of white-light holography, later known commonly as rainbow holography.[92]

Benton's holography laboratory was eventually fairly well equipped, but reliant on the financial vagaries of a struggling consumer business. Benton later described his small laboratory as 'devoting a fraction of its efforts to holography between manufacturing crises'.[93] 'You'd wait—Polaroid was very much an up-and-down business, and budgets weren't always high, especially when they were moving to production of colour film, but

[89] Benton, Stephen A. to SFJ, interview, 11 Jul. 2003, Cambridge, MA, SFJ collection.

[90] According to Stroke, Land had offered him a job in the early 1950s shortly after Stroke came to America (Aspray, William, 'George Wilhelm Stroke, Electrical Engineer, an oral history conducted in 1993 by William Aspray, IEEE History Center, Rutgers University, New Brunswick, NJ, USA.' (1993)).

[91] Benton, Stephen A. to SFJ, interview, 11 Jul. 2003, Cambridge, MA, SFJ collection.

[92] Benton, Stephen A., 'Artist files,' MIT Museum 26/584, 1978. See §10.2.

[93] Benton, Stephen A., *Holographic Imaging (draft)* http://splweb.bwh.harvard.edu:8000/courses/mas450/reading/chaptersPDF/, 2003).

he said "look, if you want three-color lasers, [conspiratorially] *this would be a good time to make a plan!*" '.[94]

Stephen Benton's work in holography thereafter seemed to follow the example set by his patron. Edwin Land had had both commercial successes and failures in a fifty-year career. The technically sophisticated SX-70 instant camera had been introduced successfully in 1975, but the Polavision instant movie system, introduced five years later, attracted few customers owing to the rising competition with video recorders. Behind both products was a marketing strategy that had been enunciated by Land in 1946:

> I believe it is pretty well established now that neither the intuition of the sales manager, nor even the first reaction of the public is a reliable measure of the value of a product to the consumer. Very often the best way to find out whether something is worth making is to make it, distribute it, and then to see, after the product has been around for a few years, whether it was worth the trouble.[95]

Benton was to pursue this strategy in holography through his career.[96]

6.5 EXPANSION IN THE EAST

Some of the most prepared workers for research in holography were those from the generation that had explored wavefront reconstruction. Leith, continuously involved in the field and a source of its explosion in 1964, had maintained momentum to exploit it. By contrast, Dennis Gabor and Gordon Rogers had found themselves rapidly outpaced by other laboratories.

Between Leith, on the one hand, and Gabor and Rogers on the other, was Yuri Denisyuk, who had set aside his wave photography as recently as 1962. For some four years his laboratory worked on other problems, beginning further research on the subject only in 1966 and in direct response to Western developments. He recalls, 'Experiments on the recording of reflection 3D image holograms were also carried out in our laboratory; however, due to a lack of imagination, coins, medals and other objects of small depth were recorded on these holograms'.[97]

By that time, papers on holography were flooding Western journals. There were at least two reasons for the revival of Denisyuk's work at this time. First, Western researchers were rediscovering, in more particular but powerful experimental implementations, ideas that Denisyuk had pursued without lasers and from an alternate theoretical

[94] Benton, Stephen A. to SFJ, interview, 11 Jul. 2003, Cambridge, MA, SFJ collection.

[95] 'Edwin Land', *Physics Today*, Jan. 1982: 35.

[96] Stephen Benton joined the faculty of MIT in 1982 forming the Spatial Imaging Group, and the new Media Laboratory there when it opened in 1984. He became Director of the Center for Advanced Visual Studies (CAVS) in 1996. In the first years of the Media Laboratory, Benton's sponsors were principally General Motors and the Defense Advanced Research Projects Agency (DARPA), providing some $500,000 per year [Brand, Stewart, *The Media Lab: Inventing the Future at MIT* (New York: Viking Penguin, 1987)].

[97] Denisyuk, Yu N., 'My way in holography', *Leonardo* 25 (1992): 425–30, p. 428.

standpoint over half a decade earlier.[98] That experience and perspective had not yet been adequately explored at the Vavilov Institute. Second, the mid-1960s was a period when assimilative repetition, or the duplication and extension of Western science and technology, was an overt Soviet technical policy. In order to surpass Western science, Soviet research since the early 1950s had been seeking to first reproduce it at home and second to take it in new directions.[99]

Thus, the explosion of holography in the West and scientific policy at home aided Denisyuk's own career. When an important Communist Party official visited the Vavilov Institute and discovered that there was no laboratory dedicated to holography, Denisyuk's 1961 Vavilov Institute laboratory was rededicated and renamed the 'holography laboratory'.[100] This comparison with the West happened more than once. When the President of the Academy of Sciences of the Soviet Union, Mstislav V. Keldysh, toured America in 1969, he asked American holographers which Soviet holographers they knew.[101] He learned that Yuri Denisyuk was well known there, but was later surprised to find that Denisyuk was little known in the Soviet Academy by his countrymen. Largely because of this Western recognition, Denisyuk was elected a corresponding member of the Academy in 1970.[102] The most positive, and repeated, support came from Petr L. Kapitsa, a former student of Lord Rutherford and prominent Soviet scientist. Kapitsa asked one of the young theoreticians working for him to write a review of Denisyuk's work. The result was a negative and rather naïve report. When Denisyuk was nominated for the Lenin Prize in 1970, Kapitsa applied pressure to the same individual to write a positive review.[103]

[98] See §7.3 on the reinvention of reflection holography in the west.

[99] Medvedev, Zhores A., *Soviet Science* (Oxford: Oxford University Press, 1979) pp. 60–64. Claims of the inherent superiority of Soviet science, based on the principle of dialectical materialism, had been promoted originally by Lenin in the early 1920s and then pursued by Stalin in the gradual sovietization of the system of science in the Soviet Union and again emphasized during the Khrushchev period 1953–64. The policy of assimilative repetition, inaugurated under Khrushchev during the 1950s, was supported by the establishment of English at all levels of the educational system and the encouragement of at least reading knowledge for postgraduates; translation of foreign literature by the Mir publishing house; and, particularly in high-technology industries, the examination of successful foreign designs and synthesis of improved versions for Soviet manufacture. This policy was replaced by a gradual integration with international science from 1971 via the establishment of the principle of *détente* in international affairs.

[100] Denisyuk, Yu N., 'The work of the State Optical Institute on holography', *Soviet Journal of Optical Technology* 34 (1967): 706–10 Translated from *Optiko Mekhanicheskaya Promyshlennost*. 34 (11): 18–23 (1967).

[101] Keldysh (1911–78) was fully capable of understanding the technical details and significance, both intellectual and social, of Denisyuk's work. Trained in mathematics and physics at the University of Moscow, he had held a variety of positions in applied mathematics and aerohydrodynamics in Moscow, working on the mathematics of atomic energy, rocketry and space flight and scientific management. He was president of the Academy of Sciences of the Soviet Union, 1961–75.

[102] Denisyuk was elected a corresponding member of the Academy of Sciences in 1970 and Academician in 1992, and awarded the Lenin Prize in 1970, The Badge of Honour (1975), the National Prizes of the Soviet Union (1982, 1989), and Orders of the Red Banner of Labour (1988).

[103] According to Zhores Medvedev, the awarding of major academic awards such as the Lenin Prize required preliminary consideration at the different Party levels and a character reference and approval by the Party Central Committee secretariat [Medvedev, Zhores A., *Soviet Science* (Oxford: Oxford University Press, 1979), p. 188]. The support from Kapitsa can therefore be assumed to have been important to Denisyuk's progress. As noted in §3.4, Kapitsa's influence is also detectable in Denisyuk's first papers: the 1962 paper on wave photography expresses his 'deep gratitude' to Kapitsa, Linnik, and I. V. Obriemov.

Denisyuk described the goals of his resumed holographic research in 1967 as 'the realization of the practical possibilities of holography which opened up as a result of the development of lasers' and 'a study of those properties of lasers which affect the hologram quality, as well as the development of those holography methods which permit using one or another feature of these new sources of radiation'.[104]

A small part of Denisyuk's own laboratory included postgraduate students, and his holographic work was extended by two of them, 'young energetic beautiful ladies: Mrs Ella G. Zemtsova (hologram recording) and Mrs Diana Zagorskaya (chemistry of holographic plates)'.[105] Alongside their work, Rebekka Protas further developed the production of holographic plates based on micro-grain silver bromide emulsion, which eventually were fabricated in a production plant.[106] Dr Yuri Usanov developed a specific developer for the plates, which continued in use thereafter. Such emulsion and processing improvements went almost unstudied in the West.

With his student Dmitri Staselko, Denisyuk had developed pulsed lasers for portraiture from 1968.[107] They first used the deep red light of a ruby laser, and later the green light of a frequency-doubled neodymium laser. Denisyuk changed his workplace in 1968, moving from the Vavilov Institute to the nearby A. F. Ioffe Physical–Technical Institute of the Academy of Sciences in Leningrad. There he began to study the applications of holography to optical computing.

During most of the 1960s, then, Yuri Denisyuk's holography research proceeded haltingly and in relative obscurity in the Soviet Union. As work in Western countries reproduced and extended his optical studies his domestic status rose accordingly. His receipt of the Lenin Prize, and membership in the Academy of Sciences, rehabilitated Denisyuk's status at home and signalled a new public awareness of holography in the

[104] Denisyuk, Yu N., 'The work of the State Optical Institute on holography', *Soviet Journal of Optical Technology* 34 (1967): 706–10, translated from *Optiko Mekhanicheskaya Promyshlennost*. 34 (11): 18–23. He cited active research areas based on optomechanical applications, image processing, viewing through heterogeneous media, optical copying, and the various other fields appearing widely in the Western literature by 1967.

[105] Denisyuk, Yu N. to SFJ, email, 16 Apr. 2003, SFJ collection; Zagorskaya, Z. A., 'High-resolution photographic emulsion for recording three-dimensional holograms', *Soviet Journal of Optical Technology* 40 (1973): 134, translated from Optiko-Mekhanicheskaya-Promyshlennost 40 (2) (Feb. 1973): 72; Zemtsova, E. G. and L. V. Lyakhovskaya, 'Recording of three-dimensional holograms on LO1-2-63 photographic plates', *Soviet Journal of Optical Technology* 41 (1974): 473–4, translated from Optiko-Mekhanicheskaya-Promyshlennost 41 (Aug. 1974): 75–6.

[106] During the 1970s, N. I. Kirillov independently developed similar plates. See, for example, Vasilieva, N. V. and N. I. Kirillov, 'Requirements to high resolution photomaterials for holography', *Tekhnika Kino i Televideniya* 7 (1972): 3–9. The Slavich factory, located some 100 miles north of Moscow, subsequently manufactured such emulsions.

[107] Staselko, D. I., A. G. Smirnov, and N. Denisyuk Yu, 'Production of high-quality holograms of three-dimensional diffuse objects with the use of single-mode ruby lasers', *Optics and Spectroscopy* 25 (1969): 505–7; Staselko, D. I., Yu N. Denisyuk and A. G. Smirnov, 'Holographic portrait of a man', *Zh nauch i prikl photogr i kinematogra*. 15 (1970): 147.

Soviet Union. Journalists on the most important newspapers, such as *Isvestia* and *Pravda*, described his work for the first time in popular terms.[108] But even fame had its price, as he later recalled:

> My struggles with conservative colleagues and my subsequent triumph had a bad influence on my investigations in the field of holography. During these long years, I was bombarded by journalists and numerous beginners who wished to become involved with holography. These were not my best years.[109]

Western observers gradually became aware of Soviet work in holography not merely through translated papers, but by direct contacts. Emmett Leith and Juris Upatnieks had their first contact from Soviet holographers Yuri Ostrovsky and Galya V. Ostrovskaya of the A. F. Ioffe Physico-Technical Institute in Leningrad that year, requesting the exchange of reprints and preprints.[110] They and their colleagues were among the first to develop holographic interferometry in the Soviet Union. Denisyuk made his first trip to a Western country in 1970 to present a paper at Besançon, France on the measurement of laser coherence by holography. His laboratory had been using domestically produced lasers since about 1965; on the advice of Prof. J. Viénot at Besançon, he obtained a Spectra-Physics laser that year. With it, and focusing further research on the chemistry of holographic processing, Soviet reflection holograms improved considerably.[111]

In the western Ukraine, early investigations of holography were also taken up under the direction of Vyacheslav K. Polyanskii at the Optical School of Chernivtsi University. The group developed referenceless holography (better known in the West as ghost-image or phantom-image holography) in their quest for a holographic associative memory. This line of work led to studies of holographic cryptography.[112]

In Tblisi, in Soviet Georgia, holography was pursued at the Institute of Cybernetics of the Georgian Academy of Sciences. There were two groups, headed by Prof. Shermazan Kakichashvili and by Prof. N. Ramishvili, respectively, from the mid-1970s. Kakichashvili (d. 2002) and his co-workers created and extended the field of polarization holography by

[108] Anonymous, *Soviet Journal*, 247 (10), 12–5. See also Denisyuk, Yu N., 'Holography at the State Optical Institute (GOI)', *Soviet Journal of Optical Technology* 56 (1989): 38–43. Translated from Optiko-Mekhanicheskaya-Promyshlennost 56, (1) (Jan. 1989): 40–4.

[109] Denisyuk, Yu N., 'My way in holography', *Leonardo* 25 (1992): 425–30, p. 428.

[110] Ostrovskaya, G. V. and Y. I. Ostrovsky to E. N. Leith and J. Upatnieks, letter, 13 Sep. 1966, Leningrad, Leith collection.

[111] Denisyuk, Yu N. to SFJ, email, 8 Aug. 2003, SFJ collection.

[112] Soskin, Marat, 'Optical holography in Chernivtsi University', unpublished manuscript, 20 Sep. 2004, SFJ collection.

developing dozens of polarization-sensitive recording materials, for which he received a Diploma of Discovery from the state.[113]

Thus, Soviet holography of this period was not merely a matter of extending Denisyuk's groundbreaking work by abstract scientific research; applications were also actively investigated, with some centres focusing on practical implementations. Among the most important of them was the Cinema and Photographic Research Institute (or NIKFI, according to the Russian acronym) in Moscow. NIKFI had been founded in 1929 as the central Soviet institute for the development of cinematography and photography. From the 1940s, S. P. Ivanov developed lenticular screens to permit stereoscopic film projection without glasses.

Gennadi Sobolev of NIKFI was one of the important investigators who explored and exploited the new medium of holography, especially through development of improved emulsions.[114] In 1967 he made the first large-format Denisyuk-type holograms in the Soviet Union, and soon thereafter began making holographic copies that allowed the image to project forward of the film plane (so-called image plane reflection holograms, discussed in §7.4). From the late 1960s, he and his colleagues recorded reflection holograms of museum objects from the Hermitage and Kremlin Museum collections, which toured the country and were later widely seen and admired internationally.[115]

During the early 1970s, with a population of some 800 scientists, NIKFI expanded its research in holography, developing holographic materials and chemical processing techniques, methods of recording, copying, and reconstruction. The stereoholography group, under I. Yu. Fedchuk, created high-quality display holograms. And under Victor G. Komar, head of the stereocinematography group at NIKFI, a prototype holographic cinema system was developed for educational purposes. A pulsed laser illuminated live

[113] Kakichashvili, S. D., 'Polarization holography: possibilities and the future', presented at *Holography '89: International Conference on Holography, Optical Recording, and Processing of Information*, Varna, Bulgaria, 1989; Akopov, Edmund to SFJ, email, 24 Feb. 2004, SFJ collection; Denisyuk, Yu N. to SFJ, email, 8 Dec. 2004, SFJ collection; Reingand, Nadya, 'Holography in Russia and other FSU states', http://www.media-security.ru/rusholo/index.htm, accessed 27 Sep. 2004. Kakichashvili and co-workers discovered over a hundred different materials, including photochromic glasses, hypersensitized silver halide emulsions, and a range of dyes in polymer matrices, having polarization sensitivities between the ultraviolet and red portions of the spectrum. This class of materials made possible new components for diffractive and coherent optics. In the Soviet system, 'Diplomas of Discovery', equivalent in status to Western patents, contrasted with 'Diplomas of Appreciation', which recognized work having less intellectual or applied importance.

[114] G. A. Sobolev (1933–2001) became Doctor of Physics and Engineering in 1969, and joined NIKFI that year. From 1970–83 Sobolev headed the Holography Laboratory at NIKFI, and with colleagues made the first widely discussed holographic movie, a thirty-sec sequence projected onto a holographic screen that he invented. From 1989 he worked on photopolymers for holographic memory and also on applications of holography for medicine. In 1996 he and his wife, Dr Svetlana Soboleva, moved to America to work at Holos ('Gennadi A. Sobolev', *Holography News*, 16 (7), Sep. 2002: 5).

[115] Similar work was done in the Ukraine by Vladimir Markov and his colleagues from 1976, when a group from the Institute of Physics of the Ukrainian SSR Academy of Sciences and the Ministry of Culture gained privileged access to Kiev museums (Markov, V. B. and G. I. Mironyuk, 'Holography in museums of the Ukraine', presented at *Three-Dimensional Holography: Science, Culture, Education*, Kiev, USSR, 1989). The large-scale reflection holograms produced in Tbilisi, Leningrad, Moscow, and Kiev had a dramatic impact on Western observers.

subjects and a holographic screen focused the three-dimensional image for four or five positions in the audience—in essence, a scaled-up version of what Conductron engineers had been seeking, but still falling short of Kip Siegel's unattainable goal. In 1976 Komar demonstrated a 20 sec holographic movie on a 60 × 80 cm. screen at a Moscow congress, and in 1984 a two-colour cartoon was displayed.[116]

Thus Soviet holography diverged in several respects from its Western implementation: it concentrated on thick, fine-grained photographic emulsions to record Denisyuk reflection holograms; the chief applications became the representation of museum objects; and display holograms, both as large-format images and short cinema clips, were widely admired. By contrast, Western holograms of the period were usually Leith–Upatnieks variants that required a monochromatic light source for viewing. The principal application of such display holograms was advertising, as illustrated by the extensive attempts by Conductron and its successor, McDonnell-Douglas Missouri, to pursue such markets. Thus, the differing economic and research environments for these activities led to distinct trajectories for American and Soviet holography.[117]

As in America and Britain, interest developed in other less experienced laboratories. Soviet scientific and technical work in holography was more centrally directed and rationalized than in the West, but there was nevertheless considerable overlap. Among the most influential groups participating in the 1971 Ulyanovsk (formerly Simbirsk) and 1975 Kiev All Union Conferences on Holography, for instance, were the Vavilov Institute in Leningrad, Moscow University workers, and those in Kiev. Work was also carried out at the capitals of other republics of the Soviet Union such as Minsk.[118]

Holographic interferometry in East Germany, for example, was developed at several separate laboratories. One such example in the German Democratic Republic (GDR) was the Institute for Optics and Spectroscopy, founded after the Second World War by Prof. Ernst Lau, which became involved in holography from 1968, when Prof. Paul Görlich, an Academician, Research Director of Zeiss Jena and Associate Director of the institute, urged East German government officials to support holographic research.[119] Like Western laboratories of the period, they sought to satisfy the diverse requirements of different end-users and found practical applications to be elusive.

[116] Komar, V. G., 'Possibility of creating a theatre holographic cinematograph with 3-D colour image', *Tekhnika Kino i Televideniya* 4 (1975): 31–9; Huff, Lloyd, 'Holography in the Soviet Union', *holosphere* 12 (1983): 4–5; Komar, V. G. and O. B. Serov, 'Work on holographic cinematography in the USSR', presented at *Holography '89: International Conference on Holography, Optical Recording, and Processing of Information*, Varna, Bulgaria, 1989.

[117] On the sociological differences between high-status scientists in the two countries, see Lubrano, Linda L. and John K. Berg, 'Academy scientists in the USA and USSR: background characteristics, institutional and regional mobility', in: J. R. Thomas and U. M. Kruse-Vaucienne (eds), *Soviet Science and Technology: Domestic and Foreign Perspectives* (Washington DC, 1977), pp.

[118] Schwider, Johannes to SFJ, email, 1 Jul. 2003, SFJ collection.

[119] As in America, there was a close association with the experimental knowledge and activities of spectroscopy and interferometry. Between 1955 and 1969 the Head of the institute was spectroscopist Rudolph Ritschl. In 1969, after merging with another academy institute working on lasers, the institute was renamed the Central Institute for Optics and Spectroscopy.

Within two years, a group in the Institute had built its first HeNe laser, so holograms could be produced with available equipment. As with research centres in the West, the Institute produced demonstration pieces—fittingly holograms of Marx and Lenin busts, illuminated by mercury arc lamps—for party officials. On the other hand, the interferometry and modern optics group under Dr Schultz investigated the prognosis of holography and suitable topics for investigation. Entering the field some two years after what they perceived as the peak activities of 1966, the group decided that the most promising field would be synthetic or computer generated holography. Seeking closer cooperation with industry, the Institute took up holography as a suitable technology for non-contact lithography. They decided rather soon, however, that non linearities of the photographic material and coherent interference made the image inferior to conventional methods. Computer-generated holograms had their own problems related to limitations in computing power and memory.

The group was producing aspheric lenses at the time, a secret project in which they were significantly ahead of other laboratories. They conceived a method of testing these optical elements with computer-generated holograms (in other words, specially designed Fresnel zone plates). However, Carl Zeiss Jena was unable to invest in an industrial plotter merely to plot such masks. Nor was the extension of computer holography to two-dimensional problems such as the lithographic application pursued, owing to high costs.[120]

Thus, Soviet laboratories sought applications as actively as Western researchers in universities, government laboratories, and commercial firms. And the first laboratories to make an impact were those with prior experience in the field.

The explosion of research in holography was not limited to America, Russia, and Britain. Researchers in other countries were taking up the new subject enthusiastically. A handful of examples can illustrate the international seeding of the subject. In France, for example, Serge Lowenthal created a holography research group at the Institut d'Optique in 1968 that began exploring spatial filtering, digital holography, and x-ray holography. The following year, Lasergruppen Konsult AB was formed in Sweden by a group of scientists working at the Royal Institute of Technology, Stockholm. The company, later renamed Lasergruppen Holovision, focused on technical consulting and patenting of hologram concepts, and later moved into display holography.[121] And in 1970, Parameswaran Hariharan at Hindustan Photo Films was developing alternatives to Kodak 649F emulsion that would be suitable for the tropical conditions of Indian photographic laboratories.[122]

[120] Despite such setbacks, the Institute was able to pursue phase-shifting interferometry analysed by computers, obtaining a working device by 1981. Significantly, the sensors used for the device had been obtained from another group's cancelled project on holographic data storage; like Leith and others at Radiation Inc., holographic memories proved elusive.

[121] Skande, Per, 'The development of commercial display holography in Sweden', *Proceedings of the International Symposium on Display Holography* 2 (1985): 47–50.

[122] Hariharan, Parameswaran, 'Photographic materials and coherent light', *Proceedings of the Indian National Science Academy* 37A (1971): 193–9. Hariharan had researched photographic emulsions in National Research Council of Canada during the early 1950s, developed interferometers at the NPL in New Delhi,

The individuals and laboratories discussed above all had privileged routes to information: either they involved those researchers who had been active for years, or they resulted from well-funded exploratory or development projects. The next most direct conduit of information was open publication, which allowed enthusiastic but relatively inexperienced researchers to take up the subject.

A notable example of early entry into the field is that of Jumpei Tsujiuchi in Japan. Tsujiuchi had first encountered the concept of wavefront reconstruction in August 1959 when attending the Fifth International Meeting of the International Commission of Optics (ICO) in Stockholm. He was, at the time, an attaché de recherche at the Centre National de la Recherche Scientifique (CNRS) in Paris studying optics, and became interested in wavefront reconstruction after listening to a paper by Hussein El-Sum, then at Lockheed Aircraft Corporation in Palo Alto, California, on 'Optical information retrieval by reconstructed wavefront techniques'. Intrigued, he looked up the papers of Gabor, El-Sum, Rogers, and Baez on wavefront reconstruction at the library of the Institut d'Optique.[123] He discovered that the technique was close to his own research subject of coherent optical filtering.[124]

The following year, upon returning to his place of work at a Japanese standards laboratory, the Government Mechanical Laboratory in Tokyo, he kept up his reading on the subject, but decided that it was impracticable owing to the conjugate image problem. He remained sceptical upon reading Leith's and Upatnieks' 1961 and 1962 papers, but recalls being 'very shocked' when he saw their 1963 paper on the reconstruction of greyscale transparencies with its 'beautifully reconstructed images'. Their 1964 paper on diffuse illumination and three-dimensional objects made him eager to take up the subject himself. He recalls that, the following year, '[w]e could get a He-Ne laser easily from Hitachi, later an Argon Laser from NEC, and [Kodak] 649F plates from the US, and we succeeded to take holograms by ourselves in 1965'.[125]

That year, three independent groups began exploring holography in Japan. One was the group of Prof. Toshihiro Kubota in the Institute of Industrial Science at the University of Tokyo, which included T. Ose, T. Asakura, and others; a second was Prof. Ryuichi Hioki's Group in the Faculty of Engineering at the University of Tokyo, including S. Tanaka, T. Suzuki, and others; and the third was Tsujiuchi's group, including K. Matsuda, at the Government Mechanical Laboratory, also in Tokyo. A fourth group, interested in

and was laboratory director of a photographic manufacturing company for a decade. He later joined the Indian Academy of Science in Bangalore, and then the Commonwealth Scientific and Industrial Research Organisation (CSIRO) in Australia from 1973.

[123] At the Institut d'Optique, Tsujiuchi worked next door to George Stroke, who was then completing his PhD on diffraction gratings there. They did not exchange ideas about wavefront reconstruction at that time, but in later years both worked on spatial filtering. They came together again at the US–Japan Seminars on Holography in 1967 and 1969.

[124] Tsujiuchi, J., 'Restitution des images aberrantes par le filtrage des frequencies spatiales I', *Optica Acta* 7 (1960): 243; Tsujiuchi, J., 'Restitution des images aberrantes par le filtrage des frequencies spatiales II', *Optica Acta* 7 (1960): 386; Tsujiuchi, J., 'Restitution des images aberrantes par le filtrage des frequencies spatiales I', *Optica Acta* 8 (1961): 161.

[125] Tsujiuchi, Jumpei to SFJ, unpublished report, 20 Jun. 2003, SFJ collection.

acoustic and microwave holography, started their investigations under Prof. Aoki at Hokkaido University in Sapporo. The Japanese research interest mirrored that developing in other countries. The first three groups initially were interested in the general properties of holograms, and soon afterwards in holographic interferometry for engineering applications. The first paper on holography published in Japan was probably a description of cylindrical holograms.[126] By the late 1960s, Japanese holographic research was pursued at a number of university and industrial centres.

The same was happening in other industrialized countries, where the promise of the medium seemed both irresistible to marketers and necessary for international competitiveness. A combination of public and private investment motivated nationally flavoured research and development. By the end of the decade, then, holography was being pursued internationally for military, aerospace, general engineering, and display applications. Interest and funding had proliferated in the industrial, commercial, and governmental sectors. There was a growing conviction that improvements in the new medium were just around the corner.

[126] Hioki, R. and T. Suzuki, 'Reconstruction of wavefronts in all directions', *Japanese Journal of Applied Physics* 4 (1965): 816.

7

Technology of the Sublime: The Versatile Hologram

During its breathless decade (1964–73), holography was an ever-expanding collection of techniques, claims, and awe-inspiring imagery. As soon as 'modern holography' seemed established, a new variant appeared to challenge definitions. This chapter discusses the features and impacts, if not the full theoretical and technical accounts, of those variants. I will resist the urge to discuss the technologies from a modern perspective, which would obscure their difficult genesis. Instead, I recommend excellent modern texts that explain the techniques according to an integrated conception.[1]

Some of these new variants of holography were technologically trivial, but visually or conceptually impressive; others were complex in theory and painstaking in their execution, but subtler for the casual observer to appreciate. Yet, like many new technologies, the sublime nature of these techniques was evanescent: the awe engendered was fleeting and demanded constant extension to maintain enthusiasm. Like an addictive drug, ever more was needed to produce its effect. Thus holography during this period was led by a search for novelty and extension, a trend that could not be sustained indefinitely.

7.1 PROSPECTS AND PROBLEMS

The explosion of interest in holography from mid-1964 led, as we have seen, to enthusiastic exploration of its possibilities in laboratories around the world. But this search for applications soon generated a list of limitations. Some appeared minor and temporary defects; others began to appear unavoidable. Technical bottlenecks appeared everywhere: they involved the light sources for recording and viewing holograms, the optical arrangements for recording them, and the photosensitive materials on which they were based.

[1] Among the most easily absorbed practical texts are Saxby, Graham, *Practical Holography* (Bristol: Institute of Physics Press, 2004) and Unterseher, Fred, Jeannene Hansen and Bob Schlesinger, *Holography Handbook: Making Holograms the Easy Way* (Berkeley, CA.: Ross Books, 1982). A good treatment including more underlying theory is Hariharan, P., *Optical Holography: Principles, Techniques and Applications* (Cambridge: Cambridge University Press, 1996). Another excellent text, incomplete on the author's death but planned for subsequent publication, is Benton, Stephen A., *Holographic Imaging (draft)* http://splweb.bwh.harvard.edu:8000/courses/mas450/reading/chaptersPDF/, 2003).

The most immediate problems centred on the laser itself. First, laser light sources, combined with the relatively insensitive high-resolution film emulsions, required exposures of minutes or, at best, seconds. This limited the subjects of holograms to small, static, and necessarily lifeless objects and made holography more suited to artistic still-lives than to a replacement for snapshot photography. The *lensless photography* and *wave photography* labels oversold the technology.

A laser was required again to reconstruct the image. Laser light produced the phenomenon of *speckle*, an ever-changing graininess caused by random interference of light reflected from diffuse surfaces. This could make holographic images distracting to view and difficult to photograph. Laser lighting arrangements were also a serious limitation. Although bright argon-ion lasers became commercially available in 1965, their intense coherent light made them potentially dangerous to unprotected eyes, and ruled them out for display purposes by the end of the decade. In 1973–74, laser use was further restricted when the Bureau of Radiological Health (BRH) imposed strict standards for the use of lasers in America, with similar legal limitations following in Europe.[2] They also demanded special power supplies, cooling water, and painstaking optical adjustment. While dimmer HeNe lasers were acceptable in terms of public safety, even they were too expensive and unreliable to consider using as display sources except by the largest corporations or laboratories. Conductron, for example, relied on filtered mercury lamps for its displays, as did the first public exhibitions in Sweden.[3] As a result, holograms were relegated to an environment of darkened rooms and arcane lighting.

These technical limitations constrained applications. The recording and display problems were worse for large-format holograms. Leith and Upatnieks moved from 4 × 5 in. high-resolution plates in 1964 to 8 × 10 in. plates for their demonstration pieces two years later, but 11 × 14 in. appeared to be a practical limit. Holographic cinema appeared even more unlikely because of the need for very short exposures.

There were also viewing limitations. The hologram was not a projection; it was more akin to a small window through which viewers peered. The restricted viewing angle made group viewing awkward, and cinema—displaying moving three-dimensional images to an audience—seemingly impossible.

And holograms were in other respects *too* good. Their storage of multiple perspectives meant that holographic television was unlikely, given the need for tremendous bandwidth beyond the capacity of existing communication networks. A way would have to be found either to reduce the information content in holographic images, or to increase the bandwidth of communication channels, or both.

There were equal difficulties in recording the hologram itself. The available photographic emulsions were extremely insensitive, because the need for high resolution

[2] In 1976 the American National Standards Institute ANSI Z136.1 further classified laser powers and specified where lasers of defined maximum powers could be directed. The British equivalent was PM19 (Health and Safety Executive, 1980).

[3] Charnetski, Clark to SFJ, interview, 3 Sep. 2003, Ann Arbor, MI, SFJ collection; Bjelkhagen, Hans I. to S. Johnston, interview, 18–19 Sep. 2002, SFJ collection.

meant that the silver grains had to be unusually small; more light was needed to produce darkening than in common photographic plates. As the Willow Run SAR developers knew well, the subsequent chemical processing was a frustratingly slow step in optically transforming radar data into an observable image, and the same was true for the growing applications of holography.

The very identification of such 'problems', of course, was conditioned both by the applications that were conceived, and the understanding of how holography related to existing technologies. As such challenges became identified during the late 1960s, laboratories either undertook open-ended research with the hope of fortuitous discovery of new and useful characteristics, or focused on goal-directed research to surmount the perceived problems. Open-minded exploration dominated technical advances in the first decade: during the rapid scramble of experimental work, investigator after investigator stumbled across new varieties of hologram. The result was an impressive but unplanned series of technical improvements over the decade. Thereafter, tightened budgets led to goal-directed research that improved the chemistry and physics of the recording medium. Display holography got bigger, brighter, and more beautiful. While that second decade generated less technical enthusiasm, its accomplishments were a necessary foundation for further expansion of the field.

7.2 INVENTING AND REINVENTING HOLOGRAPHIC INTERFEROMETRY

Some techniques were repeatedly rediscovered and privately mythologized by a series of investigators. One of the earliest extensions of holograms in Leith's and Upatnieks' laboratory also became the most widespread and commercially important application. In the first weeks of investigating holography with three-dimensional objects, they happened upon an unexpected phenomenon. Leith recalls:

> One of the objects we used was a sheet from a paper calendar, measuring about a few inches on a side. To give stability, we pasted the calendar sheet to a block of aluminum. The resulting holographic image was good except for a circular black spot about a half-inch or less in diameter, which covered up one of the numbers on the calendar. We made another hologram and yet another. It was always the same story—just one defect, and always in the same place. Examination of the object revealed that the aluminum block had a circular hole in it, just at the place where the image defect occurred. Juris remarked that this would be a good method to measure object motion.[4]

Upatnieks separately recalls the circumstances:

> One of the earliest holograms showed a dark spot where a piece of paper, cemented to an aluminum block, traversed a hole in the block. It was obvious that the paper had moved. In a later

[4] Leith, Emmett N. to SFJ, email, 5 Jun. 2003, SFJ collection. The hologram was recorded on 19 Dec. 1963.

> hologram, cardboard was cemented to a metal block with an edge extending beyond the block. This part showed fringes that are typical of moving or vibrating objects.[5]

The movement during the exposure causes a blurring out of the interference pattern, but the brightness of the reconstruction depends on the magnitude of the motion. When reconstructed, this time-averaged exposure reconstructs a bright image where the motion was a multiple of a full wavelength, and dimmer for intermediate motions. Similar observations had long been used by optical researchers studying classical interferometry.[6]

Leith and Upatnieks listed this potential application of using fringes as a measure of object motion in a patent disclosure the following April.[7] They were too busy with other work to follow up the discovery and did not mention it to their colleagues. In December 1964, though, two others in the Optics Group, Robert Powell and Karl Stetson, followed a similar path. They had used hot wax to fasten their holographic subject, another small metal train model, to a metal base. The result had been a reconstructed image obscured by black marks across part of the object.[8] After a couple of days' reflection, they realized that the metal base had cooled during the exposure causing contraction of the metal and yielding visible bands of interference. Powell later recounted the circumstances:

> By accident—fortuitously—we caused a strange, uniform set of 'visual obliterations' of the image—of the object. Our first reaction was annoyance at having screwed up an otherwise aesthetically satisfying image. On second thought, we cooled off, caught the idea to track-down, to re-introduce the 'visual obliterations' [and] found that they were *interference* fringes, that the recording apparatus must be some kind of interferometer.[9]

This was an important conclusion, but initially a confusing one. According to Stetson's later accounts, they had been exploring the limitations of their HeNe laser, which they knew produced such obliterations. The small dimension of the laser tube cavity meant that its coherence length was short, but periodic; as a result, it produced holograms that were unable to reconstruct images at certain depths. But, as they discovered, their latest holograms had *extra* dark areas:

> It was here that our magic looking glass showed something that was not there: an interference fringe between the two transverse modes of the laser that was not there in the actual laser beam. The only explanation was that the hologram had some unforeseen ability to make light

[5] Upatnieks, Juris to SFJ, email, 6 Jun. 2003, SFJ collection.

[6] See, for example, Dye, W. D., *Proceedings of the Royal Society* A 138 (1932): 1; Tolanski, S., *Surface Microtopography* (London: Longmans, 1960), Chapter 15. The use of interference fringes for metrology had been pioneered in Michelson, Albert A., and Edward W. Morley, 'On a method of making the wavelength of sodium light the actual and practical standard of length', *Philosophical Magazine* 24 (1887): 463–566.

[7] Leith, Emmett N. and Juris Upatnieks, 'Wavefront reconstruction disclosure', Patent disclosure report, University of Michigan Willow Run Laboratory, 22 Apr. 1964.

[8] Lindgren, Nilo, 'Search for a holography market', *Innovations* (1969): 16–27.

[9] Powell, Robert L. to P. Jackson, letter, 14 Oct. 1976, MIT Museum 30/850.

> beams coherent that were not originally coherent [. . .] suddenly, interferometric measurements could be made on non-mirror surfaces, that is, on structures of ordinary engineering interest.[10]

Concluding that instabilities of the object were responsible for the peculiar impact on the image, they judged that this would be a good way of measuring not merely gradual motion, but also *vibrations* of objects. They set up experiments to vibrate an object during the hologram exposure, and discovered that dark fringes occurred where the object had moved, and bright reconstructions occurred where the object had either remained motionless or moved by an integral number of wavelengths. Powell later described their work as having been ' "discovered" in a fit of serendipity'.[11] He and Stetson dubbed their technique *time-averaged holography* (see Figure 7.1).[12]

Intriguingly, this powerful method had been discovered independently twice in the same building. 'Powell and Stetson did not get the idea from us', Leith emphasized; 'they made the discovery themselves, as so many others did later'.[13] As was to be expected given the accidental nature of the discoveries, the experimental situation was not the simplest: rather than arranging for merely two distinct exposures (as would be common in classical interferometry), the two teams had stumbled upon a situation that superposed multiple images. To quantify the fringes they saw in vibrated objects, Powell and Stetson had to rely on Bessel functions, which were available from tables of scientific functions and employed in calculations via mechanical calculators or slide rules.

Yet this was not a rare case of rediscovery: it happened over and over again. Another pair of researchers in the Willow Run Laboratories (WRL) Optics Group, Bernard Percy Hildebrand and Kenneth Haines, were also working on holographic interference images during this period. The limited communication between Willow Run researchers—and conflicting benefits arising from discovery—is suggested by the subsequent priority claims of Hildebrand made when sifting through notebook records five years later. He recounted to patent attorneys that he had communicated his own ideas of holographic interferometry to Emmett Leith and others in the spring of 1965, not realizing that Stetson had recorded similar ideas in his own notebook two weeks earlier. Leith 'did not mention Karl's entry to me when he signed mine because he didn't want a conflict to arise. However, he foresaw the present problem, which of course doesn't help me now'.[14]

[10] Stetson, Karl A., 'The problems of holographic interferometry', *Experimental Mechanics* 39 (1999): 249–55, 250; Stetson, K. A., 'The origins of holographic interferometry', *Experimental Mechanics* 31 (1991): 15–8.

[11] Powell, Robert L. to S. Zussman, letter, 28 Mar. 1995, Leith collection.

[12] For example, Powell, Robert L. and Karl A. Stetson, 'Interferometric vibration analysis by wavefront reconstruction', *Journal of the Optical Society of America* 55 (1965): 1593–8; Stetson, K. A., 'A rigorous treatment of the fringes of hologram interferometry', *Optik* 29 (1969): 386–400, Powell, Robert L. and Karl A. Stetson, 'Some techniques for hologram interferometry of string instruments', presented at *79th meeting of the Acoustical Society of America*, Atlantic City, NJ, 1970.

[13] Leith, Emmett N. to SFJ, email, 5 Jun. 2003, SFJ collection.

[14] Hildebrand, B. P. to R. Washburn and Woodcock Phelan and Washburn patent attorneys, 2 Feb. 1970, Philadelphia, PA, Leith collection.

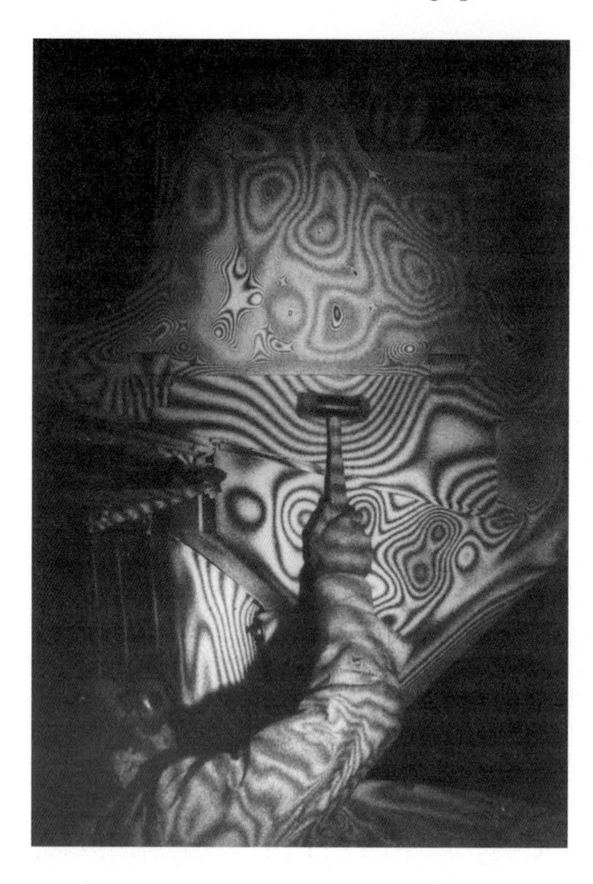

Fig. 7.1. Hammer blow, illustrating time-averaged holographic interferometry. Karl Stetson 1966 (MIT Museum MoH collection).

Stetson recalls similar contention between former WRL workers trying to apply the techniques commercially:

> Later, in 1970, when I was at the National Physical Laboratory in England, some documents came around regarding those patents which my former colleague, Robert Powell, was giving them quite a hard time over [. . .] The upshot of that argument was to separate the patents based on initial disclosures, and it is then that I learned that Emmett and Juris had filed a disclosure about six months earlier than our work proposing the use of holograms for recording the coherence of light and for recording vibration nodes of vibrating objects.[15]

As with the Leith-Upatnieks and Powell-Stetson teams, Hildebrand and Haines were discovering phenomena that they subsequently explained theoretically. They had fallen upon a distinct variant of holographic interferometry, which they called *contouring by wavefront reconstruction*.[16] In brief, they employed a laser that generated two closely

[15] Stetson, Karl A. to SFJ, email, 1 Nov. 2004, SFJ collection.

[16] Haines, Kenneth and B. P. Hildebrand, 'Contour generation by wavefront reconstruction', *Physics Letters* 19 (1965): 10–1; Hildebrand, B. P. and Kenneth Haines, 'Interferometric measurements using the wavefront reconstruction technique', *Applied Optics* 5 (1966): 172–3.

spaced wavelengths. Each wavelength produced a hologram on the same photographic plate and, when reconstructed by a *single*-frequency laser, yielded two images of subtly different scale that interfered with each other. If the object was motionless during the original exposure, those fringes mapped out spatial steps corresponding to the wavelength difference between the two laser emissions. Their method provided precisely measurable depth information about three-dimensional scenes. Haines, Hildebrand, and Stetson wrote their PhD theses on their subjects.[17]

Yet another example of the genesis of interferometric holography is that of Matt Lehmann at Stanford University:

> One of our very first holograms was of a business card with a few coins laid on it. When it was reconstructed the card had a series of black bands across it. We realized that these were caused because the card had moved during the exposure but gave no thought to the fact that we had discovered a technique of non-destructive testing. Our only thought was that the hologram was spoiled by the card movement.[18]

Other laboratories made the discovery of two further forms of hologram interferometry: the double exposure (or time-lapse) method and the real-time method. In the double exposure method, a single hologram plate recorded two *separate* exposures of an object. Between the exposures, the object would be stressed in some way: perhaps heated, inflated, or otherwise mechanically stressed to produce differential movement. When reconstructed, the two interfering images showed fringes of displacement, from which the mechanical characteristics of the object could be deduced. In the real-time method, the idea was essentially the same, except that only a single exposure was made and the holographic plate was returned to precisely its original position. When the object was subsequently perturbed with the processed plate in place, the reconstructed image would interfere with the real-time image and fringes would be seen when viewing through the hologram. A single exposure could then be used to produce an optical system to permit the real-time mechanical testing of the object. These 'live fringes' of *real-time holographic interferometry* allowed more versatile analysis than the 'frozen fringes' of the two-exposure technique, but developing and orienting the hologram was a problem.

In both these methods, the shifted object created multiple sets of interference fringes on the same hologram; when the two images were reconstructed, they interfered with each other. Like the other varieties of holographic interferometry, this was an optical situation that could not occur in classical interferometry: there, the interfering beams could be obtained only from specularly reflective objects. The new holographic

[17] Haines, Kenneth and B. P. Hildebrand, 'Contour generation by wavefront reconstruction', *Physics Letters* 19 (1965): 10–1; Haines, Kenneth, *The Analysis and Application of Hologram Interferometry*, PhD thesis, Electrical Engineering, University of Michigan (1966); Hildebrand, B. P. and Kenneth Haines, 'Interferometric measurements using the wavefront reconstruction technique', *Applied Optics* 5 (1966): 172–3; Hildebrand, B. P., *A General Analysis of Contour Holography*, PhD thesis, Electrical Engineering, University of Michigan (1967);Stetson, Karl A., *A Study of Fringe Formation in Hologram Interferometry and Image Formation in Total Internal Reflection Holograms*, PhD thesis, Institut för Optisk Forskning (1969).

[18] Lehmann, Matt, 'Holography's early days at Stanford University', *holosphere*, 11 (9) (Sep. 1982): 1, 4.

technique made any diffusely reflecting object a suitable candidate for measurements of motion. The value of such measurements was their exquisite precision and relative ease of analysis: the appearance of a dark fringe on the reconstructed image implied a movement of half a wavelength (or an integral multiple of half-wavelengths) in the direction of view—a motion of scarcely 300 nm.

The possibilities of holographic interferometry were taken up rapidly by researchers already well versed in classical interferometry.[19] At the Maser Optics section of the National Physical Laboratory (NPL) in Teddington, England, Jim Burch, Tony Ennos, and their colleagues explored holographic interferometry in these ways, after Jim Burch and a student interpreted one of their own accidents as an interferogram.[20] NPL researchers were quick to apply the effect to metrology. By recording a hologram of a prototype piston or automobile cylinder, for example, the reconstructed image could be used to interfere with subsequent production models and to reveal dimensional variations by the resulting interference fringes.[21]

The same domain was explored by John Butters at Loughborough University of Technology in England, after his group got started with help from Gordon Rogers.[22] Indeed, by 1968, British researchers organized the first truly international conference on the subject.[23] The NPL and Loughborough groups popularized holographic methods for the problems of mechanical engineering in Britain. Butters and co-workers collaborated, for instance, with Rolls Royce to develop methods of detecting defects in jet-engine turbine blades by observing anomalous fringe patterns.

Given this important application for mechanical engineering, techniques developed rapidly. At TRW Systems in Redondo Beach, California, Lee O. Heflinger, Ralph Wuerker, Robert Brooks, and their colleagues again discovered holographic interference independently, but by using a pulsed laser. Wuerker had developed the first pulsed ruby laser for holography in early 1965. At the time they had been recording stop-action holograms of bullets with a short pulse length ruby laser. During one exposure, the laser accidentally fired a double pulse, recording two closely spaced images of the bullet that

[19] In fact, it was soon realized that classical interferometry could be understood as a form of holography, with the holograms being interpreted directly rather than after reconstruction—an overcoming of contextual screening that was akin to Rogers' realization that the zone plate was a point hologram. See Bryngdahl, O. and Adolf W. Lohmann, 'Interferograms are image holograms', *Journal of the Optical Society of America* 58 (1968): 141–2.

[20] Ennos, A. E., 'Holographic techniques in engineering metrology', *Proceedings of the Institution of Mechanical Engineers* 183 (1969): 5–12; Ennos, A. E. and E. Archbold, 'Techniques of hologram interferometry for engineering inspection and vibration analysis', presented at *The Engineering Uses of Holography*, University of Strathclyde, Glasgow, 1968; Gates, J. W. C., 'Holography, industry and the rebirth of optics', *Review of Physics in Technology* 2 (1971): 173–91.

[21] Lindgren, Nilo, 'Search for a holography market', *Innovations* (1969): 16–27.

[22] Butters, J. N., 'Lasers in the visualization of surface strain and vibrations', *Proceedings of the Institution of Mechanical Engineers* 183 (1969): 67–74; Butters, J. N. and D. Denby, 'Some practical uses of laser beam photography in engineering', *Journal of Photographic Science* 18 (1970): 60–7; Denby, D. and J. N. Butters, 'Holography as an engineering tool', *New Scientist* 45 (1970): 394–6; Hockley, B. S. and J. N. Butters, 'Holography as a routine method of vibration analysis', *Journal of Mechanical Engineering Science* 12 (1970): 37–47.

[23] Robertson, Elliot R. and James M. Harvey (eds), *Symposium on Engineering Uses of Holography* (University of Strathclyde, 1968).

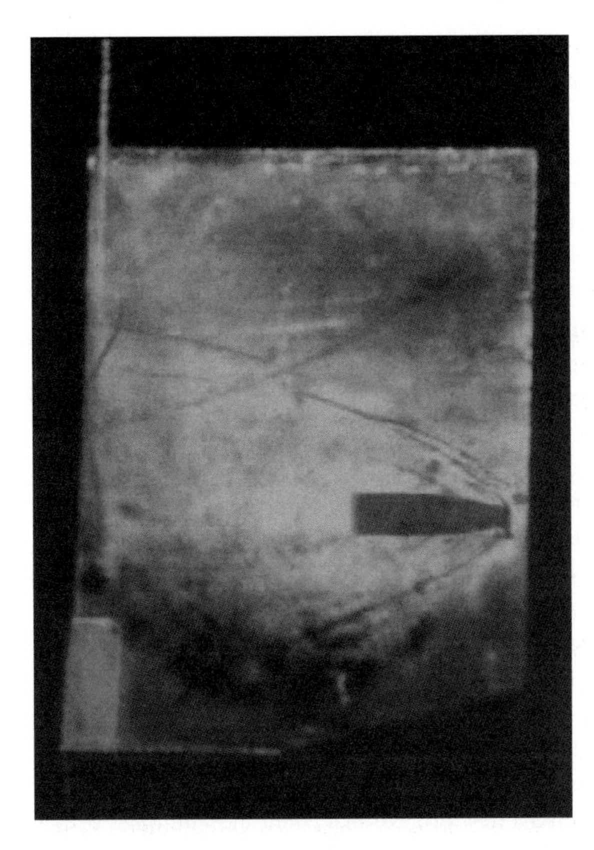

Fig. 7.2. Pulsed hologram of bullet in flight, Ralph Wuerker, 1966 (MIT Museum MoH collection).

interfered to yield dark fringes. As with Powell and Stetson, after musing about the causes they settled on optical interference as the explanation, and pursued the development of double-pulse interferometry.[24] This allowed them, for example, to capture the details of the shock waves produced by a bullet in flight (see Figure 7.2).

Wuerker also adapted the Haines-Hildebrand contouring idea, using two wavelengths of ruby to record the two holograms and replaying with one to reveal geometrical contours. Researchers at Bell Telephone Laboratories similarly pursued applications for electronics manufacturing problems.[25] In this atmosphere of intense research and institutional confidentiality, surprises were common. After the Bell researchers submitted their first paper on their techniques in August 1965, they learned that at least four other groups had papers out or in the publications pipeline.[26]

[24] E.g. Heflinger, L. O., R. F. Wuerker and R. E. Brooks, 'Holographic interferometry', *Journal of Applied Physics* 37 (1966): 642–9; Wuerker, R. F., 'Holography and holographic interferometry: industrial applications', *Annals of the New York Academy of Sciences* 168 (3) (1969): 492–505; Heflinger, L. O. and R. F. Wuerker, 'Holographic contouring via multifrequency lasers', *Applied Physics Letters* 15 (1969): 28–30.

[25] Kogelnik, H. W., 'Optics at Bell Laboratories—lasers in technology', *Applied Optics* 11 (1972): 2426–34.

[26] E.g. Nisida, M. and H. Saito, *Scientific Papers of the Institute of Physical and Chemical Research (Tokyo)* 59 (1965): 5; Collier, R. J. et al., *Applied Physics Letters* 7 (1965): 223; Stroke, George W. and A. Labeyrie, 'Two-beam interferometry by successive recording of intensities in a single hologram.' *Applied Physics Letters* 8 (1965): 42–4.

Between 1964 and 1966, then, complementary aspects of holographic interferometry were repeatedly rediscovered during exploratory experimental accidents in laboratory after laboratory. As Leith has observed, the original discovery was extremely improbable until the holography of reflecting objects became possible, but inevitable thereafter.[27] Indeed, accidental discovery was almost impossible to avoid, but to conceive of holographic interferometry from first principles in 1965 was unlikely, as suggested by the initial confusion of the various researchers.

In fact, presumptions had intimidated experimenters and hindered unpromising avenues of investigation. Would movement during exposure smear all the interference patterns on the hologram, and simply reduce the reconstruction brightness of the overall image? Would objects move as little as a few wavelengths of light, allowing them to be mapped by observable fringes? Variants of the technique were developed actively, beginning from distinct viewpoints and instrumental arrangements. Labs that had prior experience in classical interferometry, such as the NPL in England, had a head start in applying the technique to industrial applications.

The first applications, however, relied on mere detection of movement rather than on quantitative measurements. One such important and sensitive qualitative application was the detection of ply separation in pneumatic aircraft tires developed by Gordon Brown of GC Optronics in Ann Arbor (see Fig 6.6).[28] Such *holographic non-destructive testing*, or HNDT, became a rapidly growing field in its own right, and the most commercially important application of holography during the late 1960s and early 1970s.

It also created new communities of practitioners and customers. The techniques of holographic interferometry promoted a new professional connection between optical scientists and mechanical engineers, the primary users of the new technique. During the 1970s, for instance, a 'holography club' was organized by British researchers involved in interferometry at NPL, the National Engineering Laboratory in Glasgow (NEL), Rolls Royce and Loughborough University of Technology. HNDT proved particularly attractive to aircraft companies, owing to their use of light but highly stressed structures. The detection of fatigue cracks and the visualization of airflow were both amenable to the holographic method.

Not all laboratories found it sufficiently practicable, however. McDonnell Douglas dropped out of holographic testing in the early 1970s, finding that the chemical processing of photographic plates was not convenient for real-time engineering analysis. Others discovered that quantitative analysis was laborious and converting fringe patterns to

[27] Leith, Emmett N. to SFJ, email, 5 Jun. 2003, SFJ collection.

[28] For example, Brown, G. M., 'Holographic nondestructive testing (HNDT) of rubber–and plastics–containing products', presented at *22nd annual conference of electrical engineering problems in the rubber and plastics industries*, Akron, Ohio, 1970; Brown, G. M., 'A review of holographic nondestructive testing of pneumatic tires', *Materials Evaluation* 31 (1973): 37A; Grant, R. M. and G. M. Brown, 'Holographic nondestructive testing (HNDT)', *Materials Evaluation* 27 (1969): 79–84; Grant R. M. and G. M. Brown, 'Holographic nondestructive testing (HNDT) in the automobile industry', presented at *International Automotive Engineering Congress*, Detroit, Michigan, 1969; Brown, G. M., R. M. Grant, and G. W. Stroke, 'Theory of holographic interferometry', *Journal of the Acoustical Society of America* 45 (1969): 1166–79.

calculations of object motion was not straightforward.[29] As Jim Trolinger recalled concerning studies of turbine blades, 'holography was always difficult to employ in such studies and we could never seem to hand off the measurements to others'.[30] Interest continued in Japan, however, where all major car companies eventually employed HNDT. Ford, in America, and Volkswagen, in Germany, also remained faithful users, as did Peugeot and Renault in France extensively during the 1970s and 1980s.[31]

These early applications of holography had little to do with its origins in Synthetic Aperture Radar (SAR) development or with three-dimensional displays. Yet a few laboratories around the world—including the Willow Run Optics Group in the mid-1960s, the Japanese laboratories of Jumpei Tsujiuchi, R. Hioki, and T. Kubota, and the Swedish workers Nils Abramson and Hans Bjelkhagen, for example—for a time engaged in both optical measurement and display holography. The technique proved popular with physicists, chemists, and engineers of every flavour.[32] Most laboratories soon specialized in one or the other domain, however. This bifurcation of interests was the first important splitting the subject, and typical of the subsequent emergence of subcultures of holography discussed in Part III.

Although its adopters saw it as a prosaic and economically valuable application, holographic interferometry also generated at least temporary awe among seasoned optical workers. Robert Powell wrote in 1968:

> What seems to me the most exciting value of this technique [. . .] has to do with the three-dimensionality. When I first had occasion to see Dr Wuerker's holograms I was bowled over just

[29] Vest, Charles M. to SFJ, interview, 13 Jul. 2003, Cambridge, MA, SFJ collection. While double-exposure and real-time holographic interferometry were analytically simpler than the time-averaged variety, all were difficult to analyse quantitatively, because the appearance of the fringes depends on optical orientation and demands multiple viewpoints for full understanding. An attempt at automating the exposure process, at least, was the Newport Corporation HC-1000 Instant Holographic Camera introduced in 1981, which used reusable thermoplastic film to record an interferogram and to compare it with a test object via video monitor.

[30] Trolinger, James D., 'The history of the aerospace holography industry from my perspective', unpublished manuscript, May 2004, SFJ collection.

[31] Fagan, William to SFJ, interview, 19 Sep. 2002, SFJ collection; Smigielski, *Holographie Industrielle* (Paris: Teknea, 1994). Large corporations such as Aerospatiale, Renault, Office National d'Etudes et de Recherches Aerospatiales (ONERA), Michelin, Snecma, and others in France increasingly adopted holography for specific problems in structural noise, faults in composite materials, measurement of dynamic deformations, fatigue testing, and validation of computer modelling.

[32] Early examples from among thousands of conference presentations and published papers on the art and application include Tsujiuchi, J., N. Takeya, and K. Matsuda, 'Measurement of the deformation of an object by holographic interferometry', *Optica Acta* 16 (1969): 707–20; Yamaguchi, I. and H. Saito, 'Application of holographic interferometry to the measurement of Poisson's ratio', *Japanese Journal of Applied Physics*8 (1969): 768–71; Becsey, J. G., G. E. Maddux, N. R. Jackson, and J. A. Bierlein, 'Holography and holographic interferometry for thermal diffusion studies in solutions', *Journal of Physical Chemistry* 74 (1970): 1401–3; Fink, W., P. A. Buger, and L. Schepens, 'Rock probe deformation measured by holographic interferometry', *Optics Technology* 2 (1970): 146–50; Jansson, E., N. E. Molin, and H. Sundin, 'Resonances of a violin body studied by hologram interferometry and acoustical methods', *Physica Scripta* 2 (1970): 243–56; Klimenko, I. S., E. G. Matinyan, and G. I. Rukman, 'Double-exposure holographic interferometry with reconstruction in white light', *Optics and Spectroscopy* 29 (1970): 85–8; Nicholson, J. P., A. F. Hogan, and J. Irving, 'Electron-density profiles of a theta-pinched plasma by holographic interferometry', *Journal of Physics D (Applied Physics)* 3 (1970): 1387–91; Abramson, N., 'Moiré patterns and hologram interferometry', *Nature (Physical Science)* 231 (1971): 65–7.

by the aesthetic view of the engineering information and its three-dimensional display, and the extent to which one can peruse at length this stored bit of three-dimensional information; this is most dramatic.[33]

7.3 THE AMERICANIZATION OF REFLECTION HOLOGRAPHY

In the scramble to discover and publish, cases of repeated discovery were common in holography during the late 1960s. Another case is the so-called *volume hologram*—that is, the formation of interference fringes through the depth of a thick emulsion, which had been the chief characteristic of Denisyuk's technique of wave photography. Some of the underlying ideas were discussed independently by Pieter van Heerden of General Electric in 1963, but made little impact at the time. Van Heerden did not conceive a three-dimensional imaging technique and could not then demonstrate experimental confirmation of his ideas.[34]

Denisyuk's earlier work, first appearing in English translation in late 1963, went similarly unnoticed by American researchers.[35] During the subsequent two years, researchers around the world focused on the Leith–Upatnieks technique and its extensions. Not until 1965 did these experimental investigations lead workers at several sites to rediscover independently their first Denisyuk type holograms. Leith and Upatnieks themselves mused, for example, that their off-axis reference beam might be more conveniently arranged to strike the holographic plate if it came from behind it, rather than being occluded by the object itself or other components. For this pragmatic reason, they tried this geometry in October 1965. Upatnieks recalls:

> I think it was somewhat of an accident: while carrying a developed off-axis reference beam hologram from our photographic processing laboratory in one building to our optics lab in another, I noticed a reflected image generated by sunlight falling on the plate [. . .] The image was neither very bright nor sharp, so I was not impressed.[36]

[33] Powell, Robert L., 'Discussion', in: E. R. Robertson and J. M. Harvey (eds), *The Engineering Uses of Holography* (Glasgow, 1968), pp. 128–9, pp. 128–9.

[34] van Heerden, Pieter J., 'Theory of optical information storage in solids', *Applied Optics* 2 (1963): 393–400. Van Heerden discussed not a three-dimensional imaging technique, but a principle for constructing what could be called a holographic memory, a philosopher's stone for many subsequent holographers. Finding his ideas poorly received at GE, he moved to Polaroid in Cambridge, MA, and later worked on other holographic applications.

[35] However, Denisyuk's 1963 paper [Denisyuk, Yu N., 'On reflection of the optical properties of an object in wavefield of radiation scattered by it', *Optika i Spektroskopija* 15 (1963): 522–32] was cited in the Leith and Upatnieks patent disclosure (Leith, Emmett N. and Juris Upatnieks, 'Wavefront reconstruction disclosure', Patent disclosure report, University of Michigan Willow Run Laboratory, 22 Apr 1964).

[36] Upatnieks, Juris to SFJ, email, 27Jul. 03, SFJ collection.

Nevertheless, all their previous holograms had shown only a smeared spectrum of colours when viewed in sunlight or under another white light source. Moreover, those images were reconstructed by transmission of light *through* the plate, not reflected from it. After further examination, Leith recalls, 'we were all terribly excited: it was amazing. Then it dawned on me that this was exactly what Denisyuk was saying'.[37] It was nevertheless a tardy realization: their experimental discovery occurred at least eighteen months after Leith and Upatnieks had read Denisyuk's original paper in translation, and more than four years after Denisyuk had produced such holograms using a mercury arc source.

At least two other American groups discovered the same technique within weeks or months of each other. When Leith reported the new kind of hologram to one of their principal sponsors, the Holotron Corporation, he found that they already knew of the technique, but were seeking a patent and so had not reported it. Nile Hartman and C. Schwartz at the Battelle Memorial Institute (then a part-owner of Holotron) had themselves discovered the technique accidentally in August 1965.

Not all laboratories stumbled upon Denisyuk's method entirely by accident. Around the same time, French researcher Antoine Labeyrie was working in George Stroke's Electro-Optical Sciences Lab at the University of Michigan, supported by a Fulbright research fellowship from September 1965 until the following June.[38] He recalls:

> A few weeks after arriving at Ann Arbor, I remembered the Lippmann effect, which I had learned the previous years, and I wondered whether it could be combined with holography. A quick and simple reasoning convinced me that it could indeed be combined by propagating the reference beam towards the back side of the plate. In a thick emulsion, with fine grain, this obviously had to produce a Lippmann stratification within each speckle domain, and thus a reflective reconstruction of the recorded wavefront, with the same wavelength selectivity demonstrated by Lippmann. The same day, I was able to build a quick-and-dirty 'transparent' plate holder, to make a recording set-up, and to make successful exposures.[39]

Later, student Arthur Funkhouser and George Stroke told him about Denisyuk's paper. Labeyrie then realized that Denisyuk's model covered the same geometry, although not with a separately directed skew reference beam behind the photographic plate. Labeyrie concluded that Denisyuk was 'really the inventor'.[40] Stroke presented the results to his National Science Foundation sponsors in early December. In late January 1966, Stroke and Labeyrie submitted a joint paper describing these new holograms that could be viewed in ordinary sunlight or an incandescent light bulb.[41]

[37] Hecht, Jeff, 'Applications pioneer interview: Emmett Leith', *Lasers & Applications*, 5 Apr. 1986: 56–8.

[38] Antoine E. Labeyrie had completed an engineering degree at the Institut d'Optique in Paris, and the *license en physique* at the Université de Paris (Sorbonne). See also §7.3.

[39] Labeyrie, Antoine E. to SFJ, email, 2 Jun. 2003, SFJ collection.

[40] Labeyrie, Antoine E. to SFJ, email, 22 Jul. 2003, SFJ collection. Nevertheless, as the first documented inventors to describe a holographic image reconstructed with white light by recording a hologram with a large angle between the sample and reference beams, Hartman and Schwartz were awarded the American patent.

[41] Stroke, George W. and Antoine E. Labeyrie, 'White-light reconstruction of holographic images using the Lippmann-Bragg diffraction effect', *Physics Letters* 20 (1966): 368–70.

So, this striking variant of holography was rediscovered between August and October 1965 by three independent groups (although in close geographic proximity and with interacting personnel), a full two years after the first leakage of news about three-dimensional holography and eighteen months after Denisyuk's work appeared in English.

To some extent, this was more than a rediscovery of Denisyuk's technique, because it recognized that reflection holograms were viewable in white light. Denisyuk's publications had hinted at, but not highlighted, this fact. White-light reconstruction, hitherto unappreciated outside Leningrad, was exactly the new development to attract media attention, and George Stroke publicized it as a major development from his Electro-Optical Sciences Lab. His press release described his lab's white-light reflection holograms as 'comparable to a successful Apollo moon shot in this field' and 'based on a hair-thin balance between sophisticated mathematics and extremely refined experimentation'.[42] In much later recollections, he vaunted the achievement as a cold war victory, having prevented 'making holography into a Sputnik'.[43] And with Lawrence Lin and Keith Pennington at Bell Laboratories, Stroke soon extended the method to three-colour holograms, also touted as a breakthrough.[44]

The new holograms were labelled *Lippmann* or *Lippmann-Bragg* holograms by most American workers, thus denying Denisyuk a direct link in the family tree of the subject just then sprouting.[45] Thus, reflection holography became another weapon in the ongoing inter-laboratory rivalry, priority and patent disputes, and a tool in contesting the history of the new science.

7.4 IMAGE-PLANE HOLOGRAMS

The publicity surrounding reflection holograms was not sustained. Such holograms proved difficult to record because of their very fine fringe patterns, which were easily disturbed by vibration or temperature change during exposure. They were also rather dim, because Kodak 649F, the only suitable photographic emulsion, was not as thick or fine grained as the Soviet emulsions developed by Denisyuk and Protas. Nevertheless, the great convenience of white-light reconstruction, which liberated the display of holograms from expensive lasers and awkward monochromatic lamps, was commercially

[42] Stroke, George W., 'Press Release: Breakthrough in "lensless photography" sets stage for 3-dimensional home color television and a possible multi-million industrial explosion in electro-optics', 12 Mar. 1966, Leith collection.

[43] 'We beat the Russians by three months [. . .] We kept them from making holography into a Sputnik'. [Aspray, William, 'George Wilhelm Stroke, Electrical Engineer, an oral history conducted in 1993 by William Aspray, IEEE History Center, Rutgers University, New Brunswick, NJ, USA.' (1993)].

[44] Lin, R. H., K. S. Pennington, G. W. Stroke and A. Labeyrie, 'Multi-color holographic image reconstruction with white-light illumination', *Bell System Technical Journal* 45 (1966): 659–61.

[45] Somewhat later, the terms *Lippmann-Denisyuk* and *Denisyuk* holograms gained currency, but the more generic *volume hologram* or *reflection hologram* became the most common labels in the West.

exciting. Conductron, for example, began investigating reflection holograms in 1966 and was producing quantities in the hundreds for advertising clients by 1968. Conductron's enormous half-million run of Leith-Upatnieks (transmission) holograms in 1967 had made a profit, but only temporarily impressed the firm. The hologram required a red filter for viewing with white light, and the result was a dim, fuzzy image. As a result, the holography group eagerly explored reflection holograms. Wayne Dwyer worked on the development of thick, fine-grained colloidal emulsions, and Dick Zech, Stroke's former student, focused on mass production of reflection holograms. The two developed a hologram printer to record reflection holograms using images from an original master transmission hologram [46]

Before the exploration of reflection holograms had got underway, though, another variant appeared. During 1966 several groups reported forming holograms that could be viewed in white light. They were relatively easy to produce with different arrangements of mirrors and lenses, and had their own distinct collection of properties to be explored. As with the discoveries surrounding holographic interferometry a year earlier, developments occurred for several groups and the results were reported in a flurry of letters to physics journals.

From the Department of Aeronautical Engineering at the Queen's University, Belfast that February came the news that, if an image of an object illuminated with laser light were *focused* on the holographic plate, the resulting reconstruction would be rather like a photograph having a subtle third dimension. The resolution requirements were lower than in the Leith–Upatnieks type of hologram and the image could be viewed in incoherent (white) light.[47] This seemed to question many of the new orthodoxies assumed by holography researchers, but nevertheless made relatively little impact.

In March, two University of Michigan researchers reported producing a hologram that formed an image *in front* of the photographic plate (i.e. producing a focused real image) by recording a hologram of a hologram.[48] Groping their way with this new extension, they assured readers that this second-generation hologram was not significantly poorer in resolution and that an observer could interact with the suspended image in new and useful ways.[49]

The ability to make a hologram of a focused image was also reported that autumn by Lowell Rosen at the NASA Electronics Research Center in Cambridge, Massachusetts.[50]

[46] Charnetski, Clark to SFJ, interview, 3 Sep. 2003, Ann Arbor, MI, SFJ collection. Their biggest customer, La Roche Pharmaceuticals, ordered hundreds of custom reflection holograms, and others were produced to promote Conductron to investors. One reconstructed the image of a marching band, and was bound into the Nobel Prize volume alongside Gabor's printed acceptance speech (Gabor, Dennis, 'Holography, 1948–1971', in: Nobel Prize Committee (ed.), *Les Prix Nobel en 1971* (Stockholm, 1971), pp. 169–201).

[47] Tanner, L. H., 'On the holography of phase objects', *Journal of Scientific Instruments* 43 (1966): 346.

[48] Gordon Rogers had made such a hologram copy in the early 1950s, but his further research was seriously constrained by the quality of in-line holograms.

[49] Rotz, F. B. and A. A. Friesem, 'Holograms with nonpseudoscopic real images', *Applied Physics Letters* 8 (1966): 146–8.

[50] Rosen, L., 'Hologram of the aerial image of a lens', *Proceedings of the IEEE*54 (1966): 79–80. The Electronics Research Center (ERC, 1964–70), which included Space Optics among its ten labs, developed NASA expertise in holography during the Apollo era.

He had produced a focused-image hologram that June, but reconceived the technique two months later after conversations with George Stroke as a method that would allow the image to *straddle* the photographic plate, producing the unintuitive result of an image that was part real and part virtual. He portrayed this new *image-plane* hologram geometry as an improved form of lensless photography.[51] Almost in passing, the authors noted that this arrangement could allow an image to be reconstructed using white-light illumination. George Stroke published a paper on the white-light reconstruction of Gabor *in-line* holograms in late October, and within a few weeks Rosen also reported that his focused images appeared progressively sharper as they approached the holographic plate.[52] Early the following year, others extended the technique to reflection (Denisyuk) holograms,[53] but now vaunting the ability to reconstruct images with white-light extended sources. As with earlier developments, each laboratory reported a bewildering variety of new phenomena, but framed individualistically in terms of research allegiances and familiar analogies.

So 1966 was the year of (white) light: the year of realization for holographers that white-light reconstruction was possible for reflection holograms and for image-plane holograms. The new techniques considerably eased the demanding illumination needed for viewing. Holograms could leave the laboratory and be displayed in galleries, schools, and foyers, breaking free of their nurturing community to be adopted by new ones. Coming a mere two years after the Leith–Upatnieks revelation, holography appeared to be progressing rapidly.

7.5 360° HOLOGRAMS

Holographic interferometry, Denisyuk reflection holograms, and image-plane holograms were all unintuitive varieties, discovered as much by chance as by intention by Western researchers. Each variant generated excitement for the growing holography community. But not all exciting holograms were dramatic extensions of the art. Among the most compelling holograms, for instance, were those that provided all-round views.

The first one-page description of such a hologram appeared in print in a Japanese journal in 1965.[54] The idea was an extension of the Leith–Upatnieks hologram: instead of a flat-surfaced photographic emulsion, the hologram was formed on a film curved into a cylinder. The reference beam was arranged to enter the cylinder from below and to be

[51] Kock, W. E., L. Rosen, and G. W. Stroke, 'Focused-image holography—a method for restoring the third dimension in the recording of conventionally-focused photographs', *Proceedings of the IEEE* 54 (1966): 80–1.

[52] Rosen, L., 'Focused-image holography with extended sources', *Applied Physics Letters* 9 (1966): 337–9; Stroke, George W., 'White-light reconstruction of holographic images using transmission holograms recorded with conventionally-focused images and 'in-line' background', *Physics Letters* 23 (1966): 325–7.

[53] Brandt, G. B. and A. K. Rigler, 'Reflection holograms of focused images', *Physics Letters* 25A (1967): 68–9; Brandt, G. B., 'Image plane holography', *Applied Optics* 8 (1969): 1421–9.

[54] Hioki, R. and T. Suzuki, 'Reconstruction of wavefronts in all directions', *Japanese Journal of Applied Physics* 4 (1965): 816; Supertizi, E. P. and A. K. Rigler, *Journal of the Optical Society of America* 56 (1966): 524.

directed onto its inner surface by reflection from a spherical lens. Most of the light passed through the lens to strike the object sitting on it. When the processed hologram was illuminated with a reference beam having the same geometry, the reconstructed image could be viewed over a 360° angle. While this was a minor technical extension, it dramatically transformed viewers' perceptions of the hologram. Now, instead of being a small window showing only a limited amount of parallax, the hologram was transformed into cylindrical volume in which all sides of an object could be seen. That space was now accessible, and paradoxically intangible: the viewer could reach into the cylinder, but found nothing to touch.

This compelling type of hologram was further popularized by physicist Tung Jeong, working with his undergraduate students at Lake Forest College in Illinois (just as Albert Baez had studied in-line holography with his students at Redlands University in California over a decade earlier). He described simple methods of recording all-round holograms from a set of four flat photographic plates and, later, a variant of the cylindrical geometry.[55] A series of further variations appeared in print, illustrating not only the visual appeal of the technique, especially for undergraduate instruction, but also the rush by individual researchers to make a contribution to the rapidly evolving subject.[56] Here again, Conductron Corporation, quick to exploit the potential for astonishing its investors, scaled up the technique to create larger and more impressive demonstrations. In early 1967, Gary Cochran and his colleagues recorded a large cylindrical hologram about 30 cm in diameter of a Buddha statue (a humorous dig at their generously proportioned Director, Kip Siegel). This hologram, like many other Conductron products, was shown to potential investors but not to the public or scientists; no papers were written. As Cochran emphasized, 'Conductron wanted to impress people with money'.[57]

None of these variants of the 360° hologram was a dramatic extension of the state of the art. Nor were they commercially useful, because a special lighting arrangement was needed, and the enclosure was even bulkier than the object that it reconstructed. But such holograms proved unexpectedly popular with scientists and the general public, probably because they once again accentuated the difference between holography and photography. Holography was not merely three-dimensional, but potentially all-encompassing—both figuratively and literally—in ways that photography was not. 360° and image-plane holograms were both fascinating because they seemed more interactive than the original Leith–Upatnieks type: viewers could get at, and pass their hands through the images, highlighting their tenuous mystique.

[55] Jeong, T. H., P. Rudolf and J. Luckett, *Journal of the Optical Society of America* 56 (1966): 1263; Jeong, T. H., *Journal of the Optical Society of America* 57 (1967): 1396.

[56] Annulli, R. J. and J. T. Ziewacz, 'Single-beam 360 degrees holograms', *American Journal of Physics* 45 (1977): 493–4; Hsue, S. T., B. L. Parker, and M. Monahan, '360 degrees reflection holography', *American Journal of Physics* 44 (1976): 927–8; Murata, K. and K. Kunugi, 'Cone-shaped cover for 360 degrees holography', *Applied Optics* 16 (1977): 1798–800; Stirn, B. A., 'Recording 360 degrees holograms in the undergraduate laboratory', *American Journal of Physics* 43 (1975): 297–300.

[57] Cochran, Gary D. to SFJ, interview, 6 and 8 Sep. 2003, Ann Arbor, MI, SFJ collection.

7.6 PULSED HOLOGRAMS

Lasers had imposed a severe constraint on the subject matter for holograms. Objects had to remain motionless for seconds, or even minutes, during the exposure. This was a serious limitation far beyond the conditions needed for the Victorian daguerreotype, where seated positions and metal braces could keep sitters adequately still. For a holographic exposure, objects had to remain motionless to a fraction of a wavelength. A human would be recorded not as a smear, but as a silhouette.[58]

The helium–neon laser, which became ubiquitous in laboratories of the mid-1960s, generated a continuous and constant beam of light (*continuous wave* (CW) operation). The first lasers to be developed, however, were the ruby laser, which generated intense and exceptionally brief pulses of light. Such a laser could be adapted to act essentially as a flash lamp to record not just static holographic scenes, but rapidly changing or even living subjects, too. Pulsed exposures also dramatically reduced the need for optical stability. Heavy, vibration-isolated optical tables were no longer mandatory.

Pulsed lasers for holography were developed at several centres during the late 1960s. The best known worker was Larry Siebert, who joined the Conductron Corporation in 1966 and developed the technology with colleague Richard Zech.[59] McDonnell Douglas had bought the company the previous year, principally because of Conductron's work in side-looking radar. Like colleagues who had joined earlier, Siebert found that he had a free hand with holography, and was able to work unimpeded on projects that impressed visitors. The founder of Conductron, Kip Siegel, had been fascinated by the concept of holographic movies, and pressed for the development of pulsed lasers.

There were at least two distinct technical challenges: first, to develop a pulsed laser having adequate optical coherence. The first ruby lasers had a coherence length of centimetres, which limited the depth of scenes that could be recorded, because deeper parts of the scene would reflect light no longer coherently related to the reference beam.[60] The second challenge was the available recording materials, which were in some cases relatively insensitive to the deep red colour of the ruby pulse, or less sensitive to

[58] Such silhouette portraits of animate objects were recorded using continuous-wave lasers by the late 1960s (Redman, J. D., C. J. Norman, and W. P. Wolton, 'Holographic reconstruction of animate objects', *Nature* 222 (1969): 476–7), and by 1970 by artist Margaret Benyon (who referred to them as 'non-holograms' 'to express the utter lack of presence in the solid hole that is left by movement' (Benyon, Margaret to SFJ, email, 4 May 2005, SFJ collection)). In 1980 artist Rick Silberman was able to hold a complete exhibition of what he called 'shadowgrams' at the New York Museum of Holography. Such silhouettes were unlike the popular vignette of the eighteenth century in that they were three-dimensional: a brightly-lit and stable background was occluded by a three-dimensional black object.

[59] Siebert, L. D., 'Large-scene front-lighted hologram of a human subject', *Proceedings of the IEEE* 56 (1968): 1242–3; Zech, R. G. and L. D. Siebert, 'Pulsed laser reflection holograms', *Applied Physics Letters* 13 (1968): 417–18.

[60] Siebert, L. D., 'Coherence length curve for ruby oscillator-amplifier', *Applied Physics Letters* 16 (1970): 318–20; Siebert, L. D., 'Holographic coherence length of pulse laser', presented at 1970; Siebert, L. D., 'Holographic coherence length of a pulse laser', *Applied Optics* 10 (1971): 632–7.

brief pulses than to lower-power continuous illumination.[61] Siebert and Zech began by recording scenes that demanded little coherence: shallow sprays of aerosols, for example, creating 2 in. square holograms on acetate film.

Siebert spent his first six months at Conductron modifying a commercial Korad (Union Carbide) ruby laser to increase its coherence length. The original laser was far from suitable. It was capable of a single pulse every 2 to 3 min, after charging its capacitors to store sufficient energy for a strong pulse. The laser employed ⅜ or ½ in. diameter ruby rods, which were not entirely uniform and so generated optical distortions and multiple optical modes. The pulse had to be limited to a single optical mode to make it capable of interference over reasonable path lengths, so Siebert increased the power supply and had the beam pass through more ruby rods to amplify the optical power. Over a period of about six months, he added two such optical amplifiers, reduced the beam aperture to a couple of millimetres, interposed optical etalons to select one optical mode, and laboriously aligned the system. He was eventually able to tune the laser to generate some two or more joules of energy in the pulse, which was sufficient to illuminate a moderate volume and record a hologram on the insensitive Kodak and Agfa Gavaert holographic emulsions (subsequent pulsed lasers for holography have ranged from 0.2 to more than 10 joules).

One of Siebert's first holograms of a live subject was of his own hand. Several months later, after further tweaking to increase the pulse power to some 10 joules, he made the leap of recording a hologram of his face and then his upper body posing beside an oscilloscope. Although he had calculated that the pulse should not cause eye damage if properly diffused, he recalls, 'It was very late at night; I was alone and very excited'.[62] The first human portrait by pulsed laser was Siebert's self-portrait on Halloween night, 1967 (see Figure 7.3). Soon afterwards, he recorded a pulsed reflection hologram of Richard Zech,[63] and the company publicized the success for portraiture and medical imaging.[64]

Other important early centres for 'impulse laser' holography were TRW in California, the NPL in Britain by John Gates and his colleagues, and the Vavilov Institute under Yuri Denisyuk and his student Dmitri Staselko (see Figure 7.4).[65]

[61] This differing sensitivity to short and long exposures had been known as 'reciprocity failure' since its discovery by late nineteenth-century photographers.

[62] Goldberg, Shoshanah, 'Conductron and McDonnell Douglas', *holosphere* 15 (1987): 16–8; Siebert, L. D., 'Large-scene front-lighted hologram of a human subject', *Proceedings of the IEEE* 56 (1968): 1242–3.

[63] Zech, R. G. and L. D. Siebert, 'Pulsed laser reflection holograms', *Applied Physics Letters* 13 (1968): 417–18.

[64] For example, Zech, R. G., L. D. Siebert, and H. C. Henze, 'A new holographic technique for medical and biomedical applications', presented at *7th Annual Biomedical Sciences Instrumentation Symposium on Imagery in Medicine.*, Pittsburgh, 1969; Ansley, D. A. and L. D. Siebert, 'Portrait-holography by impulse lasers', *Laser* 1 (1969): 29–34.

[65] See, for example, Rogero, S., B. J. Mathews, and R. F. Wuerker, 'Pulsed laser holography—new instrumentation for use in the investigation of liquid rocket combustion', presented at *15th International ISA Aerospace Instrumentation Symposium*, Pittsburgh, 1969; Wuerker, R. F., L. O. Heflinger, R. E. Brooks, and C. Knox, 'Action holography', presented at *Northeast Electronics Research and Engineering Meeting*, Boston, 1968; Gates, J. W. C., R. G. N. Hall, and I. N. Ross, 'Holographic recording of rapid transient events and the problems of evaluation of the reconstructions', presented at *Eighth International Congress on High Speed Photography.*, Nat. Physical Lab. Teddington UK, 1968; Hall, R. G. N., J. W. C. Gates, and I. N. Ross, 'Recording rapid sequences of holograms', *Journal of Physics E (Scientific Instruments)* 3 (1970): 789–91; Staselko, D. I., V. G. Smirnov, and N. Denisyuk Yu, 'On obtaining holograms of an alive diffuse object with

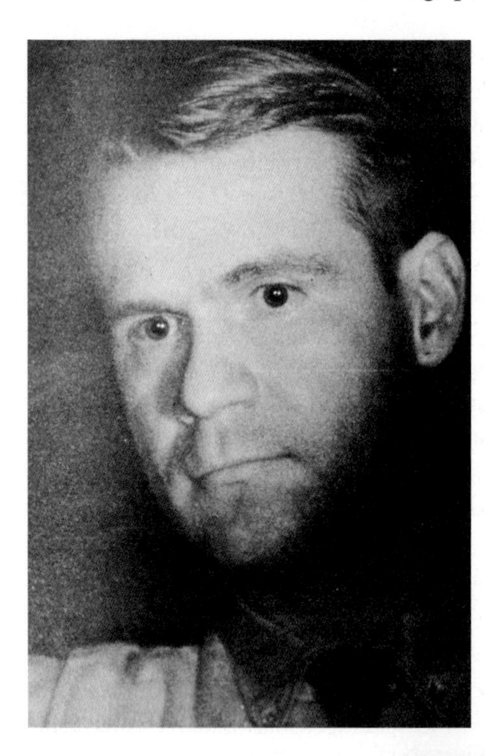

Fig. 7.3. First pulsed hologram of a human subject: Larry Siebert self-portrait, Conductron, 31 Oct. 1967 (Siebert, 'Large-scene front-lighted hologram of a human subject' (1968) 1242–3; Zech and Siebert, 'Pulsed laser reflection holograms' (1968), 417–18)

Fig. 7.4. Soviet pulsed hologram: Dmitri Staselko and A. Smirnov, 1970 (Staselko, Smirnov and Denisyuk, 'On obtaining holograms of an alive diffuse object with the help of single-mode ruby laser' (1968), 135–6).

But pulsed lasers remained expensive, uncommon, and temperamental. Only a handful of well-funded laboratories could afford them, and the early holographic community made do with continuous wave (CW) lasers for another decade. This is well illustrated by the peripatetic pulsed laser developed by Larry Siebert. Siebert had developed his laser at

the help of single mode ruby laser', *Zhurnal Nauchnoi i Prikladnoi Fotografii i Kinematografii* 13 (1968): 135–6; Staselko, D. I., A. G. Smirnov, and N. Denisyuk Yu, 'Production of high-quality holograms of three-dimensional diffuse objects with the use of single-mode ruby lasers', *Optics and Spectroscopy* 25 (1969): 505–7; Staselko, D. I., Yu N. Denisyuk and A. G. Smirnov, 'Holographic portrait of a man', *Zh nauch i prikl photogr i kinematogra.* 15 (1970): 147.

Conductron in Ann Arbor, but it was moved to Missouri (and reinstalled by Siebert) when McDonnell Douglas (MD) relocated the holography operation. Peter Nicholson, an artist interested in using pulsed holography, heard of the plans by MD to close its holography laboratory in 1974, years after Siebert's involvement with holography had ceased.[66] With David Challinor, Assistant Secretary for Science at the Smithsonian Museum in Washington, Nicholson argued that the equipment was a national heritage to be preserved. MD responded by turning over the equipment not otherwise needed by the company.

By 1975 the equipment had migrated to Brookhaven National Laboratory, NY, supported by seed money from the Kaplan Fund for artwork linked to the American bicentennial celebrations, and the cooperation of the chairman of Brookhaven.[67] Over the next two years, with the assistance of Will Walter, Bill Molteni, and Bruce Goldberg, the laser was reconstructed and used again.[68] Not until September 1977, though, was the pulsed laser refined adequately for 'push button operation'.[69]

From 1979 the laser had moved again with Nicholson to the University of Hawaii at Manoa, Honolulu, with State sponsorship. Ana Maria Nicholson, Peter's wife, began to create pulsed portraits with it. When funding ceased, the laser was returned to Brookhaven in 1980.[70] At that time, only a handful of pulsed lasers were in operation for holography—apart from Nicholson, facilities operated by Nick Phillips at Loughborough University of Technology, Hans Bjelkhagen and Per Skande at Holovision in Stockholm, and the holographers at the All-Union Center for Photographic and Cinematographic Research (NIKFI) in Moscow.[71]

7.7 RAINBOW HOLOGRAMS

Another significant innovation—although one that made only a gradual impact—was the rainbow hologram. In 1968, Stephen Benton was a part-time Polaroid employee and Assistant Professor of Physics at Harvard. He was focusing on two research topics:

[66] Peter Nicholson had met Lloyd Cross in New York in 1970, and they had obtained a joint grant from the New York State Council on the Arts to develop holography and art. Nicholson travelled with Cross to Sidonna, Arizona to set up a holography facility. There, it seems, Nicholson's ideas about pulsed holography, and Cross's on integral holography, developed. Five months later, the entourage moved to San Francisco, where Cross and Pethick formed the San Francisco School of Holography. Nicholson and his wife Ana Maria later returned to New York.

[67] Alice Kaplan, an art collector, sponsored holographic art at this time.

[68] William J. (Bill) Molteni, Jr studied physics at Monmouth College, NJ. He learned holography at the San Francisco and New York Schools of Holography, teaching at the latter for several years. Subsequently he became Director of R&D at Holographic Film Company, NYC, developing colour holograms and flat holographic stereograms with Dominick Di Bitetto. He later taught holography at Goldsmith's College, London.

[69] 'Center for Experimental Holography completes first phase of research', *holosphere*, 7 (1), 4–5; Bjelkhagen, Hans I., 'Holographic portraits made by pulse lasers', *Leonardo* 25 (1992): 443–8.

[70] 'Talk of the Town', *New Yorker*, 24 Aug. 1987; Nicholson, Ana Maria to SFJ, interview, 21 Jan. 2003, Santa Clara, CA, SFJ collection.

[71] 'Nicholson to offer pulsed holograms to advertisers/sales promoters', *holosphere* 9 (1980): 1–2.

(1) information reduction, which might allow a hologram to be transmitted electronically and thus allow 3-D television or video; and (2) image plane holograms, which appeared to be a simple implementation suited to commercial displays.

Information reduction was being pursued seriously by electronics companies, which sought a practical way of transmitting holographic still images or video. As Leith had noted publicly in 1964, the bandwidth required for holographic television was considerably greater than the entire radio spectrum. In early 1968, D. J. De Bitetto at Philips Laboratories in New York proposed strip holograms as one way of reducing the information content of a hologram significantly.[72] De Bitetto recognized that the vertical parallax of holograms was relatively unimportant, because human eyes lie in a horizontal plane. He suggested first recording a conventional hologram on a very narrow horizontal strip; he used a 1 × 100 mm strip in his experiments. This strip hologram was like an extreme version of Leith and Upatnieks' early holograms on long, narrow spectroscopic plates with which they first saw parallax effects. De Bitetto then made many copies of the strip hologram on a second photographic plate, assembling them more or less contiguously in a vertical column. The result, he reported, 'is a 100 mm square multistrip hologram which reconstructs surprisingly well the entire field of view, lacking only vertical parallax', and one that had reduced the information content by a factor of 100.[73] Even through a narrow slit hologram, the entire scene could be seen by moving one's head up and down to sight through it at different angles. De Bitetto's multistrip version made all angles visible without movement by the observer. For holographic video, he envisaged recording and transmitting the strip hologram, and then using it to synthesise a similar multistrip hologram at the receiving end.[74]

De Bitetto's paper appeared in March 1968. In August, Stephen Benton took up a similar idea. Unlike De Bitetto, he did not synthesize a new hologram, but instead tried an optical arrangement that achieved the same result in a single recording step. He used a narrow horizontal 'strip' mirror to limit the vertical parallax of the hologram. Benton had arranged the reference beam to strike the hologram from above, rather than from the side as was conventional. This had the unexpected result of allowing the image to be reconstructed from a white-light source, because the fixed vertical perspective of the image was dispersed unchanged in each colour of the light source. The hologram acted like a prism or diffraction grating, dispersing light of different colours along a vertical arc.

[72] Dominick De Bitetto (b. 1923, Vermont) had obtained degrees in Mechanical Engineering and Physics from Oklahoma and New York University, respectively, and worked at Philips Laboratories from 1956.

[73] de Bitetto, D. J., 'Bandwidth reduction of hologram transmission systems by elimination of vertical parallax', *Applied Physics Letters* 12 (1968): 176-8.

[74] Similar research was being pursued at the same time in the Soviet Union. Simon B. Gurevich's group at the Institute of Physics and Engineering in Leningrad was studying how holograms could be sent on fax channels between Leningrad and Moscow by sending them as strips with reduced grey levels (Konstantinov, B. P., S. B. Gurevich, G. A. Gavrilov, A. A. Kolesnikov, A. B. Konstantinov, V. B. Konstantinov, A. A. Rizkin, and D. F. Chernykh, 'The transmission of holograms on standard phototelegraphic channels with a limited member of sub-tones', *Zhurnal Tekhnicheskoi Fiziki*39 (1969): 374–83). This research was subsequently extended to studies of holographic television (Gurevich, S. B., 'Holographic television future', *Tekhnika Kino i Televideniya. no.* 2 (1971): 67–9).

When viewers' eyes intercepted this arc of light, they saw a three-dimensional reconstruction that appeared in different colours depending on their vertical position. By contrast, a conventional hologram would have reconstructed overlapping but displaced images of the entire object in different colours, leading to a smeared result. Few observers noticed the lack of vertical parallax, unless they turned their heads sideways, or tried to bob up and down to look over and behind a reconstructed image.

In mid-September Benton showed the *white-light transmission* hologram to Edwin Land. He later recalled, 'Land was always great to work for anyway, but came over and looked at it and said "it looks like something might happen in holography after all these years!"'[75]

Benton continued his experiments in November, using a two-step technique, creating a white-light transmission (WLT) hologram from a normal Leith–Upatnieks master hologram. In an internal Polaroid report, he emphasized the new technique as an application for reduced-bandwidth electronic transmission.

Benton first described the new technique for reducing the information content of a hologram publicly at the 1969 Annual Meeting of the Optical Society of America in Chicago. His abstract stated:

> A two-step technique for elimination of vertical parallax in hologram viewing has resulted in significant reductions of information content, allowing relaxations of the reconstruction illumination-coherence requirements. The subject for a second hologram is a real image of the scene, projected by a narrow horizontal strip of a conventional hologram. The second hologram is illuminated to reconstruct a real image of the first, so that the entire field of view becomes visible when the eye is positioned at the image of the strip, and correct horizontal parallax is presented. The original point-illumination source can then be replaced with an incoherent vertical line source to produce a continuum of vertically displaced strip images. The eyes may now move throughout a large volume while viewing an undistorted, speckle-free, three-dimensional image that displays normal changes of horizontal perspective with viewing position.[76]

This description mentioned only in passing that the hologram was viewable in white light. Instead, it emphasized that his hologram provided a useful trade-off: what the hologram gave up (vertical parallax) was acceptable; most viewers would not miss it unless they tried to view over the top of the object. The members of the audience were similarly restrained in their response: 'Nice gimmick', observed Al Friesem.[77]

Just as Denisyuk had never vaunted the white-light reconstructions from his reflection holograms, Benton's own ideas were contextually screened: his goal was information

[75] Benton, Stephen A. to SFJ, interview, 11 Jul. 2003, Cambridge, MA, SFJ collection.

[76] Benton, Stephen A., 'Hologram reconstructions with extended incoherent sources', presented at *1969 Annual Meeting of the Optical Society of America*, 1969.

[77] Benton, Stephen A., 'Artist files,' MIT Museum 26/584, 1978. Albert A. Friesem obtained his BA (1958) and PhD (1968) from the University of Michigan, the latter focusing on thick recording media and information storage. He had been employed by the Bell Aero-System Company and Bendix Corporation before joining the WRL Radar & Optics Laboratory in 1963. At the time of meeting Benton, he was with Radiation, Inc., in Ann Arbor.

reduction, not white-light viewing. The technique, producing a garish rainbow-like image, had begun with undirected lab work, and Benton continued to think of it as 'sort of a toy' or demonstration experiment. Through 1970 he turned away from holography to work on the development of a multi-lens lenticular camera and to studies of film granularity.

When published a few months later, however, his description now stressed that this was a peculiar new hologram that reconstructed the image as a rainbow of colours dispersed vertically.[78] Nevertheless, the *rainbow hologram* only became popular after Benton's collaboration with the artist Harriet Casdin-Silver.[79] Galleries, in particular, took to rainbow holograms because of the low cost and relative ease in lighting. And during the late 1970s rainbow holograms proved increasingly popular with artists because the colour dispersion allowed them to create colour effects that could be mutated. For example, by combining two or more reconstructed images, the rainbow distribution could be overlapped to create 'pseudo-colour' effects, even mimicking full-colour imagery.

7.8 HOLOGRAPHIC STEREOGRAMS

None of these innovations captivated audiences for long during the 1960s. But in combination, holograms began to turn heads. An important factor in the American acceptance of the rainbow hologram was its implementation in a separate innovation: the commercially produced animated holograms of the Multiplex Company. The product was based on the idea of creating a three-dimensional image by combining strip holograms of numerous two-dimensional photographs. As with the rainbow hologram, ideas had been circulating earlier but had not found a convincing commercial application. And like the Leith–Upatnieks, Denisyuk, image-plane, and rainbow holograms, the innovation reduced to a novel geometrical arrangement to record the hologram.

M. C. King, at Bell Laboratories in New Jersey, submitted a paper in early 1968 that described a new and more compact way of viewing all sides of an object. He noted that 360° holograms had been proposed but that all these produced bulky holograms having awkward illuminating conditions.[80] Instead, he recommended synthesizing an image from separately recorded vertical strips a few millimetres wide. By rotating the object a few degrees between the strip exposures an 'animated' hologram would be obtained with the object appearing to rotate as the viewer moved sideways.

[78] Benton, Stephen A., 'Rainbow holograms', *Journal of the Optical Society of America* 50 (1969): 1545; and Benton, Stephen A. and Herbert S. Mingace, Jr., 'Silhouette holograms without vertical parallax', *Applied Optics* 9 (1970): 2812–13.

[79] See § 10.2.

[80] King, M. C., 'Multiple exposure hologram recording of a 3-D image with a 360 degrees view', *Applied Optics* 7 (1968): 1641–2.

This idea of vertical strip exposures was extended that year by J. D. Redman at the Atomic Weapons Research Establishment (AWRE) in Aldermaston, UK, who reported that he could produce three-dimensional x-ray pictures or reconstructions of people or outdoor scenes by making strip holograms from individual photographs.[81] Because he and his co-workers also envisaged an alternate scheme in which the hologram as a whole rotated to view the successive photographic 'frames', he called the technique 'hologram multiplexing'.[82] The term 'multiplexing' had been borrowed from communication theory, where it describes the combination of signals modulated by carrier waves of different frequency. The multiplex hologram superimposed several independent holograms on the same holographic plate, but with the reference beam shifted a few degrees each time, changing the spatial frequency of the carrier wave.

Beginning the same year, these ideas were generalized by a pair of investigators at the California Institute of Technology, Pasadena, by using separate photographic images from a motion picture to form each of the strips. Although each photograph was two-dimensional, a series of photographs of a suitably rotated object could be viewed as a three-dimensional scene. They called their product a 'holographic stereogram'.[83]

The commercialization of such holograms began, however, with Lloyd Cross. From 1970, when Cross set up a holography lab with Peter and Ana Maria Nicholson in Arizona, he became increasingly committed to the idea of integrating holography with photography.[84] He recalled:

> I set up a sandbox in an old barn, a small structure in the middle of this beautiful desert in Verde Valley. I would be continually shocked every time the door was opened and looked out into this incredibly beautiful brilliant blue sky, red rock, green plants, flowers, hills and Arizona countryside, to realize that with our present technology I could not capture that incredible three dimensional scene with people and horses and so forth on a holographic plate. So, I became quite determined at that time to do something about that, and of course, one of the answers was multiplexing holography, as were a number of other people at that time. I worked out the ideas to make an integral hologram from cine film at the Verde Valley School.[85]

At the San Francisco School of Holography (discussed in §9.3) he began to gestate this idea. In the summer of 1972 he constructed a 'Mark I' printer to record *Leslie*, consisting

[81] Redman, J. D., 'The three-dimensional reconstruction of people and outdoor scenes using holographic multiplexing', presented at *Holography Seminar Proceedings*, San Francisco, 1968; Redman, J. D., 'Novel applications of holography', *Journal of Physics E (Scientific Instruments)* 1 (ser.2) (1968): 821–2.

[82] Haig, N. D., 'Three-dimensional holograms by rotational multiplexing of two-dimensional films', *Applied Optics* 12 (1973): 419–20.

[83] George, Nicholas and J. T. McCrickerd, 'Holographic stereogram from sequential component photographs', *Applied Physics Letters* 12 (1968): 10–12; George, Nicholas and J. T. McCrickerd, 'Holography and stereoscopy: the holographic stereogram', *Photographic Science and Engineering* 13 (1969): 342–50; McCrickerd, J. T., 'Comparison of stereograms: pinhole, fly's eye, and holographic types', *Journal of the Optical Society of America* 62 (1972): 64–70.

[84] Cross, Lloyd G. and Cecil Cross, 'HoloStories: Reminiscences and a prognostication on holography', *Leonardo* 25 (1992): 421–4.

[85] Cross, Lloyd, 'The Story of Multiplex', transcription from audio recording, Naeve collection, Spring 1976, p.6.

of thirty-six vertical strip holograms on a 4 × 5 in. plate from still photographs. Cross took a roll of 35 mm colour slides, moving about 60° around Leslie's face. He then set up his 2 mW laser to record vertical strip holograms onto a 4 × 5 in^2 plate, using each slide as the subject for each slit. It was a process that took all night:

> The slit was moved across the plate by typewriter platens [. . .] and I got this little postage stamp 3D picture and it worked. It worked exactly the way I knew it would, but again, the visual impact of a little head in there, especially the last frame with the smile, was really a knockout. The visual impact was more than I anticipated. That same night I took the plate, projected the image in space and made a reflection image plane hologram which also worked. I made two of those plates and they were viewable in almost any kind of white light because the image plane was essentially perfect.[86]

That winter, Selwyn Lissack showed one of the *Leslie* holograms to Salvador Dali in New York, who wanted to know whether a larger one could be made—some 12 in. high and 24 in. in diameter. He offered the group $500 to make an animated hologram of a scene that would include the rock singer Alice Cooper. Cross agreed to create the composite hologram, provided that Lissack and Dali arranged the filming. To accommodate the request, Cross and his associates conceived and built a 'Mark II' automatic system in early 1973 to sequentially record the strip exposure holograms (see Figure 7.5). He, Michael Kan, and Pam Brazier produced a successful hologram after some six months of development, just in time for the exhibition at the Knoedler Gallery in New York.[87]

Cross designed the resulting hologram as a cylinder, or portion of a cylinder, so that as viewers moved around it an animated scene was replayed. He and his colleagues called the result a Multiplex hologram (although not using the term in the sense originally adopted by the AWRE workers). This hologram required a laser or mercury arc lamp for illumination. In April 1973, at the International Symposium on Holography in Biomedical Sciences in New York, Cross first encountered one of Stephen Benton's rainbow holograms and, discerning the secret, added the recording technique to the Multiplex process. By the spring of 1973 he had developed the technique to create rainbow multiplex holograms, viewable when lit by an unfrosted light bulb at the centre of the cylinder, creating the hologram *Lilliana*. Producing a 'Mark III' printer in 1974, the company used it to print thousands of holograms over the next three years.

The Multiplex hologram proved a successful product, attracting commissions from organizations such as NASA, General Motors, Revlon, the Smithsonian Institution, Merck Sharp & Dolme pharmaceuticals, Ripley's Believe or Not Museum, and many others. Unlike the holograms produced by Conductron, the Multiplex Company holograms

[86] Cross, Lloyd, 'The Story of Multiplex', transcription from audio recording, Naeve collection, Spring 1976, p.7.

[87] Cross, 'Holostories', op. cit.; Moore, Lon to J. Ross, interview, 1980, Los Angeles, Ross collection; Unterseher, Fred to SFJ, interview, 23 Jan. 2003, Santa Clara, CA, SFJ collection. The June 1975 Knoedler hologram exhibition, *The Threshold of the Optical Renaissance* also included other pulsed and laser-lit holograms that Dali had conceived and, collaborating with Selwyn Lissack, produced at McDonnell Douglas during the early 1970s.

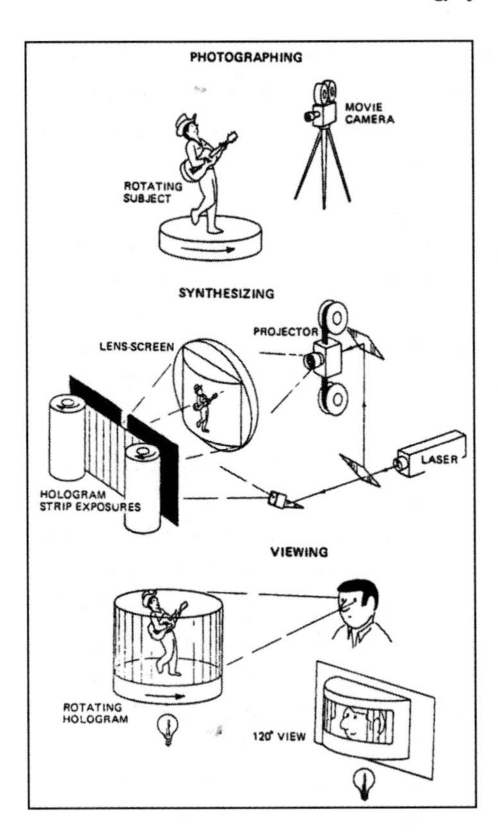

Fig. 7.5. 'Multiplex' hologram recording geometry (Industrial and Scientific Conference Management, 1976; MIT MoH collection)

also sold in large numbers directly to the public. Such holograms became the most widely seen types during the late 1970s. *The Kiss*, created in 1973 as a laser-transmission integral hologram and the following year as a rainbow-transmission integral hologram, was an animated image of Cross's girlfriend Pam Brazier wearing a broad hat.[88] As the viewer moved around the cylindrical image, the image blew a kiss and smiled. Cross described the work as 'a serious intensive effort over a period of months. I wanted to communicate the space-time event of 'blowing a kiss', as well as to explore techniques to deal with the limitations of the medium [. . .] Even after we had five final takes that were all 'perfect', I still spent days choosing the right one'.[89] A second, hatless version (*Kiss II*, 1975) was produced in thousands of copies, in two sizes (19 inch and 6 inch diameter), and sold around the world through science centres and other outlets through the 1970s, making the image iconic for the technology. Stephen Benton described the impact of Cross's integral holograms as responsible 'for this new wave of interest in holography, because they're available, they're cheap, they're flexible and they're enchanting'.[90]

[88] Provenice, Steve to J. Ross, interview, 1980, Los Angeles, Ross collection.
[89] Cross, Lloyd G. and Cecil Cross, 'HoloStories: Reminiscences and a prognostication on holography', *Leonardo* 25 (1992): 421–4, quotation p. 423.
[90] 'Variations of integral holos planned: Cross', *Holosphere* 6 (1977): 1, 4–5.

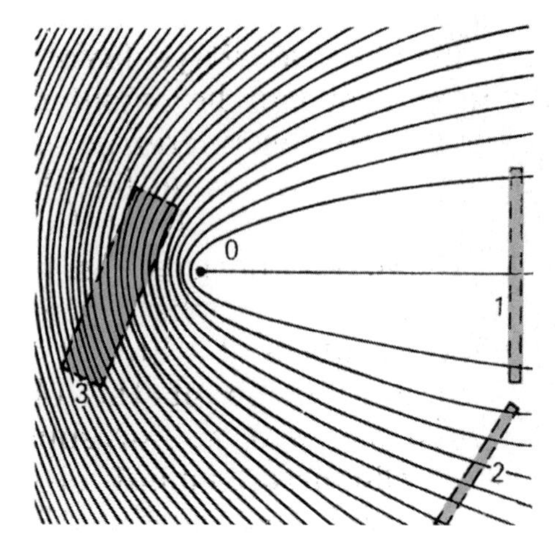

Fig. 7.6. Illustration of (1) Gabor, (2) Leith-Upatnieks, and (3) Denisyuk hologram recording geometries. The curved lines represent wavefronts produced by a collimated (parallel) reference beam and a point source object at O (adapted from Ostrovsky, Yu. I., *Holography and Its Application* (Moscow: Mir, 1977), Fig. 17(a), p. 52).

The Multiplex Company assisted other facilities in getting started (e.g. the Holographic Film Company (HFC) in New York, established in 1976 by Hart Perry Sr, William Molteni, Rudie Berkhout, Lisa Wo, and Hart Perry Jr). By 1978 HFC had produced its own version of an integral hologram printer, producing flat, as well as cylindrical, holograms.[91] During the late 1970s other workers sought to rid the Cross hologram of inherent distortions, which made the images appear tall and narrow (astigmatism), and to revert to variants of the techniques that George and McCrickerd had pioneered.[92]

Gradually, the varied characteristics of holograms (see figure 7.6) came to be described by a more generalized theoretical understanding. Yuri Ostrovsky, for example, provided a lucid view of the coalescing subject in his popular Russian text.

7.9 COMPUTER-GENERATED HOLOGRAMS

One variety of hologram that had little practical importance during the 1960s and 1970s was the computer-generated hologram. The term is ambiguous, but was first used in the sense of a complex interference pattern created by computer calculation. Such a *computer-generated hologram* (CGH, also known as a *synthetic hologram* or *digital hologram*) simulates the addition of a reference wave and object wave to produce a fringe pattern.

Digital holography had been an early area of exploration, although few saw it as a promising way forward. Adolf Lohmann, one of the earliest workers, recalls that Dennis

[91] 'New York company producing integral holos', *holosphere*, 7 (2), February 1978: 4.

[92] Aebischer, N. and C. Bainier, 'Multicolor holography of animated scenes by motion synthesis using a multiplexing technique', *Proceedings of the SPIE* 402 (1983): 51–6, Aebischer, N. and B. Carquille, 'White light holographic portraits (still or animated)', *Applied Optics* 17 (1978): 3698–700; Johnston, Sean F., 'Holographic animation apparatus', *American Journal of Physics* 47 (1979): 681–2.

Fig. 7.7. Adolf Lohmann (centre) with Byron Brown and Ronald Kay of IBM, *c.*1966 (Lohmann collection)

Gabor 'expressed his disappointment because I had used a computer to fabricate synthetic holograms. To use a computer indicated a severe lack of imagination, according to Gabor and many senior scientists. My employer at that time, IBM in San Jose, CA, felt differently'.[93]

At IBM from the early 1960s, Byron Brown (see Figure 7.7), Dieter Paris, Harald Werlich, and Lohmann comprised a small Optical Signal Processing Department.[94] Their proof-of-concept digital holograms were severely limited by the available hardware. In the early months, they used a significant portion of their budget to have the computer centre use its IBM 7094 computer to perform small (64 × 64 element) digital Fourier transformations for their first demonstration holograms.[95] Once this very crude synthetic hologram had been calculated, it had to be printed as a transparency so that it could be used to optically reconstruct the image. The available plotters, however, could print only binary black and white patterns, not shades of grey. Despite this serious limitation, they conceived a way to generate satisfactory *binary* holograms by shifting the plotted points, a technique dubbed *detour phase*.[96] The IBM central research laboratory in Yorktown, NY, hearing of the San Jose group's success, requested a three-dimensional computer hologram. Lohmann and colleagues, paying about one dollar per CPU second for calculations, correspondingly produced a demonstration hologram showing three bars, one in front of the hologram plane and two behind it.[97] These first

[93] Lohmann, Adolf W. to patent attorneys, patent declaration, Jun. 1987, Haines collection.

[94] Brown, who had first joined IBM as a summer student, was the only team member to be conversant with the theory, computer programming, and optical experimentation.

[95] Lohmann, Adolf W. to SFJ, fax, 20 May 2003, SFJ collection; Lohmann, Adolf W. to SFJ, fax, 9 Sep. 2004, SFJ collection.

[96] Lohmann, A. W. and D. P. Paris, 'Binary Fraunhofer holograms, generated by computer', *Applied Optics* 6 (1967): 1739–48 (a seminal paper, frequently referenced); Brown, B. R. and A. W. Lohmann, 'Computer-generated binary holograms', *IBM Journal of Research and Development* 13 (1969): 160–8. Detour phase was first described in Hauk, D. and A. W. Lohmann, 'Minimumstrahlkennzeichnung bei Gitterspektrographen', *Optik* 15 (1958): 275–7. As with other concepts in modern optics, this one can be related to communication engineering, via the equivalent concept of 'pulse position modulation' (PPM).

[97] Lohmann, Adolf to SFJ, fax, 9 Aug. 2005, SFJ collection

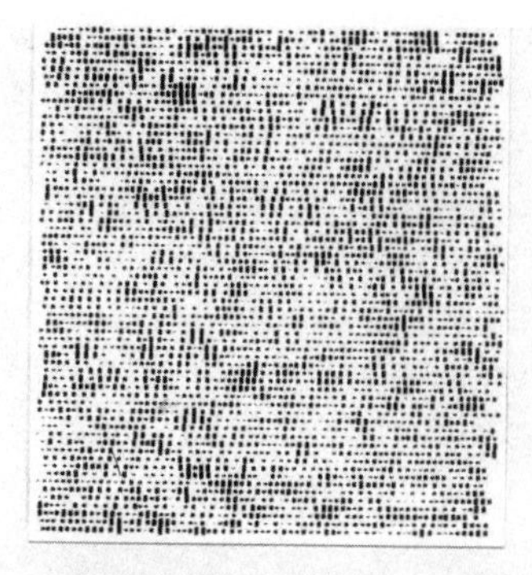

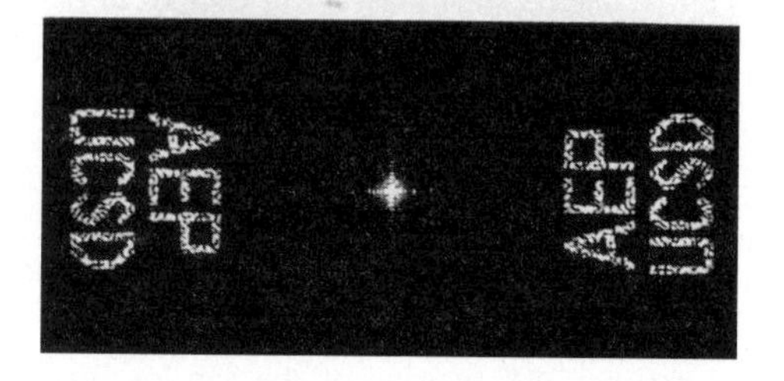

Fig. 7.8. Computer-generated binary hologram and its reconstruction, Lohmann et al., 1967 (Lohmann collection)

computer-generated holograms (see Figure 7.8) were thus digital in at least three senses: digitally computed, digitally quantized in terms of intensity, and digitally positioned as cells on the x-y plane of the hologram. While computing power and imaging quality were severely restricted, the team envisaged using these synthetic binary holograms as computer-generated optical elements such as altering the intensity profile of laser beams for making holographic optical elements or for image processing.

While such research was not popular or commercially interesting during in its early years, it was pursued elsewhere. Interest in digital holography developed in the Soviet Union about 1969. By that time, Lohmann's work on binary computer-generated holograms had become known. The USSR Academy of Sciences initiated a project to investigate the feasibility of producing a three-dimensional holographic display that used computer-generated holograms.

The work was taken up by the laboratory of Dmitry S. Lebedev at the Institute of Information Transmission Problems of the Academy of Sciences. The team, consisting principally of electronic engineers, had already worked in image processing. Leonid Yaroslavsky headed a small group studying digital image processing, and which had built image digitizing and image recording devices for two popular Soviet computers, the Minsk-22 and BESM-6.[98] A new group focusing on digital holography was formed with Yaroslavsky as group leader, Nickolay Merzlyakov responsible for optics, laser and photo-chemical equipment, and M. Kronrod for programming.

Lohmann's group at IBM had been limited by their computing and display hardware, and the Soviet team strove to improve on it. Their digital imaging system had unusually good specifications for its time: a relatively small pixel size (200 μm), recording a wide range of intensity (128 quantisation levels in a logarithmic scale—something not possible from the IBM plotters), and high resolution (image size up to 1000 × 1200 pixels). Such pixels were, however, far too large for recording the image of a hologram directly. Lebedev and Merzlyakov consequently designed an optical system to magnify or reduce the image size by a factor of 20.

For their initial experiments during 1970–71 on the digital reconstruction of holograms, for instance (essentially the inverse of what IBM workers had tried to do), the group employed an optically produced hologram of toy soldiers. The hologram was magnified, scanned with the group's image digitizer, and subsequently reconstructed by the computer. They were able to demonstrate digital processing methods that corrected nonlinear distortions of the optical hologram and finite resolving power of their digitizer. The group then used the methods that they had developed for measuring the directivity patterns of microwave antenna arrays and for reconstructing acoustic holograms.

Yaroslavsky's group also investigated the more useful reverse process, that is, creating holograms by computer for subsequent optical reconstruction. Unlike Lohmann's binary (black and white) synthesized holograms, they concentrated their efforts on greyscale holograms. By 1971 they had reconstructed images of greyscale objects from 512 × 512 pixel holograms, and combined this work with digital processing to reduce optical distortions in the recording process. Two years later, they acquired a more precise Western image digitizer purchased for processing images that were expected from the Mars-4 and Mars-5 Soviet space probes.[99] Having resolution some sixteen times better, the new equipment enabled them to fabricate computer-generated holograms without optical reduction. Within a year they were able to generate stereo holograms and a three-dimensional movie that contained about 800 holograms and displayed a 3-D model of six balls rotating pair-wise with different speeds around a vertical axis. By 1975 they had mastered the technique of bleaching photographic film and applied it to their computer-generated holograms. In 1977 Yaroslavsky and Merzlyakov collaborated on a Russian

[98] Yaroslavsky, Leonid P., 'Recollections on early works on digital holography in the former Soviet Union', unpublished report, 14 Jun. 2003, SFJ collection. Yaroslavsky, for example, had received a *Kandidat* degree in Radio Engineering.

[99] Optronics International Photomation P-1700 photo-densitometer and image writer.

book on their methods, which, on the recommendation of George Stroke, was published in English.[100]

Unlike Western laboratories, where a combination of readily-available NASA, NSF, DoD, and other government funding jostled with corporate-sponsored research, the Soviet work was more rationally organized. Yaroslavsky's group was one of several widely distributed teams in the Soviet Union studying digital holography. For example, W. Koronkevich headed a group at the Institute of Autometry of the Siberian Branch of the USSR Academy of Sciences in Novosibirsk; Victor Soifer headed another group located at the Aviation Institute in Kuybyshev; and a third was headed by I. N. Sisakyan at the General Institute of Physics of the USSR Academy of Sciences in Moscow. They focused on the synthesis of computer-generated optical elements such as axicons and beam formers.[101]

Nevertheless, digital holograms became a significant area of research and commercialization only when computers, displays, and printers became capable of the prodigious calculation and resolution necessary for recording fringe patterns that can diffract light. A seemingly unpromising avenue during the 1960s, it was to assume commercial importance during the 1990s, as discussed in §12.4.[102]

7.10 PHOTOSENSITIVE MATERIALS

While lasers and hologram recording geometries were the focus of technical innovation during the 1960s and early 1970s, further progress, in a commercial and technical sense, proved to be elusive. Attention began to shift to subtleties of the recording medium. Improving the recording material could transform applications: more sensitive emulsions

[100] Yaroslavsky, Leonid P. and N. S. Merzlyakov, *Methods of Digital Holography* (New York: Consultants Bureau, 1980).

[101] Yaroslavsky, L. P., 'Digital holography: 30 years later', *SPIE Proceedings* 4659 (2002): 1–11; Yaroslavsky, Leonid P., 'Recollections on early works on digital holography in the former Soviet Union', unpublished report, 14 Jun. 2003, SFJ collection.

[102] As late as the early 2000s, computing power remained a limitation for computer-generated holograms. Attempts to reconstruct synthetically generated optical images in real time attained limited success. For example, the Spatial Imaging Group at MIT and Qinetiq (formerly the Royal Signals and Radar Establishment, Malvern, UK), developed high-capacity parallel computer/display systems to calculate the synthetic hologram and to display it on acousto-optic modulators, which then serve to diffract a laser wavefront into a three-dimensional image. However, a different species of computer-generated hologram is the *dot-matrix hologram*, a type common since the late 1980s, in which a computer-directed optical system records tiny 'diffractive pixels', each of which is a diffraction grating made up of straight fringes of fixed spacing and orientation (Iwata, Fujio and Kazuhiko Ohnuma, 'Grating images', *Optical Security Systems Proceedings* (1988): 12–4, an idea followed up quickly by Frank Davis and Craig Newswanger in America). Each such micro-grating pixel diffracts light of a particular colour to the viewer's eye and combines collectively to create a multicoloured image or several images visible at particular angles. This type of digital hologram, also sometimes known as an *elemental hologram*, has become common in increasingly elaborate variations for security applications, but much less so for high-quality imaging.

would reduce the need for powerful and expensive lasers or could lessen the requirements for mechanical stability by reducing the exposure time; an instant recording process devoid of chemicals would make holographic interferometry easier and quicker; new recording media might also allow more efficient reconstruction and so produce images that could be viewed with ambient lighting or in brighter surroundings. Researchers therefore sought replacements for high-resolution photographic emulsions that would be more sensitive, better adapted to specific laser wavelengths, easier or faster to process, and yielding brighter holographic reconstructions. Given this more goal-directed technical emphasis from the mid-1970s, recording materials steadily improved.

Until the late 1960s, the recording of holograms had relied on a wide variety of silver-halide photographic emulsions. Soviet investigators, unconstrained by the need to satisfy commercial markets in the way that the large Western manufacturers of photographic media were, developed emulsions optimized for reflection holography. During the late 1960s, Rebekka Protas of the Vavilov Institute developed the LOI-2 photoplate, a thick, fine-grained emulsion for Denisyuk holograms. Another popular formulation for Soviet holographers was the PE-2 and PFG-03 photoplates developed independently by N. I. Kirillov working with Gennadi A. Sobolev at NIKFI (see Figure 7.9) in Moscow.[103] Although the characteristics were less stable over time, the PE-2 type was some ten times more sensitive than the LOI-2 plates. Kirillov also developed two-layer emulsions sensitive to red and to green for recording colour holograms, an approach also used later at Ilford Ltd in Britain.

The first holographic recording materials available in the West were high-resolution emulsions manufactured by Eastman Kodak in America, Ilford in England, and Agfa Gavaert in Belgium, with researchers showing a strong regional preference.[104]

Kodak had originally manufactured its very-high resolution 649F emulsion for use in spectroscopy and was the exclusive choice for holograms during the mid-1960s.[105] This emulsion was the highest resolution that Kodak produced, and was the only choice for recording the fine fringe patterns generated by reflection holography. The lower resolution SO-253 emulsion, developed for the American space program, had been available since the mid-1960s. The Kodak product line was extended specifically for holography during the early 1970s with type 120, SO-173 and SO-131 emulsions, but remained substantially unaltered over the following thirty years. Despite the expansion of holography

[103] His wife, Svetlana Soboleva, had worked in Denisyuk's lab, developing the first dichlorinated, gelatin-based holographic emulsion. She and her husband later developed emulsions for the HOLOS Corporation in New Hampshire.

[104] Cox, M. E. and R. G. Buckles, 'Evaluation of selected films for holography', presented at *Optical Society of America Spring Meeting*, Philadelphia, PA, 1970. Dennis Gabor and Gordon Rogers appear to have relied on Ilford materials during the 1950s and American investigators of the mid-1960s used Kodak materials almost exclusively. Given their importance for the Denisyuk method as well as distribution problems, Soviet researchers used the emulsions produced there.

[105] See Goodman, J. W., 'Film-grain noise in holography', presented at *Proceedings of the Symposium on Modern Optics*, Stanford University. CA USA, 1967 and Couture, J. J. A. and R. A. Lessard, 'Diffraction efficiency of specular multiplexed holograms recorded on Kodak 649F plates', *Applied Optics* 21 (1979): 3652–60.

Fig. 7.9. Victor Komar, Deputy Director of NIKFI, with holographic movie film, 1986 (Novosti Press Agency, MIT MoH collection).

as a research subject, the market for photographic materials remained small. From the late 1970s, Kodak devoted few resources to further technical development, and declining support for their commercially available holographic materials, but targeted products developed for other applications to the particular needs of holography.[106]

The major British supplier of materials was Ilford, a photographic firm founded in Essex, England in 1879, which moved into holography in the 1970s.[107] Ilford produced 'He-Ne 1' film in the early 1970s, a moderate-resolution emulsion that was most sensitive to the red light from the HeNe laser. From 1984, encouraged by Kodak's gradual retirement from the market, Ilford introduced a line of holographic films and processing chemicals. This coincided with a marketing push to support anticipated sales from a high-volume hologram printing machine co-designed with the British firm Applied Holographics.[108]

The third, and most important, Western supplier of silver-halide materials was Agfa Gavaert. Agfa, an acronym for the dye and chemicals manufacturing firm *Aktiengesellschaft für Anilinfabrikation*, was founded in 1888 in Berlin. The Gavaert firm, manufacturing photographic products, was founded in Belgium the same year. From dyes, the firm extended products to photographic developers in 1891. The Agfa and Gavaert companies, which merged in 1964, had each been producing photographic films since the mid-1920s.

Given the relatively poor availability and high cost of Kodak holographic materials in Europe, European researchers turned increasingly to Agfa Gavaert materials from the

[106] See Eastman Kodak Company, *Kodak materials for holography* (Rochester, NY: Eastman Kodak, 1976); Eastman Kodak Company., *Two new Kodak materials for holography: Kodak holographic plate, type 120–02: Kodak holographic film (Estar base) SO-173* (Rochester, NY: Eastman Kodak Co., 1973); James, Dorothy, 'Report from Rochester', *holosphere*, 13 (1) (1985): 13–5; Eastman Kodak Company, 'Kodak High-speed Holographic Glass Plates 131CX and 131PX', Kodak, Jan. 2002.

[107] Hercock, R. J. and George Alan Jones, *Silver by the Ton: The History of Ilford Limited, 1879–1979* (London: McGraw-Hill Book Co., 1979).

[108] Brentnall, Roger, Alan Hodgson, Nigel Briggs, and Mary B. Dentschuk to SFJ, interviews, 20 Sep. 2002, Mobberley, Cheshire, SFJ collection. See also § 12.3.

early 1970s. The firm thereafter began to actively market its existing high-resolution materials—originally designed for the microcircuit industry—for holographic use.[109] Its *Scientia* materials were rebadged as *Holotest*, with the emulsion reformulated to produce the higher resolution 'HD' variants, in 1976. In 1986 the company moved from a polyester to triacetate base, which provided a more stable foundation for their high resolution products. As with Kodak and Ilford, the rise in silver costs in 1980 caused Agfa to double its prices.[110] More responsive than Kodak to the small but growing holography market, Agfa Gavaert had a market share of about 85% by the late 1980s, but this still represented a very small income; even within their Diagnostic Imaging Systems Division, holographic products amounted to less than 2% of sales in the mid-1980s, probably the peak of their popularity. Holography attained scarcely 0.1% of all the sales of the firm in Britain at that time. [111] As the name suggests, the manufacture of holographic emulsions shared coating facilities with x-ray and scientific film production for scarcely two half-days per year, and generated negligible profits.[112]

So holographers clung desperately to diffident Western manufacturers of silver-halide emulsions. Even with the routine requirements of holographic interferometry and a growing market for display holograms, sales were insignificant compared to regular photographic applications. This small market also meant that the emulsions could not be made available cheaply or reliably on demand. As Duncan Croucher of Agfa Gavaert confided, the yearly production of holographic materials required only a few hundred metres of film stock, which meant that the coating machines in the Belgian factory scarcely got up to full speed. As a result, product uniformity was difficult to achieve.[113] Owing partly to the small production runs consistency of the emulsion and coating was difficult to maintain, which dissatisfied their customers. Agfa Gavaert ceased production of its Holotest materials in 1997, citing ever-declining markets.

Similarly, Ilford workers, who prepared the special fine-grained emulsions as a batch process, could not mix a new batch for coating more frequently than every few months even though, like beer and cheese, stocks aged and altered their characteristics. And Ilford resisted supplying their holographic emulsions on glass plates, because converting their coating facilities was uneconomic.[114]

On the other hand, the concoction of optimized emulsions had been a strength of Soviet research since the work of Protas and Denisyuk around 1960. They benefited from

[109] Frecska, S. A., 'Characteristics of the Agfa-Gevaert type 10E70 holographic film', *Applied Optics* 7 (1968): 2315–17; Croucher, Duncan, 'Agfa-Gevaert photographic materials for holography', *Proceedings of the International Symposium on Display Holography* 1 (1983): 71–8; Frecska, S. A. to SFJ, telephone interview, 16 Aug. 2002, SFJ collection.

[110] Prices rose to nearly $15 US per 8 × 10 in. plate. Prices doubled again in 1993, with a 10 m roll of 114 cm film costing $1231.

[111] Pepper, Andrew, 'Expansion of holographic film products', *holosphere* 15 (1987): 12–3.

[112] Boone, Pierre M., 'Report of the nations: Belgium', presented at *International Symposium on Display Holography*, Lake Forest, Illinois, 1991.

[113] Croucher, Duncan to S. Johnston, telephone interview, 16 Aug. 2002, SFJ collection.

[114] Brentnall, Roger, Alan Hodgson, Nigel Briggs, and Mary B. Dentschuk to SFJ, interviews, 20 Sep. 2002, Mobberley, Cheshire, SFJ collection.

the close association there between research in optical design and photographic chemistry. An important characteristic of these emulsions was their unmarketability: the very fine emulsions gave them a limited shelf-life and the unusually thick emulsions made them difficult to coat reliably in large batches. As a result, Soviet emulsions were unavailable in the West. Only the finished product could be seen. From the mid-1970s, Soviet holograms were increasingly judged by Western researchers to be technically superior to Western products: they were bright and had little 'noise', or scattering of light, because of the uniform and fine-grained emulsion, careful coating, and special processing chemistry.[115]

Given the unresponsiveness of major photographic manufacturers and the incentive of bright high-quality Soviet holograms, Western researchers turned to film chemistry. Nick Phillips at Loughborough University of Technology became the leading investigator of photographic processing techniques in the West during the 1970s.[116] There was considerable scope for experimentation. Although photographic emulsions had by then been studied scientifically for over a century, the requirements for holographic imaging were distinct. Gordon Rogers had made 'phase holograms'—holograms having little or no absorption, but that modulated light by causing a phase shift, or optical retardation of the wave—in the early 1950s. Bright holograms could be produced by chemical bleaching and had become more popular once public displays grew in popularity at the end of the 1960s.[117] But all bleaches degraded the hologram to some degree. By removing the absorbing silver salts the emulsion itself was distorted and random distortions introduced noise in the reconstructed image. Such distortion could also be produced by aggressive developers or harsh 'fixers', used to make the resulting fringe pattern permanent in the emulsion. And without adequate fixing, holograms were likely to discolour with time. The investigation of hologram chemistry was therefore a multidimensional problem only weakly supported by theoretical understandings. The work of Phillips and others was hard-won; a cult of secrecy pervaded the techniques, particularly when they could be used to generate commercially important improvements in holograms.

[115] After the end of the Soviet Union, such emulsions became available in the West. The Lithuanian firm Slavich, for example, founded in 1931 to serve the Russian photographic market, began producing silver-halide and dichromated gelatin holographic materials in the early 1980s, and selling them worldwide in the late 1990s. These were characterized by extremely fine grain, allowing high resolution recording of fringes and hence suited for reflection (Denisyuk) holography (Barachevsky, V. A., 'Russian advances in holographic recording media', *Proceedings of the SPIE The International Society for Optical Engineering* 3417 (1998): 142–53).

[116] E.g. Phillips, N. J. and D. Porter, 'An advance in the processing of holograms', *Journal of Physics E (Scientific-Instruments)* 9 (1976): 631–4; Phillips, N. J. and D. Porter, 'Organically accelerated bleaches: their role in holographic image formation', *Journal of Physics E (Scientific-Instruments)* 10 (1977): 96–8; Phillips, N. J., 'Modes of holographic image formation-major areas of progress in the various regimes', presented at *Colloquium on Holographic Displays*, Loughborough UK, 1978.

[117] For example, Latta, J. N., 'The bleaching of holographic diffraction gratings for maximum efficiency', *Applied Optics* 7 (1968): 2409–16; Buschmann, H. T., 'The production of low noise, bright-phase holograms by bleaching', *Optik* 34 (1971): 242–55; Hariharan, P., 'Bleached reflection holograms', *Optics Communications* 6 (1972): 377–9; Eastman Kodak Company., *Reversal bleach process for producing phase holograms on Kodak spectroscopic plates, type 649-F* (Rochester, NY: Eastman Kodak Co., 1974).

Despite achieving hologram qualities that approached the Soviet exemplars but with Agfa, Ilford, or Kodak materials, Western researchers also investigated other photosensitive materials. From the late 1960s silver-halide emulsions began to compete with dichromated gelatin or DCG. As a photosensitive material that was not commercially available and yet relatively simple in fabrication, DCG offered considerably more scope for optimization than did silver-halide emulsions.

DCG was, in fact, an old technology, used since the early twentieth century for some photographic applications. The material, then referred to as 'bi-chromated gelatin' had first been applied to wavefront reconstruction by Gordon Rogers in the early 1950s, but long forgotten. It was rediscovered in the late 1960s, though, by holographers seeking bright holograms. T. A. Shankoff used DCG to make phase holograms from 1968 at Bell Laboratories in New Jersey. His colleagues L. Lin and others soon investigated the optical and chemical properties of the material.[118] At IBM, Keith Pennington and J. Harper used DCG made from Kodak 649F plates and sensitized by the silver grains.[119] Milton Chang at Northrop and Meyerhofer at RCA further studied the chemistry and structure of DCG, and Don Close and others at Hughes Research Laboratories used it for producing holographic optical elements, which was to become its most commercially important application.[120]

DCG does not darken like alkali-halide materials; it cross-links after exposure to laser light, changing its refractive index and acting as a phase hologram. It reconstructs very bright images from a seemingly transparent plate. It is, however, extremely insensitive to laser light, particularly to red lasers, the most inexpensive and widely available type. The quality also depends critically on a wide range of poorly understood chemical and environmental factors, requiring the resulting hologram to be sealed to prevent moisture from swelling.

DCG as a sensitive medium therefore had some distinct advantages, as well as serious disadvantages. It was extremely transparent and, consisting of a homogeneous emulsion free of suspended particles, generated very bright images having extremely low light scattering. But its very short shelf life prevented photographic manufacturers from

[118] Shankoff, T. A., 'Phase holograms in dichromated gelatin', *Applied Optics* 7 (1968): 2101–5; Brandes, R. G., E. E. Francois and T. A. Shankoff, 'Preparation of dichromated gelatin films for holography', *Applied Optics* 8 (1969): 2346–8; Lin, L. H., 'Hologram formation in hardened dichromated gelatin films', *Applied Optics* 8 (1969): 963–6; Curran, R. K. and T. A. Shankoff, 'The mechanism of hologram formation in dichromated gelatin', *Applied Optics* 9 (1970): 1651–7.

[119] Pennington, K. S. and J. S. Harper, 'Techniques for producing low-noise, improved efficiency holograms', *Applied Optics* 9 (1970): 1643–50; Pennington, K. S., J. S. Harper, and G. Kappel, 'Forming high efficiency holograms', *IBM Technical Disclosure Bulletin* 13 (1971): 2282–3.

[120] Chang, M., 'Improved dichromated gelatin for holographic recording', presented at *Optical Society of America Annual Conference*, 1970; Chang, M., 'Dichromated gelatin of improved optical quality', *Applied Optics* 10 (1971): 2550–1; Meyerhofer, D., 'Spatial resolution of relief holograms in dichromated gelatin', *Applied Optics* 10 (1971): 416–21; Meyerhofer, D., 'Phase holograms in dichromated gelatin', *RCA Review* 33 (1972): 110–30; Close, D. H. and A. Graube, 'Holographic lens for pilots head-up display', report, Hughes Research Laboratories, Aug. 1974.

producing DCG material directly, although initially such plates were produced starting from Kodak or Agfa emulsions.[121]

These factors altered the balance between producers and users and promoted the production of holographic materials by practitioners themselves. The development of cottage industry holography in the late 1970s went hand in hand with local production of DCG. As a result, a number of laboratories began to carefully explore the processing conditions. While seemingly simpler than silver-halide materials, dichromated gelatin had its own collection of experimental variables. Don Broadbent at Westech Development Laboratories, for example, studied methods of coating plates with dichromated gelatin: spin coating, dip coating, and scalpel coating.[122] Others, both in scientific laboratories and start-up firms, examined its chemistry and environmental dependencies (DCG proved notoriously dependent not only on processing conditions, but on moisture absorption, which can cause a colour shift or complete fading of the image). During the mid-1970s, Richard Rallison discovered reliable methods of coating commercial gelatin onto glass surfaces, tuning the colour of the holograms recorded on it, and reliably sealing them in a glass sandwich. The result was an inexpensive and high-quality holographic medium that was ideal for reflective pendants.[123] A significant disadvantage for uptake by amateurs was that 'dichromates' were most sensitive to short-wavelength light, and required the blue–green wavelengths from relatively expensive and unreliable argon-ion lasers. Sensitization techniques were eventually found to improve sensitivity to HeNe exposures.[124]

Yet another medium that had its own distinct pattern of distribution and usage was the photopolymer. Photopolymers used in holography are transparent long-chain molecular materials that cross link when exposed to short-wavelength light. Unlike DCG, which had been nurtured by entrepreneurs and cottage industry after corporate interest waned, photopolymers began as, and remained, a product of large companies. From the late 1980s, the Polaroid Corporation marketed 'DMP-128', a grainless photopolymer recording material requiring a carefully controlled wet process.[125] Bruce L. Booth at Du Pont, Wilmington Delaware, developed photopolymer materials there in 1975, and Du Pont introduced its own 'HRF' photopolymer materials in 1989.[126] The Du Pont materials,

[121] For example, Oliva, J., P. G. Boj, and M. Pardo, 'Dichromated gelatin holograms derived from Agfa 8E75 HD plates', *Applied Optics* 23 (1984): 196–7; Hariharan, P., 'Silver handle sensitized gelatin holograms: mechanism of hologram formation', *Applied Optics* 25 (1986): 2040–2.

[122] DiBiase, Vincent to S. Johnston, unpublished report, 24 Jan. 2003, Santa Clara, California, SFJ collection.

[123] Rallison, Richard D., 'The history of dichromates', http://www.xmission.com/~ralcon/dichrohist.html, accessed 31 oct. 2002; Rallison, Richard D. to SFJ, email, 22 Mar. 2004, SFJ collection. On the commercial impact of DCG, see § 12.2.

[124] Kubota, T., T. Ose, M. Sasaki, and K. Honda, 'Hologram formation with red light in methylene blue sensitized dichromated gelatin', *Applied Optics* 15 (1976): 556–8; Rallison, R. D., 'Control of DCG and nonsilver holographic materials', presented at *International Symposium on Display Holography*, Lake Forest, Illinois, 1991.

[125] Ingwall, R. T. and H. L. Fielding, 'Hologram recording with a new photopolymer system', *Optical Engineering* 24 (1985): 808–11; Hay, W. C. and B. D. Guenther, 'Characterization of Polaroid's DMP-128 holographic recording photopolymer', *Proceedings of the SPIE* 883 (1988): 102–5.

[126] Anon., 'Heat is on with new Du Pont photopolymers', *Holographics International* (1989): 26; Salter, J. L. and M. F. Loeffler, 'Comparison of dichromated gelatin and Dupont HRF-700 photopolymer as media

processed by curing with ultraviolet light followed by heat treatment in a low-temperature oven, proved better adapted to amateur use, but the company licensed the process to a handful of producers. The impetus to study such materials increased after the withdrawal of Ilford and Agfa from the holographic emulsion market, but had little impact for amateurs or cottage industry.[127]

7.11 THE MEDIUM AND ITS MESSAGE

As these species of holograms show, technical evolution had occurred in waves of innovation having distinct intellectual and social features. Between 1964 and 1974 holography repeatedly hit the headlines with a series of achievements that seemed to bring it ever closer to a practical future: three-dimensional imaging itself (1964); holographic interferometry (1965); reflection holograms and image-plane holograms for white-light reconstruction (1966); cylindrical holograms (1966); pulsed holograms (1967); rainbow holograms (1972); and, holographic stereograms (1974). This publicity sustained further development work and encouraged a continuous stream of corporate and governmental sponsors and entrepreneurs. Most of these developments were unplanned and unpredictable; even emulsion and process development was often a hit-or-miss affair dominated by cut-and-try methods. Incongruously, these developments also encouraged the view that continued progress was both inevitable and tangible, promoting a wider interest from new communities, as discussed in Part III.

Alongside these technical advances, though, lay recognized technical obstructions and, even more seriously, a tentative and fickle public engagement. During the relatively fallow period for new technical developments, 1967–1972, McDonnell Douglas and other serious players desperately sought commercial applications (discussed further in Part IV). But, from 1975 onwards, technical innovations seemed to slow to a crawl.

From about 1968 to 1978, the most rapid improvement in technical abilities occurred in the recording materials for holography. The media were distinct in production requirements, recording characteristics, and adoption by their users. Silver-halide materials were produced in the West by large photographic manufacturers and became successively more expensive and less available. Those practitioners who learned how to control the experimental variations in processing, opening possibilities for sales of

for holographic notch filters', *Proceedings of the SPIE The International Society for Optical Engineering* 1555 (1991): 268–78; Gambogi, W. J., A. M. Weber, and T. J. Trout, 'Advances and applications of DuPont holographic photopolymers', *Proceedings of the SPIE The International Society for Optical Engineering* 2043 (1994): 2–13; Stevenson, S. H., 'DuPont multicolor holographic recording films', *Proceedings of the SPIE The International Society for Optical Engineering* 3011 (1997): 231–41.

[127] For excellent technical accounts and bibliographies of these materials and processes, see Bjelkhagen, Hans I., *Silver-Halide Recording Materials for Holography and Their Processing* (Berlin; New York: Springer-Verlag, 1993) and Barachevsky, V. A., 'Holographic recording media: modern trends', *Proceedings of the SPIE The International Society for Optical Engineering* 3011 (1997): 306–18.

commercial holograms, remained reticent about the details. In the Soviet Union, by contrast, silver-halide materials were made close to the point of use and never offered for commercial sale; as a consequence, a clear East–West separation developed in both the media and the performance of holograms. Differently again, dichromate materials became most accessible in the West to amateur and military producers of holograms over the space of a decade, while photopolymer materials reinstituted a more hierarchical division by restricting sales to major businesses. In short, the materials and their users evolved together.

PART III
Creating an Identity

As holography expanded during the 1960s and beyond, its practitioners became distinct. Their backgrounds, working environments, skills, and goals divided them. The following four chapters make explicit this social dimension hinted at in the first half of the book. This section illustrates how practitioners nurturing common values created a scientific working culture (Chapter 8), how it was riven by cultural and political currents (Chapter 9), how the subject acquired an aesthetic dimension through the contributions of artists (Chapter 10), and how stable but heterogeneous communities of holographers developed by the 1980s (Chapter 11).

The technology was instrumental in this process, transforming a would-be discipline (optical engineering) and spawning limited artist–scientist collaborations by the late 1960s. But a distinct artisanal community shaped a countercultural form of holography via a radical technology: the sandbox optical table. Their equipment catalysed a group solidarity that undermined the privileged exclusivity of scientists. The three principal communities—scientist–engineer holographers, artisanal holographers, and aesthetic holographers—developed traditions defined by their backgrounds, tools, sponsorship, products, literature, and engagement with wider culture, but maintained a limited technological trade.

Holographic technology, which shaped, and was shaped by, these new communities, provides a striking illustration of the co-evolution of new technology along with highly dissimilar user groups, none of which fostered the secure establishment of a profession or discipline. In short, I focus on how this new breed of technical worker, the holographer, came to be, and how different tools and environments shaped that identity.

8

Defining the Scientific Holographer

From Gabor's work in the late 1940s until the late 1970s, the young subject of holography spawned distinct communities. These coalesced not just because of their shared expertise but also because of changes in wider culture. Alongside questions about the content and applications of the subject were important questions of identity: How did holographers differ from other technical experts and from each other? The first practitioners gravitated towards a disciplinary home that combined physical optics with electrical engineering theory to yield a revitalized subject, optical engineering. But the period around 1970 was a turning point for practitioners and events illustrate how the identity of the holographer was shaped by deeper cultural influences. Both the invention and the user communities co-evolved to support a growing differentiation of products and practices. While such interaction and divergence is a common feature of new technologies, holography provides an extreme case of this process: the communities remained highly distinct.[1]

This chapter focuses on the first holographers, exploring how they came to differ from their contemporaries in occupational, disciplinary, and professional respects. The period demonstrates how the political dimensions of a technology can be important, and yet evanescent, in the growth of such communities. I argue that four factors were crucial in generating contrasting identities for holographers: first, the adoption of a philosophy having political undercurrents; second, a radical reconceptualization of equipment; third, the establishment of semi-formal training; and, fourth, the promotion of the subject and its practitioners through new communication channels.

8.1 RESHAPING OPTICAL ENGINEERING FOR HOLOGRAPHERS

The experiences of Gabor, Denisyuk, Leith, and their respective co-workers differed in many respects. The administrative structures, resourcing, and interplay of disciplines in

[1] Of course, many technologies foster new communities, for example, users' groups for computer hardware and software. As argued here, though, holography generated an unusually broad spectrum of technical communities divided by tools, products, and philosophies.

post-war Britain, America, and the Soviet Union contrasted sharply. Nevertheless, these environments melded a union between two previously separate scientific fields. As the milieus of Gabor at BTH and Leith and Upatnieks at Willow Run indicate, the foundations of holography were created by two mutually unrecognizing technical cultures, those of physical optics and electrical engineering. Electrical engineering, in which signal processing was becoming a routine tool, seemed to share few intellectual principles or practical skills with optics.

Words and the concepts that they represented separated practitioners. As is typical across disciplinary divides, jargon has two purposes: it consolidates a sense of identity of practitioners by helping them to communicate precisely and economically about their subject and it also excludes outsiders.

Adolf Lohmann, for instance, recalls the mutual incomprehensibility and lack of respect between disciplines at his German university:

> In 1954 two radio-and-TV engineers came to our optics lab in Braunschweig. They asked: 'How many dB has a typical lens?' [I replied], 'A lens has aberrations and a bit of diffraction, but no dB' (and I thought to myself you'd better learn what aberrations are before you come back and steal our time).[2]

Decibels and lenses had never coexisted. The disciplines were of unequal and shifting status: optics was an established, but staid, scientific subject; radio engineering was a young but rapidly rising upstart.[3] In this area of overlap, a merging of their vocabularies was beginning to occur. In a memo a decade later, discussing an optical plate used to correct a wavefront, Emmett Leith wrote of an 'insertion loss of about 10 or 13 dB' and a '5 or 10 percent' proportion of radiation being diffracted, unconsciously combining the mindsets of electronic engineers and optical physicists.[4] Technical specialists in isolated domains of the two fields remained separated by their vocabularies, however. One editor, listening to papers on coherent optical information processing at a 1966 meeting, complained that he 'found the verbalized papers almost unintelligible. Many computermen attending the session left at intermission, never to return. Those that weathered the storm of technical jargon were heard to mutter, when are those birds going to become as coherent as a laser's output?'[5]

Given its various origins in electron microscopy, in the optical processing of radar data and in optical filtering, holography did not graft neatly onto the science of optics. It is

[2] Lohmann, Adolf W. to SFJ, fax, 13 May 2003, SFJ collection.

[3] Radio engineering, boosted by the First World War, became the home for low-powered electrical designers and the spark for the new discipline of electronics during the Second World War. The youthfulness of the subject is suggested by its professional affiliations: in America, the Institute of Radio Engineers (IRE) had been founded in 1912 as a response to the conservative and nationally oriented American Institute of Electrical Engineers (AIEE), which focused on electric power engineering. The IRE and AIEE negotiated merger from 1957, combining to form the Institute of Electrical and Electronics Engineers (IEEE) only in 1963.

[4] Leith, Emmett N. and Juris Upatnieks, 'Wavefront reconstruction disclosure', Patent disclosure report, University of Michigan Willow Run Laboratory, 22 Apr. 1964.

[5] 'Computermen hear five laser papers—a critique', *Laser Focus*, 1 May 1966.

significant that Dennis Gabor himself never applied for membership in the Optical Society of America (OSA), the largest such society internationally, as the directors found to their chagrin when he was awarded the Nobel Prize. Indeed, while late nineteenth-century optics had made important forays into interference phenomena, the subject during the first half of the twentieth century had been dominated by geometrical optics for lens design and by spectroscopy, which proved to be a valuable tool in the development of modern physics and chemistry. The new subject of holography consequently was developed by a variety of workers in electrical engineering, optical engineering, and general physics and applied to an ever-wider variety of interdisciplinary problems.

This competition for the field between electrical engineers and physicists was also occurring for other variants of the subject and other locales. Stuart Leslie recounts how the Microwave Laboratory at Stanford University, staffed by electrical engineers during the early 1960s, sought to prevent the creation of a new department of Applied Physics and rival research in quantum electronics, plasmas, lasers, and related fields. The Provost of Stanford, Frederick Terman, argued that scientists would move into interesting new fields if engineers did not move aggressively, and would relegate engineers to dull trade school subjects. Nevertheless, Stuart Leslie credits the synergism of physics and electrical engineering as the foundation for much of Stanford's post-war success in both fields.[6]

Such synergism proved crucial for the emergence of holography. George Stroke's post as Professor of Electro-Optics was one of the first to be established in the subject, signalling recognition by the University of Michigan's Institute of Science and Technology and Electrical Engineering Department that a new discipline was in the offing. Such appointments can be crucial in institutionalizing a subject, and this clearly was the intention of U-M administrators and others further afield. The establishment of such a post seems to have been widely supported by optical scientists. Just as Leith had experienced in the cosseted environment of Willow Run, Stroke recalls a gradual awareness of the codependence of electrical and optical viewpoints.

> In 1962 I was at MIT, but I wanted to become a professor. I was a consultant at Perkin Elmer at that time. I knew Dick Perkin. I could see that there was no modern optics being taught in universities. I talked with Brian Thompson and Bill Browers and I tried to convince Harvey Brooks at Harvard and Charlie Townes who at that time was already provost at MIT. Everybody said optics should be taught—modern optics, not the Arthur Hardy-type geometrical optics, although I admired the man. We did this for three, four, or five years—writing papers mobilizing the Optical Society and so on—without success. At that time nuclear physics was the big thing, and the realities are that sometimes you have to travel. This seems to be an understandable phenomenon. From a distance you see things clearly. I was in yet another conference in 1961, maybe in Paris again, and was fascinated by yet another meeting of classical optics, of famous people. I went out with somebody else, an Italian friend of mine, Giuliano Toraldo di Francia. I spontaneously said to Guiliano, 'I'm getting into electrical engineering.' And he said, 'That's a great idea! That's the right thing.' I went back to MIT and walked into Peter Elias's office. Peter Elias, a Harvard Junior Fellow, had also

[6] See Leslie, Stuart W., *The Cold War and American Science: The Military–Industrial–Academic Complex at MIT and Stanford* (New York: Columbia University Press, 1993), pp. 175–81.

written an early paper on Fourier transforms in optics with Robinson, who was then the Science Advisor to one of the presidents, or chief scientist, I forget. I walked into Peter Elias's office and said, 'Peter, I am now an electrical engineer.' He said, 'Great! Then I can appoint you.'[7]

Stroke was citing researchers who influenced the institutional and cognitive scope of optics. Brooks was Dean of Engineering and Applied Physics at Harvard University; Townes of MIT was the inventor of the ruby laser; physicist Arthur Hardy of MIT had made notable contributions to spectrophotometry and was long-time Secretary of the OSA and officer in the Commission Internationale de L'Éclairage (CIE); Toraldo di Francia was one of the founders of the International Commission for Optics (ICO) established after the Second World War; Elias was later Professor of Electrical Engineering at MIT and Nobel Prize winner in information theory. A new mood in optics was definitely in the air.

After joining the University of Michigan, Stroke vaunted this intellectual melding in his summer courses:

Many of the most dramatic advances in the field of optics in the last decade or two were directly stimulated or originated by advances in electrical engineering, in its various branches of communication sciences, microwave electronics and radio-astronomy. The operational Fourier transform treatment of optical image forming processes and spectroscopy, the introduction of resonant structures and of optical feedback control, the remarkable simplicity of optical computing, communication systems, and coherent background (heterodyne) detection, the exploitation of the statistical and coherence properties of electromagnetic signals and radiations, as well as polarization in interferometry and astronomy, the dramatic development of light amplification and control in optical masers, and more recently, the newly dramatic achievements in 'lensless' photography and 'automatic' character recognition, and non-linear optics, are some of the more well known examples of the interdependence of theory and techniques throughout broad ranges of the electromagnetic domain, in astronomy, radio-astronomy, physics and electrical engineering. Skilful recognition and exploitation of basic similarities in pursuits throughout the entire electromagnetic domain is proving most useful in pinpointing new areas of research and of industrial applications in what may be called 'electro-optical science and engineering'.[8]

Modern optics at the University of Michigan came to be taught in the Electrical Engineering Department, electrical engineering itself having been a spin-off of the Physics department a half century earlier.[9] Emmett Leith described the special characteristics of the University of Michigan graduates:

The engineer working in these areas must not only have training in both optics and electrical science, but in addition, must understand how these areas relate so that he can treat such

[7] Aspray, William, 'George Wilhelm Stroke, Electrical Engineer, an oral history conducted in 1993 by William Aspray, IEEE History Center, Rutgers University, New Brunswick, NJ, USA', 1993.

[8] Stroke, George W., 'An Introduction to Optics of Coherent and Non-Coherent Electromagnetic Radiations', Course notes, Summer School, University of Michigan Ann Arbor, March 1965 (viewed August 2003).

[9] The U-M Department of Engineering was first established in 1857, focusing on mechanical engineering. In 1889 the first electrical engineering course was offered in the physics department, and in 1895 the Electrical Engineering Department was formed. The Department became part of the College of Engineering in 1915.

> hybrid systems in a unifying way. Our program is designed to meet this need. On the other hand, the program does not produce conventional optical engineers, trained in the design of lenses, cameras or other optical instruments in which the electrical aspect is absent or minor, or readily separated from the optical function. Indeed, such an aim is probably not appropriate to an electrical engineering department. The optics courses include physical optics, presented from the Fourier viewpoint, modern optics, infrared technology, incoherent optics, holography and optical data processing, electro-optical devices, physiological optics, nonlinear optics, and geometrical optics, the latter emphasizing the use of computer techniques.[10]

There had been a decided shift in the centre of balance from relatively straightforward classical optics taught in physics departments to the modern optics taught in electrical engineering departments. Leith later suggested that holography had emerged as a successful field closely related to electrical engineering because of a one-way cross-fertilization between radar science and conventional optics: all published work in optics was available to radar scientists, but the corresponding work in synthetic aperture radar was all classified and unavailable to optical scientists. He cites the example of the holographic complex spatial filter as a case of optical concepts infusing into communication theory.[11]

The examples mentioned by Stroke and Leith nevertheless covered an unwieldy expanse of intellectual territory. Few institutions recognized anything approaching a *discipline* of electro-optics. As late as 1978 the borders were disputed. Nicholas George of the University of Rochester, NY, observed that his was the only American institution that awarded a bachelor's degree in Optics. He denoted its graduates 'opto-electronic engineers' in contrast to 'electro-optical engineers' he claimed to be produced by Cal Tech, the University of Arizona, and a few other institutions. George noted 'it's a new breed of engineer, reflecting the close coupling of electronics and optics', witnessing 'a five fold increase' in numbers since the early 1950s. The editor of *New Engineer* magazine deplored the 'salami-slicing of the profession' into so many specialities, though, because demand changed more quickly than university engineering departments could adapt their teaching. He urged greater breadth and generalization.[12]

8.2 CARVING A NICHE WITH JOURNALS

Schismatic professions were a particular problem for journals, most of which traditionally served existing disciplines. The first papers on holography necessarily were submitted to the journals that seemed either closest to the authors' professional allegiances (e.g. the organs of societies of which they were members) or which seemed most apt for the

[10] Leith, Emmett N., 'Electro-optics and how it grew in Ann Arbor', *Optical Spectra*, May 1971: 25–6, quotation p. 26.

[11] Leith, Emmett N., 'The legacy of Dennis Gabor', *Optical Engineering* 19 (1980): 633–5.

[12] George, Nicholas to R. H. Jackson, 1978, MIT Museum 42/1278. Rochester, NY, had been the American centre for traditional optics since the late 1920s. In 1929, the University of Rochester opened an Institute of Optics, with funding provided by Eastman Kodak and Bausch & Lomb and faculty recruited from England. By the end of the century, the Institute had spawned over 100 optical companies in the city.

particular information to be communicated. In the process, the scope and meaning of holography were constricted as much as the new subject stretched disciplinary boundaries. Different practitioners learned of holography either as a new optical science, as an expression of information theory, or as an application for microscopy, acoustics, metrology, or mechanical engineering. Kendall Preston, for example, an optics research manager for Perkin Elmer, complained that

> the understanding you need to be really effective in holography must come from an understanding of electromagnetic radiation, not silver halide emulsions. Therefore, you need a concentration of people whose background lies in electric field theory, and this invariably means electrical engineers and physicists rather than chemists.[13]

These new specialists were also defined by their workplaces. A key organizational event broadening the subject was the appropriation by optical engineers of the intellectual perspectives and potential dividends of holographic technology. In return, the working context of the engineers directed and, in some ways, constrained the scope and content of the technology. This complementary co-evolution gave stability to both the subject and its workers. As discussed earlier, the early community of holographers practiced their art at a narrow but expanding range of sites. Initially, industrial laboratories highly dependent on military contracts (e.g. TRW Defense and Space Systems, Lockheed, Hughes Aircraft, Aerodyne Research, Rockwell, Grumman Aerospace, and Harris Electronic Division) pursued holographic applications under DoD and NASA contracts, while other large industrial laboratories (e.g. Bell Telephone, RCA, Texas Instruments, CBS, and McDonnell Douglas) explored commercial applications such as optical data storage, communications, and imaging. Most such sites were well funded but largely invisible in wider culture: holographers were isolated from public interaction even if they were becoming an increasingly self-aware community.

Journals helped optical engineers to become a visible specialist group from the mid-1960s. The journals of publication shifted as holography was redefined as a subject and budding specialty. The early publications on wavefront reconstruction were in journals of general science or physics (e.g. *Nature, Journal of Applied Physics, Applied Physics Letters, Oyo Buturi*). During the holography boom of the 1960s, publications appeared as frequently in these as in a wide range of journals of modern optics (e.g. *Journal of the Optical Society of America, Optica Acta, Applied Optics, Optics & Spectroscopy, Optics Communications, Optik*) and electronics (e.g. *Bell Systems Journal, Proceedings of the IEEE, Radiotechnika i Elektronika*). By the early 1970s the profession of optical engineering, considerably boosted by military funding, usurped these prior domains of publication and in America *Applied Optics* and *Optical Engineering* became increasingly important journals for papers on holography.[14]

[13] Wolff, Michael, 'The birth of holography: a new process creates an industry', *Innovations* (1969): 4–15, quotation p. 14.

[14] *Applied Optics* first appeared in 1962. Deane Judd, the editor of Journal of the Optical Society of America (*JOSA*), had hoped that the new journal would publish the more mundane applications-related

The process of publication can be portrayed in ecological terms. The development of the new knowledge of holography was akin to the pressure of a new alien species on a stable ecological system. Research on holography altered the distribution of resources for the existing species (seen as research teams, subject areas, or journals), thereby altering their distribution.[15] Existing specialties jostled for position and adapted in the process, often by accepting holography as a variant of themselves for a time. Eventually, optical engineering—a young specialty that adapted rapidly to the newcomer—seemed to offer the most suitable home to holography.

A decade earlier, however, optical engineering had been a very different subject indeed. Optical engineers until the mid- to late-1960s were versed principally in geometrical optics and mechanical integration. The original name of the professional organization for American practitioners, the Society of Photographic Instrumentation Engineers (SPIE), reflects their dominant orientation. A group of seventy-four engineers founded the SPIE in July 1955. It invaded an environment already crowded with similar but narrowly conceived professional societies: The Society of Photographic Engineers (SPE), concerned with photographic emulsions, exposure, and processing; the Society of Motion Picture and Television Engineers (SMPTE), dealing with standards used in the entertainment industry; the Instrument Society of America (ISA), which focused on transducers and telemetry; the American Society of Photogrammetry (ASP), concentrating on accurate aerial mapping from stable aircraft; and, the Optical Society of America (OSA), a catch-all organization concerned with optical science.[16]

Even so, there was a niche for a new technical organization for the test engineers then working on military development. One of the founders recalled that, 'all the societies had an aversion to discussions about hardware of any sort, nor did they properly recognize the need for new equipment'.[17] In the post-war environment, the important and unsatisfactory applications of optical engineering were rapid-exposure still and cine cameras for recording and measuring transitory incidents such as nuclear explosions and tracking equipment to follow rapid events such as missile and rocket launches.

These concerns dominated the early members of the SPIE, who cast themselves as the first society for optical engineers. The December 1964/January 1965 issue of *SPIE*

papers, but John Howard, the new editor of *Applied Optics*, aimed to publish subjects on lasers, space optics, and optical engineering with rigorous refereeing standards (Howard, John M., 'The early years of *Applied Optics*', *Optics & Photonics News*, 14 (9), Sep. 2003: 22–3).

[15] See, for example, Abbott, Andrew Delano, *The System of Professions: An Essay on the Division of Expert Labor* (Chicago: University of Chicago Press, 1988) for a theoretical discussion focusing on professional psychology; and Divall, Colin and Sean F. Johnston, *Scaling Up: The Institution of Chemical Engineers and the Rise of a New Profession* (Dordrecht: Kluwer Academic, 2000) for analysis of an emerging engineering profession. On the growth of other engineering professions, see Layton, Edwin, *Revolt of the Engineers: Social Responsibility and the American Engineering Profession* (Cleveland: Case Western Reserve University Press, 1971).

[16] For a discussion of the transition from 'narrow-niche' technical groups to a 'generic' or research-technology perspective in America, see Shinn, Terry, 'Strange cooperations: the U.S. Research-Technology perspective, 1900–1955', in: B. Joerges and T. Shinn (eds), *Instrumentation: Between Science, State and Industry* (Dordrecht: Kluwer Academic Press, 2001), pp. 69–95.

[17] Woltz, Robert L., 'Before the beginning of SPIE', *SPIE Journal* 5 (1967): 1–7, quotation p.2

Journal, for example, had articles on photographic studies of convection patterns, image motion and resolution for air-to-air photography, and optical techniques of visual simulation. Advertisements were principally for special cameras, developing machines, projectors, and tracking devices.

A paradigm for such conventional optical engineering was the CORONA surveillance satellite programme. Inaugurated in 1958 by the CIA and American Air Force as a short-term project to develop satellite reconnaissance, CORONA developed into a series of satellite-borne cameras that would return a large canister of exposed high-resolution film to earth. The project, initially pursued under the cover story of a scientific and engineering satellite programme called Discoverer, proved more successful than an Air Force alternative that relied on a television camera and telemetered data. Up to 1972, 145 launches of five generations of the satellite took place and the programme was fed into succeeding classified surveillance satellite programmes and the civilian LANDSAT earth resources satellite series.[18] The CORONA programme enrolled optical engineers from traditional American centres of expertise. An experienced reconnaissance camera design team from Boston University formed the Itek Corporation in the late 1950s, which was to define the camera concept and design and manufacture the lenses; Fairchild Camera designed and built the cameras. The Boston University Optical Research Laboratory (BUORL) had a genealogy extending back to military funding during the Second World War. From 1941, aerial camera lenses were designed in the basement of Harvard College Observatory, in Cambridge, across the Charles River from Boston and manufactured by the Perkin Elmer Company in Connecticut. In 1943, this was replaced by a brick facility on the Harvard University campus known as the Harvard Optical Laboratory. At the end of the war the chief of astronomy at Harvard, Harlow Shapley, and its President, James B. Conant, decided that the university must not profit from wartime research and that it would return to peacetime activities. So its Optical Laboratory was razed to the ground and the staff transferred to Boston University to become BUORL in 1946. In this way the network of universities, colleges, and firms along the Charles River was strengthened during the war and supported the post-war establishment of optical engineering and, eventually, holography research.

The Lockheed Missiles and Space Company in Palo Alto, California, managed the CORONA project as a whole and another group at Stanford University (also in Palo Alto) investigated aspects of camera design during the early 1960s. Important advisors for the project and other intelligence agency programmes included Edwin Land of Polaroid and Carl F. J. Overhage, Chief of Eastman Kodak's Color Laboratory.

Several of these players also became important in early holography, but optical engineering of the time was a very different subject. Despite its exotic platform, for

[18] Yet, as a military space analyst and historian notes, the repercussions of this important application of optical engineering are still murky: 'the role that the CORONA [. . .] played in early civilian earth resources policy and technology development is still largely unknown' (Day, Dwayne A., 'The development and improvement of the CORONA satellite', in: D. A. Day, J. M. Logsdon, and B. Latell (eds), *Eye in the Sky: The Story of the Corona Spy Satellites* (Washington DC, 1998), pp. 48–85, quotation p.(82). See also Lewis, Jonathan E., *Spy Capitalism: Itek and the CIA* (New Haven: Yale University Press, 2002).

example, the CORONA optical design was largely conventional. Over the life of the programme the cameras and satellite stabilization gradually improved the resolution of images from 24 to 6 ft and to allow for different satellite heights (satellite motion, which smeared the imagery, had to be compensated). Eastman Kodak developed high-resolution film on a polyester base to make the most of the available camera lenses and the necessarily tortuous and high-vacuum film path through the cramped camera and the camera manufacturer Hasselblad adapted its own designs for space-based operations.

These activities, while supported by military funding and motivations much like those supporting Project MICHIGAN, were far removed intellectually and practically from the physical-optics concerns of the Willow Run Laboratories (the same division of labour occurred in the Soviet Union; while Denisyuk's wave photography and synthetic aperture radar research was carried out at the Vavilov Institute, the *Zenit* Soviet surveillance satellite cameras were designed under Dr Yuri Frumkin and manufactured at the Krasnogorsk Optical-Mechanical plant near Moscow).[19] At WRL, opto-mechanical design was relatively low in the hierarchy of necessary skills; research was dominated by problems of diffraction, interference, and optical coherence. The Optics Group at Willow Run, and the commercial research and development activities at nearby Conductron Corporation, had genealogical connections with other specialties: with spectroscopy, for example, through their shared need for high-resolution photographic plates and with electrical engineering through communication theory.

Gary Cochran recalls that most of the people he hired for Conductron's research and development of holography had no conventional optics experience at all. The available optics specialists were dominated by lens designers, bereft of the kind of experience he was looking for.

> I don't think I could have worked with a pure optical person; because they were interested in rays, they are interested in analysis [. . .] There is nothing in what I was doing that had to do with incoherent light; that's not what we did, because with coherent light, you had to have an entirely different lens and mirror structure. Coherent light was all we were into; we weren't into lens or camera design, or movies, or anything else.[20]

The same shift of requirements was beginning in the aerospace industry, which gradually broadened the domain of optical engineers. The study of airflow in wind tunnels and shock tubes, for example, benefited from Schlieren photography, a method of visualizing variations in air density by optical shadowing techniques. Interferometers provided a still more sophisticated way of imaging airflow. So, at least for qualitative uses refined optical methods grew in value. Government-supported institutions supported this gradual transition. For instance, the Arnold Engineering Development Center (AEDC), opened in Tullahoma, Tennessee in 1951 and became the primary US Air Force test centre for aircraft and spacecraft, operating wind tunnels for flight simulation, rocket stands for testing, engine cells for simulating jet engine flight, vacuum cells for simulating space, and launchers to simulate ballistics. Some of

[19] Gorin, Peter A., 'ZENIT: The Soviet response to CORONA', in: D. A. Day, J. M. Logsdon, and B. Latell (eds), *Eye in the Sky: The Story of the Corona Spy Satellite* (Washington DC, 1998), pp. 157–70.

[20] Cochran, Gary D. to SFJ, interview, 6 and 8 Sep. 2003, Ann Arbor, MI, SFJ collection.

the facilities consisted of German hardware captured at the end of the Second World War, but by the late 1950s the Center was gearing up for missile research. The Air Force contracted the nearby University of Tennessee to instigate a graduate study programme for AEDC workers in 1956, and the University of Tennessee Space Institute (UTSI) was founded in 1964 with the continued close support of the Air Force and NASA. Indeed, the NASA Marshal Space Flight Center in Huntsville, Alabama, was a mere 70 miles southwest of AEDC.

This environment of institutions led to a revitalized role for optics and a receptive context for scientific holography. Jim Trolinger recalls:

> Optical instrumentation groups in the aerospace industry were some of the first people sufficiently funded to be able to afford lasers. The two major fronts were powered by NASA, in a race to put satellites into orbit and to go to the moon and beyond, and the military, who were trying to build defenses as well as offensive weapons in a race against the Soviets. Funds were fairly easy to come by for new research, and especially my friends in NASA had practically unlimited materials budgets, so much that they sometimes had difficulty spending the money.[21]

In 1968, UTSI developed a short course on holography to support the growth of a holography group at the University. It included Trolinger, Juris Upatnieks, John DeVelis (author of one of the first books on scientific holography)[22] and others and provided an academic background in the subject for the AEDC engineers. There was also a limited amount of exportation of concepts and applications: the university helped to organize an aerospace conference at the Von Karmen Institute in Brussels, Belgium, through NATO's AGARD (Advisory Group for Aerospace Research and Development), with a heavy American presence. Umbrella organizations furthered this dissemination of information and research culture to clusters of individuals. For example, an AGARD review of the use of optical diagnostics in NATO countries allowed a further sharing of information between centres such as Rolls Royce in England (where Rick Parker, Bernard Hockley, and Peter Bryanston Cross were using holography to study jet engine interiors to observe flow fields and blade vibration), Loughborough University (where John Tyrer was using holographic interferometry for vibration analysis and Nick Phillips was developing advanced bleaching methods), The French-German Research Institute of Saint-Louis (where Paul Smigielski, Gunter Smeets, Bernard Kock, and Hans Pfeifer were studying particle velocimetry), and the Office National d'Etudes et de Recherches Aérospatiales (ONERA) in France (where Claude Veret was doing holographic interferometry of flow fields). On the other hand, plans to create a major holography institute at the University of Tennessee did not come to fruition owing to personnel changes.[23]

[21] Trolinger, James D., 'The history of the aerospace holography industry from my perspective', unpublished manuscript, May 2004, SFJ collection. Trolinger (b. 1940) had studied physics at the University of Tennessee while working at AEDC as an intern. After gaining a PhD there, he worked at the UT Space Institute with colleagues Joe O'Hare, Mike Farmer, and Ronald Belz on holographic applications.

[22] DeVelis, John B. and George O. Reynolds, *Theory and Applications of Holography* (Reading, MA: Addison-Wesley Pub. Co., 1967).

[23] Trolinger, James D., 'The history of the aerospace holography industry from my perspective', unpublished manuscript, May 2004, SFJ collection.

In this way, changing technical requirements altered the culture of optical engineering. The *SPIE Journal* responded by inaugurating a new topic for articles in 1964: a tutorial on physical optics, with which the editor hoped to provide 'a complete notebook of physical optics principles',[24] and thereby provide a new base of knowledge for practicing optical engineers. That year, the name of the organization also changed, from the Society of *Photographic* Instrumentation Engineers to the Society of *Photo-Optical* Instrumentation Engineers. But even this name change was too little, too late.

Owing to developments such as coherent optical processing and aerospace optical diagnostics, optical engineering began to broaden its intellectual scope during the 1960s and 1970s, remaining an important outlet for military-funded research especially in the new field of digital image processing. The source of project funding and applications were subtly reinforced by the journal illustrations: images of tanks and jeeps frequently served as examples for image-processing algorithms. The society journal *Optical Engineering* replaced the *SPIE Journal* in 1972, and in 1981 the SPIE submerged its title beneath an acronym and the descriptor 'The International Society for Optical Engineering', thereby consolidating a new profession and de-emphasizing its narrower intellectual origins. Until the end of the cold war period, some SPIE meetings had entry restricted to American citizens or North American residents, intended to slow the dissemination of information deemed to be of potential military usefulness. The profile of SPIE members was shaped by their employment: a significant fraction worked on defence-related research. In 1964, for instance, at least one-third of the local chapters of the SPIE were located on military testing ranges or bases (including the Naval Ordnance Test Station (NOTS) at China Lake, California; Vandenberg Air Force Base, California; the Kwajalein launch facility, Marshall Islands; and White Sands, New Mexico).

This association between military research and optical engineering was to persist. The Strategic Defense Initiative (SDI, or the 'Star Wars program') was founded on optical engineering research for lasers, optical components, and sensors. Similarly, following the Sep 11, 2001 attacks and the George W. Bush administration's establishment of a Department of Homeland Security, the SPIE formed a 'Homeland Security Special Interest Group' in 2003.[25] Despite the close connections with American military and aerospace funding, the SPIE at the end of the cold war began to fulfil its claim to international coverage by opening chapters in the former Soviet Union. Through this disciplinary and geographical expansion, the membership of the organization rose from 1000 in 1961, to 3000 in 1980 and to 14,000 in 1999. Thus the environment of sponsorship that had produced synthetic aperture radar processors and off-axis holography seeded the revitalized profession of optical engineering.

[24] 'Physical Optics Notebook', *SPIE Journal* 3 (1964).

[25] The name was altered almost immediately to the '*Global* Homeland Security *Technical* Group' to reflect the Society's international aims, which its President described as a 'complicated task' involving learning how to 'operate efficiently and effectively under a broad range of political, economic and social/cultural conditions' (Bilbro, James W., 'Letter from the President', *SPIE Member Guide*, (2004), 1).

Despite the predominance of military and governmental funding for American holography, there was a degree of cultural and political resistance that reduced the coherence of the budding subject, and which was to grow, as discussed in the next chapter. One such divergence was the 'Subversive Optical Society', or SOS, which grew in the Boston area *c.*1955–70. According to Adolf Lohmann, the members of this unofficial club were young, anti-establishment, humorous, and oriented away from the staid geometrical optics and Rochester, NY, focus of the Optical Society of America. They included Ed O'Neill of Boston University and Itek, Larry Mertz of Block Associates, and Emil Wolf. Significantly, though, both Itek and Block Associates were funded predominantly by remote surveillance projects for the CIA and military.[26]

8.3 MEETINGS AS SOCIAL NUCLEI

Meetings on holography were the clearest strategy for promoting the would-be profession and its activities. One reason for this is the dramatic influence of demonstration: seeing examples of new imaging techniques that could not be illustrated adequately in journals. Dennis Gabor had cited his conference presentations at the London Conference of the Electron Microscope Group in April 1948 and the BAAS meeting that July as more important than papers in publicizing his message. The same was true to a much greater extent with Upatnieks' and Leith's OSA presentation in March 1964. As holographers repeatedly emphasized, seeing is believing.

Meetings were concerned not merely with presenting new results, however: they were also important for this fledgling subject in defining nomenclature, techniques, and a growing sense of community. Such activities are a necessary precursor to the formation of a sub-discipline and, eventually, to a profession based upon it. Interestingly, the first venues for conferences dedicated to holography appeared outside America. The first dedicated conferences had an engineering slant and focused on applications. As early as 1967, for instance, a symposium of holography was held in the spring meeting of the Japanese Society of Applied Physics, and the papers presented were from a combination of universities and large firms. They included presentations on microwave holography by Y. Aoki of Hokkaido University; on the measurement of lens characteristics by holographic techniques by T. Ose of the University of Tokyo; on the characterization of film resolution by N. Nishida of NEC; and, on holographic interferometry by K. Matsumoto of Canon, J. Tsujiuchi of the Government Mechanical Laboratory, and T. Tsuruta of Nikon.[27]

[26] Block Engineering focused on spacecraft-borne spectrometers, particularly Fourier-transform spectrometers, for which Lawrence Mertz made important conceptual improvements. Mertz opposed the narrow publication criteria of the *Journal of the Optical Society of America* during the early 1960s by publishing technical notes in it as paid advertisements.

[27] Tsujiuchi, Jumpei to SFJ, unpublished report, 20 June 2003, SFJ collection.

The first truly international conference on engineering applications was organized at Strathclyde University in Glasgow in September 1968.[28] Held over a period of four days, it brought 255 participants from eighteen countries including 167 from around Britain, twenty-six from America, twenty-three from Germany, ten from France, nine from Belgium, four from Sweden, three from Switzerland, two each from Japan, Denmark, and Italy, and one each from South Africa, Czechoslovakia, Luxemburg, Austria, Ireland, Canada, and Monaco. The meeting was accompanied by a commercial exhibition that included the major laser manufacturers. The organizers vaunted the conference not merely as an internationalization of the holography community, but as a key element in the very creation of such a grouping:

> This acceleration has come about because the physicist and the engineer (the academic scientist and the practical man of affairs), are, at last, coming to understand each other better. It is not so long ago that they had entirely different outlooks with little understanding of each other's problems.[29]

While this portrayal of holographers as the merging of what the speaker described anachronistically as 'string and sealing wax abstract sort of persons' and 'rule of thumb' types was inaccurate, it illustrated his awareness of a new breed of worker and one characterized more recently by Shinn's term *research technologist*.

Another early attempt at internationalization and consolidation of holographers was a series of three US–Japan Holography Seminars, in October 1967 (Tokyo), October 1969 (Washington DC), and in January 1973 (Hawaii). Jointly sponsored by the Japanese Society for the Promotion of Sciences and by the National Science Foundation, USA, the seminar series was first proposed by George Stroke, who had just moved to SUNY (Stony Brook) from Ann Arbor.[30] The subject of the first and second seminars was principally optical holography; the third concerned acoustical holography and imaging. The first two therefore attracted similar participants—six from each side in 1967 and ten each in 1969 and 1973.[31] This two-country grouping again encouraged the common interests of scientific and engineering research.

Holography conferences did more than merely demonstrate new techniques and foster collaboration, though; they also linked disparate fields by the technology they both used. This is illustrated by the Gordon Research Conferences (GRCs) on coherent optics

[28] Robertson, Elliot R. and James M. Harvey (eds), *Symposium on Engineering Uses of Holography* (University of Strathclyde, 1968); Fagan, William to SFJ, interview, 19 Sep. 2002, SFJ collection. The conference was repeated at the same venue through the 1970s.

[29] Sayce, L. A., 'Closing address', in: E. R. Robertson and J. M. Harvey (eds), *The Engineering Uses of Holography* (Glasgow, 1968), p. 560.

[30] Details of the seminars were reported in Stroke, George W., 'U.S.–Japan seminar on holography, Tokyo and Kyoto, 2–6 October 1967', *Applied Optics* 7 (1968): 622; Barrekette, Euval S., *Applications of Holography: Proceedings of the United States-Japan Seminar on Information Processing by Holography, held in Washington, D.C., 13–18 October 1969* (New York; London: Plenum Press, 1971); and Stroke, George W., 'Ultrasonic imaging and holography: medical, sonar, and optical applications', presented at *United States-Japan Science Cooperation Seminar on Pattern Information Processing in Ultrasonic Imaging, 1973*, University of Hawaii, 1974.

[31] Tsujiuchi, Jumpei to SFJ, unpublished report, 20 Jun. 2003, SFJ collection.

and holography held biennially from 1972 until 1997 (see Figure 8.1).[32] The Gordon conferences had been founded in 1931 as an informal venue for discussing advances in science, particularly chemistry. After the Second World War, their role mutated to promote closer communication within the American chemical industry and academic chemistry. The peculiar environment of the Gordon Conferences promoted a distinct sense of professional identity. For example, in order to ensure an atmosphere that encouraged camaraderie and the sharing of ideas and results, the organizers prohibited the publication of presentations and discussions. A typical attendee praised the conferences for providing 'intimate peer review with immediate feedback'.[33] These were also very much nationally oriented meetings. Celebrating a quarter century of activity in 1956, one article referred to attendees as 'Gordon Frontiersmen', emphasizing both the perceived post-war research frontier and American frontier spirit.[34]

By the 1960s, the GRCs had expanded to include meetings in a wide range of burgeoning scientific fields, but retained its national homogeneity. H. John Caulfield, then a researcher at Sperry Rand Research Center in Massachusetts, proposed a Gordon Conference on holography for 1972, seconded by Emmett Leith and Brian J. Thompson, Director of the Institute of Optics in Rochester, NY.[35] Caufield argued that the Gordon conferences could shape the new discipline and its professional community:

> Holography is a field in which the activity is so widespread and the interest so great that simultaneous invention of a new technique by several groups is quite common. Typical meetings of the Optical Society of America devote about one-third of their sessions to holography and allied areas. Over 100 industries and over 50 universities in the United States alone carry out research in this area. Interest seems to be increasing rapidly in this whole area. The very rapid growth of this field has been achieved at some expense in terms of commonly-propagated errors, ill-defined concepts, and inadequate consensus in terminology. No meeting has been either held or scheduled which provides the opportunity for most of the serious investigators to meet and work out these problems.[36]

Thompson, seconding the proposal, suggested that the conference 'would provide an excellent format for thrashing out some of the more important aspects of the subject and

[32] The titles of the GRC conferences varied, but included 'holography' and 'optical processing' or 'image processing'.

[33] The meetings were held in association with the American Association for the Advancement of Science (AAAS). In 1948 the Chemical Research Conferences were renamed Gordon Research Conferences in honour of chemist Neil E. Gordon, their founder (who committed suicide a year later) (Wedlin, Randy, 'Scientific utopia: the Gordon Research Conferences', *Chemistry* (Aug. 2002): 25–9; quotation p. 26). See also Storm, Carlyle B., Jimmie C. Oxley, and Alexander M. Cruickshank, 'The Gordon Research Conferences: a brief history', presented at *GRC 50 Years in New Hampshire*, New Hampshire, 1997.

[34] Parks, W. George, 'Doctor Gordon's serious thinkers', *Saturday Review*, 4 (Aug. 1956): 42–9. On the portrayal of American research as a drive to extend frontiers, see Zachary, G. Pascal, *Endless Frontier: Vannevar Bush, Engineer of the American Century* (New York: London: Free Press, 1997).

[35] H. John Caulfield (b. 1936 in Halletsville, Texas), having obtained his PhD in physics from Iowa State University, was Senior Scientist at Texas Instruments 1962–7, Principal Scientist at Raytheon Co. Night Division Dept in Massachusetts during 1968, and then moved to Sperry Rand. When the first Gordon Conference was held in 1972, he had moved to Block Engineering. Brian J. Thompson worked in many aspects of holography, specializing in its application to particle measurement.

[36] Caulfield, H. John to A. M. Cruickshank, by letter, 15 May 1970, Sudbury, MA, Chemical Heritage Foundation.

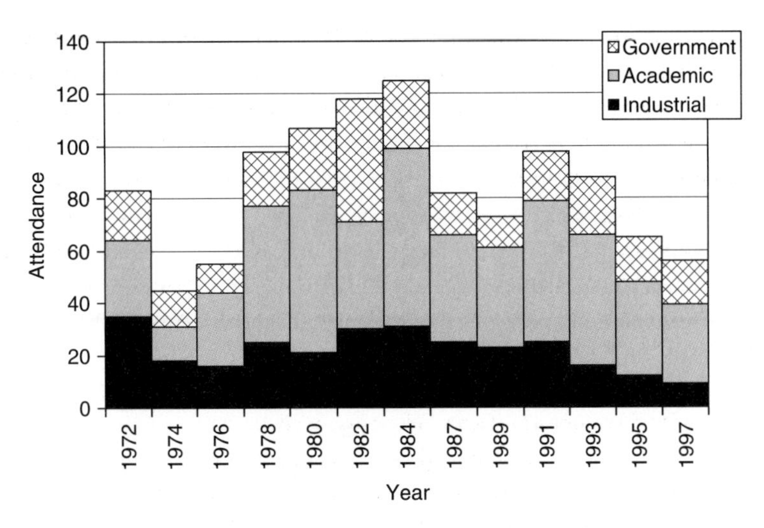

Fig. 8.1. Attendance at the thirteen Gordon Research Conferences on Holography and Optical Information Processing.

perhaps be able to guide uses of the science in a more meaningful way'.[37] Joseph Goodman of Stanford Electronics Laboratories added his support:

> Such a gathering would be particularly important since, for the most part, holography has reached the stage of development where some concentrated attention to the various proposed applications would help to outline those areas which have the greatest potential benefit to society.[38]

The Gordon Conferences not only organized the new discipline; they also self-consciously defined its professional community. The *chairman* traditionally chose Discussion Leaders, Principal Speakers, Invited Participants, and discussion topics (in fact, the original organizer referred to the chairman's right to invite participants as a means of keeping out 'disruptive elements').[39] Continuity was assured by electing the Chairman and Vice-Chairman for the next conference by the attendees of the current conference. The remainder of attendees were drawn to the Conference for the informal discussions:

> I anticipate about a 3:2 ratio of academic to industrial conferees on the basis of attendance at other meetings. We will probably draw a few Canadians and Europeans. Most of those attending will be doing research in holography. About 10% will be from highly diverse fields. Their only common interest will be the use of holography as a tool in their fields.[40]

[37] Thompson, Brian J. to A. M. Cruickshank, by letter, 15 May 1970, Rochester, NY, Chemical Heritage Foundation.

[38] Goodman, Joseph W. to A. M. Cruickshank, letter, 18 May 1970, Stanford CA, Chemical Heritage Foundation.

[39] Caulfield, H. John to SFJ, email, 12 May 2003, SFJ collection.

[40] Caulfield, H. John to A. M. Cruickshank, by letter, 15 May 1970, Sudbury, MA, Chemical Heritage Foundation.

Caulfield's projection proved accurate. The 1972 conference, held in New Hampshire, drew 83 attendees, 35 of them from industry (nearly all from large industrial research labs) 19 from academe and the remainder from government laboratories. Even so, mobility of personnel and overlap of funding made such categories imprecise. For example, a number of the industrial laboratories represented (e.g. TRW Defense and Space Systems, Lockheed, Hughes Aircraft, Aerodyne Research, Rockwell, Grumman Aerospace, Harris Electronic Division, and ERIM) were funded primarily by military contracts and several of the government laboratories (e.g. Sandia National Laboratories, Naval Research Laboratory, Brookhaven Laboratory, and Defense Advanced Research Projects Agency) were also focused on military research.

As mentioned above, the preponderance of attendees were from America, largely because of the direct invitations by the conference Chair. This national perspective did not go unnoticed; Howard M. Smith of the Eastman Kodak Research Laboratories in Rochester suggested that the Gordon Conferences would correct a previous imbalance, observing, 'I think that a need definitely exists as most of the large conferences devoted solely to holography have been in Europe'.[41] Emmett Leith sized up the American holography community as he saw it:

> There are, in the United States, several hundred persons who have had some experience in this field, as evidenced by the publication of one or more papers. There are, perhaps, two or three hundred people who are seriously engaged in holography as a full-time activity. Their major aims are varied; they are primarily concerned with the application of holography in such fields as non-destructive testing, biology, display devices, data storage, and pattern recognition.[42]

Of the 1972 attendees, 95% were from American firms, industry, or government (see Figure 8.2). Subsequent attendances through the 1970s were even more tightly knit: of the 45 attendees in 1974, all but one was American; 1976 garnered 55 attendees, 51 of them American; and the 1978 conference attracted 87, and still included only five from outside the America.[43] The holography conferences also attracted few non-scientists, further consolidating a sense of community (see Figures 8.3, 8.4, 8.5, 8.6 and 8.7).[44]

[41] Smith, Howard M. to A. M. Cruickshank, by letter, 19 Jun. 1970, Rochester, NY, Chemical Heritage Foundation.

[42] Leith, Emmett N. to A. M. Cruickshank, letter, 20 May 1970, Chemical Heritage Foundation.

[43] Non-American attendances at GRCs rose gradually and by the last holography conference in 1997 amounted to 21. The American dominance was not universal across all fields studied, though: some Gordon Conferences in other subjects were held outside the USA from the 1990s (Storm, Carlyle B., Jimmie C. Oxley and Alexander M. Cruickshank, 'The Gordon Research Conferences: a brief history', presented at *GRC 50 Years in New Hampshire*, New Hampshire, 1997).

[44] There were two exceptions. The 1976 conference brought Lloyd Cross and members of his San Francisco School of Holography in a bus to the Santa Barbara Biltmore Hotel to display multiplex holograms. The culture shock was later recalled by several parties (Caulfield, H. John to SFJ, interview, 21 Jan. 2003, Santa Clara, CA, SFJ collection; Leith, Emmett N. to SFJ, interview, 22 Jan. 2003, Santa Clara, CA, SFJ collection; Unterseher, Fred to SFJ, interview, 23 Jan. 2003, Santa Clara, CA, SFJ collection). At the next Gordon Conference in 1978, artist Anaït Stephens attended, and Posy Jackson spoke on the work of the New York Museum of Holography, which again stressed artistic, rather than scientific, applications. For more on artisans and artists, see the following two chapters, Chapters 9 and 10.

Fig. 8.2. Scientific holographers: attendees at the first Gordon Research Conference on Holography and Coherent Optics, July 1972 (Trolinger collection).

Fig. 8.3. Hans Bjelkhagen recording a reflection hologram of the Swedish crown at the Royal Institute of Technology, Stockholm, 1974 (MIT MoH collection).

The meetings proved popular with participants. The GRCs were often held in isolated locations in New England, with mornings and evenings devoted to lectures and papers, late evenings open for conversations and debates, and afternoons free for recreation and informal discussions. Adolf Lohmann, who attended most of the thirteen meetings, recalls, 'I remember a lively discussion about evanescent waves, on a sailboat. Or between tennis matches we debated the virtue of pre-exposing the photographic plates before using them for holographic recording'.[45]

[45] Lohmann, Adolf W. to SFJ, fax, 18 Aug. 2003, SFJ collection.

Fig. 8.4. Juris Upatnieks at ERIM, Ann Arbor, MI, 1979. The photograph shows the hologram recording setup, without the film holder, of a 12 in. diameter 360° hologram recorded for the Museum of Holography (MIT MoH collection).

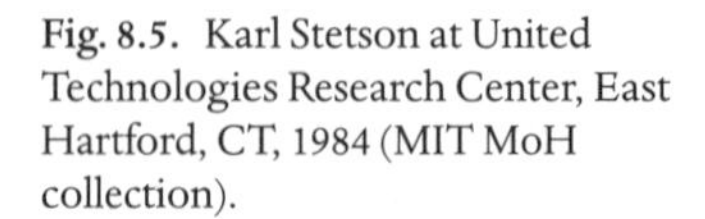

Fig. 8.5. Karl Stetson at United Technologies Research Center, East Hartford, CT, 1984 (MIT MoH collection).

Fig. 8.6. Experimental laser for hologram recording, NIKFI, Moscow, *c.*1985 (Novosti photo; MIT MoH collection).

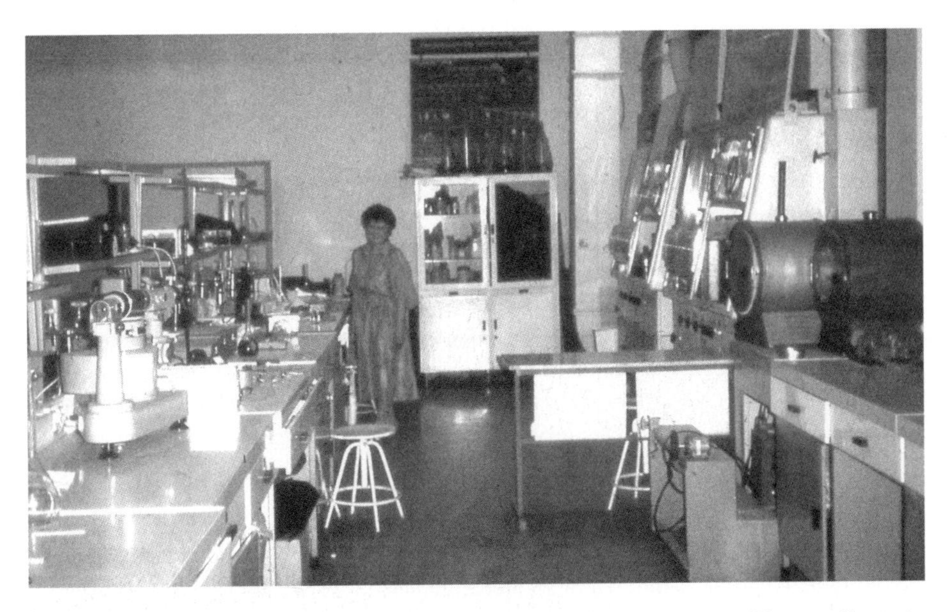

Fig. 8.7. Svetlana Soboleva in Moscow laboratory, 1989 (H. Bjelkhagen photo, Bjelkhagen collection).

The opportunities for socializing were superior to the semi-anonymity and brief sessions at more general optical conferences, breeding a sense of community among the participants. This rolling group of mainly American scientists and technologists was simultaneously exclusive and liberating for its attendees. Its invited attendees bolstered common perspectives and interests and the absence of published proceedings allowed frank and free-ranging discussions between peers. Several of the elements of a proto-profession were in place: an occupational homogeneity, a common intellectual base, and mutual recognition by peer selection for attendance. It is all the more interesting, then, that this embryonic profession failed to prosper.

One complication was the increasing competition from other meetings. By the 1970s, holography meetings were appearing in an ever-more complex ecology of applications. Caulfield had observed in 1970 that

> The only regularly-scheduled meetings of great interest to holographers are the Spring and Fall meetings of the Optical Society of America. In the spring of 1971, the Society of Photographic Scientists and Engineers will hold a major conference which might be competitive with a 1971 Gordon Conference. This is our reason for suggesting the 1972 date.[46]

The conference terrain was also being crowded by a burgeoning number of offerings from the Society of Photo-optical Instrumentation Engineers (SPIE). Recognizing an opportunity for expansion, the SPIE had begun offering its own symposia and

[46] Caulfield, H. John to A. M. Cruickshank, by letter, 15 May 1970, Sudbury, MA, Chemical Heritage Foundation.

conferences on aspects of holography from 1968; by 2005, it had organized some ninety conferences devoted to the subject, with as many as nine in a single year.[47] The longest-lived series of SPIE conferences were the *Practical Holography* conferences held annually in California and chaired by Stephen Benton between 1986–2002 and thereafter by Tung H. Jeong. Benton, a perennial campaigner for the subject, recalls that a particular vision of the holographer was promoted by the series:

> I could see that for people to get interested in these meetings, they would have to see holograms, they would have to be presented respectably [. . .] I had been active in the SPIE—a vice president, and invested in the thing on the grounds of previous research on film granularity. They grew it up from nothing to a substantial organisation and service. Not a scientific society in the terms of the Optical Society, frankly, but an important enterprise. SPIE, OSA, and then at a distance the IS&T [Society for Imaging Science and Technology], which is more chemically oriented. SPIE was more entrepreneurially oriented—anyone wanting a conference could sell it to a couple of people, or if they had confidence in you.[48]

Where the Gordon Research Conferences promoted an elite grouping of professional holographers (namely senior scientists and engineers in American corporate, academic, and government laboratories), the SPIE conferences cast their net widely to represent the field of workers comprehensively. But here again, the Chair played a role in defining the profile of speakers. Benton recalls:

> One of the things that gave it mass was, I said (I have a lot of friends in Japan) 'why don't you guys come over and give papers?', and up to half started showing up and giving interesting papers. Also sponsorship—their companies were eager to have them get well known. I would use that money to bring in [. . .] whatever scientists were making things happen in the States.[49]

8.4 DEFINING THE HOLOGRAPHER

Within the broad but receptive disciplinary category of optical engineering, and buttressed by journals and targeted conferences, the identity of the scientific holographer was born. But there was disagreement about precisely what skills and attributes the holographer should have.

By wedding physical optics with electrical engineering, the early subject clearly required disparate skills seemingly best suited to a complementary group of investigators. Such small teams had been a feature of the earliest work. Albert Baez and his undergraduate students at Redlands University in California worked in small groups to

[47] SPIE, 'Seminar In-Depth on Holography', presented at San Francisco, 1968. A peak of conferences and conference papers was reached about 1992, but had evinced a threefold drop by 2000. See Appendix for a graph of conference activity.

[48] Benton, Stephen A. to SFJ, interview, 11 Jul. 2003, Cambridge, MA, SFJ collection.

[49] Ibid.

reproduce and extend published work, modelling his arrangements on the laboratory culture of Paul Kirkpatrick, his former supervisor.[50] The team approach was also the organizational basis for the Willow Run Optics Laboratory, where an investigator or two, a technician, and occasionally other support staff had worked together on projects. Mixed teams brought complementary experience to this still unfamiliar subject. Emmett Leith observed how the Optics Lab had worked well because of the combination of different kinds of specialist: Willow Run hired mainly electrical engineers, but Leith's optics background and one or two conventional optical engineers yielded a fruitful balance. The mixture of an academic and skills-based approach also worked well. Many of those in the Optics Laboratory were pursuing graduate degrees.

The small-team model was applied elsewhere, even if the precise mix was disputed. By 1973, with about five years' experience of forming numerous research teams in holography, a NASA survey could recommend:

> Persons with experience and training in optical theory and laser systems are probably best qualified to perform holographic analysis and testing, although some electrical and mechanical engineers have completed university and industrial courses in the fundamentals of holography. The ideal holographic team is an optical engineer or scientist and a mechanical and/or an electrical engineer. The optical specialist should know some practical photography and laser technology, since film is the usual medium for recording laser beam holograms. If acoustic or microwave holography is to be pursued, then an acoustic engineer, electrical engineer, or physical scientist is highly desirable. The field of holographic data storage, processing and retrieval is even more demanding and is recommended only for an expert. A solid state physicist or engineer with a sound background in optical theory and experimentation would also be a valuable addition to a team.[51]

Trying to build his own research team for holography at CBS Laboratories, Dennis Gabor wrote to his friend Peter Goldmark concerning holography research:

> The ideal thing would be if you could provide two optical laboratories, similar to the one in existence, side by side, one of them with a dark room. Also, next to it, an office for the optical workers. In one of these two labs, I would like to work myself, with suitable assistance. If need be, I could do my work with one or two good technicians, of the type of Roland Muehleisen, but evidently it would be desirable to have also a trained physicist in the team. After the disappointing experiences we have had, I would suggest a young first degree physicist, say 22–26 who would be willing to learn, and whom I could train myself. If he had a course in modern optics, all the better, but it is not absolutely necessary.[52]

As suggested by the quotation, Gabor saw the holographer as a physicist trained on the job, having a smattering of modern optics, and probably working in a small team with varied skills—indeed, rather similar to his own background.

[50] Baez, Albert V., 'Anecdotes about the early days of x-ray optics', *Journal of X-Ray Science and Technology* 7 (1997): 90–7; Baez, Albert V., to SFJ, email, 13 Mar. 2003, SFJ collection.

[51] Dudley, David D. and Computer Sciences Corporation, *Holography: A Survey* (Washington, DC: Technology Utilization Office National Aeronautics and Space Administration, 1973), p. 91.

[52] Gabor, Dennis to P. Goldmark, letter, 3 Jun. 1967, IC GABOR LA/9.

While everyone seemed to agree by the end of the 1960s that holography was best practised in small teams, there was some dispute about the intellectual expertise the members should embody, citing a conflicting mixture of skills, training, and goals. Ralph Wuerker at TRW suggested that the occupation of holography was defined at least partly by its dangers. Inadvertent exposure to laser beams was becoming recognized as a potential danger; Wuerker suggested that 20 mW was considered 'a level of safety', and that 'the second thing on the line of safety is glass plates. I don't know how many of you have scars, but that will also be the mark of the holographer!'[53] He also defined the holographer in opposition to conventional optical scientists, by the attributes of their new subject.

> If you are an old man and have cut your teeth on classical interferometers, interferograms shouldn't focus. Yet when you illuminate a hologram from behind with a ground glass screen, these holographic interference fringes do focus and locate.[54]

Jim Burch, heading the Laser Optics Section of the National Physical Laboratory and its research in holographic interferometry, referred to the practising holographer as a hands-on investigator concerned with engineering analysis.[55] Alexander Metherell, active in acoustical holography, even identified a sub-class, the seismic holographer, who was concerned with reconstructing hidden geophysical features for prospecting.[56]

By the late 1960s, then, holographers had colonized and reshaped optical engineering and were strongly represented in the new academic departments of electro-optics. Their papers filled specialist journals and dedicated conferences around the world. The holographer was becoming established as a specialist technical worker and appeared to be on course as a professional in the making. The collection of practical skills, and the occupation they engendered, were becoming self-recognized. But this was also the very time at which the identity of the holographer suddenly fractured.

[53] Wuerker, Ralph, 'Discussion', in: E. R. Robertson and J. M. Harvey (eds), *The Engineering Uses of Holography* (Cambridge: Cambridge University Press, 1968), p. 72. Other participants at that 1968 conference suggested values between 1 mW and 50 μW entering the pupil of the eye as dangerous.

[54] Wuerker, Ralph, 'Discussion', in: E. R. Robertson and J. M. Harvey (eds), *The Engineering Uses of Holography* (Glasgow, 1968), p. 129.

[55] Burch, J. M., 'Laser speckle metrology', *Proceedings of the SPIE* 25 (1971): 149–56.

[56] Metherell, A. F., 'The present status of acoustical holography', *Proceedings of the SPIE* 25 (1971): 137–46.

9

Culture and Counterculture: The Artisan Holographer

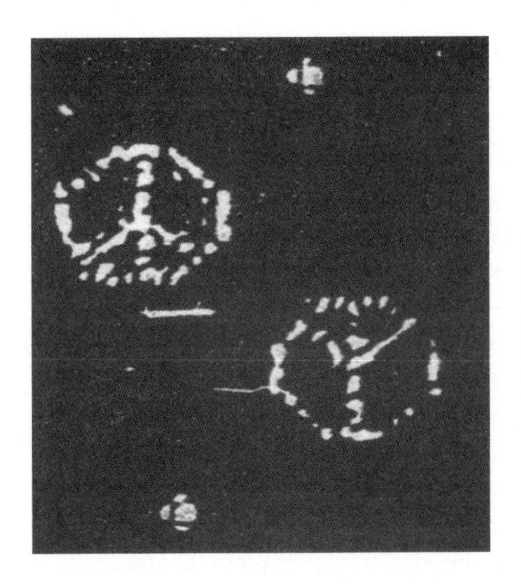

Fig. 9.1. The peace symbol as the subject of a computer-generated Fourier-transform hologram, an atypical illustration appearing in a scientific journal (Dallas, 'Phase quantization in holograms—a few illustrations' *Applied Optics* 10 (1971), 674–6).

9.1 INTRODUCTION

Despite their momentum, holographers did not, after all, evolve into clearly defined scientific professionals. Instead, their definition bifurcated during the 1970s. The source of this schism was the entry of new practitioners who infused wider culture into the subject. The politics of technology drew artisans to holography. They brought craft knowledge: optical ingenuity, renewed curiosity about materials, and an alchemical mystique to processing, and with those skills a new gaze for the subject.[1]

[1] The term *artisan*, used to label a craftsperson skilled in a manual art, sometimes suggests outdated hierarchies. Here, however, I use it to distinguish holographers who have developed wide-ranging competences underlain by tacit knowledge, and trained outside universities or art schools, making them distinct from scientists, engineers, and artists. They are not merely practical tradesmen of a stable art, but rather innovators who saw beyond the contextual channels of earlier workers.

The dominant culture of holography during the 1960s had been that of post-war science, largely allied to contract research in university, government, or corporate laboratories. A popular understanding had developed about what modern science entailed: it was an esoteric, intellectually progressive, elitist and well-funded activity, and having wide economic and intellectual value. Newsreel and television stories presented this culture as one of neck-tied and disciplined male physicists working in a clean laboratory environment among powerful lasers and expensive optical equipment.

Yet this conventional public image did not capture the enthusiasm engendered by modern optics. Holograms, and the lasers that made them, were part of a new kind of optics that physicists and engineers found exciting and intriguingly open-ended. Chapter 7 hinted at the intimate connection between exploration and the invention of new kinds of hologram. Speaking during the question period at one of the early conferences, Ralph Wuerker of TRW suggested the playfulness these workers felt towards their studies:

> It is fascinating to shine a laser on all sorts of things. If you shine it on inanimate objects you see a speckle pattern. If you shine it on a person you do not see a speckle pattern. If you shine it on fruit and other organic things you will find out that the speckle pattern rotates and vibrates. You can calibrate how fresh your fruit is that way![2]

That fascination was largely shielded from the public, though. Young post-war audiences developed a negative evaluation of modern science. The growing distance between 'big-science' holographers and wider culture is illustrated by the student protests of the late 1960s. Ann Arbor was home not only to the Willow Run Laboratory and classified optical research, but also to the Students for a Democratic Society (SDS). While these two events had no initial correlation, their proximity soon became significant.

SDS had been established in 1959 by Alan Haber, a sometime Ann Arbor student, from the youth branch of an older organization for socialist education, the League for Industrial Democracy. Fifty-nine founding members held the first meeting in Ann Arbor in 1960, and two years later the fledgling group adopted the 'Port Huron Statement', a political manifesto written principally by Tom Hayden, former editor of the University of Michigan (U-M) student newspaper. The group called for a more participatory form of democracy to address the social problems of racism, poverty, materialism, and militarism. As early as October 1963—when Leith and Upatnieks were about to begin their first successful experiments with three-dimensional holography—a large student rally on campus protested American intervention in Indo-China.

Ann Arbor's SDS chapter was the largest one in the country during those early years.[3] SDS and the 'Free Speech Movement' (formed at the University of California at Berkeley

[2] Wuerker, Ralph, 'Discussion', in: E. R. Robertson and J. M. Harvey (ed.), *The Engineering Uses of Holography* (Glasgow, 1968), pp. 501, quotation p. 501.

[3] Unger, Irwin and Debi Unger, *The Movement: a History of the American New Left, 1959–1972* (New York: Dodd Mead, 1974). On the youth movement and the role of students at Ann Arbor, see also Breines, Wini, *Community and Organization in the New Left, 1962–1968: The Great Refusal* (New York, NY: Praeger; J.F. Bergin, 1982) and Bunzel, John H., *New Force on the Left: Tom Hayden and the Campaign Against Corporate America* (Stanford, CA: Hoover Institution Press Stanford University, 1983).

in 1964 to protest against heavy-handed actions by their university administrators) became the core of the New Left, part of the wider youth movement dubbed the 'counterculture' by social analysts of the 1960s. In March 1965, Liberal Arts faculty members at U-M organized the first 'teach-ins' in conjunction with students' 'sit-ins' to discuss issues surrounding the war, an activity soon taken up and repeated at dozens of other campuses.[4] The SDS opposition to militarism became more focused on protest against the Vietnam War from January 1966, when the Johnson administration ended automatic student deferments for the draft.

SDS membership mushroomed when the National SDS Convention was held in Ann Arbor that year. The tempo of protest increased year by year, broadening its philosophy and further politicizing its stance. In 1969 some twenty thousand persons protested the war at the city's Michigan Stadium.[5] An extreme faction, the Weathermen, developed from the splintering of SDS that year, going underground and adopting more militant tactics against establishment targets and specifically activities supporting the Vietnam War. Ann Arbor had become not only a major centre for classified research, but also a focus for political activism.

In this way the U-M at Ann Arbor, and more specifically the Willow Run Laboratories (WRL) and the Institute for Science and Technology (IST), became a focus for student protests through the 1960s. In 1967 a sit-in at the U-M Administration building protested the university's classified research at Willow Run and on North Campus; this helped to spearhead protests by students at over a hundred American campuses against the war and against local militarily funded research over the following year.[6] Even more directly, the IST building was bombed one night in the autumn of 1968, destroying the door and windows of the Radar and Optics Laboratory along its east wing.[7]

In response to such protests, the university administrators debated whether to absorb WRL into the College of Engineering, or to affiliate it with an independent non-profit organization such as the Battelle Memorial Institute. Research funding from military sponsors had fallen from a peak of $13 to $9 million in 1969. The Director of WRL, Rune Evaldson, and spokesmen for individual laboratories at Willow Run reported to the press that they saw 'campus unrest' concerning classified research as a major cause for uncertainty about future financing of contracts by the Defense Department, citing

[4] 'The President [of the University of Michigan] said they had reflected the tenseness of the time in which we live. The members of the teaching staff, many of them highly dedicated social scientists, had been told by the President that there was a time and a place and a method to express one's views and to discuss them, but that it was inappropriate to cancel classes for such a purpose' (U-M Board of Regents Proceedings, March 1965).

[5] Feldkamp, John C., 'Student life since 1945', in: W. B. Shaw (ed.), *The University of Michigan: An Encyclopedic Survey* (Ann Arbor, 1974), pp. 411–412.

[6] For example, Al Razutis, later an art holographer, recalls resigning from the PhD programme in nuclear chemistry at University of California, Davis in response to the anti-war movement in 1968.

[7] The dynamite sticks and fuse were detonated at 11:45 p.m. on 14 October 1968, according to the *Michigan Daily* student newspaper, which noted that the east wing housed some of the Great Lakes research units, a holography darkroom, radar laboratory, and classified research, including the Willow Run Radar and Optics Laboratory. No perpetrator was identified or claimed responsibility.

specifically the SDS and the Radical Caucus, 'which are campaigning against performance of classified research at universities'.[8] A Detroit newspaper reported:

> Willow Run labs, which thrived during the 1950s on open ended research grants, have been compelled to seek contracts for specific projects whose aims are defined in advance. Moreover, there seems to be a serious morale problem at the labs, stemming mainly from the classified research controversy which began in 1967 and culminated in a report recommending guidelines for secret research and the establishment of a classified research committee to review contract proposals submitted by lab researchers [. . .] It has been distressing beyond words for the researchers to find themselves looked down upon as being involved in an 'evil' business.[9]

Holographic research in Ann Arbor became a fugitive activity. By 1972, the student opposition convinced the university to opt for the extreme solution: the Willow Run Laboratory as a whole was to be reorganized as a not-for-profit company called the Environmental Research Institute of Michigan (ERIM), and moved at the end of the year to the former Bendix building a half-mile from North Campus in Ann Arbor, where the Apollo Lunar Rover had been developed.[10] Carl Aleksoff, who had worked at Willow Run as an undergraduate summer student, and completing his PhD at U-M and joining ERIM a few months after the move, recalls:

> It turned out to be a surprise to everybody that it was called the Environmental Research Institute of Michigan because it was supposed to be called the *Research* Institute of Michigan, and the signs were getting ready with R-I-M—'Rim'. But we needed an endorsement from the State of Michigan [. . .], and there was one state senator that was pushing the bill through the state legislature to form the company, and he decided at that time that 'environmental' was a very good thing to have, it could help pass a bill very quickly; it was the 'in' word to use [laughing], a popular term, so 'environmental' got stuck on the front and it passed, and to everybody's surprise we were the Environmental Research Institute of Michigan! [11]

Some staff retained dual roles: Emmett Leith, for example, remained a professor in the Electrical Engineering Department while also consulting at ERIM as Chief Scientist, an association that was later to draw further protests.[12] Juris Upatnieks remembers:

> Initially [at North Campus] we could pursue any work we liked but during the Vietnam era war protests began to hamper our choice of projects. Moving to ERIM removed this hindrance and we could proceed as before. Around 1970 US Congress prohibited the Defense Department

[8] 'Labs' fate linked to ROTC', *Ann Arbor News*, 10 Dec. 1969, p.1.

[9] 'Willow Run future', *Detroit Daily News*, 4 Dec. 1969, p.2.

[10] Leonard, Carl to SFJ, email, 15 May 2003, SFJ collection. At the decisive U-M Senate meeting, WRL had no representatives present, and no one supporting them present from the College of Engineering. The dissociation of the university from classified research was a popular one (Leonard, Carl to SFJ, email, 15 May 2003, SFJ collection).

[11] Aleksoff, Carl to SFJ, interview, 9 Sep. 2003, Ann Arbor, SFJ collection.

[12] Gates, Max, 'Holography a "re-creation of reality"—defense debate', *Ann Arbor News*, 18 Mar. 1982, p.2. A group called Science For the People protested Leith's involvement at ERIM when he took part in the university's highest honour, the Henry Russel lecture. Over the next couple of years, the Radiation Laboratory at Willow Run was also the focus of student protests about military contracts.

from funding research that was not of direct interest to the military. Also, NSF funded basic research only at educational institutions. These events limited what we could do at ERIM.[13]

Thus the researchers at WRL/ERIM found their relatively unfettered research style of the early 1960s increasingly constrained by Congress on the one hand, and student protests about this classified research on the other.[14] This conflict between sponsors and interest groups, and their separate perceptions of the purpose and application of research in holography, was an important factor encouraging the growth of distinct communities in specific locales. Ann Arbor's unusual situation, with its concentration of holography researchers, on the one hand, and students opposing militarily related technologies, on the other, was bracketed by two other American centres: the Bay Area of California, and Boston on the opposite coast.

The area encompassing San Francisco Bay and cities and towns to the south had been economically dependent on higher education, research, and technological development since the Second World War. San Francisco and its environs also had long shared liberal activities; the Berkeley campus across the Bay from San Francisco, for example, had been the source of the Free-Speech movement, one of the early triggers for student activism.

Technologically, the region was also ripe. The peninsula south of the city had prospered with fruit orchards in the early twentieth century, but by the 1950s was transforming into the urban sprawl later dubbed Silicon Valley.[15] Stanford University, near Palo Alto, was the initial focus of this expansion. It was also the earliest American centre of 'wavefront reconstruction' and later holography, having been seeded at the university in the late 1940s with the work of Kirkpatrick, El-Sum, and Baez.[16] During the same period, the university fostered post-war expansion. The Stanford Research Institute (SRI) was founded in 1946 to engage in non-traditional university research, which, like the WRL in Ann Arbor, was founded on classified research. The university also formed the Stanford Industrial Park in 1951, offering long-term leases to companies on university-owned land. The first tenant was Varian Associates, the founder of which had provided

[13] Upatnieks, Juris to SFJ, email, 15 May 2003, SFJ collection.

[14] This was not solely an American phenomenon. When Dennis Gabor lectured in Munich in 1969, he was reproached for his participation by students protesting the military–industrial complex and its visions of the future. Unlike other speakers, however, an observer later explained, 'Gabor is clean. He is as pessimistic about the future as the students' (Edson, Lee, 'A Gabor named Dennis seeks Utopia', *Think*, (January-February 1970): 23–7]. Gabor's books on science and society [Gabor, Dennis, *Inventing the Future* (London: Secker & Warburg, 1963; Gabor, Dennis, *The Mature Society* (London: Secker and Warburg, 1972)) unfashionably questioned the widespread confidence in technological progress and technocracy, a theme taken up more overtly by cultural historian Theodore Roszak (Roszak, Theodore, *The Making of a Counter-Culture: Reflections on Technocratic Society and Its Youthful Opposition* (London: Faber 1970)]. Nevertheless, Gabor espoused an elitist intellectual view of society at odds with what he characterized as the permissiveness of the counterculture.

[15] The term first appeared in print in 1971 by a trade journalist, Don C. Hoefler, to describe the proliferation of electronics firms in Santa Clara County as a distinct technical grouping.

[16] One link between both the Bay area and the Boston region as a focus of holography and the counterculture is Albert Baez's daughter Joan, who as a folk singer became a well-known voice of the youth movement. The Baez family lived in California until 1958, and then in Cambridge, Massachusetts, home of MIT, until 1961.

the initial idea for the optical processing of synthetic aperture radar data. Through the 1950s, other research operations were founded there to conduct research and development in electronics or optics, including Eastman Kodak, General Electric, Admiral Corporation, Shockley Transistor Laboratory of Beckman Instruments, Lockheed, and Hewlett-Packard.

Stanford Research Institute was one element in the redefinition and confrontation of cultures in the Bay Area. And just as the CORONA surveillance satellite programme had been developed in the Boston and Bay areas, student activism nucleated at those centres, too.[17] Student protests in the spring of 1969 centred on the SRI research. Several hundred students occupied the Applied Electronics Laboratory at Stanford. During the occupation, some 8000 faculty and students met and agreed almost unanimously that classified research at the university should end. Two weeks later, the university Trustees voted to sever Stanford University's ties with SRI. As was to happen with at Willow Run three years later, the classified research was not strictly controlled as the students urged, but merely dissociated from the university campus. Joe Goodman recalls that his Stanford holography research group, long supported by the Air Force and Office of Naval Research (ONR) disbanded when Stanford decided to leave the classified research arena.[18]

Student opposition in Ann Arbor, Boston, and the Bay Area fostered a counterculture that had direct repercussions for holography. A subculture or counterculture of holography can be identified that contrasts with the 'orthodox' representation of the holographer, that is, the professional, government-agency-funded, or academic researcher within the revitalized field of optical engineering. Cultural historian Elizabeth Nelson, who has studied the British underground press of the late 1960s and early 1970s, has noted the difficulty of attempting to define the characteristics of the wider counterculture commonly identified with that period, and argues that defining a counterculture must begin with a definition of the dominant culture. Various analysts have discussed the dominant culture in terms of class, ideology, or power.[19] One of the earliest of these, Milton Yinger, identified a 'contra-culture' as one in conflict with the values of the total society. In a similar way, a counterculture is at struggle against the powerful or the orthodox.[20] Thus far, I have argued that the youth movement directly generated themes that promoted a recasting of holography. The student protests against classified research and, more broadly, against establishment technologies and assumptions, provided a critique of holography itself. Their stance attacked particular centres such as the U-M,

[17] For more on Stanford, see Leslie, Stuart W., *The Cold War and American Science: The Military–Industrial–Academic Complex at MIT and Stanford* (New York: Columbia University Press, 1993). On MIT, see Eskowitz, Henry, 'The making of an entrepreneurial university: the traffic among MIT, industry, and the military, 1860–1960', in: E. Mendelsohn, M. R. Smith and P. Weingart (eds), *Science, Technology, and the Military* (Dordrecht; London, 1988), pp. 515–40.

[18] Goodman, Joseph W., 'Research in Holography at Stanford: 1960's 1970's and 1980's, unpublished report, SFJ collection, 30 Aug. 2005; Goodman, Joseph W. to SFJ, email, 31 Aug. 2005, SFJ collection.

[19] Roszak, Theodore, *The Making of a Counter-Culture: Reflections on Technocratic Society and Its Youthful Opposition* (London: Faber 1970).

[20] Yinger, Milton, 'Contraculture and subculture', *American Sociological Review* 25 (1960): 629; Nelson, Elizabeth, *The British Counter-Culture, 1966–1973* (Basingstoke: MacMillan, 1989).

Stanford University, and MIT, their funding, and the nature of the research itself. More subtly, the anti-technological perspective and esoteric philosophies attaching to the youth movement urged a re-evaluation of the uses of holography.

While there were relatively small numbers of people involved in the developing counterculture of holography, the activities, products, and writings of these practitioners created a new artisanal breed of holographer and opened the subject to non-professionals.

Defining our counterculture as one conflicting with, or confronting, the orthodox view, it is apparent that a new way of doing holography blossomed during the early 1970s. The new group of practitioners rejected methods and values of orthodox holography, along with a rejection of some social values as well. Ann Arbor and the Bay Area were linked not only by a subject (holography) and a force of opposition (the youth movement), but also by individuals who straddled the cultural divide: Lloyd Cross (see Figure 9.2) and Jerry Pethick.

Lloyd Cross, introduced already as a product of the WRL and Ann Arbor companies, was seminal in synthesizing this new technical counterculture. While working at KMS Industries in Ann Arbor, Cross counted many artists and 'crafters' among his personal friends, and operated a gallery and a print and framing shop in town. He met Canadian sculptor Jerry Pethick in early 1967.[21] Pethick was creating sculptural works that incorporated two- and three-dimensional components, and was experimenting with optical effects based on multiple lens arrays, a technology that had been pursued by Gabriel Lippmann at the turn of the century. Having returned to Canada in 1966 after art school in Britain and seeing his first hologram the following year, he became interested in holography. Pethick learned some of the practicalities of the subject from Dr George W. Jull, a scientist at the Communications Research Centre in Ottawa and former student of Dennis Gabor.[22] With a Canada Council Arts Bursary, and casting around for others who could help him explore holography, he travelled south to Ann Arbor and met Lloyd Cross. During early 1968 Pethick and Cross collaborated in setting up a basic holographic arrangement in London, England, on the floor of Pethick's studio, yielding what Cross recalls as 'a couple of very tenuous sort of holograms'.[23] Later that year they both returned to Ann Arbor and, using a krypton laser that Cross had obtained from KMS Industries, spent 'lots of time working together trying to simplify and justify what were then the formidable requirements of holography'.[24]

Pethick's first successful holograms in Ann Arbor were of small, simple sculptural objects of plastic, his favoured artistic medium; Cross provided the technical expertise in

[21] Pethick (1935–2003) had been educated in Ontario and England, studying at the Chelsea School of Art and the sculpture school of the Royal College of Art in London 1957–64 [Pethick, Jerry, 'Animals Dream', in: R. Amos (ed.), *Collection* (Victoria, BC: 1999), pp. 2].

[22] Jull had been Gabor's PhD student (1951–1955), and had spent 1966 at Imperial College with him exploring holography, Gabor's first experimental foray into the field since the mid-1950s.

[23] Cross, Lloyd, 'The Story of Multiplex', transcription from audio recording, Naeve collection, Spring 1976, p.1.

[24] Cross, Lloyd G. and Cecil Cross, 'HoloStories: Reminiscences and a prognostication on holography', *Leonardo* 25 (1992): 421–4.

Fig. 9.2. Lloyd Cross, from left to right, at Trion Instruments, Ann Arbor (1961), Editions Inc., Ann Arbor (1969), N-Dimensional Space exhibition, New York (1970), and the Multiplex Company, San Francisco (1973) (Broadbent collection).

holography, which initially was more theoretical than practical.[25] Their explorations convinced Cross that his technical background in lasers provided a fertile outlet. He recounts crossing the U-M campus one day during 1968,

> thinking in my head that I had finally found something that I wanted to put my time and energy into for a number of years. I wanted to become a holographer; it wasn't as much of a decision as a revelation to me. I could really feel my interest and my energy going into this craft, art, trade.[26]

Conceiving holography as a craft, an art or a trade was unconventional in 1968. Holographers saw themselves as scientific professionals, allied with optical engineering, and part of well-funded science.

That year, Cross joined with Pethick, Allyn Lite, and Peter Van Riper, an artist/musician, to explore laser light and holography in art, one of the first scientist–artist collaborations. Lite, an artist and U-M alumnus with a Master of Fine Arts degree from Rutger's University, had moved from sculpting and painting to holography after meeting Cross. Van Riper, interested in combining light, music, and installations, was similarly drawn to Cross and his expertise in laser technology.[27]

While still employed at KMS Industries in 1968, Cross set up another firm, Sonovision Inc., to promote the use of laser lighting effects in entertainment. He exhibited the system—the first laser light show—at a local Optical Society of America meeting in the autumn of 1969, and also set it up as pre-movie entertainment at an Ann Arbor cinema.[28]

[25] Pethick, Jerry, 'On sculpture and laser holography: a statement', *Arts Canada* (1968): 70–1.

[26] Cross, 'The Story of Multiplex, op. cit., p. 4.

[27] Peter van Riper (1942–1998) obtained a BA in Art History from the University of North Carolina in 1965, and an MA in Arts at Tokyo University in 1967, finally studying for a Master of Fine Arts at the University of Michigan in 1968, where he encountered Cross (van Riper, Peter, MIT Museum 31/930, 1976).

[28] Charnetski, Clark to SFJ, interview, 3 Sep. 2003, Ann Arbor, MI, SFJ collection. Cross sought a patent for the device, offering a version based on a 2 mW HeNe laser for $1095, and another based on a Krypton laser that employed red, blue, green, and yellow 'sonic deflectors'. The company lasted about a year after Cross left KMS Industries. Edmund Scientific Ltd in Barrington, NJ, was selling a variant of Sonovision in 1969, but Sonovision itself was untraceable by the art-science journal *Leonardo* in 1972.

This relatively high-technology version of the light shows that were increasingly ubiquitous at rock concerts in the late 1960s and early 1970s impressed the young public and scientists alike.[29] Unlike most light shows of the period, which employed overhead projectors, coloured liquids, strobe lights, moiré transparencies, and slides and film, Cross's Sonovision was expensive but self-contained and automatic, responding to music by reflecting the laser beam from mirrors fastened to a reflective membrane over a loudspeaker. The group showed it in *The Laser: Visual Applications*, an exhibition of laser art and holography at the nearby Cranbrook Academy of Art in suburban Detroit in November 1969.[30] Clark Charnetski, holographer at Conductron Corporation in Ann Arbor, recalls encountering a new kind of holography there:

> Things were dim where the holograms were shown, because the lasers were dim. He took the L shaped gallery, and filled one arm with an inflatable building. You crawled through a slit—like a rebirth experience—where a laser light show was running with music and beanbag chairs. Then into the holograms, once your eyes were accommodated—one series was of a pie, with bites successively taken out of it. Another was a holographic jigsaw puzzle; really neat things. The subjects were interesting, more artistic in nature, and he had a whole environment these were displayed in.[31]

Charnetski went back and told his Conductron associates, 'we have to hire an artist, at least part time', now realizing the need to concentrate on the display environment instead of the toys, models, and technical tricks that had preoccupied their efforts to promote holography.[32]

Cross, too, had plans to market holography, telling a magazine reporter:

> Within a year or so, I think there will be hundreds of little holographic studios all over the country with people exploring holography the way photography was explored. Commercial holography is now where photography was in the mid-19th century, and the next step will be to develop a simple, cheap pulsed laser—this will do for popular holography what the flash bulb did for photography.[33]

Pethick, Cross, Lite, and Van Riper formed Editions, Inc. in Ann Arbor that year, creating art holograms, and producing light and sound shows travelling around New York's small

[29] Other (non-laser) light show producer/artists active from 1967, and who later were influential in holography, were Richard Rallison, who set up 'Electric Umbrella' while he was a student at Ohio State University and Fred Unterseher who worked in San Francisco for Gerry Abrams of 'Headlights' and others while a student in 1968–9.

[30] 'Be-it-yourself works of art: laser exhibit at Cranbrook', *Detroit News*, 20 Nov. 1969, 21–2; Cranbrook Academy of Art, 'The Laser: Visual Applications,' Detroit, Michigan, 1969; Hakanson, Joy, 'They create art that isn't there', *The Sunday News Magazine*, 8 Feb. 1970: 18–20, 44. The exhibition was in two parts: the *Sound, Light and Air* portion was directed by Peter Van Riper, and a contiguous but separate exhibit entitled *Holograms* was directed by Cross, Pethick, and Lite (Cross, Lloyd to SFJ, email, 26 Oct. 2003, SFJ collection). Local Ann Arbor craftspeople and/or artists included Don Broadbent, Gene Gilliam, Margaret Pater Bennett, and Carl Goldenberg.

[31] Charnetski to SFJ, op. cit. The Lemon Meringue Pie (a series of four holograms), Champagne Glass, and some two dozen other holograms generated queues of curious viewers (Cross, 'The Story of Multiplex', op. cit.).

[32] Charnetski to SFJ, op. cit.

[33] Wolff, Michael, 'The birth of holography: a new process creates an industry', *Innovations* (1969): 4–15.

theatres during 1969–71.[34] This culminated in a 'Laser Ball' in Detroit in May 1970, described by a reporter as a high-society 'happening' of 'vivid laser beams in ever-changing patterns, somehow controlled by the music' and 'a demonstration of holograms [. . .] three-dimensional reproductions of objects so convincing you are certain that they are real, but they aren't'.[35] Their holograms and laser light effects appeared again at the Finch College Museum of Art, after its curator, Elayne Varian, saw the Cranbrook show, again in the first holography exhibition in New York City, *N-Dimensional Space*, and then toured museums in Rochester, Syracuse and Albany, New York through 1970. 'At that point', reflected Cross, 'we were definitely into holography and we decided to move to New York'.[36] After further peregrinations, he was to reach San Francisco.

Cross slipped between three subcultures: the militarily funded Willow Run research environment, to commercial laser development, and finally to the expanding youth culture that had, thus far, been firmly anti-technological.

9.2 CHALLENGING THE ORTHODOX OPTICAL LABORATORY: MATERIAL CULTURE AND COMMUNITY IDENTITY

We can see the clash between these subcultures not only in the profiles of the holographers, but also in their working environments and in the equipment they used—the material culture. Technology itself became a significant force in giving an identity to practitioners and in liberating the subject for new communities. In effect, the orthodox optical lab was transformed into an environment that was intellectually subversive and socially cathartic.

Through the 1960s, given the linkage between generous project grants and the laboratory equipment purchased with them, the professional practitioners had developed a clear definition of appropriate apparatus. The implicit hierarchy of equipment is suggested by Michael Michalak of the Goddard Space Flight Center in Maryland, who wrote:

> At first, we started making holograms on a laboratory bench using rather crude apparatus. New equipment and a granite slab soon put transmission holograms on a scientific rather than an artistic basis.[37]

Repeatable 'scientific' results, he and colleagues argued, required stability. Leith and Upatnieks had filled their train model with epoxy and glued it to a steel track to ensure

[34] Burns, Jody, 'Messages to the future http://www.holonet.khm.de/Holographers/Burns_Jody/text, consulted 11 Aug. 2005; Cross, Lloyd, MIT Museum 30/843, 1977.

[35] Whittaker, Jeanne, 'On the beam . . . in color!' *Detroit Free Press*, 10 May 1970, 4-D. See also Wilfong, Joan, 'Conductron—Ann Arbor laser photography is 'Belle of Ball', *Conductron antenna* 4 (1970): 7.

[36] Cross, 'The Story of Multiplex', op. cit.

[37] Michalak, Michael W., 'Holography in the Test and Evaluation Division at Goddard Space Flight Center', presented at *Holographic Instrumentation Applications*, Ames Research Center, Moffett Field, CA, 1970.

adequate stability over the exposure time of several minutes. The optical arrangement of mirrors, object, and film had to remain motionless to less than a wavelength of light during the exposure to properly record the interference fringes comprising the hologram. The solution was a heavy and rigid table on which to mount the apparatus. Leith echoed Michalak, observing that the increasing sophistication of holography led to increased requirements for stability, requiring that 'holography moved from its original place of performance on ordinary optical rails onto massive granite tables'.[38]

However, a rigid mass also transmits high-frequency vibrations efficiently, which could perturb the lighter optical elements mounted on it. So it was also necessary to isolate this massive table from the environment. Frank Denton in George Stroke's laboratory had mounted his table on solid foundations sunk into the dirt floor of the IST building, unattached to its the walls. A similar arrangement was built by Norm Barnett across the street in the basement of the Electrical Engineering building. This had also been the approach adopted by researchers in astronomy and interference optics from the late Victorian period.[39] By the late 1960s, though, other labs without access to solid foundations employed more modern methods of vibration control such as 'floating' the heavy marble or granite slab on compressed nitrogen pistons.

But more was required: mere weight and vibration isolation were not enough. Smooth, firm surfaces were needed to position optical elements precisely, as were solid and precisely controllable optical mounts. Following the tradition of optical rails and mounts, the positioning equipment generally featured screw or even micrometer adjustments. Individual optical mounts in later commercial tables were commonly stabilized by being screwed into tapped holes. Such heavy, isolated tables and mounting fixtures became *de facto* criteria for a serious holography laboratory.[40] Workers without such equipment were either apologetic or admitted themselves to be second rank.

[38] Leith, Emmett N., 'A short history of the Optics Group of the Willow Run Laboratories', in: A. Consortini (ed.), *Trends in Optics* (New York: Academic Press, 1996), pp. 1–26, quotation p.23. Leith recalls that during the late 1950s the Optics Group had used readily available and inexpensive Cenco optical benches, and then movedas to more versatile Gaertner equipment. Finally, they dedicated a fabricator at WRL to design and build equipment full time; thereafter, when he formed his own firm, they bought from him and from Jodon, another spin-off company of a former WRL employee. All the WRL optical processing and holography until the late 1960s used very massive granite tables some 2 × 2 × 8 ft (Leith, Emmett N. to SFJ, interviews, 29 Aug.–12 Sep. 2003, Ann Arbor, MI, SFJ collection). For a 1966 collaboration with photographer Fritz Goro, for example, they employed a 7 t table floated on tire inner tubes to make the 5 min exposures for 11 × 14 in. holograms.

[39] Denton, Frank to SFJ, telephone, 1 May 2003, SFJ collection; Barnett, Norm to SFJ, interview, 11 Sep. 2003, Ann Arbor, SFJ collection. This was true, for example, of Albert A. Michelson at the University of Chicago, who had developed interference spectroscopy and high-precision diffraction grating ruling engines. Michelson's early interferometer experiments in Berlin during the early 1870s had been swamped by vibrations from traffic. At his later Chicago laboratory he employed a granite table resting on a block of wood that floated in a tank of mercury.

[40] The Newport Research Corporation, founded by Milton Chang (a former graduate student of Nick George at Caltech) in 1969, soon dominated the market by developing optical tables a relatively lightweight honeycomb structure. The company purchased GC-Optronics in the early 1970s to pursue holographic non-destructive testing, and marketed instant 'holocameras' based on thermoplastic recording materials in 1980. The company's involvement in the field provided a welcome visibility for holographic testing during the 1970s and 1980s, when the technique was being supplanted by simpler methods and declining as an engineering tool.

Just as crucial as vibration isolation was a clean working environment. Dust-free conditions were essential to produce diffraction patterns free of rings and blotches, as Dennis Gabor and Hussein El-Sum had found in battling against optical imperfections to yield 'clean' and 'noise-free' images.[41] An example of this is the specification of optical components. When Gary Cochran at Conductron obtained a commercial Kodak lens to construct an optical processor in 1963, he was appalled to discover that it contained bubbles. Arthur Ingalls, the optical engineer hired before him, dismissed his concerns with the comment that even good lenses have bubbles and that they would not perceptibly affect imaging. In incoherent light the optical imperfections added only a tiny amount of light scatter or noise. But as both soon realized, *coherent* light processing could not use such lenses because the bubbles caused diffraction and optical interference: undesired diffraction rings introduced excessive noise. Conductron's lens specifications were thus unorthodox and troubling: they required reducing the content of bubbles and specular reflection. Ingalls and Cochran had to learn about the peculiar requirements of coherent optics and lens specification, as did their suppliers at Kodak and Goodyear Aircraft. This crucial reliance on clean optics was yet another factor distinguishing the new optical engineering from the old.[42]

Such requirements went largely unformalized. Nor were many other practical necessities documented: how to mount, clean and process photosensitive plates; how to employ spatial filters effectively to remove imperfections of laser illumination; and how to ensure bright hologram recordings. This *tacit knowledge* acquired by a generation of researchers defined the orthodox laboratory of coherent optics.[43] Seemingly incontrovertible technical demands called for high-quality, spotless optical components, and a flat and massive table that was isolated from vibrations. Nevertheless, Jerry Pethick and Lloyd Cross challenged these conventions by developing a radically different set of skills and tools. Unlike Cross's colleagues at Willow Run and KMS Industries, they had little money for equipment. Pethick and Cross planned initially to use a scaled-down table based on tombstone slabs:

> The initial concept was manifested during a long day's pondering how we were going to move a granite slab into the low windowed basement studio we had leased for a holographic studio. It had one small door down three steps and around a tight corner—and after hours of measurements and figuring we had to admit it couldn't be done without busting a huge hole in the brick wall of the building or smashing the granite slab into pieces that could be manhandled through the restrictive passage. It was this last idea that we focused on as our patience with the

[41] The term 'noise-free' referring to an optical image is an interesting carry-over from communication theory. It is today synonymous with the visual term 'clean', but was not used in this sense until the rise of coherent optics and optical engineers. Obtaining such a clean beam of laser light was accomplished from the 1960s using a *spatial filter*—essentially a pinhole positioned after a lens that produces a sharp focus of the laser beam, used to strip off irregularities of the beam (or, more precisely, to occlude the higher spatial frequency components of the Fourier transform of the beam profile, retaining only the lowest frequency component, which yielded a bell-shaped (Gaussian) intensity profile after the filter).

[42] As mentioned in §2.7, during the 1950s, El-Sum at Stanford had bypassed conventional lens fabrication by rotating his optical components to smear out optical imperfections. During the 1970s, Lloyd Cross constructed large clean lenses from bags of oil sandwiched between Plexiglas sheets. Both technical solutions were abrupt changes of approach from conventional optical methods.

[43] On tacit (craft-based, undocumentable) knowledge as an ever-present feature of science see Polanyi, Michael, *Personal Knowledge: Towards a POST-Critical Philosophy* (London: Routledge & Kegan Paul, 1958).

> problem and the day grew short. After a few beers and a smoke or two to relieve our frustrations, we began to get into the image of smashing the granite slab to bits with sledge hammers—a few more beers and we were visualizing bagging up the pulverized granite and then—the jump to—why not start with the starting material—sand! Continuing the thought we realized that a box of loose sand would allow a new way of mounting and adjusting the many optical elements required in holography—on plastic or metal tubes stuck in the sand!—allowing a smooth control of the laser beams by hand and eliminating many expensive, delicate and often jiggly conventional optical components.[44]

In an earlier account, Cross reflected:

> [We spent] many hours and days spending a lot of time sitting around trying to get to the bottom of the complications of holography, which at that time seemed enormous. Huge granite slabs required for stability with expensive three point metal components. Neither of us knew much about the process itself; I knew it from the theoretical point of view but not from a practical point of view. It was during one of our evening conversations that Jerry came up with using boxes of sand as a base for making holograms. And after going through a few ideas, we tried it once and then a couple of times and wound up with using tubes to hold the components in the sand. It of course worked incredibly well. We were both amazed at the amount of stability and the incredible ease of setting up and making holograms [. . .] Everyone we showed the holograms to were more than amazed—they were overwhelmed by the simplicity of it all.[45]

Pethick conceived the idea of building a large plywood sandbox (see Figure 9.3) filled with washed sand and, 'stuffing nagging thoughts of stability, dust, the alliance of gritty

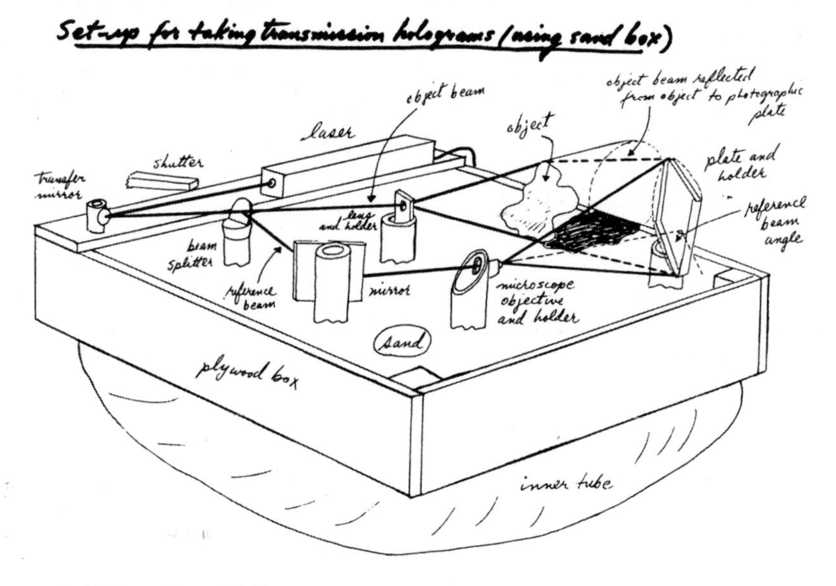

Fig. 9.3. Pethick sandbox table[46]

[44] Cross, Lloyd to SFJ, email, 25 Oct. 2003, SFJ collection.

[45] Cross, 'The Story of Multiplex', op. cit.

[46] Pethick, J., *On Holography and a Way to Make Holograms* (Burlington, Ontario: Belltower Enterprises, 1971), 25.

sand and front surface mirrors and the potential ridicule of the scientific/technical community into the furthest recesses of our minds,' they had tried the idea and found it worked.[47] The entire box was to be mounted on a semi-inflated inner tube to 'float' the mass and isolate it from vibration. Optical components were then mounted on the ends of poly vinyl chloride (PVC) tubing that could be sunk into the sand. Fine adjustments of angle and position were possible with this arrangement, and the sand kept the components in place and damped vibrations.[48] Provided that the apparatus was allowed to settle for a few minutes after being positioned, holographic exposures of several minutes were possible. Cross recalled:

> We even applied for a patent, which for all I know may have been issued; but we decided not to attempt to keep the information a secret, which seemed a ludicrous thing to try to do with such a simple and basic concept. We quickly found out that even though the means were available to acquire and produce the technology, the art was still totally arcane to most people, except to those who were trained in it.[49]

Pethick and Cross never did finish that first basement studio, but instead set up briefly in the building occupied by Cross's Sonovision company, and then moved to the basement of the Print Shop building in downtown Ann Arbor, where they set up the Editions Inc. studio and later made the holograms for the Cranbrook and Finch College exhibitions. Thus was born the 'Sand table Holographic Camera' first shown to the general public and the Ann Arbor scientific and technical community.[50]

As suggested by Cross's accounts, the free flow of information was a guiding principle of their activities, and a contrast to the restricted publication practices of classified and commercial research. Pethick and Cross shared their ideas on sandbox holography widely. They chatted and demonstrated their techniques openly to the wide assortment of visitors, and a Californian MA student made the sand table the subject of his thesis at about the same time.[51] Alongside free information were practicality and thrift. The sand table and its methods reduced the cost of a holography laboratory from some $20,000 to about $3000.[52] They tried to reduce costs even further: about three years after their invention of the sand table, and during a time when they were very low on funds in San Francisco, Cross and Pethick 'drove to the city beach across the great highway from the old Playland amusement park and began bagging sand for our new holographic studio',[53] but, Cross recalls, 'it didn't work out at all—it was full of debris, dusty and continued to

[47] Cross, Lloyd to SFJ, email, 25 Oct. 2003, SFJ collection.

[48] Pethick, *On Holography and a Way to Make Holograms*, op. cit.

[49] Cross, Lloyd G. and Cecil Cross, 'HoloStories: Reminiscences and a prognostication on holography', *Leonardo* 25 (1992): 421–4.

[50] Cross, Lloyd to SFJ, email, 26 Oct. 2003, SFJ collection.

[51] Reeves, Daniel Paul, *Holography with a sandbox optical bench system*, MA thesis, Loma Linda University (1970).

[52] Billings, Loren, Vince DiBiase, Jerry Fox, Mark Holzbach, Chris Outwater and Gary Zellerbach, 'Bay Area holography: An historical view', *L.A.S.E.R. News* 2 (1985): 10–1.

[53] Cross, Lloyd to SFJ, email, 25 Oct. 2003, SFJ collection.

develop odors until we threw it out. We went back to washed silica sand which was very clean and uniform'.[54]

Cross later simplified the arrangement by replacing the plywood box with one made from heavy particleboard clamped in a tension structure. The particleboard was inexpensive and, being a glued composite, was nearly free of mechanical resonances, helping to damp vibrations further.[55] It also made the entire table transportable: after removing the sand, unscrewing the boards and deflating the inner tubes, the component parts could be moved readily and reassembled at the next rented basement or garage.[56]

The apparatus liberated its users from constraints of funding, location, and social stability. The technology was also intellectually subversive: no one—particularly a researcher in a well-funded lab—would previously have considered that abrasive sand and delicate optics could coexist in the same working environment. Gary Cochran, his colleague at KMS Industries, recalls Cross showing him an early version: a large square area in his basement with a 9 × 9 ft sandbox completely filled with sand. In the foot-deep sand were long steel poles with optics sitting on top. Cochran cites the unorthodox arrangement as an example of Cross's nonconformist nature. 'This is the kind of experimentalist he was; it didn't bother him at all about the expensive lenses falling in the sand'.[57]

This style of fabricating equipment from found materials was a philosophy consciously embraced and extended by Cross and his colleagues (see Figure 9.4). In 1973–4, they devised a holographic camera for 'multiplex', or integral holography.[58] The device, used for exposing individual frames of a movie film as strip holograms, employed door springs as gears, cams made from particleboard, and large cylindrical lenses constructed from bent Plexiglas sheets sandwiching a plastic bag filled with mineral oil. The patience required to design and use these mechanical improvizations is suggested by John Fairstein's recollections:

> Tuning the lenses required considerable skill as I later learned first hand. To the eye, the lens might appear fine, but invisible phase changes played havoc on the laser image. Each 'warp' had to be carefully tuned out with tiny adjustments to the lens curvature. Between adjustments, the lens required a 'settling' time to come to equilibrium.
>
> The 9 in. high holographic film was placed in a cylindrical film platen made of plexiglass on an indexing table. The film extended over a 120-degree semi-cylinder. The indexing table, also made of particleboard, used a door spring as the ring gear. Lloyd shimmed it along its length to exactly adjust the travel. An off the shelf worm gear drove the door spring gear. The entire mechanism

[54] Cross, Lloyd to SFJ, email, 26 Oct. 2003, SFJ collection.

[55] Others employed variants such as a poured concrete base with cinder block sides and multiple inner tubes.

[56] 'Bay Area holography: An historical view', *L.A.S.E.R. News* 2 (1985): 10–1.

[57] Cochran, Gary D. to SFJ, interview, 6 and 8 Sep. 2003, Ann Arbor, MI, SFJ collection. The table designs continued to evolve during the early 1970s to improve stability, convenience, and portability (Unterseher, Fred to SFJ, telephone interview, 11 Aug. 2005, SFJ collection).

[58] On the genesis of multiplex holograms, see §7.8.

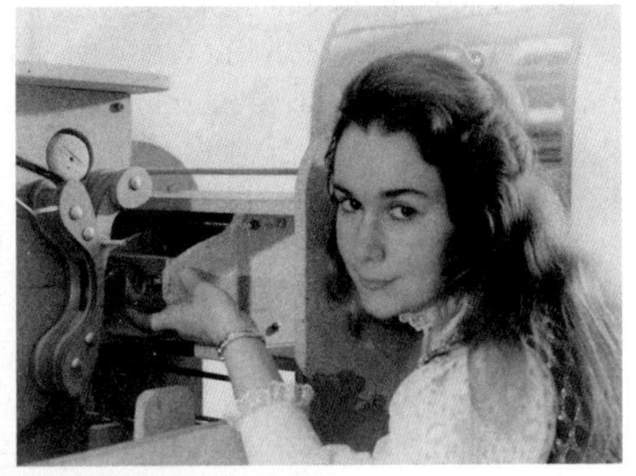

Fig. 9.4. Artisans and proselytizers (clockwise from top left): Lloyd Cross, Sharon McCormack, and Michael Kan of the School of Holography / Multiplex, mid-1970s (Sharon McCormack photos; Naeve collection).

> from projector to film platen was controlled by a system of solenoids and motors controlled by a bank of rotating particleboard cams with switches. To adjust timing required changing the speed of the DC motor with the rheostat and / or adjusting the duration of the cam.[59]

And artist Nancy Gorglione, an early student of Cross, remembers, 'Expensive lab equipment was shunned; we were taught to explore refuse containers behind industrial parks for our components. People found lasers this way'.[60]

In their ingenious improvization of equipment, Fred Unterseher recalls 'a very strong anti-technology consciousness', particularly on the part of Cross.[61] Cross counters that his orientation 'was not so much anti-technology as against the process and procedures of technical innovation which separate and isolate the technical specialities'.[62] Thus his goal was to mutate technology for new purposes and new audiences, and had the effect of transcending disciplinary boundaries and reducing the distance between expert and layperson.[63]

[59] Fairstein, John, 'The San Francisco School of Holography', http://www.jfairstein.com/SOH.html, accessed 28 Feb. 2003.

[60] Gorglione, Nancy, 'Lloyd Cross', *Holographics International* 1 (1987): 17, 29.

[61] Unterseher, Fred to SFJ, interview, 23 Jan. 2003, Santa Clara, CA, SFJ collection.

[62] Cross, Lloyd to SFJ, email, 25 Oct. 2003, SFJ collection.

[63] Cross's development of low-cost solar optics with Ambjörn Naeve and Sharon McCormack is documented in the film Cort, David, 'Focusing the Sun', Electronic Arts Intermix, 1977.

This design philosophy had much in common with west-coast youth culture. It dovetailed, for example, with the *Whole Earth Catalog*, a counterculture collection of tips, sources, and views published yearly from 1968.[64] The small organization and its shop were located in Menlo Park, some fifteen miles southeast of San Francisco and 2 miles from Stanford University. The publishers described the purpose of *Whole Earth* as supporting the development of 'a realm of intimate, personal power—power of the individual to conduct his own education, find his own inspiration, shape his own environment, and share his adventure with whoever is interested'. This individualistic, self-sufficient slant was allied with a mistrust of the large scale, because 'so far remotely done power and glory—as via government, big business, formal education, church—has succeeded to the point where gross defects obscure actual gains'.[65] This did not translate into a rejection of technology *per se*; indeed, the publishers of *Whole Earth* cited the insights of the freethinking technologist Buckminster Fuller as a source of inspiration for their publication. In fact, Fuller's emphasis on tension and compression—famously embodied in his geodesic domes—was taken up by Pethick in his use of curved plywood sheets in the design of the lab dividers for the School of Holography, and by Cross in his designs of holographic printers.[66] The *Catalog* was filled with an eclectic assortment of tools, book reviews, poetry, and observations on science, technology, philosophy, sociology, politics, and more. It reflected the youth movement's growing themes of individualism, alternative technologies, holistic perspectives, and opposition to authority. The paths of Fred Unterseher and others at the School had crossed those of the Whole Earth people and other alternative groups from the late 1960s.[67] The illustrations and layout of his later popular holography do-it-yourself book, the *Holography Handbook*, stressed these aspects while creating a modern folk tale about the Bay Area holographers.

> How did they do this while being so poor that they often ate their food stamps before they could be redeemed? The secret lay simply in understanding some basic principles of holography, and using a little common sense. It was possible to build a holographic lab, in many ways superior to those costing many thousands of dollars, *out of scrap materials*! (Technocrats watch out! Do you suppose there is a hidden lesson in all this?[68]

These perspectives were also nurtured in a specifically visual form by interactions with a growing Bay Area concentration of video artists and artisans. Lloyd Cross and Jerry

[64] Interestingly, Stewart Brand, the editor of the original *Whole Earth Catalog* wrote a book two decades later lauding the high-tech future promised by MIT researchers, including the holography research of Stephen Benton. Brand, Stewart, *The Media Lab: Inventing the Future at MIT* (New York: Viking Penguin, 1987)–a reversion to the ethic of expert designers and uneducated users.

[65] Brand, Stewart, *Whole Earth Catalog–Access to Tools* (Menlo Park, CA: Portola Institute Inc., 1970), quotation from inner cover.

[66] Unterseher, Fred to SFJ, telephone interview, 11 Aug. 2005, SFJ collection.

[67] For example, Unterseher first encountered Stewart Brand and his colleagues through the 1969 project Life: Raft: Earth, and via the Ant Farm artists' collective. The School members also knew Jay Baldwin, industrial designer, former student of Fuller and design editor of the *Whole Earth Catalog*, and the subversive artists/producers of the underground *Zap Comix*.

[68] Unterseher, Fred, Jeannene Hansen, and Bob Schlesinger, *Holography Handbook: Making Holograms the Easy Way* (Berkeley, CA.: Ross Books, 1982), quotation p. 18.

Pethick contributed information and articles for *Radical Software*, a journal that sought to alter both culture and the future via communications technologies. The journal had been founded in 1970 by a collection of artists, writers, musicians, and filmmakers. They argued that the dissemination of information outside the usual commercial media channels could transform social power structures. The subjective, homemade style of *Radical Software* mirrored that of The *Whole Earth Catalog* and the ethos of the San Francisco holographers. As David A. Ross later summarized their motivations,

> technology might have brought us to the brink of global destruction, may have enabled the alignment of power and money that kept us on the verge of devastation, yet technology was not our enemy. In fact, if properly developed and humanely managed, the new communications technologies held within them the power to unleash something truly revolutionary.[69]

The Bay Area was just the place for establishing such views and communities. Besides the video activists emerging at the beginning of the decade, its editors cited a burgeoning music scene and subversive group of comic book illustrators and cartoonists, sustained by the laid-back Marin county lifestyle and the participatory democracy constructed at the Berkeley campus across the Bay.[70] The video artists found the developing vision of Cross and Pethick consistent with their own. In his first interactions with the journal while still living in New York, Cross had touted laser projectors for video more than he did holograms:

> Considering such things as holographic television, mass transference via laser beam, projection in free space without screens and stuff like that, either forget it forever as a totally fucked up idea or maybe wait ten or fifteen years, if we last that long, for some kind of holographic 3-D video.[71]

But two years later, echoing the optimistic forecasts of Kip Siegel, Cross reported their developments in recording animated scenes, natural-colour holograms and, more optimistically, the prospects of three-dimensional television and movies. Even more excitingly, he suggested forecasts such as the use of holograms on skyscrapers' window glass to create huge pictures for people inside.[72]

Thus the freewheeling practical technologies espoused by Pethick and Cross were not unique, but a particularly attractive implementation of countercultural ideals, the emerging technological art movement, and a direct reaction to 'orthodox' holographers. The simple ideas embodied in the sand table promoted a new constituency for holography: the non-scientific amateur. Their brand of amateurism was distinctly unlike the 1960s model of the scientific amateur, however. Although professional holography was usually

[69] Ross, David A., 'Radical Software redux', http://www.radicalsoftware.org/e/ross.html, accessed 11 Nov. 2004.

[70] Gigliotti, Davidson and Ira Schneider, 'Videocity: Summer 1973', http://www.radicalsoftware.org/e/volume2nr3.html, accessed 11 Nov. 2004. On the widely shared countercultural and counter technological elements espoused by other journals such as *Mother Earth News*, *New Alchemy*, and *Domebook*, see Aquarius Project, 'Revolutionary engineering: towards a 'counter-technology'', *Radical Software* 1 (1970): 7.

[71] Cross, Lloyd, 'The potential impact of the laser on the video medium', *Radical Software* 1 (1970): 6.

[72] Albright, Thomas, 'School of holography', *Radical Software* 2 (1972): 56–7.

portrayed as the work of small teams during the 1960s, an individualistic, non-professional dimension to the subject nevertheless began during this time. This was promoted by popular articles in *Scientific American*, which was then a magazine targeted at the educated classes.[73] A regular feature of the magazine had been the 'Amateur Scientist' section, which described scientific construction and experimental projects.[74] Such articles appealed to lone experimenters and the occasional adolescent at ease with workshop skills. The methods and apparatus they adopted were generally scaled-down or simplified models of professional equipment. They advocated the creative use of 'found' materials more out of practical need than any underlying philosophy, however. These were usually more locally or regionally based than the 'disciplined' holography practised by scientists as an extension of the disciplines of optics, electronics, or information theory, because the amateurs studied their craft near home and usually had fewer resources to travel or interact with either professionals or each other. There are striking parallels between philosophy and practice of the Cross–Pethick group and of the 'Sidewalk Astronomers' organized in San Francisco by John Dobson from 1968 as a democratic and empowering movement, even if no direct links can be demonstrated. Both cases are qualitatively different from more traditional amateur–professional interactions such as sky surveys in astronomy or bird surveys in ornithology, in which there is a clearer consensual division into complementary activities and a shared philosophy of purpose.

9.3 TRAINING ARTISANS: THE BIRTH OF SCHOOLS

Cross, Pethick, and their followers broke away from the WRL model of the holographer in two important ways: they adopted a counterculture philosophy that challenged the goals and funding of conventional holographers, and they promoted appropriate technologies that contrasted sharply with orthodox equipment. Their role in shaping

[73] Leith, Emmett N. and Juris Upatnieks, 'Photography by laser', *Scientific American* 224 (1965): 24–36; Pennington, K. S., 'Advances in holography', *Scientific American* 218 (1968): 40–8; Metherell, A. F., 'Acoustical holography', *Scientific American* 221 (1969): 36–53.

[74] C. L. Stong, the editor of the 'Amateur Scientist' section since 1955, followed up the professionally written scientific articles of the magazine with informal descriptions of build-it-yourself projects, such as practical laser construction and hologram recording (Dickson, Leroy D. and C. L. Stong, 'The Amateur Scientist: Stability of the apparatus: insuring a good hologram by controlling vibration and exposure', *Scientific American* 223 (Jul. 1971): 110–2; Heumann, Sylvain M. and C. L. Stong, 'The Amateur Scientist: How to make holograms and experiment with them or with ready-made holograms', *Scientific American* 219 (Feb. 1967): 122–8; Further 'Amateur Scientist' articles followed during the next decade under the editorship of Jearl Walker (Walker, Jearl, 'Amateur Scientist: Easy way to make holograms', *Scientific American* 232 (Feb. 1980): 158; Walker, Jearl., 'Amateur Scientist: Rainbow holograms', *Scientific American* 238 (Sep. 1986): 114; Walker, Jearl, 'Amateur Scientist: How to stop worrying about vibrations and make holograms viewable in white light', *Scientific American* 241 (May 1989: 134). A more personal example in the spirit of Cross is Johnston, Sean F., and C. L. Stong, 'Amateur Scientist: A high school student builds a recording spectrophotometer' *Scientific American* 237 (Jan. 1975): 118.

identity went further, however: they created a way of sharing their perspective and building a community on it. Schools of holography promoted a third form of socialization that was based neither on small scientific teams nor solitary experimenters. The short courses fostered the growth of a loosely interacting constituency of enthusiasts, particularly in America during the early 1970s. The best known were those in San Francisco, New York, and Lake Forest, Illinois, each established between 1971 and 1973.

Following their exhibitions, Pethick had moved from New York to the Art Institute of San Francisco in early 1971, turning from holography back to his earlier medium of vacuum-formed plastic for his three-dimensional explorations and teaching. Cross's and Pethick's aspirations in holographic art had peaked in New York and they had set up a studio first on Broom Street and then on Prince Street, finding them plagued by vibrations. They still shared a common interest in holography, but had no independent means of support. At the time there were a few holograms to be seen publicly in the Bay Area. Matt Lehmann of Stanford University was displaying his holograms at the College of Arts and Crafts. Impressed by enthusiastic reports of these holograms from other artists, Pethick decided to continue to pursue holography as an art form and, more immediately, to generate money. With Fred Unterseher he started to prepare a holography laboratory in borrowed space in the basement of Project One, an artists' cooperative in San Francisco.[75]

Lloyd Cross moved to San Francisco to join them later in 1970, via a sojourn in Arizona, during which his entourage had grown. He spent three months teaching optics and holography to eight or ten high school students in Verde Valley, Arizona, and found this first experience teaching the subject to people with no prior technical experience using the sandbox idea to be very positive.[76]

When Cross joined Pethick in San Francisco, their holographic interests proved magnetic in attracting acolytes and enthusiasts. One source of this magnetism was the excitement of seeing well-produced and artistic holograms, still a rare experience both for scientists and the general public. Cross and Pethick displayed their holograms at the Exploratorium in San Francisco during the summer of 1971, where Cross worked part-time for a few months and from it attracted students. The interactive exhibits at this novel centre were designed, like Cross's own holographic equipment, from low-cost 'unprofessional' materials.

After brief misgivings about spending time and energy teaching others their liberating techniques of holography, Cross agreed that he and Pethick needed a baseline income from teaching to support their holographic activities. Hugh Brady provided some $10,000 to start a school, and let the group live out of his large Victorian house on Masonic Street, San Francisco.[77] This public access to the mysteries of the subject began

[75] Unterseher (b. 1946), having lived in California since his early teens, studied at the San Francisco Art Institute 1967–70, obtaining a Bachelor of Fine Arts in sculpture. Between 1969 and 1971 he was a member of the Ant Farm art/media collective. Unterseher's career in holography was continuous from 1970, when he met Pethick, who was teaching a vacuum-forming class at the Art Institute: he formed Holografix, a teaching studio 1973–81, was Director of Educational Services at the New York Museum of Holography 1982–4, Technical Director of Holocom in Hamburg, Germany for the following three years, and subsequently and between times worked as an engineer, lecturer, and consultant on holography.

[76] Cross, 'The Story of Multiplex', op. cit., p. 4.

[77] Brady, Hugh to SFJ, letter, 23–24 Sep. 2003, SFJ collection.

when Cross taught a short course on holography at the Exploratorium and simultaneously began teaching students at the Project One studio in October 1971. The striking holograms made an impact, recalled years after by individuals who were drawn to the subject and to the courses that Cross and Pethick offered.[78] Lon Moore, their first student, had been impressed by the projected-image hologram of a champagne glass that Cross had made at KMS Industries, and that had been shown in Cross and Pethick's Cranbrook Museum and Exploratorium shows.

They set up four sand tables at Project One, and hand-printed posters on surplus computer paper.[79] Pethick also devised attractive three-dimensional advertisements from vacuum-formed plastic, but they quickly disappeared from shops where they had been displayed. Peter van Riper also moved West, teaching 'graphics, laser/holography and conceptual performance' at the California Institute of the Arts for its first three years from 1970. In this new environment, and with new colleagues, their Editions Inc. collaboration transmuted into the first School of Holography.

In early 1972 the group looked for permanent premises, scouring the Bay Area, and found that their special requirements for a quiet laboratory did not seem well suited to their demeanour. Unterseher recalls, 'They had to "check the vibrations in your building"—and owners thought they were totally nuts. One guy said, "You guys aren't going to make drugs are you?" And of course we looked the part'.[80]

They moved to 454 Shotwell Street, a cavernous warehouse and former bread bakery in the Mission district of San Francisco, that spring. John Fairstein described it as dimly lit and atmospheric, 'divided roughly down the middle—the left half consisting of a workshop, darkrooms, and a small classroom area in the back. The right side was used as a display area, meeting room, and living space for some of the SOH denizens'. Others recall a parachute hanging from the ceiling tent-like, under which Cross and others could live and discuss.[81]

The San Francisco School of Holography trained several hundred practitioners between 1971–74, offering its first advanced projects course in the summer of 1972 to a number of students, 'who were more or less totally interested in holography and eventually became assistants and began helping us with teaching and then instructors in the school'.[82] By that summer, 130 students had taken the $85, eight-week course.[83] Teaching varied with the changing instructors and audiences: by 1976 a 16 h course cost $120, and

[78] Moore, Lon to J. Ross, Interview, 1980, Los Angeles, Ross collection. Gary Zellerbach remembers it as the only hologram still displayed at the Exploratorium in San Francisco during the late 1970s (Zellerbach, Gary to SFJ, email, 2 Aug. 2003, SFJ collection).

[79] Cross, Lloyd G. and Cecil Cross, 'HoloStories: Reminiscences and a prognostication on holography', *Leonardo* 25 (1992): 421–4.

[80] Unterseher, Fred to SFJ, interview, 23 Jan. 2003, Santa Clara, CA, SFJ collection.

[81] Fairstein, op. cit.; Gorglione, Nancy, 'Forms of light: a personal history in holography', *Leonardo* 25 (1992): 473–80; Gillespie, Donald to SFJ, interview, 29 Aug. and 4–6 Sep., 2003, Ann Arbor, MI, SFJ collection.

[82] Cross, 'The Story of Multiplex', op. cit., p.5. Many of those students became active in art and commercial holography (Anon., 'School of holography flourishes on west coast', *holosphere* 2 (1973): 1, 5–6; Billings, Loren, Vince DiBiase, Jerry Fox, Mark Holzbach, Chris Outwater, and Gary Zellerbach, 'Bay Area holography: An historical view', *L.A.S.E.R. News* 2 (1985): 10–1).

[83] Albright, Thomas, 'Holography's accelerating impact', *S. F. Sunday Examiner and Chronicle*, 11 Jun. 1972, 30–1.

the School's 1979 brochure listed a six-week basic holography course consisting of either six 4 h lessons, or an intensive three-day course for $250, or a one-day seminar for $50.[84] From 1974 until 1977, teaching activities came to a halt as the group members turned their attention to producing the Cross 'Multiplex' holograms and organizing the Multiplex Company, but resumed again from January 1978.[85]

As their products were more widely displayed and a commercial market began to develop, the potential income from such holograms caused the School of Holography to metamorphose into the Multiplex Company in the autumn of 1973, eventually filling 15,000 sq. ft of the warehouse.[86] Over the previous two years the School had expanded from Cross and Pethick to include several of their former students, and operated as a loose collective (see Figure 9.5). The School remained as a separate entity, but most of the energy went into developing multiplex holograms.[87] The Multiplex Company merged the School of Holography group and others connected with cinematography.[88] Developments at the chronically money-poor company were driven by unconventional financing and commissions from customers. Cinematographic equipment, including four Mitchell cameras, became available from a former pornographic film studio, and cinematic expertise from Peter Claudius and David Schmidt.[89] Cross hoped to patent the Multiplex technique, but did not pursue the paperwork. Holosonics, the major patent-holder in holography, was interested in buying the Multiplex Company, but decided that it was too unconventional and unstable to be worth the risk.[90]

The Multiplex Company was a further indication of Cross's desire for commercial success, even if it was an unconventional firm. Through a wide collection of contacts, the company found distributors and custom orders. Edmund Scientific, the major American distributor of consumer scientific goods, listed it in their 1977 catalogue as 'Multiplex Holography—Moving Images of the Future [. . .] A major breakthrough, our "Moving Kiss" hologram may be a forerunner of future holographic motion pictures'. Multiplex and other suppliers offered a wide range of images available for between $18 and $75,

[84] Crane, Alison to P. Jackson, letter, 25 Apr. 1979, MIT Museum 29/822.

[85] The School of Holography reopened in 1978 and, under the direction of Sharon McCormack, endured until her move to Oregon in 1990. Pethick himself, becoming frustrated by the tedium and limitations of holography, left the School of Holography in 1973, and moved with his wife and son to Hornby Island in British Columbia in 1975. His subsequent eclectic artwork combined optics (particularly multiple lenses) with sculptural and other elements.

[86] Gorglione, Nancy, 'Lloyd Cross', *Holographics International* 1 (1987): 17, 29. The members of the School received shares in Multiplex, but the partners were in a constant state of flux. Gorglione lists early members of the Multiplex Company as Lloyd G. Cross, Michael Kan, Gary Adams, Michael Fischer, Pam Brazier, David Schmidt, Peter Claudius, Bob Taunton, David Harell, and Rufus Friedman. By the early 1980s, others recall Claudius as the sole operator in one half of the Mission Street warehouse, which subsequently closed.

[87] Unterseher, Fred to SFJ, interview, 23 Jan. 2003, Santa Clara, CA, SFJ collection.

[88] The SoH group included Cross, Michael Fischer, Gary Addams, Pam Brazier, George Dowbenko, and Fred Unterseher; the cinematography group included David Schmidt, Michael Kan, Peter Claudius, Bob Tauton, and Carl Extine.

[89] In common with other imaging media, some of the best-selling Multiplex holograms had erotic themes. See §13.1.

[90] Haines, Kenneth to SFJ, interview, 21 Jan. 2003, Santa Clara, CA, SFJ collection.

Fig. 9.5. Entrepreneurs—members of the Multiplex Company 1973. Top: Pam Brazier; middle: Michael Fisher, Lloyd Cross, Michael Kan; bottom: Peter Claudius, Dave Schmidt (Multiplex Co.; MIT Museum MoH collection).

with custom holograms made for under $1000. Multiplex holograms became ubiquitous in gift shops at Science centers, and museums. They were also widely seen in two films, both released in 1976: *Logan's Run* and *The Man Who Fell to Earth*, starring David Bowie. Cross recalls being 'immersed in a kaleidoscope of images from around the world, from commerce, art, famous people, each with a new challenge in this unexplored medium [. . .] The expansion of the technology diverted energy from the solution of these difficult basic problems'.[91] He left the Multiplex Company in 1976, moving to set up another studio in a building a block away with Sharon McCormack and Michael Kan to pursue more advanced printer designs, holographic movies, and computer calculation of the required optics. He freelanced, teaching for the School, later working on improved stereogram cameras in Los Angeles and Japan, and organizing exhibitions in Australia. Between times, Cross worked in a secluded home in northern California, inspiring a cachet of mystery and a burgeoning mythology.[92]

Reminiscences are fragmentary and impressionistic, but ubiquitous among practitioners: a wide range of American holographers of the 1970s became tangentially or intimately associated with the School as students or holographers (e.g. Lon Moore, Fred Unterseher, Ana Maria Nicholson, Pam Brazier, Michael Fischer, Gary Adams, Jody Burns, Larry Lieberman, Nancy Gorglione, Sam Moree, Rudie Berkhout, Anaït Arutunoff Stephens, Sharon McCormack, Rufus Friedman, Michael Kan, John Fairstein, Randy James, David Schmidt, and Peter Claudius), production workers for the Multiplex Company (e.g. Steve Provence and Vince DiBiase), friends, business associates (e.g. Steve

[91] Cross, Lloyd G. and Cecil Cross, 'HoloStories: Reminiscences and a prognostication on holography', *Leonardo* 25 (1992): 421–4.

[92] From the late 1970s, Cross lived on Hugh Brady's land in Point Arena, California, sharing in the activities of organic gardening while pursuing his interests in alternative-technology optics.

McGrew, Don Broadbent, and Linda Lane) or merely hangers-on, in a loose-knit association cum collective.[93] Jason Sapan summarizes them as

> a loose knit group of free souls who had slowly gathered around a concept of what they saw as the future of visual images. The key players were a mix of countercultural gypsies and hermits who for a brief moment wove together into a fabric that glowed brightly. It was a mix of science, pornography, business, and a 'crash pad' for wayward souls.[94]

Recalls Gorglione,

> People carved out cubby-holes to live in; under a tie-dyed parachute tent, Lloyd lived and perfected his system of multiplexing stereograms.[95]

Jerry Pethick's 1971 booklet summed up their zeal:

> The application of holography to communications and the human environment could soon have a very great and far-reaching effect on our society. Using holography, the physical environment could be anything that man can conceive. Holography can create the future.
>
> Those interested in the medium, either as a purely aesthetic statement or for its numerous commercial applications, need not worry unduly about the economic and technical problems, as the majority of these are temporary and solvable. Holography is simple. Anyone with interest, basic information and minimum equipment can make a hologram.[96]

Many remember it as a special, if transient, time:

> The idea of holography was then, far more than now, imbued with a breathless sense of being on the threshold of some new form of consciousness. (Randy James, 1979)[97]
>
> It was an amazing period; quite different than it is now. Everyone was full of hopes and dreams, and everyone was going to get rich; when everyone was down and no one could pay utility bills, suddenly someone would call from New York and ask 'can you make a hologram of Alice Cooper?' And all the wheels would start turning and the lights would come back on; it was a magical time; you could never recreate it. (Linda Lane, 1980)[98]
>
> The atmosphere, the vibes, at the School were so intense; holography was what we had all been looking for. Not just holography as the end product; all creative paths to holography were embraced. Suddenly we were all artists, using holography. Lloyd taught and encouraged us all. We all felt at the time that history was being made. (Nancy Gorglione, 1987)[99]
>
> It was all going to happen any minute. It was a magical time; everybody talked. (Vince DiBiase, 2003)[100]

[93] Rufus Friedman had worked on television commercials and film as an actor and technical consultant. At the 1969 Woodstock Festival he had built concession stands and was an assistant in the filming (the principal cinematographer for *Woodstock* was Hart Perry, also associated with Selwyn Lissack and holography during the 1970s). Friedman first met Pethick and Cross in New York City in 1970 and had been 'been hooked ever since' (MoH, 'Exhibits: Holography '75,' MIT Museum 33/1022, 1975).

[94] Sapan, Jason to SFJ, email, 17 Mar. 2004, SFJ collection.

[95] Gorglione, Nancy, 'Lloyd Cross', *Holographics International* 1 (1987): 17, 29.

[96] Pethick, *On Holography and a Way to Make Holograms*, op. cit., quotation pp. 6–7.

[97] James, Randy, 'Off the wall', *holosphere*, 8 (12), Dec. 1979: 3.

[98] Lane, Linda to J. Ross, interview, 1980, San Francisco, Ross collection.

[99] Gorglione, Nancy, 'Lloyd Cross', *Holographics International* 1 (1987): 17, 29.

[100] DiBiase, Vince to SFJ, interview, 21 Jan. 2003, Santa Clara, CA, SFJ collection.

We thought this was the coolest thing—thought that there would be 3D movies and TV, everything. (Fred Unterseher, 2003)[101]

In 1976 Cross recorded reminiscences about the formation of the School and Multiplex, threading his conversation with the themes of openness of information and ecological sensitivity. Both organizations had unusually fluid membership, providing technical information openly—even eagerly—to visitors, seeking to create holograms without polluting the local environment with chemicals, and hoping to make them viewable with low-voltage or even solar-powered light sources.[102]

This unconventional group and its holograms also attracted widespread attention. Steve McGrew, later a major contributor to commercial holography, heard about their first multiplex holograms and, visiting in 1972, 'was completely shocked; I'd never seen anybody like that before. I was extremely impressed by their creativity'.[103] Tung Jeong visited in 1973, and left a few days later converted to the methods of sandbox holography. For Kenneth Haines, who visited the Multiplex Company with an eye to its acquisition for Holosonics, Lloyd Cross was 'a strange, hippy guy' with whom he did not dare do business.[104] Emmett Leith described Cross as 'remarkable in his way, a free-spirit scientist [. . .] working with artists in a kind of communal society, and making some fantastic holograms'.[105]

Besides low-cost self-sufficiency in the style of the *Whole Earth Catalog*, the San Francisco School of Holography absorbed wider meanings for holography. Since the late 1960s physicist David Bohm had mused publicly about an analogy between holography, human perception, and physical reality itself. Psycho-physiologist Karl Pribram similarly had promoted the analogy of human memory as holographic.[106] These links between holism and holography resonated with eastern and mystical elements in counterculture thinking. Rather than stressing holography's theoretical basis in communication theory and image processing as had optical engineers, artisanal holographers stressed that holograms resisted a reductionist analysis: they were intensely non-intuitive and yet mind-expanding. Writing about the School and its vision of holography in 1973, *Rolling Stone* magazine suggested how it resonated with counterculture themes.

The hologram is as likely as anything technological to push your subliminal awe and wonder button and leave an ancient message flashing somewhere below the surface of consciousness: Here we have some Powerful Magic.

Perhaps because of the changing loose coalition of interests that impelled the School and the Multiplex Company, it also proved unusually fertile in generating independent holographers. In 1973, for instance, some of the first individuals associated with the

[101] Unterseher, Fred to SFJ, interview, 23 Jan. 2003, Santa Clara, CA, SFJ collection.
[102] Cross, 'The Story of Multiplex', op. cit.
[103] McGrew, Steve to J. Ross, interview, Dec. 1980, Los Angeles, Ross collection.
[104] Haines, Kenneth to SFJ, interview, 21 Jan. 2003, Santa Clara, CA, SFJ collection.
[105] Leith, Emmett N. to SFJ, interview, 22 Jan. 2003, Santa Clara, CA, SFJ collection.
[106] On these wider understandings of the subject, see Chapter 13.

School (Lon Moore, John Fairstein and, later, Nancy Gorglione and Randy James) formed Celestial Holograms, an independent holography studio. In this respect, the San Francisco School of Holography equalled the Optics Group of the WRL a decade earlier in spawning a generation of active holographers (see Figure 9.6).

It also spawned other teachers of holography. Fred Unterseher set up his own school, Holografix, in 1974 in Emeryville California and later taught a course at the California College of Arts and Crafts, which, by the late 1970s, had a profile of students shifting to photographers and artists.[107] Lon Moore taught privately from the holography laboratory in his home. Randy James and Steve McGrew became full-time commercial holographers at Light Impressions.

Other, more conventional, schools of holography began teaching in other cities. Lake Forest College, Illinois, began offering week-long summer classes in 1972 under Tung Jeong, a professor of physics there. The summer course began to attract some two dozen students per year, increasingly from long distances.[108] Similarly, the New York School of Holography was established in October 1973 by Joseph R. ('Jody') Burns, Cecile Ruchin, and Selwyn Lissack.[109] Burns had founded the New York Art Alliance Inc., a commercial and fine arts business representing artists interested in new processes, in 1970, after four years as a photo-interpreter with the US Navy. Even such early adopters, though, could be counted a second generation. Burns studied holography via workshops with Stephen Benton, Tung Jeong in the summer of 1972 and Lloyd Cross in the summer of 1973. In October 1973, Burns turned his business completely over to holography and the New York School of Holography opened under his direction, with Gary Adams from the San Francisco School teaching the first cohort. In the first eighteen months the school taught some 300 students, and had developed a curriculum of nine courses ranging from basic skills to history and aesthetics. Each course was 20 h long and cost $180.[110] Like the San Francisco School of Holography, the New York School (see Figure 9.7) attracted a growing group of holographers and diversified into other activities, the most significant of which was the New York Museum of Holography, discussed in §11.4.

Other early teaching was conducted at Fringe Research in Toronto from 1974, formed by David Hlynsky and Michael Sowdon, and the School of Holography at Chicago's Fine Arts Research and Holography Center, opened in 1978.[111]

[107] DiBiase, Vince to SFJ, interview, 21 Jan. 2003, Santa Clara, CA, SFJ collection.

[108] Jeong, Tung H. to SFJ, interview, 21 Jan. 2003, Santa Clara, CA, SFJ collection. During the 1980s specialized courses were provided on colour holography, interferometry, portraits and photopolymers, and taught by specialists. Thus Hans Bjelkhagen taught a silver-halide course and Nils Abramson taught interferometry, as did Charles Vest over four summers.

[109] Burns Jr, Joseph, 'Update on the New York School of Holography', *holosphere* 4 (1975): 3–4. Burns (b. 1946, California) saw his first hologram at a small exhibit in the lobby of the Time-Life building in 1966. Artist/musician Selwyn Lissack (b. 1942, South Africa) had been involved in production of holographic products and art work since about 1970, having formed with Ruchin the Holographic Communications Corporation of America and later liaising with McDonnell Douglas, Quaker Oats, Timex, Salvador Dali, and others. Burns founded the Museum of Holography (MoH) in New York in 1974 (see ß11.4 for more on the MoH). Lissack left the New York School in 1974.

[110] 'Update on the New York School of Holography and the Museum of Holography' *holosphere* 4 (4) (1975) 3–4.

[111] The Fine Arts Research and Holographic Center comprised a Museum of Holography, School of Holography, and the commercial firm Holographic Design Systems. Operated by Loren Billings, the name

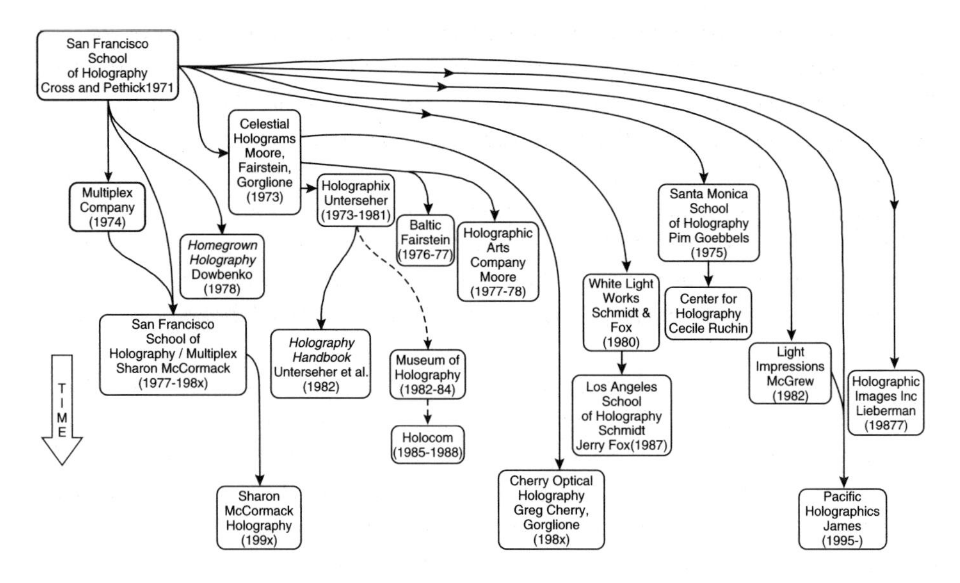

Fig. 9.6. Genealogy of San Francisco holography.

Fig. 9.7. New York School of Holography advertisement, *c.*1976 (MIT Museum MoH collection)

The students for their courses, recalls Unterseher, came 'from everywhere—some biker guys, to little old ladies, housewives and stuff. A Marin County lady rubbing shoulders with a hippy, and a biker kid'.[112] Teaching activities aimed at the general public received frequent newspaper coverage. A typical item from the period vaunted modernity, exclusivity, and accessibility:

> If you like the idea of being the first kid on your block to do something, take up holography. With an investment of under $1000—and it can be way under depending upon the kind of

was shortened to The Holographic Center in 1995. During the 1980s schools flourished, such as l'Atelier Holographique in Paris; Ideecentrum Foundation in Eindhoven, Netherlands; Richmond Holographic Studios in Richmond, UK; and Goldsmith's College, University of London.

[112] Unterseher, Fred to SFJ, interview, 23 Jan. 2003, Santa Clara, CA, SFJ collection.

> equipment you use—you can set up an holography studio in your basement or spare room. When you start making holograms you will join an elite group of artists and researchers working in this new field. Maybe one or two thousand people in the world.[113]

There was a true cultural shift in these activities. Social observers such as E. P. Thompson have defined culture by its practices and function, by 'what [it] *does* (or fails to do)'.[114] Indeed, the diverging occupational identities of holographers were elaborated by additional strategies of promotion and expansion. The new breed of artisanal holographers began to portray themselves as such by their tools (sand tables), their products (display holograms), and by their social milieu, and underlying philosophies (an eclectic mix of counterculture themes, some of which were directly opposed to the conventions of optical engineers). In these ways artisanal holographers, seeking wide popular applications, began to distinguish themselves from engineer/scientists, who were more concerned with goal-directed research. These diverging occupational identities were elaborated by different strategies of promotion and expansion. Artisanal holographers began to present themselves as such via publications, exhibitions, and conferences. This differentiation was at the level of occupational practice rather than representing deep intellectual schisms. And no overt professional distinction was pursued, except insofar as some existing differences were maintained (e.g. artisans were seldom found in engineering research laboratories and engineers seldom worked in artists' studios).[115]

9.4 TRANSMITTING THE COUNTERCULTURE: PRACTICAL PUBLICATIONS

Cross and Pethick thus founded a new community of holographers based on a counterculture philosophy, unorthodox equipment, and a freewheeling school. Cross's multiplex holograms and unconventional company brought holography to an even wider audience. But the growing community also promoted its identity through a handful of publications. This was nevertheless a distinct contrast with scientific holographers, who communicated almost exclusively via conference presentations, published papers, and a growing number of books. It is ironic that the most coherent communities of holography—the early Willow Run optical processing group and the San Francisco school—were the least focused on publication.

[113] Schultz, Catherine, 'Holography: they do it all with lasers', *Berkeley Gazette*, 21 (Oct. 1977): 3.

[114] Thompson, E. P., 'The long revolution—II', *New Left Review*, (9) (May-Jun. 1961): 32.

[115] But there were a number of exceptions as holographers sought to carve out careers. Fred Unterseher, and later Melissa Crenshaw, Ana MacArthur, Pearl John, and Andrea Robertson, for example, oscillated between technical applications and aesthetic work.

'Orthodox' holographers had amassed a record of successful research and applications through technical publications. Unlike the peer-reviewed papers by holographic scientists, however, the Bay Area holographers published little aside from ephemeral publicity information. Cross and his followers attended few conferences and published few papers during the 1970s. Nor did they self-publicize in print, apart from flyers for courses and brochures for their Multiplex products. Yet within a few years, the Bay Area holography community nurtured a new contingent of holographers through the writings of former students. These promoted ideals espoused by the wider youth culture, and reacted against the 'orthodox' practices defined by optical engineers to spawn a distinct subculture. As suggested by Lloyd Cross's career trajectory from Willow Run to San Francisco, the writings of his followers consciously rebelled against centrally managed, government-funded research laboratories and sought to liberate the subject for non-scientist artisanal practitioners.

Practical instruction books began to appear in the wake of the schools of holography, and had a straightforward genealogy. The first publication was Pethick's booklet on sand table holography, which served as the handbook for the first classes at the San Francisco School of Holography.[116] It contrasted with Tung Jeong's *Study Guide on Holography*, similarly intended as a course reference but in a more traditional vein.[117] While non-mathematical, Jeong's diagrams and analogies were conventional developments of scientific teaching; it devoted considerably less space to sandboxes than to the lab-stands found in the typical college physics laboratory. Neither was widely disseminated, but Outwater's 1974 *Guide to Practical Holography* sold well.[118] Two subsequent books adopted an overtly countercultural style: *Homegrown Holography*[119] and the more carefully produced *Holography Handbook*.[120] George Dowbenko, an early student and worker at the San Francisco School, promoted its early ideals, interspersing free-hand sketches of bearded holographers with admonitions to reject 'science hoodoo':

> science hoodoo's whatchacall those textbooks
> making common knowledge inaccessible
> removing it from people's mind grasp
> through specialized and secret code words
> number symbols without reference
> created by the science priestcraft
> to confuse what is and to ensure the chaos.[121]

[116] Pethick, *On Holography and a Way to Make Holograms*, op. cit.

[117] Jeong, Tung H., *A Study Guide on Holography (Draft)* (Lake Forest, Illinois: Lake Forest College, 1975).

[118] Outwater, Chris and Eric Van Hamersveld, *Guide to Practical Holography* (Beverly Hills, CA.: Pentangle Press, 1974).

[119] Dowbenko, George, *Homegrown Holography* (Garden City, NY: Amphoto, 1978).

[120] Unterseher, Fred, Jeannene Hansen and Bob Schlesinger, *Holography Handbook: Making Holograms the Easy Way* (Berkeley, CA: Ross Books, 1981).

[121] Dowbenko, op. cit., Preface.

Both books incorporated free-drawn figures illustrating a wide gamut of holographic techniques, but the book by Unterseher et al. was in many respects the culmination of the San Francisco school of holography philosophy. It carefully described the alternative technologies (adopting, in places, the eclectic style of the *Whole Earth Catalog*), the necessary science, and a chapter on the wider philosophical connections. Within four years some 22,000 copies had been sold. The publisher, Franz Ross, suggested that the intent of the book was to 'introduce the practical possibilities of holography to the general public. We had in mind people that were in the middle of nowhere'.[122] Its championing of low-tech holography transformed a counterculturally inspired theme of the early 1970s into a more modern Californian dream in the 1980s: public access holography (see Figure 9.8).

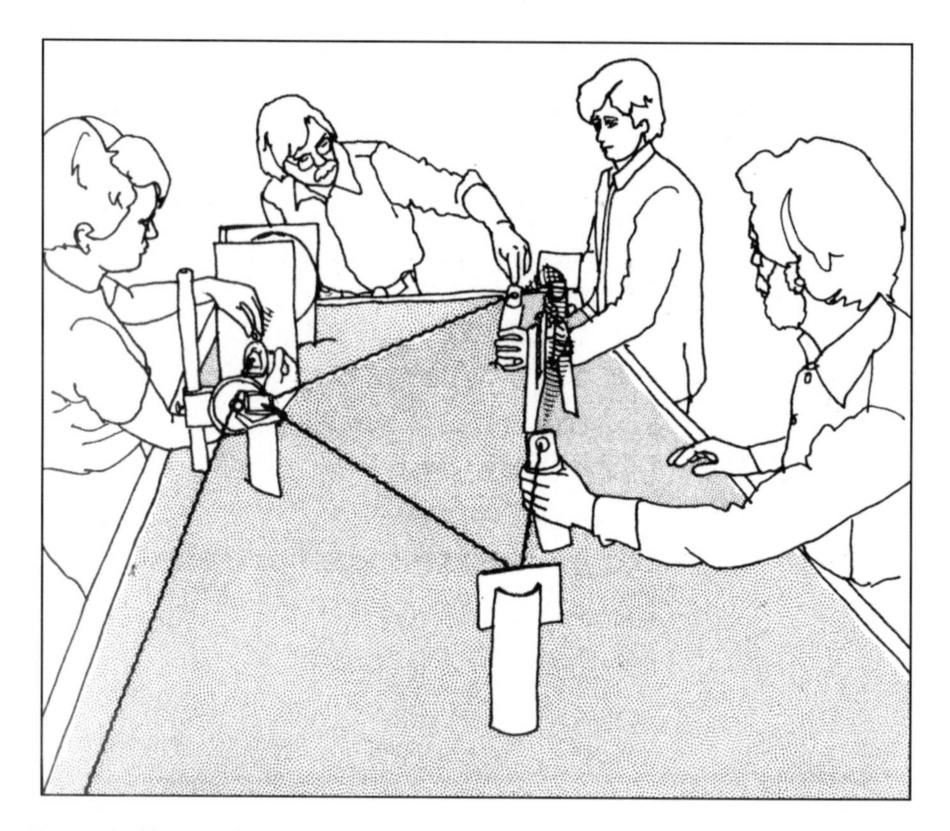

Fig. 9.8. 'Public access holography' via the sand table, 1982.[123]

[122] 'A book for all seasons: Holography Handbook', *L.A.S.E.R. News*, 4 (2) (1985), 6–7.

[123] Unterseher, Fred, Jeannene Hansen, and Bob Schlesinger, *Holography Handbook: Making Holograms the Easy Way* (Berkeley, CA.: Ross Books, 1982).

As Nancy Gorglione put it, 'The stable table took it out of the physicists' laboratories and into the hands of the people'.[124] That personal liberation was supported by certificates of completion offered by the publisher to readers who sent in their holograms. It appealed to a latent need for identity in professional holographers as much as in amateurs: the publisher accumulated a box full of holograms sent in from around the world and Unterseher recounts finding framed certificates at laboratories he visited. Unterseher saw it as a mark of broadening expertise.

> The Brits all knew reflection holography, but didn't know from rainbow holography; the east coast knew rainbow holography, but didn't know from reflection holography so much; and our courses, which Jerry [Pethick] and Lloyd [Cross] taught, that I passed on, was that you get it all; and then whatever comes along you integrate as fast as possible, so in a sense a certificate from the west coast tradition meant that you were learning at least from the broadest spectrum.[125]

The book, and its certificates, clearly filled a need: as practitioners of an undisciplined subject, holographers sought to certify their special knowledge, and to communicate their expertise to others, either novice or professional.[126] It was followed by a growing number of how-to guides.[127]

9.5 SHAPING AND RESHAPING AN IDENTITY

So the creation of the scientific holographer during the 1960s, and the splitting and remoulding of identity during the 1970s to yield an artisanal practitioner, involved both conscious tactics and evanescent contextual factors. They catalysed a coalescence of identities among marginal and variously defined groups. Research interests, economic pressures, amateur enthusiasms, and artistic concerns motivated the practitioners and gave them group solidarity. But the absence of a stable occupational and disciplinary niche was crucial in limiting the growth of communities and in damping any spark of professionalism that may have been smouldering.

Holography, then, in some respects typical of post-war endeavours in science and technology, was very unusual in aspects concerning the emergence and stabilization of its

[124] Gorglione, Nancy, 'Lloyd Cross', *Holographics International* 1 (1987): 17, 29.

[125] Unterseher, Fred to SFJ, interview, 23 Jan. 2003, Santa Clara, CA, SFJ collection.

[126] By 'undisciplined' I mean unrestricted by intellectual and professional constraints, as developed in Johnston, Sean F., *A History of Light and Colour Measurement: Science in the Shadows* (Bristol: Institute of Physics Publishing, 2001). Practical holography (and photometry) demanded extreme *experimental* discipline, of course.

[127] Other books appearing during this period included the influential British publications of Graham Saxby and other, less commercially successful, books (Saxby, Graham, *Holograms: How to Make and Display Them* (London; New York: Focal Press, 1980); Saxby, Graham, *Practical Holography* (New York; London: Prentice-Hall International, 1987); Saxby, Graham, *Manual of Practical Holography* (Oxford: Focal Press, an imprint of Butterworth-Heinemann Ltd. 1991); Wenyon, Michael, *Understanding Holography* (Newton Abbot: David and Charles, 1978); McNair, Don, *How To Make Holograms* (Blue Ridge Summit, PA.: TAB, 1983); Kasper, Joseph Emil and Steven A. Feller, *The Hologram Book* (Englewood Cliffs, NJ: Prentice-Hall, 1985)).

technical communities. The young subject first fostered the formation of a new occupation (the holographer) in the mid-1960s by radically expanding the cognitive content of optical engineering. During a period of rapid transition from military to commercial sponsorship, its practitioners defined the character and tools of their subject based on the subculture of classified research at the University of Michigan and a handful of similar centres. This first community of holographers evolved its practice by adapting and merging pre-existing disciplinary models.

By contrast, the emergence of a community of artisanal holographers was triggered by wider cultural changes. Beginning from two specific cultural events of the late 1960s—the reaction of university students to classified academic research and the rise of the technological art movement—a holographic community sensitive to countercultural themes appeared during the early 1970s. This bifurcation of communities was mediated by the technologies themselves. For the first generation of aesthetic holographers, limited access to the tools and tacit knowledge of holography had hindered their ability to apply it creatively to aesthetic projects. As explored in §10.2, they consequently sought collaborations with scientists, interpreting and applying the technology in ways prescribed by the science–technology community. By adopting inexpensive materials and methods, however, Bay Area artisanal holographers not only shaped the technology to make it accessible to a wider public, but also linked it to deeper countercultural themes. And by redefining the purpose and goals of holography, they valorized different practices, technological solutions, and cultural products. This eventually reshaped the public interaction with holography. In contrast to the corporate attempts to identify markets during the 1960s, which had involved the optical engineering community, the artisanal holographers of the 1970s fostered a cottage industry.

The historical trajectory of holography and its technical communities cannot, then, be represented as a natural sequence of development or a paradigm of technical progress. In the emergence of two of the earliest and most important technical communities in holography, optical engineers and artisanal holographers, both groups self-consciously employed the technology to augment their social stability, but in distinctly different ways. Optical engineers, who defined the orthodoxy of holographers, expanded their cognitive domain through new theoretical perspectives borrowed from electrical engineers, which channelled the applications and meaning of the new subject. They courted DoD sponsorship and developed apparatus and procedures that promoted privileged exclusivity. Artisanal holographers, by contrast, marshalled the sand table to both liberate holographic practice and to extend its meanings and products.

Their equipment mirrored their social structures. The heavy granite tables employed by 'orthodox' holographers were immovable without marshalling major resources of time, labour, and money; they required considerable social stability. In effect, the inertia of the table itself echoed a certain rigidity and resistance to change in the administration and projects studied by their users. In a complementary way, the sand table embodied the qualities of its developers: malleable and versatile, its components and materials could be improvized with a fluidity that reflected the social interactions of its artisanal users.

This technological segregation was enmeshed with political factors. Orthodox holography was buoyed up by post-war military research, while the appropriation of the subject by aesthetic holographers was founded on the critiques and social norms of the wider youth culture of the late 1960s. Indeed, this history illustrates how a technology itself can become, in an *evanescent context*, intensely political in the sense described by historian of technology Langdon Winner: redistributing hierarchies of power and enabling, or requiring, alternate social arrangements. Like Winner, I argue that artefacts can configure environments, and furthermore bring into existence new communities and new practices. The sand table defined a new kind of stability, both technical and social. During a brief period, it mediated a new user community and occupation (the artisanal holographer); it enabled new products (art holograms); and, those products promoted a new interpretation of the subject itself (holography as an expression of holism or an altered epistemology). The transitory political and cultural effects of this technological artefact contrast with the more stable examples cited by Winner.[128]

The technology of holography redistributed hierarchies of power in one other important respect. The equipment that had promoted the segregation of technical communities played a subsequent role as an independent agent or quasi-agent in the socio-technical network, in the sense developed by science studies scholars Bruno Latour and Michel Callon.[129] The equipment of artisanal holography eventually became divorced from its early schools and countercultural context. It transcended its Bay Area community and was taken up by individualistic hobbyists, artists, and entrepreneurs around the world. It is an example of Latour's contention that objects can themselves be understood as actors, or quasi-actors, in a socio-technical network. By the 1980s, the optical table represented such an autonomous actor, enabling creative expression for individuals in a range of social contexts no longer mediated by any dominant user community. The equipment of holography had developed a life of its own.

Despite buttressing technical communities, the technology of holography did not entirely stabilize them. These emergent communities of holographers struggled with an elusive disciplinary identity. Incongruously, optical engineering had been redefined but had not provided the occupational, disciplinary, and professional foundations to support holography as a recognizably stable intellectual contour within it. As introduced earlier, such 'undisciplined' technical occupations straddling industry, academe, and government sponsorship—and yet resisting professionalization and sometimes even occupational identification—have been discussed analytically as research-technologies. Sociologist of the sciences Terry Shinn has argued that open-ended technologies that potentially fit many disciplinary niches engender *research-technologists* who resist occupational

[128] Winner, Langdon, 'Do artifacts have politics?' in: (ed.), *The Whale and the Reactor: A Search for Limits in an Age of High Technology* (Chicago: University of Chicago Press, 1986), pp. 19–39.

[129] A pertinent example from a large body of work is Latour, Bruno, 'Where are the missing masses? The sociology of a few mundane artifacts', in: E. Bijker Wiebe and J. Law (eds), *Shaping Technology/Building Society* (Cambridge MA, MIT Press, 1992), pp. 225–58. For recent reviews of actor-network theory, see Law, J. and J. Hassard (eds), *Actor Network Theory and After* (Oxford: Oxford University Press, 1999).

categorization. These workers form interstitial communities and pursue hybrid careers.[130] As discussed in another context, the incongruity of the holistic subject of holography, like other research-technology specialties that rely on the integration of disciplines, is that its specialists have remained socially dispersed through industry, government, and academia.[131] Artisanal holographers, defined in response to, and in combination with, wider cultural trends, proved even more fluid and difficult to define. The technique-based perspective, which discarded military, industrial, and academic connections, thus challenges the breadth of the definition of research-technology.

Holography during the 1970s, then, offered a cogent illustration of the interactions between a new technology, its emergent communities, and wider culture. The experiences illustrate the importance of political context, and how a seemingly neutral technology can be applied in distinct ways to establish the identities of embryonic technical groups. And it demonstrates that this co-evolution of technology and community is particularly sensitive for fields in which uncontentious success is elusive. Although they lasted scarcely a decade, the Bay Area holographers liberated holography for new creators and new audiences.

[130] Joerges, Bernward and Terry Shinn (eds), *Instrumentation: Between Science, State and Industry* (Dordrecht: Kluwer Academic Press, 2001). pp. 1–11. Examples of such communities are common in, but transcend, the field of optics and instrumentation.

[131] Johnston, Sean F., 'In search of space: Fourier spectroscopy 1950–1970', in: Joerges, B. and T. Shinn (ed.), *Instrumentation: Between Science, State and Industry* (Dordrecht: Kluwer Academic Press, 2001), pp. 121–41.

10

Aesthetic Holographers and Their Art

10.1 INTRODUCTION

The experience of viewing a hologram—the strongest motivation for the entire field—has been only touched upon thus far. The words of Dennis Gabor, Emmett Leith, Juris Upatnieks, and other early holographers merely hint at the magical effect of that first view of the intangible three-dimensional image, a response described by art historian Jonathan Benthall as 'a specialized and esoteric form of exhilaration'.[1] For a growing number of viewers, this sense of delight provoked a desire to explore the mystery and to create their own imagery.

At about the same time that artisanal holographers were reacting against professional optical engineers as models, a separate and more diverse group was emerging. I will refer to them as *aesthetic holographers*, because existing terms are too imprecise.[2] 'Holographic artists' and their products, 'holographic art', proved contentious to define and to evaluate. So, this chapter does not seek to judge quality by an absolute criterion of acceptability or value, but rather to recount and identify the social, technical, and aesthetic factors that guided activities in this domain.

The terms are, of course, variously defined and imbued with subtle understandings. I will use the term *aesthetics* according to its philosophical definition, namely the study of what is beautiful or valued, and consequently as a dimension of human thought and creation distinct from another branch of philosophical study pursued in the first part of this book, epistemology.[3] More broadly, I will use the term *aesthetic holographers* for

[1] Benthall, Jonathan, *Science and Technology in Art Today* (New York: Praeger Publishers, 1972), p. 38.

[2] Andrew Pepper defined the domain of *art holographers* as *creative* holography (Pepper, Andrew, 'Creative Holography Index,' 1992–95), but this common term is contentious: how are art, artists, and creativity to be defined? The label *fine art holographer* has had some currency, but a more popular and inclusive term—although too wide for my purposes—is *display holographer*, a label used to describe, for instance, the participants at the Lake Forest *Symposia on Display Holography* discussed in the next chapter

[3] Philosophy traditionally is divided into the four subjects of aesthetics, epistemology, ethics, and logic. Two of those branches are directly addressed in this book. Epistemology, the study of structures of knowledge, is concerned with the methodologies of discovery and organization of facts, and is thus a

practitioners concerned with producing holograms that have a purpose transcending mere utilitarian value. It is this *intentionality* that separates aesthetic products from goal-driven commodities. Thus a holographic interferogram was usually recorded to provide data for an engineering study, not to evoke a sense of wonder or beauty, even if these were common by-products. And an aesthetic practitioner seeks explicitly to create a piece having transcendent qualities, while an optical engineer or artisanal holographer does not.[4]

Nevertheless, some analysts would argue that the worlds of art history and science studies are closely related. Sociologist of science Bruno Latour proposes, for example, that concepts of 'aesthetics', 'truth' and 'evidence' are all socially constructed and may be contrastingly defined by different social groups. Indeed, he suggests that science studies often amount to an 'aesthetization of science'—a matter of being sensitized to the subtle historical, social, and political forces that cannot be separated from intellectual developments—and that the two fields could learn from each other. But he also cites a deep division between art history and science studies: art historians explore and value the multiple interpretations that artistic works attract; on the other hand, science historians tend to mistrust claims of 'truth' or 'evidence' that attract more than a single interpretation. Latour argues that claims about Truth are weakened by considering their social construction, while claims about Beauty may be strengthened.[5] The aim of this chapter is, nevertheless, to focus on those multiple interpretations to argue that there is no objective and neutral story to be told about holographic art.

The terms holographic *art* and *artist* are, of course, imbued unavoidably with socially constructed meanings. New artistic movements, by definition, redefine meanings and judgements about artistic products. The Australian art theorist and critic Donald Brook, for example, in 1981 defined art as experimental modelling, in contrast to a craft that relies on full knowledge of its materials.[6]

The roles of creator and technician have also repeatedly been redrawn. From the standpoint of 1972, Jonathan Benthall suggested the less emotive terms *designer* or *art worker* for aesthetic activities in what was then known as the 'Art and Technology' or 'Science in Art' movements. Perhaps the most important proponent of this movement (although, it seems, not connected directly with holography) was the electrical engineer Billy Klüver (1928–2004). A Swedish citizen with interests in film as well as engineering, Klüver came to America in 1954 and, during the early 1960s, collaborated with artists such as Robert Rauschenberg, Jasper Johns, and Andy Warhol on artworks that incorporated new technologies. In 1966 he co-founded Experiments in Art and Technology (E.A.T.), a service

central concern of Part I, which dealt with the validation and categorization of new methods of imaging. Ethics is broached indirectly in Chapters 5, 6, 11, and 12.

[4] On the aesthetics of holography, see Benyon, Margaret, 'Do we need an aesthetics of holography?' *Leonardo* 25 (1992): 411–6 and Maline, Sarah Radley, 'The aesthetic problem of figural holography', *Proceedings of the SPIE the International Society for Optical Engineering* 1732 (1992): 438–42.

[5] Latour, Bruno, 'How to be iconophilic in Art, Science, and Religion?' in: C. A. Jones and P. Galison (eds), *Picturing Science, Producing Art* (New York, 1998), pp. 418–40, esp. pp. 422–4.

[6] Brook, Donald, *The Social Role of Art* (Adelaide: Experimental Art Foundation, 1981).

organization for artists and engineers that, by the 1980s, undertook projects around the world. Elayne Varian, the Director of the Contemporary Wing of the Finch College Museum of Art and organizer of *N-Dimensional Space* (1970), had been an early New York presenter of artists such as Andy Warhol, Jasper Johns, Robert Rauschenberg, and others and she made comparisons between the holograms she showed and the works of those contemporary artists. And Nancy Gorglione has suggested that part of the appeal of Lloyd Cross and Jerry Pethick's San Francisco School of Holography was not merely the artisanal skills they promoted but also the *found-art* aesthetic that had been fostered by Rauschenberg.[7] As discussed in §9.2, there were also links between holography and emerging video-art, again in the Bay Area of California. Aesthetic holographers have tended to cast their origins further back, though, more frequently citing connections with early European Modernism.[8] Cultural barriers also played a role in the relationship between holography, art, and science. While the small Art and Technology movement of the late 1960s promoted cross-fertilization, Modernism, beginning a century earlier, had emphasized the contrast and separation between the domains of science and art. Conflicting tensions attracted and separated scientists and artists.

Writing at the beginning of the aesthetic holography movement, Benthall suggested that the key attributes of an artist are a sense of creative imagination combined with skills in a technical medium with which he transmits his or her perceptions.[9] This visionary quality conveniently distinguishes the aesthetic holographer from the artisan for our purposes, even though the two were to be closely associated for some time. Holography and other forms of reputed 'technological art' pose some problems for Benthall's definition, however, because the creative and skills-based traits were not always combined in aesthetic holographers. A significant number, especially in the early days, relied on technicians to translate their concepts into holograms, transforming a solitary act of creation into a team effort.[10]

[7] Gorglione, Nancy, 'Forms of light: a personal history in holography', *Leonardo* 25 (1992): 473–80. Robert Rauschenberg (b. 1925), an American painter who had pioneered the use of 'found' objects in creating sculpture and collage, adopted methods subsequently pursued by members of the San Francisco School, who sought opto-mechanical junk from the rubbish discarded by nearby Silicon Valley companies. Jerry Pethick's works after his move to British Columbia in 1973 continued this inspired magpie approach in producing optical art.

[8] See, for example, Benyon, Margaret, 'Holography as an art medium', *Leonardo* 6 (1973): 1–9; Tyler, Douglas E., 'Holography: roots of a new vision', *Proceedings of the International Symposium on Display Holography* 2 (1985): 379–86 and Casdin-Silver, Harriet, 'Holographic installations: sculpting with light', *Sculpture* 10 (1991): 50–5.

[9] Benthall, Jonathan, *Science and Technology in Art Today* (New York: Praeger Publishers, 1972), pp. 12–13.

[10] The creation of art has often involved a division of labour into creative and artisanal components, of course, from the art works completed by students in medieval schools of painting, to the production of prints in Andy Warhol's Factory. However, such examples usually involve a master artist who *in principle* is capable of performing all the skills-based tasks. A prominent example of an artist who relied on intermediaries to execute his holographic concepts is Salvador Dali, who collaborated (via intermediaries) with Conductron engineers to produce pulsed laser holograms and with the Multiplex Company to produce holographic stereograms. The Knoedler Gallery in New York City exhibited the resulting holograms in April 1972 and again in 1973. Among Dali's holograms was a work, *Holos! Holos! Velázquez! Gabor!*, inspired by a Valazquez painting, and another founded on the novelty of the multiplex hologram, *First Cylindric Chromo-Hologram Portrait of Alice Cooper's Brain*.

What, then, constitutes a *fine-art holographer, holographic artist* or *aesthetic holographer*? I focus on intentionality, that is, on those individuals who identified themselves as being concerned with the aesthetic dimensions of holography and on the holograms that they produced.

10.2 ARTIST–SCIENTIST COLLABORATIONS

While the early scientific demonstrations of holograms produced a frisson of visual excitement, their aesthetic potential appears not to have been broached until the late 1960s. In 1966, photographer/artist Fritz Goro collaborated with Emmett Leith and Juris Upatnieks to devise an aesthetically pleasing hologram for an article to be published in *Life* magazine (see Figure 10.1).[11] Neither Goro nor the *Life* editors classed the result as art, but the striking hologram was intended to evoke a sense of the sublime while illustrating the three-dimensionality of the medium. It was widely reproduced by the Time/Life Corporation in photographic collections that highlighted visual patterns and colours evoking unfamiliar perspectives on the natural world and modern science.

Perhaps the earliest suggestion focusing on the artistic possibilities of the medium was that proffered two years later by physicist Hans Wilhelmsson in *Leonardo*, the newly founded 'International Journal of the Contemporary Artist', which promoted an association between art, science, and technology.[12] Wilhelmsson did little more than describe the Leith–Upatnieks method, and left open how holography might be used by artists.

Even the well-read Dennis Gabor hesitated to invade aesthetic territory. Prompted by queries from an art historian about potential art applications, Gabor suggested a single direction: his ongoing, but slowly progressing, work at CBS Labs on 'panoramic holograms' in which a small model, imaged through a special lens, would appear as if it extended to infinity, giving 'an unlimited third dimension to the artist'.[13] Gabor described the possibilities in utilitarian terms to his friend Peter Goldmark:

> I have an idea, and I would like to have your reaction, because you have an instinct for what will sell and what will not. Holography has so far reproduced chessboards and suchlike tedious objects in three dimensions. A portrait may be just within its reach, with the tricks which I wish to try out in October. But there is something more interesting. One could put a real, deep three-dimensional landscape in the hologram, at which one could look within a wide angle. Of course the coherence length required for a real landscape is out of the question; we have only 10 inches or so, even with the best gas lasers. My trick is similar to that in my 3D invention: we make a strongly flattened and reduced-scale model of the landscape, and

[11] *Life*, Dec. 23, 1966. On Goro (1901–86), see Smith, C. Zoe, 'Fritz Goro: emigre photojournalist', *American Journalism* 3 (1986): 206–21.

[12] Wilhelmsson, Hans, 'Holography: a new scientific technique of possible use to artists', *Leonardo* 1 (1968): 161–9.

[13] Gabor, Dennis to D. S. Strong, letter, 18 Jan. 1972, London, IC Gabor ML/16.

Fig. 10.1 Fritz Goro, with Juris Upatnieks and Emmett Leith, hologram for *Life* magazine, 1966 (Leith collection).

> put it just inside the focal surface of a wide angle lens, which must be somewhat larger than the hologram. The plate will then see the large, real-scale landscape, and the viewer will see the same. Technically the problem is much easier than portraiture, because we can use gas lasers, and with a neon plus argon-ion laser one can get quite nice natural colours. Now the question to the man with the instinct: would you buy a picture on the wall which is like a window into a sunny landscape? It would be one up on the 'false daylight' windows, such as they had on some ships, where people in the below-water dining rooms had the impression that there was brilliant daylight outside. What is more, it would be a new art, as the models would have to be artist's work.[14]

This was at best a commercial art allied with home decoration. Holographer–scientists seemed largely bereft of aesthetic ideas for their new medium.

The first artists to explore holography faced their own obstacles. During the late 1960s holography was still a professional semi-discipline allied to optical engineering; access to its skills and hardware required collaboration with practicing scientists. The transition of Lloyd Cross from classified researcher to commercial engineer and then to artisanal holographer has been discussed earlier. Cross was a transitional figure, but was publicized as a physicist, not an artist or artisan, until the early 1970s and the formation of the San Francisco School of Holography. The handful of artists who became involved with holography in the late 1960s generally approached scientists without prior interactions, and then negotiated arrangements for collaboration or assistance.

Bruce Nauman, for instance, worked with Conductron in Ann Arbor in 1968 making the first holographic portraits by an artist.[15] His series of self-portraits, *Making Faces*

[14] Gabor, Dennis to P. Goldmark, letter, 26 Jul. 1967, IC GABOR LA/9.

[15] Bruce Nauman (b. 1941 in Fort Wayne, Indiana) studied Fine Arts at the University of Wisconsin, Madison and the University of California, Davis.

(see Figure 10.2), was recorded scarcely a year after Larry Siebert at Conductron had first attempted portraiture using a pulsed laser. As Siebert gradually coaxed more power from his ruby laser system, Nauman produced further series of images, beginning with his face and eventually including most of his body. Nauman exhibited his holograms at the Nicholas Wilder Gallery in Los Angeles and the Knoedler Gallery in New York in late 1968.[16]

During the same year, another of the handful of artists to begin working with holography was Carl Fredrik Reutersward in Sweden. Reutersward, born in 1934 in Stockholm, studied art in France and Sweden in the early 1950s and first saw a laser beam at Bell Laboratories in New Jersey in 1965.[17] Beginning to learn about holography from 1967, he began studying the medium as an art form in 1969. He was assisted technically by a network of Swedish scientists: Nils Robert Nilsson, at the University of Uppsala, Sweden; Nils Abramson and Hans Bjelkhagen of the Institute of Technology, Stockholm; Hans Hammarskjold in Stockholm; and Eric Sigg in Lausanne, Switzerland.[18]

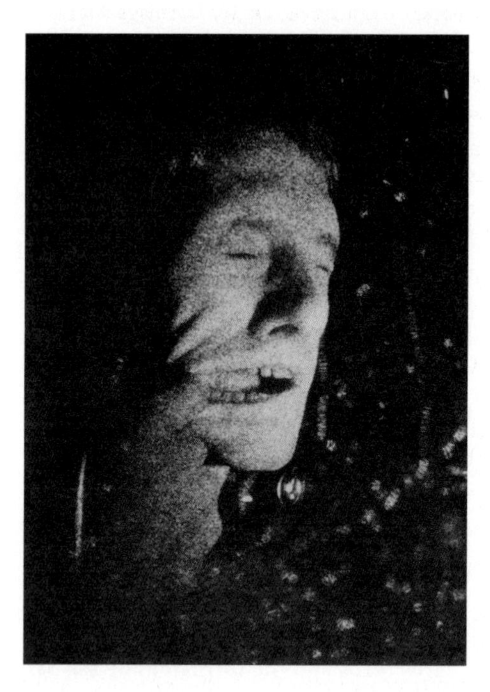

Fig. 10.2. Bruce Nauman, *Making Faces*, 1968 (MIT Museum MoH collection).

[16] Nauman also exhibited in other locations, including the Leo Castelli Gallery in New York City in May–June 1969, as indicated by an exhibition postcard announcing 'Holograms, Video Tapes and Other Works'. His holograms were shown in three European locations in 1970: the Stedelijk Van Abbe Museum in Eindhoven, the Netherlands; the Museum Am Ostwell, Dortmund, Germany; and, the Kunsthalle, Berne, Switzerland.

[17] Reutersward, Carl Fredrik, 'Rubies and rubbish: an artist's notes on lasers and holography', *Leonardo* 22 (1989): 343–56.

[18] Reutersward, Carl Fredrik, 'Artist files,' MIT Museum 30/859-860.

Also from 1968, Richard Wilt, a professor of Art at the University of Michigan, began to work with Conductron engineers to produce holograms of moon globes, a subject that he had been exploring previously in painting and sculpture.[19] For a popular audience, Wilt drew parallels between holography and Impressionism, and mused about the uses and limitations of the medium:

> One might ask, 'What can holography have to do with art?' The most obvious answer is that sculpture can be translated into the more convenient two dimensional plane; all of the sensory attributes contained within the three dimensional form—those nuances of aesthetic relationship—would be possible by moving one's head instead of one's feet. Similarly all the illusions of space sought by the graphic artist would be obtainable in reality—a true, let us say, séance of time and space.

Wilt became concerned with the artist–technician relationship in creation:

> Do the ideas stem from holographic potentialities or do the potentials prod holographic capabilities? In the case of the artist, the answer should be simple if one considers holography to be merely a photographic reproduction of that 'masterpiece' produced by an artist's ego. Such is hardly the case because holography is a re-creation of a subject rather than a reproduction of the visible [. . .] Because this is true, holographic art requires a very special definition of the artist's role [. . .] The holographer is the equalizer of the 'aesthetic' expression. If the artist's idea, and model, are plausible, the holographic 'artist' can bring it to 'reality'; if not, then only a scientific capability has been realized.[20]

Nauman, Reutersward, and Wilt worked with holographic technicians and scientists to create their works. But two artists who began exploring the medium in 1968 defined the subject in a new way. Both were female; both pursued long and relatively successful careers in the medium; and both were the first artists to learn the skills of the medium for themselves. And, despite the individuality of artists and their creative products, Margaret Benyon and Harriet Casdin-Silver also shared some career experiences and perspectives.

In Britain, and quite disconnected from happenings in America, Margaret Benyon took up holography systematically as an artistic medium. As perhaps the most articulate and self-reflective artist exploring holography extensively, her experiences and views are particularly relevant. During the late 1960s, she was a painter preoccupied with where her art was going.[21] She recalls, '1968 was a revolutionary year politically, and art language was broadening; artists going out in the countryside and walking—that was art; putting texts up on the wall; trying to break out of their own discipline into the world'.[22]

Benyon remembers reading about holography in a newspaper at the time and being struck by the connections with her own art, which was concerned with using graphic

[19] Charnetski, Clark and Richard Wilt, 'Interviewed on holography and art by Ed Burroughs', audio recording, Jan. 1969, WUOM, 'Eleventh Hour'.

[20] Wilt, Richard, 'Holograms have a message', report, 1969.

[21] Margaret Benyon (b. 1940) spent her youth in Kenya and studied painting at the Birmingham College of Art and at the Slade School of Fine Art in London. She taught art and (later) holography in Britain from the 1960s to 1990s, with an intervening half-decade in Australia. She was the recipient of an MBE, a Royal honour, for services to art in 2000, and returned to Australia in 2005.

[22] Benyon, Margaret to SFJ, interview, 21 Jan. 2003, Santa Clara, CA, SFJ collection.

interference patterns (such as moiré patterns, a graphic idea employed in op art) to modulate the picture plane so that it no longer looked flat, and provided the illusion of depth.[23] Some of her works were two-colour 'anaglyph' paintings intended for binocular viewing with coloured glasses to create the illusion of three dimensionality. Immersed in a sense of the avant-garde and progression, she recalls feeling that holography could be the foundation of a movement, just as Impressionist, Abstract, Pop Art, and other movements had succeeded each other.

As a part-time art lecturer in the British midlands, Benyon (see Figure 10.3) was isolated from new technologies; she had never seen a hologram and knew no holographers. She located and contacted Gordon Rogers, who had the longest continuing interest in holography in Britain, but—perhaps because of his own activities or particular perspective—he was uninterested in meeting her at the time.

Benyon won a Fellowship at the University of Nottingham in 1968 with few responsibilities other than teaching Life Drawing once a week and mounting one show per year. The Art History Department was able to locate a suitable shared laboratory in the Production Engineering Department at Nottingham, where she worked nights and a BSc student worked during the days. During periods when the lab was not available, she pursued her concurrent interests in anaglyph paintings. With the basic laboratory facilities available, and already experienced in photographic processing, Benyon was able to teach herself holography largely on her own by arranging optics as illustrated in papers. She produced her first hologram, of a shiny steel component from the lab, on a 9 × 12 cm (about 3.5 × 5 in.) plate without much difficulty. Other small holograms followed, exploring the various characteristics of the medium's imagery.

Margaret Benyon was able to use the facilities of several scientific labs. Larger holograms—all 5 × 7 in. plates for her first exhibition—were made in 1969 with the assistance of Peter Spicer at the British Aircraft Corporation in Bristol. And while

Figure 10.3. Margaret Benyon, 1970 (Central Office of Information, London)

[23] Benyon distanced herself from the term *op art*, which she categorized as a superficial journalistic category largely ignored by artists (Benyon, Margaret, *First Nottingham exhibition* (Nottingham: Nottingham University Art Gallery, 1969)).

working at Nottingham that year, she also spent time in the laboratory of Tony Ennos at the National Physical Laboratory (NPL) in Teddington, some 150 miles away.[24] They both were pursuing applications of holography, albeit from different perspectives. Jim Burch of the NPL, who had pioneered holographic interferometry, felt that Benyon's art holography was yet another application worth pursuing. But there were more pertinent reasons for the NPL interest: The 1970 World's Fair in Osaka, Japan, was looming, and Britain's national laboratory was interested in producing an attention-grabbing hologram. Expo '70, like previous international fairs stretching back to London's Great Exhibition of 1851, was concerned with emphasising national prestige and extending export markets for new technologies.

After her first exhibition of holograms at the Nottingham University Art Gallery in the spring of 1969, Benyon was also able to borrow Prof John Butter's lab at Loughborough University during six weeks of the university vacation to make 8 × 10 in. holograms.

For that first exhibition, Benyon was forced to improvise. As lasers were scarce, she arranged her 9 × 12 cm holograms in a vertical circular turntable turned manually by the visitors. In another room, she mounted five 5 × 7 in. holograms in a cylindrical arrangement on a horizontal motorized turntable so that viewers could press a button rotate each hologram into the path of the expanded laser beam for viewing. In subsequent years (at the Lisson gallery in 1970 (billed as the first London exhibition of holograms) and Nottingham shows in 1970 and 1971), she used more turntables to display the rising number of holograms (see Figure 10.4), with a laser, filtered mercury arc and sodium lamp producing progressively fuzzier images in red, green, and yellow. At the 1969 show, Jerry Pethick excitedly informed her that her solo exhibition was the first in holography, although she later learned that she had been preceded by a few months by Bruce Nauman, exhibiting at the Knoedler Gallery in New York.[25]

After her three years at Nottingham, Benyon gained a fellowship at the University of Strathclyde in the Architecture Department. She got advice on building her optical table from engineer Bill Fagan, who had just completed his PhD in holographic interferometry, and contacted Matt Lehmann for further technical information.[26]

As an initially isolated and pioneering British artist experimenting with holography, Benyon thus faced several hurdles: to gain proficiency in the medium; to present holograms to a largely unfamiliar public; and to explore and eventually define an artistic meaning for her work.

[24] Ennos had been an early worker in the use of holography for industrial measurement (Ennos, A. E., 'Holographic techniques in engineering metrology', *Proceedings of the Institution of Mechanical Engineers* 183 (1969): 5–12; Ennos, A. E. and E. Archbold, 'Vibrating surface viewed in real time by interference holography', *Laser Focus* 4 (1968): 58–9).

[25] Benyon knew Pethick from their student days, when he was studying sculpture at the Royal College of Art and she was studying painting at the Slade.

[26] As a graduate student, William Fagan (b. *c.*1943) worked with Peter Waddell, who had begun holography at the university in 1967. Like many holographers, Fagan subsequently had an international career, working in holographic interferometry research and development at firms around Europe, Asia, and America (Fagan, William to SFJ, interview, 19 Sep. 2002, SFJ collection). Benyon had been using practical information from Lehmann, Matt, *Holography: Technique and Practice* (London; New York: Focal Press, 1970).

Fig. 10.4. Benyon exhibition, Nottingham University Art Gallery, 1971, with laser transmission holograms on a carousel (Benyon collection).

She recalls her hologram *Bird in a Box* (1973) as her first holographic art piece having an emotional and personal undercurrent: 'although officially it was a hologram of a complete closed surface, unofficially it was about my feelings as a scapegoated female artist in the architecture department, a boxed-in "bird" '.[27]

Benyon was truly a rare bird constrained by her environment: in Britain of the early 1970s, female students of mechanical engineering were sparse; non-scientist holographers remained marginalized; and female holographers were unknown: Benyon found herself to be the only woman attending a meeting of the Holography Club in London in 1970, and met her first female scientist, Liz McLaughlan, in East Kilbride three years later. Benyon described her subsequent status as perpetually peripheral: 'I work by ducking in and out of institutions, sometimes pooling whatever equipment was there with my own extremely modest homemade equipment, for much shorter working sessions than I need to carry out experiments'.[28]

By the late 1970s, having emigrated to Australia with her husband and family, she worked in the Physics Department at the Australian National University (ANU) in Canberra on a fellowship. At the nearby Royal Military College in Duntroon from 1979, which was linked to University of New South Wales by central government funding, she gained access to an argon-ion laser. One series of holograms that she produced there, 'with cadets stomping up and down outside', was an anti-war statement critiquing the international arms build-up.[29] Despite the context, a relaxed atmosphere prevailed in

[27] Benyon, Margaret, 'DEFINING TRADITIONS 1969–96. Living and working with holography', presented at *Art in Holography*, Nottingham, 1996. Benyon produced the hologram after having been contacted by Gordon Rogers to collaborate as an experimentalist/artisan while at Strathclyde, sculpting a mathematically derived surface for holography (Rogers, G. L. and M. Benyon, 'Holographic recording of a complete closed surface', *Applied Optics* 12 (1973): 886–7). He had not recalled her first request to visit him five years earlier, but in the interim had seen her Nottingham shows.

[28] 'An interview with Margaret Benyon', *holosphere* 9 (1980): 1–4.

[29] Benyon, Margaret to SFJ, interview, 21 Jan. 2003, Santa Clara, CA, SFJ collection.

both institutions, promoted by creative funding: when her laser tube failed, the ANU was able to find money to replace it from the light fixtures and fittings budget. In return, the ANU Physics Department received demonstration holograms to attract students and a report on Benyon's findings. And at the Royal Military College, Benyon was charged only for holograms taken away, giving her the unusual luxury of being allowed to make mistakes without penalty. She subsequently returned to Britain and collaborated with John Webster, a physicist at the Central Electricity Generating Board, to make pulsed laser holographic artworks from 1981.[30] Webster had developed a pulsed laser for recording holograms of nuclear fuel rods, so that images of the rods could later be carefully and minutely inspected for faults.

Throughout her career in holography, Benyon theorized about criteria for defining aesthetic values in holographic fine art. Indeed, her 1994 PhD thesis for the Royal College of Art was entitled *How is Holography Art?*[31] She argued that every fine-art holographer theorized, in effect, about the practice and the product, in common with the beginning of art movements through the twentieth century.[32] Benyon's writings suggest diffidence about the aesthetic holographer, whom she wishes to rank from the practitioner of fine art (i.e. a non-utilitarian practice that has no direct association with earning a living) to a craft worker or artisan (who is paid or commissioned and identified by technical virtuosity). While Benyon herself arguably became adept according to both these criteria, her entry into holography was not by an artisanal route; indeed, she allies fine-art holographers and scientists, on the one hand, and artisanal holographers and craft workers on the other.

Benyon is also diffident about her criteria of art and the desirability of an explicit aesthetic of holography. She suggests a balance between social construction and objective classification. For example, while she expresses the definition that I adopted at the beginning of this chapter linking art with intentionality and judgement ('If the work is to be classified as art, it should be intended as art by its maker or recognized as art by the art world'), she also suggests that holograms can be classified as bad art or good art, even if the contemporary art world had not yet expressed an interest in doing so.[33] This tenuous and variously defined connection between intentionality, critical attention, aesthetic criteria and objectivity were debated by other fine-art holographers and their critics without resolution over the following decades.

[30] Benyon, Margaret, 'Pulsed holographic art practice', presented at *Practical Holography*, Los Angeles, California, 1986; Benyon, Margaret and J. Webster, 'Pulsed Holography as Art', *Leonardo* 19 (1986): 185–91. Webster has been responsible for extending the use of pulsed laser facilities around the world.

[31] Benyon, Margaret, 'Holography as an art medium', *Leonardo* 6 (1973): 1–9; Benyon, 'Do we need an aesthetics of holography?', op. cit.; Benyon, Margaret, 'Holography as art: cornucopia', presented at *Fifth International Symposium on Display Holography*, Lake Forest, Illinois, 1994; Benyon, Margaret, *How is Holography Art?*, PhD thesis, Royal College of Art, London (1994).

[32] Benyon, Margaret, 'DEFINING TRADITIONS 1969–96. Living and working with holography', presented at *Art in Holography*, Nottingham, 1996; Benyon, Margaret to SFJ, interview, 21 Jan. 2003, Santa Clara, CA, SFJ collection.

[33] Benyon, 'Do we need an aesthetics of holography?', op. cit., quotation p. 415.

Such concerns were to influence most aesthetic holographers. Another important early worker was Harriet Casdin-Silver, who benefited from a series of successful collaborations with scientists in the American northeast. Born in 1925, Casdin-Silver had painted and produced installation art from the late 1950s before taking up holography. Her 1968 piece *Exhausts*, for example, combined sound recordings, exhaust pipes and variable lighting to encourage audience interaction within a 10 m steel cube. Having viewed a laser beam that year, she considered adding one to her installation:

> I was looking for more sophisticated lighting for that stainless steel piece, and I went off to American Optical Company because I knew they were doing holography there, but I wanted a laser—I wanted to borrow a laser in my stainless steel. They invited me in to learn this new crazy thing they had already done. They were doing 4 × 5" holograms of rabbits, and I was really intrigued, and thought it would give me more to integrate with my other stuff.[34]

At the American Optical (AO) Research Laboratories, located in Framingham, Massachusetts, she collaborated with Raoul Van Ligten, head of the Coherent Optics department, for her first holograms.[35] Casdin-Silver had recently had a show in the area, and van Ligten initially was curious to see what an artist could do with holography. She came to Ligten's lab for some two days every couple of months, learning the techniques from him and proposing subjects to be 'holographed'. There she 'would run around looking for little objects—they had to be little and they had to be *objects*—because nobody knew that you could just take laser light'.[36]

Her first 4 × 5 in. hologram was of a dishwasher basket filled with silverware. She recalls that the constraints of the medium soon became apparent. The coherence length of the 50 mW Spectra-Physics laser limited her to a volume of about a cubic foot, much smaller than her previous 10 ft steel cube, so her 'artist friends jeered'.[37] Nevertheless, she discovered beauty in the accidental meanderings of her reflected laser beam, patterning the walls and ceiling in complex interference patterns. She recalls this as a crossroads: 'Marshall McLuhan's idea that the medium is the message was prevalent then, and I was enveloped constantly in the laboratory by constant laser light shows'.[38]

Casdin-Silver began to explore these special attributes of the new medium in directions that her scientist patrons had not foreseen. After a couple of years, Van Ligten's work expanded to several laboratories, at which point Casdin-Silver was given

[34] Casdin-Silver, Harriet to SFJ, interview, 3 Jul. 2003, Boston, MA, SFJ collection. American Optical, one of the three largest optical companies in America, had specialized in eyeglasses and microscopes until the 1950s when it developed the Todd-AO widescreen cinema system, and was investigating fiber optics, as well as holography, during the early 1970s.

[35] Van Ligten had completed an MSc in Physics at the Technological University of Delft, Netherlands, in 1957. Before joining AO, he had worked at Itek Corporation on lens design and evaluation.

[36] Casdin-Silver to SFJ, op. cit.

[37] Casdin-Silver, Harriet, 'My first ten years as artist/holographer (1968–77)', *Leonardo* 22 (1989): 317–26, quotation p. 317.

[38] Ibid., quotation p. 318.

access to an entire lab when she wished. As a result, Casdin-Silver worked at AO from 1968 to 1972:

> They asked me for nothing. They gave me plates and everything else; they did not pay me! But they funded the materials and stuff, and the space. The lab was like no other lab you would see at this point. The tables were marble top; just beautiful. I knew so little I didn't know how unusual that was.[39]

However, as an artist interested in expressing feminist statements in her products, Casdin-Silver was aware of her own unusual status. Her privileged access to the medium was all the more unusual in that she was not merely a non-scientist, but female:

> In 1969 the AO scientists, from the optical physics and other laboratories, came to study the strange intruder bulding a weird construction—which they thought would never culminate in a hologram. Moreover, the strange creature was not like them at all: it was female. There were no others like it in any of the labs.[40]

During 1970, Casdin-Silver exhibited holograms at venues in Kansas, Illinois, and Massachusetts. At a demonstration evening of her holograms in Worcester, Massachusetts, one of the visitors invited her to lecture and display her work at Polaroid Corporation in nearby Cambridge. There she met Stephen Benton in April 1972 at a Polaroid Research and Development workshop.[41] Benton, then a 31-year-old part-time researcher, sixteen years Casdin-Silver's junior, had been doing research in display holography there since his undergraduate days, as discussed in §6.4. Casdin-Silver recalls:

> I did not know him at all until I was invited to do a show at Polaroid. Someone at Polaroid had seen my work somewhere else. That person did not know there was a Steve Benton at Polaroid doing work in holography. When I went there, somebody said 'oh, there's somebody here doing this, he has a little lab over here somewhere'. They went to get Steve; I looked at his work, and he looked at my work, and we tried to do something together. I invited him to be part of my little show at Polaroid.[42]

For his part, Steve Benton remembered their collaboration as crucial for popularizing his technique of rainbow holography:

> It wouldn't have gone anywhere, frankly, if I hadn't run into the artist Harriet Casdin-Silver [. . .] I brought her downstairs and told her 'this is what I do!' and she said, '*Do you know*

[39] Casdin-Silver to SFJ, op. cit.

[40] Casdin-Silver, 'My first ten years as artist/holographer (1968–77)', op. cit., quotation p. 318.

[41] Benton, Stephen A., 'Artist files,' MIT Museum 26/584, 1978. Stephen Benton (b. 1941, San Francisco; d. 2003, Boston) obtained his BS in Electrical Engineering from MIT and PhD in Applied Physics from Harvard. He remained at Harvard until 1973 as an Assistant Professor of Optical Physics. From his student days, he had been working at the Polaroid Research Laboratory in Cambridge, also across the Charles river from Boston, and within three miles of both Harvard and MIT. With Herb Mingace and Will Walter he formed the Cambridge Stereographic Society in the 1970s. Benton was the founder of the Spatial Imaging Group at MIT in 1982, and a member of the founding faculty at the MIT Media Lab in 1984. In 1996, he became Director of the Centre for Advanced Visual Studies there. Benton was also active in professional societies associated with holography, being an office bearer in the Optical Society of America from the late 1970s, the SPIE from the early 1990s, and of the Society of Imaging Science & Technology from 1999. Between 1987 and 1992 he was also on the Board of Trustees of the Museum of Holography in New York, and was the principal organizer of the purchase of the MoH collection for the MIT Museum.

[42] Casdin-Silver to SFJ, op. cit.

> *what you have here?* White light holograms that are nice and bright! We could make holograms together!'[43]

Although Polaroid had a history of supporting artists (usually by supplying free materials and working space in return for getting first choice on purchasing works of photographic art), Casdin-Silver gained no special status there. She continued to work at her AO lab, with Benton joining her at AO about once a week from May 1972.

Casdin-Silver had both a growing reputation as an artist, and attention in the local art community around Boston. Benton had solid technical expertise and a new but little known method for displaying holograms in white light. In their first month together, they created a series of holograms from scattered light, a series they dubbed *Cobweb Space* (1972), and inspired by accident:

> I was in that AO lab, and the laser would bounce off the walls, and I thought 'instant art!' So when I got to collaborate with Steve, which is what we were doing at that time—we were a team—I showed him these images, and how could I get this into a hologram, they were so exciting [. . .] He worked out the way, and I worked a few things in. This is Steve's thing—imagine, we could see it with the white light! The physicists at AO knew nothing about the white light—and I was sworn to secrecy *not to tell the system!* God bless Raoul van Ligten, who made it possible for Steve from another company, another corporation, to come into American Optical, because he was a physicist with an artist's soul.[44]

She recalls the AO physicists as being astonished by the brightness of the rainbow holograms when they first saw them. The large 11 × 14 in. holograms also enabled Benton to demonstrate to his employer, Edwin Land, the potential of rainbow holography. They attracted local interest, too: Fellows of the Center for Advanced Visual Studies (CAVS) and from the Massachusetts Institute of Technology (MIT) visited, both organizations based, like Benton's Polaroid laboratory, a few blocks apart in Cambridge, Massachusetts.

Yet the new technology did not transform art. Neither Casdin-Silver nor Benton were initially enthusiastic about white-light transmission (rainbow) holograms. Casdin-Silver characterizes herself as ambivalent about the piece; on the one hand, it was dramatic work that they hoped would startle the small holographic community and the art world; on the other, she missed the special mystique of laser light, and resolved that future holograms would be designed for viewing one colour at a time.[45] She recalls:

> I don't know if I really thought it was such a big deal, the white light system, because I loved my laser light better! But museums and galleries were very loath to play with the laser.[46]

Benton concurred:

> Not everyone agreed! I was making holograms with people at Polaroid, too, who *liked* the blurriness of ordinary white-lit full parallax—they'd say 'well *that's* art . . .' [The rainbow

[43] Benton, Stephen A. to SFJ, interview, 11 Jul. 2003, Cambridge, MA, SFJ collection.
[44] Casdin-Silver to SFJ, op. cit.
[45] Casdin-Silver, 'My first ten years as artist/holographer (1968–77)', op. cit., p. 319.
[46] Casdin-Silver to SFJ, op. cit.

hologram was less appealing] also because it was image plane, which was *not* a big deal at the time; most images were quite far out of the plane.[47]

Like Margaret Benyon, Harriet Casdin-Silver was effective in discovering laboratory space. Her access to the AO laboratory eventually ended when van Ligten departed for another post in Indonesia in 1972. She then came to an arrangement with Henrik Gerritson, a Dutch professor of physics at Brown University, whom she characterized as 'a kind of patron'.[48] He was able to purchase and give her access to a steel optical table, 50 mW laser, and the requisite supplies. Like van Ligten at AO, Gerritson also gave Stephen Benton access to the Brown laboratory to continue his collaboration with Casdin-Silver.

But even in these first forays of collaboration between artists and scientists, there was a world between them. Emmett Leith, for example, first learned of rainbow holograms when he saw one exhibited in New York.[49] And early artist-holographers were less intimidated by presumed constraints than were scientists. Margaret Benyon, who was beginning to make holograms of light patterns at about the same time, recalls, 'we took on the rigid rules of holography as a challenge. If an artist was told not to make a hologram of any light source that would act as a second reference beam, what do you think we would do?'.[50]

The collaboration between Benton, the inventor/scientist, and Casdin-Silver, the artist, lasted about a year. Benton found the fifty-mile trip from Polaroid (in Cambridge, Massachusetts) to Brown University (in Providence, Rhode Island) awkward. And, according to both, there were differences in the goals and products of the collaboration. Casdin-Silver recounts:

> I started to do phallic images [notably the *Phalli* series, 1973]. Henrik [Gerritson] didn't care about that; he was from the Netherlands, more free. Stephen thought it was *sick*. He didn't want his name on it, but it was okay to say he was the consultant [. . .] *Cobweb Space* and *Skeletal Hand* were the two pieces where we were really a team, and then he became more of a consultant because he just couldn't be around.[51]

Benton adds:

> That's when it really stopped, when she really just wanted people to make holograms the way *she* liked them. I said 'Well, that's really not much fun for me. [. . .] I was really just interested in inventions, when it came right down to it. Harriet was great to work for that first year, but [. . .] artists just love to become independent.[52]

[47] Benton, Stephen A. to SFJ, interview, 11 Jul. 2003, Cambridge, MA, SFJ collection.

[48] Casdin-Silver, 'My first ten years as artist/holographer (1968–77)', op. cit., p. 321.

[49] Leith, Emmett N. to SFJ, interviews, 29 Aug.–12 Sep. 2003, Ann Arbor, MI, SFJ collection.

[50] Benyon, Margaret to SFJ, email, 4 May 2005, SFJ collection.

[51] Casdin-Silver to SFJ, op. cit. The *Phalli* holograms were recorded with a continuous wave HeNe laser, using commercially purchased dildos, and were intended as a feminist statement. From 1980, her figure studies were of live nude subjects using a pulsed laser.

[52] Benton, Stephen A. to SFJ, interview, 11 Jul. 2003, Cambridge, MA, SFJ collection.

Thereafter, their collaboration transmuted into less frequent telephone consultation. Casdin-Silver's first piece done 'without the physicists' was *Glass Balls*, a laser transmission hologram done in the AO laboratory that marked her independence. Benton produced his own independent pieces such as the widely reproduced *Engine No. 9* (1975), *Pum III* (1976), and *Aphrodite* (1977). Intended more as high-quality demonstration pieces than art, Benton's holograms nevertheless found their way into exhibitions and collections.

Just as independence from scientists became a recurring theme, Harriet Casdin-Silver was increasingly independent of the wider community of art. The reception received by her holograms was not an issue, though, in the local environment of Boston/Cambridge, dominated by MIT and technological firms that valorized linkages between art, science, and technology. In particular, CAVS at MIT allowed Casdin-Silver to benefit from resources without restricting her freedom (by contrast, Stephen Benton was not formally associated with CAVS until becoming its Director in 1996).[53] In 1975 she participated in ARTTRANSITION at CAVS, and its Director, Otto Peine, made her the first CAVS Fellow with a commitment to holography the following year. For a time she commuted between the Brown University and CAVS laboratories, and from 1977 focussed on teaching for the MIT programme for a Master of Science in Visual Studies, for which CAVS was one of five participants (see Figure 10.5). She taught her course, 'Holography as an Art Medium', for the following nine years. In 1978, she returned to earlier themes by creating holograms that were illuminated by a sun-tracked beam, a 144 ft long environmental outdoor sculpture called *Centerbeam* exhibited in the mall near the Smithsonian Air and Space Museum in Washington, DC.

Fig. 10.5. Harriet Casdin-Silver with *Equivocal Forks*, *c.*1977 (MIT Museum MoH collection)

[53] Casdin-Silver to SFJ, op. cit.; Connors, Betsy to SFJ, interview, 10 Jul. 2003, Cambridge, MA, SFJ collection. CAVS was founded in 1967 by Gyorgy Kepes to combine activities in art and science.

10.3 ARTISTS AND ARTISANS

Despite relatively successful stories such as those of Benyon, Casdin-Silver, and Reuterswärd, collaborations between scientists and artists were uncommon, especially after the 1970s.[54] As Steve McGrew, an American holographer-entrepreneur observed,

> Collaboration between scientists and artists in holography is mostly a myth. A few members of the scientific community, tossing out tidbits to hungry artists, have done their reputations a lot of good while not doing more than momentary good for art holography. Real collaboration must take the form of open communication and mutual education. We cannot build a language of art holography when every impulse to communicate is damped by the fear of giving away something of value.[55]

Even Harriet Casdin-Silver was concerned about the 'second class citizenship of the artists in this supposed union of scientists and artists. Everyone working together to send holography into space. The artists know this is untrue'.[56]

Despite this disdain from the perspective of the 1980s, art and science, particularly when allied to commerce, were the driving forces over the decade. The environment that brought these elements together was greater accessibility of holographic equipment, short courses to impart the necessary skills, and growing public awareness and interest.

Relatively few artists were able to interact directly with scientists, because this required enthusiasm from both sides and mutual interests to be fulfilled.[57] Instead, and in parallel with the handful of artists who negotiated such arrangements with professional holographers, most artists became associated with artisans who were developing more accessible environments. The earliest of these artist–artisan collaborations nucleated around Lloyd Cross and Jerry Pethick in Ann Arbor. Their first group show of aesthetic holograms took place at the Cranbrook School of Art in Detroit, soon followed by the Finch College Museum of Art and then *N-Dimensional Space*, in New York City, all during 1969–70.[58]

[54] Andrew Pepper has explored the various forms taken by collaborations, and identifies the early collaborators that I have discussed, a second generation (including Setsuko Ishii, Dieter Jung, John Kaufman, Sam Moree, Ruben Nuñez, Dan Schweitzer, Rick Silberman, and Fred Unterseher who 'either worked with, brushed against or collaborated with technical experts', and subsequent workers for whom collaborations were uncommon. See Pepper, Andrew T. and E. P. Krantz, 'Art of collaboration: a conflict of disciplines or constructive relationship', presented at *Fifth International Symposium on Display Holography*, Lake Forest, Illinois, 1994. As argued below, the growth of artisanal holographers and their techniques mediated this growing independence of artists.

[55] Steve McGrew, quoted in Jackson, Rosemary H., 'Off the wall', *holosphere* 9 (1980): 4–7, p. 6.

[56] Casdin-Silver, Harriet, 'Of holography and art and artists', *Proceedings of the International Symposium on Display Holography* 2 (1985): 403–10.

[57] Besides Dali, Nauman, Benyon, Reutersward, Casdin-Silver and Richard Wilt, another early exception was Ruben Nuñez (b. 1930, and working with physicists Maurice Françon and Jean Sagaut at the Institute D'Optique, Paris in 1974).

[58] McBurnett, Ted, *N Dimensional Space* (New York: Finch Museum of Art, 1970). The *Village Voice* reported, 'sixteen examples of the real thing are on view at the Finch College Museum of Art, 62 East 78th Street. These were created by seven avant-garde artists—Lloyd G. Cross, Robert Indiana, E. N. Leith, Allyn Lite, Bruce Nauman, George Ortman, and Gerald Pethick—working in collaboration with trained laser technicians' (Anon., 'N-Dimensional Space exhibition', *Village Voice*, 14 May 1970). All were products of Michigan: Ortman was at the time artist in residence at the Cranbrook Academy; Nauman's holograms had

Cross was then beginning a transition between technical subcultures. In the catalogue for *N-Dimensional Space* he was portrayed as a scientist and PhD candidate at the University of Michigan, member of the Optical Society of America and author of 'numerous technical papers and several patents pending in the laser field', in contrast to his associates, who were identified as artists.[59] Yet, as discussed in §6.3 and §9.1, Cross's methods and aims were then shifting to artisanal and aesthetic concerns, as suggested by his partnership in the Editions Inc gallery.

Indeed, from the early 1970s, most of the second crop of holographic artists were trained not by scientists, but by artisans. Their early learning environments were the schools of holography that were beginning to appear. Such environments gave them not only the requisite skills, but also the freedom to define holography and its methods in their own terms.

Thus, from the San Francisco School of Holography came artists Fred Unterseher (1971), Anaït Arutunoff Stephens (1971) and Nancy Gorglione (1972).[60] From the Lake Forest College classes of Tung Jeong came Jody Burns and Posy Jackson (1973), Ana Maria Nicholson (1983) and Ana MacArthur (1987–8). And from the New York School of Holography (later transformed into the New York Holographic Laboratories) came Dan Schweitzer and Sam Moree (see Figure 10.6) (1974), Rudie Berkhout (1975) (see Figure 10.7), Mark Diamond (1976), Dieter Jung (1977) (see Figure 10.7), Edwina Orr (1979), Arthur Fornari (1980), Marie-Christiane Mathieu and Andrew Pepper (1981), Claudette Abrams (1982), and Georges Dyens (1984).[61]

Artisans and artists had become not merely students, but also teachers, of holography soon after the first schools opened. Sculptor Jerry Pethick, co-founder of the San

Fig. 10.6. Sam Moree at the New York School of Holography (MIT Museum MoH collection)

been produced at the Conductron Corporation in Ann Arbor; Lite, Cross and Pethick were co-directors of Editions Inc; and Indiana had made a hologram in their studios.

[59] McBurnett, Ted, *N Dimensional Space* (New York: Finch Museum of Art, 1970).

[60] Anaït Arutunoff Stephens (1922–98), born in Berlin, received her art education in Mexico and America (Anait, 'My art in the domain of reflection holography', *Leonardo* 11 (1978): 306–7). She also collaborated with scientist Hans Bjelkhagen in producing colour reflection holograms during the 1980s.

[61] Daniel K. Schweitzer (1946–2001) studied drama at Penn State University during the late 1960s. He met Sam Moree at a summer stock company and the two became involved in video art and then holography. In 1998 he co-founded the Center for the Holographic Arts in Long Island with Ana Maria Nicholson; on his early career, see Bush, Edward A., 'A conversation with Dan Schweitzer', *holosphere* 10 (1981): 3–6. Sam

Fig. 10.7. Rudie Berkhout, 1979 (MIT Museum MoH collection)

Francisco School of Holography, is an example. And former students became teachers. Thus Fred Unterseher taught at the San Francisco School and others in California from the early 1970s; Dan Schweitzer and Sam Moree taught at the New York School of Holography; Edwina Orr and David Traynor, taught in New York, opened Richmond Holographic Studios in London to teach a generation of British holographers. This feeding back of knowledge promoted a maturation of skills and stability of location for an emerging community of aesthetic holographers.

10.4 FORMALIZING THE ART: ACCREDITED SCHOOLS

While short courses on holography became available through the 1970s and inspired would-be holographic artists, more formal training in aesthetic holography at the

Moree (b. 1946) co-founded the New York Holographic Laboratories in 1977, and taught Visual Arts at the New York School of Holography 1990–95. Rudie Berkhout (b. 1946, Amsterdam) studied electronics in the early 1960s. He became one of the most successful and respected holographic artists, claiming to obtain a full-time living from the art from 1979. Dieter Jung (b. 1941) studied theology and art in Berlin and at the École Nationale Supérieure in Paris during the 1960s, and film in Berlin during the early 1970s. He was appointed professor of the newly founded Academy of Media Arts in Cologne in 1991.

undergraduate or postgraduate level began to appear a decade later.[62] Interestingly, this specific form of academic training was limited to fine art rather than the sciences, for which postgraduate studies defined holography merely as a tool to solve a problem.

The appearance of aesthetic holographers was in part a consequence of such formal training. Unlike those trained in the artisanal schools, holographers graduating from an academic programme found that they had the credentials that made exhibitions more likely, although a large fraction turned to commercial pursuits. Thus Marie-Andrée Cossette obtained an MA (Fine Art) in Holography and Photography from the Université du Québec à Montréal in 1983. Obtaining an MA (Holography) from the Royal College of Art in London, for example, Jean Bailey, Susan Cowles, Melinda Menning, and Jeffrey Robb continued to work in holography; Patrick Boyd and Pearl John taught holography classes; Graham Tunnadine and Matthew Andrews ran a commercial holography studio in East London for some years; and Kevin Baumber became the Creative Director of Light Fantastic and then Applied Optical Technologies, responsible for hologram design. Andrew Pepper obtained his PhD in Fine Art Holography from the University of Reading in 1989. Martin Richardson obtained his PhD in Display Holography at the RCA in 1988, and Margaret Benyon and David Pizzanelli completed their PhDs there in 1994.

Few comparably focused art schools or universities offered training in art holography. Exceptions included the Center for Advanced Visual Studies at MIT from the late 1970s; the School of Art and Design at Goldsmith's College, London, from 1980; the MIT programme in Media Arts and Sciences founded by Stephen Benton in 1987 at the MIT Media Lab (offering technology-related projects rather than art *per se*, but passively sympathetic to the Art and Technology perspective of MIT's Center for Advanced Visual Studies);[63] the School of the Art Institute of Chicago, offering Bachelors and Masters degrees in Fine Arts; St. Mary's College, Notre Dame, Indiana; the Academy of Media Arts in Cologne, Germany, instituted by Dieter Jung, taught holography to students of media art and launched a Fellowship programme for artists and engineers; the University of Tsukuba, Tokyo, which included holography in its Plastic Arts and Mixed Media Course from the 1980s; the Ontario College of Art Holography Department, operating from the early 1990s; and, in Australia during the 1980s and early 1990s, holography laboratories at Sydney College of the Arts, the Design School of the University of

[62] Scientific training in holography had, of course, been available at postgraduate and sometimes undergraduate levels at many universities since the mid-1960s. On its Soviet implementation, see Larkin, A. I., 'Holography and education in the USSR', *International Symposium on Display Holography* 1600 (1991): 412–7.

[63] Stephen Benton, who directed the programme, reflected, 'My students here are not classical scientists; it only mattered to a few of them if they had papers in the Optical Society of America. For most of them having SPIE papers was just fine. They are mostly non-mainstream science students, a lot of them with art-type background and art interest. And a lot of stuff had to be pounded from metal to build equipment. Whereas most of the Media Laboratories was programming; they didn't know how to use a tool' (Benton, Stephen A. to SFJ, interview, 11 Jul. 2003, Cambridge, MA, SFJ collection).

Technology, and at the College of Arts of the University of New South Wales.[64] Even for fine-art holographers, commercial sponsorship was important. Richardson's PhD studies were sponsored by Pilkington PE, a large British optical firm then pursuing display holography alongside its much larger military HOE markets.[65] And Benyon's PhD was sponsored by Ilford Ltd, which was then actively marketing holographic media and chemicals for numerous commercial applications.

The factors determining the origin and stability of formal training programmes in holography are illustrated by the experiences of the Royal College of Art (RCA). Founded in 1837, the RCA established a Holography Unit in the Photography Department of its School of Fine Arts and Communications in 1985. Conceived by John Hedgecoe, Professor of Photography, and Michael Langford, Head of the Photography department, technical aspects of holography were taught at the college from 1984 by Jeff Blyth and author-teacher Graham Saxby.[66] Beginning as a short course in the Cromwell Road premises of the RCA opposite the Natural History Museum, it was based initially on a sand table and a 5 mw laser, one studio, and a darkroom. In 1996, the Unit moved to purpose-built facilities in the main RCA premises, located next to the Albert Hall in South Kensington.

From 1985, Nicholas Phillips headed the Unit as Senior Visiting Tutor and Peter Miller, followed by Mike Burridge, and then Rod Murray, as Tutor.[67] Miller invited contributions from a pool of visiting lecturers including Andy Pepper, Margaret Benyon, John Kaufman, and Sally Weber. Rob Munday, the Unit technician, subsequently became the *de facto* tutor, and later ran a successful large-format company in Surrey.[68]

[64] The Goldsmith's course, intended for public access, was organized by Peter Cresswell and Paul Watson and taught by Michael Wenyon (and later Susan Gamble), receiving funding from the Arts Council of Great Britain towards equipment costs (MoH, 'Workshops: Goldsmith's College,' MIT Museum 42/1301, 1980). The holography programme at the School of the Art Institute of Chicago was an outgrowth of Fine Art Photography teaching at the School, as a series of workshops in the late 1970s taught by Tung Jeong, a physicist at Lake Forest College some 40 miles away. Teaching was taken over by Ed Dietrich during the late 1970s, and thereafter continued under Ed Wesley until the early 1990s. The fifteen-week courses attracted some 100 students per year to one or more classes from the early 1980s (Dietrich, Edward, 'The development of the holography program at the School of the Art Institute of Chicago', *Proceedings of the International Symposium on Display Holography* 2 (1985): 435–40; Dietrich, Ed to SFJ, interview, 19 Nov. 2003, Vancouver, SFJ collection). At Tsukuba, a CAVS Fellow, Donald Thornton, provided initial training in holography. By 2000, when Prof. Mitamura left the University, six undergraduate, eight masters, and one doctoral degree had been conferred for holographic art (Kamata, Yasumasa to SFJ, interview, 19 Nov. 2003, Vancouver, SFJ collection; Kamata, Yasumasa to SFJ, email, 17 Dec. 2003, SFJ collection). The Australian labs were taught successively by Paula Dawson.

[65] Owen, Harry to SFJ, interview, 9 Sep. 2003, Ann Arbor, MI, SFJ collection.

[66] Saxby, Graham to SFJ, interview, 16–17 Sep., 2002, SFJ collection; Munday, Rob to SFJ, interview, 31 Mar. 2004, Richmond, UK, SFJ collection; Miller, Peter, 'An educator with some light at the end of the tunnel!' *L.A.S.E.R. News*, 3 (3) (Fall/Winter) 1986: 8–9. On Blyth, see §12.2.

[67] Lloyd, Scott, 'The Holography Unit at the Royal College of Art', *holosphere*, 14 (3) (Summer 1986): 15–6; Murray, R., 'Holography at the Royal College of Art, London', *International Symposium on Display Holography* 1600 (1991): 237–9. Rod Murray had taught holography within the Fine Art course in the Art and Design Faculty at Liverpool Polytechnic, which had closed before the RCA operation. He was external examiner of the RCA holography course in 1988, and joined the staff in 1989. In 1991 Murray moved to the RCA to train its final cohorts of graduate students in holography, remaining there until 1996 (Murray, Rod to SFJ, emails, 4 Jul. and 10 Sep. 2002, SFJ collection).

[68] Munday saw his first hologram in 1981 while pursuing a Photography, Design, Illustration, and Computer Graphics degree at Plymouth Polytechnic in Cornwall. From 1983, he worked for Edwina Orr

The Holography Unit was well resourced, with equipment that included a pulsed ruby laser and krypton laser. The facilities expanded to include four CW laser studios, a pulsed laser studio, two darkrooms, a course room, and office. A multiplex camera was constructed, and a computer interface was devised for creating holograms of computer-generated images. Some thirty students, many subsequently attempting to pursue careers in holography, were trained at the RCA. The supply of students gradually fell, however. One reason for this was that the RCA granted graduate degrees (MA and PhD) and required students to possess a Bachelor's degree at entry.[69] Another problem was maintaining an adequate supply of prepared students. During the mid-1980s, the RCA was the only art college granting MA and PhD degrees in art holography, although the MA degrees were never solely directly towards fine art. As part of the Photography Department, the teaching was also oriented towards commercial design and visual communication. Yet most British universities had ceased to operate undergraduate holography laboratories by the late 1980s. One of the last British institutions providing undergraduate teaching was Wolverhampton Polytechnic, where Graham Saxby taught holography.[70]

In most cases, the decision to close holography units reduced to a matter of economics. The demise of the RCA holography programme in 1994, after eight academic years, reflected this dual constraint of dwindling students and high programme cost: lasers and holographic emulsions were more expensive than easels and paint.

10.5 DISTINGUISHING SUBCULTURES

The disparate collection of aesthetic holographers came from different backgrounds. As we have seen, some were trained traditionally in art, while others migrated to aesthetic interests from other fields. Some developed their skills in orthodox professional laboratories, and others alongside artisans. Some were self-taught, and others acquired professional qualifications. These qualities differentiated and bound together clusters of aesthetic holographers. While remaining an incongruent grouping, aesthetic holographers nevertheless became clearly distinguished from both professional scientists and their artisanal counterparts. Geography, sociology, and intellectual motivations distinguished them further.

and David Traynor at Richmond Holographic Studios. Both Orr and Traynor had studied sculpture at the RCA. From 1985, Munday was Senior Technician for the RCA Holography Unit. He left in 1991 to form Munday Spatial Imaging (Spatial Imaging from 1994) to produce and sell holograms, purchased another firm, Laza Holograms, and subsequently moved into the space fomerly occupied by Richmond Holographic Studios (Munday, Rob to SFJ, email, 11 Jul. 2005, SFJ collection).

[69] Munday, Rob to SFJ, interview, 31 Mar. 2004, Richmond, UK, SFJ collection; Benyon, Margaret to SFJ, interview, 30 Apr. 2005, London, SFJ collection.

[70] Saxby (b. 1925), after obtaining his bachelor's degree in Physics and Mathematics in 1945, trained as a photographer in the RAF, where he served eighteen years and rose to Chief Technician. From the mid-1960s, he taught at Wolverhampton Teacher's College (later Wolverhampton Polytechnic) and wrote freelance on photography.

Attempts to survey aesthetic holographers have always been inadequate, in part because of the ambiguous definitions that categorize artists and artisans. The *Creative Holography Index*, for example, collated the works and biographies of holographic artists active in the early 1990s.[71] It found fine-art holographers to be unevenly distributed, with some fifteen in America,[72] a dozen in Britain,[73] eleven in Canada,[74] four in Germany[75] and a one or two each in Japan, France, Spain, Sweden, Denmark, and Australia.[76] This distribution does not wholly reflect the production of native-born holographers, because America, in particular, had an influx from other countries. It reveals, though, a difference in age distribution of holographers presumably because the short courses that first became available in America generated some of the first aesthetic holographers; the postgraduate courses established in Britain produced a younger crop of trained holographers a decade later. Intriguingly, a significant cluster—Rudie Berkhout, Brigitte Burgmer, Marie-Andrée Cossette, Setsuko Ishii, John Kaufman, Sam Moree, Al Razutis, Dan Schweitzer, and Fred Unterseher—was born in 1946, post-war products just like Gabor's concept of holography itself.

Other compilations of fine-artist holographers have yielded different numbers, partly because of the instability of the occupation and its career paths, and partly because of differing categorizations of aesthetic holography. The New York Museum of Holography kept files on some 280 holographic 'artists' during its existence, although not all would

[71] Pepper, Andrew, 'Creative Holography Index,' 1992–1995.

[72] In America, Harriet Casdin-Silver was the earliest active artist. Others included Betsy Connors (b. 1950), Rebecca Deem (b. 1949), Nancy Gorglione (b. 1949), Eduardo Kac (b. 1962), John Kaufman (b. 1946), Shu Min Lin (b. 1963) Ana MacArthur (b. 1955), Sharon McCormack, John Perry (b. 1944), Rick Silberman (b. 1951), Douglas Tyler (b. 1949), Doris Vila (b. 1950), and Fred Unterseher (b. 1946). The largest congregations of holographers were in California and New York City.

[73] In Britain, Margaret Benyon (b. 1940), Matthew Andrews (b. 1959), Susan Cowles (b. 1962), Susan Gamble (b. 1957), Adrian Lines (1960–85), Jon Mitton (b. 1954), Paul R.Newman (b. 1961), Edwina Orr (b. 1954), Andrew Pepper (b. 1955), David Pizzanelli (b. 1955), Martin Richardson (b. 1958), and Michael Wenyon (b. 1955) exhibited widely.

[74] Art holographers were active in three Canadian regions: on the West coast, Al Razutis (b. 1946) began in the early 1970s, Melissa Crenshaw (b. 1952) from the early 1980s, and Mary Harman (b. 1944) somewhat later. Razutis had worked in experimental film and video from the 1960s, and took up holography from 1973 in his Visual Alchemy studio in Vancouver, with the support of two Senior Arts Grants from the Canada Council for the Arts. He later co-founded *Wavefront*, a holography periodical, and made the documentary *West Coast Artists in Light* (1997). In southern Ontario, Michael Page (b. 1951), Claudette Abrams (b. 1959), Michael Sowdon (b. 1948), and David Hlynsky practised. Sowdon and Hlynsky set up Fringe Research in Toronto in the mid 1970s. And, in Quebec, Philippe Boissonnet (b. 1957), Marie-Andrée Cossette (b. 1946), Georges Dyens (b. 1932), and Marie-Christiane Mathieu (b. 1953) were notable holographers.

[75] In Germany, Brigitte Burgmer (b. 1946), Dieter Jung (b. 1941), Vito Orazem (b. 1959), and Thomas Luck (b. 1964).

[76] In France, Pascal Gauchet (b. 1956) and Francois Mazzero (b. 1952); in Japan, Setsuko Ishii (b. 1946) and Shunsuke Mitamura (b. 1936); in Denmark, Frithioff Johansen (b. 1939); in Australia, Paula Dawson (b. 1954); in Spain, Máximo Alda (b. 1958), and Julio Ruiz (b. 1957). Others listed included cosmopolitan Ruben Nuñez (b. 1930) who lived and worked in several countries. Dawson (b. 1954), who took up holography in 1974 and subsequently studied Fine Art at the Victorian College of Arts and the University of New South Wales, later obtained a PhD in the subject. The British artist Alexander (b. 1927), who trained at St Martin's School of Art and later studied holography at Goldsmith's College in London, also collaborated with physicist Parameswaran Hariharan in Australia. See Lucie-Smith, Edward, *Alexander* (London: Art Books International, 1992).

have classified themselves as aesthetically oriented.[77] The comprehensive *Holonet* website, created by Urs Fries during the late 1990s, listed over two hundred names. And the 2002 work *Art Holography: The Real-Virtual 3D Images* lists 102.[78]

The subculture of aesthetic holographers was also distinguished by its gender balance: unlike Western scientists, aesthetic holographers were often women.[79] Although Benyon and Casdin-Silver had seen themselves as isolated pioneers in a male profession, schools of holography later produced comparable numbers of female and male holographers.[80] Indeed, women were more frequently employed than men at museums of Holography as effective and dynamic administrators as well as artists.

Even distinct styles of aesthetic holography became discernable to some. Posy Jackson, the first and founding Director of the New York Museum of Holography, mused that, by the early 1990s, it was becoming possible to categorize holographic artists according to visual styles. She suggested that artists and teachers in New York such as Dan Schweitzer, Sam Moree, and Rudie Berkhout promoted certain design elements such as two-dimensional imagery, shadowgrams, and optical effects and that these, in turn, influenced Rick Silberman, Abe Rezny, Bill Molteni, Jody Burns, and Hart Perry. She also mused that the environment of New York precluded much serious work in reflection holography, which was more demanding of stability during exposures.[81]

Jackson had a point about geographical focus. While holography as a subject developed internationally, artists, more than scientists, often had fewer opportunities for funded travel than did many holographic scientists who travelled regularly to conferences. This exacerbated the already prominent national focus that had developed for scientific and artisanal holographers. For artists, regional differences were eclipsed by larger divisions that developed between countries.

Location and working environment became important factors in shaping the aesthetic sensibilities of early holographers. The most profound difference was between Soviet and developing Western views of holography and art. Owing to cold war limitations on travel, Soviet and Western holographers interacted little until the mid-1970s. Most Soviet holographers first saw holograms from Britain and America at the Ulyanovsk Conference in 1979. Denisyuk recalls that they found the Western holograms unimpressive: the technique of rainbow holography seemed unnecessarily complicated compared to their single-beam method of recording white-light viewable reflection holograms.[82] And while they were much brighter than reflection holograms, the rainbow variety provided an

[77] MIT Museum, MoH archive series 3, boxes 26–31.

[78] http://www.holonet.khm.de/; New York Museum of Holography Artists' Files, MIT Museum MoH Series III; Dyens, Georges, 'Art Holography—The Real Virtual 3D Images,' CD-Rom, Montreal, 2002.

[79] There were rather more female scientist-holographers in the Soviet Union and among artisans in the West, but still a distinct minority.

[80] Pepper's compilation lists sixty-five art holographers, twenty-five of whom (38%) are women (Pepper, Andrew, 'Creative Holography Index,' 1992–1995). Dyens identifies 22% of his fine-art holographers as women (Dyens, Georges, 'Art Holography—The Real Virtual 3D Images,' CD-Rom, Montreal, 2002).

[81] Jackson, Rosemary H. to M. Tomko, letter, 1991, New York, MIT Museum 12/265.

[82] Denisyuk, Yu N. to SFJ, email, 16 Apr. 2003, SFJ collection.

unappealing colour shift and lack of vertical parallax as the viewer moved. Moreover, the principal application of such holograms seemed to be in abstract art, a function held in low esteem by Soviet scientists, who sought above all the faithful reproduction of three-dimensional imagery.[83] Publicly, Denisyuk vaunted this as a demonstration of the primacy of science:

> In the USA and other western countries research lays too much stress on technology—it is driven by practical applications. In the USSR research has been more concerned with pure science; we encourage bright ideas. This shows in holography: we have approached it as a science, a very beautiful result of science. The rainbow hologram, invented in the west, is subjectively attractive and so it is used for advertising and by artists, while my holograms are much more precise and accurately record space.[84]

On the other side, Western observers were shocked to discover the large number of Soviet workers—some 300, covering the gamut of scientific applications—and the high quality of their holograms. The fine-grained emulsions yielded exquisite imagery that could not be rivalled in the West and which gave an unequalled sense of realism to the images. While Denisyuk's first holograms recorded with a mercury lamp had been mere feasibility demonstrations, Soviet holograms of the 1970s were unequalled in clarity, brightness, and contrast. When Soviet holograms were displayed at Strasbourg, France, in 1976, and Ulyanovsk in 1979, Western observers were awestruck by their superior technical quality. Attendees such as Stephen Benton of Polaroid and Nicholas Phillips of Loughborough University in England were chastened to observe the technical lead that had been achieved with straightforward recording conditions combined with superior emulsions and chemistry. Benton reported, 'they're on a completely different track, concentrating on materials and developing' and admitted, 'it's one of the few times in my life when I've seen a hologram and it seemed like a glass case full of objects themselves'.[85] Thus workers were separated by their choice of technology.

The sense of Soviet superiority was underlined by the subjects of Russian and Ukrainian holograms. The Denisyuk method of reflection holography had been widely applied for the recording of museum objects. Its elegance of recording and simplicity for illumination, often by using an arc lamp at the centre of the room and reflection holograms mounted on the walls, contrasted with the more artificial-looking Western product.

But despite this superficial connection between Soviet holograms and the recording of art—typically museum items such as precious gold objects and religious icons—art holography never developed in the same form in the Soviet Union.[86] The developing

[83] Denisyuk, Yu N. to SFJ, email, 3 May 2003, SFJ collection.

[84] Denisyuk, Yu N., 'Denisyuk on holography in the USSR', *Holography News* 4 (1990): 2.

[85] 'Soviet holography conference shows high amount of research activity', *holosphere*, 7 (9) (Sep. 1978) 1–2. See also Phillips, N. J., 'Bridging the gap between Soviet and Western holography', in: P. Greguss and T. H. Jeong (eds), *Holography: Commemorating the 90th Anniversary of the Birth of Dennis Gabor* (Bellingham, 1990), pp. 206–14.

[86] See Markov, Vladimir B., 'Display and applied holography in culture development', in: Greguss, P and T. H. Jeong (eds.), *Holography: Commemorating the 90th Anniversary of the Birth of Dennis Gabor* (Bellingham: SPIE, 1990), pp. 268–304. After the fall of the Soviet Union, a handful of holographic artists appeared, but

visual vocabulary of aesthetic holography was limited geographically and sociologically: in the Soviet Union, a non-official artistic community developed little; instead, the centrally organized Soviet institutes established programs to holographically record valuable art objects and to make them available to audiences in the provinces.[87] Where some Westerners saw holograms as art, Soviets (at least those Soviet scientists having access to holography) defined them as a medium to *record* art, a mere technology of transmission. And whereas art holography became identified with non-scientific perspectives in the West, it promoted the state culture in the Soviet Union.

Alongside such distinguishing features of their demography and practice, the growing numbers of aesthetic holographers produced a certain cohesion of purpose and outlook. Margaret Benyon identified a truly interactive social network of practitioners that helped to define aesthetic criteria:

> The artists in holography have their own networks and are in regular communication with each other through group exhibitions, correspondence and personal visits to each other and to the holographic centres that exist in each country. Not enough recognition is given to the potential power of these networks. Ideas and visions are important, but their success and permanence rest on organizational suppport, not on their intrinsic worth. Whether a work is 'lasting' is the major criterion by which people recognize great art.[88]

10.6 ENCHANTING AUDIENCES THROUGH EXHIBITIONS

Developing such criteria was a matter not just for artists themselves, but for their audiences, too. Just as scientific holographers were defined by their research papers, engineering holographers by their applications, and artisanal holographers by their equipment, aesthetic holographers were defined by their exhibitions. As the first generation of holographic artists discovered, exhibitions were crucially important in raising their profile and status among the general public and critics. For a decade from the mid-1970s, some 500 shows brought the medium to the attention of critics and the public, and provoked judgements of its achievements and prospects as an aesthetic

working in figurative, rather than abstract, holography and seemingly as a commercial rather than an aesthetic venture. Artist Aleksander Prostev, for example, studied painting at the Leningrad Academy of Fine Arts 1976–1984, and holographer V. Bryskin, who had worked under Yuri Denisyuk at the Vavilov State Optical Institute, produced holograms later sold in the west ('Art holography from Russia', *L.A.S.E.R. News*, 13 (1), 9).

[87] Holographic recording of precious objects and artworks did not become popular in the west, despite initiatives in a few centres (notably Ralph Wuerker of TRW (Westlake, D., R. F. Wuerker, and J. F. Asmus, 'The use of holography in the conservation, preservation and historical recording of art', *Journal of the Society of Motion Picture and Television Engineers* 85 (1976): 84–9 and Wuerker, Ralph, 'Holography of art objects', *Proceedings of the SPIE* 215 (1980): 167–71), the Smithsonian Institution in Washington, and proposals made to the Victoria and Albert Museum in London) to promote the idea.

[88] Benyon, 'Do we need an aesthetics of holography?', op. cit., p. 414.

medium. They allowed critics and the public to evaluate the aesthetic claims of holographic practitioners.

Some of the earliest shows have been mentioned above: Bruce Nauman's holograms displayed at galleries in Los Angeles and New York in late 1968; Margaret Benyon's exhibition at the Nottingham University Art Gallery in May 1969; the Editions, Inc. exhibition *The Laser: Visual Applications* at the Cranbrook Academy of Art in Detroit in November 1969, and *N-Dimensional Space* at the Finch Gallery, New York, in April 1970. These works and others were exhibited at subsequent art exhibitions and festivals and attracted attention as the first of their kind.

But the first major American exhibition of art holography was a 1975 show, *Holography 1975: The First Decade*, held at the International Center for Photography (ICP) in New York. Organized by Jody Burns and Posy Jackson, who were then involved with a school of holography in the city,[89] the exhibition included holograms from a wide range of scientists, artists, and amateurs. The exhibition was important in bringing holography to public awareness in the largest Western city and centre for art.

A series of major exhibitions followed. Burns and Jackson organized *Holografi: Det 3-Dimensionel Mediet* (*Holography: the 3-Dimensional Medium*) at the Kulturhuset in Stockholm in March 1976, drawing attendance of 60,000 over its two-week run. The same year, they launched *Through the Looking Glass* (Figure 10.8), the inaugural show at the new Museum of Holography in New York. Many cities had major holography shows from this period that introduced holography to enthusiastic local audiences: Vancouver (*Whole Message*, 1976); London (*Light Fantastic* and *Light Fantastic 2*, 1977 and 1978); Tokyo (*Alice in the Light World*, 1978); Berlin (*Holographie Dreidimensionale Bilder*, 1979); Rome (*Olografia*, 1979); Groningen (*Hololight '79*, 1979); as well as small exhibitions.

In the Soviet Union, hologram exhibitions made a similar impact. Reflection (Denisyuk) holograms were increasingly used to record historic treasures by teams in Moscow and Kiev. The first exhibition of Ukrainian holograms took place in the Crimea in 1976, attracting long queues of visitors and reports in *Pravda* and *Isvestia*. Throngs were so much higher than typical museum attendance that museum staff initially complained, but were eventually placated by the increased returns from ticket sales. The Ministry of Culture in the Ukraine supported a holographic laboratory under the direction of Vladimir Markov from 1979, with holograms recorded at museums and displayed in permanent exhibitions in Kiev, L'vov, Sebastopol, Yalta, and other centres. Travelling holographic exhibitions were also installed in buses to reach smaller population centres. An estimated one million visitors saw the Ukrainian exhibitions every year, and travelling exhibitions were taken, for example, to London (1980) (Figure 10.9), Yugoslavia (1984), Italy (1985), and Korea (1989).[90]

[89] Discussed in Chapter 11.

[90] Markov, Vladimir B., 'Display and applied holography in culture development', in: Caulfield, H. J. (ed.) *Holography: Commemorating the 90th Anniversary of the Birth of Dennis Gabor* (Bellingham: SPIE, 1990), pp. 268–304; Markov, Vladimir B. to SFJ, interview, 26 May 2004, Kiev, SFJ collection.

Fig. 10.8. Exhibition poster, Museum of Holography, 1976 (MIT Museum MoH collection).

Fig. 10.9. Mobile hologram display, Ukraine, early 1980s (*Russian Holograms: Treasures Trapped in Light* (London: Light Fantastic, 1992; V. Markov photo)).

New York and London became centres for exhibitions during this period. In London, the first *Light Fantastic* exhibition, held at the Royal Academy of Arts in 1977, attracted some 250,000 visitors over 28 days. Its successor, *Light Fantastic 2*, ran for six weeks the following year. These two large shows, in quick succession, suggested great progress: the second displayed reflection holograms and pulsed portraits, and the organizers forecast that full-colour reflection holograms would be produced by the end of the year, and that 'before the end of 1984 it will be commonplace to have holograms in the same way that we have photographs in our homes today'.[91] Both exhibitions had a large impact on British artisans and entrepreneurs, if less on artists themselves. These were not professionally curated exhibitions: the London shows exhibited principally the work of Holoco, a company based on the technology developed by physicist Nick Phillips of Loughborough University of Technology, and was sponsored by Guinness, the brewers. Nor were most of the organizers and promoters of exhibitions artists themselves. Journalist Itsuo Sakane, for example, produced a series of exhibitions in Tokyo. Eve Ritscher, a London publicist, Linda Lane in Los Angeles, and Jim Finlay in Paris, organizers of some of the largest shows of the period, sought to bring dramatic, rather than necessarily aesthetically pleasing, holograms to the public.

Nevertheless, the role of a second wave of exhibitions (see Figures 10.10 and 10.11) in creating a certain solidarity among holographers is illustrated by another influential British exhibition, *Light Dimensions: The Exhibition of the Evolution of Holography*, held at the National Photographic Centre of the Royal Photographic Society (RPS) in Bath from June to September, 1983. Like the earlier British shows, it attracted large audiences, estimated at a half million visitors over its one-year run. The RPS, the principal organizer, became an institutional supporter of holography following the formation of its Holography Group that year, which brought together British scientists, artists, and amateur holographers.

In the Preface to the exhibition catalogue, the RPS president drew strong parallels between *Light Dimensions* and the first major photographic exhibition, held in 1852–53 in London. That exhibition, supported by the Society of Arts, had displayed 800 prints and had triggered the formation of the Photographic Society in 1853, which had held annual exhibitions ever since. With an eye to history and the Society's own traditions, the RPS supported the holography exhibition which was meant 'to evoke all the excitement of those early photographic exhibitions of the 1850s because, as they were, it is involved with a developing technology giving to the artist new possibilities of self-expression, to the technologist new applications, and to the uninitiated viewer, a new experience'.[92] Similarly, the organiser, Eve Ritscher, drew analogies between the history and impact of photography and holography. Twenty-five firms, principally British and American, contributed displays.

[91] Boyle, Sandy, *Light Fantastic 2: A New Exhibition of Holograms by Holoco* (London: Bergstrom and Boyle, 1978), 1.

[92] Roberts, Christopher, 'Preface', in: E. Ritscher, J. Reilly, J. Lambe, and R. MacArthur (eds), *Light Dimensions: The Exhibition of the Evolution of Holography* (Bath, UK, Bergströmt Boyle Books, 1983), p. 4.

Fig. 10.10. Exhibition poster melding art with technological wonder, based on *The Kiss* multiplex hologram by Lloyd Cross, Renaissance Center, Detroit, 1980s (Renaissance Center, Detroit, MI; MIT Museum MoH collection)

Fig. 10.11. bookmark for the Hologram Gallery, Amsterdam (Hologrammen Galerie; MIT Museum MoH collection)

Other large-scale exhibitions had impact through press coverage and large numbers of viewers. *Alice in the Light World* (Isetani Museum, Tokyo, 1978), *SPACE-LIGHT* (a touring Australian show, 1982), *LICHT-BLICKE* (Deutsches Filmmuseum, Frankfurt, 1984), and *Images in Time and Space* (Montreal, 1987, and thereafter a travelling show through North America) continued this format of exhibiting large collections of historic, scientific, and aesthetic holograms to the public.

10.7 CRITIQUES FROM MAINSTREAM ART

The exhibitions of holograms brought the medium to growing audiences, but not always to critical acclaim. Exhibitions proved ambivalent in promoting aesthetic holographers and their holograms, partly because the shows vaunted the versatility and wonder of the medium more than its aesthetic content.

Nevertheless, the holographers who participated in the exhibitions often were consciously attempting to extend or reinvent contemporary schools of art. The coalescence of a community, trade, or profession seldom occurs in isolation. As discussed for scientific holographers in Chapter 8, such groups are part of a shifting ecology, in which battles for territory, status, and even survival are continuous. The community of aesthetic holographers experienced this process as they challenged conventions of art and struggled to find a place within it.

Initial scouting for territory appeared hopeful. During the late 1960s and early 1970s, holography was a new medium like installation art and video and with seemingly equal potential.[93] In some respects, as discussed above, it was an outgrowth of Op (optical) Art and its focus on visual perception and illusion, which both Margaret Benyon and Jerry Pethick had been exploring since the early 1960s. And from the early 1970s, holograms were also combined with other media to create environmental pieces.[94]

Yet in other respects, the hologram was a medium without precedent. The early public and critical reception to aesthetic holograms was mixed. Emerging from the secluded environs of universities and government laboratories, the origins of the new art form seemed suspect to some viewers. Margaret Benyon recalls that when she took part in a Colloquium with David Bohm and Jonathan Benthall at the Institute of Contemporary Art, London she got questions afterwards: 'how do you feel about using a laser that was an instrument of war?' She responded that anyone using pencil and paper or going in a

[93] Benthall, Jonathan, *Science and Technology in Art Today* (New York: Praeger Publishers, 1972).

[94] To cite a few aesthetic holographers, Benyon, for example, included holograms as key components in mixed-media pieces, placing drawings in ink, gouache and/or feather under them; Burgmer, Dyens and MacArthur produced sculptures incorporating holograms; Caodin Silver, Doris Vila, Paula Dawson, Phillip Boissonnet and others created large environmental pieces; collages of holograms were created by Douglas Tyler, Jeff Robb, Sally Webber and others; geometrical abstract images were made, for example, by Benton, Andy Pepper, and Dieter Jung; and more organic abstracts were created by Berkhout and Pizzanelli.

plane is using technology that is applicable to war.[95] Similarly, for *N-Dimensional Space*, the first New York exhibition, Lloyd Cross and his associates felt compelled to write:

> This exhibition has been so designed, by individuals with skill and experience in the handling of all types of laser, that it will not be possible for the viewer to come into direct contact with a laser beam. One may view and enjoy the exhibition, as laser beams of the low power used will not burn or harm human flesh in any way.[96]

Audiences were also unprepared for how to view holograms. When Benyon gave her first solo show in Nottingham, she found the audience reaction muted:

> There was a motorized turntable with a little perspex sign, edge-lit, in three languages, saying 'to move turntable on, press switch once, look close to the lit glass plate and through it'. Arrows on the floor and turntable. And unfortunately the holograms were at table level as well, so people had to bend down to catch first sight of the images. The art professor was quite a conservative guy, and the idea of this dangerous object—the laser—in the exhibition raised safety concerns. James Bond (Goldfinger) had been on, so suddenly these laser caution notices had to be put up in the exhibition. He asked 'will it disturb the traffic on the Derby Rd?'. I found people scared stiff, standing with their backs to the wall in the gallery not knowing what they were there to see.
>
> I always put down that exhibition as a complete failure, because most people never got to see the holograms unless I was there to show them, even though I had put arrows etc, because they didn't know what to expect; they had no idea. I had to teach them how to see the holograms. I had to go right back to basics, from abstract images to familiar stuff they could get a grip on. They would be looking, and I'd say 'can you see the banana there?'. . . . and they'd see colours! 'Yeah, I can see that, it's yellow' (no [interpolates Benyon] it's red actually), 'no, I can see it!'[97]

A similar teaching effort was mounted at the first shows of the Museum of Holography in New York. Mark Diamond recalls, 'we did a lot of placing of little stickers on the ground [. . .] like those games where they show you how to tango, and where to put your feet; it was instructions on how to view the hologram, because people were unfamiliar with how to view the space'.[98]

Thus the medium and its trappings were unfamiliar and disquieting. A decade later, Benyon dismissed her first show of holograms:

> The very first phase was a false start. I thought I could continue in holography the preoccupations as a painter which led me into it. I quickly discovered after the failure of my first show in 1969 that I could not do this and that I should have to go back to square one. So initially I was

[95] Benyon, Margaret to SFJ, interview, 21 Jan. 2003, Santa Clara, CA, SFJ collection.

[96] McBurnett, Ted, *N Dimensional Space* (New York: Finch Museum of Art, 1970). The danger of lasers had been raised by at least one holographer: 'holographically reconstructed 'lensless' images were not only generally single colored, but they could only be viewed in laser light (a dangerous practice at best)' (Stroke, George W., 'Press Release: Breakthrough in "lensless photography" sets stage for 3-dimensional home color television and a possible multi-million industrial explosion in electro-optics', 12 Mar. 1966, Leith collection). The constraints on laser use became more severe in America when the Bureau of Radiological Health set more restrictive exposure standards in 1974.

[97] Benyon, Margaret to SFJ, interview, 21 Jan. 2003, Santa Clara, CA, SFJ collection.

[98] Diamond, Mark, 'Holotalk interview', Internet radio interview to F. DeFreitas, 2003, www.holoworld.com.

concerned to use only those aspects that were exclusive to holography, introducing people to unfamiliar notions about space with time-reversed imagery or double-exposures in which solids seem to share the same space, or non-holograms, which play havoc with received notions of surface, volume, part and whole.[99]

In a similar vein, Al Razutis describes the early 1970s as a period when 'there were no established technical precedents, no presumptive aesthetics to follow. Everything was created in the spirit of 'invention' and discovery'.[100] But art critics, as well as the public, remained ambivalent about the aesthetic content and uses of the medium. Categorizing holography was a central difficulty. A *Christian Science Monitor* review noted that a hologram is 'neither painting nor sculpture but a curious, intangible distillation of the two'.[101]

A crucial event was the first major holography group exhibition in New York: *Holography 1975: The First Decade*. The show, held at the ICP, was the most visible display yet of holograms in an artistic venue, and in a city that prided itself on sophistication and art criticism. It attracted mixed reviews. *The Village Voice*, for instance, noted, 'So far the medium is more entertaining than it is artistically expressive. But while holography has yet to find its Stieglitz or Steichen, it is plain to see that the medium will continue to lure new and devoted devotees'.[102] As the references to Stieglitz and Steichen suggest, holography was being portrayed as analogous to early photography. The location of the exhibition in the ICP was the strongest indication of this often-voiced claim.[103]

Nevertheless, a firm rebuff for the fledgling subject appeared in the *New York Times* in 1975. Art critic Hilton Kramer (b. 1928) archly identified the culture of holography as one defined by second-rate subjects and dubious artists:

It is, to judge by the present exhibition, a gadget culture, strictly concerned with and immensely pleased by its bag of illusionistic tricks and completely mindless about what, if any, expressive possibilities may lie hidden in its technological resources. There are, to be sure, a few artistic attempts here at abstraction and pop art and the familiar neo-dada repertory, but these are even more laughable than the outright examples of kitsch. Much of the work has, I gather, been produced not by artists but by physicists professionally involved in holographic technology. The physicists appear to favor objects out of the local gift shop, whereas the artists do their shopping in provincial art galleries, and both, it seems, are much taken with television commercials. It is difficult to know which is the more repugnant: the abysmal level of taste or the awful air of solemnity that supports it.[104]

[99] 'An interview with Margaret Benyon', *holosphere* 9 (1980): 1–4.

[100] Razutis, Al, 'Detailed history of holography at Visual Alchemy', http://www.alchemists.com/visual_alchemy/holo_hist.html, accessed 6 Oct. 2004.

[101] Artner, Alan G., 'Rhetoric, not results, at holography show', *Chicago Tribume*, 12 Jun. 1977.

[102] Bourdon, David, *Village Voice*, 21 July. 1975.

[103] Alfred Stieglitz (1864–1946) and Edward Steichen (1879–1973) were American photographers who strongly influenced the medium as an art form. Both contributed prominently to schools of photographic representation such as Pictorialism and the Photo-Secession, and developed media (such as gum bichromate and glycerine printing), styles of representation, and new subject matter (such as landscape photography and fashion portraiture). Steichen was curator of photography at the Museum of Modern Art in New York for fifteen years. During the early 1980s, the Royal Photographic Society in Britain, citing the analogy between early photography and contemporary holography, also created a holography group and mounted major exhibitions.

[104] Kramer, H., 'Holography: a technical stunt', *New York Times*, 20 Jul. 1975, 1–2.

The creators were as significant as their creations. As Sarah Maline notes, 'Kramer's discovery that most of the holograms were produced by physicists helped him to summarily dismiss the medium as a technical stunt unfit for public display'.[105] There was a tension between the need to form an alliance between holographers broadly defined, and the contrasting requirement of defining art holography in terms that could satisfy its critics. Fine-art holographers needed allies from other technical communities to swell their numbers, to encourage them and to share experiences, but they also needed to coalesce with communities of artists, and to downplay their focus on technique and medium. These two aims, like the separate communities themselves, could not readily be reconciled.

Contributors to the exhibition disagreed with each other, too, at least in retrospect. Harriet Casdin-Silver later recalled:

> I wanted to take my art out of that show [. . .] I thought it stank to high heaven. There was so little real art in it [. . .] But from the holographic artists' point of view, the art critics stood in the doorway, looked in and really saw nothing [. . .] I never saw them at my installation. The show should not have happened, it was too early on. Typical people who had six weeks of holography called themselves artists.[106]

There was, indeed, a danger in the organizers' inadequately differentiating 'art' from 'clever imagery' for critical consumption, but as some artists noted then and afterwards, such distinctions were not trivial to judge. The Executive Director of the ICP, Cornell Capa, wrote a response to Kramer in which he largely concurred:

> Mr Kramer is completely right about holography. He criticizes holography for its overly-complete depiction of reality [. . .] To this we can only plead pictorial poverty; there just isn't enough holography around to permit the same kind of critical judgments that go into an exhibition of pictures in which images, rather than their processes, are the subject-matter. And suppose all of us—exhibitors and critics alike—just happen to be wrong. Suppose that among the several thousand people who have already experienced the 'esthetic kick of a postcard from Montauk' there is just one person who has formed one idea that might make meaningful holography feasible. It isn't impossible, and that's why we're here.[107]

And rebuttals to the criticism drew the now familiar analogies with early photography. Jody Burns, the co-organizer of the exhibition, wrote:

> I believe that one of the important functions of the International Center of Photography is to expose possibilities, such as this new visual experience, holography, just as it was important to introduce photography in 1839, to potential artists who may transform and advance a technology ever closer to greater artistic expression.[108]

[105] Maline, Sarah Radley, 'Eluding the aegis of science: art holography on its own', *International Symposium on Display Holography* 1600 (1991): 215–9, quotation p. 217.

[106] Casdin-Silver to SFJ, op. cit.

[107] Capa, Cornell to P. Jackson, by letter, New York, MIT Museum.

[108] Burns, Jody to C. Capa, by letter, New York, MIT Museum 33/1022.

Physicist Tung Jeong added:

> Photography began as a highly technical process, with contributions from many individuals. It suffered through a period of immense technical difficulties, having unclear and monochromatic images; it then acquired motion, sound, and color; finally, it became simple. Holography, after the initial discovery by Gabor in 1947, lay dormant until the sixties. It then went through a period of development by physicists and engineers, with their multi-ton granite stones and mysterious laser beams. Within the last decade, it has evolved into new formats, developed new and simple techniques, and incorporated motion. We can soon expect true color [. . .] If you marvel at how photography has arrived, wait and see how far holography will go![109]

While such arguments were widely voiced, they presumed a standard trajectory of improvement despite dramatic differences in social and economic contexts between the mid-nineteenth century and late twentieth century. Even if technology followed a familiar path—itself a doubtful assumption—art could not be strapped willingly to this notion of progress.

Writing a review of the exhibition in an optics journal, Stephen Benton, Casdin-Silver's sometime collaborator, mounted a different defence. Blaming the recession of 1972, which had led to declining support for holographic research, he argued that there were too few holographers to support a major exhibition. Benton, part of the corporate and scientific culture of holography, saw the field in decline during the mid-1970s because of the withdrawal of blue-sky research funding. The scientific community had already seen such optimistic support evaporate. For artists, however, the subject was just opening up. As a result, he denied the existence of numbers large enough to support a culture, at least thus far.

> If you thought that things seemed quiet for holography of late, you were right. Since the collapse of support for three-dimensional laser photography, only a few image makers have managed to keep working with the new medium. But they have made some progress, and are catalyzing a new and growing interest, so that more people may be seeing holograms in the next two years than in the last ten combined.[110]

Holographer David Hlynsky of Fringe Research complained in the same vein, 'The economy is taking on water like a sieve and the first ballast to be dumped is culture'.[111] There was a problem, too, inherent in what the *Times* critic saw as premature exposure and immaturity of the subject: art holographers had a limited time in which to attract funding and retain corporate interest. Unlike the cultural reception to photography in the nineteenth century, the window of opportunity for holography in the overcrowded cultural context of the 1970s was brief. Exhibitions were necessary to sustain momentum.

Several holographers interviewed for this book recalled the *New York Times* review as a serious blow. Two years after the event, Posy Jackson wrote back to Hilton Kramer upon the opening of a new show:

> While we objected to the tone of the piece and many of the conclusions it drew, it might interest you to know that we shared your feelings about holography as an art form at that point.

[109] Jeong, Tung H. to P. Jackson, letter, 1975, New York, MIT Museum 33/1022.
[110] Benton, Stephen A., 'From the inside looking out', *Applied Optics* 14 (1975): 2795.
[111] Hlynsky, David to P. Jackson, letter, Nov. 1979, MIT Museum 27/682.

> For this reason, we have not contacted you until now when we feel we have something to show which should be considered holographic art work.[112]

But it did not elicit another review from the *New York Times*, nor were other newspapers much more supportive. Alan G. Artner, an opinionated *Chicago Tribune* critic reviewing a hologram exhibition at Gallery 1134 in Chicago in 1978, observed that the pieces were 'accompanied by the usual extravagant claims', adding:

> There is, of course, an implicit glorification of technology which doubtless has accounted for part of the acceptance in Soviet countries. Furthermore, my limited imagination does not yield many holographic possibilities that would effectively transcend that 'miracle of science' status [. . .] Until it is firmly established, holography will continue to perform its one spatial variation on some of the most traditional art forms of all. That is hardly enough to justify the revolutionary rhetoric.[113]

One result of the damning critiques was a loss of confidence. Jackson, as first Director of the New York Museum of Holography, later mused, 'I though I had an idea what art was all about, and where holography fitted into that model when I started the museum. Now I feel I have no concrete idea of what art is at all, or how holography fits into that category now or in the future'.[114] More frequently, however, holographers criticized what they interpreted as the prejudices of the New York art world.[115]

Two decades after the ICA exhibition, British art critic Edward Lucie-Smith characterized the holography community as imbued with both enthusiasm and paranoia. Noting that curators of avant-garde exhibitions and critics alike saw holography as 'irredeemably kitsch', he argued that holographic art had been a victim of art trends since its emergence in the early 1970s.

> Even then critics were rather sniffy. Op Art was just going out of fashion—and anything 'optical' was suspect. Three-dimensional holographic images were disturbingly there but not there. They hovered like ectoplasm both before and behind the transparent or reflective plates to which clear celluloid sheets bearing laser-generated interference patterns were attached. Sometimes, even in those days, two incompatible forms seemed to occupy precisely the same space.
>
> [. . .] Another problem with holography was quite simply that it was highly technological, and, in its early manifestations at least, extremely expensive and difficult to do. Holographers who aspired to be artists boasted of the fact when they managed to produce six or seven small images a year. This was entirely contrary to one important trend in the art of the 1980s and 1990s, which valued scale (vast environmental works were and are the height of fashion), untrammeled spontaneity, and the use of so-called 'poor' materials. Arte Povera, as it came to be called, eventually saw off a number of other fashionable manifestations, such as Neo-Expressionist painting, and established itself as the dominant mode of the day, especially among European artists.[116]

[112] Jackson, Rosemary H. to H. Kramer, letter, Apr. 25 1977, MIT 2/25.

[113] Artner, Alan G., 'Rhetoric, not results, at holography show', *Chicago Tribume*, 12 Jun. 1977.

[114] Jackson, Rosemary H., 'An accounting, 1976–1978', *holosphere* 8 (1979): 3, 5–6, 8–10, quotation p. 10.

[115] An interesting survey highlighting the lack of interest by New York curators, gallery owners, and art publishers appears in Lightfoot, D. Tulla, 'Contemporary art-world bias in regard to display holography: New York City', *Leonardo* 22 (1989): 419–24.

[116] Lucie-Smith, Edward, 'A 3D triumph for the future', *The Spectator* 277 (1996): 57–8.

Many of these criticisms had been triggered by the first aesthetic holograms. As early as 1969, gallery owner Leo Castelli, speaking to an interviewer about Bruce Naumann's portrait holograms and their reception, had observed:

> I had the other show with the holograms in it, which to me was a great show. Here nobody except Jasper [Johns, the artist], Andy Warhol, and Philip Leer understood that. They were very involved in it [. . .] They really understood [Bruce Nauman]. I also liked the show very much and [was] very pleased with it. There was no really . . . Nothing was sold . . . People didn't . . . New York is difficult. It's strange why people don't realize that that is the probably the last place (except for a few individuals) that really is pioneering. This is not an easy place to make one's mark in.[117]

Despite a growing number of exhibitions during the late 1970s and early 1980s, reviews by art critics remained sparse. Aesthetic holographer Nancy Gorglione, like several other observers and participants, argued that holographers could be compared to early photographers and to Impressionist artists. Yet she cited a major difference with Impressionists: that disaffiliated group had gained frequent reviews and commentaries in contemporary French newspapers and magazines but, by contrast, holographers had few critiques, even negative ones, to keep their subject in the public eye.[118] Margaret Benyon, arguably the most publicly visible fine-art holographer and gaining State recognition with her award of the Member of the Order of the British Empire in 2000 'For services to Art', felt isolated by the lack of reception from the art world:

> Despite my favourable reception in the early days, I have not had a solo show in the United Kingdom in a serious art gallery since 1972. This lack of dialogue is bound to affect one's work, in the same way that female artists' exclusion from the life classes affected their careers in former days.[119]

From the 1980s and 1990s, subsequent critiques in the form of books and dissertations have taken a more analytical internal view of holographic art.[120] Peter Zec, for example, interprets aesthetic holography as a medium unlike those of conventional art, and one that must be divorced from any association with photography and faithful imagery.[121] Exhibitions that he curated in Germany during the 1980s involving work by holographers such as Dieter Jung reinforce his theme of the hologram as an artwork of light alone rather than as a mimetic medium.

Art historian Sarah Maline, on the other hand, argues that the early association with science had doomed aesthetic holography to a marginalized, minority existence. The

[117] Castelli, Leo to P. Cummings, interview, 14 May 1969, New York, Smithsonian Archives of American Art.

[118] Gorglione, Nancy, 'A partial view of a three-dimensional world', *Leonardo* 25 (1992): 407–9.

[119] Benyon, 'Do we need an aesthetics of holography?', op. cit., p.413.

[120] See, for example, Berner, Jeff, *The Holography Book* (New York: Avon Books, 1980); Burgmer, Brigitte, *Holographic Art—Perception, Evolution, Future* (La Coruña, Spain: Daniel Weiss, 1987); Zec, Peter, *Holographie: Geschichte, Technik, Kunst* (Köln: DuMont, 1987); Hayward, Philip, *Culture, Technology & Creativity in the Late Twentieth Century* (London: J. Libbey, 1990); and Maline, Sarah Radley, *Art Holography, 1968–1993: A Theatre of the Absurd*, PhD thesis, Art History, University of Texas at Austin (1995); Dyens, Georges, 'Art Holography—The Real Virtual 3D Images,' CD-Rom, Montreal, 2002.

[121] Zec, P., 'The Aesthetic Message of Holography', *Leonardo* 22 (1989): 425–30.

emphasis on technical accomplishment and progress obscured the artistic merit of holograms, making them 'a monument within the medium's history, not necessarily its art history' and perpetuated the notion that art holography was as yet 'incompletely formed'. She suggested that it is dangerous, in the history of art, to assume that the best art is still to come; much better is to critique aesthetic quality in terms of how the artist works within the constraints of the medium.[122]

These assessments transcended artistic judgments: holographers found themselves trapped at the nexus of marketing and technological, as well as creative, concerns. While an aesthetic, or aesthetics, may have been emerging, the community of fine-art holographers remained marginalized in the wider art world and peripheral to the issues that arose at technical conferences of holography and in the marketplace. Strengthening and uniting the communities of holographers was one of the tactics that promised a vital role for the medium during the 1980s.

[122] Maline, Sarah Radley, 'Eluding the aegis of science: art holography on its own', *International Symposium on Display Holography* 1600 (1991): 215–9, pp. 215–6.

11

Building Holographic Communities

11.1 UNCERTAIN IDENTITIES

Holographer scientists, artisans, and artists developed distinct traditions. Opposing tensions simultaneously divided and united these groups, keeping them from achieving a critical mass. On the one hand, the tribes of holographers sought separate identities, and groomed them at distinct venues: scientific conferences, galleries, and exhibitions. They distinguished themselves by the equipment they adopted and by the holograms they made, and trained a new generation either through graduate schools, artisanal workshops, or academies of art. These competing forces sapped consensus and helped keep the subject on the margins during the 1970s. The small field of holography was riven just as profoundly as the two cultures described by C. P. Snow during the 1950s.[1] And its practices were more distinct than the division between engineers and scientists first characterized by historian of science Derek de Solla Price during the 1960s and revived by recent scholarship such as Peter Galison's study of microphysics.[2]

A by-product of such divisions was a cult of secrecy. Improvements in holograms had been painstaking and incremental. Holographer Ronald Erickson reported, 'secrecy in its various forms is part of the heritage of holography, and I take it as a personal and professional challenge to work against this counter-productive practice in the holographic community'.[3] Nick Phillips (see Figure 11.1), responsible for many of the improvements in processing chemistry from the 1970s, also reflected that he and others had been secretive.

[1] Snow, C. P., *The Two Cultures; and, A Second Look—An Expanded Version of the Two Cultures and the Scientific Revolution* (Cambridge: Cambridge University Press, 1964).

[2] See, for example, de Solla Price, Derek J., 'Is technology historically independent of science? A study in statistical historiography', *Technology and Culture* 6 (1965): 553–68. He argues that scientific culture is *papyrocentric*, or based on writing papers and communicating news orally and at conferences, while engineering culture eschews publication but promotes avid reading to stay ahead of competition. For instance, Emmett Leith in later years described himself as a 'paper writer' who did not wish to 'reduce ideas to practice' (Leith, Emmett N. to SFJ, interviews, 29 Aug.–12 Sep. 2003, Ann Arbor, MI, SFJ collection). On the engineer–physicist divide in mid-twentieth century science, see Galison, Peter, *Image and Logic: A Material Culture of Microphysics* (Chicago: University of Chicago Press, 1997), especially Chapter 4.

[3] Erickson, Ronald R., 'There is this "attitude" in holography', *holosphere* 16 (1989): 4.

Fig. 11.1. Nick Phillips in Bulgaria with Russian holographic plates, 1990 (H. Bjelkhagen photo, Bjelkhagen collection).

> [I]n the early days, everyone was absolutely paranoid, and I include myself. It wasn't very nice, because you were absolutely shattered from so many hours in the lab [. . .] Over the years, the question of confidentiality has been a big issue, and leads to a kind of feeling of paranoia. In the end you feel got at, and you can't sleep because you worry about whether somebody is going to walk off with an idea, and yet you want the idea to be published, and that's awful; that is all because of the overblown nature of the subject.[4]

The level of secrecy varied from conference to conference, too. Artists, amateurs, teachers, and scientists had shared information freely since the days of the Multiplex Company, although there were also short courses available for those willing to pay to learn the subtleties—the *tacit knowledge*—of the latest state of the art. On the other hand, other venues were distrustful affairs: the series of conferences held from the 1980s by the emerging embossing and packaging industry were unlike Lake Forest. Indeed, the Holopack-Holoprint conferences organized by Reconnaissance International felt it necessary to caution presenters about ethics: they were not to use their time to over-promote their products, and were to present useful and accurate information for other delegates. The market intelligence firm also stressed the damaging effects for the industry of rumour mongering and pessimism.[5] Even so, presentations were more than

[4] Phillips, Nicholas J. to SFJ, interview, 18 Sep. 2002, Leicester, UK, SFJ collection.

[5] Lancaster, Ian M., Holopack-Holoprint 2003 presenters' documentation, Holopack; Lancaster, Ian M., 'An apologetic industry', *Holography News*, 8 (7) (Sep. 1994): 2; Lancaster, Ian M., 'Warning: rumours can seriously damage your health', *Holography News*, 9 (1), (Feb. 1995): 2; Lancaster, Ian M., 'Bad news is bad news for the holography industry', *Holography News*, 9 (9/10) (Dec. 1995): 2.

once interrupted by a member of the audience claiming intellectual theft.[6] As Nick Phillips noted, 'they won't talk to each other [. . .] At conferences they show things, and boast to each other; that's the way they stay alive'.[7] The problem was endemic: for holographic entrepreneurs, information meant money.

Secrecy had, of course, been a feature of holography from the 1950s, when Emmett Leith at Willow Run and Yuri Denisyuk at the Vavilov Institute had begun their research. There, however, security concerns gradually declined. In its place, a growing sense of community can be illustrated by the convergence of those two seminal workers. Leith and Denisyuk had first come into contact at the Novosibirsk meeting on holography in the winter of 1973, but Leith, presumably for security reasons, did not meet many Russians.[8] Six years later, Leith was invited to visit the Soviet Union as the guest of the A. F. Ioffe Institute, visiting important centres of Soviet holographic research: the Ioffe Institute itself and the Institute of Nuclear Researches, both in Leningrad, the P. N. Lebedev Physical Institute, the Research Kino-Foto Institute (NIKFI) and the Institute for Information Transmission Problems, both in Moscow, the Physico-Technical Institutes in Kiev and Riga, and the Institute of Automatics and Electrometry in Novosibirsk.[9] Leith and Denisyuk met again at a Soviet holography conference in Leningrad, with Leith and his two daughters visiting Denisyuk's family and the Vavilov Institute, although not Denisyuk's laboratory owing to security concerns. Denisyuk himself made his first trip to America in August 1989, after the fall of the Soviet Union, and subsequently made more frequent visits to Western countries. During his second visit to Ann Arbor in 1989, he travelled by car with Emmett Leith on an 800-mile trip to a Gordon Research Conference in New Hampshire, with stops at Niagara Falls and the Adirondack mountains.[10] The two, having travelled the same road half a world apart, could do so together at last.

11.2 STRENGTHENING NETWORKS

Such individuals became symbolic figures in what remained a subject of restrained scientific interest, frustrated artistic acceptance and discouraging commercial penetration. But just as scientists on opposite sides during the cold war could be reconciled, so, too, other holographic communities could find mutual interests, especially in promoting a wider awareness of holography and an interest in developing markets for holograms. The heterogeneous groups found themselves intersecting in specific arenas.

Holographers' periodicals were important in nurturing community visions. Just as Elizabeth Nelson has observed that, 'from the early days of the counterculture its only

[6] Kontnik, Lewis T. to SFJ, interview, 20 Nov. 2003, Vancouver, SFJ collection.

[7] Phillips, Nicholas J. to SFJ, interview, 18 Sep. 2002, Leicester, UK, SFJ collection.

[8] '5th All-Union School on the Physical Principles of Holography', Novosibirsk, 29 Jan.–3 Feb. 1973; Denisyuk, Yu N. to SFJ, email, 16 Apr. 2003, SFJ collection.

[9] Tuchkevich, V. M. to E. N. Leith, letter, 12 Mar. 1979, Leningrad, Leith collection.

[10] Denisyuk, Yu N. to SFJ, email, 16 Apr. 2003, SFJ collection; Leith, Emmett N. to SFJ, interview, 30 Aug. 2003, Ann Arbor, SFJ collection.

viable "institution" had been the underground press', which served the counterculture 'as a communications and advisory medium [. . .] promoting the ideas current in the counter-culture',[11] holographers, too, increasingly attempted to define themselves through publications. As San Francisco of the late 1960s and beyond had specific communities whose views, thoughts, and philosophies were represented by various local underground papers, holographers a decade later launched periodicals seeking to represent, bind together or, indeed, create communities.

For a wide range of holographers, the best-known periodical was *holosphere*, a news-cum-technical notes magazine published between August 1972 and 1990. For its first five years, it was published as a newsletter for the electro-optical industry, thereafter shifting its focus to aesthetic holography when taken over by the New York Museum of Holography (MoH). It had been acquired from International Laser Systems of Orlando, Florida, its publisher since 1972. Originally cast as 'the newsletter of holographic science and technology', it was soon rebadged by the MoH as 'the newsletter of holographic science, technology *and art*', and gradually expanded to a magazine.

Less widely available periodicals followed, each pursuing a distinct vision of a sociotechnical community, and seldom exceeding circulations of a few hundred. They included *Image Plane: A Journal of Holographic Art* (*c.*1980, Rick Silberman and Judith Parker, Brown University), a fleeting journal described as 'a tool for those working and interested in exploring the art of holography [. . .] an open, unbiased space for free expression [. . .] a magnet for provocative and controversial thinking, contributing toward creating a healthy and vital holographic community';[12] *L.A.S.E.R. News* (of the San Francisco-based Laser Arts Society for Education and Research, 1980–), specializing in practical details of holography and eventually outgrowing a parochial perspective on the field; *The Holo-gram* (Frank DeFreitas, Allentown, PA, 1983–), one hobbyist's newsletter; *Wavefront* (Al Razutis et al., Vancouver, Canada, 1985–87), focusing on critical reviews of art and business, a stance that soon alienated its sponsors in the arts and industry; *Holographics International* (Sunny Bains, UK, 1987–90), with editorial staff principally from Imperial College, London, seeking to be an independent and objective voice for the holography industry;[13] *Holography News* (Reconnaissance International, 1987–), a business newsletter for that emerging industry; *Interferenzen* (1990–97), a quarterly periodical of the Deutsche Gesellschaft für Holographie e.V. (German Society of Holography); *Holography & 3-D Software*, (Klaus Unbehaun, 1990–92), a German newsletter for enthusiasts of holograms and other three-dimensional medial; *Creative Holography Index* (Andy Pepper, UK, 1991–94), focusing on art holography and its aesthetic criteria; and, short-lived newsletters from museums, galleries, and practitioners' collectives, some of which moved to web-based dissemination in the late 1990s. None of them succeeded even as well as *holosphere* in fostering commmunity identity.

[11] Nelson, Elizabeth, *The British Counter-Culture, 1966–73* (Basingstoke: MacMillan, 1989), quotation p. 103.
[12] Silberman, Rick, 'Artist files,' New York Museum of Holography Collection, MIT Museum 30/889, 1980
[13] Bains, Sunny, 'Editorial', *Holographics International* 1 (1987): 4.

Some of those periodicals were the voices of small organizations. The Laser Arts Society for Education and Research (L.A.S.E.R.), for example, was formed as a non-profit institution in Northern California in 1980, by members of Laser Affiliates, another non-profit group belonging to the Haight-Ashbury Arts Workshop (HAAW). The founders of L.A.S.E.R. argued that the greatly increased interest in laser art forms merited the creation of this separate group.[14] By August 1987 it had gained some 150 members in California and beyond.

Smaller groups came and went. The Philadelphia Society of Holographers, for example, based on a group of students at the Tyle School of Art at Temple University, appeared during the mid-1980s. The Chicago Holographic Cooperative, a non-profit group of artists, holographers and interested individuals formed in the mid-1980s to hold monthly meetings, disseminate information and stage exhibitions. In France, GREPA—the Groupe de Recherche et D'Expérimentation en Photonique Appliquée, Strasbourg—appeared at roughly the same time.[15] And in Japan, the Holographic Display Artists and Engineers Club (HODIC) was formed in 1979 as a research subgroup of the Optical Society of Japan for the development of holographic 3D displays. This unusual society comprised both artists and scientists/engineers interested in the theory and practice of holographic display. By the early 2000s, membership had expanded to about 250, and to all applications of holography.[16] In this respect the group was similar to the SPIE Holography Applications Working Group formed in 1976, and especially its rebadged Technical Interest Group on Holography revived during the 1980s, which also attracted non-scientist holographers.

Early computer networks combined some of the functions of holographers' newsletters and associations. A 'Holography Network' began in August 1985 as one of several groups using the New York Institute of Technology computer conferencing system, using software known as Participate. This was discontinued a year later owing to policy changes at NYIT, but was restarted on a system called UNISON in Denver, Colorado soon afterward, allowing up to thirty-two people to conference simultaneously. By 1989 a New York-based holography computer bulletin board was being advertised in *holosphere*, although it is doubtful whether such real-time and message-posting systems gained much use.[17]

With the invention of the World Wide Web in 1991 and its availability to a wider public from about 1996, software began to facilitate social activities for holographers. The new medium permitted bulletin boards, chat rooms, the display of images, and the publicizing of holography firms and products. Some notable and relatively long-lived websites included holoworld.com, independently operated by Frank DeFreitas to promote

[14] L.A.S.E.R., 1980–88

[15] MoH, 'Holography: societies and associations,' 1980–88.

[16] Tsujiuchi, Jumpei to SFJ, unpublished report, 20 Jun. 2003, SFJ collection.

[17] The expense of dial-up access to the bulletin board and its limited promotion limited take-up. By comparison, the archives of two holography discussion groups available on the web in the early 2000s reveal that they seldom attracted as many as a dozen different contributors in a *week*, although accessible around the world.

holography by amateurs, holonet.khm.de, an extensive site for aesthetic holographers developed by Urs Fries and partially supported by the Shearwater Foundation, and holography.ru, managed for hobbyists by holographers in Russian firms and academic institutions. The web allowed widely dispersed amateur holographers—who previously had congregated around schools or the occasional local conference—to interact usefully for the first time. One of the more successful online discussion groups, holographyforum.org, registered over two hundred members in its first two years of operation.[18]

But consolidation of holographers' common interests had to be nurtured in other ways besides newsletters, local organizations, and communication channels. It took something special to provide a sense of self-identity for holographers: special events, such as meetings, and special places, such as museums and galleries. These brought holographers together from the mid-1970s to mid-1980s out of enlightened self-interest, if not a desire to adopt a wider sense of community. Their function is consistent with Peter Galison's description of 'trading zones' between technical subcultures. They became venues for sharing information across boundaries, proselytizing for the subject, and providing the nuclei on which to build more cohesive communities. The special activities promoted a degree of coalescence of scientific, industrial, artisanal, aesthetic, and commercial interests. They swelled the numbers of practising holographers, making it possible to identify holographic communities for the first time. While the disparate groups still represented a modest number of practitioners, their numbers rose from hundreds to a few thousand.[19]

11.3 SPECIAL CONGREGATIONS: SYMPOSIA

As discussed in Chapters 8 and 10, scientists and artists disseminated their products for distinct audiences in different venues: conferences and exhibitions, respectively. There was a degree of conflict that introduced conflicting pressures for segregation and unification of these two groups. On the one hand, artists sought to gain respectability and

[18] About half the members of this English-speaking list who registered their locations were from North America, one-third from Europe, and the remainder from Pacific Rim countries. The commitment and activity of the members is comparable to face-to-face clubs: while four members of this holography forum site had posted over 1000 messages at the time of viewing, only fourteen had posted over hundred, and fifty had posted more than ten messages. And despite physical separation, the virtual environment permitted an amateurs' esprit de corps: its Porcelain Cat Group shared technical expertise by posting reconstructions of that typically mundane subject of amateur holograms, the ubiquitous knick-knack. As this suggests, the discussion group became a communication channel for amateurs, although a handful of professional artisanal holographers participated.

[19] INTEGRAF, a holography firm operated by Tung Jeong from the early 1970s until 2004 and his son thereafter, lists some 1200 individuals in America and 1500 elsewhere in the world associated with holography, compiled from the late 1980s (Jeong, Tung H., 'HOLOGRAF.INTERNATIONAL Directory of Holographic Companies, Institutes and Individuals—Worldwide', report, Photics Corp, Sep. 2002). The list excludes some amateurs, many artists, and most professionals, including some no longer actively publishing or working in the field.

Fig. 11.2. Tung Jeong at first Summer School on Holography, Lake Forest, Illinois, 1972 (Jody Burns photo, MIT Museum MoH collection).

acceptance from the wider art world by dissociating themselves from scientist-holographers, and to make their exhibitions representative of a higher aesthetic sensibility. On the other hand, and particularly when interacting with artisanal holographers, both scientific and artistic practitioners engaged in a limited technological trade, picking up techniques and news from each other.

By the early 1980s, mutual interest began to overcome these isolationist tendencies: conferences and exhibitions were becoming more similar. Holography conferences had developed a tradition of including a display of holograms; and exhibitions frequently included catalogues or workshops to explain the principles of holography.

These ideas were effectively merged by physicist Tung Hon Jeong during the early 1980s. His role in holography relates to his teaching background. Jeong (b. 1936, and known as 'TJ' to colleagues) immigrated to America from China as a child, settling initially with relatives in Texas and learning English. He obtained his BS at Yale in the late 1950s, and a PhD in nuclear physics from the University of Minnesota in 1962. Jeong had found optics unappealing as an undergraduate, because the subject was taught primarily as geometrical optics.[20] He took a post at Lake Forest College, a small liberal arts institution in Illinois, in the early 1960s (see Figure 11.2). Finding it difficult to pursue his original research interests in nuclear physics there, he spent summers at Oak Ridge, Tennessee where, as a junior researcher, he frequently was relegated to midnight shifts for experiments on the accelerator. He found that the research limited his time with his young family even when they came to Tennessee with him. But, while working at Oak Ridge, Jeong got to know spectroscopists and, through them, about lasers and modern optics.

While searching for a suitable laboratory subject for teaching undergraduates, he encountered articles by Emmett Leith on holography. Jeong wrote to, and later visited, Leith to discuss how the subject could be taught. The subject enchanted him as 'so visual, tactile, easy enough to do'—a subject quite unlike his own nuclear physics, which was inaccessible and unsuitable for encouraging enthusiasm for science among liberal arts students.[21]

[20] Feldman, Ruth Duskin, 'Trapping the light fantastic', *Dynamic Years* (1986): 66–70; Jeong, Tung H. to SFJ, interview, 21 Jan. 2003, Santa Clara, CA, SFJ collection; Kontnik, Lewis T., 'A teacher of holography (Tung Jeong)', *holosphere* 15 (1987): 31.

[21] Jeong, Tung H. to SFJ, interview, 21 Jan. 2003, Santa Clara, CA, SFJ collection.

Jeong consequently made a mark as a popularizer and teacher of holography. He devised, for instance, a portable holography demonstration kit for Gaertner Scientific in 1968, and later marketed his own equipment suitable for high school or undergraduate students.[22] Jeong's undergraduate teaching experiences parallel those of Albert Baez in California some fifteen years earlier. Indeed, the two collaborated on a 17 min *Encyclopaedia Britannica* film on the subject in 1972.[23]

Jeong began offering short courses on holography at Lake Forest College during the early 1970s, and produced sound recordings and instruction booklets to supplement them.[24]

As a popularizer, Jeong had crossed the borders that were then developing between communities of holographers. His undergraduates were students of non-technical humanities subjects more often than science. He was also exposed to a broad spectrum of students taking his summer courses, many of whom had aesthetic interests. And seeking low-cost equipment, Jeong was drawn to the sand table and other tricks of the west coast artisans.

Jeong was thus well placed to inaugurate an activity having wider appeal. His *International Symposium on Display Holography*, held at Lake Forest College in 1982, was intended to bring together scientists, entrepreneurs, artisanal, and aesthetic holographers (see Figure 11.3). A summer event, it incorporated the kind of extra-conference activities that had made the Gordon Research Conferences popular a decade earlier with scientists. It also had an associated exhibition where artists could display their work; and the sessions themselves were open, interactive, and accessible to all. The relatively isolated environment of Lake Forest College provided an intimate and accommodating locus for display holographers, and proved to be a memorable experience for many of the attendees who previously had felt marginalized or isolated.

[22] Jeong, Tung H., *Gaertner-Jeong Holography Manual* (Chicago: Gaertner Scientific Corp., 1968); Jeong, T. H., 'Holography comes out of the cellar', *Optical Spectra*, 2 (6): 59–65.

[23] Baez, Albert V., Tung H. Jeong, and Encyclopaedia Britannica Educational Corporation., 'Introduction to Holography', video recording, Encyclopaedia Britannica Educational Corp. In 1969, Albert Baez designed a $10 'holographic teaching unit' with David Bond of Holovision in Torrance, California, intended for primary school students to learn about coherent optics, light refraction, and holography. Baez was, at the time, an advisor to UNESCO on science education (Lindgren, Nilo, 'Search for a holography market', *Innovations* (1969): 16–27). The use of holography as a medium for teaching science was sporadically popular. In 1978, for example, New Mexico State University had a summer programme to bring in good high school students to study holography; in 1981, the National Science Foundation funded a similar programme at Erskine College in South Carolina; and during the 1990s Pearl John, having obtained an MA from the RCA Holography Unit in 1992, exhibited her art through the 1990s and went on to teach photonics in a Missouri vocational college and later at the University of Southampton in a secondary school outreach programme from the early 2000s. Similarly artist Ana MacArthur, having established a studio for dichromate hologram production in Colorado with partner August Muth during the 1980s, later taught at a 'Photonics Academy' for American high school and middle school students.

[24] Jeong, Tung H., *A Study Guide on Holography (Draft)* (Lake Forest, Illinois: Lake Forest College, 1975); Jeong, Tung H., 'A demonstrated lecture on holography', sound recording, Instant Replay; Jeong, Tung H. and Francis E. Lodge, *Holography Using a Helium-Neon Laser* (Bellmawr, NJ: Metrologic Instruments, 1978); Laser Institute of America and Tung H. Jeong, *Holography: Principles and Applications* (Toledo, Ohio: Laser Institute of America, 1975).

Fig. 11.3. Attendees at the first *International Symposium on Display Holography*, Lake Forest College, 1982 (MIT Museum MoH collection).

The Symposia became special short-lived trading zones (to use Peter Galison's metaphor) that united the disparate groups around a single product, the hologram.[25]

Artists had been the most diffident and loosely knit community among holographers. Aesthetic holographers defined themselves, in part, by the problems they faced collectively for funding and recognition. Their backgrounds, as sketched in §10.5, were various: graduates of conventional art schools, or part of the Art and Technology movement, supplemented by others who had drifted towards the subject as they perceived it as a commercial art from, for collecting, or due to technological enthusiasm. The Lake Forest Symposia opened the borders developing between the groups and brought artists into the fold.

Individuals such as Tung Jeong played key roles in redefining the community of holographers and enthusing the next generation through proselytizing. Others also adopted roles as synthesizers and broad analysts of information, thereby helping establish the community. Stephen Benton (see Figure 11.4), of the Polaroid Research Laboratories and later the MIT Media Lab, was the most visible of them, providing regular reviews of display holography[26] and chairing the annual SPIE *Practical Holography* series of conferences.[27] Besides rainbow holograms, he also devised improvements such as

[25] Galison, Peter, *Image and Logic: A Material Culture of Microphysics* (Chicago: University of Chicago Press, 1997). There, Symposia attendees exchanged techniques, information about their respective national activities, and holograms themselves.

[26] E.g. Benton, Stephen A., 'Holographic displays—a review', *Optical Engineering* 14 (1975): 402–7; Benton, Stephen A., 'Holographic displays 1975–80', *Optical Engineering* 19 (1980): 686–90.

[27] The series was founded in 1986 by Tung Jeong and Jack Ludman, and co-chaired by Benton, Jeong and / or Hans Bjelkhagen until Benton's death in 2003 and Jeong's retirement in 2005.

Fig. 11.4. Stephen Benton with *Crystal Beginnings*, *c.*1980. (Jeanne Benton; MIT MoH collection)

achromatic holograms (which reconstructed black and white images from rainbow holograms).[28] At the Media Lab from its foundation in 1984, his group researched further improvements to display holography, devising, for example, improved edge-illuminated holograms (which allowed rainbow holograms to be lit from a light source concealed in the base),[29] 'ultragrams' (very wide-angle holographic stereograms),[30] a holographic video system (which generate three-dimensional images in real-time),[31] and computer-generated 'haptic' interfaces for holograms (which allow a viewer to 'touch' the virtual image by means of force transducers).[32] Benton was known and respected by scientists and artists alike, and probably represented the least contentious spokesperson that the field had produced. Perhaps as a consequence, he was also important in attracting a particular coterie of students that straddled the divisions between science, art, and technology. Benton reflected on the unique type of graduates of the Media Lab:

> It is highly international lab; almost half our graduate students are non-US. And certainly the students in this group are not what you would consider strong theoretical students; they have skills, but some of them have been from art backgrounds who have learned a bit of holography. I'm very interested—Land built Polaroid mostly around people who had not been to

[28] Benton, Stephen A., 'Achromatic images from white light transmission holograms', *Journal of the Optical Society of America* 68 (1978): 1441; Benton, Stephen A., 'Achromatic holographic stereogram', *Journal of the Optical Society of America* 71 (1981): 1568.

[29] Lin, L. H., 'Edge-illuminated hologram', *Journal of the Optical Society of America A* 60 (1970): 714A; Benton, S. A., S. M. Birner, and A. Shirakura, 'Edge-lit rainbow holograms', *Proceedings of the SPIE The International Society for Optical Engineering* 1212 (1990): 149–57.

[30] Halle, M.W., S. A. Benton, M. A. Klug, and J. S. Underkoffler, 'Ultragram: a generalized holographic stereogram', *Practical Holography V* 1461 (1991): 132–41.

[31] St.-Hilaire, P., S. A. Benton, M. E. Lucente, J. D. Sutter, and W. J. Plesniak, 'Advances in holographic video', presented at *Practical Holography VII: Imaging and Materials*, San Jose, California, 1993. The principal sponsors were the US Navy, which envisaged using such a display for battlefield tactics, and Honda, for visualizing car design.

[32] Pappu, R. and W. Plesniak, 'Haptic interaction with holographic video images', *Proceedings of the SPIE The International Society for Optical Engineering* 3293 (1998): 38–45; Plesniak, Wendy J. and Massachusetts Institute of Technology. Dept. of Architecture. Program in Media Arts and Sciences., *Haptic Holography: An Early Computational Plastic*, PhD thesis (2001).

college. It was really toward the end, when the company had to grow quite quickly, that you might find some PhDs and Master's students. I grew up there at that time. That's what you want: people who are really committed to it, and sort of self-trained within the culture. And that's what *we* had: sort of unconventional students.[33]

11.4 SPECIAL PLACES: MUSEUMS

In the same way that symposia became important venues for holographers, museums and permanent galleries of holograms played a crucial role in promoting a wider sense of community. Exhibits of holograms blossomed into exhibitions for wider audiences, first in science departments, then art galleries, both commercial and public, and subsequently displays for wider publics. In the same way, temporary exhibitions led quite naturally to permanent displays and to 'museums' of holography. The very awkwardness of holographic set-ups inspired stability of gallery space. Holograms demanded special illuminants and carefully positioned lighting arrangements: the labour and cost of preparing such facilities required shows of long duration. And audiences were easy to find during the 1970s; a well-produced exhibition yielded a proportionately large audience. Thus permanent exhibitions seemed likely to increase public interest, to refine aesthetic sensibilities, to publicize scientific developments and to encourage commercial applications.

The first and most influential museum of holography was bootstrapped into existence with little pre-planning by its organizers. The organization and its founding Director, Rosemary H. (Posy) Jackson, proved central to the organization of holographic communities from the mid-1970s to the early 1990s. In February 1974, the New York *Village Voice* reported

> **scenes:** America's first holographic museum has opened at 120 West 20th Street in conjunction with the New York School of Holography. From noon to 5 pm from Monday through Friday, Jody Burns will demonstrate the six kinds of three-dimensional still and motion picture holography now in use today. There is even a small vibration isolation table at the second floor gallery complete with its own laser, so you can fool around with your own holograms. Also on display are the works of old masters of this new art, including a couple of three-dees by Salvador Dali.[34]

The journalist had come to interview Jody Burns and, convinced that his gallery of holograms constituted a museum, wrote an article to that effect. Although Burns protested that it was, in fact, a commercial showroom for an ongoing business in holography, his New York Art Alliance Inc. took the decision to formally found the Museum of Holography (MoH) on October 2 of that year as a profit-making subsidiary. Jackson, a

[33] Johnston, Sean F., 'Stephen Benton on holography, Polaroid and MIT', *Optics & Photonics News*, 15 (8) (Aug. 2004): 32–5.

[34] Smith, Howard and Brian Van der Horst, 'Scenes', *Village Voice*, 14 Feb. 1974.

Fig. 11.5. Rosemary (Posy) Jackson at the New York Museum of Holography, 1976 (MIT Museum MoH collection).

graphic designer, became closely involved with the Museum during that year, initially hired to make frames and design graphics.[35]

Through 1975, the Museum grew rapidly in size and stature, showing holograms at various venues in the New York region, with Burns giving lectures at more than thirty institutions. In July 1975, Burns and Jackson co-produced the *Holography '75* exhibition at the International Center for Photography (ICP). As discussed earlier, the show broke ICP attendance records but generated critical reviews.[36]

As the demands of the Museum continued to escalate, Posy Jackson became Director in January 1976 and began proceedings to make it a non-profit, tax exempt institution (see Figure 11.5). Jody Burns sold the New York School of Holography operation to Sam Moree and Dan Schweitzer in 1977, who taught smaller groups and used the facilities for their holographic art. The purpose of Jackson's revamped institution was 'to create and administrate the Museum of Holography; to create an environment in which the public becomes acquainted with and educated about holography as an art form; to provide an exhibition of historic holographic works; to regularly exhibit state-of-the-art holography; to provide—through the MoH Bookstore—a retail outlet for the sale of holograms and related products'[37] or, as expressed a few years later, to 'provide a focal point for the growing public interest in holography', and to act, in effect, as an institution promoting the subject.[38] Her initial plan was to retain the link with the New York Art Alliance Inc. by renting space, and providing free membership and exhibition space to the School of Holography students, but within a month decided to go it alone. A new location for the

[35] Jackson, Rosemary H. to S. Bains, letter, 14 Sep. 1989, MIT Museum 28/743. Rosemary H. Jackson obtained a Bachelor of Fine Arts degree from the New York/Parsons School of Design in graphics and photography in 1970. From 1970–75, she headed Jackson Graphics, a NYC based design firm. During this time, she took a certificate course in holography from Tung Jeong at Lake Forest and, in 1975, she completed a second course under Dan Schweitzer and Bill Molteni Jr at the New York School of Holography.

[36] Jackson, Rosemary H., 'History of the Museum of Holography October 2, 1974–May 1, 1976', report, MIT Museum 8/191.

[37] Jackson, Rosemary H., 'A proposal for the maintenance and development of the Museum of Holography under the Directorship of Rosemary H. Jackson', proposal, MIT Museum 8/191, 1 Feb. 1976.

[38] 'A holographic interface', *Laser Focus*, Jul. 1979: 12–22.

Fig. 11.6 Preparing for the opening of the New York Museum of Holography, 1976 (MIT Museum MoH collection).

Museum at 11 Mercer St, in the Soho district of New York City, was leased in May exclusively by the MoH. However, the collection of some 400 holograms was transferred directly from the New York Art Alliance.[39]

The holograms appeared in three shows in the American northeast. And in March 1976, in the midst of reorganization, the New York Museum of Holography (see Figure 11.6) put on its second major exhibition, this time in Stockholm, Sweden. The foreign exhibitions promoted not only holography, but America, too: Jackson and Burns gave lecture demonstrations on holography in four European cities on behalf of the MoH and sponsored by the US Information Agency.

The Museum opened with a staff of two on Dec 12, 1976 at its new premises, with the mayor of New York, Abraham Beame, cutting the ribbon, Tung Jeong as official host, and some 500 attendees. The first exhibition, *Through the Looking Glass*, displayed 75 holograms.[40]

Jackson was a tireless and charismatic administrator who quickly became a facilitator for holographic activities uniting the growing art, commercial, and scientific spheres. Making contacts initially through exhibitions, she proved empathetic and approachable, commiserating with struggling artists, cajoling and coercing prominent participants on the Board of Trustees, Advisory Board, judging panels and other bodies, and urging corporate donations of 'historic' holograms and sponsorship. Her activities swelled the hologram collection by yielding long-term loans and direct donations of holograms, both historic and contemporary, and principally from scientists. The museum adopted a policy, however, of paying artists the full price for their works, 'so that prices could be set for work in a field where basically no one else was buying'.[41] A 1980 performance

[39] The first organizational meeting was held in Sep. 13, 1976. Original Board Members were Jackson, Joseph R. (Jody) Burns, Peter Nicholson (then Director of the holography program at the Cooper-Hewitt Museum of Design, Smithsonian Institution), Fernanda K. and Frank S. Gilligan (on the Board of the Lincoln Center), and Tung H. Jeong of Lake Forest College.

[40] Jackson, Rosemary H., 'An accounting, 1976–1978', *holosphere* 8 (1979): 3, 5–6, 8–10.

[41] Jackson, Rosemary H. to S. Bains, letter, 14 Sep. 1989, MIT Museum 28/743.

appraisal judged Jackson to have

> Director syndrome—the energy / enthusiasm which developed or sustained the museum causes problems now and then. Tends to make decisions on intuition instead of gathering all facts. Spreads self too thin. Doesn't delegate enough. Very persuasive and articulate, well spoken, but slight tendency to hyperbole; tends to lengthiness. Terrific capacity to meet the public / convince the unbelievers.[42]

In the frenetic first two years of the new museum, Jackson and the growing staff coped with continual change. She reflected:

> I think the establishment of holography in the consciousness of the American psyche will take a lot longer than I thought it would. And it will not happen, as I felt two years ago, by any one person or action. It will be a slow, indirect process fed by everyone involved in the medium at every conceivable level. Convincing people that holography is an art is not something you can do by ramming it down their throats. It has to be absorbed almost without knowledge of the process by those who 'make' art 'arrive', and those who pass the word along, and by those who believe things just because they hear and read them.[43]

The MoH provided direct sponsorship of holographic art (still scarcely half a decade old when the Museum was founded) by purchasing holograms and by establishing an Artists In Residence (AIR) programme, which operated between 1980 and 1989. This was modelled on a similar programme operated by Hart Perry of the Holographic Film Company, also in New York. Other AIR programmes in holography were run, for example, by Fringe Research in Toronto and The Holographic Studio in Vancouver. AIR programmes provided a new way of circulating the skills and perspectives of aesthetic holographers, encouraging the budding culture.

Nevertheless, the aspirations and image of the Museum of Holography were never entirely clear, nor was its place secure in the rapidly shifting sands of holography in wider culture. The claims and reality of internationalism, for example, jostled through the Museum's life. The MoH had begun as the offshoot of a private New York business, and developed a new image through the organization of the 1976 New York and Stockholm holography exhibitions by its founders. Its position was neither recognized nor respected for some time. The McDonnell Douglas Corporation, for instance, declined Posy Jackson's request that they donate their large collection of pulsed holograms to the Museum because of their 'desire to use a more internationally known institution'.[44] By organizing international shows, however, the MoH staff encountered active artists, and sought connections with practising scientists and engineers.

In satisfying those ambitions, Posy Jackson had toured England in late 1978 in the wake of the two large *Light Fantastic* exhibitions, meeting an enthusiastic collection of holographers and holographic firms there for the first time: Nick Phillips (Holoco), Jonathan

[42] Harvey, Robin to P. Jackson, memo, 1980, MIT Museum 12/265.
[43] Jackson, Rosemary H., 'An accounting, 1976–1978', *holosphere* 8 (1979): 3, 5–6, 8–10.
[44] Baxter, K. D. to R. H. Jackson, letter, 8 Oct. 1976, MIT Museum 29/804.

Ross (The Hologram Place), Eve Ritscher, Edward Lucie-Smith (art critic), Paul Walton (See 3) Michael Wenyon, Andy Pollard, Nigel Abraham (The Hologram Place and See 3), Peter Saabor (Carlton Cleves), Ian Lancaster (Gulbenkian Foundation, which had instituted Holographic Awards for creative holography), Andy Harris (Holographic Developments Ltd) and, not least, Dennis and Marjorie Gabor. Jackson consulted on obtaining Rockefeller Foundation funding for a holography course at Goldsmith's College, London. She recommended that course administrators and lecturers be trained in America, and that some teaching at Goldsmith's should be performed by American holographers, thereby exporting the American model abroad.[45]

Such cultural and technical colonization embodied the MoH view of the subject. These views were supported by the MoH Board of Trustees, most of whom were American holographers or east coast business people with arts connections. At times, the MoH desire to speak internationally for holographers became overt, such as when a British documentary appeared to exaggerate the status of holography in Britain.[46] The following year, the MoH proposed an International Congress for the Art of Holography. The programme proposal stated the Museum's *de facto* understanding of the geographic distribution of the subject:

> Holography is now a worldwide practice, and its practitioners in art have not had the benefit of dialogue through established scientific channels [. . .] The Museum's goal for this congress is nothing less than attendance by all holographers in the US, where holographic activity is currently concentrated, and the rest of the international holographic art / science community.[47]

Such a Congress, held in the US, would have been a means of consolidating the organization of holography by American workers in general, and by the Museum of Holography in particular, but it never took place.

Similarly, during the early 1980s, the New York Museum of Holography had attempted to claim trademark on the name 'Museum of Holography'. Other

[45] Jackson recommended that the Goldsmith's administrator be trained in holography at the New York Holographic Laboratories, that the assistant be trained at the School of Holography in Chicago, and that the teacher take a one-week course at Lake Forest College, because 'none of the proposed staff have administered or taught this medium. In order to learn the necessary details, each should go to a different holography center and take a course suited to their needs' (Jackson, Rosemary H., 'Workshops: Goldsmith's College', proposal, Jan. 1979).

[46] Correspondence from the late 1970s suggests a tension between Benton and Nick Phillips in Britain founded in part on national rivalries. For example, Benton (who was to serve as Board member of the MoH 1987–92) complained to the producer of Horizon / Nova, a joint UK/US television programme about 'an obvious and chauvinistic misdirection of credit' as compromising 'the program's credibility and unnecessarily strain[ing] British scientific ties' (Benton, Stephen A. to T. Edwards, by letter, 15 May 1978, MIT Museum 26 / 584). Jackson, upon receiving a copy, replied, 'I have now seen the program twice [. . .] I was appalled at the geographical focus of the tape, and even more disturbed at the choice of film images used' (Jackson, Rosemary H. to S. A. Benton, letter, 19 May 1978, New York, MIT Museum 26 / 584). Behind such jostling for attention was the technical rivalry of techniques of white-light holography: the bright rainbow holograms of Benton versus the high-quality reflection holograms of Phillips, both in competition with the superb reflection holograms produced in the Soviet Union.

[47] MoH, 'First International Congress for the Art of Holography—Program Proposal', Program Proposal, Museum of Holography, MIT 3 / 56, 1979.

organizations—particularly the Museum für Holographie und neue visuelle Medien in Pulheim, Germany (founded 1979), the Musée de l'Holographie in Paris (founded 1980) and the Fine Arts Research and Holographic Center in Chicago (founded 1982)—resisted, responding that the words 'museum' and 'holography' were hardly unique.[48] It was clear, however, that the New York based organization was under-financed and ultimately limited in its connections, unable to keep abreast of developments in Western Europe, the Soviet Union, and Japan. By 1990 there were organizations offering similar attractions in Chicago, Cologne, Paris, and Leningrad, and a scattering of private shops and galleries in other locations. Where in the late 1970s the MoH had been a welcome and unique focus for activities in holography, by the mid-1980s it had lost its international cachet.

Despite its initially unique status as a permanent gallery for holography with claims to be a rallying point for holographers internationally, the MoH had organizational problems that increased through the 1980s. Posy Jackson sent her resignation letter to the Board on 15 September 1982, to be effective 31 December 1983. She argued that the Museum was entering a new phase of development requiring extra skills, and had been operating at a deficit, with herself as sole fund-raiser. The Board of Trustees was unable to provide the necessary detailed management to support her.[49]

Jackson's successors assumed Directorship of the MoH with different qualities, a new operational context, and distinct management styles. She was succeeded by David Katzive (1984–86), Ian Lancaster (1986–88), and Martha Tomko (1988–92).[50]

When Katzive succeeded Jackson, he inherited eighteen staff and a complicated octopus of an administrative hierarchy that included the Board of Trustees, Board Executive Committee, Director, Management Committee and Advisory Board, and separate managers for the Holography Laboratory, Exhibition Services Dept, Membership Dept, Educational Services Dept, Information Services Dept, Bookstore, and Administration. Having come to the MoH with a background in museums and video art, Katzive considered changing the museum's focus from holography to a media-arts research institute that would include computer and video art. Aesthetic holographers generally opposed the change, interpreting it as a step backwards for what they saw as an art form on the cusp of wider acceptance. Within a year the Board expressed concern about communications, management, and Katzive's de-emphasis of holography. Katzive resigned that summer, and Posy Jackson publicly criticized the Museum's operations under her successor's regime.[51]

[48] MoH, 'Museums,' MIT Museum 41/1263–1264, 1976–92.

[49] MoH, 'MoH Board of Trustees,' 1976–92, MIT 8/191.

[50] David H. Katzive taught Art and Technology to Humanities students at Cooper Union from 1987, and was President of the Visual Technology Group, the new media and broadcast division of Ruder Finn.

[51] Law, Linda, 'David Katzive interview', *Wavefront*, Fall 1986.

In 1986, the new Consulting Director, Ian Lancaster,[52] noted the changing terrain for the Museum in a report to the Board of Trustees:

> Ten years ago holography was a stripling. It had moved forward rapidly since the pioneering work which followed the invention of the laser yet no-one was quite sure whether it was a technique, a tool, an end-product, a science, a technology, an art-form or a medium. Of course, it was and is all those things, but then and now different aspects of holography had achieved various degrees of development. There were high expectations of it as a major new industry and art medium. One symptom of this high expectation was the involvement of a number of major companies, while another was the number of specialist companies developing holographic products for the perceived mass market. At the same time there were a surprising number of enterprising people from a variety of backgrounds who were actively pursuing their creative and artistic aims in the medium. For the first five years of the Museum's life putative commercial holographic companies came and went like Monopoly properties, some of their personnel reincarnating in new situations, others disappearing from the holographic scene. The major companies originally involved produced several significant works, but then became wary of pumping money into their researchers' promises with no immediate return and closed their holography facilities.[53]

Despite these market realities, Lancaster emphasized an international role for the institution, a vision reiterated by his successor and final director, Martha Tomko, who described the Museum of Holography as 'one of the keepers of the culture'.[54] The Museum's accounts of the period emphasized its seminal role and anticipated a positive future.[55] Yet when Tomko assumed direction of the faltering MoH in 1988, she reported to the Board that taxes had been unpaid for two years, its publication *holosphere* was delaying publication and the identified sources of income were inadequate. Within six months she had replaced all staff except the Education Director.[56]

MoH Board meetings discussed a series of devastating realities. The curator reported that the archive of holograms, photographs, and books was deteriorating, and that even Gabor's Nobel Prize, enclosed in Plexiglas since its donation to the Museum years earlier, was

[52] Ian Lancaster (b. 1948) obtained a degree in American Studies and Drama at the University of Hull and a diploma in Arts Administration at the Polytechnic of Central London. He served as administrator of the Gulbenkian Foundation in Britain for a decade before his appointment at the MoH, and was also a principal in Third Dimension Ltd, a hologram marketing business. He subsequently joined Reconnaissance International, a market analysis and exhibition organizing firm specializing in the holographic market (Kontnik, Lewis T., 'Personality profile: Ian M. Lancaster—Bend your knees!' *holosphere* 15 (1987): 27), becoming its sole director in 2003.

[53] Lancaster, Ian M., 'The Museum of Holography: Its role and policies in a changing environment', report for Board of Trustees, MIT Museum 8/191, 1986.

[54] Tomko, Martha, 'Letter to the editor—what's going on at the MoH?' *holosphere*, 17 (1) (Spring 1990): 4–5.

[55] See Cowles, Susan, 'Museum of Holography: past, present and future', *Holographics International* (1989): 26–9; Jackson, Rosemary H., 'Posy writes . . .' *Holographics International* (1989): 4–5.

[56] Tomko, Martha, 'Letter to the editor—what's going on at the MoH?' *holosphere*, 17 (1) (Spring 1990): 4–5. Martha L. Tomko (1947–92, born Bellevue, Ohio) had obtained a Bachelor of Fine Arts at Ohio University in 1969 and did postgraduate studies at the University of Nebraska 1977–8. She was an Assistant librarian and later Research Assistant at the Columbus Ohio Museum of Art, then General Manager of the B.C. Pops in New York.

becoming mildewed and permanently stained. Just as significantly, she reported that 'the collection is stagnant and does not represent the latter half of the 1980s' and that 'it is tremendously embarrassing that the Museum has not kept pace with the accelerating development in the field. No one is interested in making donations of artworks to the Museum'.[57]

The public face of the Museum also suffered. Attendance was declining steadily. Its target audience, defined as out-of-town college educated people aged between 18 and 35 with an interest in both art and science, proved difficult to sustain even when supplemented by visits from local schools. Annual paid attendance had been almost static at about 55,000 to 60,000 from the late 1970s, but then dropping steadily to 47,000 in 1983, 44,000 in 1989 and 26,000 in both 1990 and 1991, exacerbated by shorter opening hours.[58] Membership in 1991 was about 350, down from a maximum of 500 a few years earlier. The location itself had deteriorated since Posy Jackson's upbeat description in 1978:

> We chose Soho because it is the center of avant garde art in New York, which is to say it is the right place for us. Since we have had over one hundred thousand people through the museum in a year and a half I would not say we gave up a well traveled place for low rent. The rent is substantial, and the traffic is likewise.[59]

A decade later, the neighbourhood was increasingly shabby, owing to a rise in local prostitution, homeless persons, and sidewalk sellers. There was also little foot traffic, inadequate parking, and few retail businesses of like character nearby.[60] British visitors recalled uneasily leaving the safety of their taxi to stand on the step of the unwelcoming front door.[61]

The periodical *holosphere*, published by the Museum since November 1977, also faltered. Having attracted about 200 subscribers before the Museum took over its publication and then climbing to some three to four hundred, the frequency of publication gradually diminished from about one issue per month until 1982 to quarterly thereafter. After a series of five subsequent editors, it struggled to produce two issues during 1990 and was then suspended while the MoH Board negotiated unsuccessfully with the American Institute of Physics (AIP) to take over publication of the periodical. But holographers were now too diverse: the AIP administrators found the profile of the readers a poor match to its membership, and were unimpressed with what appeared to be a declining market for the magazine.[62]

Tomko, and members of the Board of Trustees such as Stephen Benton, wrote increasingly desperate letters to potential sponsors of the MoH. Behind the scenes, Posy Jackson

[57] Dinsmore, Sydney, 'on MoH collection from curator', report, MIT Museum 10/192, 18 Oct. 1990 and 22 Jan. 1991. Dinsmore had previously curated/directed the Interference gallery in Toronto.

[58] Director, 'MoH attendance information', report, New York Museum of Holography, 1983; 'Board minutes', report, New York Museum of Holography, 1986; Dinsmore, Sydney, 'Strategic plan', report, New York Museum of Holography, Jul. 1990.

[59] Jackson, Rosemary H. to G. Zellerbach, letter, 16 Oct. 1978, New York, MIT Museum 28/743.

[60] MoH, 'Strategic plan', Museum of Holography, MIT 10/192, Jul. 1990.

[61] Briggs, Nigel to SFJ, interview, 20 Sep. 2002, SFJ collection.

[62] MoH, 'Strategic plan', Museum of Holography, MIT 10/192, Jul. 1990.

and her husband Bailey Smith provided substantial loans and donations directly and via the Shearwater Foundation, for which she had disbursed monies for art holography since 1987.[63] Spending the following eighteen months fire-fighting to impose new administrative procedures and clarify the museum's financial status, Martha Tomko gained the appreciation of the holography community but made few friends with the cost-cutting and abrasive style necessary in the circumstances.

By the end of 1990 bankruptcy loomed, despite paying half of the outstanding debts. At the end of 1991 the Museum was challenging an eviction notice for unpaid rent in the courts while the MoH administrators vacillated over purchasing a new premises outright. The Board considered letter-writing appeals for New York City and State support, radio appeals, local television news items, and newspaper advertisements which were to read 'Death stalks the Museum of Holography'.[64] The last-minute fund-raising and organizational activities proved inadequate, however, and the Museum of Holography finally closed its doors in 1992.

The following year, when the holdings of the Museum were auctioned, Stephen Benton of the MIT Media Laboratory and Warren Seamans, Director of the MIT Museum, were able to raise $180,000 from a handful of private donors and MIT funding via President Charles Vest to purchase the entire lot, saving it from fragmentation through private purchases.[65]

The New York MoH was not the only museum to be dedicated to holography, even if it was the the most ambitious in terms of aims, collection, and programmes. In Pulheim, near Cologne, Germany, the first European museum of holography, the Museum für Holographie und neue visuelle Medien, opened in December 1979. It was privately owned and directed by Matthias Lauk, who had first had contact with holography when he encountered the Cartier jewellery store hologram in 1973. Over the next few years he represented American firms Holoconcepts Corporation of America, and Brown, Boveri & Co. to seek orders of laser-lit holograms, but sales were low. When McDonnell Douglas and Brown, Boveri & Co ceased production of holograms, Lauk acquired a number of them, which was the beginning of a large collection.[66] In 1979 he represented the British firm Holoco in Germany and, in conjunction with a custom order, arranged a small exhibition of fourteen holograms from holographers Rudie Berkhout, Harriet Casdin-Silver, Dieter Jung, Sam Moree, Carl Fredrick Reutersward, Dan Schweitzer, and Rick Silberman, sponsored by the tobacco company Roth-Händle. Seeking funding for a

[63] MoH, 'Correspondence XVII–XXII,' MIT Museum 2/39–44, 1988–92.

[64] MoH, 'Board,' MIT Museum 8/163, 1989–92.

[65] The MIT Museum became the repository for the collection, although there was limited display space and insufficient money for a curator and (at least initially) for cataloguing.

[66] Few collectors of holograms appeared. Most, like holographer Hans Bjelkhagen and promoter Eve Ritscher, were either holographers or exhibition organizers and collected privately; Jonathan Ross was a rare exception who, after accumulating holograms as an owner of the production company See 3 Holograms, made his collection accessible from the late 1990s through his Gallery 286 in London.

museum from large industries, cultural organizations, and local government without success, Lauk self-financed the museum.

When Holoco, the British firm behind the major *Light Fantastic* exhibitions, also ceased production in 1981, Lauk purchased many of their holograms to expand the gallery and its opening hours. Some ninety holograms from a collection of several hundred could be displayed in two uncluttered galleries. Travelling exhibitions were also staged through Germany. Although the location on the outskirts of Cologne was not propitious, the museum later claimed attendances considerably higher than those of the New York MoH. During the late 1980s Lauk began producing travelling hologram shows, displayed in banks and town halls, cultural centres, trade fairs, and museums, especially in Germany. One such show in Denmark attracted some 80,000 visitors in 1985. Like a number of other small holography businesses, the Museum für Holographie supplemented income from visitors and hologram sales with holography classes from 1984.[67]

Lauk's Museum further promoted holography by instituting the first European prize for art holography in 1990. Its first curator, Peter Zec, founded the Deutsche Gessellschaft für Holographie (German Society for Holography).[68] That year, Matthias Lauk's private collection of holograms, still the basis for the Museum, was sold to the newly founded Zentrum fur Kunst und Medientechnologie (Centre for Art and Media Technology). This German state institution was almost unique in conserving and collecting holograms, being paralleled on a much smaller scale by the Victoria and Albert Museum in London.[69]

The first French museum of holography, the Musée de l'Holographie, was active longer than the others. It opened in Paris in March 1980 in rue Beaubourg, near the Pompidou Centre. The museum, under the direction of Anne-Marie Christakis, attracted some 100,000 visitors per year with its large collection of holograms and associated activities. Even more than the museums in New York and Pulheim, the Musée de l'Holographie organized numerous exhibitions at home and abroad. Some 150 were held in France during its first fifteen years, and foreign exhibitions included Canberra, Jakarta, Budapest, Madrid, Buenos Aires, and Peking. It offered courses from its first year for the public, artists, and scientific specialists. Like holographers in the Ukraine, the Musée de l'Holographie outfitted a bus as a travelling laboratory for the Ministry of Research, touring holiday villages with a travelling exhibition and opportunity for visitors to create holograms. The Musée also promoted careers in holography, becoming associated with l'Association pour le Développement des Arts et des Techniques Holographiques (L'ADATH), which was founded in 1983. The two had different remits: the Association had scientific and artistic members and a mission of contributing to scientific research,

[67] Misselbeck, Reinhold, 'The Museum für Holographie und neue visuelle Medien and its influence on holography in Germany', *Leonardo* 25 (1992): 457–8, Misselbeck, Reinhold and Matthias Lauk, 'Holography in Germany', *L.A.S.E.R. News* 10 (1990): 8; MoH; 'Holography: museum fur Holographie & Neue Visuelle Medien,' MIT Museum 41/1264, 1981; Museum für Holographie & neue visuelle Medien., *Museum für Holographie & neue visuelle Medien* (Pulheim: The Museum, 1989); Pepper, Andrew, 'German museum celebrates 5th anniversary', *holosphere* 13 (1985): 9–10

[68] Öhlmann, Dietmar and Niklas Möller, 'The new goals of the German Association for Holography, DGH', *Proceedings of the SPIE* 3358 (1989): 152–4.

[69] On the archiving of hologram collections and documentation, see §14.3.

particularly in industrial metrology, while the Museum was a meeting place between scientists, artists, and the public.[70]

Even in France, museums of holography supported the careers of aesthetic holographers more than those of scientific holographers. Fine-art holographers remained arguably the most marginalized of holographers internationally, with no single representative expressing their common interests, a matter of some concern at conference congregations.[71] Their continued visibility owed a great deal to the activities of a single sponsor: the Shearwater Foundation.

Based in Ft. Lauderdale, Florida, Shearwater proved to be the most reliable and generous sponsor of artistic holography. While its 1987 press release noted that the award recipients were selected by an anonymous Board of Trustees and 'by a confidential group of advisors hired by the Foundation to make recommendations from the worldwide field of holographers', its origins were linked to those of the New York Museum of Holography. In fact, the Foundation's grants to holography were underpinned by the involvement of Posy Jackson, who administered the programme from February 1987 until May 1998, and subsequently administered by fine-art holographer Andrew Pepper.[72] The Foundation closed in November 2004 upon the death of the last of its Board of Trustees. Over its seventeen years, the organization disbursed some $1.5 million dollars to individual artists and organizations (including the New York MoH, MIT Museum, and Center for the Holographic Arts, New York) to support publications, conference, and exhibition attendance and organization, operating funds for museums, AIR programmes, the distribution of holographic film to artists, the purchase of holograms for collections, and other activities to promote creative holography around the world.

There were many tactics, then—both conscious and unconscious—of assimilation. Scientists, engineers, artisans, entrepreneurs, and artists active in holography were brought together and sustained by periodicals such as *holosphere*, symposia such as Lake Forest, spokespersons and enthusiasts such as Steve Benton, organizations such as the New York MoH, and sponsors such as the Shearwater Foundation. To varying degrees, such protagonists united scientists, engineers, and artists, helping to construct a wider community despite the absence of the common trappings of a definitive technical journal, viable professional society, and dedicated university departments. But each was unique and difficult to replicate, unlike the wider social underpinning of more obviously successful subjects. And one further feature was crucial for sustaining a viable holographic community: a market for holograms.

[70] Christakis, A., 'Musee de l'Holographie of Paris and its activities: 1980–94', *Fifth International Symposium on Display Holography* 2333 (1994): 245–7.

[71] For example the aesthetic holographers attending the 2003 *Practical Holography* conference considered banding together in an association to seek funding and provide mutual support, a proposal that produced a muted response.

[72] The operating officers listed on the 1987 tax return were Rosemary Jackson Smith (President), Bailey Smith (Vice President), and Alexander Nixon (Secretary). Jackson described the Shearwater funds as not her own, but 'a rather delicate balance within my family [. . .]. Suffice it to say that I am able to direct about $75,000 a year to holography for as long as they choose to have me in that position' (Jackson, Rosemary H. to E. O'Neill, letter, 1990, MIT Museum MoH Box 12).

PART IV
Creating a Market

During the late 1970s and beyond, holography was expanding its markets and popular appeal. While new technical capabilities were appearing less frequently, holographers were consolidating a stronger sense of community. Exhibitions and cottage industries—outcomes of that growing confidence—enrolled an increasing number of businesses (Chapter 12). But despite the growing economic importance of a narrow range of holographic products, popular awareness of the subject was fostered as much by fictional depictions in films and television as by consumer products (Chapter 13). Holograms became relevant to new audiences for new purposes.

By the late 1980s, then, holography was a field with well-explored intellectual territory, a cluster of interacting communities and an economically viable medium for a handful of applications. Holographers and other observers were beginning to recount histories of their subject and judgments of their success. But these accounts and verdicts varied between communities, illustrating the differing visions at their heart (Chapter 14).

12

Commercialization and Ubiquity

For most observers, the commercial future for holography and holographers appeared obvious. As groups emerged, jostled and differentiated, the identity of holographers shifted and new opportunities beckoned. But while commercialization was a constant environmental pressure, it did not yield steadily expanding terrain. Instead, the process was chaotic and dispiriting: entrepreneurs and firms struggled to define, develop, and exploit markets.

The patterns and expectations of commercialization developed early on. As an industrial scientist, Dennis Gabor had been impelled to invent, conceiving his two-step imaging process specifically for microscopy, and was surprised when manufacturers remained unimpressed. Leith and Upatnieks were urged by their superiors to patent and seek applications as soon as their discoveries became known, and stretched their minds to envisage how lensless photography might be applied. And when patents were granted during the 1970s, the most important patent holders were the military, corporate contractors, film manufacturers, and opto-electronic firms.[1] But the imaginative ideas and hard-won patents proved difficult to translate into new markets and profits.

12.1 MAKING HOLOGRAPHY PAY

The proliferation and commercial failure of holography's numerous small firms may be a statistic that defeats historical documentation. So, too, is the expenditure on blue-sky research funding, exploratory grants, feasibility studies, venture capital, and government

[1] The most important patents were No 3,506,327 'Wavefront Reconstruction using a Coherent Reference Beam' (1970, filed 1964); No. 3,838,903 'Wavefront Reconstruction' (1974, filed 1970), and No. 3,894,787; and 'Holograms' (1975, filed 1971), with Leith and Upatnieks as listed inventors. Some eight years after the laser-induced renaissance of holography, the major American patent holders were IBM (36 patents), Bell Telephone Laboratories (26), TRW Inc (21), RCA Corporation (20), Holotron Corporation (18), Battelle Development Corporation (13), Texas Instruments (12), AGFA-Gavaert and the British Aircraft Corporation (8 each), and the UK Atomic Energy Authority, US Secretary of the Navy, NASA and the US Philips Corporation (6 each). See Kallard, Thomas, *Holography; state of the art review 1971–1972* (New York: Optosonic Press, 1972) and McLuskie, Caroline, 'Holography Community', *Wavefront* 2 (1987): 4. By 1978, Holotron was the most important patent holder, having acquired some 150. See also Pizzanelli, David, 'Public or patented', *Holographics International* 1 (1990): 33

seed money for promising ideas. But there were two clear characteristics of these activities: first, expectations of commercial success were wide-ranging and ubiquitous and, second, those visions proved a poor match to consumer expectations.

As discussed in §6.3, the businesses of the 1960s growing in Ann Arbor had been among the first to explore the commercial possibilities of holography. The most important of these, Conductron, KMS Industries and then McDonnell Douglas (Missouri), actively sought commercial applications during the late 1960s and early 1970s, and anticipated advertising as a likely use of holograms. Conductron methodically explored markets, supported by imaginative engineers, a marketing team, and its entrepreneurial director Kip Siegel. But despite a handful of successes such as the *Science Year* hologram of 1967, the Conductron team attracted advertisers ineffectually.

By 1970, Clark Charnetski, a physicist on the holography team, could summarize what the company had already learned in its attempts to promote holography. His message was sobering: holography had not met their expectations. Consumers were still largely unaware of holograms or bemused by what they saw.

But although far more people had *read* of holograms than had actually *seen* them, there were genuine achievements to publicize: the first five years of commercial holography had led to significant technical improvements. The maximum size of holograms had grown from 4 × 5 in. to 18 × 24 in.; mass production had generated nearly a million holograms; the development of pulsed lasers having a long coherence length had enlarged the scope of subjects from toy locomotives to groups of people and even, according to unofficial reports, dancing nudes. He noted, however, that the spectacular predictions made by Siegel and others had provided large investment capital but relatively few commercially viable applications. To the surprise of Conductron engineers, media or display holography had been less important than other uses such as holographic non-destructive testing. Charnetski ascribed this to the cost and difficulty of producing holograms of large scenes and in developing copying processes. But the problem was also a cultural one.

> Holography was and still is largely done by scientists and engineers, and these people are as a whole more interested in the scientific things and therefore lend their talents more readily to scientific applications than toward breaking into the advertising game.[2]

The enthusiasms of technologists did not necessarily translate to those of wider culture. Nor did the Conductron engineers fully appreciate how their holograms were perceived by the public. Charnetski noted, for example, 'we have always shied away from lasers for reconstructing holograms because of [. . .] adverse public reaction to them as death rays'.[3] Marketing holograms proved unexpectedly difficult. The half-million transmission holograms that had been produced for the *1967 Science Year* were well received by the publishers, but the impact on the readers was limited. The hologram had

[2] Charnetski, Clark, 'The impact of holography on the consumer', presented at *EASCON Convention*, Washington DC, 1970, quotation p.2.

[3] Ibid., quotation p.3.

to be cut out of the book, mounted in a cardboard arrangement with a supplied red plastic filter (which limited the bandwidth of the light) to yield an unspectacular but recognizable image of a chessboard. Many copies of the hologram were never viewed.

Conductron engineers consequently developed reflection holograms as products that could reconstruct images in white light. One advantage of the reflection hologram was that it could be pasted into a book page or brochure. Even this more convenient format demanded some familiarity and enthusiasm from viewers, though. In 1969 the pharmaceutical manufacturer Hoffmann La Roche ordered some 35,000 reflection holograms of a model of the colon, to be sent to gastrointestinal specialists across America along with a penlight to view it. The Conductron engineers learned from anecdotal reports that many of the doctors receiving the package had kept the penlight but not bothered to view the hologram.[4]

A similar technical success and consumer failure was the creation of a hologram display for General Motors' New York headquarters, consisting of large 18 × 24 in. transmission holograms in a square arrangement, so that the image—a carriage and, with a change of lighting angle, a modern car body—could be viewed from all sides.[5] These large-format holograms represented the state of the art: they incorporated two separate images (by employing two separate reference beam angles) which could be viewed independently; their size was the maximum then achieved; and the recording conditions were scrupulously controlled (the cloth upholstery of the car model demanded absolutely no air currents during the long exposures). For reconstruction, sequenced lighting replaced one image with the other and then with illumination to reveal an empty box. Nevertheless, the impact of the display was disappointing. Most passers-by did not pause long enough to realize that the arrangement was anything more than a glass case displaying a model car.

After this experience, the Conductron team insisted that their commercial products be freestanding holograms, not images contained within a box. The viewers' interaction with holograms—and so their perception of the illusion—would be promoted by generating real image or image-plane holograms, in which all or some of the image appears in front of the hologram plate. The action of seeking, and failing to touch, this ghost-like apparition retained the interest of viewers. Even so, it remained difficult to capture viewers' attention in the first place.

The marketing attempts made by Conductron explored still other techniques. In 1969 the company offered the advertisers and the public a five-dollar Conductron hologram kit, including two reflection holograms, a penlight to view them, and an explanatory article.[6] The following year their holograms were displayed at a Laser Ball at the Detroit Institute of Arts.[7] As noted earlier, Conductron also promoted holograms for art, and

[4] Charnetski, Clark to SFJ, interview, 3 Sep. 2003, Ann Arbor, MI, SFJ collection.

[5] Wilfong, Joan, 'Hologram product display announced', *Conductron antenna* 2 (Dec. 1968) 4.

[6] Wilfong, Joan, 'Holography use increases', *Conductron antenna* 3 (Dec. 1969) 4.

[7] Wilfong, Joan, 'Conductron—Ann Arbor laser photography is "Belle of Ball" ', *Conductron antenna* 4 (1970): 7.

collaborated with artists Richard Wilt and Bruce Nauman. But Charnetski noted that most artists who approached the company were dissuaded by the high costs involved and the technical limitations.

Efforts continued after the move of the Conductron holography facility to McDonnell Douglas Missouri, despite a more ponderous management structure there. In 1972, for example, after Bob Schinella joined the marketing team and became interested in marketing holograms, he oversaw the production of *Hand and Jewels*, a hologram 'depicting a woman's outstretched hand wearing a stunning diamond ring and holding a diamond bracelet'.[8] The bejewelled hand was displayed in the window of Cartier's New York Fifth Avenue shop, illuminated by a mercury arc lamp to seemingly project the real image 14 in. beyond the window (see Figure 12.1). While the hologram briefly attracted small crowds, most of the attention came from advertisers curious about the new medium, and investors anticipating new growth markets.[9]

Like Conductron before it, McDonnell Douglas Missouri dabbled in mass production, manufacturing some one hundred thousand holograms for the Quaker Oats Company, which promoted its King Vitaman cereal in 1972 with a reflection hologram recorded on film and mounted in a plastic finger ring. The cereal promotion did not perceptibly boost sales and was not repeated.[10] Customers—adults and children—remained difficult to find and to convince. After three years of promotion, M-D withdrew from the unpromising market.

The efforts of Conductron and its successor, McDonnell Douglas Missouri, to promote transmission, reflection, image plane, and pulsed holograms for advertising between 1965 and 1973 sank without a trace. Other hologram producers, and other investors, picked up where they had left off, motivated by improvements in the technology and inspired by a fresh crop of promoters. Scarcely four years after the company closed its holography division, the *San Diego Union* reported that hologram marketers were testing the water with white-light holograms. In a southern California shopping mall, Sears Roebuck & Co showed holograms to test market their use in advertising. Multiplex holograms appeared the most promising, as they were visible from multiple angles, and raised interest by the rainbow colour and animation of their images, still an unfamiliar novelty to the public.[11] A handful of advertisers in major American industries—principally for cigarettes, hamburgers, and health products—investigated the possibilities of holograms in marketing anew during the late 1970s, once rainbow holograms had made lighting less critical.[12] Yet such test marketing revealed that shoppers rapidly tired of the images, expecting new examples on a weekly, rather than monthly, basis.

[8] 'A touch of the future', *Display World*, Jan. 1973: 3.

[9] Goldberg, Shoshanah, 'Conductron and McDonnell Douglas', *holosphere* 15 (1987): 16–8.

[10] Kontnik, Lewis T., 'Commercial holography: a decade of emergence', *holosphere* 14 (1986): 12–6. The film holograms were produced by McDonnell Douglas Missouri via the intermediary Selwyn Lissack. There are no contemporary accounts of customer reaction and only a handful of the plastic rings can be traced today.

[11] 'Shoppers get the picture: Hologram marketers test the water', *Union*, 20 Jan. 1977: 2.

[12] Anon., 'National advertisers test display holograms throughout the country', *holosphere* 6 (1977). The firms included Proctor & Gamble, McDonalds, and R. J. Reynolds Tobacco Corp. Multiplex-type holograms

Fig. 12.1. An 18 × 24 in. pulsed ruby laser hologram for the Cartier New York shop window by McDonnell Douglas Missouri, 1972 (Boeing Inc; MIT Museum MoH collection).

Aggravating the difficulties of selling holograms for advertising purposes, over-eager and naïve promoters again widely misrepresented their capabilities. Citing Multiplex holograms, the *Red Bluff News* in California exaggerated:

> It's done with a process called holography—laser photography. The same process can put a 10-foot tall gyrating rock singer beside a highway or hanging in space over an intersection. It can put a three dimensional cowboy over a toy counter. It would be animated by the viewer walking by.[13]

Such expectations, piqued during the 1960s and peaking through the 1970s, continued to fill newspaper stories despite the earlier doubts of government and corporate sponsors about the likelihood of identifying commercially viable applications and markets.

The reality of commercial holography was that it remained a field strongly supported by government programmes until the mid-1980s.[14] By that time, the most important commercial market was not display holograms for popular advertising, but holographic optical elements (HOEs) for two niche applications: Head Up Displays (HUDs), which

were then being investigated by several major companies. For example, Westwood Pharmaceuticals displayed a scene of models of a steroid molecule, and scales on a human hand, at a dermatology meeting in 1977; General Tire showed a hologram of a woman swinging a tennis racquet at a Sportsman's show in 1978; ICI Americas sponsored Jason Sapan in 1978 to record a hologram showing the effects of rodent-killing chemicals on a rat and mouse at a National Pest Control Association meeting; and, Revlon Cosmetics commissioned Lon Moore in 1980 to create a stereogram for a lipstick advertisement (which was not used). See James, Dorothy H., *Holographic Illustration: A Handbook*, Master of Fine Arts thesis, Fine Arts, Syracuse University (1983).

[13] 'Holography-Three Dimension Photos', *Red Bluff News*, 20 Aug. 1976.

[14] See, for example, Kontnik, Lewis T., 'Governments underwrite holography industry', *Holography News* 1 (1987): 1. For 1987, he identified 34 projects accounting for $7.2 million, mainly with Kaiser Optical Systems, which had contracts to manufacture HUDs for the F16, F15, F14, and A6 fighter aircraft, as well as the Italian Airmachi. During 1980, on the other hand, Bosco, Mary C., *What Ever Happened to Holography?*, MA thesis, John F. Kennedy School of Government, Harvard (1981) found, based on data from the Smithsonian Science Information Exchange, that total American government funding for holography projects was $16–22 million for 177 projects of which 105 were under the DoD and twenty-three from other agencies.

combined a view of cockpit instruments with a direct view out the window of a military aircraft,[15] and holographic scanners for supermarket check-outs, which rapidly swept a laser beam across a product package to read the barcode.[16]

In Britain, for example, Marconi and Pilkington, heavily supported by military contracts, independently developed HOE applications for aircraft displays. Thereafter, as military expenditure fell at the end of the cold war, companies that initially had explored holography for military sponsors gradually turned their attention to commercial markets. Pilkington PE, for instance, which had produced a production facility to concentrate on the production of HUDs for fighter aircraft, sought to fill in the troughs of the production cycles by marketing commercially oriented products such as display holograms and holographic awards. The two applications were difficult to support, though: military-spec production teams coexisted uncomfortably with table-top consumer items. Similarly, Kaiser Optical Systems in Ann Arbor moved from the production of HUDs as a major income stream to the production of holographic optical elements for commercial Raman spectrometers during the 1990s, a development that transformed the instruments' technical specifications and engineering elegance.[17]

Thus, the markets for holography changed. The most important applications envisaged for holography during its first decade had been conceived on the grand scale: corporate visions of information storage and retrieval, information processing (particularly image filtering and recognition),[18] and improved imaging and communication. Among the most long-lived and widely anticipated application of holography was information storage, which was expected to be compact and virtually instantaneous to read. From the late 1960s onwards, the development of holographic memories was announced regularly, but nevertheless had not been introduced commercially at the end of the century.[19] Holographic cinema, television or video, on the other hand, were also widely predicted both in specialized journals and in the popular press, but widely judged to be unfeasible by the mid-1970s. Like holographic memories, though, a variety of schemes were touted as a means of securing venture capital or entertainment headlines.[20] Less familiar to

[15] For example, Lindvold, Lars, 'New uses for HOEs', *Holographics International* 1 (1988): 33.

[16] See, for example, 'IBM supermarket scanner uses holographic optics', *holosphere* 9 (1980): 1–2.

[17] Leith, Emmett N., 'The evolution of information optics', *IEEE Journal of Selected Topics in Quantum Electronics* 6 (2000): 1297–304; Owen, Harry to SFJ, interview, 9 Sep. 2003, Ann Arbor, MI, SFJ collection.

[18] For example, Winzer, G. and N. Douklias, 'Improved holographic matched filter systems for pattern recognition using a correlation method', *Optics and Laser Technology* 4 (1972): 222–7.

[19] For example, of this longstanding interest include Weitzman, C., 'Optical technologies for future computer system design', *Computer Design* 9 (1970): 169–75; Johnson, C. and E. Briggs, 'Holography as applied to information storage and retrieval systems', *Journal of the American Society for Information Science* 22 (1971): 187–92; Frateschi, G. and E. Gatier, 'Gigamemories based on recyclable holograms', *L'Antenna* 45 (1973): 446–52; Lang, M. and H. Eschler, 'Gigabyte capacities for holographic memories', *Optics and Laser Technology* 6 (1974): 219–24; Mikaelyan, A. L., 'The present state and future prospects for the development of optical memory systems', *Soviet Journal of Communications Technology & Electronics* 35 (1990): 21–39.

[20] The range of publications is suggested by Mariotti, D., '3D television', *Radio Industria* 36 (1969): 33–4; Jacobson, A. D., V. Evtuhov, and J. K. Neeland, 'Motion picture holography', *IEEE Journal of Quantum Electronics* QE-5 (1969): 334–5; Youngblood, Gene, *Expanded Cinema* (London: Studio Vista, 1970); Gurevich, S. B., 'Holographic television future', *Tekhnika Kino i Televideniya*. 2 (1971): 67–9; Leith, Emmett N., D. B.

non-specialists were the engineering applications of mechanical testing, which proved to be a growing and durable market during the 1970s.[21]

Yet none of these applications had generated the anticipated sales. One market survey of the mid-1970s, based on forty American companies and organizations, suggested that the four expected growth areas (holographic non-destructive testing (HNDT), optical memories, holographic optical elements, and display holography) would amount to scarcely $2–4 million and that, by 1980, the collective market opportunities would be annually in the $10–25 million range, or barely the income of a single moderate-sized firm.[22]

Nevertheless, holography attracted an unusually wide assortment of proponents and potential investors to colonize what they saw as virgin territory. Expectations of growth and profit, whether expressed publicly or not, united the various communities of holographers. Engineering applications and the consequences for increased export trade motivated metrologists at national laboratories. The Bay Area holographers, catalysed by Lloyd Cross and Jerry Pethick, dreamed of an explosion of public interest in their holograms and corresponding sales. The Multiplex Company and its cylindrical holograms dominated the small market during the mid-1970s, although it proved too disorganized to pursue its commercial lead or even to preserve a clear account of its accomplishments.[23] But, as corporate dreams of commercial holography were declining, a sparse network of commercial or semi-commercial galleries, museums, and sales outlets grew to support a small hologram production industry that anticipated expansion. They included Laser Light Concepts in New York and Third Eye in Paris (1976); Fine Arts Research in Chicago (1977); The Hologram Place in London (1978); the Museum für Holographie und Neue Visuelle Medien near Cologne; and the Holos Gallery in San Francisco (1979). By the late 1970s, then, a cottage industry was developing and seemed ripe for expansion.

Brumm, and S. S. H. Hsiao, 'Holographic cinematography', *Applied Optics* 11 (1972): 2016–23; Denisyuk, Yu N., 'Holography motion pictures', *Soviet Physics—Technical Physics* 18 (1974): 1549–51; Komar, V. G., 'Possibility of creating a theatre holographic cinematograph with 3-D colour image', *Tekhnika Kino i Televideniya* 4 (1975): 31–9; Fujio, T., 'Optical communication and image technology, the new future', *Journal of the Institute of Television Engineers of Japan* 32 (1978): 349–54; Domanski, G., 'Three-dimensional television', *Funkschau*, (25–26), 60–4; Ostria, V., 'The Beginnings of Cinematic Holography', *Cahiers Du Cinema* (1982): R11; Remis, X., 'Holography and the Cinema', *Revue Du Cinema* 343 (1983): 102–3; Halas, John, *Graphics in Motion: From the Special Effects Film to Holographics* (New York: Van Nostrand Reinhold, 1984); Salmon, Pippa and Kamala Sen, 'Holography in Hollywood?' *Holographics International* 1 (1988): 26–7; Kompanets, I. N. and A. P. Yakimovich, 'Three-dimensional television (a review)', *Telecommunications and Radio Engineering, Part 2 (Radio Engineering)* 46 (1991): 40–7; Palais, J. C. and M. E. Miller, 'Holographic movies', *Optical Engineering* 35 (1996): 2578–82; Winston, Brian, *Technologies of Seeing: Photography, Cinematography and Television* (London: British Film Institute, 1996).

[21] See, for example, Ennos, A. E., 'Holographic techniques in engineering metrology', *Proceedings of the Institution of Mechanical Engineers* 183 (1969): 5–12; Butters, J. N., 'Measurement techniques using laser holography', *Electronics and Power. Journal of the Institution of Electrical Engineers* 20 (1974): 585–8.

[22] Harvey, Thomas W., 'Holotron reports results of holography market survey', *holosphere* 4 (1975): 1–3.

[23] The holographic stereogram nevertheless was widely publicized during the late 1970s. Multiplex claimed in 1978 to have one system capable of producing 2,500 holograms in 8 hrs. The same year, the Holographic Arts Co. in Chicago claimed a system capable of producing 1,800 per day, with advertisers paying $25 to $135 per day (Schmidt, David to P. Jackson, 1978, MIT Museum 40-1209: holography advertising). Standard commercial versions were widely available in the gift shops of science centres and museums.

12.2 ENTREPRENEURS AND COTTAGE INDUSTRY

Unlike accounts of the slow research, investment and marketing cycles of corporations, tracing a path through holography's small firms is chaotic. The story, repeated city by city and country by country, is one of incestuous relationships between investors, technologists, and promoters. I will evoke the flavour of this period, if not its multiple threads, by focusing on the careers and products of some of the individuals who maintained an enduring association with the field.

Even during the 1960s, when corporations were lavish in exploratory research and development, small firms were setting up to exploit holography's commercial potential. As discussed in earlier chapters, several of the scientists, engineers, and technicians who explored holography in Ann Arbor had become entrepreneurs, defecting to new start-up firms: Jodon Engineering, GC Optronics, and Photo-Technical Research, for example. Most holographers, at most periods, proved to be entrepreneurs of sorts, spinning-off companies, then schools and museums, and eventually hologram shops. Visions of holography as a profit-making activity even encouraged franchisers: one entrepreneur of the 1970s sought to sell hologram franchises to would-be local producers.[24]

The same was true in other centres, notably Stanford University and firms in the San Francisco Bay Area. At Stanford, Matt Lehmann and J. W. Goodman, who had pursued studies of holographic compensation of distorting atmospheres, for example, later consulted on holographic storage of information for credit card verification. Lehmann also became involved with the Santa Clara Company Creative Imagery, which tried to market the production of holograms of art works for travelling exhibits, a field pursued with greater success in the Soviet Union from the late 1970s. Lehmann outfitted a holographic laboratory in a trailer truck, intending to visit museum collections for on-site recording of holograms.[25] He also consulted with advertising firms interested in promoting holograms for use in domains as diverse as displays for jewellery shops or stage plays.[26]

By the mid-1970s, some forty small firms were selling holographic products, but none had significant sales. One of the more successful entrepreneurs was Richard Rallison who, having built light show equipment while in college during the late 1960s, also began experimenting with holograms, and then went to work at Hughes Aircraft in 1973, where he designed and built lasers. He also began to specialize in dichromated gelatin (DCG)

[24] 'The most exciting photographic medium since TV is reaching out for you!', trumpeted an advertisement in the May 1976 issue of *Industrial Photography*, described in Anon., 'Entrepreneurs urged to set up holographic centers', *holosphere* 5 (1976): 4, 8.

[25] The same application has been pursued more recently by Yves Gentet, whose full-colour reflection holograms are well suited to museum objects.

[26] Lindgren, Nilo, 'Search for a holography market', *Innovations* (1969): 16–27. *Theatre Design & Technology* 14 (1978) devoted an entire issue to holography and its potential uses as a display medium on theatre sets. The *Detroit Free Press* reported in 1980, for example, that a Czech production of Macbeth projected a holographic real image onto clouds of steam.

holograms or 'dichromates', an appealingly crystal clear and bright hologram.[27] By the late 1970s, after moving back to Utah, he was specializing in the design of optical scanning systems based on rotating DCG hologram disks. He began producing holographic pendants in 1977, supplying dichromate products to Holex Corporation, in Norristown, Pennsylvania, and the Holotron Corporation in Richland, Washington.[28]

His pendant products proved so popular that Rallison's Fire Diamond Company became a division of Holotron in 1978 and a template for its commercial liaisons. Rallison continued to sell dichromate products through his original company, Electric Umbrella, which changed name in 1979 to the more respectable International Dichromates, and again to Dikrotek in 1984. After selling the firm, it spun off Krystal Holographics, which later became Dupont Holographics. Along the way, Rallison was involved in setting up DCG production for a number of other firms, including Raven Holographics in London, UK, and firms in Minneapolis, Kansas City, and Seattle.[29]

Another seminal figure during this period was Steve McGrew. While completing an MS degree in physics at the University of Washington, McGrew (b. 1945) pursued interests in electrical engineering and became intrigued by optics. When teaching a physics laboratory class for non-science majors during his first year of graduate school there in 1970, he helped them make holograms as a class project. He recollections, decades later, hint at the artisanal, inspirational, and subversive aspects of holography that were uniting the Bay Area coterie of practitioners eight hundred miles to the south:

> I was attracted to holography because it seemed so clear and understandable that it bordered on obvious; and at the same time it seemed like magic. I loved the garish colors of holograms. I loved the mathematics that described what a hologram is and does. I loved the *accessibility* of holography: if you could afford a few hundred dollars for a HeNe laser and some holographic plates, and if you were willing to learn to use a hacksaw and a drill and build components out of PVC pipes, scrap metal and a few precious lenses and mirrors purchased from the Edmund Scientific catalog, then you could become a holographer. I loved the fact that even *non*-technical people were attracted to holography and, in the process of building their own equipment, *became* expert in things that university physics professors perhaps should have known (but didn't).[30]

To make the class hologram, he consulted available textbooks, and improvised equipment by using a piece of marble on an inner tube and mirror mounts constructed out of plywood and screws. The result of McGrew's exposure to this new kind of hands-on optics convinced him that the field was not 'encompassed in the lens-maker's formula'.[31] Over the next couple of years, he conceived an optical analogue computer to allow him to

[27] On DCG, see §7.10.

[28] Holex, set up in 1973 by Larry and Jerry Goldberg, initially manufactured silver-halide holograms sold through the Edmund Scientific Company, a major mail-order supplier in Barrington, New Jersey.

[29] Production of dichromate holograms demands low-humidity conditions and several of the American laboratories were founded in the dry climate of Utah and Colorado.

[30] McGrew, Steve, 'My reflections on holography', unpublished manuscript, 5 Dec. 2004, SFJ collection.

[31] The formula expresses the relationship between the focal length of a lens and the image and object positions, and is part of the geometrical optical approximation that holographers disparage.

two-dimensional calculations to find solutions to field equations.[32] And having seen a multiplex hologram in Seattle, he spent a few weeks at the San Francisco School of Holography and learned more advanced techniques, 'soaking up everything that Michael Kan would teach me'.[33] In 1975 he formed Europlex Holographics B. V. in Amsterdam with Barrie Boulton and two investors, making stereograms, rainbow holograms, and DCG holograms for what proved to be a non-existent market. McGrew recalls selling only a handful of holograms, despite filming several scenes to be used in the multiplex process. The company folded about a year later.[34]

McGrew joined the Holex Corporation (Norristown, PA) in early 1977 where, like Richard Rallison, he developed processes for DCG hologram production.[35] Because he was using ultraviolet (UV) light to cure optical cement for hologram cover glasses (to protect the hygroscopic DCG), McGrew conceived the idea of a UV embossing process. He also explored other technologies of hologram recording, including thermal embossing processes, photopolymer materials, and tanning bleaches for silver-halide emulsions.[36]

When Rallison's Fire Diamond Company was sold to Holotron, McGrew again switched companies and developed a method of mastering DCG reflection holograms there that was compatible with their equipment. Holotron moved from Washington State down to Sunnyvale, California, in 1979, and McGrew, as its Head of Research and Development, hired Steve Provence as assistant.

Provence (b. 1954) had met McGrew when visiting Europlex in Amsterdam. The first hologram he had seen was a Conductron reflection piece brought back by his father from a convention in 1968. Provence had studied holography in New York in 1975, at what was then the interconnected New York School of Holography—Museum of Holography—New York Art Alliance. A year later he had joined the Multiplex Company in San Francisco.[37] At Holotron in Sunnyvale, McGrew and Provence worked in the back of a room of a games company, where pinball machines were being assembled, with McGrew developing better methods of using photoresist techniques to produce hologram masters based on surface relief, with the intention of making hologram copies by embossing. After only a few months, though, Holotron abandoned its holography plans. The move

[32] With it, McGrew hoped to study a theoretical expression for the structure of elementary particles, using 'a 3-dimensional analog computer (2 spatial dimensions plus time) using coherent light, solving my equations by generating a movie of the solution' (McGrew, Steve, 'My reflections on holography', unpublished manuscript, 5 Dec. 2004, SFJ collection). His scientific sources included a decade's worth of *Applied Optics* and the seminal text Preston, Kendall, *Coherent Optical Computers* (New York: McGraw-Hill, 1972).

[33] McGrew, Steve to SFJ, email, 28 Oct. 2004, SFJ collection.

[34] McGrew, Steve, 'Steve McGrew comments', unpublished notes, 26 Oct. 2004, SFJ collection.

[35] Holex was incorporated in 1972 by brothers Larry and Gerry Goldberg as one of the first firms focusing on educational and hobby markets for holography. Like nearly all the small holography firms of the period, it eventually manufactured and sold custom and standard white-light, transmission, reflection, dichromate, and integral (stereogram) holograms, but foundered and closed a decade later.

[36] Ultraviolet (UV) embossing selectively hardens exposed parts of the material (often a photoresist or photopolymer), allowing the unexposed material to be removed and to leave a thickness-modulated surface corresponding to the hologram fringe pattern that can then selectively delay light to reconstruct the holographic image. Thermal embossing impresses a pattern into a transparent material softened by heat. A tanning bleach for silver-halide emulsions produces a similar result by altering the emulsion thickness.

[37] Provence, Steve to J. Ross, interview, 1980, Los Angeles, Ross collection.

to Sunnyvale, which had carried along only half the staff, ruptured the company; within months it was bankrupt. McGrew, ever adaptable, set up a laboratory in his garage to improve his DCG and thermal embossing processes, and Provence took a job with Spectra-Physics as a laser technician.

McGrew continued to develop and disseminate technology. Using borrowed money and the income from DCG hologram sales to buy a HeCd laser (and other holography equipment at low prices from the bankruptcy sales of Holotron/Holosonics and from Holex), he set up a clean room and better facilities. By 1980 he envisaged embossed rainbow holograms cheap enough to use everywhere—as a transparent product stuck on a car window, or as greeting cards and book jackets if treated with reflective coating.[38]

Because his method of embossing holograms had technical similarities with methods used to create replica diffraction gratings, he contacted the Diffraction Company in Baltimore, Maryland, which had been producing such gratings for several years, and began supplying them with photoresist master holograms. He approached Atari to suggest applications for games and, when he discovered that Atari had already been discussing possibilities with another company, Eidetic Images, he joined the firm briefly as a consultant.

In common with a handful of other firms of the late 1970s, Eidetic Images Inc., led by Ken Haines, had developed an improved version of Lloyd Cross's cylindrical multiplex hologram. The company had also been investigating how such holograms could be replicated inexpensively, and had developed a pressure-moulding process. McGrew decided to pursue embossed holograms as a commercial venture, and named his garage company Light Impressions Inc. He formed Light Impressions Europe in 1981 with John Brown. As described at greater length below, both McGrew and Haines were key figures in the commercial development of embossed holography and its patent battles.

Besides producers of holograms, exhibitors and sellers increasingly appeared. By the late 1970s short courses, gallery exhibitions with gift shops, and other organizations and sales outlets were becoming more visible to the public, but nearly all had short, troubled lives. Perhaps the best example is the first hologram shop on the American west coast, the Holos Gallery in San Francisco. It was launched by Gary Zellerbach in February 1979, intending to take advantage of the Bay area holography culture spawned by the School of Holography, the Multiplex Company, and its associated holographers, arguably the most widely based community anywhere at that time. After university in Boston, Zellerbach had played guitar in Los Angeles but, with an interest in science and art and seeing his first hologram there, was drawn to holograms as a subject for a dedicated art gallery. The grandson of Harold Zellerbach, a prominent San Francisco art patron, he hoped to sell a range of holograms ranging from signed artworks to inexpensive trinkets.[39]

Zellerbach took courses from Lloyd Cross and from Fred Unterseher and met a wide range of holographers who visited his gallery. But despite a good location on Haight

[38] McGrew, Steve to J. Ross, interview, Dec. 1980, Los Angeles, Ross collection; McGrew, Steve to SFJ, email, 28 Oct. 2004, SFJ collection.

[39] 'It's Spring, and holo galleries are popping up from coast to coast', *holosphere*, 8 (3) (Mar. 1979).

Street, frequent newspaper coverage and the support of local holographers, Zellerbach found selling holograms difficult. During the early 1980s, some purchasers acquired holograms as investments, but without a reliable means of gauging the perceived value:

> There were no auction records as a basis to compare and justify prices. There were precious few reviews and critiques to help establish artistic merit and intent, and most of the criticism that did exist was quite negative anyway. There was no historical basis for predicting future increases in value for those concerned with the appreciation of their 'investment,' and there was no perceptible resale market that could offer potential customers any type of liquidity.[40]

He felt, too, that the there was too strong a reliance on the novelty of the imagery, and too little on its aesthetic content.

> People go into holography and think they're going to get rich quick at it. Then the reality sets in which is that, for some reason, it's slow going and it's hard to convince the public of the worth of what you have. I see it as being a continual educational process for a few more years [. . .] For now, it's really hard to sell something when 50% of the people walking in the door don't know what it is, have never seen it or have never even heard of it. So the first step towards the future is a little more education.[41]

Like other entrepreneurs of the period, Zellerbach found himself drawn into educational activities, joining with other Bay Area practitioners to form the Laser Arts Society for Education and Research (L.A.S.E.R.).[42] He had conceived the gallery as an outlet for art holograms, but found that there was little or no market for holographic fine art. The company instead evolved into the largest American distributor of 'holo trinkets, mostly watches, pendants, small framed or matted holograms'.[43]

To further develop the sales volume he needed, Zellerbach's company opened up a wholesale market by targeting science museums. The Smithsonian eventually became his largest customer. The firm also developed lucrative distribution agreements for Japan and Australia. This market grew, largely owing to novelty and little competition, but was becoming saturated by the late 1980s. The company was drawn into the distribution of a wider range of inexpensive novelty goods, but only rare sales of art holograms. By then, sales at Holos peaked at about $1million per year, with probably less than $10,000 contributed by the sale of art holograms. Zellerbach's products were forced to follow a downwards spiral of lower prices and kitsch appeal that he had eschewed a decade earlier.

[40] Zellerbach, Gary A., 'Selling holographic art: an analysis of past, present and future markets', *Proceedings of the SPIE The International Society for Optical Engineering* 1600 (1991): 446–54. This aim of making holograms into collectable art was shared with holographer Larry Lieberman and his business partner Frank Millman in Miami, who in 1995 undertook the 'C Project', producing large-format holograms by some twenty-one well-known artists to break down barriers for holography exhibitions in important galleries and museums. However, given artists' disagreements and the commercial, rather than aesthetic orientation, a group show was never mounted and most of the limited-edition holograms were never reproduced or sold, sending the project into bankruptcy (Maline, Sarah, 'State of the art', *Interferenzen*, 7 (4), (1997): 6–7; Munday, Rob to SFJ, interview, 31 Mar. 2004, Richmond, UK, SFJ collection).

[41] Zellerbach, Gary to J. Ross, interview, 1980, Los Angeles, Ross collection.

[42] For more on L.A.S.E.R., see §11.2.

[43] Zellerbach, Gary to SFJ, email, 9 Aug. 2003, SFJ collection.

The Holos Gallery was sold to A. H. Prismatics (Brighton, England) in 1990, which had operated the gift shop of the New York Museum of Holography. While this gave them the principal mail-order outlet for gift holograms in America, A. H. Prismatics, also finding that the novelty market was declining, closed Holos Galleries in San Francisco soon after its purchase and pulled out of the business entirely during the mid-1990s.[44] Zellerbach subsequently focused on a manufacturing firm, The DZ Company, co-owned with Dan Cifelli. They manufactured a range of novelty products including the Dazer—a 4 in. diameter flat metal disc with a dimple in the middle and covered with embossed holographic foil. When spun, the disc reflected a rapidly changing pattern of colour and shape. Some three million were sold, but the easily copied product led once more to rapidly declining sales. The DZ Company also marketed holographic bookmarks and an increasing fraction of even less holographically inspired products such as rainbow sparklers (a wand with slit foil on the end, to be waved in the air). The DZ Company was acquired by US Holographics in 1992, but within two years it, too, was bankrupt. Zellerbach revived an online version of the Holos Gallery in 1994, as the Internet was just entering public consciousness. Over the following decade, however, the website generated only a single sale of an art hologram for $500.[45]

In Britain, too, a wave of hologram companies was founded during this period, triggered by the successful *Light Fantastic* exhibitions of 1977–8 at the Royal Academy of Arts, London. The company behind *Light Fantastic* was Holoco, founded in the early 1970s by John Wolff, Nick Phillips of Loughborough University of Technology, and Anton Furst.[46] The rock group The Who, via their lighting and effects manager Wolff, supported the fledgling company by supplying lasers that had been used in their laser light shows during concerts. Eve Ritscher, who later organized the 1983 *Light Dimensions* holography exhibition at the Royal Photographic Society in Bath and the Science Museum, London, had previously done promotion work for The Who. The group withdrew its financial backing in 1980, and the company was reformed as Advanced Holographics Ltd in 1981, employing Phillips and two Loughborough University colleagues, D.W. Tomkins and L. Holden. When other firms joined the market, Advanced Holographics eventually became part of Markem Systems, a European manufacturer of stamped-foil embossed holograms, to supply hologram masters.[47]

[44] One news item described the company as the world leader in the small field of holographic gift items ('A. H. Prismatics', *Holography News*, 7 (1), (1993): 1).

[45] Zellerbach, Gary to SFJ, email, 2 Aug. 2003, SFJ collection.

[46] Nicholas Phillips (b. 1933) studied physics at Imperial College, and then worked in nuclear fusion research until the early 1960s. He joined Loughborough University in 1965 and rose to Professor of Applied Optics there in 1991. Phillips was also Senior Visiting Tutor in Holography at the Royal College of Art from 1986, and later Professor of Imaging Science at de Montfort University in Leicester. He was awarded the Institute of Physics Thomas Young Medal and Prize in recognition of contributions to holography in 1981, particularly the development of high-quality holograms for visual display. Anton Furst (1944–1991), having graduated from the Royal College of Art in 1969, worked in laser special effects, and later set design and art direction for major Hollywood films.

[47] Saxby, Graham, 'Display holography in Britain: 1991', presented at *International Symposium on Display Holography*, Lake Forest, Illinois, 1991.

The Holoco holograms shown at *Light Fantastic* unleashed a wave of interest in British holography. Paralleling Zellerbach's Holos Gallery experiences, the Light Fantastic gallery was founded by Peter Woodd (the accountant for Holoco), at Covent Garden, London, in 1981, supporting it with a production facility in Leicestershire from 1989.[48] The firm also opened a hologram attraction in the Trocadero at Piccadilly Circus in 1983 but closed five years later when the lease was purchased, opening a production facility in Leicestershire in 1989. The Covent Garden gallery closed in 1991, and the firm abandoned exhibitions to concentrate on sales of embossed holograms, which included holographic trading cards and book stickers.

Holography firms in Britain, America, France, Germany, Holland, and Sweden struggled to generate income with small investment capital, promising technical ideas and untested markets. The international intermingling of creative individuals, schemes, and companies can be suggested by tracing the career of Jeff Blyth. Born in 1938, Blyth obtained a degree in applied chemistry from Brighton Polytechnic and then worked during the early 1970s for two firms making dry transfer lettering, which was then based on pigmented dichromated gelatin. He found himself enchanted by holograms via a cluster of events: the *Light Fantastic* exhibition of 1977; a laser light show presented in the Victoria Palace Cinema, London; and the dichromate holograms made by Richard Rallison, and sold there by John Brown and Andy Harris. Convinced that his experience with dichromated gelatin would allow him to develop a process for mass-producing holograms, he tried recording one using the Victoria Palace laser with Brown and Harris after the last show one night—without luck.

Spurred on by his failure, Blyth began to collaborate with the Medical Physics department of the Royal Sussex Hospital in Brighton. Two medical physicists there, Brian Keane and Laurie Wright, had been developing a technique to produce a holographic stereogram from individual CAT scan images. While Blyth learned more about holographic processes, his chemistry experience helped Keane and Wright improve the brightness of their images. Blyth's graphic arts background also provided the idea of recording a reflection hologram by scanning the laser beam rather like a photocopier. The advantage was that the beam could be more intense and the exposure shorter than if a fanned-out laser beam were used. Together, these techniques allowed the creation of a master reflection hologram that could be copied straightforwardly.

In 1977 John Brown formed Holographic Developments Ltd, and Blyth joined as Technical Director. Via Brown, he met other entrepreneur/technologists such as Steve McGrew and, travelling to America in early 1979, a number of holographers in Boston and New York. Blyth later joined Laser Point, a Cambridge firm, where he set up a production system for his reflection master technique. Although his holograms were sold via the New York Museum of Holography shop and others, income was low. The owners of one hologram outlet, the Parallax boutique in Covent Garden that opened in 1979,

[48] Saxby, Graham, 'British holography at Light Fantastic', *British Journal of Photography* (1982): 476–7.

became interested in Blyth's manufacturing process and, obtaining venture capital from the City, were able to found Applied Holographics four years later.[49]

Blyth himself moved to another small firm, Hollusions, formed in 1980 with Gil Searl, Ray Mumford, and David Pizzanelli to produce silver-halide holograms for customers.[50] It lasted scarcely a year, and wiped out Searl's savings. Blyth and Pizzanelli, however, remained involved in holography. Blyth taught holography classes at the Royal College of Art in London during 1981 and 1982, and then joined another small firm, Third Dimension Ltd, founded by Ian Lancaster. Lancaster hired a small factory unit in West London where Blyth once more set up the contact copying system for reflection holograms, but again the returns from sales proved unsustainable. Lancaster later joined the New York Museum of Holography as Director, while Blyth pursued ideas for reflection holograms as anti-forgery devices, without commercial success, and new formulations of silver-halide plates that contributed to the basis of a commercial product, BB plates.[51]

For his part, David Pizzanelli first got into holography in 1979, and saw one of Steve McGrew's early embossed holograms in 1981. He made contact with a Royston company, Pat Centre (later PA Technologies), which was then developing laser discs. He and holographer Nigel Abraham, who had collaborated with Hollusions, spent a weekend at the Centre making a rainbow hologram using the firm's photoresist technology, absorbing in a few hours what had taken others months to develop.

After wasting several months on another attempt to create a hologram company in Sunningdale, Gloucestershire, they set up a laboratory in the house of Jonathan Ross in Clapham. Ross been drawn to holography in 1978, following the *Light Fantastic 2* show in London. They founded SEE3 to produce multiplex holograms and had survived on a handful of commissions and a sales outlet, The Hologram Place, in north London, to sell multiplexes and dichromate pendants.

Beginning with information they had gained about methods of photoresist, electroforming, and embossing and with searches of the published literature, Abraham and Pizzanelli developed a method of creating embossed holograms from silver-halide masters: they discovered that by swelling emulsions and casting from them, and then metallizing the casting to make a nickel shim, good quality embossed holograms could be produced. By 1983, they had succeeded in producing hologram masters in Clapham

[49] Blyth, Jeff to SFJ, letter, 7 May 2005, SFJ collection.

[50] David Pizzanelli had been completing an MA in Literature at Cambridge, but had a background in art and initially created solid models as holographic subjects. Gil Searle subsequently worked for ABNH in America (Pizzanelli, David to SFJ, email, 4 May 2005, SFJ collection).

[51] BB plates, an acronym for Birenheide / Blyth, were developed during the early 1990s by Richard Birenheide after a brief collaboration with Blyth, and produced from 1996 in a small factory some 50 miles from Frankfurt. He sold the technology to Mike Medora of Colourholographics in 2001, who was a former co-worker at Third Dimension and holography instructor at the RCA. Third Dimension went into receivership in 1990, and was purchased successively by Light Impressions Europe that year, Martin Richardson of The Holography Image Studio (T.H.I.S) in 1994, and the German firm Laser Trend in 1995.

and having them embossed as polyvinyl chloride (PVC) copies at another site. Pizzanelli stresses the commercial value of the technique:

> It was kept as secret as possible. To a large extent it still is. If holographers have developed a system or methodology that works well, they will go to great lengths to keep it to themselves [. . .] Although the basics are widely known, it's much like cooking—the basic ingredients of a baked Alaska may be known, but each master chef has their own secret method for making it.[52]

In 1984, Abraham began negotiations with Applied Holographics (see below) to sell them the SEE3 embossing technology, but instead joined Applied Holographics as employee. Now competing with the much larger firm using their techniques, SEE3 continued to produce embossed and silver-halide holograms for a variety of customers, but closed by 1990.[53]

Another British example of the collision of innovative technology with impervious markets is New Holographic Design Ltd, founded in 1982 by Michael Waller-Bridge. With others from the Physics Department of Imperial College, the firm produced holograms for sale and marketed an achromatic viewer for holograms. Developed by Kaveh Bazargan, the viewer allowed conventional full-parallax transmission holograms to be viewed with a white-light source by interposing a holographic screen to compensate for spectral dispersion. Reformed as Icon Holographics Ltd, the company struggled to interest other firms in its use as a point-of-sale display or viewer for educational holograms (producing, for example, an 'Evolution of Man' education pack of 11 in. square holograms and viewer). Adding to their business tribulations, their patent application was restrained for a time by a 'gagging order' from the UK Ministry of Defence.[54] Stephen Hart and Michael Dalton, two astronomers at Imperial, worked on the computer imaging aspects. By 1986, having explored schemes for producing integral holograms of synthesized tomographic images with little commercial success, Icon went into receivership, and Hart and Dalton bought the firm's intellectual assets, continuing development for two further years with funding from the UK Science and Engineering Research Council. In 1988 Dr Allan M. Wolfe, an American cardiologist, invested in the invention. The resulting company, Voxel, produced a holographic camera to go with the viewer, which multiplexed 200 MRI or CAT scan images to create a 14 × 17 in. three-dimensional image. However, the Voxcam and Voxgram (priced at about $100,000 and $1000, respectively, by 1992) found increasingly difficult competition from improving computer display technologies.

As these cases suggest, entrepreneurial activities multiplied during the 1980s with little market research. The American firm Reconnaissance, launched in 1984 by Lewis Kontnik

[52] Pizzanelli, David to SFJ, email, 4 May 2005, SFJ collection.

[53] Ross, Jonathan to SFJ, interview, 3 Apr. 2003, London, SFJ collection. Ross himself subsequently became a prominent collector of holograms, and Pizzanelli joined their original competitor Light Impressions, obtaining a PhD in 1994 from the RCA Holography Unit.

[54] Waller-Bridge, Michael to SFJ, internal reports and publicity documents, 9 Aug. 2005, SFJ collection; Kennard, A. H. W. and The Patent Office to New Holographic Design Ltd, letter, 22 Feb. 1983, London, Waller-Bridge collection; 'Voxel', *Holography News* 6 (1992): 8.

as Travels Bounty to market high-technology art and collectibles, further illustrates the rapid evolution of companies to chase an ephemeral international market. Kontnik initially focused on selling retail holograms via display carts in a Washington DC shopping mall. Attention from the press allowed the firm to reach a broader market and to begin representing artists and holographers.[55] Kontnik started Reconnaissance with the catch phrase 'Technology IS art, 'Technology AS art'. Meeting European holographers during the mid-1980s, he sold holograms to the Smithsonian Museum and other outlets. Kontnik joined forces with Ian Lancaster, who had represented the Gulbenkian Foundation in Britain, and was then operating Third Dimension in London.

Seeing the emergence of a market for holography, Kontnik started *Holography News* in 1987 to provide commercial information for investors and entrepreneurs, and was joined again by Lancaster after Directorship of the New York Museum of Holography. Their newsletter attracted some thirty-five subscribers in its first year and tripled after three or four years, 'just enough to keep it barely scraping along'.[56] By the late 1980s, the company refocused to provide not just market information, but conference venues for the commercial holography industry, too.[57] With the shift of focus of Reconnaissance to Lancaster in England, the company inaugurated the *Holopack-Holoprint* annual conferences on commercial holography in 1989. The original focus was intended to be holographic packaging and printing, but rapidly shifted to the burgeoning security and embossed hologram industry.

Artists, too, became entrepreneurs. For instance, while Margaret Benyon's early exploration of holography in Britain and Australia gained critical and popular attention, getting decent prices for her work was another matter. She returned to Britain in 1981 and set up a holography studio in her garage, confiding to Posy Jackson:

> Have been trying to make myself into a v. small scale production holographer, making 4"x5"s commercially available (in Britain only so far), although it's becoming obvious that this is going to be a fairly dodgy way to survive [. . .] I'm going to have difficulty in finding time to do any serious work, since this is taking all my time and I'm not getting much back so far.[58]

Other artists adopted a mixed economy, creating fine-art holograms for exhibition and sale, as well as teaching, and operating small shops and galleries. A striking number of them, given their art rather than science backgrounds, found employment as technical consultants and science educators. Fred Unterseher, for example, having been an early member of the San Francisco School of Holography, founded Holografix in the late 1970s and Zone Holografix in 1995 with Rebecca Deem. Along the way, he was employed

[55] Lawrence, Pamela J., 'Retailing holography in the U.S.,' *holosphere* 13 (3), (Summer 1985): 8–9. Kontnik studied math and chemistry at the University of Denver, and obtained a MS in Environmental Engineering from University of North Carolina. After obtaining a Law degree, he practiced law until 1984.

[56] Kontnik, Lewis T. to SFJ, interview, 20 Nov. 2003, Vancouver, SFJ collection.

[57] See, for example, Lancaster, I. M. and L. T. Kontnik, 'Market conditions for display holography in the USA and Europe', presented at *Practical Holography IV*, Los Angeles CA, 1990 and Lancaster, I. M. and L. T. Kontnik, 'Holography industry: market size and trends in commercial activity', presented at *Fifth International Symposium on Display Holography*, Lake Forest, Illinois, 1994.

[58] Benyon, Margaret to P. Jackson, letter, 10 May 1983, MIT Museum 26/587.

as a college instructor in Pasadena, California and in Columbia, Missouri, and consulted on holography and imaging projects for the Jet Propulsion Laboratory and NASA.

Such peripatetic careers were impelled both by professional invisibility and the lack of a viable market for fine-art holograms. A later example is that of Melissa Crenshaw who, after gaining a bachelors degrees in social work, later studied holography at the Fine Arts Research and Holographic Center in Chicago and obtained an arts degree in the early 1980s. She says of her contemporaries,

> There were no rigid standards then for who could be there and who couldn't be there, and I really liked that. Probably most of the people were there because of that. Not necessarily misfits, but outside the mainstream of arts, science, engineering and everything [. . .] People were pretty eclectic; like an artist who was a closet scientist, or a scientist who was a closet artist.[59]

Working subsequently for Chromagem in Chicago and The Holography Studio in Vancouver, Crenshaw collaborated with artist Sydney Dinsmore in Toronto for seven years, and subsequently found full-time employment as a designer of holographic optics for a Vancouver lighting manufacturer.[60]

Just as individual careers were unsettled, only a handful of firms survived, and few established before the mid-1980s grew substantially. Holographics North, for example, founded by physicist John Perry in 1984, began with a contract to produce large-format holograms for Pratt & Whitney.[61] Perry had earlier taught astronomy at Vermont University (see Figure 12.2), and moved into freelance photography and directorship of a gallery. His 30 × 40 in. laser transmission holograms of commercial aircraft engine models proved successful and orders for large format holograms continued to appear. Perry

Fig. 12.2. John Perry at the Living and Teaching Center, University of Vermont, late 1980s (Kevin Eaton photo; MIT Museum MoH collection).

[59] Crenshaw, Melissa to SFJ, interview, 19 Nov. 2003, Vancouver, SFJ collection.

[60] Also in Toronto was Fringe Research, founded in 1974 by Michael Sowdon and David Hlynsky, and its associated Interference Gallery.

[61] Perry, J. F., 'Design of large format commercial display holograms', *Proceedings of the SPIE* 1051 (1989): 2–5; 'John Perry: doyen of large format holography', *Holography News* 7 (8) (1993): 69–70; 'Large format holography: a major producer of large format display holograms explains his trade', *Holography Marketplace* 8 (1999): 30–6.

produced white-light transmission holograms for major exhibitions, trade shows, institutional displays, and artists such as Harriet Casdin-Silver and Michael Snow. The commercial stability of this small one-man operation, however, was atypical. Most surviving firms mutated substantially in personnel, products, and financing over their operating lives.[62]

12.3 OPTIMISTIC INVESTMENT: THE ILFORD STORY

Like the Holy Grail, a market for holography during the 1980s seemed tangible, worthy of the trouble and expense of seeking it, and yet perpetually elusive. The case of Ilford Photographic Ltd., mirrored by dozens of smaller firms, illustrates the painful gestation of holographic markets.

Founded in Essex, England, in 1879, Ilford was a long-established photographic firm looking to new markets, and had seen potential for holography sales during the 1970s.[63] The company marketed *He-Ne 1* film in the early 1970s, a straightforward adaptation of one of its existing photographic emulsions for recording high-resolution fringes produced by helium–neon laser light. The original 1970s product was too coarse-grained for reflection holography, however, and produced moderate light scatter.

Alongside the cottage industry springing up in Britain after the *Light Fantastic* exhibitions, a handful of major firms became interested in developing their sales via holography. Glenn Wood, a marketing manager at Ilford, examined the potential holography market and concluded that it was small and would remain focused on glass plates for display work. As production at Ilford had long been dedicated to film production, the firm could not justify the cost of setting up a production plant for glass holographic plates.[64] But through the 1980s, Ilford's Science and Technology Imaging (STI) Group undertook feasibility studies and research and development projects in silver-halide technology in the quest for new markets. The formation of the Group had been triggered by the attempt in about 1980 to corner the silver market by Nelson Bunker-Hunt.[65] As a result, silver-halide materials costs doubled, and Ilford saw its markets diminish. One

[62] Besides the companies discussed, notable among the 1970s wave of hologram firms were the New York Art Alliance (incorporated 1973, New York), Holovision AB (1974, Stockholm), the Holographic Film Company (1976, New York), Third Eye Productions (1976, Paris), and the Holographic Arts Company (1977, Chicago), all of which claimed annual sales of at least $50,000 (*Who's Who in Display Holography* (New York: New York Museum of Holography, 1978)).

[63] Hercock, R. J. and George Alan Jones, *Silver by the Ton: The History of Ilford Limited, 1879–1979* (London; New York: McGraw-Hill Book Co., 1979).

[64] Pepper, Andrew, 'Expansion of holographic film products', *holosphere* 15 (1987): 12–3

[65] Texas billionaires Nelson Bunker-Hunt and his brother William Hunt had begun buying silver in the mid-1970s, acquiring some one-third of the world supply by 1980.

result was Ilford's withdrawal from marginal markets such as x-ray products while investigating new opportunities.[66]

By the early 1980s, a commercial market for photographic emulsions optimized for holography appeared more likely. This was initially a matter of both technological push and commercial pull: optical data storage and holography were two growth areas identified by Ilford, both of which seemed particularly attractive opportunities because they shared the same silver-halide emulsion technology and a requirement for very high resolution. The rising output power and falling price of lasers also made it appear to its development staff that slow (i.e. light-insensitive), high-resolution emulsions were becoming more marketable.[67]

The firm identified some potential high-volume customers. One was the American military, seeking to store maintenance manuals and other bulky documents in compact form. The solution was optical data storage on a form of videodisc dubbed 'LaserFilm', the trademark of McDonnell Douglas Electronics; another was the same technology in the form of a credit card sized memory card. Both were non-holographic optical technologies with digital storage and readout, employing dots for binary code, and reproduced at low cost by contact printing.[68] This very specific application is typical of those that had been identified for holography between the late 1960s and early 1970s. The expectation, as with previous narrow market niches, was that the particular development could engender large profits that would allow this risky ledge to be broadened into a more reliable platform for commercial ventures.

A second major application and customer was high-volume hologram production by Applied Holographics, already mentioned in the context of the small firms Laser Point and SEE3. The company had been founded in 1983 when Hamish Shearer, who ran the Parallax hologram shop in London, looked into manufacturing large numbers of copies for customers.

With physicist Larry Daniels, Shearer designed a holographic printer, which incorporated a pulsed laser to record reflection holograms on silver-halide film. When they met Simon Rogers, an entrepreneur, the three formed Applied Holographics in April 1983, floated by merchant bankers. While many earlier firms had won large research and development contracts from government and private sponsors, Applied Holographics

[66] Brentnall, Roger, Alan Hodgson, Nigel Briggs, and Mary B. Dentschuk to SFJ, interviews, 20 Sep. 2002, Mobberley, Cheshire, SFJ collection.

[67] For further background on Ilford's involvement in holography, see Pierattini, G. and G. C. Righini, 'Sensitometric and holographic data of Kodak 120–02 and Ilford He-Ne 1 plates', presented at 1976; Saxby, Graham, 'Ilford's new holographic emulsions', *British Journal of Photography* (1985): 1286–7 and Wood, Glenn P., 'New silver halide holographic materials from Ilford Ltd', presented at *Proceedings of the International Symposium on Display Holography*, Lake Forest College, Illinois, 1985; Wood, Glenn P., 'Ilford holography—1986 progress report', *Proceedings of the SPIE* 747 (1986): 62–6; Wood, Glenn P., 'Ilford holographic consumables world strategy', presented at *Proceedings of the Third International Symposium on Display Holography*, 1988; Dentschuk, Mary B., 'Visit report', Ilford Ltd, Aug. 1988. See also §7.10.

[68] Photographic storage of information had long been in use, beginning with microfilm for storing records of transactions (1920s), particularly Emanuel Goldberg's 1927 system combining microfilm records with data retrieval cards, or 'Statistical Machine'. Ilford's implementation was an analogue opto-electronic system of storing data in non-image form.

was the first holography company to raise substantial funding from the public. The small firm moved into Braxted Park in Essex later that year and hired James Copp as resident holographer. When Agfa Gavaert proved unreceptive to engaging in research and development for their holoprinter development, the firm approached Ilford Ltd.[69]

From 1984 there was a close collaboration between Ilford and Applied Holographics to produce a commercially viable product for high-volume users. The two firms worked together to develop an automatic hologram printer (the HoloCopier)[70] and its associated expendable materials for about four years, with some two years of shared commercial activity. As a venture-capital-supported business, with an imported City manager and run by an artist and by a businessman, Applied Holographics was uncomfortably dissimilar to the long-established and hierarchical Ilford Ltd.

As a consequence of this technological push and commercial pull, Glenn Wood became responsible for holography market development at Ilford. The firm marketed a line of holographic films, and later processing chemicals, from March 1984.

Ilford research and development set out to produce finer-grained emulsions with good stability, as well as processing chemicals specifically for amateur or semi-professional markets: these had low toxicity, good shelf-life, and were non-corrosive and replenishable (i.e. their operating life could be extended by the addition of other chemicals). A number of specialized films were produced and marketed as Holofilm. Two emulsion types and three substrate types were developed: a red and a blue/green sensitive formula, coated either on sheet film having a triacetate base (a cast polymer to avoid problems of birefringence when the film was mechanically stressed in film holders, which would lead to streaks or blotches on processed holograms), a polyester base, or glass plates.[71] Ilford hoped to garner a substantial part of the existing market and to expand it further. The first Western firm to develop materials specifically for holography, Ilford was able to promote itself as a leader and innovator in the emerging industry.[72]

[69] Saxby, Graham, 'A visit to Applied Holographics plc', *British Journal of Photography* (1984): 667–8, 81. Applied Holographics, one of only two publicly traded companies specializing in holography in the early 1990s, later became part of Applied Optical Technologies plc, and included OpSec. The other company offering stock options was HoloPak in America.

[70] Agfa Gavaert could not support the research and development for the holoprinter, so Applied Holographics approached Ilford. The HoloCopier produced 480 exposures on a roll of film, and could produce holograms either of real objects or of transmission masters. The firm initially saw a market for at least a dozen HoloCopiers, but by 1987 only two had been sold (Salmon, Pippa, 'Commercial survey', *Holographics International* 1 (1987): 4).

[71] SP672T green sensitive film; SP673T red sensitive film; red sensitive and SP695T blue/green sensitive plates, where 'SP' denoted 'Special Products'. The red and green emulsions were later renamed *HOTEC R* and *HOTEC G* when Ilford sold HOTEC processing chemicals for holography. Its advertisements stressed safety and low toxicity of the processing chemistry, and encouraged educational applications.

[72] Ilford Ltd, 'ILFORD: The leader in specialised holographic film, plates and chemistry', *Holographics International* 1 (1) (Autumn 1987): 9. Later advertisements stressed its 'integrated approach' to the medium [Ilford Ltd, 'The integrated approach to holography giving you greater profitability with reduced labour and wastage. ILFORD', *Holographics International* 1 (4) (Summer 1988): 4), its continuing research program (Ilford Ltd, 'ILFORD HOLOGRAPHY—In the last two years alone ILFORD research has created . . .' *Holographics International* 1 (7) (Winter 1989): 11), and product safety (Ilford Ltd, 'ILFORD HOLOGRAPHY—HOTEC—A new family of holographic products to make holography simpler and safer—HOTEC designed with teaching in mind', *Holographics International* 1 (8) (Summer 1990): 30).

As a relatively small firm with a growing team of enthusiastic holographers, Ilford was responsive to its potential customers. Field trials at various holographic facilities were fed back to adapt the films, with the red-sensitive emulsion eventually undergoing four different specifications.[73] An experimental, unmarketed panchromatic silver-halide emulsion was reported in 1990 and used by Paul Hubel for experimental work on his DPhil, and supplied its emulsions in a wide variety of sizes.[74] These products were not identical to the Agfa or Kodak products: they appeared well-suited to colour holography because of lower haze levels than competing products, but were less sensitive (slower) than the Agfa material.

During 1986, Ilford and Applied Holographics began to supply product for new markets. They produced reflection holograms for a Dungeons and Dragons game with characters having holographic panels. Some one million images were recorded on 100,000 m^2 of holographic film (recorded on 240 mm wide film, with some four to eight images across, then die-stamped). The process was prone to wastage as processing chemicals became exhausted or optical parts shifted, so larger quantities were produced. Despite this success, for Ilford it amounted to, at best, a moderate-sized order for film, and one that recurred only sporadically. Similar special orders included hologram promotions for biscuits, Ghost Busters cereal prizes, and Tonka Toys SuperNaturals action figures.[75] Ilford worked with a handful of other large purchasers, including Edwina Orr and David Traynor of Richmond Holographics and Larry Liebermann in Florida.

Wood hoped to supplement these customer orders with production for an in-house product dubbed *FlashPrint*. This was based on the concept of using a transmission hologram as a packaging window so that customers walking past a product display would see a bright flash of diffracted light from the holographic reconstruction at one position, thus grabbing their attention. The Ilford Research and Development team participated in the development of another automated exposure machine, based on a scanned HeNe laser optics, specifically for the FlashPrint product. They also made up various mock-ups of soap packaging, Olympics security badges, Ilford visitor badges, and so on. They were unable, though, to attract advertising firms or other customers, a failure they attributed at least in part to an unavoidable slight yellowish cast to the holograms. Another in-house development was the use of HOEs, particularly for fenestration products: wavelength-selective glazing film for use on windows for architectural and automotive applications. These never reached market, either, owing again to the yellowish cast.

Much like the attempts fifteen years earlier at Conductron, the marketing group at Ilford creatively, or desperately, sought other conceivable avenues for sales: Wood launched a programme for students at a handful of secondary schools to produce

[73] Pepper, Andrew, 'Expansion of holographic film products', *holosphere* 15 (1987): 12–3, p. 13.

[74] Anon., 'Three colours from one Ilford emulsion', *Holographics International* 1 (1990): 14. Plate sizes: 4 × 5 in., 8 × 10 in., 30 × 40 cm, 50 × 60 cm^2; film sizes: 4 × 5 in., 8 × 10 in., 20 × 24 in. and the continuous roll 9.5 in. × 4 ft for the Applied Holographics Holocopier.

[75] Holograms for toys were made on alkali-halide emulsions, photopolymers, and then as embossed holograms. On the market near its peak, see Sadowski, Patrick, 'Holographic child's play', *holosphere* 15 (1987): 17.

holograms to develop both publicity and future holographers. He undertook discussions with Rolls Royce at Derby to use Ilford materials for non-destructive testing. HUDs and HOEs for the military aircraft and cars were discussed with British and foreign manufacturers. Ilford engineers investigated anti-counterfeiting holographic applications, before the market was exploited successfully by the embossed holograms of American Bank Note from 1983 (discussed below). And the concept of hallmarking expensive and genuine products—initially envisaged as perfume bottles and whiskey, but eventually shifting to lower-cost products such as toys—also consumed the time of the development staff. Equally important to long-scale plans to attract display holographers was the production of coated glass plates, which proved a particular problem at Ilford owing to the high purity of the gelatin and other emulsion ingredients required, which could fail to adhere to the glass plates.[76]

The number of potential but unreliable market niches grew, as did the rise in Ilford resources devoted to holography. Some two dozen people were employed at the peak of development,[77] with one internal product meeting bringing together twenty people. Managers judged the group too large, by Ilford standards, to sustain.

By the late 1980s, the other markets of hoped-for major customers were proving capricious. The laser videodisc project ended in 1987; Ilford had found increasing competition from CDs, and from other varieties of videodisc. Applied Holographics also turned increasingly from reflection holograms recorded on Ilford's silver-halide emulsions to embossed holograms as it responded to its customers' demands for less expensive products. What had begun as a creative endeavour for producing high-quality display holograms mutated within half a decade into a high-volume, low-cost production business. Like the smaller-scale experiences of Gary Zellerbach and his Holos shop, low-quality alternatives inexorably squeezed out the potential for Ilford's technological contributions and profit.

This clash of cultures was fatal to Ilford's hopes. By 1989, the market for Ilford was declining steadily. The development team had little market information; according to one staff member, Ilford policy preferred that scientists not talk to customers. As a result, customers' desires were not always well matched to Ilford's capabilities.

As Alan Waller, Manager of Scientific Products, explained, Ilford could not afford to produce less than 10,000 m^2 of film in a batch, but 'the market was too small'.[78] Having invested over £1million, the Ilford holography business was effectively closed in 1991, with marketing ceasing but production activity continuing through 1992. Surplus film stocks were sold to bulk purchasers and still remained available from these sources over a decade later.

[76] Ilford emulsions were prepared in 250 litre batches and stored until coating runs, irregularly scheduled according to market demand. The specifications of the emulsions were constrained by this detail, preventing the commercial production of finer-grained emulsions such as the superior, but slower, Soviet products of the period.

[77] Anon., 'Mobberley pioneers 3-D film', *Focus* (8) (Jun. 1984): 1.

[78] 'No more Ilford Holo Film', *Holography News* 5 (5) (1991) 1, 7.

Marketing manager Glenn Wood later reflected bitterly on the experience:

> There can be few product user groups in the world more particular about their materials than holographers and few who are more commercially naive. The money invested by Agfa and Ilford in improving emulsion quality and processing techniques has not been rewarded by the sales that might have resulted from a willingness to bring exposure, processing and marketing techniques into the twentieth century. Ilford's venture into holography was based on the assumption that, given the right materials, holographers would build businesses based on volume production, thus reducing the costs of their products to the consumer [. . .] History may never record the effort expended by Agfa and Ilford in trying to create a bigger market for their holographic products and so create the profits which would have paid for the art [. . .] Those who make the most excessive demands on their materials suppliers are unsuccessful as commercial suppliers themselves, being cushioned for the most part by universities, financial institutions or state grants. Silver halide holography must either pay its way and live up to the promises it has made for so long or risk the rejection from others that it has now received from Ilford.[79]

12.4 EMBOSSED HOLOGRAMS AND PROFITABILITY

The Ilford experiences illustrate how the firm was overtaken by an unanticipated commercial trend: the availability of inexpensive embossed holograms. Embossing transformed the market for holograms, increasing their availability by orders of magnitude and saturating public engagement with the medium.

In embossed holograms, the fine fringe patterns are recorded as surface relief—variations in the thickness of the underlying medium on which reflective film is deposited, rather than as variations in transmission as in photographic plates. From the early 1980s embossed holograms were almost always made reflective (usually by hot stamping with aluminium or gold foil), although transmissive types were also possible. This allowed embossed holograms to be mounted on opaque media such as magazine pages, postage stamps, credit cards, or walls. Such phase holograms yield bright reconstructions because of the lack of light absorption, and are inexpensive compared to photographic emulsions or the relatively complex handling procedures required for dichromated gelatin or photopolymers.[80]

[79] Wood, Glenn P., 'Silver halide film supply: follow-up—Silver halide users let down the suppliers', *Holography News* 7 (5) (1991) 3.

[80] On photopolymers, Stephen Benton recalled, 'We wanted to make reflection holograms [. . . that] got Du Pont back in the business where they had been before, when they saw that Polaroid was interested. We had a lot of good chemists at Polaroid, who came up with chemistry for photopolymers. But it was the way that a company like Polaroid is, with tens of thousands of employees, it's very hard to convince anybody that you're making any money. Sure, it's novel; sure it's fun, but all of a sudden you've got forty new employees who didn't have anything else to do. There was always [. . .] a conflict between Cambridge [Massachusetts], the intellectual, inventive side, and the part out in Waltham, that's trying to make a living on a large scale. You can look at the history of the company: trying to manufacture cameras was just a

A handful of individuals already mentioned were important in the development of the embossing industry, notably Kenneth Haines, Michael Foster, and Steve McGrew. Ken Haines had begun his career with four years at Willow Run. In 1966, Haines joined Holotron to develop commercial applications of holography. There, he and his colleagues developed a videotape player that employed a laser to read holograms recorded on photopolymer tape. The tape was based on photopolymer material that had been developed by their parent company, Du Pont. In effect, the recording medium was equivalent to photographic film, and was exposed by a laser beam. The photopolymer's refractive index is altered by exposure, which causes cross-linking of the polymer material. During the same period, however, engineers at RCA were developing a very similar system, but based on an embossed plastic tape. The conception of an embossing process—literally stamping the holographic information into soft vinyl plastic—permitted mass production and low costs for the system, dubbed Holotape.[81]

Thus both Holotron and RCA developed video storage systems based on holography. Holotron had the patents, and RCA initially agreed to unite to share their methods, which proved almost identical. However, the system was never commercialized. The video storage scheme was intended only for playback of pre-recorded films by consumers. RCA decided to pursue magnetic-tape video recording and playback using patents that it held for magnetic recording. Holotron did not enter the market because they deemed the RCA magnetic video tape recorder to be a better and less expensive product.

Nevertheless, the technical capabilities of producing transmission holograms by embossing had been explored. Mike Foster, who had operated light shows in Utah during the late 1960s and took up holography during the early 1970s, further developed the technology. Foster began investigating dichromate processes and in 1974 also developed a novel method for transferring holographic information from holographic plates to nickel embossing shims, which could then be used to stamp holograms onto plastic. He was soon trumpeting the commercial possibilities of holographic gift-wrapping, clothing, embossed LP records, and children's games for Mattel Inc.[82] Steve McGrew—who, as described above, had already had an eventful four-year career at Europlex Holographics, Holex, and Holotronics—made the method commercially viable in 1979 via his new company Light Impressions(see Figure 12.3).[83] Foster's own embossed hologram company, Spectratek, was founded a year later, and became a major producer of holographic foils for packaging.

mistake for a company like Polaroid. They *invent* technologies and then go to other companies and try to get them to build them" (Benton, Stephen A. to SFJ, interview, 11 Jul. 2003, Cambridge, MA, SFJ collection).

[81] See, for example, Bartolini, R., W. Hannan, D. Karlsons and M. Lurie, 'Embossed hologram motion pictures for television playback', *Applied Optics* 9 (1970): 2283–90.

[82] Rolfe, Lionel and Nigey Lennon, 'Holography—new light on a new dimension', *Delta Airlines Sky*, Dec. 1974.

[83] McGrew, Steve, 'Mass produced holograms for the entertainment industry', *Proceedings of the SPIE* 391 (1983): 19–20; McGrew, Steve, 'Custom embossed holograms', presented at *Proceedings of the International Symposium on Display Holography*, Lake Forest College, Illinois, 1983

Fig. 12.3. Steve McGrew at Holex Corporation, 1978 (Holex Corp.; MIT Museum MoH collection).

McGrew developed embossing techniques suitable to the graphics industries, and particularly for applications such as packaging and colourful stickers. His so-called 2D–3D embossed holograms were synthesized from flat artwork, and reconstructed an image in two planes, one a few millimetres behind the surface of the hologram and another at the surface itself. These holograms, producing bright colour from diffraction grating effects produced by the hologram surface corrugations, proved inexpensive and appealing to customers. His crude but colourful *Holographic Sunrise*, a six-inch square embossed 2D souvenir image, sold quickly at the Holos Gallery in San Francisco, and encouraged McGrew to expand production and marketing. His embossed products were much more widely distributed as promotional stickers for the movie *ET* in 1982, and became widely used during the 1980s.

In this way, the technology of embossed holograms was developing satisfactorily by the early 1980s. The medium was expanded more dramatically, however, by the development of a niche application and effective marketing. As mentioned above, in 1980 Ken Haines' small firm Eidetic Images was producing commercial holograms, including an improved form of integral hologram that Haines was able to mass produce by embossing. The director of American Bank Note (ABN), Ed Weitzen, decided that embossed holograms could be used for security purposes. That year, ABN acquired a limited license from Holosonics for security applications of holographic patents.[84]

ABN developed its capability in embossed holograms by acquisitions. It purchased Haines' company, and Haines researched and developed embossing techniques for the application. ABN then bought an embossing company, Old Dominion Foils, and created a new company, American Bank Note Holographics (ABNH). Weitzen approached MasterCard to promote the idea of small embossed holograms as anti-counterfeiting

[84] The principal business of Holosonics was acoustic holography. The firm had developed a method of using high-frequency sound waves to illuminate objects underwater, mix the diffracted sound waves with a reference wave on a liquid/air interface, and to diffract light off the surface ripples to form an optical image of the underwater objects.

devices on their credit cards. Such holograms carried a complex optical image that could not be reproduced or simulated and defacement of the image would reveal attempts at tampering.

By the end of 1982, ABNH was beginning to manufacture holograms for MasterCard, and the distribution of cards to customers began a year later.[85] When Weitzen sold the same idea to Visa for their own credit cards, the largest ongoing production of embossed holograms had been established. The company reported that counterfeit losses in 1985 were 8% lower than in 1984; losses for the first five months of 1986 were 52% lower than the same period a year earlier.[86] By mid-1986, all MasterCard cards incorporated a hologram.

As hologram embossing and coating processes improved, however, reflective embossed rainbow holograms began to appear more widely. The acceleration of production and marketing indicates that the subject attained international proportions from this time. Embossed holograms also reached wider audiences as graphics featured on magazine covers during the 1980s. These were more striking, and considerably less expensive, than the few transmission holograms that had appeared previously in books (e.g. the *Science Year* of 1967) or the hologram bound into a Random House psychology textbook in 1981.[87] Perhaps the last example of the use of transmission holograms in publishing was a transmission rainbow hologram that appeared in an issue of the *Journal of the Society of Instrument and Control Engineers* in 1981.[88] The new reflective foils swamped the market.

In 1982, the Toppan Printing Company in Tokyo prepared a twenty-five-page book, *Holograms for Displaying 3D Images*, which included a 3 × 4 in. embossed hologram of toys. The same year, 100,000 copies of a 6 × 6 in. hologram appeared on a record album cover for the British reggae group UB40, produced by Optec Design Ltd in Leeds. Wide circulation magazines soon began to incorporate holograms, with the June 1983 issue of the British magazine *Amateur Photographer* among the first.[89] A more widely publicized and influential hologram, however, was the March 1984 *National Geographic Magazine* hologram of an eagle, produced by Ken Haines, for which 11 million copies were printed; Ed Weitzen had convinced the magazine publishers to use a hologram that already had been produced and used for the 1982 annual report of the MasterCard International Corporation. In 1985, the journal *Artforum International* included a cover hologram, publicizing the new version of the medium for artists.[90] Cover holograms marked a design fashion of the late 1980s, when many magazines sought to boost their circulations by incorporating them. The November 1985 and December 1988 *National Geographic*

[85] Vuotto, Joe, 'Holography—a new security device', *holosphere* 12 (1983): 1, 3.

[86] Bender, David C. to E. Weitzen, letter, 12 Jun. 1986, Haines collection.

[87] The book was *Elements of Psychology*. Another example was the rainbow transmission hologram bound into Unterseher, Fred, Jeannene Hansen and Bob Schlesinger, *Holography Handbook: Making Holograms the Easy Way* (Berkeley, CA: Ross Books, 1981).

[88] *Journal of the Society of Instrument and Control Engineers* 20 (1981). The 3 × 4 in. hologram of the society logo, produced by the Dai Nippon Printing Company, was supplied in an envelope in the journal.

[89] *Amateur Photographer*, 167 (26), 25 Jun. 1983: cover.

[90] *Artforum International* 55 (1985): cover.

magazines again extended the impressiveness of embossed reflection holograms: the 1986 was a large hologram of a primitive skull; the 1988 cover was in fact a full front and back gold-embossed cover, with the front showing a double-image hologram of a crystal globe exploding, and the back cover a holographic advertisement for McDonald's hamburgers.[91] The February 1988 issue of the British *New Scientist* magazine brought the medium to the attention of readers of popular science.[92]

Then the floodgates opened: that year, Australia issued the first banknotes incorporating small holograms as an anti-counterfeiting measure. Unfortunately the holograms cracked and rubbed off the first versions.[93] In October 1988, Austria issued the first postage stamp with an embossed hologram on it (a 3D image of a shipping crate), designed by Wolfgang Stocker, with hologram design by 3D AG in Switzerland and Gunther Dausmann at Holtronics GmbH.[94]

American Bank Note Holographics and other embossing firms encouraged further development of markets for embossed holograms. Among them were holograms on Tonka Toys Go-Bots, a holographic label on Liberta Italian wine and holographic logos on romance novels.[95] The first wave of these holograms added product distinctiveness, and provided an argument for expanding use during the 1990s to promote products such as toothpaste packaging.

In contrast to a decade earlier, industry forecasts of the mid-1980s were cautiously optimistic.[96] A 1987 commercial survey reported expanding markets in the toy and security industries, and growing demand for holographic novelty items, all intended either to promote, decorate, sell, or protect.[97] By 1993, a decade after commercial exploitation had got underway, embossed holograms represented nearly 95% of a holography market estimated at $150 million, and then dominated by American firms.[98]

[91] The project generated losses of some $2.2million for the magazine: it rather unnecessarily employed a pulsed laser to create a hologram of a glass globe as it was exploded by a gunshot. Unusually, the holograms themselves were cast, not embossed, which further complicated production and raised the unit cost (Haines, Kenneth to SFJ, interview, 21 Jan. 2003, Santa Clara, CA, SFJ collection).

[92] *New Scientist* 117 (1988): cover.

[93] Pizzanelli, David, 'Captain Cook in half-baked hologram', *Holographics International* 1 (1988): 7–8.

[94] On the proliferation of philatelic applications, see Bjelkhagen, H. I., 'Holography and philately: postage stamps with embossed holograms', *Proceedings of the SPIE The International Society for Optical Engineering* 4149 (2000): 12–31.

[95] Erickson, Ronald R., 'American Bank Note Holographics: 8 years of innovation', *holosphere* 17 (1990): 14–5. The Zebra Books line of historical romance novels, introduced in 1985, included a small embossed hologram logo label 1 × 1.5 in., displaying 'A Zebra Romance' and alternately an image of two lovers embracing inside a rainbow-coloured heart. Walter Zacharius is chairman of Zebra Books. The company press release says 40% of all books sold (in America) are romances, a $583 million market. Titles include 'Golden Ecstasy', 'Stolen Ecstasy', 'Texas Torment', 'Rapture's Tempest' and 'Passion's Dawn' (MoH, 'Large editions', MIT Museum 41/1253, 1985). Other publishers, including Penguin and Hodder & Stoughton, used holograms on softback covers during the late 1980s.

[96] See Edelstein, J. Y., 'Technology Forecast—Commercial Applications of Holography', *Proceedings of the Society of Photo-Optical Instrumentation Engineers* 523 (1985): 343–6 and Kontnik, Lewis T., 'Commercial holography: a decade of emergence', *holosphere* 14 (1986): 12–6. For a review of earlier commercial history, see Lawrence, Pamela J., 'Retailing holography in the U.S.', *holosphere* 13 (3)(1985): 8–10.

[97] Salmon, Pippa, 'Commercial survey', *Holographics International* 1 (1987): 4.

[98] Lancaster, I. M. and L. T. Kontnik, 'Holography industry: market size and trends in commercial activity', presented at *Fifth International Symposium on Display Holography*, Lake Forest, Illinois, 1994.

Holographers nevertheless received embossed holograms ambivalently. The inexpensive products made holograms widely available, but reduced their technical gamut and visual impact. One commentator interpreted these technical limitations as creating a new social divide:

> Especially with the cheapo candy-bag-sized item, the image remains constrained within a small effective viewing angle, and is difficult for the untrained eye to see. For purists, this creates a nice cultural sifting effect: shiny magazine and magazine covers go in one eye and out the other with the hoi polloi, allowing true believers to strut their stuff by getting the message that 'lesser' folks miss. For pragmatists, such elitist smugness—necessary to keep the faith alive in the bad old days of struggling obscurity—has become a serious impediment. Now that there actually *is* an audience, concrete reforms must be effected soon to hold it.[99]

Such reforms were not immediately forthcoming. For artists, struggling to sell their fine-art holograms, the drawbacks of embossed holograms were apparent: instead of deep three-dimensional images, embossed holograms collapsed images into a shallow region of space; instead of a high contrast image in a pure colour, they provided a shimmering, gaudy spectrum of bright colour; because they were commonly mounted on magazine covers, the illumination was unpredictable and the surface was not flat; and instead of seemingly solid but unreachable images, their thin reflective surfaces reconstructed difficult-to-see and distorted pictures. And, perhaps most cutting of all, embossed holograms moved hologram production out of small laboratories and into commercial printing factories, thereby removing their mystique, value, and income from the original image creators (see Figure 12.4).

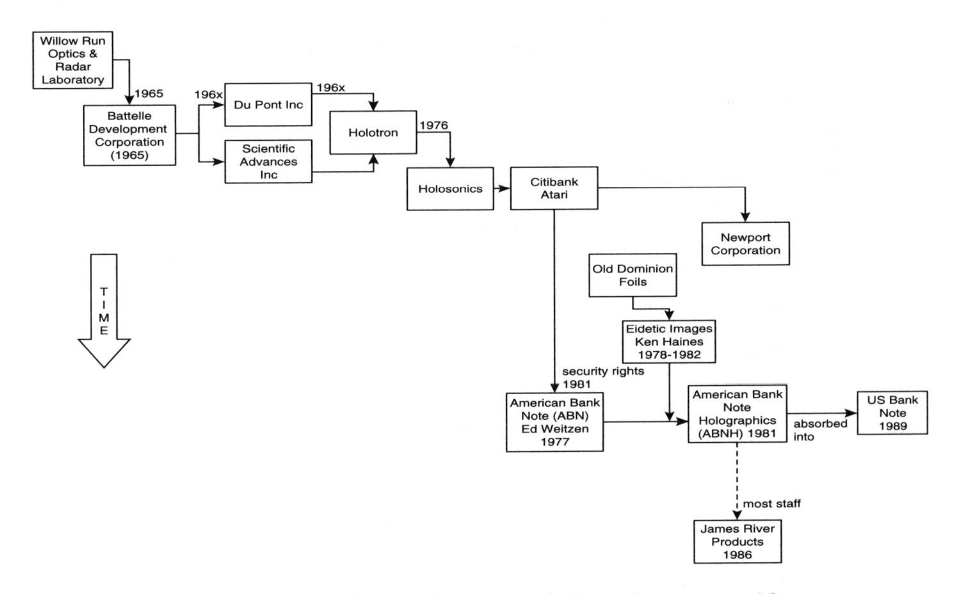

Fig. 12.4 Migration of some of the primary holography patents and firms.

[99] De Marrais, Robert, 'Holography in the future tense', *holosphere* 12 (1984): 4, 6.

12.5 PATENTS AND COMMERCIAL HOLOGRAPHY

In opposition to this optimistic headlong expansion was a force of restraint: patents. Patents were significant in formalizing the priority claims and official history of the young field of holography, as discussed in §5.4, and also captured early perceptions of technical applications and market niches. But even more significantly, holography patents played an important role in constraining commercial development. The patent problem had grown little noticed alongside commercial optimism, until it threatened to stifle the young industry.

Commercial expectations, and patent claims, had both developed early for holography. The first patent granted for applications of wavefront reconstruction to microscopy, Dennis Gabor's UK 685,286: 'Improvements in and relating to microscopy' and US #2,492,738: 'Method of obtaining enlarged images', filed at the end of 1947 and 1948, respectively, attracted no challenge and little interest before their expiry in the mid-1960s. Gabor later reflected,

> it was so novel that it was granted in the record time of two months. This was rather unfortunate, because it lapsed in 1966, just when holography was getting into its stride. If the U.S. Patent Office had waited, as is more usual, for a few years before granting the patent, it is well possible that the Courts would have declared it a master patent, valid far beyond the wording of its too restricted claims.[100]

In the interim, Gabor filed other patents on more sophisticated versions of his optical system for compensating the conjugate image in wavefront reconstruction, and for other ideas he developed while consulting at CBS Laboratories, but these, too, garnered no interest.[101]

The situation was very different fifteen years later, however. Emmett Leith and Juris Upatnieks were advised by their patent attorney in early 1964—shortly after news of three-dimensional imagery had been spread by press releases—to provide a disclosure of

[100] Gabor, Dennis to *Sunday Times*, letter, 14 Apr. 1968, IC B/GABOR/11/111.

[101] Gabor's first inventions filed at Imperial College were British patent 727,893: 'Optical apparatus'; American and Canadian versions were also filed. During his final years at Imperial College, he filed patents for IBM on a holographic information retrieval system (British patent 1,143,086: 'Information correlating'), and for CBS on deep holographic imaging (US patents #3,561,838: 'Holographic imaging', filed 24 Mar. 1967, and #3,545,836: 'Holographic pictures', filed 11 Dec. 1967) and acoustic holography (US patents #3,506,952: 'Sonar system employing holographic techniques', filed 30 Aug. 1967, and #3,745,814: 'Sonoradiography system', filed 13 Sep. 1971). After retirement, he filed further patents for IBM on information storage (US patents #3,600,054: 'Holographic association memory permitting conversion of a pattern to machine readable form', filed 11 Jun. 1970; #3,899,240: 'Distinguishing similar subjects using discriminating holograms', filed 9 Dec. 1970; and #3,764,979: 'Holographic system for subject recognition permitting conversion of a pattern into machine readable form', filed 3 Aug. 1971). Of Gabor's 172 patents (German, British, Canadian, and American, sometimes duplicated in separate countries), 10% were for nine distinct holographic inventions. For a list of his patents, see Tanner, P.G. and T. E. Allibone, 'The patent literature of Nobel laureate Dennis Gabor (1900–79)', *Notes and Records of the Royal Society of London* 51 (1997): 105–20.

all the applications that they envisaged for wavefront reconstruction. They collaborated on a document in late April that combined earlier memos to file and newer ideas.[102] They discussed eight ideas and possible inventions:

(1) lensless microscopes;
(2) lensless imagery, particularly with wavelengths other than the visible;
(3) production of phase plates to correct optical systems aberrations;
(4) techniques for production of reconstructions in colour;
(5) generalized holograms;
(6) motion picture and television applications;
(7) electro-optical holograms;
(8) measurement of motion.

The document, much more than published papers, provides a snapshot of their knowledge of previous work and their research plans as well as their ideas for inventions. For example, item (4), concerning colour holograms, shows their awareness of Denisyuk's concept of Lippmann reflections, but suggests that they had not realized that reconstruction could be achieved with white light instead of a laser of the same type used in the recording. The experimental confirmation of such white-light reconstructions in colour occurred two years later in a variety of laboratories. Similarly, item (8), detailing the measurement of motion via interference fringes in the reconstructed image, was ascribed to Juris Upatnieks. This was rediscovered and publicly confirmed by their Optics Group colleagues and others from 1965, as discussed in §7.2. Thus the eight concepts gradually appeared in print over the following three years.

Leith and Upatnieks filed a patent application on 23 April 1964, three weeks after their dramatic OSA meeting demonstration.[103] That year, Fred Llewellyn, then Director of Willow Run, approached Battelle Development Corporation (BDC), the development division of the Battelle Memorial Institute for commercial collaboration. Over the next nine years (until the Willow Run Laboratory mutated into ERIM), the Optics Group filed numerous patents for basic holographic geometries, holographic interferometry, contour generation, aberration reduction, image processing, data reduction, chemical processing, and applications. The University consequently assigned to BDC all 'rights, title and interest' to the holography inventions of Leith, Upatnieks, and other members of the Radar and Optics Laboratory. In return, BDC provided 15% of income from the patents back to the inventors via the university, with the university and BDC sharing any remaining royalties after expenses.[104]

[102] Leith, Emmett N. and Juris Upatnieks, 'Wavefront reconstruction disclosure', Patent disclosure report, University of Michigan Willow Run Laboratory, 22 Apr. 1964.

[103] A 'continuation', or further detail, was submitted six months later.

[104] Wolff, Michael, 'The birth of holography: a new process creates an industry', *Innovations* (1969): 4–15. On the reaction of George Stroke to the potential windfall for his rivals, see §5.4. Leith has estimated his personal income from the patents to have been some $100,000 (Leith, Emmett N. to SFJ, interviews, 29 Aug.–12 Sep. 2003, Ann Arbor, MI, SFJ collection).

In turn, Battelle Development Corporation approached Du Pont de Nemours Inc. and Scientific Advances, a wholly owned subsidiary of Battelle tasked with pursuing commercially risky technical products. The Du Pont connection was based on the company's experience with photographic films, its strong technical base, and freedom from preconceived ideas and commitments.

At Du Pont, chemist Russell Peterson oversaw the new venture.[105] In 1967, Du Pont and Scientific Advances established its new subsidiary, Holotron, in Wilmington Delaware, headed by Dan St. John, a chemist formerly in Du Pont's Development Department. The new company had several broad and influential patents. For example, the 1964 patent application was finally issued as US Pat #3,580,655 'Wavefront Reconstruction' issued 25 May 1971. Others included: #3,506,327 'Wavefront Reconstruction using a Coherent Reference Beam' (filed 23 Apr. 1964; issued 14 Apr. 1970; expired 14 Apr. 1987); # 3,838,903 'Wavefront Reconstruction' (filed 7 Dec. 1970; issued 1 Oct. 1974; expired 1 Oct. 1991); #3,838,787 'Holograms' (expired July 15, 1992). Most underwent varying degrees of patent litigation, such as the crucial patent #3,894,787—for the holography of diffuse objects—which was not granted until 1975.

Thus Holotron had considerable power to shape the embryonic industry. Its strategy for exploitation was nevertheless unfocused and fickle. In 1971, Holotron granted the first license for producing holograms. The company made little, if any, money by licensing the various patents during the 1970s, however, because the large companies that had explored the technology in the 1960s generated few viable products, and only a rudimentary cottage industry developed during the following decade. In December 1976, Holotron was acquired by Holosonics Inc., a maker of ion lasers and equipment for optical and acoustical holography, which paid a reputed $1.7 million, funded by Peoples' National Bank of Seattle and Citibank, for its collection of some 150 holography patents. The new administrators of Holotron judged that holography would not blossom as soon as anticipated, and that the company could not prosper from licensing alone; instead, it would concentrate on promoting the marketing and manufacturing of products based on holography.

Holotron planned mass production of holograms for commercial markets, focusing on the production of dichromate holograms and on embossed holograms, both of which were promising new technologies that offered low-cost production. Both technologies were absorbed by the acquisition of what had hitherto been cottage industries. Holotron Corporation's Fire Diamond division, founded on Richard Rallison's dichromated hologram capability, acquired the skills of one of the most active entrepreneurs of the period and consequently inherited the reputation of being the major American manufacturer of dichromated holograms for jewellery.[106] By 1978 orders of tens of thousands of

[105] Peterson was Republican governor of Delaware 1969–74.

[106] Anon., 'Fire Diamond, hologram maker, moved to Richland, Washington in reorganization program', *holosphere* 7 (1978): 1–2; Anon., 'Holotron licensing outlined by Cravens', *holosphere* 7 (1978): 3, 6. Another major producer of dichromate jewellery was Gary Cullen (b. 1954), a photographic technician who took up holography in 1976 and formed Canadian Holographic Developments (Holocrafts) in British Columbia in 1979. Specializing in thin DCG emulsions that yielded gold-coloured images, the firm eventually succumbed after 22 successful years to competition from photopolymer and DCG holograms produced in the Far East.

pieces could be filled. Similarly Holotron benefited from the embossing technology of Steve McGrew when he became its manager of research and development.

Not until the late 1970s did the company more aggressively seek to control the market through patents. In 1978, the company decided to generate more income by forcefully licensing the patented technologies. Its chairman, Bill Cravens, summarized the patent situation for hologram producers by stating 'if they are made by the off-axis method, they're ours'.[107] Holographers, most of whom were eking a marginal living from the sale of holograms, or were part of struggling small firms in the late 1970s, for the most part ignored the licensing requirements. For small shops and suppliers of holograms, the Holotron decision nevertheless raised anxieties about retroactive license fees.

Holosonics hired Harry Fondiller as Chief Executive in June 1978. He confidently predicted sales increasing from $2 to $25 million in the first year; instead, they declined. Fondiller was castigated by *Business Week* as a 'smooth talking name-dropper', who had previously claimed to reinvigorate companies but often led them to bankruptcy.[108] Holosonics defaulted on $3.5 million in bank loans, 80% to Citibank and 20% to People's, in May 1978. In rapid succession, Fondiller purchased the firm and moved it from Richland Washington to Sunnyvale, CA, as described above. Holosonics could not deliver its $2.8 million in orders, however, and new orders dried up. The company was declared bankrupt in December 1979 when the banks demanded full payment of their $4.3 million loans. At that point its patent portfolio was estimated at between $2 and $10 million. The patents reverted to the People's Bank of Seattle and Citibank. Sources close to the banks noted that 'bidding on the patents has not been especially active'.[109] Nevertheless, Citibank found that it could parcel out the patents by granting patents for specific domains of use. Citibank was able to negotiate exclusive licenses for a handful of firms that saw a bright future in holography. These included Atari, which envisaged applications in toys, games, and advertising (see Figure 12.5); IBM, for data storage; Newport Corporation, for industrial applications of non-destructive testing; and International Banknote, for security products. These had varying degrees of success.

Atari's experience was not atypical. In 1980 Atari bought the patent license for consumer holography. In 1981 it announced Cosmos, a holographic game system. The tabletop game turned out to be Atari's sole commercial experiment in holographic (or, as Atari dubbed it, Holoptic) technology. The games relied on flashing light-emitting diodes to reconstruct two different images from a semi-transparent rainbow hologram. One of eight different holographic backdrops could be inserted to play games such as Space Invaders, Football, and Basketball. The Cosmos, designed by Atari engineers Allan Alcorn, Harry Jenkins and Roger Hector, and with the optical layout designed by Steve McGrew of Light Impressions, was introduced at the New York Toy Fair in 1981. Reviewers were unimpressed by the concept, noting that the use of three-dimensional

[107] 'Holotron licensing outlined by Cravens', *holosphere* 7 (11) (Nov. 1978).

[108] 'On Holosonics and H. Fondiller', *Business Week*, 28 Jan. 1980: 96–7.

[109] Anon., 'Atari buys patent license for consumer holography', *holosphere* 9 (1980): 1–2.

Fig. 12.5. Advertisement for the Atari Cosmos game system, 1981 (Atari Interactive Inc).

background images was not important to the game itself. Nevertheless, the Toy Fair demonstrations generated some 8000 pre-orders. A pilot run of 250 units was ordered, with Steve Provence (then working for Light Impressions) producing the holograms for the unit. Like Holotron before it, Atari decided to set up a holographic laboratory in Sunnyvale, which Provence was to man. The company also began development of a second game, known as Spector, which was to have employed holograms more innovatively. The Spector tabletop game would incorporate a cylindrical hologram spinning to produce a holographic playing field. In the centre of the cylinder, an array of LEDs would reconstruct players and opponents.

In an uncertain commercial climate, however, Atari decided not to launch the Cosmos game as planned for the Christmas 1981 market, so as not to affect sales of its more expensive home video game system, deciding instead to delay Cosmos for a year.[110] In fact, the game was not marketed at all, and the company never proceeded past the concept and prototype mock-up stage for the Spector game. Provence recalls, 'While working at the Atari Holoptics lab, there was a wall stacked with Cosmos cases and hardware, perhaps as

[110] 'Atari unveils Cosmos, a holographic game system', *holosphere* 10 (1981): 1.

many as 1,000. When the money dried up and the lab was closing, some people from Atari came in and hauled all of the hardware away and then the lab closed.'[111]

In 1980, the same year that Atari entered the commercial hologram market, ABN purchased an exclusive license for three of the principal holography patents, intending to manufacture holograms as greeting cards and as security devices.

For all firms, commercial expansion was constrained nevertheless by threats, or at least perceptions, of forthcoming patent disputes. The unclear patent situation came to a head by the mid-1980s, when commercial applications were beginning to become successful and cottage industry was growing.

Polaroid, for example, claimed in 1984 that ABN's holographic processes infringed their patent on rainbow holography and sought a license agreement.[112] In response, ABN argued that the Polaroid patent concerned bandwidth reduction rather than rainbow holography and that, in any event, Stephen Benton had acknowledged the concept as originating with DeBitetto in his first paper on the subject.[113] Polaroid subsequently re-evaluated the validity of its own patents for rainbow holograms, finding that several of its claims were not defensible as being novel. It issued a disclaimer giving up rights to claim license fees or royalties for eight of the patent claims.[114] The retraction permitted small hologram producers to manufacture rainbow holograms with more freedom.

However, the validity and breadth of other patents threatened to choke the embryonic industry. In March 1986, ABN sued Steve McGrew of Light Impressions Inc. and United States Banknote Corporation (USBC) for infringement of three patents. McGrew, and his company Light Impressions, had worked with USBC, then the second largest security printer in America, to develop security applications of embossed holograms during the early 1980s, when Ken Haines played a similar role at ABNH. In 1981, USBC had been unsuccessful in negotiations to obtain a license to use Holotron patents for security applications; ABN had obtained the rights instead. However, because Steve McGrew had a sub-license from Atari to manufacture holograms 'for the purpose of games and security as it relates to advertising', USBC believed that their employment of McGrew as a holographer would cover them. USBC and ABN both bid for MasterCard and Visa hologram contracts, although Terry Gallagher, on behalf of ABN, had filed patents for

[111] http://www.atarimuseum.com/videogames/dedicated/cosmos/cosmos.html, consulted June 2003.

[112] Benton, Stephen A., Pat. No. 3,633,989 'Method for making reduced bandwidth holograms' (1972), assigned to Polaroid Corp., Peck, R. F. to G. P. Parsons, by letter, 27 Jul. 1974, Cambridge, MA, Haines collection.

[113] Benton, Stephen A. and Herbert S. Mingace, Jr., 'Silhouette holograms without vertical parallax', *Applied Optics* 9 (1970): 2812–13; DeBitetto, D. J., 'Bandwidth reduction of hologram transmission systems by elimination of vertical parallax', *Applied Physics Letters* 12 (1968): 176–8; Parsons, G. P. to S. F. D'Amato, letter, 21 May 1986, Haines collection.

[114] The disclaimer was issued 27 Jul. 1984 by Polaroid Corporation re: patent of Stephen Benton No. 3,633,989 issued 11 Jan. 1972 for 'Method for making reduced bandwidth holograms': Polaroid Corporation [...] has reason to believe that claims 1, 2, 4, 5, 9,10,11 and 14 of said Letter patent are too broad or invalid, and that it has no reason to believe that said claims were presented with any deceptive intention.' Polaroid, which had employed 13,000 at its peak, declared bankruptcy in October 2001 after posting debts of nearly $950 million, principally because of competition with digital photography.

the process of hot-stamping holograms onto a plastic substrate.[115] McGrew found his Atari contracts and sub-license abruptly cancelled, effectively removing him from the commercial holography business. He claimed that his process for mastering holograms did not infringe the Holotron patents, and counter-sued ABNH. In addition, McGrew sued Atari, Citibank, and the Holosonics bankruptcy trustee for restraint of trade. McGrew also continued to file patent applications for new processes or variants, such as a method of casting the microrelief pattern by using ultraviolet-curing resins.

The hologram patents themselves rested on the priority of Leith's and Upatnieks' inventions. McGrew therefore wrote to Emmett Leith, seeking to end the patent constraints by convincing Leith to withdraw claims:

> The Holotron trustees and ABN are seeking an injunction to prevent me and my company from making any holograms at all. The lawsuit is depressing the market for holograms. Many of our potential clients are afraid to buy holograms due to the threat of being sued themselves! One only needs to keep an ear open at an SPIE meeting or at Lake Forest to know that the threat of lawsuits over these patents is affecting a lot of people.
>
> [. . .] It is apparently the position of the Holotron trustee and ABN that display holograms of all types infringe various of these claims.
>
> I believe that these claims were drafted too broadly and that they cover holograms described by yourself, Denisyuk, Rogers and Gabor more than a year before the patent application was filed. I also believe that the methods we use and the products we make are not even covered by the claims.
>
> [. . .] Hopefully at our meeting we can clarify the fundamental technical questions and come up with a statement that will end the current litigation and let people use holograms the way they would like to for a change![116]

As with the priority disputes of the mid-1960s, historiography again was marshalled as an effective tool. McGrew publicized brief historical accounts to support his claims. He argued that Leith and Upatnieks had simply used a convenient light source to discover rather obvious phenomena, and that the edge of a Gabor in-line hologram is the same as a Leith-Upatnieks off-axis hologram. Indeed, McGrew contended that off-axis holography had actually been invented by Lawrence Mertz and N. Young, because they had proposed using zone plates to form images of x-ray stars.[117]

For the first time, the history of the subject assumed an importance beyond individual career trajectories: it was now of crucial commercial importance. Adolf Lohmann was one of the technical experts consulted to provide a dispassionate opinion. According to Lohmann's analysis,

> The history of holography from Gabor's original invention up to the work of Leith and Upatnieks is predominantly a history of the battle against the twin image problem. The most

[115] Colgate, Gilbert, 'Patents—in the history of security embossed holography', *Holography News* 10 (1996): 3–5.

[116] McGrew, Steve to E. N. Leith, letter, 26 Mar. 1986.

[117] McGrew, Steve, 'Holography before the laser', presented at *Proceedings of the International Symposium on Display Holography*, Lake Forest College, Illinois, 1985; Mertz, Lawrence and N. Young, in: K. J. Habell (ed.), *Proceedings of the International Conference on Optical Instruments* (1961): 305–12.

popular explanation of the twin image problem was put forward by Rogers, who alluded to the diffraction of light by a so-called Fresnel zone plate. Gabor himself, Rogers and myself all fully understood and tried to solve the problem so that the quality of the reconstructed image could be improved. But it was Leith and Upatnieks' off-axis techniques which provided the breakthrough that solved the problem and revived holography.

Gabor tried to suppress the twin image by a technique called 'double hologram in quadrature' (1956/1966) and by a 'spatial filtering method' (1952). Rogers tried the quadrature method and Lohmann tried 'single sideband holography'.

Denisyuk's idea led to impressive holograms only much later (not before 1965). He used a 180-degree angle between object and reference beams. He used COLLIMATED light, not diffuse light. He also used reflecting objects. He used a reflecting sphere and a reflecting micrometer scale,[118] which were specularly reflecting. There is no information about rough objects.[119]

Lohmann partitioned the developments of holographic techniques according to angle, mode of observation, mode of illumination, thickness of the hologram, and type of object. His categories argued that Leith and Upatnieks had indeed provided the information crucial for commercial exploitation.

While these patent disputes between major players dominated concerns for more than a decade, they also influenced holographers engaged in cottage industry. Richard Rallison recalled an unstable business environment mired in conflicting patent specifications:

Originally all holographers that made any money at it faced suits from Holotron and later Holosonics and I paid some royalties and eventually traded dichromate skills for stock in Holosonics which is now worthless [. . .]. Most patents are worthless anyway. Some of the stories are really bizarre; my single beam method of making HUD elements for Pilkington of England was patented 4 years later as the 'air gate ' method by Bob something of Portland Oregon and I thought it was too obvious to patent. One of my head mounted display patents was re-patented by a Hughes employee years after my patent had issued and Dupont, Polaroid and I were all issued separate patents for the same process in the same year, the only difference was in the words chosen by each of us to describe the process.[120]

Steve Smith, then of Holographic Imaging Studios in Chicago, argued that fear of lawsuits inhibited potential clients for holograms, such as Hallmark greetings cards, and that cut-throat competition was dividing the budding industry:

It isn't just ABN doing this. This involves many people who I thought were decent people, who got hungry. We used to sit side by side in places like Lake Forest College and talk about the excitement of doing holographic images. Now we know so much junk about each other, it's amazing we aren't all writing biographies.[121]

Smith lost his Chicago business and recovered a patent only after years of litigation. Tung Jeong also recounts how an Illinois holography company employed his students and

[118] Fig. 4 in his third paper, 'On the reproduction of the optical properties of an object by the wave field of its scattered radiation. II', *Optics & Spectroscopy* 18 (1965), 152–7.

[119] Lohmann, Adolf W. to patent attorneys, patent declaration, Jun. 1987, Haines collection.

[120] Rallison, Richard D. to SFJ, email, 22 Mar. 2004, SFJ collection.

[121] McLuskie, Caroline, 'Holography Community', *Wavefront* 2 (1987): 4.

reaped patentable ideas in the process but excluded Jeong himself.[122] In the same vein, Wasy d'Cruz, who became President of ABNH in 1990, recalled that he was appalled by what he found when he joined ABNH as president in 1990: 'It was a bunch of cowboys, the Wild West. No one respected anyone's patents or proprietary technology'.[123]

The ABNH/USBC lawsuits eventually were settled out of court, with the two companies cross-licensing technologies and with USBC absorbing ABN in 1989. McGrew's own lawsuits continued until 1994–5. They found in favour of the Leith–Upatnieks patent (although deeming two claims to be invalid because they were a case of double patenting, i.e. a claim already valid in a previous expired patent). ABNH was obliged to license McGrew's 2D–3D patents and ultraviolet embossing patents.

The years of patent litigation surrounding holography thus had a profound impact on the growth of the industry. For over a decade, they constrained the formation of new firms and applications and created work for corporate lawyers in the legal manoeuvring. Ironically, the intellectual property represented by the patents was not only difficult to define, but had its origins in another time and context.[124]

12.6 SECURITY AND ITS INFLUENCES

As discussed earlier, the security market drove a handful of commercial applications of holography during the 1980s. Credit cards almost universally incorporated embossed holograms by the late 1980s; paper currency increasingly included holograms from the 1990s, beginning with high-denomination banknotes and gradually extending to other forms of monetary items such as cheque guarantee cards, debit cards, and travellers' cheques, with a handful of large producers such as De La Rue Holographics in Basingstoke, UK.[125] Identity cards, visas, and drivers' licenses followed, and a rising market during the late 1990s was the labelling of items that could be counterfeited by low-standard substitutes, such as pharmaceuticals and concert tickets, or items subject to high taxation, such as alcohol and cigarettes, particularly in the Far East.

In an industry still riven by patent disputes, commercial secrecy, and competitive niche markets, security hologram manufacturers competed aggressively for an edge. One response was the formation of the International Hologram Manufacturers' Association (IHMA) in 1992. The Association adopted a code of practice in 1993, attracting thirty international members that year.

[122] Jeong, Tung H. to SFJ, email, 11 Jan. 2005, SFJ collection.

[123] Beeching, Dave, 'Holopack-Holoprint Special', *Holography News* 15 (2001): 1.

[124] Patents remained significant to industry activities. In 2000, the Dai Nippon Printing Company purchased some 150 holography patents from Raytheon, the US military manufacturer, most of which had been bought from the Hughes Aircraft Company. Dai Nippon had amassed nearly 1200 holography-related patents to expand its global share of holographic products.

[125] On security applications for currency, see Bloom Murray, Teigh, *The Brotherhood of Money: The Secret World of Bank Note Printers* (Port Clinton, OH: BNR Press, 1983).

Commercial applications also shaped the technology itself. Holographic anti-counterfeiting products came to acquire several discrete (and sometimes conflicting) characteristics:

(1) identifiability by consumers that the product is genuine;

(2) identifiability by the producer that the product is genuine;

(3) forensic information to allow experts to identify the origin of a product and prosecute counterfeiters;

(4) accentuation of the visual appeal of the product.

While security holograms could not be reproduced by photographic or photocopying processes, they could be copied. Indeed, within a year of the distribution of hologram-based credit cards, Jeff Blyth in England demonstrated that simple, if mediocre, copies could be recorded merely by placing holographic film in contact with the reflective hologram on the card and illuminating with a laser, generating a Denisyuk-type reflection hologram (see Figure 12.6).[126]

Holographers Steve McGrew and David Pizzanelli of Light Impressions challenged the future of the young industry given the ease of counterfeiting.[127] And by 1991, the first true security breach was trumpeted: the counterfeiting of security holograms used to label the packaging for Microsoft's MS-DOS 5 software.

As a result, the production of master holograms, from which embossed copies were made, became gradually more sophisticated. A technique that came to dominate production of security products was the dot-matrix hologram. In this type, conceived by Fujio Iwata and Kazuhiko Ohnuma of the Toppan Printing Company, artwork from a two-dimensional original is translated into a series of dots, each of which consists of a diffraction grating, a series of linear fringes having an orientation and frequency that generate a desired colour at a desired viewing angle.[128] In effect, the collection of gratings

[126] Blyth, Jeff, 'Holographic security', *holosphere* 13 (1985): 16. A. D. Little in Boston, technical consultants to MasterCard, nevertheless decided that his demonstration of a copy on 100 μm photographic film was sufficiently unlike the much thinner embossed foil to be readily discernible (Blyth, Jeff to SFJ, letter, 7 May 2005, SFJ collection; Cross, Michael, 'Fake hologram, that will do nicely', *New Scientist* (10) (25 Oct. 1984): 24).

[127] McGrew, Steve, 'Countermeasures against hologram counterfeiting', presented at Zurich, 1987; Pizzanelli, David, 'Countering counterfeiting', *Holographics International* (1988): 18–9. Nevertheless, the security provided by holograms remained of chronic concern for the industry. An article fifteen years later, for example, argued that many anti-counterfeiting holograms could be copied readily and undetectably for use on fake merchandise. The author proposed standards to ensure that manufacturers and public were aware of the types of hologram that are 'safe' (for now) and 'unsafe' ('Holograms don't work . . .' *Holography News* 16 (2002): 1–2). More recently still, McGrew concludes that, 'Holography is still highly regarded as a security feature, but holograms are widely counterfeited. Hologram copying techniques are so widely known, and high-quality holographic mastering equipment so widely spread, that holograms *per se* do not offer any more security than ordinary security printing features'. He suggests, though, that further innovations may provide another two decades of further utility for holographic anti-counterfeiting measures (McGrew, Steve, 'My reflections on holography', unpublished manuscript, 5 Dec. 2004, SFJ collection).

[128] Iwata, Fujio and Kazuhiko Ohnuma, 'Grating images', *Optical Security Systems Proceedings* (1988): 12–4. As background see the discussion of computer-generated holograms in Chapter 7, and Iwata, F., 'History of holography display in Japan', *Proceedings of the SPIE* 747 (1987): 116–8.

Fig. 12.6. *New Scientist* response to news of methods of counterfeiting security holograms (M. Peyton illustration; *New Scientist* (10), 25 Oct 1984, p. 24).

creates an image as a digital display does, with the pixels in this case producing bright and high-contrast images that are visible in most ambient lighting.

The new technology led to the rapid evolution of security holograms. The original dot-matrix holograms produced bright two-dimensional images, but with the unique characteristic of being animated when rotated. During the early 1990s Toppan produced three-dimensional imagery by the dot-matrix method, and other companies further extended the faithfulness of colour reproduction, image depth, and fine-scale encoding in the images. In all these variations, the technique is inherently digital, and so dot-matrix holograms are generated by computers coupled to the hologram printer, which writes each pixel individually at high speed. Steve McGrew's 2D–3D type of imaging, introduced in 1982, can also be rendered via a computer; both he and Ken Haines (early claimants to processes of creating embossed holograms) filed patent claims for computer-aided holography. And, direct production of holograms from digital images displayed on an LCD screen (a Digital-Holographic, or *Di-Ho* system) was first developed by Rob Munday of Spatial Imaging in 1994.[129]

Increasingly, security holograms combined techniques. Thus some incorporated microtexts having dimensions between 0.01 and 0.1 mm, invisible to the naked eye but discernable under a microscope; hidden images that can be reconstructed only with laser light or carefully oriented illumination; modulated pixels, in which the pixel position encodes information; or selective demetallization to produce regions that are transparent and amenable to other forms of optical encoding. Each of these variations enabled forensic analysis to verify genuine products and to identify counterfeits.

[129] The holographic printer is an example of technology developed along several lines by different workers. Walter Spierings of Dutch Holographic Laboratory, for example, also used an LCD screen as the source of images to record digital holograms, but using a one-step process (Spierings, W. and E. van Nuland, 'Development of an office holoprinter. II', *Proceedings of the SPIE The International Society for Optical Engineering* 1667 (1992): 52–62). By contrast, both the Toppan and Dai Nippon Printing Companies in Japan developed methods of writing computer-generated gratings and holograms directly by electron beam during the late 1990s.

As 2D–3D holograms gave way to sophisticated dot-matrix types, the essence of the hologram itself became more difficult to distinguish. In the *kinegram*, patented by Landis & Gyr and introduced in 1984, the two-dimensional computer-generated image is created from individual dots and lines made up of light-diffracting microstructures. While an essential feature of the device is the diffraction of light, holographic principles are less apparent. As with earlier patent disputes, the use of computers for digital holography created contention concerning priority and licensing.[130] By the early 2000s, these generic security devices were increasingly classified as Diffractive Optical Variable Image Devices (DOVIDs). The rising market requirements had mutated the product from a three-dimensional medium to one satisfying diverse producers and consumers. But just as importantly, patent disputes created minor nuances in the production technologies, allowing competing companies to operate.

Another influence of security applications concerned the commercial importance and power provided by such technologies. Security holograms became big business not only for government agencies and commercial firms, but for counterfeiters as well. In countries newly exposed to the methods of market capitalism, such as in the former Soviet Union, holographers found themselves in high demand. Engineer Sergey Poleyatev, for example, born in 1963, set up the Ukraine's first hologram company, Holograma, in 1996 to supply the burgeoning market for security devices on government documents. In 2003 he was shot dead in Kiev. Another Kiev holographer, Vladimir Markov, has recounted the growing need for protection from the pressures of organized crime in the security hologram industry since Ukrainian independence in 1991.[131]

The Ukrainian situation was an illustration of the gradual shift of focus for holographic producers and consumers owing largely to developing markets. With the collapse of the Soviet Union, traditional funding disappeared during the 1990s. Holographers there found financial support from the Russian Foundation for Basic Studies, but also from the US military and European grants organizations. Research institutes began to manufacture stock and custom holograms, and new businesses sprouted; a decade later about a dozen organizations were involved in holography in the Moscow area, six around St Petersburg and in the Ukraine, and another dozen sprinkled across other former Soviet republics.[132] From America during the 1960s and early 1970s, to an American–European axis during the 1980s, and to Eastern Europe and the Far East during the 1990s, commercial holograms served different purposes and markets.

[130] Kontnik, Lewis T., 'Computers, pixels and holograms: the patents', *Holography News* 8 (1994): 3–5 and Pizzanelli, David, 'Direct-write digital holography', presented at *Holography, Art and Design*, Royal College of Art, London, 2002. The incorporation of diffraction grating as picture elements in computer-generated holograms was patented in America by Takahashi, Toda and Iwata for the Toppan Printing Co. Ltd in 1991 and 1992 (#5,058,992 and #5,132,812); Frank Davis for Dimensional Arts Inc. (#5,262,879) in 1993; and Applied Holographics Corp. in 1994 (#5,291,317). The fine distinctions between the claims embroiled the companies in another wave of patent litigation during the mid-1990s.

[131] Markov, Vladimir B. to SFJ, interview, 26 May 2004, Kiev, SFJ collection.

[132] Reingand, Nadya, 'Holography in Russia and other FSU states', http://www.media-security.ru/rusholo/index.htm, accessed 27 Sep 2004. A number of former Soviet holographers also emigrated to take up holography in the West.

In Japan, commercial firms had maintained a continued interest in potential applications of holography. Japanese interest initially mirrored that in the West, namely for holographic interferometry, information storage, and display products. Photographic and printing companies were particularly active. At Canon, Fuji Photo Optical Company and Nikon, for example, holographic interferometry were researched; at NEC and Fujitu, holographic optical elements (HOEs) were studied; and at Toppan Printing and Dai Nippon Printing, holographic optical security devices and 3D displays were the topic of research.[133] Research and development continued on hologram printers for consumer applications.[134] Holographic stereograms were also refined by the Toppan Printing Company during the early 1980s.

With the rise of the security industry, however, anti-counterfeiting holograms attained a dramatically increased importance, particularly in the burgeoning Chinese economy from the late 1990s, where the hologram industry mushroomed. China became the most rapidly expanding market for holograms and hologram companies. The first public exhibition of holograms had been organized by the Musée de l'Holographie (Paris) in 1983, attracting some 6000 people daily. In 1987, Qingao Qimei Images Ltd became the first Chinese holography company to be founded, producing and selling embossed and DCG holograms. Western firms played an important role in the early development, but Chinese companies soon took over. For instance, Light Impressions set up ten holography factories in China between 1988 and 1994.[135] In 1993, the first national hologram industry conference in the country attracted ninety firms. China's production of holograms also became a major concern for other countries, because the rapid development of technologies sometimes ignored the licensing restrictions that had bottled up the industry in the West. Moreover, China was a vast new market for such holograms. Many food and packaged products were produced and by state or regionally owned concerns. The resulting territorial monopolies were prone to counterfeiting and so holograms proved a popular solution for tax papers, identification cards, and packaged goods. Cigarette packages, for example, expanded the Chinese market for holograms by orders of magnitude. As a result, embossing machines—and counterfeiting, too—proliferated during the 1990s.[136]

[133] Anon., 'Dichromate and Integral holography in Japan', *holosphere* 7 (1978): 2.

[134] Yamaguchi, M., N. Ohyama, and T. Honda, 'Holographic three-dimensional printer: new method', *Applied Optics* 31 (1992): 217–22.

[135] Nevertheless, Light Impressions California closed in 1996 with cash-flow problems, transferring its accounts to Light Impressions International in London. McGrew himself diversified his interests during the 1990s, working on government contracts for holographic solar concentrators, genetic algorithms for logistical planning, spectroscopy, and magnetics.

[136] See, for example, Hsu, D., J. Jiao, H. Tao, and P. Long, 'Recent developments on holography in China', *Three-Dimensional Holography: Science, Culture, Education* 1238 (1989): 13–7; 'Rapid expansion in Chinese hologram production', *Holography News*, 7 (2) (1993) 13–4; Hsu, Dahsiung, 'Anti-counterfeiting holograms and government anti-piracy activities in China', *Proceedings of the SPIE The International Society for Optical Engineering* 3358 (1998): 318–21.

Despite the hopes, scams, disputes, and hard work, commercial holography remained a small industry and so was susceptible to closures of some of the larger businesses. Just as the early 1970s had witnessed the establishment of initiatives in teaching holography, signs of decline in display holography raised comment two decades later. In 1990, for example, the two periodicals having the widest circulation, *holosphere* and *Holographics International*, both ceased publication; in 1991, Ilford stopped all production of holographic film materials; in 1992, the New York Museum of Holography filed for bankruptcy and *Wavefront* magazine suspended publication; in 1994, the Museum für Holographie und Neue Visuel Medien in Germany closed, and the Canada Council ceased funding for aesthetic holography; in 1995, Light Impressions Inc. in California closed operations owing to inadequate sales; in 1996, Agfa Gavaert, long the dominant supplier, announced that it would end production of silver-halide holographic emulsions; and in 1997, Hughes Power Products, a division of Hughes aircraft, ceased photopolymer hologram production, used principally for Head Up Displays, owing to the slowing of military contracts in the post-Soviet era. And over that decade, holography exhibitions declined twofold.[137]

Nevertheless, commercial holography continued to intrigue and enthuse investors as well as holographers. The continuing confrontation of technical zeal and conservative market niches is exemplified by the case of Dimensional Foods, which developed the ideas of Eric Begleiter from the early 1990s. Begleiter, an MIT alumnus, developed the idea of producing transmissive or reflective holograms in foods.[138] Initially he envisaged imprinting chocolate with the holographic fringe pattern, but then explored candies, starchy materials, and other food substances that could be embossed, ablated, or otherwise imprinted on a surface. The concept proved remarkably versatile, being applicable to a variety of edible materials and production methods. While there remained technical constraints (chocolate, for example, would lose its fine embossed pattern in a matter of weeks unless refrigerated or adapted to embody denser material, which proved to have an unappealing 'mouth feel') he established that the optical effects could be combined with palatable foods.

Obtaining a broad patent for holograms produced in foods, Begleiter found his time divided between investigating problems of food processing and trying to develop markets for the idea. Initially, he operated as a consultant in the food industry for firms interested in the possibilities of the process. With Mike Wodke and Paul Graham, two former employees of the candy manufacturers M&M-Mars, he formed Dimensional Foods during the early 1990s. The firm set up a pilot plant in Cincinnati to make demonstration pieces and began to produce limited runs of holographic candies for trade shows and

[137] See Appendix, Figure A-1D.

[138] Steve McGrew had made holographic candy for his associates at Holex in 1978 by pouring hot syrup onto a silicone rubber mould made from a surface relief hologram. The relief hologram had begun as a conventional silver-halide hologram developed in a tanning bleach to accentuate its surface variations and rinsed in hot water. McGrew filed a disclosure on the method to the US Patent office, but did not pursue the idea further.

corporate promotions. In the process, they investigated new markets for the products, and the technical complexities of modifying existing food processing equipment.

Begleiter discovered that, while the food industry was willing to fund feasibility studies and consultancy contracts, enthusiasm was difficult to sustain. Large producers were conservative, hesitant to lose their market share in a highly competitive field, and reluctant to invest in development of new food manufacturing equipment. Similarly, focus groups found the idea of candy or other foods with holograms intriguing, but disconcerting as well. Consumers sometimes assumed, for instance, that the rainbow colours produced by the holographic reconstruction were due to dyes, which suggested toxic materials. Others assumed that the bright sparkles produced by the foods were due to high sugar content, because sugar crystals sparkle similarly in sunlight. In response, Begleiter vaunted the purity of his holographic foods, which can look colourful while being completely free of colorants, and attempted to explain in his product literature that the diffractive effects of the holograms were due to 'tiny prisms'.

Searching further afield for product niches—as had Kip Siegel of Conductron, Glenn Wood of Ilford, Ed Weitzen of American Bank Note, and dozens of small entrepreneurs before—Dimensional Foods produced holographic communion wafers, matzoh, and components of sushi (specifically seaweed, which has a hygroscopic inversion, changing it from soft to hard and allowing a holographic impression to be cast). While such products had an unlikely appeal, Begleiter found the company drawn into an educative process for both potential manufacturers and their customers. The image was not enough; indeed, the optical image was accompanied by other undesired connotations that proved difficult to shake.

As so many others had found, the enthusiasms of holographers proved difficult to export to others. Begleiter, after more than a decade of solving a stream of technical problems and tirelessly seeking new markets, concluded:

> In some ways it is frustrating, because I see some of the applications as being absolutely huge [. . .] and at the same time, there is a sense—and this might be an illusion—but it's *just around the corner*. I always have enough big companies feeding me enough money doing enough research that I just *know* that it's nearly there [. . .] but it's like if you're digging for gold; you must feel that the next day you're going to find it![139]

The cases of Dimensional Foods, Ilford, Conductron, and numerous smaller firms were typical in commercial holography. From the 1960s to the 1990s, commercial applications, or at least perceptions of them, remained tantalizingly close and yet exasperatingly beyond grasp.

[139] Begleiter, Eric to SFJ, interview, 10 Jul. 2003, Cambridge, MA, SFJ collection.

13

The Hologram and Popular Culture

As businesses sprang up to produce holograms, one after another learned that public interest was fickle and unfathomable. The cultural history of the hologram is unusual for a technological artefact, because the medium was successively introduced to new audiences, and became imbued with new purposes and meanings at every step. During the 1960s, as we have seen, holograms were both created and used by scientists and engineers. The wider public was excluded from access to the medium, which retained an aura of sophistication and elitism. During the 1970s, artisans and artists brought the hologram to a wider audience. And from the 1980s, holograms were commercialized to an extent that made some forms ubiquitous. But perceptions of holograms mutated in the process. Ubiquity had a price.

13.1 SHOCK AND AWE

The phrase 'shock and awe' used fleetingly to describe the planned use of intense, high-technology weaponry in the second Iraq war could be co-opted for holograms.[1] In a similar, if distinctly less militaristic, way, the cultural impact of holograms has always relied on their ability to shock viewers and produce awe.

Awe was an emotion commonly voiced by viewers of holograms during the 1960s. The notion of a 'technological sublime', explored by historian of technology David Nye, is relevant to the experience.[2] Nye focuses on the effects of technology in creating a sublime experience: large-scale architecture, transportation, and engineering (dams, bridges, railways, and the atomic bomb, for example) illustrate how the American context shaped technology and its reception. Holograms are distinctly limited in scale, but their impact was similarly profound, if ephemeral.

[1] For the origins of the American policy, similar in concept to the *blitzkrieg* employed by the German army during the Second World War, see Harlan K. Ullman and James P. Wade, *Shock and Awe: Achieving Rapid Dominance* (Newport: NDU Press, Oct. 1996).

[2] Nye, David E., *American Technological Sublime* (Cambridge, MA: MIT Press, 1994). Nye does not define the technological sublime precisely, but offers examples. In his introduction, he describes the sublime object as one that 'cannot be comprehended through words and images alone' (p. xi). From his examples, he suggests that the sublime 'taps into fundamental hopes and fears' and that it 'can weld society together' (p. xiii).

Holograms entered the realm of the sublime through their connection with the viewing environment and natural optical phenomena. Consider the rainbow appearing after a severe storm, and dividing the sky into bright and dark regions by its colourful arc; ice haloes appearing in the fog of a cold morning, and noticed from an isolated hilltop; bright lightning bolts in an otherwise inky sky; the *heiligenschein*, a halo of light surrounding the shadow of one's head reflected from the dewy grass of morning; or the *glory*, a circular rainbow around the shadow of an airplane as it flies above a cloud. One reason for the impact of these optical phenomena is that each is experienced in a peculiar context that isolates the observer from the everyday world.[3]

The same transcendent experience can be captured in laboratory or gallery demonstrations of holograms (see Figure 13.1). Like natural phenomena, the experience of viewing a hologram has amazed or awed observers, because of the profoundly unfamiliar imagery and disorienting environment. The holographic images of the 1960s were isolated from the viewer, appearing behind the hologram surface, and were illuminated by the unnatural speckled light of the laser. Rooms were dimly lit, and sometimes shrouded in curtains; the hologram was illuminated by a hidden source and appeared bright and disembodied from its environment. Later hologram displays were in less isolated environments but were equally unworldly and unsettling: the image itself straddled or hung disembodied in front of the hologram, offering interaction with the viewer but disappearing entirely when the observer moved too far. The colours were rainbow-like (for Benton holograms) or an unfamiliar monochrome green or yellow (for Denisyuk reflection holograms), disorienting the viewer. And, most importantly, the image was three-dimensional, realistic, and yet untouchable. Even when, like the first

Fig. 13.1. Holographers François Mazzero and Pascal Barr viewing hologram (H. Bjelkhagen photo, Bjelkhagen collection).

[3] A classic study of optical phenomena in nature is Minnaert, M., *The Nature of Light and Colour in the Open Air* (New York: Dover, 1954).

generation of holograms made by scientists, the images were mundane—tools, trains, or chessmen—the context wrapped them in an aura of ghost-like unfamiliarity. The captivating qualities of holograms point to incongruities: we juxtapose our expectations of photographs onto holograms, and are surprised when the analogy does not match our common sense. Thus photographs that we can look around, and holograms of lenses that really work, have the capacity to awe, at least temporarily, new generations of viewers.

Nevertheless, for many viewers, this sublime experience is brief. The awe generated by the hologram, to the dismay of its proponents and promoters, proved capricious and evanescent. Historians of technology have cited other examples of the speed with which the public absorbs new, awe-inspiring capabilities: electric lighting, radio, and jet transport rapidly became mundane and unremarkable.[4] The same was true for holography. A natural response of holographers was to continually seek to extend the sublime experience by augmenting the technical capabilities of the medium or, when that was not possible, to claim that such advances were in train. Thus a medium dependent on generating a sublime experience was inevitably harnessed to the notion of technological progress.

The first consumers for holography were scientists and engineers, but Jim Trolinger suggests the difficulties of promoting holograms even in the technical sphere:

> In some ways it was misleading since the limitations were not always apparent, and the holograms were often irrelevant to the application under discussion. Many charlatans who had never made a hologram themselves marketed with the holograms produced by others. We could show a potential customer a hologram of a few chess pieces and then explain how we would make similar holograms of ice crystals in clouds. This gave the impression that one would be able to look into a hologram and see an ice crystal microscopically in 3-D. There are several reasons why this is not so straightforward that were not obvious. To see microscopic detail required looking through a microscope. If direct light was used, this was not easy. If diffused light was used, then speckle led to a very noisy image. Most scientific studies employed TV cameras to examine the particles so this took away some of the excitement in viewing the holographic image. [. . .] After many failed promises holography actually took on a stigma of charlatanism during some periods and the word holography in a proposal could mean instant death. This state of affairs has come and gone several times in my career as new breakthroughs offer to breathe new life into holography.[5]

Beyond the emotion of wonder and awe, holograms provoked a degree of shock, a consequence of surprise at events that do not match expectations. One source of a shocking, if not sublime, experience is sexual imagery. For any imaging medium, the allure of the prurient and the attraction of lucrative markets eventually finds connections with erotic imagery. Unsurprisingly for such a realistic medium, this connection developed early for holograms. Perhaps less expectedly, however, the application failed to develop.

[4] See, for example, Marvin, Carolyn, *When Old Technologies Were New: Thinking About Electric Communication in the Late Nineteenth Century* (Oxford. Oxford University Press, 1988) and Douglas, Susan J., *Inventing American Broadcasting: 1899–1922* (Baltimore: Johns Hopkins University Press, 1987).

[5] Trolinger, James D., 'The history of the aerospace holography industry from my perspective', unpublished manuscript, May 2004, SFJ collection.

The history of erotic holograms is understandably less visible than other applications. Nor can the subject readily be classified, ranging from aesthetic nude studies, to erotica, and to pornography. As mentioned in §10.2, among the earliest examples of such contentious art—conceived as a feminist statement—was *Phalli* (1974) by artist Harriet Casdin-Silver, an image-plane hologram of disembodied penises. The uncongenial environment required for pulsed holography was not conducive to erotic portraiture, although such images could be, and perhaps were, recorded as early as 1968 by firms such as Conductron (later McDonnell Douglas (MD) Missouri), according to unconfirmed folklore. McDonnell Douglas certainly sought Playboy as a customer, but the liaison remained unfulfilled when the pulsed holography facility was closed in 1973. During the 1980s Harriet Casdin-Silver at the Royal College of Art in London, Fred Unterseher in Germany, and Jason Sapan, portrait holographer for the Museum of Holography, also produced nude portraits as artworks, although Casdin-Silver found one of her exhibition catalogues censored by the gallery owner.[6] And pulsed holograms by Ed Wesley described by a reviewer as 'tasteless . . . sado-masochistic erotica' were shown in a 1986 solo exhibition in Chicago entitled *Recent Pulsed Stuff and Other Delights*, and decried by Dennis Gabor's surviving brother André and friend T. E. Allibone as 'bastardized art'.[7] A series of more commercially oriented pulsed portraits of Playboy Playmate models and similar subjects were produced from 1990 by Ron and Bernadette Olson of Laser Reflections in San Francisco.[8] Their customers could purchase lifelike, if unnervingly frozen, large-format reflection images of voluptuous models for $1895 each.

A lower-resolution but more lifelike portrayal had been exploited earlier through holographic stereograms. The Multiplex Company in San Francisco produced a series of erotic short subjects, typically 15 s long or less, in their ubiquitous cylindrical format during the mid-1970s. The first of these was *Pam and Helen* (see Figure 13.2), one of the earliest commercial multiplex holograms produced by Pam Brazier and Lloyd Cross in 1974. Such examples were followed by more explicit holograms by Peter Claudius.[9] The most highly publicized example was the *Holoblo* (see Figure 13.3), a coin-operated viewing machine constructed in 1976 that consisted of a rotating multiplex hologram positioned on a horizontal axis at waist level. One magazine review summarized:

> *Holoblo* features an attractive young Eurasian girl fellating a penis; *Quickie* depicts close ups of a penis and vagina grinding away (and similarly oriented in the machine). They have been installed in two San Francisco theatres, and reaped hundreds of dollars of quarters a night.

[6] Unterseher, Fred to SFJ, interview, 23 Jan. 2003, Santa Clara, CA, SFJ collection; Sapan, Jason to SFJ, email, 18 Mar. 2004, SFJ collection; Casdin-Silver, Harriet to SFJ, interview, 3 Jul. 2003, Boston, MA, SFJ collection. The censored catalogue was from Gallery Naga, Boston, 1995.

[7] Lightfoot, D. Tulla, ' "Recent Pulsed Stuff and Other Delights"—Ed Wesley's show at Benny's Casino', *holosphere* 15 (1987): 28–9; Gabor, André and T. Edward Allibone, 'Letter to the editor', *holosphere* 15 (1988): 5.

[8] Olson, Ron and Bernadette Olson, 'Playmates to primates', *Proceedings of the SPIE* 3358 (1989): 251–6.

[9] Claudius, with an MA in English Literature, had worked in screenwriting, kinetic sculpture, music, and photography in the early 1970s, for a time known to the photographic world by the pseudonym 'F. Stop Fitzgerald', according to his Museum of Holography biography. He was a General Partner in the Multiplex Company, and was the final partner during the early 1980s (Claudius, Peter, MIT Museum 33/1028, 1976).

Fig. 13.2. Erotic hologram: *Pam and Helen*, Multiplex Company, 1974 (cylindrical white-light transmission holographic stereogram) (Pamela Brazier and Lloyd Cross; MIT Museum MoH collection).

Fig. 13.3. Marketing erotica: Peter Claudius and the 'Holoblo', 1976 (Dave Patrick Photo; MIT Museum MoH collection).

> Claudius also takes them to rock concerts. 'The machine was originally in the Ophanall sex theatre and is now at home in a punk rock club in San Francisco, where it is considered a novelty'.[10]

Not coincidentally, the photographic equipment at the Multiplex Company had been acquired from a defunct pornographic studio. The long-term sales from these modern

[10] 'Lunch Box Review,' MIT Museum MoH 26/619, 1978.

updates to 'What the Butler Saw' are not known, but erotic subjects comprised one-fifth of the first Multiplex hologram catalogue.[11]

The risqué connections and lifelike possibilities of three-dimensional erotica proved an appealing attraction for the viewing public. During 1978 the New York Museum of Holography mounted a two-week exhibition in their newly opened basement gallery (held over for three weeks) of a range of holograms by Peter Claudius of Multiplex. On seeing the opening lecture by Claudius and the 'Lunch Box Review' exhibition itself, science fiction writer Isaac Asimov penned a limerick:

> Said the public to Dennis Gabor,
> 'Just what is holography for?'
> 'With pictures in three
> Dimensions', said he
> 'The love-stuff is really hard-core.'[12]

The potential for publicity from holographic erotica continued. In the early 1990s John Mitton ('John F', b. 1964, Amersham, England), a light-show designer and graphic artist, completed an MA at the Royal College of Art in London. Using the College's equipment for producing flat integral holograms, he exhibited works with titles such as 'Testosterone' and 'Orgone Accelerator' at his RCA Degree show in 1992. Organizing F.UK, an artists' collective, after his degree, he mounted an exhibition entitled *Hardcore Holography* two years later at the John Moore University Hope Street Gallery, displaying a collection of integral holograms displaying about 10 s of animation depicting sexual subjects. The Merseyside police closed the exhibition after three days owing to a complaint, although some of the holograms were subsequently shown at a Festival of Erotic Art held at the Coventry Gallery. Holographer Rob Munday suggests that many, if not most, commercial holographers have dabbled in holographic erotica.[13]

13.2 FROM THE SUBLIME TO KITSCH

Holographic erotica remained a marginal market, and one seldom emphasized by more than a few promoters of the medium. Indeed, it may be that the intended effect demands covertness; the shock value—or more generally, the sublime experience of any hologram—requires a certain surprise, and hence promotional constraint.

By contrast, the mass-marketing of holograms, as exemplified by the histories of the Holos Gallery in San Francisco and Light Fantastic in London (both discussed in Chapter 12), produced a trend towards lower-priced products. These were dependent on novelty

[11] Among the holograms produced by Claudius were *Celeste Undressed, Peek-A-Boo, Flasher, Garden of Eden*, and *Banana Lady*.

[12] Asimov, Isaac, 'erotic limerick', 26 Sep. 1978, MIT Museum MoH 26/619.

[13] Munday, Rob to SFJ, interview, 31 Mar. 2004, Richmond, UK, SFJ collection.

value—a kind of low-order surprise—for their sales. When one audience became saturated and jaded, losing its capacity for surprise at the medium, marketers moved to new audiences. Thus the sublime and other-worldly holograms of the 1960s—experienced in the disorienting environments of the laboratory or gallery—were replaced by holograms that surprised or pleased narrower audiences in private environments—connoisseurs of aesthetic holograms or consumers of erotic holograms—and then by wider audiences intrigued, amused, or dazzled by optical tricks. This extension to mass consumption automatically entailed different viewing environments and purposes.

By the late 1980s, inexpensive holograms for the wider public were being described as 'holokitsch'.[14] The definition of tawdry and vulgar holograms was, however, a matter of taste, and one that evolved gradually. The first example of holographic jewellery—apart from the King Vitaman cereal box promotion of 1972—may have been the kind of decorative ring given to Dennis Gabor, which he claimed his wife would 'wear proudly on festive occasions'.[15] And the dichromate holograms produced at the end of the decade by Richard Rallison, Holocrafts, and others, found their most frequent use in pendants, earrings, and belt-buckles sported by enthusiasts. A few years later, though, Sunny Baines, editor of *Holographics International*, complained that much holographic jewellery had become 'twee and uninteresting', was poorly made, and devoid of style and cachet.[16] Such evaluations were an extension of the aesthetic criticisms levelled at holograms by an earlier generation of art critics.

The cultural history of holograms has important parallels to that of other imaging technologies, notably the stereoscope. Invented by Sir David Brewster, the stereoscope was first exhibited at the British Association meeting in 1849, and a number of the instruments were produced by a Paris optician, M. Duboscq-Soleil. When stereoscopic daguerreotypes were shown at the Great Exhibition of 1851, Queen Victoria ordered viewers, and photographers began to experiment with the new medium. Supported by such patronage, hand stereoscopes and thousands of stereoscopic views were available to the middle classes within a decade, and patents peaked during the 1870s. Regional practitioners produced images for limited audiences, but national firms began to record striking images to extend sales and give their viewers a sense of worldly sophistication. The subjects were wide-ranging: from landscapes, to monuments, to educational, cultural, or humorous topics. Companies around the world grew to exploit the fad. It is estimated that by 1862—about a decade after its introduction into America—there were over a thousand commercial photographers in America producing stereographs, or roughly the number of holographers active during the 1980s.[17]

[14] Peter Zec, quoted in Smith, P. J., 'Between the Lines', *Leonardo* 22 (1989): 291–4, p. 291.

[15] Gabor, Dennis to L. Cross, letter, 27 Apr. 1973, MIT Museum MoH 40 Correspondence XVIII.

[16] Bains, Sunny, 'Editorial', *Holographics International* 1 (1989): 4. For a more industry-centred account, see 'Holograms as giftware: the commercialization of artistic holography', *Holography Marketplace*, 8 (1999): 37–43. At the upper end of the market, and contrasting with garish costume jewellery, were the pieces by August Muth of Lasart Ltd, founded in Colorado in 1983 and New Mexico in 1991, and the associated DCG lab/studio shared with Ana MacArthur and Steve Siegel, Aurorean (1988–2005).

[17] Green, Harvey, '"Pasteboard masks": the stereograph in American culture 1865–1910', in: E. Earle (ed.), *Points of View: The Stereograph in America—A Cultural History* (New York: Visual Studies Workshop Press, 1979), pp. 109–15.

The introduction of the stereoscope during the 1850s mirrored aspects of the introduction of holography over a century later. Stereoscopic images had, in their day, been as awe-inspiring and shocking as holograms were later. Historian Harvey Green describes them as provoking 'at once wonder, exuberance, hesitation and confusion'. Moreover,

> the three-dimensional quality produced when an obviously two-dimensional image was viewed through the proper apparatus occasioned an intensified debate on the nature of reality and truth, art and science. In both Europe and America, analysts wrote often contradictory explanations of the nature, possibilities and power of the new photographic process. For some, the stereoscopic image was truth, fact.[18]

Stereo photographs were vaunted for their honesty, because they could not readily be retouched by photographers (the retouching would stand out from the plane of the photographic subjects, making artifice detectable). Pulsed hologram portraits proved difficult to market for precisely this reason.[19] Moreover, stereo photographs were described as more complete and beautiful than the two-dimensional photographs that had been available for more than a decade, and provided imagery that conventional paintings could not provide. Indeed, these debates of the 1860s employed arguments rehearsed for holography a century later:

> Form is henceforth divorced from matter. In fact, matter as a visible object is of no great use to us any longer, except as a mould in which form is shaped [. . .] Matter must always be fixed and dear; form is cheap and transportable.[20]

But most relevantly to holography, stereography did not endure. The historical attraction of stereo photography noted by Howard Becker applies equally to holography:

> We usually study successful artistic innovations, those which not only developed a national or international culture around the production and use of their typical products, but also persisted and became part of the main stream of work in that medium or genre. But stereography [. . .] eventually declined, and turned into a dead end.[21]

While stereography diffused into popular culture more successfully than did holography during the following century, ultimately it was a victim of changing fashions and competition from other media.

By the 1890s stereoscopic views had lost popular appeal. Publishers sought to extend the market, producing colour stereographs by three-colour lithographic printing, but by the First World War the experience of stereoscopic viewing had become distinctly

[18] Ibid., quotation p. 109.

[19] Munday, Rob to SFJ, interview, 31 Mar. 2004, Richmond, UK, SFJ collection.

[20] Holmes, Oliver Wendell, 'The stereoscope and the stereograph', *The Atlantic Monthly* III (1859): 738–48, p. 747.

[21] Becker, Howard S., 'Stereographs: local, national and international art worlds', in: E. Earle (ed.), *Points of View: The Stereograph in America—A Cultural History* (New York: Visual Studies Workshop Press, 1979), pp. 89–96, p. 89.

unfashionable. The market was nevertheless recaptured a generation later using a repackaged form of the technology. The View-Master, a moulded Bakelite stereoscope used to view seven pairs of colour transparencies mounted on a cardboard disc, was introduced at the New York World's Fair in 1939 and proved popular, particularly when it was retargeted for children from the 1960s. The new, robust packaging of the technology made it suitable for less discriminating audiences.

New adult audiences were also cultivated for the stereo photograph. *Amateur* stereography became possible with the introduction of the Stereo-Realist and competing stereo cameras from about 1948, and endured as a niche hobby until the mid-1950s. While stereo box cameras and folding cameras for amateurs had been available since the turn of the century, the Stereo-Realist cameras (Milwaukee, 1947–72) usually were used to produce stereo slide transparencies. This allowed more straightforward commercial processing, and was free from the démodé associations of cardboard-mounted stereogram photographs.

At about the same time (1952–56), 3D cinema had a boom of popularity as commercial studios sought to compete with early television. The live three-dimensional action added an exciting dimension to stereo cinematography, but 3D filmmakers overused the technology for its shock value. Thus typical film subjects (*Hondo* (1953)), starring John Wayne or Hitchcock's *Dial M For Murder* (1954)) gave way to a declining series of horror films (e.g. *Creature From the Black Lagoon* (1954) and *13 Ghosts* (1960)). Not surprisingly, the technology soon became literally stereotypical. A brief revival during the 1970s was limited to subjects enhanced by three dimensionality (e.g. a series of soft-core porn films such as *Prison Girls in 3D* (1972), *International Stewardesses* (1974), and *Wildcat Women* (1975), or mainstream thrillers like *Friday the Thirteenth Part 3* (1982) and *Jaws 3D* (1983)). During the 1990s, stereoscopic cinema was reintroduced in two settings: either for awe-inspiring large-screen high-resolution cinema such as IMAX theatres or for shocking attractions at theme parks such as Disney World.[22]

Two other three-dimensional media rising in popularity during the 1950s were the anaglyphic (2-colour) comic (1953–4) for children, and the Xograph, or lenticular-screen image, from 1952. Subsequent interest in 3D imaging was more muted. A notable later example is the Nimslo camera system, a stereo camera combined with commercial processing to yield lenticular photographs, which proved to have a limited market during the 1980s.[23] Most of these applications evinced a trajectory of popular acclaim, declining interest, niche novelty, and a transformation for the children's' market.[24] Holograms mirrored this trajectory.

[22] At Disney World, the *Honey, I Shrunk the Audience* attraction combined stereoscopic computer-generated animation with live action and tactile effects such as puffs of air and water sprays to generate realistic sensations.

[23] See Anon., '3D photography coming to Europe, inventor pair says', *holosphere* 7 (1978): 2.

[24] See Speer, L., 'Before Holography—a Call for Visual Literacy', *Leonardo* 22 (1989): 299–306; Darrah, William C., *The World of Stereographs* (Gettysburg, PA: Times & News Publishing Co., 1977).

13.3 PERSISTENT IDEAS

The evolution of holograms in popular culture was accompanied by a growing assortment of ideas concerning them. Beyond their visual enchantment, viewers were intrigued by more abstract properties of holograms. The notion of the full picture contained within a fragment was unintuitive and mysterious, and was rehearsed in many contexts as a general property of holograms, even for image-plane holograms (the most common variety since the 1970s) for which it scarcely applies.

The spread of ideas concerning holograms can be discussed in terms of Richard Dawkins' notion of the *meme*, an idea or intellectual construct rather analogous to the *gene* in biology.[25] According to Dawkins, the meme is a unit of intellectual information—an idea fragment—that resists decomposition and can be recombined to yield new and pervasive ideas. Like biological genes, memes are transmitted by being copied by their hosts: popular ideas are passed on and so propagate through culture and may be replicated in new variant forms. Such elementary ideas may or may not be objectively true; may or may not be advantageous to the individuals who harbour or propagate them; and may or may not prove culturally infectious (i.e. highly popular) or long lived.

The cultural perception and uses of holograms can be represented at least superficially in this way. Popular beliefs about holograms have arisen at different periods and proven more or less tenacious. A few of them, listed in roughly chronological order of appearance, are:

(1) holograms are a special, advanced form of photograph;
(2) any portion of a hologram includes the entire image;
(3) holograms project an image into space;
(4) holograms are improving steadily in technical quality;
(5) holograms are any form of three-dimensional image;
(6) holograms are recreations of an object from the past;
(7) holograms are three-dimensional objects generated by a computer;
(8) holograms flicker, buzz, and fluctuate;
(9) holograms can be walked around and viewed from any angle, exactly like a real object;
(10) holograms are an example of the interconnectedness of the natural universe.

Not all these beliefs are consistent with each other or even judged to be correct by expert audiences. Some (such as (1) and (2)) came from early holographers and popularizers. Others (such as (3), (4), and (9)) were misunderstood or created more cynically by

[25] See Dawkins, Richard, *The Selfish Gene* (Oxford: Oxford University Press, 1976) and Blackmore, Susan, *The Meme Machine* (Oxford: Oxford University Press, 1999). *Memetics*, or the study of the propagation and evolution of such *cultural replicators* of information, expanded rapidly from the early 1990s and with the rise of the Internet.

promoters and some (such as (6), (7), and (8)) by fictional cinematic or textual portrayals. And still others (such as (5)) have no identifiable point of origin at all. Yet each of them nevertheless found a durable niche in the form of a community to propagate it. Some of the ideas coexist, supporting each other and clustering together in what Susan Blackmore and other theorists of memetics dub *memeplexes*.

The value of the memetic approach is that it helps to explain the cultural trajectory of the hologram in ways that a more traditional technological history normally does not. Economic historians might argue that the history of the hologram can be understood through the market forces influencing the demand for particular applications. Deterministic historians of technology may interpret the history in terms of progressively improving technological capabilities. And philosophers of science might well be dismayed by the persistence of irrational or poorly substantiated claims about the medium. The public perception of holograms consistently has confounded the promoters of holography, too. As successive waves of engineers, holographers, advertisers, and museum managers discovered, some ideas about holograms were potent; others resisted transmission despite energetic attempts to disseminate them.

The sources and chronology of some of these memes can be identified. The idea of a close correlation between holograms and photographs, for instance, arose with little conscious construction from the early 1960s work of Denisyuk, Leith, and Upatnieks: both photographs and holograms were, after all, recorded on light-sensitive emulsions and both produced an image when suitably viewed. In fact, the initial cultural response to Leith–Upatnieks holograms closely mirrored that following the announcement of daguerreotype photographs to the French Académie des Sciences and Académie des Arts in August 1839:

> There was a great deal of excitement, and the crowd's reaction to the announcement was intense and immediate: within hours every optician in town was besieged with people trying to obtain cameras in order to share in the wonder of the new art-science.[26]

Other memes also spread widely. Public attitudes towards, and beliefs about, holography were shaped by expert predictions. Kip Siegel of Conductron, discussed in Chapter 6, was effective in creating memes and encouraging their replication in culture. His promotional activities—interviews, speeches, demonstrations, and samples—spread the idea that progress in holography was unstoppable and autonomous, that holograms would supersede other forms of imaging, and that they could record a faithful, if intangible, copy of any scene. Extravagant claims generated an understandable response from potential consumers. Arthur Funkhouser recalls a not-unusual case:

> A fellow from the public roads department came wanting us to provide expertise on a scheme he had for putting up holograms on the expressways. He wanted to have them project a 3-D brick wall for people wanting to drive up the wrong exit ramps the wrong way. I tried to

[26] Barger, M. Susan and William B. White, *The Daguerreotype: Nineteenth-Century Technology and Modern Science* (Washington: Smithsonian Institution Press, 1991), p.1. See also Rudisill, Richard, *Mirror Image: The Influence of the Daguerreotype on American Society* (Albuquerque: University of New Mexico Press, 1971).

convince him that it would take a hologram the size of a billboard and a very bright light source to make that really effective. I said that holograms of that size had not been made, but that perhaps one could put together a mosaic of them and do it that way. Although I found the scheme very impractical and tried to tell him so, he would have none of it. He had gone out on a limb, so to speak, in his department and his career there was at stake. They had even already put it out for bids and one company in California (where else?) had submitted a paper saying they were willing to study the question (for a price, of course). The fellow went ahead and convened a panel of experts which included me and a group of physicists from the Goddard Space Flight Center and we all tried as gently as we could to tell him his scheme would not work. He cried. We suggested other things he might think of doing to improve the roads, but he was not able to listen to us.[27]

Inflated claims about the capabilities of holography plagued its early years. A typical example was the touting of holographic media in a Los Angeles newspaper in 1970 to attract investment capital:

Strick Will Make Laser Beam Film

Joseph Strick, director of 'Ulysses' and 'Tropic of Cancer' has obtained a license to make the first movie in which actors will be seen in the round, and will even jump into the laps of the audience, without anyone wearing trick glasses.

The film will be a hologram motion picture, shot with the use of laser rays. Holograms can be projected so that the audience can see completely around the image presented—and a hologram actor can even mingle with the audience, like an incorporeal ghost [. . .]

The hologram was discovered by Drs. Dennis Gabor and Ernest Lieth [*sic*] in the early 1960s [. . .] at present audiences of only 50 to 100 persons can be seated in a hologram movie, but Strick says work is already under way on a different projection system that can be seen by 400 to 600 people [. . .]

Lasers, Strick says, are harmless. 'The worst you can get is a headache from it, because you are staring at pure energy—but then you can get a headache from gazing at an electric light,' he explains [. . .] His first hologram picture should be ready next year, a movie-in-the-round that may make the Woodstock generation feel slightly threatened by the pace of technological change.[28]

Given such extravagant claims, it is not surprising that public perceptions of holography were influenced. The New York Museum of Holography probably fielded more enquiries from the public than any other organization during the late 1970s and 1980s. Besides the face-to-face enquiries faced by the MoH staff at their numerous lectures and exhibitions, bulging correspondence files illustrate how holograms were a continued source of puzzlement and curiosity for the general public. The letters ranged from young enthusiasts seeking answers for practical projects, to bewildered enquiries about the modern world. For example:

I am a widow of 82 years whose neighbor has been trying to scare me for 4 years to sell my house for small price so they can build a garage for the son's antique cars. There appeared at

[27] Funkhouser, Arthur to SFJ, email, 9 Apr. 2003, SFJ collection. Similar anecdotes about holographic traffic signs have attained the status of urban myth, and were recounted by Matt Lehmann and Gary Cochran (Lehmann, Matt, 'Holography's early days at Stanford University', *holosphere*, 11 (9) (Sep. 1982): 1,4; Cochran, Gary D. to SFJ, interview, 6 and 8 Sep. 2003, Ann Arbor, MI, SFJ collection).

[28] 'Strick will make laser film', *Los Angeles Times*, 21 Aug. 1970, 14.

> first white vapor—plain. Now, comes also in shapes as circles, half circles, squares with perfect straight lines, horses head, chicken on nest, back of man with arms outstretched etc [. . .] Please answer—would appreciate! Does this sound as holography?[29]

A member of the museum staff replied patiently that such a phenomenon did not seem to be holography and that perhaps she should ask her neighbours about it.

By the late 1980s such ad hoc correspondence and educational work were sapping the museum's time. Even worse, newspaper items showed little evidence of a gradual transfer of information from holographers to the public. Each new article between the early 1970s and early 1990s repeated substantially the same information for readers who were geographically distributed or culturally constrained. Indeed, the appearance of a newspaper story or article on holography now demanded something special: a new exhibition, a new claim, or a new achievement. Moreover, the flow of information was increasingly impeded by new public understandings of holograms—or, in Dawkins' terms, the circulation and competition with more successful and persistent memes—based on movie and television portrayals.

These appealing ideas, initially passed on by enthusiastic scientists, technology journalists, and would-be entrepreneurs, mutated to become autonomous memes that developed a life of their own. Assertions of cynical promoters became detached from them and spread as an urban legend (itself an example of a meme) that resisted correction. The notion of the inherent progress of holography required no further promotion: it became an accepted fact underlain by persistent myths. With the rise of the World Wide Web, a repository for such memes became established, making them readily accessible and transmissible. The Web also promoted the dissociation of information from contemporary commentators, regurgitating old and inaccurate information to form persistent new combinations. Such mixed sources of dubious reliability diluted analyses and predictions and tended to produce anodyne and progressivist accounts.[30]

While enthusiasts and promoters more or less consciously emphasized the progressive and appealing ideas about holograms, others transmitted views inspired by other sources. Science fiction writers, in particular, absorbed prevailing ideas (the circulating memes) about holograms and extrapolated them to embodiments and applications of the future. These altered conceptions in two important respects: first, they built upon popular ideas of holograms, rather than the technical realities; and, second, they applied the expectations of technological progressivism. In effect, science fiction writers converted Kip Siegel's cynical marketing promotions into naïve predictions of technical achievement.

Fiction created and motivated an imagined future, as suggested by Yuri Denisyuk's recollections (§3.2) of having been influenced by the science fiction stories of Yuri Yefremov in 1957. While some of Yefremov's stories hinted at optical devices rather like holograms and Lippmann photographs, he was describing a hypothetical object when

[29] Abraham, G. to MoH, letter, 13 Aug. 1983, MIT Museum 2/27.

[30] See, for example, Pescovitz, David, 'The future of holography', *Wired*, Jul. 1995: 50.

writing during the late 1940s. From the 1970s onwards, science fiction writers employed holograms more frequently as a futuristic technology that, in combination with newspaper stories, seemed to represent an inevitable future.

Some fiction related the hologram to the wider world, using it merely as a metaphor rather than centrally as an artefact of the future. The first published short story to use a hologram as a plot device was William Gibson's 'Fragments of a hologram rose', published in 1977. The story, his first published work, concerns the musings of the narrator about a holographic postcard (a white light reflection hologram of a rose) sent by his girlfriend. Gibson's narrator compares people to the pieces of the laminated plastic hologram, which he has shredded into pieces:

> Parker lies in darkness, recalling the thousand fragments of the hologram rose. A hologram has this quality: Recovered and illuminated, each fragment will reveal the whole image of the rose. Falling toward delta, he sees himself the rose, each of his scattered fragments revealing a whole he'll never know.[31]

Thus each person represents the fragments of another through the limitations of their relationship, just as the fragments of the hologram show the whole image from a different angle. But in combination with this restrained and realistic analogy, Gibson also makes passing reference to the (fictional) decline of the popularity of holographic amusement arcades, 'the block-wide Fuller domes that had been the holo temples of Parker's childhood'.[32] For Gibson, two visions of the hologram coexist: it is simultaneously a simple object like the 1970s hologram as well as a powerful future technology for producing a simulated three-dimensional environment. Atypically of subsequent writers, the hologram in Gibson's world was an expression of decline, and even dystopia.

Other science fiction writers were beginning to employ holograms in equally extreme extrapolations. Films from the mid-1970s began to represent the hologram as a synthetic three-dimensional *projection*, an attribute neither claimed nor predicted by contemporary holographers. In *Logan's Run* (1976), Multiplex holograms were used to represent disembodied heads rather than holograms *per se*, and can perhaps be discounted as directly influencing public perceptions of the medium (although they did produce publicity for holography through movie trailers and interviews).[33] *Star Wars* (1977) was the first film to portray holograms of the future directly: the robot R2D2 (see Figure 13.4) displays a recorded three-dimensional, miniaturized, and animated image of Princess Leia. The hologram first appears in the second draft script of 1975, and gets refined in subsequent drafts.

[31] Gibson, William, *Burning Chrome* (New York: Ace Books, 1987).

[32] Gibson (b. 1948), a Canadian writer who became known as the originator of the *cyberpunk* science fiction style, characterized by bleak, dystopic depictions of technology (in contrast, for example, to the politically idealistic and technologically progressive work of Yefremov), was also noted as the conceiver of *cyberspace*, or computer-generated environments, recognition that appeared during the 1990s birth of the Internet and its computer-mediated environments of chatrooms, distributed databases, and simulated environments.

[33] One such trailer highlights Chris Outwater, described as master holographer, as the source of these modernistic special effects.

Fig. 13.4. Iconic representations: a hologram reconstructed by robot R2D2 in the film *Star Wars* (Lucasfilms Ltd).

> The fragment breaks loose with a snap, sending Luke tumbling head over heels. He sits up and sees a twelve-inch three-dimensional hologram of Leia Organa, the Rebel senator, being projected from the face of little Artoo. The image is a rainbow of colors as it flickers and jiggles in the dimly lit garage. Luke's mouth hangs open in awe.[34]

The eventual cinematic representation distanced itself from what sounds like a Multiplex hologram. The film showed the holographic image as monochrome, rather than rainbow-tinged. But George Lucas's depiction added another element to cinematic holograms: its flickering and stuttering image suggested intrinsic links with electronic communication media like a television having poor reception. Through this unwitting fictional depiction, the archetypal hologram for cinema audiences inadvertently reunified optics and electronics, the fields that had merged to generate holography decades earlier.[35]

Television science fiction began to employ holograms as plot devices soon after *Star Wars*, but with distinctly different attributes. The British comedy television series *Red Dwarf* (pilot written in 1983, series broadcast in Britain 1988–94, and set onboard a spaceship of the distant future) extended the capabilities of fictional holograms far beyond contemporary expectations. One of the characters was a hologram, calculated in real time by the spaceship's computer, his form reconstructed by a 'light-bee', which rapidly painted his three-dimensional image on the fly. In early episodes, the character was described as being generated by 'Soft Light', and could not be touched; he was later said to be rewired with a 'Hard Light drive' to make him solid, thus extending the notion of a hologram from a mere optical effect to one imbued with mechanical, as well as electronic, underpinnings.[36]

[34] Lucas, George, 'Star Wars script (revised fourth draft)', 15 Jan. 1976.

[35] Subsequent films in the *Star Wars* series depicted holograms as static displays like paintings or lobby displays; as control panel imaging devices; and as superior playback devices that showed recordings in magnified or reduced form.

[36] http://www.reddwarf.nildram.co.uk/rimmer_fr.htm.

A similar idea was developed in the second *Star Trek* television series *The Next Generation* (broadcast 1987–94) with its stories of a 'holodeck', or holographic visualization room, in which an entire environment can be reconstructed by computer (and, indeed, interacted with by the human characters). As in *Red Dwarf*, the fictional optical technology was gradually augmented by other capabilities: some objects in the holodeck were described as being 'replicated' (that is, material objects manufactured from individual atoms) and animated with 'weak tractor beams' and 'shaped force fields'. In the succeeding series, *Deep Space Nine* (1993–99), the technology becomes commercialized: 'holosuites', or holodecks of various capacities, are rented out for private use. And in the *Voyager* series (1995–2001) these technologies are used to implement an Emergency Medical Hologram, or virtual doctor. This virtual character, and his tribulations as a sentient computer program embodied in a computer-generated hologram, spawned a popular book.[37] Thus the ontology of popular belief was extended: the hologram became a staple of science fiction plots, alongside time travel, robots, black holes, and interplanetary travel. Indeed, one online science fiction magazine adopted the title *Hologram Tales* not because of a central interest in these plot devices or the underlying technology, but merely because of the ubiquity of the concept in contemporary science fiction. Most importantly, the science fiction holograms were paradigms of progress, like the stories in which they were immersed. From the late 1970s, then, there was an increasingly obvious bifurcation of technologists' and science fiction writers' perceptions of the hologram.[38]

As a result of these factors, technical realities diverged from popular expectations. Entrepreneurs bemoaned the impossibility of satisfying an increasingly demanding public.[39] Rob Munday of Spatial Imaging Ltd complained, 'We have people calling us up all the time, asking us for something they've seen on *Star Trek*—almost like holography isn't good enough any more'.[40] The fictional representations offered much more than real holographers could hope to provide, as Fred Unterseher reflected:

> The novelty was the attraction; isn't that always the hidden level of the anticipation/expectation of the three-dimensionality, [. . .] always so far ahead of the reality? I have to start my classes with what holography *isn't*. So many of my students assume that if they've seen it in a movie, if they've seen it on television, then that's reality. And there's this lack of context in what is really real, and it's profound; they have this *expectation*.[41]

[37] Picardo, Robert, *Star Trek Voyager: The Hologram's Handbook* (New York: Pocket Books, 2002).

[38] See also Pizzanelli, David, 'Evolution of the mythical hologram', *Proceedings of the SPIE The International Society for Optical Engineering* 1732 (1992): 430–7. A similar observation has been made about popular understandings of cold fusion, which passed from 'science' to 'an object of science fiction or, worse, the pathology of science' (Simon, Bart, *Undead Science: Science Studies and the Afterlife of Cold Fusion* (New Brunswick, NJ: Rutgers University Press, 2002), quotation p. 95).

[39] Alexander, Kent, 'Seller beware', *holosphere* 13 (1985): 10.

[40] Munday, Rob to SFJ, interview, 31 Mar. 2004, Richmond, UK, SFJ collection.

[41] Unterseher, Fred to SFJ, interview, 23 Jan. 2003, Santa Clara, CA, SFJ collection.

13.4 HOLOGRAMS AS ANALOGY AND PARADIGM

Unterseher's challenges in teaching practical holography were a consequence of incompletely overlapping cultural ideas and competing memes. The holographic television characters threatened to submerge the visual excitement of real holograms. Holographers, though, were extending their own understandings in other, non-technical, directions. Unterseher's practical workshops, for instance, were imbued with a wider sense of what holography and the hologram represented. During the 1980s, he and others increasingly portrayed the subject as an expression of what had been, a decade earlier, hinted at with countercultural themes. One newspaper story entitled 'A laser-man's way to altered reality' reported of him:

> What could have passed for the set of a B-grade horror movie was in fact the dimly lit Studio Thirty-Five of Fred Unterseher. Standing amid floating images and red laser light, Unterseher told his audience 'In at least nine ways, I can alter your perception of reality in a clear, objective state of mind'. Unterseher was lecturing on the art of holography and how to make your own holographs [*sic*] inexpensively. Holography is a process in which laser light is reflected off a subject and onto a specially treated plate of glass. After a series of development steps, a three-dimensional image of the original object is created, appearing to sit in mid-air. 'It's only as complex as we make it', Unterseher said. 'There is no need for lengthy mathematical formulas to enjoy holography'. The holograph [*sic*] is a three-dimensional 'photograph or slide'. As eerie as it may look, Unterseher insists it is no trick or illusion. 'We live in a world of light as a fish lives in a world of water'. 'There is interaction and defraction [*sic*] of light all around us if we could perceive it as such', he said [. . .] Unterseher's job, as he sees it, is to simplify and demystify holography and make it available to the curious. He offers classes as well as free introductory lectures the first Saturday of every month.[42]

Such heady ideas coexisted uneasily with the growing orthodoxy of science-fiction portrayals of holograms. They fit well, however, with expanding scientific views. From the late 1960s, holograms came to be used as an analogy of other processes, particularly human memory and the interconnectedness of physical phenomena.[43] By the 1980s, it extended to speculative cognitive science, cosmology, and even interpersonal relations.[44]

The idea that the hologram could be conceived as an analogy for the human brain was explored from the mid-1960s. It was provoked mainly by the characteristic that the holographic image (at least in the Leith–Upatnieks diffusely-illuminated hologram) was non-localized: one perspective of the reconstructed object was available from every

[42] 'A laser-man's way to altered reality', *Daily Californian*, 17 Oct. 1977, 31.

[43] 'Computer men see laser trick', *Times*, 7 Aug. 1968, 2.

[44] These were most developed in the works of physicist David Bohm and psychophysiologist Karl Pribram, discussed below. The analogy between holography and self-actualization and interpersonal relationships was expounded in, for example, Levenson, Edgar A., 'A holographic model of psychoanalytic change', *Contemporary Psychoanalysis* 12 (1976): 1–20, and even Miller, Larry, *Holographic Golf: Uniting the Mind and Body to Improve Your Game* (New York: HarperCollins 1993).

portion of the plate. Neurophysiological experiments investigating the storage of human memories also indicated that memories were not localized in any one region of the brain. An early example of such speculations came from G. H. McLaughlin, a lecturer in Human Communication at the City University, London:

> Dear Professor Gabor,
>
> Can you please spare a couple of minutes to consider a Partly-Baked Idea? I should explain that my specialisation is the psychology of language, but my PBI involves neurophysiology and holography. Because the latter is more crucial I am writing to you.
>
> As I understand it, a hologram can represent an object as a series of interference fringes between coherent light reflected from that object and a reference beam. I also believe that information about the entire object can be present in all parts of a hologram. Ablation studies suggest that information stored in the brain is similarly diffused.
>
> Could it then be that, on entering the brain, neural impulses initiated by the stimulation of a sense organ modify the 'reference' alpha-rhythm, caused by the spontaneous synchronised discharge of cortical cells, to produce 'beats', analogously to the reference beams producing interference fringes in holography? It could be assumed that the pattern of beats is projected (along paths made available by more or less random processes) throughout the association areas of the cortex, thus distributing the information as in a hologram. If we further assume that the mere use of these pathways would facilitate their future use, then any later presentation of a stimulus similar to the first would tend to be channelled along many of the same pathways, thereby evoking reactions similar to those evoked by the first stimulus.
>
> Thus holography might be considered to be a model of memory. The model could be made to cover visual and other types of memory by taking into account the topological characteristics of the various kinds of stimuli and their appropriate sense organs.
>
> The catch may be that alpha rhythm has a frequency of only 10 c.p.s [cycles per second, or Hertz]. Is this adequate to cope with the colossal rate of information intake through the visual system? Perhaps I have not taken sufficient account of the requirement of 'coherence' in the sense organs . . .
>
> In any case I would be deeply obliged if you would let me know whether this PBI deserves the oven or the garbage pail.[45]

McLaughlin's speculations were even then beginning to circulate as a meme. Gabor replied that he already had heard of the idea from neuroscientist Jack Cowan, 'but it must have got about quite a bit, because the other day somebody from the B.B.C. rang me up about it'.[46]

Gabor himself published attempts at mathematical models two years later,[47] but the idea was more publicly explored by psychophysiologist Karl Pribram. These new ways of thinking about holography also entailed new ways of categorizing its practitioners or

[45] McLaughlin, G. Harry to D. Gabor, letters, 22–24 Nov. 1966, London, IC GABOR MM/3.

[46] Gabor, Dennis to G. H. McLaughlin, letter, 24 Nov. 1966, London, IC GABOR MM/3. Cowan, later a professor of mathematical neuroscience at the University of Chicago, was at the Imperial College Engineering Department at the time.

[47] Gabor, Dennis, 'Holographic model of temporal recall', *Nature* 217 (1968): 584 and Gabor, Dennis, 'Improved holographic model of temporal recall', *Nature* 217 (1968): 1288–9, both of which were responses to Longuet-Higgins, H. C., 'Holographic model of temporal recall', *Nature* 217 (1968): 104.

adepts. A more inclusive, if less skills-based, grouping was proposed. Thus Karl Pribram suggested that the adherents of the predominant neurophysiological theory differed from supporters of his own ideas in both intellectual and social respects. The former, concentrated on the American East coast, he termed 'feature creatures', as opposed to members of Pribram's own ideological camp which he labelled 'frequency freaks'.[48] As the facile terms suggest, most of Pribram's writing on the subject appeared in semi-popular accounts and conference proceedings.[49]

During the same period, David Bohm, a professor of Theoretical Physics at Birkbeck College, London, elaborated a different analogy between holography and wider aspects of perception and, indeed, the physical world. His scientific views had been influenced by the Indian philosopher J. Krishnamurti, with whom he engaged in extended conversations or dialogues as a route to new knowledge. Bohm found strong correlations between his ideas on quantum mechanics and Krishnamurti's thinking. First meeting in 1961, they had numerous dialogues over the subsequent thirty years.[50] Bohm had corresponded with Gabor as early as 1952, responding to Gabor's request for information about quantum mechanical calculations (relating to Gabor's own work on electron gases in electronic devices) and providing a paper on his ideas concerning causal interpretations of quantum mechanics.[51] Their next communication was seventeen years later, this time concerning how holography might provide insights into quantum mechanics. Bohm wrote to him in 1969:

> Dear Professor Gabor,
>
> I have been thinking about holograms for a long time. It has struck me very forcibly that the hologram has an even greater significance for philosophy and for language than it has for technology. In short, as in the time of Galileo, the lens, under the relationship of image and object, became a paradigm of human thinking, especially in science and out, so it seems that the hologram can now lead us to a new paradigm. The lone order in the older paradigm was image–object correspondence. In the newer paradigm, it is a relationship of the whole to the whole.
>
> It seems to me that the hologram helps lead us to a new informal language for discussing quantum theory, relativity, and other subjects. I would very much like to discuss this with you some time. Also, I would like actually to see a hologram in operation.[52]

[48] Anon., 'Holographic model of brain function explicated by Pribram at conference', *holosphere* 7 (1978): 1–2.

[49] See, for example, Pribram, Karl H., 'The neurophysiology of remembering', *Scientific American* 220 (1969): 73–86; Anon., 'Holographic model of brain function explicated by Pribram at conference', *holosphere* 7 (1978): 1–2; Goleman, Daniel and Karl Pribram, 'Holographic memory', *Psychology Today*, 13 Feb. 1979: 71–84; Pribram, Karl, H., 'Rethinking neural networks: quantum fields and biological data', presented at *First Appalachian Conference on Behavioral Neurodynamics*, Hillsdale, New Jersey, 1992,

[50] David Joseph Bohm (1917–92) was born and studied physics in Pennsylvania, studying and working under Robert Oppenheimer during the Manhattan Project to develop the first atomic bomb. Investigated by the House Un-American Activities Committee in 1951, he took posts in Brazil, Israel, and finally Britain, where he was Professor of Theoretical Physics at Birkbeck College of the University of London from 1961 to 1987 (Hiley, B. J., 'David Joseph Bohm', *Biographical Memoirs of Fellows of the Royal Society* 43 (1997): 105–31).

[51] Bohm was best known for his interpretation of quantum mechanics as a 'hidden variable', or deterministic, theory, in opposition to the Copenhagen interpretation of Neils Bohr and others, which some physicists, notably Einstein, felt represented an unsatisfactory metaphysical turn.

[52] Bohm, David to D. Gabor, letter, 14 Mar. 1969, London, IC GABOR MB/9.

Although Gabor's archives illustrate his lifelong eclectic interests, the extant papers show no further correspondence with Bohm.[53] Bohm explored these ideas at length, but published relatively little on the subject. His approach to philosophy and physics are expressed in two books: *Wholeness and the Implicate Order*, and *Science, Order and Creativity*.[54]

Alongside the expanded scientific analogies concerning holograms, scholars in the humanities began to note their cultural ramifications. Literary critics cited the hologram as metaphor for the zeitgeist.[55] The literary theorist and novelist Umberto Eco, for example, extending ideas framed by social philosopher Jean Baudrillard, discussed holograms as a technology of hyper-reality. He argued that the hologram exemplifies a characteristic of the information age, presenting a simulation of reality that is 'more real than real'. As a result, there is a reversal of reality and simulation, with 'real' things made to copy images portrayed by the media. One consequence is that the boundaries between culture and its media representation are dissolved, with the latter dominating. He notes, 'Holography could prosper only in America, a country obsessed with realism, where, if a reconstruction is to be credible, it must be absolutely iconic, a perfect likeness, a 'real' copy of the reality being represented'.[56]

Popular engagement with holography shifted further during the early 1980s when, owing largely to the writings of Pribram, Bohm and popularizer Ken Wilber, holograms were increasingly touted as a fruitful analogy or even model of human consciousness and the physical universe and emblematic of a trend towards holistic thinking in science. The 'holographic paradigm', an imprecise cluster of notions popularized by Ken Wilber, was used to cite links between psychology, cosmology, and fiction.[57]

Among optical scientists, holography was also gradually extending its meaning. Adolf Lohmann argued, for example, that the principles underlying holography had imbued all of optics, and were the most significant organizing concept in the field.

[53] Neither the Gabor files at Imperial College, London and the MIT Museum, Cambridge, MA, nor the Bohm archive (Birkbeck College, London) have any further correspondence between Bohm and Gabor or on the subject.

[54] Bohm, David, *Wholeness and the Implicate Order* (London: Routledge, 1980) and Bohm, David, and F. David Peat, *Science, Order, and Creativity* (New York: Bantam Books, 1987). See also Bohm, David, 'Quantum theory as an indication of a new order in physics. II. Implicate and explicate order in physical law', *Foundations of Physics* 3 (1973): 139–68.

[55] See, for example, a discussion of Baudrillard's 1981 *Simulacres et Simulations* and Kac, Eduardo, 'On Baudrillard's text "Hologrammes" ', *holosphere* 17 (1990): 25–6.

[56] Eco, Umberto and William Weaver, *Travels in Hyper-Reality: Essays* (London: Picador, 1986), p. 1.

[57] For some of these extensions to the holographic analogy, see Wilber, Ken, *The Holographic Paradigm and Other Paradoxes: Exploring the Leading Edge of Science* (Boulder: Shambhala, 1982); its application to literature (Martinez, Zulma Nelly, 'From a Mimetic to a Holographic Paradigm in Fiction: Toward a Definition of Feminist Writing', *Women's Studies: An Interdisciplinary Journal* 14 (1988): 225–45); to psychology (Vandervert, L. R., 'Chaos Theory and the Evolution of Consciousness and Mind—a Thermodynamic–Holographic Resolution to the Mind–Body Problem', *New Ideas in Psychology* 13 (1995): 107–27); and, to of a deeper connection with cosmology (Gibbons, G. W., 'Holography and the future tube', *Classical and quantum gravity* 17 (2000): 1071–80; Tipler, F. J., 'The ultimate future of the universe, black hole event horizon topologies, holography, and the value of the cosmological constant', *AIP Conference Proceedings* (2001): 769–74).

Indeed, holography could be applied to subjects as disparate as optical processing, acoustic imaging, and super-resolution.[58]

13.5 CONFLICTING VALUES AND COMPETING MEANINGS

While they bolstered the relevance of holograms in wider culture, these wider interpretations caused overt dissension among holographers. Stephen Benton and Posy Jackson, for instance, began exchanging examples of what Benton dubbed 'holo-litter' in the late 1970s.[59] One such example he provided was a book blurb from a University of Cincinnati assistant professor's guide on biofeedback:

> A hologram is the fascinating and mysterious new medium that materializes four-dimensional light sculpture (it moves) in the center of a ring of interference patterns. A hologram is the model for the human energy rings designed by Dorelle Markley Heisel, during which each person makes himself apparent—that is, turns her present self into the nurturing source and protector of his Artist, the kind and unified combination of the sane, logical and free creative Child s/he lights into life over and over again.[60]

Similarly, physicist John Caulfield publicly rebuked the extrapolations of the holographic paradigm, but was preaching mainly to the converted at an SPIE conference in 2002.[61] Some aesthetic holographers, too, riled against 'hippies babbling cosmic nonsense' about a 'wildly speculative and half baked philosophy, appealing but unsubstantial'; for Michael Sowdon, this was a problem of separating 'the wheat from the chaff'.[62] In effect, Caulfield's as well as Sowdon's criticisms were myopic identifications of the different subcultures of scientific, artisanal, and aesthetic holographers. But no one community had authority over the meaning of the hologram or holography any longer.

By the early 2000s, the understandings of holograms and holography were fractured, like Gibson's hologram rose, into limited perspectives for different audiences. An illustration of the lack of impact was the production of a hologram of Queen Elizabeth II in 2004 to mark 800 years of the island of Jersey's loyalty to the Crown (Figure 13.5). The hologram itself—a 3 × 4 ft rainbow transmission holographic stereogram based on a sequence of 205 digital photographs and lit with the light from blue LEDs—represented the state of the art. But evoking echoes of prior controversies, the holographers involved

[58] Lohmann, Adolf W., 'The holographic principle', in: W.-M. Boerner (ed.), *Inverse Methods in Electromagnetic Imaging* (Boston: Reidel, 1985), pp. 1033–42; Lohmann, Adolf W. to SFJ, fax, 1 Dec. 2003, SFJ collection.

[59] Benton, Stephen A., 'Artist files,' MIT Museum 26/584, 1978.

[60] Heisel, Dorelle, *Human Holograms and the Biofeedback Guide* (Cincinnati: Metropolitan Printing, 1978)

[61] Caulfield, H. John, 'Holographic brain: a good analogy gone bad', *Proceedings of the SPIE The International Society for Optical Engineering* 4737 (2002): 124–30.

[62] Sowdon, Michael, 'The sixties revisited', *Wavefront* 1(3) (1986) 4.

Fig. 13.5. A mark of establishment: hologram of Queen Elizabeth II, by Rob Munday, Chris Levine, John Perry, and Jeffrey Robb, 2004 (white-light transmission holographic stereogram) (J. Perry photo, Munday Collection).

in the production, Rob Munday and Jeff Robb of Spatial Imaging in Richmond, UK, and John Perry of Holographics North, Vermont, USA, garnered little publicity and struggled to share credit with the conceiver of the project, Chris Levine. Few practicing holographers learned of the piece except through Internet bulletin boards. Even fewer news stories tracked the event; those that did either transmitted the lacklustre press release verbatim or used it as an opportunity to rehearse the ambivalent public engagement with the medium.[63]

Some sixty years after its invention, then, the hologram evoked distinct visions for different users. For some communities it was a mere tool, and one increasingly replaced by simpler alternatives. For others, it was a transcendent expression of holism and the murky frontiers of cosmological science. For yet others, it was a technology that was destined to blossom in the near or distant future. For the wider public, however, it was more often regarded as an expression of low-order novelty and a medium unequal to its promises.

[63] Duffy, Jonathan, 'Holograms: high art of just a gimmick?' *BBC News*, 23 Jun 2004, webpage.

14

Conclusion: Creative Visions

Over the span of half a century, ideas, practices, and communities co-evolved to create a new science. Holography was interwoven with distinct theoretical threads; its technologies evolved to yield exotic species of hologram adapted to distinct environments; and its communities of practitioners, critics, and audiences developed contrasting perspectives.

This chapter highlights those different visions of holography to emphasize a central theme of this book, that the subject cannot be separated from its communities. And, given the limited coalescence of those communities, no synoptic view can fully encapsulate the subject. These multiple perspectives—both optical and social—are a defining feature of holography and the dominant theme of this book.

14.1 THROUGH A GLASS DARKLY: VISIONS OF SCIENTIFIC GENESIS

Just as I have identified the emergence of broad types of holographer having distinct visions of their field, so too there are different outsiders' perspectives. This book has enrolled several disciplinary viewpoints to study how the new field of holography came to be. The range of such approaches has multiplied over the past century as historians, philosophers, social theorists, and scientists discarded simpler but unsatisfactory explanations of how new subjects evolve. Until recent decades, in fact, two mutually supporting understandings have suggested that such study is largely unnecessary.

The first basis for devaluing historical study came from philosophy of science, which was dominated by positivism during a century of rapid technological and scientific expansion (*c.*1860–1960). Although philosophers and scientists largely parted ways from the beginning of the twentieth century, it is fair to say that most practising scientists during that period, and beyond, subscribed implicitly to the tenets of positivism. And the confidence of positivism spread beyond the physical sciences to the social sciences, medicine, economics, and wider culture. As defined by its originator, Auguste Comte, during the 1830s (and extended as *logical positivism* or *logical empiricism* from the 1920s), positivism concerns itself with observable quantities, rather than with theoretical constructions, to yield verifiable 'positive knowledge'. It argues that such empirical knowledge increases incrementally and inexorably over time. Its claim about the universality of expanding knowledge implies that all new science follows a reliable trajectory, beginning with imprecise speculations and even metaphysical descriptions and ascending to mathematical formulations and reliable predictions. A positivistic understanding of scientific knowledge suggests, then, that the detailed analysis of emerging subjects is largely pointless: any subject of worth will progress naturally over time in a regular fashion. And, because this process of observation is seen as universal, no one investigator or team is crucial, in the long run. According to this view, writing the history of science need be little more than recording who added a new fact or quantum of information to the growing pyramid of knowledge, if indeed it is worth acknowledging such a regular process at all. As physicist and historian Abraham Pais concluded regretfully, 'physics is an ahistoric discipline'.[1]

A related understanding that became popular over the same century was *technological progressivism* and its more radical form, *technological determinism*. Based on the empirical observation of progress—the growing power, efficiency, economy, and complexity of machines, for example—it argues that this is a natural and irresistible process. Like positivism, progressivism assumes that applied knowledge will increase inevitably over time in a regular and natural manner. Simple mechanisms will be the basis of better, more sophisticated mechanisms, and so there is an inbuilt tendency for improvement. Technological progressivists generalize and extend their claims considerably, though: they argue that the technical expansion that got underway during the nineteenth century is a template for all human societies. Indeed, Victorian writers were fond of citing the Industrial Revolution, and its prominent engineers, as a moral example of how hard work leads to more rapid inventive progress.[2] So-called *Whig history*, in which Britain became the exemplar of social, economic, and political advance, extended such notions of progress beyond science and technology to wider culture. Like positivists, progressivists argue for the universality of their claims, that is, that societies inevitably follow a trajectory of technological expansion and dependence, and that this technological development is a feature of an 'advanced' society.

Even more contentiously, progressivism can be extended to determinism, which argues that this process is largely beyond human control, and that societies naturally adapt to the inevitable progress of technology. According to determinists, societies will evolve in lockstep with their technologies, carried along on a relentless wave. As summarized by the motto of the Century of Progress International Exposition held in Chicago in 1933, the rather depressing determinist claim is that 'Science Finds—Industry Applies—Man Conforms'. Here again, the writing of technological history reduces to a list of heroes and dates. It also suggests that forecasting the future is a straightforward

[1] Pais, Abraham, *Niels Bohr's Times: in physics, philosophy, and polity* (Oxford: Clarendon Press, 1991), p. 29.

[2] See, for example, Smiles, Samuel, *Self-Help: With Illustrations of Character and Conduct* (London: John Murray, 1859) and his *Lives of the Engineers* (London: John Murray, 1861, four volumes).

matter: inventions and technologies will continue to improve in an upward spiral, forever altering and benefiting society.

From the 1950s, however, philosophers and historians of technology began to turn away from such simplistic accounts of change. Thomas Kuhn, for example, argued that there have been many occasions in intellectual history when the corpus of knowledge changed abruptly rather than by following a smoothly ascending slope.[3] Rather than focusing on observation and experiment as the key ingredients to new knowledge, Kuhn and others stressed the seminal role of conceptual schemes. For Kuhn, the elaboration of *paradigms* is the central activity of new science: these conceptual frameworks could be tested, tweaked, and even terminated by empirical observations. Yet Kuhn's historical studies indicated that paradigms are not universal: they are constructed by groups of researchers, and can differ between them, or be discarded following the *Gestalt* shift of a scientific revolution. His work opened the door to the consideration of social factors in the creation of new knowledge. Historians of technology, too, began to interpret 'progress' as a culturally defined concept. Since Kuhn, the social and cultural aspects of science and technology have attracted an increasingly critical gaze.

Nevertheless, such new anti-positivist perspectives had little impact on practicing scientists, entrepreneurs, and the wider public, who had long been conditioned to expect continued scientific and technological change. The understanding of real-world sciences was pigeon-holed in two ways: either by being mapped onto the reassuring linear model of progress, and hastened by great men and their breakthroughs, or else relegated to the class of failed subjects, undeserving of attention.

The case of holography does not follow these models well. As I have argued, there was no breakthrough moment, but instead characteristically tentative evolution of ideas. The subject of holography emerged after some twenty years of research in the subjects of wavefront reconstruction, wave photography, and lensless photography in dissimilar intellectual contexts. Holography was not positivistic: rather than building on past knowledge, the subject was reinvented afresh in variant forms in different environments. And, in contrast to Comte's expectation, holography accumulated layers of metaphysical interpretation as it aged, rather than shedding them.

Nor does holography conform well to anti-positivist models of intellectual development introduced during the 1960s. It was not a neat case of theoretical generalization, or merely fortuitous and experimental discoveries that mapped onto a conceptual scheme. Instead, we can identify periods during which the subject was shaped by different dominant influences. Between about 1947 and 1963, as discussed in Chapters 2 to 4, the subject was largely theory-driven. The work of Gabor, Rogers, El-Sum, Baez, Lohmann, Leith, and Upatnieks was impelled by slowly developing insights and generalizations, and rewarded by gradually improving technical capabilities. Nevertheless, this was not a matter of testing competing theories, but instead separate, and seemingly disconnected, goal-directed searches

[3] Kuhn, Thomas S., *The Structure of Scientific Revolutions* (Chicago: University of Chicago Press, 1962).

Between about 1964 and 1972, by contrast, the subject expanded primarily by unguided experimental explorations. This is typified by the multiple occurrences of rediscovery and the proliferation of techniques and terminology for products discussed in Chapters 6 and 7. And between about 1972 and 1983 the subject was reshaped to suit the new communities that appropriated it. On the other hand, the dominant forces influencing the subject from the early 1980s were economic markets, increasingly steered by entrepreneurs having less technological commitment to the art of holography.

According to this periodization, holography sprouted successively as a 'subject' (an intellectual domain with theoretical underpinnings), a 'medium' (a collection of experimental techniques and products), an 'identity' (a social locus for holographers), and a 'market' (a collection of economically viable commodities, consumers, and application niches). For that reason, this roughly chronological series of forces also has been the basis for the organization of the chapters of this book.

14.2 ASSESSING PROGRESS: VISIONS OF SUCCESSFUL TECHNOLOGY

While contemporary historians had a diffident engagement with holography, holographers were vocal in assessing its progress. Practitioners are often the principal narrators of the evolution of a young technical subject, and the first judges of its significance and potential. As hinted above, while the validity of technological progressivism has been eroded over recent decades within such communities of technologists and scientists, it has continued to inform judgments of success. Historians of science and technology can inadvertently sustain such viewpoints by omission, overlooking subjects that do not demonstrate commonly recognized indicators of achievement.[4] Such criteria are usually taken to include the intellectual, cultural, and economic impact of new sciences and technologies. Other, sociological, indicators may include the emergence of a disciplinary presence in academic curricula; a professional identity; and, the growth of occupations related to the new subject. Yet the absence of some of these characteristics excludes a wide range of subjects in science and technology from consideration, and indeed recent studies by historians and sociologists argue, as mentioned in earlier chapters, that such fields represent a distinct class they dub *research-technologies*.[5] These unstable subjects

[4] By contrast, perceptive analyses of 'failure' include, for example, Terry Shinn, 'Failure or success? Interpretations of 20th century French physics,' *Historical Studies in the Physical and Biological Sciences* 16 (1986): 353–69; Elzen, Boelie 'The failure of a successful artifact: the Svedberg ultracentrifuge,' in: S. Lindqvist (ed.), *Center on the Periphery: Historical Aspects of 20th-Century Swedish Physics* (Canton, MA: Science History Publications, 1993), pp. 347–77; Kunkle, Gregory C. 'Technology in the seamless web: "success" and "failure" in the history of the electron microscope,' *Technology and Culture* 36 (1995): 80–103; Graeme Gooday, 'Re-writing the "book of blots": critical reflections on histories of technical failure,' *History and Technology* 14 (1998): 265–91; and McCray, Patrick 'What makes a failure? Designing a new national telescope, 1975–1984,' *Technology and Culture* 42 (2001): 265–91.

[5] See Joerges, Bernward, and Terry Shinn (eds), *Instrumentation: Between Science, State and Industry* (Dordrecht: Kluwer Academic Press 2001) for contributions on research-technologies.

resist recent sociological explanations of consensus. They may not, for example, show convincing closure of technical and intellectual debates.[6] By exploring the evolution of holography—a subject that does not follow simple models of development—this book confronts beliefs about progress, historical change, and technical identity.

The case of holography demonstrates how interest groups can differently assess 'success' and 'failure', and thereby influence the fate of the technology and its subsequent historical evaluation. But a more sensitive approach than studying these two alternate end-points is to study attributions of *progress* for a subject-in-the-making. During the evolution and lifetime of a technology, outright success and failure are seldom judged; instead, practitioners and adopters evaluate progress so as to apply corrective measures, make decisions about adoption, or revise forecasts. Only in retrospect does the subject acquire the totalizing label 'success' or 'failure'. By observing how progress is evaluated group by group and case by case, we can gain a clearer understanding of their effects on the technological trajectory and ultimate judgment of a subject, and how they relate to historians' assessments. Such analysis can reveal the overgeneralizations and unbalanced perspectives that promote faith in technological determinism.

A study of this kind is perhaps easiest for a young subject that has attracted several technical constituencies and for which consensus has been elusive. From its conception in 1947, the field that became holography has mutated repeatedly and successively enrolled new communities of practitioners, sponsors, and consumers. While in some respects a typical post–Second World War technical subject, holography has been unusually wide-ranging in the applications and social groups that it encompassed. As we have seen, the subject found relatively stable niches as a scientific specialty, technical solution, and art form, but attracted ambivalent assessments of progress. Holography has been both vaunted and criticized based on the contrasting criteria of its unusually broad range of technical communities. As a result, it is a rich historical case for exploring attributions of progress, success, and failure.

How do the backgrounds of different communities, and changing scientific, economic, and political environments, influence the reception of a new technology? Context is crucial. This book has situated the new science of holography within the peculiar late twentieth-century environment that melded the military, commercial, and popular engagement with scientific and technological subjects. This section revisits the narrative of earlier chapters to examine the competing criteria by which they assessed their products, and how notions of success and failure became tied to expansion and progress.

14.2.1 Origins and forecasts

As we have seen, during the first two decades of holography (1947–1966) concepts clarified but forecasts shifted dramatically. The young subject was shaped in three

[6] On the social factors in the closure of debates surrounding technological options, see Pinch, Trevor J. and Wiebe E. Bijker, 'The social construction of facts and artifacts: or how the sociology of science and the

intellectual environments, and became tied to existing concepts, inventions, and metaphors, each of which shaped perceptions of its prospects and defined its criteria of success.

As discussed in Chapter 2, Dennis Gabor's broad background and research interests led him towards a new microscopic imaging technique, but no further. Despite his experience in physics, electronics, information theory, and stereoscopic cinema he was *contextually screened* by his working environment and commercial goals; he was *contextually channelled* to develop an awkward, if theoretically intriguing, variant of microscopy. But by merging electron microscopy with visible optics, wavefront reconstruction had aspects that appeared retrograde rather than progressive. Instead of the immediacy of seeing an image on a fluorescent screen (as some electron microscopes then produced), the reconstructed image was to be obtained more painstakingly via a half-hour exposure, followed by conventional photographic processing, unintuitive optical transformation, and observation through a conventional microscope eyepiece. For Gabor, his AEI colleagues and practicing microscopists, the meagre practical accomplishments amounted to a failure. Nevertheless, they recognized *different* failures: Gabor blamed lack of enthusiasm from industrial collaborators and microscopic manufacturers; the AEI team held instabilities of the electron apparatus to be responsible; Gordon Rogers and the Californian investigators were stymied by the twin-image problem; and microscopists dismissed a hybrid and unfamiliar technique.

The 'failed' subject nevertheless was rehabilitated posthumously; indeed, Gabor was to be awarded the Nobel Prize in Physics for holography in 1971. Subsequent evaluations of this work were recast, converting it from a white elephant into an example of the inevitable progress of science.

As explored in Chapter 3, Yuri Denisyuk's research contrasted with Gabor's, and supported this rehabilitated view of technical progress. His wave photographs could reconstruct the image of a three-dimensional surface, but without the necessity of a focusing lens. The technique was different in concept and implementation from Gabor's. It could reconstruct three-dimensional images by reflection from the hologram in white light and had no conspicuous link with either microscopy or information theory. Instead, its more demanding recording conditions made it rather analogous to a nineteenth-century daguerreotype, but with the addition of depth and (potentially) colour. Denisyuk portrayed his technique as a superior and generalized form of Lippmann photography for a limited class of objects, or as a colour-dependent optical element. But because of his unimaginative portrayal of applications and lack of an influential early mentor, his Soviet contemporaries largely ignored Denisyuk's research.

sociology of technology might benefit each other,' in: Bijker, W. E., T. P. Hughes and T. J. Pinch (eds), *The Social Construction of Technological Systems* (Cambridge, MA: MIT Press, 1987), pp. 17–50. My discussion of holography follows excellent studies of the early history of technologies, such as Bijker's study of the bicycle (ibid.), and Susan Douglas's account of radio (Douglas, Susan J. *Inventing American broadcasting: 1899–1922* (Baltimore, MD: Johns Hopkins University Press, 1987)). However, it differs from such cases in dealing with a technology that has not achieved a consensual evaluation by any user group for more than a brief period.

By merging physical optics with communication theory, as discussed in Chapter 4, Leith and Upatnieks found another method of sidestepping the technical disadvantages of the twin image problem; and by restricting their work to optics, they avoided the complexities and contextual screening of microscopy. Impressive results followed from their research: first, the ability to produce clean reconstructed images of line drawings in early 1961; second, high-quality greyscale images at the end of 1962; and finally, with the use of the newly available laser as a coherent light source in late 1963, an astonishing form of three-dimensional imagery in which the reconstructed images exhibited depth and parallax with unprecedented realism. When viewed in the light of Gabor's work, the achievements seemed to represent a rapidly rising ladder of accomplishment.

Another source of this redrafted ascent was the new link with the evolution of photography: when the imaging achievements of the 1960s were linked with those of the late 1830s, a soaring rise was obvious to all. As Adolf Lohmann recalled, the idea of 3D lensless photography was 'effective for our reputation and for the funding of our projects'.[7] But the subject strained to support this perceived link with the photograph. The Leith–Upatnieks hologram was a kind of transparency, but the image was observed by looking *through* the hologram as through a window. Its featureless surface was described as storing the image for later reconstitution. The copy of a hologram yielded not a negative image, but another positive. And unlike a photograph, the hologram could recreate a view of the entire image from any part; the pieces of a broken hologram still worked. The technique was also restrictive: only small laboratory scenes could be recorded. And the transmission hologram was tied to the laser as a light source, not just for its initial recording but also for subsequent reconstruction. The unfamiliar attributes of this 'window with a memory' were difficult to reconcile with concepts of photography but, despite the imperfect correspondence, photography was to be a convenient guide to understanding the new medium and in forecasting its future development.

Thus the work of Gabor, Leith–Upatnieks, and Denisyuk created at least three versions of an intellectual concept and its associated technologies: either an instrument for improved microscopy; a type of three-dimensional, lensless photograph in the form of a transmissive window; or, a method of recording the complete optical properties of a shallow object on a reflective plate. These divergent conceptions, arising from different technical and occupational contexts, profoundly shaped the early forecasts of the subject that became holography. Moreover, their respective 'successes' were differently evaluated. Gabor's narrow portrayal of wavefront reconstruction during the 1950s had yielded few forecasts beyond improved microscopy. His concept was self-limiting and of interest principally to workers interested in ultra-microscopy and the then limited field of physical optics. Denisyuk's self-assessment was similarly derided or ignored. By contrast, the Leith–Upatnieks conception excited great interest far beyond the domain of

[7] Lohmann, Adolf W. to SFJ, fax, 28 Jul. 2005, SFJ collection. In a complementary way, holographic research was submerged under different labels during the 1970s, when it was widely seen by corporations as having been over-sold.

physicists and engineers. The corresponding potential for career building and patenting (and consequential rivalries and history-painting) were examined in Chapter 5.

For most American observers, the Leith–Upatnieks technique was framed in terms of a potential success story, especially when linked with the early work of Gabor or, even more convincingly, with early photography. These historical associations acted as a lever for the slope of progress: with them, the accomplishments of the 1960s were cast upwards, and made to represent an exponential improvement that seemed bound to continue. Forecasts were tied to genesis myths: the predictions of the future were crucially dependent on particular claims about origins.

Predictions made between 1964 and 1971—bracketed by the announcement of the Leith–Upatnieks hologram and Gabor's Nobel Prize—were uniformly expansionist, making optimistic extrapolations based on laboratory demonstrations or anticipated applications.[8] Gabor himself predicted incautiously in 1969 that by 1976 his brainchild would become a billion-dollar industry.[9] Edwin H. Land, Director of the Polaroid Corporation, provided a typical prophecy when he declared around the same time that ' "hologram" will be a household word in 25 years'.[10] In fact, the word became ubiquitous sooner, but in the context of science fiction rather than consumer products, as discussed in Chapter 13. During the same period, Kip Siegel sold holography to investors by drawing upon their assumptions that the technology was ripe with latent potential for inevitable expansion. Conductron's approach of producing proof-of-concept demonstrations, discussed in Chapter 6, was an extension of methods pursued in military research contracts. Underlying this approach was the expectation of an inevitable technical pay-off.

Siegel proselytized a message seeded with positivist and progressivist philosophy, to which military and commercial sponsors were receptive. Yet the success of Conductron's holography development was ambivalent. Its engineers worked to realize Siegel's expectations of progress, advancing the technical possibilities such as producing progressively larger holograms for trade show displays. While such holograms attracted interest and exemplified the technical progress being made in image reproduction, their display requirements (such as the two or three carefully aligned and power-hungry lasers necessary for colour holograms) made them too unwieldy to be sold or even displayed outside the laboratory.

[8] For reviews of the technical capabilities of the medium, see, for example, Leith, Emmett N. and Juris Upatnieks, 'Photography by laser', *Scientific American* 224 (1965): 24–36; Gabor, Dennis and George W. Stroke, 'Holography and its applications', *Endeavour* 28 (1969): 40–7; Mikaelyan, A. L., 'The present state and future prospects of holography', *Telecommunications and Radio Engineering, Part 2 (Radio Engineering)* 25 (1970): 66–74; Leith, Emmett N. and Juris Upatnieks, 'Progress in holography', *Physics Today* 25 (1972): 28–34. Reviews emphasizing the commercial potential include Gates, J. W. C., 'Holography, industry and the rebirth of optics', *Review of Physics in Technology* 2 (1971): 173–91; Hammond, A. L., 'Holography: beginnings of a new art form or at least of an advertising bonanza', *Science* 180 (1973): 484–5; Dolgoff, E., 'Commercial holography', *Optical-Spectra* 9 (1975): 26–31.

[9] 'The incredible hologram', *Newsweek*,29 Dec. 1969: 29.

[10] Benton, Stephen A., 'Edwin Land, 3-D, and holography', *Optics & Photonics News*, 5(10), 41–46, p.43.

Similarly, the development of pulsed-laser capabilities illustrated clear technical improvement in a limited domain over only a short period, and was company-funded and commercially sterile. Conductron engineers realized that producing practical holographic movies would require a high-powered pulsed laser to record moving objects—such as people—before they could move enough to smear the interference fringes on the photosensitive plate, and that such a laser for recording outdoor scenes would certainly have dangerous and unattainable power requirements.[11] Nevertheless, they developed a pulsed ruby laser that produced a sufficiently intense beam to record a human portrait in a darkened room. When further technical improvement proved impossible, the Conductron marketing staff sought to redefine the goals, turning from holographic movies to three-dimensional human scenes.[12]

Progress was also touted in terms of production range and capacity. Between 1965 and 1970, the firm's promiscuous creation of over a thousand custom holograms for clients ranging from pharmaceutical manufacturers to artists culminated in the unprecedented achievement of manufacturing half a million holograms for the *Science Year*. Nevertheless this, too, suggested uncertain commercial success: production costs were covered, and the publication proved the best-selling edition of the series, but no orders of comparable size followed.

So the notion of progress was problematic from the outset: displays became more impressive, but the necessary equipment multiplied in cost, complexity, and unreliability. And unlike other commercial ventures, which commonly have such a discouraging and resource-draining development phase, Conductron's holography operations were never satisfactorily 'black boxed'.[13]

Like a series of holography companies that followed it, Conductron was therefore a messenger of a particular view of progress. Its holograms were an embodiment of all that was new and valued in that technologically optimistic decade, melding the laser, high science, and awe-inspiring imagery into an example of seemingly inevitable technical advance. Its engineers were habituated to exploring applications creatively in a classified context and with relatively abundant funding, and were imbued with confidence in the very notion of progress. They arguably were less sensitive to commercial pressures than were typical workers in industry, however, because they envisaged holography as a dramatic and inevitable extension of photography. Conductron consequently sought display applications that highlighted its three dimensionality and visual impact. Less convincingly, its engineers forecast and pursued the extension of holography to colour imagery, movies, and television based on the technical trajectories of those earlier imaging media. And most misleadingly, again using the analogy of early photography,

[11] This disparity between the technical requirements and plausibly achievable 'progress' is reminiscent of the later Strategic Defense Initiative ('Star Wars' project) promoted by physicist Edward Teller, which also relied on pulsed lasers.

[12] Charnetski, Clark 'The impact of holography on the consumer', presented at *EASCON Convention*, Washington DC, 1970.

[13] On the social and technological aspects of 'black-boxing' (creating a well-established fact or unproblematic product), see Latour, Bruno, *Science in Action: How to Follow Scientists and Engineers Through Society* (Milton Keynes: Open University Press, 1987), esp. pp. 131–2.

they predicted a rising public appeal and inevitably growing market like its antecedents. This divergence between marketplace reality and commercial claims is not unique, of course: it was a feature in other new technological fields such as nuclear power during the 1950s, biotechnology from the 1980s, and nanotechnology from the 1990s.[14]

In a more restrained fashion, other firms, too, cited practical indicators of progress for the new science. From the late 1960s, for instance, holographic interferometry (or *holographic non-destructive testing*, HNDT) became popular with metrologists and mechanical engineers and found a market niche. The commercial holographic tire-tester developed by Ann Arbor's GC Optronics, for example, allowed the lamination of airplane tires to be verified rapidly. For this application, the criteria of success were economic (lower costs), technical (better testing reliability), and social (improved customer safety).

But the often-vain expectation of steady technical progress led a number of large firms to withdraw quietly from the field by the early 1970s. The CBS Laboratories, for instance, which had long employed Dennis Gabor as a consultant, generated patents but devised no promising holographic products. RCA developed a prototype consumer video playback system (Selectavision Holotape) in 1969, but cancelled the project when the firm was in financial difficulties and facing competition from more versatile magnetic recording technologies. McDonnell Douglas closed its pulsed holography operation in 1973 owing to lukewarm interest from the advertising industry and corporate customers. IBM's early enthusiasm for holographic computer memories did not culminate in products. And Polaroid Corporation, while developing recording techniques and new photosensitive media during the 1970s, did not effectively market them.

Thus the first flush of scientific and technical confidence in holography's progress was eroded within a decade. And the enthusiasms of technologists did not necessarily translate to those of wider culture. Marketing holograms proved unexpectedly difficult, and there was increasing disjunction between technical forecasts and economic reality. But, as argued in Chapters 8, 9, 10, and 11, new constituencies of holographers defined new goals.

For artists, success in holography was evaluated according to different criteria. The medium had to have adequate technical versatility to support aesthetic expression, and the new art form required the acceptance of art critics and a receptive public. While this was relatively free of positivist underpinnings, these criteria did embody implicitly progressive ideas, namely the assumption that the capabilities and audience for the medium would expand.

Artists applied their own distinct criteria during the early 1970s, too. Their definition of progress was a combination of growing audiences and prices for fine-art holograms, alongside the development of new techniques of production, especially cost-effective or simplified methods. It is noteworthy that the community of artisans, like aesthetic or fine-art holographers, often supported this implicit assumption of technical progress and consequent mass popularity even while embracing countercultural themes.

[14] See, for example, del Sesto, Steven L. 'Wasn't the future of nuclear engineering wonderful?', in: Corn J. J. (ed.), *Imagining Tomorrow: History, Technology, and the American Future* (Cambridge MA: MIT Press, 1986), pp. 58–76.

For each of these communities of holographers, success was defined in terms of *expansion*, which amounted to anticipation of continual progressive increase. And for all three communities—scientists, artists, and artisans—stagnant conditions, measured in terms of income generation, technical abilities, and acceptance by critics, consumers, students, or the wider public, equated to failure for the subject.

By the 1980s, despite its exposure to hundreds of thousands of viewers through public exhibitions of holograms and the growing ubiquity of mass-produced holograms, the subject could not be characterized reliably by these criteria and appeared different to each constituency. The impalpable state of progress can be illustrated by bibliometric indicators: by the mid-1990s, while the annual publication rates of papers and patents were rising, those of books and theses were falling, and the number of scientific conferences and hologram art exhibitions had diminished to half their value of a decade earlier.[15] For artists and artisans, the field was declining; for scientists, it had periodic ups and downs; but for inventors and investors, it continued to look promising. Economic, rather than population, indicators consequently became a widely accepted mark of success of the medium. For different communities then, holography was a subject that either evinced obvious success, or remained latent with potential, or had outlived its promise.

14.2.2 Segregating communities: judgments of successful imagery

The contrasting judgment of emerging constituencies is further illustrated by the goals of holographic imaging. The occupational specialists of holography variously identified the strengths and weaknesses of holographic technology.[16] They argued that a collection of limitations surrounding the hologram prevented the expansion of the technology of holography in wider culture. This was a two-way process: their mutually incompatible criteria encouraged the holographic communities to differentiate further. The divergence of holograms and their associated communities is consistent with the framework of social construction of technology (SCOT) and its concepts of relevant social groups and interpretative flexibility.[17] However, for most of these groups consensus and closure were not attained. Instead, their technical goals diverged while they continued to seek improvement for their medium and stability was elusive.

The gaze of these distinct communities identified contrasting limitations for the medium, some of which were discussed in Chapter 7. The first to be noted were problems with the laser itself. Holograms illuminated by lasers were obscured by laser speckle making the reconstructed image and, indeed, any surface illuminated by the laser

[15] See the appendix.

[16] This consumers' assessment of hologram technology contrasts with the assessment by its inventors, discussed in Chapter 7.

[17] Pinch, Trevor J. and Wiebe E. Bijker, 'The social construction of facts and artifacts: or how the sociology of science and the sociology of technology might benefit each other,' in: Bijker, W. E., T. P. Hughes and T. J. Pinch (ed.), *The Social Construction of Technological Systems* (Cambridge, MA: MIT Press 1987), pp. 17–50.

shimmer and sparkle with a graininess that depends critically on the position of the observer's eyes. This artefact was a complaint of scientists and engineers more than aesthetic holographers: speckle was a particular problem when *photographing* holograms, an activity common in scientific and engineering studies, but relatively unimportant for casual viewers.[18]

On the other hand, the cost of the laser was a crucial constraint for artists and advertisers but of relatively little importance to well-funded scientists. Lasers were also unfamiliar and intimidating for non-scientists, and their use was curtailed severely with the introduction in the early 1970s of safety legislation concerning eye exposure. But lasers were also relatively dim light sources for reconstructing the holographic image—adequate to illuminate a single hologram well in a normally lit room, but not in a daylight-flooded shop window. If any other form of light were used to reconstruct a hologram, the image would be unacceptably blurred. Neither of these restrictions was a particular problem for scientific applications such as holographic interferometry, but judged to be a severe limitation for public displays.

A third characteristic constrained holography as a medium for portraits. The very monochromaticity of laser light provided eerily unworldly images akin to the street illumination from sodium lamps, or the orthochromatic (blue- or green-sensitive) images of early photographic and cinematographic films. The contrast and tonal gradations of reconstructed images appeared unfamiliar, and was inferior to the panchromatic black-and-white films that had been used universally since the Second World War.

This effect was exacerbated in the portraits made with pulsed ruby lasers. Because human skin is slightly transparent to the deep-red colour of a ruby laser, portraits made subjects look waxy-skinned, blotchy, and disturbingly morbid. Holograms produced with pulsed lasers, acting like a fast flash camera, captured unsettlingly frozen facial expressions of their subjects who had been sitting in near darkness, often accentuating the unfamiliarity by showing the unusually wide irises of the dark-adapted eye. Artists who adopted pulsed lasers most successfully employed them for figure studies rather than facial depictions.[19] For photographers, then, holographic portraiture represented problems, not progress.

In sum, these limitations restricted holography to a narrow class of subjects and applications, paradoxically in opposition to its highly realistic perspective. The problems and putative solutions were ranked differently by different communities.

The most pressing problem for aesthetic and commercial users (but irrelevant for scientists) was the need for a laser to display the hologram. Denisyuk's holograms of the

[18] Speckle is most noticeable from a fixed position and using a small camera aperture to photograph an image having a large depth of field. For observers bobbing their heads to see the parallax of the image, though, the speckle is smeared out.

[19] Only a handful of pulsed laser portrait studios appeared, notably Conductron (Michigan, late 1960s), McDonnell Douglas (Missouri, early 1970s), the Museum of Holography (New York, 1980s), Musée de l'Holographie (Paris, 1980s), Laser Reflections (California, 1990s), Spatial Imaging (England, 1990s), Holocom (Illinois, 1990s), and Russia (1990s).

early 1960s offered a solution, by providing a clear green image when viewed in sunlight or room lighting. But creating such holograms demanded extremely high-resolution photographic emulsions and very stable conditions during the exposure. They had a low uptake in the West because suitable emulsions and chemistry were not readily available and because artists perceived them to offer limited options for creativity.

A solution for some audiences was the image plane hologram, which produced little colour smearing of reconstructed images for points near the plate, so a white-light source was adequate to view holograms of shallow objects. A secondary advantage was that such images were even more striking than conventional holograms: the image appeared to pass through the hologram plate. This appealing attribute became ubiquitous in commercial and art holograms by the mid-1970s, but was of little interest to scientists.

During the early 1970s, the rainbow hologram also became widespread. With it, a sharp image could be viewed in white light, although cast in a spectrum of colours that shifted with the viewing position. Its developer, Stephen Benton, as a well-known intermediary between scientific and artistic communities, developed variants when he joined the new MIT Media Laboratory. The laboratory united a collection of enthusiastic engineers and scientists and sought to transform culture via new media technologies, aiming, as one breathless account put it, to invent the future.[20]

Rainbow holograms importantly reduced the cost and complexity of display, but had an uneven popularity that further divided supporters. East-coast American holographers, close to Benton's Massachusetts laboratory, adopted rainbow holograms more enthusiastically than did their west coast and European counterparts. And artists championed the technique, discovering that, by overlaying several exposures, a single hologram could display multicoloured images. But, as discussed in Chapter 10, rainbow holography was rejected by Soviet practitioners, who saw the technique as complex and poorer in quality than their own reflection holograms, and by most American scientists, who were concerned with the accurate recording and metrological analysis of three-dimensional objects or transitory events. Benton's later variants, such as the ultragram and edge-lit rainbow, were little adopted even for display purposes.

From the 1980s, as discussed in Chapter 12, embossed holograms provided new audiences and opportunities for technical judgments and forecasts. Manufactured by the millions on metal foil, they became ubiquitous in packaging, graphic arts, and security applications. While this brought holograms to a much wider audience, it generated dramatically divergent judgments. Unlike the previous varieties of holograms, this new type generated not just indifference from different communities, but outright animosity.

Embossed holograms were inexpensive, reducing the cost of copies by a hundredfold. They could be mass produced reliably using a number of proprietary techniques. And they were chemically and mechanically stable, unlike most previous hologram materials that were susceptible to breakage, humidity, or aging. Together, these technical advantages promoted the widespread application of embossed holograms.

[20] Brand, Stewart *The Media Lab: Inventing the Future at MIT* (New York: Viking Penguin, 1987).

On the other hand, connoisseurs of imaging—the self-styled display holographers made up of artists and artisans—derided embossed holograms. Their flexibility, particularly on magazine covers, caused colour shifts and image distortion. And because the holograms were usually viewed in uncontrolled lighting, images could appear fuzzy or dim. In response to these limitations, their producers progressively simplified the imagery to incorporate shallow, eye-catching patterns, a product that some in the industry dubbed contemptuously 'shiny shit'.[21]

By moving towards less ambitious images, embossed holograms evolved to minimize their perceived weaknesses and to exploit new markets. While applications such as magazine illustrations declined, others expanded to suit new industries and adopters. Visual appeal was redefined. Their image characteristics made embossed holograms particularly suitable for attention-grabbing product packaging (a profitable and growing industry from the early 1990s) and for security applications, where any defect in the complex pattern could indicate tampering or counterfeiting.[22]

There was an irony in this market success: this technical mutation arguably amounted to a reversal of the original aims of the medium. Yet consensus about success could not be defined in utilitarian terms. Embossed holograms promoted low-cost mass production but had relatively poor image quality; they brought three-dimensional imagery to vastly increased audiences, but simultaneously reduced the sublime characteristics of depth, parallax, and image clarity. Security applications exploited the complex colour shifts and angle dependence of embossed holograms, making the forgery of credit cards and bank notes more difficult. But for imaging purposes, these characteristics were deemed to be a serious defect. Fine-art holograms declined in popularity, with artists complaining that embossed holograms irreparably devalued the aesthetic attraction of the medium.[23] This expansion of holography into the mass market was thus judged by its initial supporters to be a failure, because it had deviated from their forecast trajectory. The 'shiny shit' had defiled the utopian predictions.

In this transmutation of meaning, the hologram itself began to disappear. Embossed versions of silver-halide display holograms shaded by imperceptible steps into grating-pixel kinegrams for passports. So, too, did holographers, transforming from solitary shrouded workers in quiet laboratories to computer graphics technicians and press operators.[24]

[21] Kontnik, Lewis T. to SFJ, interview, 20 Nov. 2003, Vancouver, SFJ collection; Munday, Rob to SFJ, interview, 31 Mar. 2004, Richmond, UK, SFJ collection.

[22] From the introduction of credit card holograms in 1983, the hologram industry was dominated by packaging and security applications, and represented by a periodical (*Holography News*, published by Reconnaissance International from 1987), by annual conferences (Holopack-Holoprint, from 1989), and a trade body (International Hologram Manufacturers' Organization, 1992) seeking to monitor and regulate an industry growing most rapidly in the Far East.

[23] Benyon, Margaret to SFJ, interview 21 Jan 2003, Santa Clara, CA, SFJ collection.

[24] The economics of embossed holograms did not improve the professional situation of holographers: embossing processes were taken over by commercial printing companies using fairly conventional equipment and relied on holographers only for the production of the original master hologram. Further developments by firms such as Dai Nippon threaten to embody holographers' expertise in automated machines.

The evolving techniques for producing bright, white-light holograms thus both liberated the growing field of display holography in the 1970s for commercial use and constrained its acceptance in the 1980s, particularly for artists. The tribulations of display holographers were not faced by most scientific and engineering users, who continued to employ laser-viewable holograms; nor were they recognized by marketers of packaging and anti-counterfeiting holograms. Thus, the applications of holograms supported the growing segregation of practitioners and conflicting definitions of success.

14.2.3 Wider judgments: critics and consumers

Despite the discordant assessments of display holography, expectations of progress remained strong. Forecasts during holography's first active decade—extending from 1965 to 1974—had been uniformly expansionist, making unrealistically optimistic extrapolations based on laboratory demonstrations or even speculative applications.[25] Indeed, some forecasts, oft rejuvenated, seemed impervious to attributions of failure despite the continued lack of demonstrable viability.[26]

Nevertheless, holography was also criticized for having failed in expanding enthusiasm, garnering audiences, and developing markets—a failure, in effect, to conform to wider expectations of technological progressivism. Artists responded with dismay, for instance, to negative reviews of *Holography '75* in New York, which curtailed their expectations of aesthetic acceptance and growing markets. Both critics and artists (to their chagrin) portrayed holography as immature and in a state of early aesthetic and technical development; both, indeed, were imbued with a similar definition of progress.

In order to sustain continued confidence, predictions mutated. History-in-the-making demands repeated re-evaluations and changes of course. Dennis Gabor's 1971 Nobel Prize for holography provided a convenient perspective from which to evaluate the emerging subject's past and future.[27] So, too, did the mid-1970s (the end of the 'first decade'), the mid-1980s (the end of the 'second decade'), and the 1990s and onwards, when early workers in the subject began to reflect on their careers and their subject's

[25] See, for example, reviews by Gabor, Dennis 'The outlook for holography', *Optik* 28 (1969): 437–41; Gabor, Dennis 'Holography, past, present and future', *Proceedings of the SPIE—The International Society for Optical Engineering* 25 (1971): 129–34; Hammond, A. L. 'Holography: beginnings of a new art form or at least of an advertising bonanza,' *Science* 180 (1973): 484–5; Dolgoff, E. 'Commercial holography,' *Optical-Spectra* 9 (1975): 26–31; Denisyuk, Yu N. 'Holography and its prospects (review),' *Journal of Applied Spectroscopy* 33 (1980): 901–15; de Marrais, Robert 'Holography in the future tense,' *holosphere* 12 (1984): 4, 6–7; Jeong, Tung H., 'Future holography', *holosphere* 12 (1984): 18.

[26] The most tenacious *technical* forecast concerned holographic memories for graphic or digital storage (Chapter 12). Popular forecasts, on the other hand, ranged from holographic television during the 1970s (Chap 6) to holographic computer-generated personalities, as explored by science fiction from the 1980s (Chapter 13).

[27] Chapter 5 focused on the creation of an 'official' history validated by the Nobel Prize. Examples of the first wave of evaluation include Shushurin, S. S., 'On the history of holography', *Soviet-Physics-Uspekhi* 14 (1972): 655–7; and, Gabor, Dennis, 'The history of holography', *Fizikai-Szemle* 24 (1974): 289–303.

achievements.[28] Although these accounts varied in their predictions, all cast the development of their field as an historical narrative linked with latent or manifest progress.[29] This perspective of imminent growth suffused newspaper and popular magazine accounts even more pervasively.[30]

When, after one decade, two decades, or a quarter century, material achievements were not obvious to all, the original commentators and others—notably Stephen Benton, who became the most prominent conference organizer and holography pundit—recast the development of their field as an historical narrative either still linked with progress or portrayed simplistically as a classic tale of market failure.[31] When compared to its optimistic forecasts, the subject seemed periodically to pause or stumble, if not decline.

So progressivist accounts of holography coexisted with attributions of its failed potential. Yet, as discussed in Chapter 13, none of these later depictions dominated public consciousness of holography. Instead, understandings became shaped by fictional portrayals. Popular anticipation, supported by faith in progress, threatened to outstrip reality. This splitting of real and imagined futures, evident in the earlier commercial forecasts as well as later science fictional accounts, is a theme common to many new technologies. It has parallels with the account that Colin Milburn has given of nanotechnology, for instance. Milburn argues that popular and professional writing about nanotechnology amounts to a 'teleological narrative' that transforms a dream into something that is inevitable. He suggests that promotion of the subject has transgressed a line between 'speculative science' (an extrapolation of current scientific thinking, describing what could be) and 'fictional science' (an account of what, inevitably, will be, in some world to come).[32] According to this view, Kip Siegel's forecasts could be characterized as fictional science that influenced science fiction writers a decade later. Such incredible extrapolations may not require the disorienting qualities of the hologram,

[28] Examples of these waves of historiography include Benton, Stephen A., 'Holography: the second decade', *Optics News* (1977): 16–21 and Benton, Stephen A., 'Ten years of white-light holography', presented at *Electro-optics/Laser International 1980*, London, UK, 1980; Leith, Emmett N., 'Some highlights in the history of display holography', *Proceedings of the International Symposium on Display Holography* 1 (1983): 1–4; Jackson, Rosemary H., 'A thirty-five year account of the development of holography—Part I', *holosphere* 12 (4) (Summer 1983): 5–12; Jackson, Rosemary H., 'A thirty-five year account of the development of holography—Part II', *holosphere*, 12 (5) (Fall 1983): 13–7; Jackson, Rosemary H., 'A thirty-five year account of the development of holography—Part III', *holosphere*, 12 (6) (Winter 1984): 19–23; Wesly, Ed, 'Silver anniversaries', *holosphere* 15 (1988): 12–3.

[29] For typical examples of forecasts, see de Marrais, Robert, 'Holography in the future tense', *holosphere* 12 (1984): 4, 6–7 and Jeong, Tung H., 'Future holography', *holosphere* 12 (1984): 18. Greguss, P., 'Thoughts on the future of holography in biology and medicine', *Optics and Laser Technology* 7 (1975): 253–7; Denisyuk, Yu N., 'Holography and its prospects (review)', *Journal of Applied Spectroscopy* 33 (1980): 901–15; Cross, Lloyd G. and Cecil Cross, 'HoloStories: Reminiscences and a prognostication on holography', *Leonardo* 25 (1992): 421–4.

[30] E.g. Albright, Thomas, 'Holography's accelerating impact', *S. F. Sunday Examiner and Chronicle*, 11 Jun. 1972: 30–1; Hammond, A. L., 'Holography: beginnings of a new art form or at least of an advertising bonanza', *Science* 180 (1973): 484–5.

[31] For example, Bringolf, Peter H. 'Holography: a medium in the making,' *Proceedings of the SPIE—The International Society for Optical Engineering* 2043 (1994): 319–321 and Bosco, Mary C., *What Ever Happened to Holography?*, MA thesis, John F. Kennedy School of Government, Harvard (1981).

[32] Milburn, Colin 'Nanotechnology in an age of posthuman engineering: science fiction as science,' *Configurations* 10 (2002): 261–95, quotation p. 263.

though. The near-utopian predictions for holography, promoted by its fantastic early commercial claims, have been made of other, more mundane, technologies.[33]

In any case, such fictional diversions increased expectations, and adversely affected the cottage industries of holography that appeared during the 1980s. The small firms selling holograms for home viewing, which had sprung up in large cities after major exhibitions, did not thrive. Small commercial galleries, such as the Holos Gallery in San Francisco, gradually discovered that sales of holograms could sustain them only if their businesses were transformed into wholesaling operations for distributing holographic trinkets to museums of science and technology.[34] The sale of holographic art, always marginal, declined as holographic kitsch in the form of embossed foils for children's stickers and magazine covers began to flood the market from the mid-1980s. As discussed above, it is significant that artisanal and artistic holographers identified this trajectory as non-progressive and hence an indicator of failure. They commonly characterized the altered focus of public interest as a descent similar to the history of earlier three-dimensional media, transforming them from a sublime technological experience to mere children's products having lower intrinsic value. The criteria of success become more abstract: they judged the type of audience to be more important than its size.

The limited public acceptance of commercial holograms meant that real-world holographers continued to struggle for occupational status and acceptance of their products. The subject, its communities and their aspirations of progress were closely interlinked. The technical groups associated with holography proved unstable partly because public engagement and employment were themselves uncertain.

Holography did not develop applications that generated a stable occupation supported by university-taught courses. The growth of long-lived occupations and accredited teaching programs, usually deemed crucial for the consolidation of a new profession and a new discipline, could not be sustained by the applications of holography. Instead, the subject spawned several marginal constituencies, along with distinct forecasts and criteria of success. Even the best supported of these, the broad field of optical engineering and scientific holography, found its military and corporate funding difficult to sustain after the cold war. Artists and artisans found their exhibitions and income reduced by the expansion of embossed holograms and changing public expectations. Colleagues in other fields consequently interpreted the relative social invisibility of holographers as a failure of the subject.

As discussed in Chapter 11, during the late 1980s, when holography was at a peak of visibility, practicing display holographers comprised an active community of about a thousand individuals ranging from scientists, to artisans, artists, and entrepreneurs. The

[33] A range of cases is discussed in Corn, Joseph (ed.), *Imagining Tomorrow: History, Technology and the American Future* (Cambridge MA: MIT Press, 1986) and Sturken, Marita, Douglas Thomas, and Ball-Rokeach (ed.), *Technological Visions: The Hopes and Fears That Shape New Technology* (Philadelphia: Temple University Press, 2004).

[34] Zellerbach, Gary to Jonathan Ross, interview Dec 1980, Ross collection; Zellerbach, Gary to SFJ, email, 2 Aug 2003, SFJ collection.

New York Museum of Holography, founded in 1976 to serve not just the disparate subcultures of holography but also the general public, discovered that holographers' sense of community was ephemeral and inward looking. As discovered by the schools and cottage industry that appeared during the 1970s and early 1980s, the Museum found that the general public absorbed the ideas and enthusiasms of holographers with difficulty. In response, these budding organizations mounted education campaigns that sapped more traditional profit-making activities. But these oft-repeated initiatives appear to have had only a local and transient impact. While the early 1970s had witnessed sustained growth in the constituencies of holography, signs of decline in institutional support of display holography became apparent during the 1990s.[35]

14.2.4 Evaluating progress, success, and failure

As the cognitive boundaries of a technical subject shift, so, too, do its applications and users, and their criteria of success. Examples abound in holography of how 'failures' and 'successes' were interpreted inconsistently by shifting audiences. The various advocates of holography had distinctive aspirations, employed contrasting criteria to evaluate its goals, problems, and solutions, and buttressed their own differentiation in the process. Thus, Gabor's wavefront reconstruction was typecast as a technically constrained, and even backward-looking, microscopy during the 1950s, unworthy of forecasts. During the 1960s, the revitalized subject was widely understood in terms of photography, an analogy that directed predictions in ways that were difficult to sustain. Scientists, artists, and artisans portrayed their subject as potential filled, and judged it by its expansion, especially by the number of adopters. They had divergent definitions of good imagery, however, and so judged progress in conflicting ways. Was the technique developing towards metrological accuracy in a laboratory environment, colourful displays in shop windows, aesthetically nuanced fine art, the recording of public events, or an ubiquitous anti-forgery product? Given the multiple constituencies, no consensus was possible, nor can any generally agreed attribution of progress be made. For the same reason, we cannot identify straightforward technological failure here. There was, however, a failure of technological forecasting, owing to over-confidence in short-term achievements made in an over-inflated funding environment.

Judging progress, success, and failure is further complicated by the altering prominence of these marginal technical communities. Embossed holography could be represented as an unalloyed success in the late 1990s not only because of its commercial profitability, but also because there were then fewer holographic artists to criticize its imaging characteristics than two decades earlier. Holography as a concept and technique

[35] The New York Museum of Holography closed in 1992, and its holdings were auctioned and transferred to the MIT Museum a year later; the Museum für Holographie und Neue visuelle Medien in Pulheim, Germany, founded in late 1979, closed in 1994, as did Le Musée de l'Holographie in Paris, founded in 1980; The Holography Unit of the Royal College of Art, an important source of postgraduate fine-art holographers, closed in 1994; the Canada Council ceased funding for holography in 1995; and, the final Gordon Research Conference of scientist-holographers, initiated in 1972, was held in 1997.

was categorized discordantly by its users, successively rejected, resurrected, and relegated to vulnerable commercial niches.

The history of the assessments and forecasts of holography has implications for other studies in the history of science and technology. As this subject demonstrates, historical evaluations of progress can be critically sensitive to appraisals made by different communities, particularly for unstable technologies that are adopted by distinct social groups. Each of them—such as scientists, the military, artists, businesspeople, and the public—may employ different criteria in judging the subject. And while we expect attributions of progress to depend on established or enunciated criteria, the case of holography shows that judgments may be based almost entirely on implicit assumptions and superficial analyses.[36]

Expectations for the trajectory of holography were supported by faith in both philosophical positivism and technological progressivism and fuelled by the expansive funding environment of 1960s America. The predictions of progress relied on little-examined assumptions and short-term forecasting, and its monitoring flavoured subsequent judgments of success and failure. But re-examination of such assessments is difficult for such insecure subjects: lack of market penetration or professionalization can hinder the documentation of a field. Would-be sciences like holography must be tracked by the historian as they evolve, not from scanty archival records. There will be a tendency to under-represent subjects that have not been judged progressive and successful by its contemporary practitioners and critics.

Holography further illustrates how closure is not an inevitable outcome for debates in scientific subjects that do not reach disciplinary status, or for technologies that do not achieve commercial viability. It suggests caution surrounding uncritical assumptions about the evolution of technological subjects: the inconsistent assessments of progress and success cannot be attributed merely to the youth of a subject or to inchoate relevant social groups. The notion of the 'maturity' of a field is a problematic one, and must be divorced from scholars' own expectations of progress towards consensus. Not all technologies become black-boxed; some merely lose their supporters and relevance and are forgotten.

14.3 PRESERVING AND PREDICTING: VISIONS OF THE PAST AND FUTURE

How, then, can the historical trajectory of holography be explained? As we have seen, the subject came to be represented in historical terms as early as 1971, when Dennis Gabor won the Nobel Prize for Physics. Historical accounts were essential to pursue early priority claims and to resolve patent disputes. Subsequent accounts were less personally

[36] Another example of such superficiality is the case of the New National Telescope, which astronomers widely judged a failure because it was never built (McCray, Patrick 'What makes a failure? Designing a new national telescope, 1975–1984', *Technology and Culture* 42 (2001): 265–91). McCray, by contrast, suggests that the project could be deemed a success because of its liberating effect on telescope design, on promotion of

directed, but promoted the prevailing expectations of technological progress and the expansion of scientific knowledge. As late as the 2000s, however, many practitioners resisted the historicity of their subject, perhaps seeing this as an unfavourable way of judging its trajectory alongside its prognostications, or of relegating it to the past. It is equally difficult for those still active in the field to recognize their activities in a historical sense. But, of course, histories do inevitably get constructed, often without the direct intervention of the historical actors and frequently in a simplified form that serves particular agendas. Direct interaction of the historian with those practitioners is ambivalent: on the one hand, holographers provide direct (if occasionally conflicting) personal accounts and interpretations of episodes; on the other, they may resent the interference of an outsider seeking to explain events in ways that may not actively support existing interpretations or promote the subject as they would themselves. The historian's account may conflict with others that inevitably suffer from selective recollection and reshaping and rehearsing of events to satisfy simplified chronologies and accounts. The origin stories that have circulated in popular accounts of holography often deviate dramatically from the account given in this book.

The role of artefacts can be significant in embodying or reifying a sense of history. Hologram exhibitions frequently have been used to make the evolution of holography tangible. Nevertheless, the desire to locate missing links can misrepresent, too, as Emmett Leith reflected concerning the preservation of early holograms:

> People ask 'well, what was your first hologram, which is the first hologram?' And museums around the country and private collections and so on, people have enough of the 'very first hologram' around [like pieces of the real cross during the Middle Ages, when] you could find, in crypts and grottos, enough pieces of the real cross to start a lumber yard. And some of these holograms that people claim to be the original might be the thousandth.[37]

Attributing a relic-like identity to holograms deemed to be historically important began during the late 1970s, when a historical perspective was becoming established. The flurry of large public exhibitions and retrospectives during that period sought to chronicle a clear history of the young field. The Museum of Holography in New York, which organized some of the first large exhibitions, became, for a time, the repository for these significant objects. Religious parallels can be suggested: the identification of relics (carefully transported from one temporary place of veneration to another)—indeed, tales about transporting important holograms to exhibitions (shrines?) are recounted that assume the dimensions of Chaucer's *Canterbury Tales*; the multiplication of holographic relics as Leith describes, and their rapid escalation in value; their display in carefully oriented reliquaries; and their home in a dark and respected sanctuary.

international cooperation, and on public promotion of astronomy. Similarly, Elzen argues that the Svedberg ultracentrifuge was seen as a successful artefact by his contemporaries despite its lack of influence on present-day designs (Elzen, Boelie 'The failure of a successful artifact: the Svedberg ultracentrifuge', in: Lindqvist, S. (ed.), *Center on the Periphery: Historical Aspects of 20th-Century Swedish Physics* (Canton, MA: Science History Publications, 1993), pp. 347–377).

[37] Leith, Emmett N. to SFJ, interview, 22 Jan. 2003, Santa Clara, CA, SFJ collection.

The analogy to the cult of relics can be taken further, with a handful of individuals—those associated with the production of the holograms—identified in nearly saint-like terms. Thus, for a constant stream of conference delegates, to be photographed with Emmett Leith or Yuri Denisyuk was akin to receiving special grace, anchoring the photographee in history and implicitly validating their own work. This concerns more than mere celebrity: their acolytes express their admiration with anecdotes (parables?) about good deeds and character (modesty and honesty most frequently) and their charismatic influence. The elaboration of these analogies represents a rich oral folklore for holography. But even if the experience of viewing holograms can be evocative of the sublime, the analogy of the hologram as relic cult is imperfect: few observers suggest that holograms, and their creators, are imbued with powers beyond their ability to evoke a connection with beauty, meditation, or perhaps holism.

Such musings provoke the question of the purpose and future of historical collections. Museums and galleries actively construct popular history. With the perception of holograms as historical objects, and a material culture to be preserved, a relationship grew between holograms, museum curators, and their representation of history.[38] However, the uneven preservation of the documentary and material culture of holography illustrates the peripheral status of the field in wider culture.

The papers of Dennis Gabor were collected and archived by Imperial College largely because of the status he achieved late in life with the award of his Nobel Prize. The papers of other early practitioners such as Hussein El-Sum have not been preserved.[39] Similarly, a survey of the holdings in Ann Arbor, the crucible of development of the subject academically, commercially, and artistically during the 1960s, shows that there are historically important documents, equipment, and holograms scattered around the small city, but no historical collections or exhibits focusing on them. The Bentley Historical Library in Ann Arbor holds some of the administrative records of Willow Run, for example, but does not identify holography as a particular collecting category, nor does it presently hold much archival material specifically on the subject. Most documents remain in the hands of individuals—participants as students, entrepreneurs, classified research workers, or commercial engineers—still living in the area.

Nor have firms and institutions made more than a casual attempt to preserve their past. Carl Aleksoff, for instance, at Willow Run Laboratories (WRL) as a student and then with its successors ERIM (Environmental Research Institute of Michigan), ERIM International, Veridian, and General Dynamics, recalls that optical processing equipment reaching the end of its working life was revealed to visitors, but ultimately neglected:

> For a number of years these things were displayed in the lobby—but it's all been thrown away. You can only do it for so long. For a bottom line profit motive, you've got to account

[38] Material culture as an intellectual concept owes its origins to anthropology and archaeology, which, from the late nineteenth century, drew object lessons from ethnographic studies of artefacts. Collections of illustrative artefacts go back, in turn, to the 1851 Great Exhibition, which sought specifically to demonstrate Victorian industrial progress, and still earlier to eighteenth-century *cabinets of curiosities*, intended to reveal the hidden or unusual aspects of the natural world to educated audiences.

[39] At the time of writing, negotiations are under way for the papers of Albert Baez to be donated to the Neils Bohr Library of the American Institute of Physics. The papers of Gordon Rogers have been archived at the Science Museum, London.

for how much area you have—so many square feet—what are you doing with it? How productive is it?[40]

Despite the relatively long-term stability of organizations funded principally by military contracts, WRL / ERIM / Veridian / General Dynamics suffered from the demands of secrecy, which are incompatible with the preservation of open history. This is equally true of the more commercially oriented Ann Arbor firms such as the Conductron Corporation, KMS Industries, and GC Optronics. The companies were simultaneously constrained by classified contracts and by the desire to control commercially useful proprietary knowledge, on the one hand, and the desire to vaunt technologies that they hoped would become major income-generating streams, on the other.

This patchy preservation is not restricted merely to commercial firms and classified-research organizations. From 1993, the MIT Museum in Cambridge, Massachusetts, held the largest collection of publicly accessible holograms in the world. Founded in 1971, the MIT Museum shares a former radio factory situated on the northern edge of the MIT campus with a number of other MIT tenants, and preserves, displays, and collects artefacts in five disparate subject domains significant for the institution's history.[41] The holography collection includes some 1500 holograms acquired after the demise of the New York Museum of Holography (MoH), and covering a period from the early 1960s to the late 1980s. Only a few dozen holograms, at most, can be displayed owing to space restrictions, and the MIT Museum attracts somewhat fewer visitors than did the New York Museum of Holography.[42]

The collection is disproportionately distributed, with more holograms from the mid-1970s to 1980s when the MoH was most active, and few after the mid-1980s when the MoH encountered more serious financial and administrative difficulties. The associated archives of the MoH held at the MIT Museum provide an excellent snapshot of holography's most fertile and expansive period as a cottage industry and would-be art form,

[40] Aleksoff, Carl to SFJ, interview, 9 Sep. 2003, Ann Arbor, SFJ collection. The ambivalent relationship between classified research and historical preservation complicates historiography. Another example is the CORONA surveillance satellite, which employed dozens of optical engineers from the mid-1950s (see Chapter 8). In 1972, when the programme ended, a 'classified museum display' of a CORONA spacecraft was installed in the headquarters of the National Photographic Interpretation Center in Washington DC. The existence of surveillance satellites was not admitted publicly until a speech by President Carter in 1978 and the satellite was not displayed publicly until 1995 at CIA headquarters in Virginia. A few years later it finally achieved true public display at the National Air and Space Museum in Washington (Day, Dwayne A., 'The development and improvement of the CORONA satellite', in: Day, D. A. J. M. Logsdon, and B. Latell (ed.), *Eye in the Sky: The Story of the Corona Spy Satellites* (Washington DC: Smithsonian Institution Press, 1998), pp. 48–85, pp. 83–4). By contrast, most details of the Willow Run SAR and holography work, now mainly declassified, have not been disseminated publicly.

[41] Besides holography, the MIT Museum collection categories include architecture, nautical design, science and technology, and MIT ephemera (Connors, B. A., S. A. Benton, and W. Seamans, 'Report from the MIT Museum', *Fifth International Symposium on Display Holography* 2333 (1994): 146–51).

[42] When the holography collection was acquired, the museum opened an exhibition of holograms in the main gallery in 1994 and attendance figures increased dramatically. Nevertheless, the gallery space was later subdivided, with the hologram display reduced and moved behind a more popular exhibit on Artificial Intelligence.

1975–85. This leaves, though, a substantial period little represented in archival collections. Despite the commitment to preserve and make available these resources, the MIT Museum has not had the luxury of a permanent curator of holography, controlled-environment storage conditions for the collection, nor an explicit collection policy that enabled it to continue to acquire representative examples of holograms or documentary records.[43] This is perhaps understandable for a university museum that has a remit primarily to document the institution itself. However, MIT has been a major participant in holographic research through the Media Laboratories, so the holography collection arguably conforms to the Mission Statement, aiming to 'document, interpret and communicate, to a diverse audience, the activities and achievements of the Massachusetts Institute of Technology and the worldwide impact of its innovation, particularly in the field of science and technology'.[44] Interestingly, relatively few of the Media Laboratories' holograms were displayed at the Museum during Stephen Benton's life, although there was a limited exchange of examples serving as demonstration items for courses. A collecting policy was drafted in 2001, but limited funding and curatorial resources restricted new acquisitions.[45]

The MIT Museum has attempted consciously to make best use of its holography collection while serving other requirements. Its limited resources are not unusual, however. Larger national museums—the Smithsonian Museum in Washington, the Science Museum and Victoria & Albert Museum in London, the National Museum of Photography in Bradford, UK, and the Deutsches Museum in Munich, for example—each of which has mounted holographic displays, commissioned, or acquired holograms—have not established collecting policies to preserve the ephemeral material culture of holography.[46] This too, may be understandable, if one assumes the remit of technology or cultural museums to be the recording and valorizing of technologies perceived to be successful, relevant, or influential. Historians increasingly question such asymmetrical representation of the past, however. It biases the historical record to suggest that progress is natural and straightforward, and that subjects declining in economic or popular importance are unworthy of attention. As discussed above, a balanced treatment of perceived successes and failures is necessary not only to understand past events, but also to learn from them. Yet museums, defined by their sponsors, remits and audiences, are often compelled to present stories of progress. By being

[43] Leen, Mary and Joan Whitlow to SFJ, discussions, 2–14 Jul. 2003, Cambridge, MA, SFJ collection; Connors, Betsy to SFJ, interview, 10 Jul. 2003, Cambridge, MA, SFJ collection. Betsy Connors (b. 1950) obtained a BA in Art at the University of Massachusetts and an MS at MIT. She worked in video and was a Fellow of the Center for Advanced Visual Studies (CAVS) at MIT, later teaching holography at Tufts University and the Media Laboratories, and served as curator of the first exhibition mounted from the collection in 1994.

[44] MIT Museum, 'Strategic Plan, Nov. 2000–Jun. 2005', report, MIT.

[45] MIT Museum, 'DRAFT MIT Museum collecting guidelines for holography', MIT, Aug. 2001.

[46] Chris Titterington, while Assistant Curator of photographs at the Victoria & Albert Museum in London from the 1980s until 1995, established a hologram collection policy, but this did not outlive his tenure.

pigeonholed in this way, the history of holography is pared down to an unfaithful representation. As a result, there has also been an understandable dissonance between the stories told for different audiences.

14.4 CAREERS IN LIGHT: VISIONS OF TECHNICAL IDENTITY

For individual holographers, historical storytelling and judgements of success always had a personal dimension. As with many new technologies, reconciling achievements with forecasts could be dispiriting. Fred Unterseher reflected:

> People got tired after several years of starving and not making any money; you kind of end up having to go other ways [. . .] What has always suffered is an easy, accessible way to expand the development of this, because there really were immediate plug-ins to where people could go to learn holography, so we've always had all these students but *what the hell do you do with them?* That was always the difficulty.[47]

Individual careers in the field differed dramatically, but often were surprisingly tenacious. Unterseher's own long and varied career in holography samples many of the significant places and events, and illustrates the difficulty in establishing a stable niche. Beginning with his involvement in the original School of Holography in San Francisco, he set up his own studio cum cottage industry to create and sell aesthetic holograms. He conceived and co-wrote the *Holography Handbook* and, after a period as Director of Education at the New York Museum of Holography, worked as a teacher and researcher of holography around Europe and America. A similar example of a seminal contributor with a varied career is Ken Haines, who began holography at WRL, moved to Battelle to pursue commercial holography, taught modern optics in New Zealand, played a major role in embossed holography for magazines, credit cards, and more, and continued to explore new niches in the early 2000s. By contrast, Emmett Leith worked at the same institution, engaged in substantially the same discipline, for over half a century (see Figure 14.1).

For others, the experiences of arduous darkroom work, priority disputes, chemical exposure, commercial secrecy, and challenging markets were ambivalent. Robert Powell, one of the original group of holographers at WRL, mused, 'the entire experience has (and continues to) cost me too much, emotionally and financially. So I am inclined to avoid remembering details of the past.'[48] Similarly Michael Kan, a key innovator at the Multiplex Company of San Francisco during the 1970s, discouraged recollections, noting that, 'though it was a somewhat educational chapter in my life, it had ended in a rather painful and sour note'.[49]

[47] Unterseher, Fred to SFJ, interview, 23 Jan. 2003, Santa Clara, CA, SFJ collection

[48] Powell, Robert L. to P. Jackson, letter, 14 Oct. 1976, MIT Museum 30/850.

[49] Kan, Michael to SFJ, email, 29 Jan. 2004, SFJ collection

Fig. 14.1. Emmett Leith with 1966 Fritz Goro model, 2003 (S. Johnston photo). He died aged 78 in December 2005, 8 days before his official retirement.

As the careers sketched in this book indicate, the development of a professional identity was fraught for holographers. The troubled emergence of their occupation was not unique, but rather is an underreported feature of modern science and technology. The hinterland between these varieties of knowledge accommodates many such in-between subjects, which appear to have characteristic attributes. Some such specialties disappear in the face of culturally stronger occupations. Others like holography endure, their technical expertise being appropriated or mutated to serve the needs of other professional groups.[50]

Terry Shinn's and Bernward Joerges' characterization of these interstitial subjects as *research-technologies*, a category closely related to my own independently developed and slightly broader definition of *peripheral science*, is apt.[51] Both formulations identify

[50] Optical sciences are characteristically of this form. See, for example, studies of photometry (Johnston, Sean F., 'Making light work: practices and practitioners of light measurement', *History of Science* 34 (1996): 273–302), colorimetry (Johnston, Sean F., 'The construction of colorimetry by committee', *Science in Context* 9 (1996): 387–420), and Fourier spectroscopy (Johnston, Sean F., 'In search of space: Fourier spectroscopy 1950–1970', in: B. Joerges and T. Shinn (eds), *Instrumentation: Between Science, State and Industry* (Dordrecht: Kluwer Academic Press, 2001), pp. 121–41 and Johnston, Sean F., 'An unconvincing transformation? Michelson's interferential spectroscopy', *Nuncius* 18 (fasc. 2) (2003): 803–23). On spectroscopy from the early nineteenth to late twentieth centuries, see also Jackson, Myles, 'From theodolite to spectral apparatus: Joseph Von Fraunhofer and the invention of a German optical research-technology', in: B. Joerges and T. Shinn (eds.), *Instrumentation: Between Science, State and Industry* (Dordrecht: Kluwer Academic Press, 2001), pp. 17–28; Mallard, Alexandre, 'From the laboratory to the market: the metrological arenas of research-technology', in: B. Joerges and T. Shinn (eds), *Instrumentation: Between Science, State and Industry* (Dordrecht: Kluwer Academic Press 2001), pp. 219–40; Hentschel, Klaus, *Mapping the Spectrum: Techniques of Visual Representation in Research and Teaching* (Oxford: Oxford University Press, 2002).

[51] See Joerges, Bernward and Terry Shinn (eds), *Instrumentation: Between Science, State and Industry* (Dordrecht: Kluwer Academic Press, 2001) for case studies of 'research-technology' and Johnston, Sean F.,

technical communities that straddle the more commonly recognized terrains of science and technology, and recognize a characteristic interchange of knowledge and practice between technology, applied science, and fundamental research. Research-technologists, according to Shinn and Joerges, comprise 'a distinctive (but never distinct) transverse science and technology culture that generates a species of pragmatic universality, which in turn provides multiple and diversified audiences with a common repertory of vocabularies, notational systems, images, and perhaps even paradigms'.[52] By peripheral science I also highlight the social cohesiveness of a subject that does not achieve the status of an intellectual discipline or a profession, and in which there is a long-lived lack of autonomy and authority over the subject by any one group of practitioners.

Research-technology is, in effect, a bridge between more widely recognized academic and technological spheres. This bridging is supported by a variety of tactics ranging from knowledge, language, concepts, and artefacts, all of which act to increase the stability of its practitioners in the interstices. Shinn and Joerges argue that a key ingredient in this fluid interplay is a 'generic scientific instrument', which is transformed for and by distinct user groups. Such a generic device is characterized by an openness or availability for new applications. I would extend this fertile categorization in two respects: first, to incorporate generic concepts and *practices* that could assume the same role as a seminal instrument and, second, to include explicitly the non-rational *aesthetic* dimension of technical engagement.[53]

This slight generalization can be envisaged by the analogy of a planetary system. The concepts of holography can, in effect, be seen to have exerted an attractive force on disparate technical communities, some of which were deviated in their practices, others perturbed in their social composition, and yet others wholly captured in its orbit. The characteristics of research-technologies or peripheral sciences, however, make them less stable and deterministic than the more conventionally identified academic sciences or engineering professions. It is important to emphasize that, like an unstable planetary system, the concepts, practices, and instruments of holography attract but also interact with their technical communities, creating a metastable but largely unpredictable system of cognitive and social forces. Shinn argues that three modes of interaction between science and technology coexist: discipline-based science and technology (where stable subjects and profession become established); transitory science and technology (where workers spend time alternately in 'pure' and then 'applied' activities); and research-technologies, in which there is a continuing transfer of workers between scientific, engineering, and other activities.

A History of Light and Colour Measurement: Science in the Shadows (Bristol: Institute of Physics Publishing, 2001), Chapter 10 for a discussion of peripheral science.

[52] Joerges, and Shinn, op. cit., Chapter 1.

[53] The fluidity of occupational identity is an important feature of research-technologists or peripheral scientists. A few examples, among many, illustrate hybrid careers, and the not-uncommon shifting of holographers from artisanal to research or educative roles: Fred Unterseher, Pearl John, and Ana MacArthur, who established careers as holographic artists, also have worked as educators during the early 2000s to promote the burgeoning subject of *photonics* for high school children. And Melissa Crenshaw, like Unterseher during the 1990s, combined contract research and development with artistic work.

This analytical framework, with holography identified as a research-technology, is more satisfactory for explaining the trajectory of the subject than are other recent alternatives. Attempts to portray holography as a discipline, for example, would taint the subject with questionable assessments of failure, as we have seen. And by shoehorning the subject into disciplinary niches, its cohesiveness and boundary-crossing character would escape notice. By contrast, research-technology recognizes that holography has always been a moving target, historically and sociologically speaking. Its concepts, practitioners, and products have been fleeting, and demand greater-than-usual attention to trace their fragmentary trails.

The research-technology approach also contrasts with the radical constructivist analyses championed, for example, by Bruno Latour, which explain knowledge as a hybrid of both nature and culture, and in which there is a continual interplay between human actors (researchers, sponsors, promoters, adopters, and so on) and inanimate actors (e.g. instruments and physical resources). Sociologist Ivan Tchalakov's study of a Bulgarian holography laboratory employed this approach fruitfully.[54] While this intrinsic connectedness between the social and physical worlds is unquestionably a feature of the history of holography, the research-technology perspective seeks a deeper understanding of local context. The division of labour in the field was crucial to its socio-cognitive development. Holography, I argue, has been characterized by a series of fluid groupings of research-technologists, each of which has appropriated and reshaped the subject, and themselves, in the process. By contrast, the 'seamless web' approach in socio-historical studies can be frustratingly imprecise in ascribing causes and effects to scientific developments, and so is less amenable to developing either rich understanding of local circumstances or generalisations that are helpful for predictive purposes. In short, these two analytical approaches apply different insights and vaunt distinct goals.

14.5 IMAGINED AND INVENTED FUTURES

While holographers struggled to build stable careers, they also drew their own, often incompatible, conclusions from their experiences in the field, and developed distinct explanations for the deviations from forecasts. Such practitioners' explanations understandably can be more insightful and candid than commercial rationalizations proffered

[54] His ethnographic study of the Bulgarian Central Laboratory of Optical Storage and Processing of Information (CLOSPI), which began the development of holographic memories in 1974, was carried out between 1993 and 1998. The most accessible publications currently available in English are: Tchalakov, Ivan, 'Innovating in Bulgaria—two cases in the life of a laboratory before and after 1989', *Research Policy* 30 (2001): 391–402, and Tchalakov, Ivan, 'The object and the other in holographic research—approaching passivity and responsibility of human actors', *Science, Technology and Human Values* 29 (2002): 64–7, which are drawn from his book Tchalakov, Ivan, *Da napravish holograma: kniga za svetlinata, uchenite if vsichko ostanalo (Making a Hologram: a Book about the light, about the scientists and their world. (Development of opto-electronics in Bulgaria—1969–1998))* (Sofia: Marin Drinov Publishing House, 1998).

for public consumption. One common observation, for example, was the elegance, but economic non-competitiveness, of holographic applications. Emmett Leith and H. John Caulfield propounded 'folk theorems' in holography, which boiled down to one: 'Holography is the best way of doing anything in optics, but also the most expensive'.[55] An even more basic theme was the oft-recited aphorism 'holography is a solution in search of a problem': there were fewer applications—or at least less money to be made—than most forecasts suggested.

Holographers have proffered their own divergent evaluations and explanations of its disputed successes, and the corresponding effect on their careers, combining technical pessimism with market optimism. In a 2002 discussion, for instance, Hans Bjelkhagen and Bill Fagan—both holographers since the late 1960s—offered contrasting views about the tribulations and future of holography, Bjelkhagen (with his later work dominated by display holography) suggesting that the faithfulness of image quality was the key factor, and Fagan (with a career in engineering interferometry) that the medium inexplicably had failed to capture popular appeal:

> Fagan: Once the honeymoon period was over, they'd say 'it's one of those hologram things'. Holograms must go beyond being a novelty to being an information carrier, like the photogravure, the first class photographic plate.
> Bjelkhagen: No doubt in my mind that there is a huge, huge market for holography if it is perfect; but as long as it is not perfect, and the illumination of the hologram has not been solved, there is no market.[56]

Don Gillespie, who had supplied commercial holographers with equipment through his businesses Jodon and Eldon from the mid-1960s, echoed and extended Bjelkhagen's evaluation with an economic tinge:

> The future requires a good 3-colour laser (which doesn't exist)—in the right portion of the spectrum with the right intensity. Other criteria are (1) you have to be able to illuminate the hologram in the right way, without shining lights on it. Steve Benton came up with one answer: edge illumination on glass, but that won't make it. And (2) it has to be relatively inexpensive and (3) it needs acceptance in the art community as art.[57]
>
> When you get pulsed three-color holography so you can take a portrait, and you can light it from the base, I think holography will take off like you would not believe, but until you can do that it will not take off in the market; It'll go when you've got a studio, and can drop in for a portrait of the family, or your wife.
>
> Emmett Leith: Without dropping three thousand dollars?
>
> Gillespie: Perhaps several hundred—3 to 5 hundred.[58]

[55] Caulfield, H. John to SFJ, interview, 21 Jan. 2003, Santa Clara, CA, SFJ collection.

[56] Fagan, William and Hans Bjelkhagen to SFJ, interview, 19 Sep. 2002, SFJ collection. Bjelkhagan (b. 1945) completed an MSc degree in 1969 and PhD in 1978 and worked with Nils Abramson at the Royal Institute of Technology, Stockholm. His peripatetic career in Sweden, America, and Britain ranged from colour display holography to emulsion development, holographic interferometry, bubble-chamber holography, medical diagnostics, and pulsed portraits (including co-production of one of former President Ronald Reagan).

[57] Gillespie, Donald to SFJ, interview, 29 Aug. and 4–6 Sep. 2003, Ann Arbor, MI, SFJ collection.

[58] Leith, Emmett N., Donald Gillespie and Brian Athey to SFJ, interview, 11 Sep. 200003, Ann Arbor, SFJ collection.

Others saw the challenges in deeper terms, in the very categorization of the subject. Glenn Wood, for instance, former marketing manager of Ilford Ltd, suggested that holography had to merge with up-and-coming technologies to retain popular interest in the early 1990s. He advised Martha Tomko, the final Director of the New York Museum of Holography, to drop holography as a name and to link the museum instead with virtual reality:

> If you could somehow hook onto it pointing out that the hard copy version has been around for some time and is called holography maybe you could get something going that way. There are also many similarities between the Los Angeles crowd involved in VR and holographers (like they are all crazy people). [. . .] Change the MOH to the MOVR and relaunch. I don't think anyone gets excited about holography anymore, at least not the corporations or institutions you need to stay viable.[59]

Along different lines, Fred Unterseher, reflecting on the various audiences for holography, suggested that its value was not intrinsically economic:

> You get three different kinds of people: you get the science guys who think that they're going to come into some company, and find some way to invent something [. . .] you get the artists, who are intimidated by the technology, and feel threatened by it all, and then you get, of course, the people who think they are going to make a million dollars in the next week; it tends to break down into these sorts of categories [. . .] and I tell them if they work really, really, really hard it's really an uphill battle to achieve something, but that's not the virtue of this. The virtue of a holography class and the holography experience is to transform the way you see, to transform the way you experience; *that's* the value of this, the value of this is to see light and to experience in new ways—*that's* what it's about. And when it's all said and done, there ain't many people who are *rich* doing this. There aren't many famous artists doing this, and the science people, to some extent it's not really been . . . But people's *lives* have been changed, and the dynamic of the experience has continued.[60]

Steve McGrew evinced a similar commitment, but from an entrepreneurial perspective:

> There are still some significant inventions yet to be made in holography. Holography hasn't yet been truly integrated into the printing process, but I think it can be done. Holography hasn't yet been integrated into everyday photography, but I think it can be done. There aren't yet holographic portraits on passports and ID cards, but I think it can be done.
>
> Holography is far more than a graphical medium, though. I think holography may have signaled the birth of a new paradigm in communications, computation, and the control of matter and processes. Before holography came into the common consciousness, information and control processes were largely conceived of as serial, one-dimensional things. Holography has taught thousands of people to think in parallel, in 2D and 3D, in terms of wavefronts and wave fields. I believe that holography is, at the very least, one aspect of a major shift in the way that engineers and scientists think. Adaptive optics, coherent control of atomic and molecular quantum states, passive radar, passive sonar, synthetic aperture radar, photonic crystals, quantum computation, systems biology and many other emerging fields are other aspects of the same new way of thinking.[61]

[59] Wood, Glenn P. to M. Tomko, letter, 13 Jun. 1991, MIT Museum 2/38.
[60] Unterseher, Fred to SFJ, interview, 23 Jan 2003, Santa Clara, CA, SFJ collection.
[61] McGrew, Steve, 'My reflections on holography', unpublished manuscript, 5 Dec. 2004, SFJ collection.

In a similar vein, Stephen Benton reflected three months before his death:

> It's the display aspect that keeps people excited. And unfortunately it's not a big business, or even a *small* business. You do it if you can manage to hold on to it.
>
> [Holography is] still the only way to do that kind of 3-D, and one of my goals all along has been to invent other kinds. But up to that point most 3-D just looked like crap, people making post-cards. I think it did up the ante quite a bit. That was the one benefit of it: to make it serious, to make it respectable. But *it* came, and then *electronics* came in and stole away the bright young people. I think they're starting to come back. All of sudden there's a little more interest, since holographic video was demonstrated. [. . .] You know, *somebody's doing it!* [chuckles] Maybe they'll build it up and get it another step or two.[62]

This handful of holographers, despite contrasting backgrounds and perspectives, suggest both the inexpressible zeal that drew them to their subject and the career frustrations they experienced. They variously cited high cost, lack of applications, constraints on image quality, public misunderstanding, and rising cultural expectations as the principal reasons for the subject's perceived tribulations. As all had discovered, holography by the 1990s was no longer an effective term to generate donations, feasibility studies, or corporate interest. But it continued to capture the enthusiasm of new practitioners and provided attractions beyond commercial profitability. That intangible attraction, released by Dennis Gabor (see Figure 14.2) a half-century earlier in his BTH laboratory, had spread haltingly to suffuse group after group of technical workers.

Gabor's last months were spent in a nursing home in London, incapacitated by a series of strokes that he had suffered from the mid-1970s. Posy Jackson, then Director of the New York Museum of Holography, visited him in January 1979, shortly before his death. Writing afterwards to Gabor's old friend and colleague, T. E. Allibone, she recalled:

> I brought him the first in-line hologram that had been recently completed as the beginning of the technical explanation of our historic exhibition. It was a white light version of the very same type he had first made in 1947, and it pleased him so much he would not let Marjorie take it home, but set it up right there in his room to look at. It has always seemed somewhat ironic to me that that hologram was probably the last he saw before he died, and it was a representation of the very first he had ever made.[63]

Gabor, the originator of holograms, was also a humanist in the Renaissance sense. An early member of the Club of Rome, he had written and lectured in his later years on science and society and their entwined futures. One of his books, *Inventing the Future*, argued that while technological societies find themselves unable to predict the future, they can invent it for themselves.[64] Most today would agree that this aim, echoed by the engineers at Conductron and the MIT Media Laboratories, was not reached. But, in altered form, the claim can be applied to predictions about the subject that Gabor initiated: the

[62] Benton, Stephen A. to SFJ, interview, 11 Jul. 2003, Cambridge, MA, SFJ collection.
[63] Jackson, Rosemary H. to T. E. Allibone, letter, 17 Dec. 1979, MIT Museum 2/40.
[64] Gabor, Dennis, *Inventing the Future* (London: Secker & Warburg, 1963).

Fig. 14.2. Dennis and Marjorie Gabor preparing for his portrait at the New York Museum of Holography, 1977 (MIT Museum MoH collection).

imagined future for holography has been recast repeatedly by successive waves of holographers and continues to be reinvented by its subsequent communities and adopters.

A dual perspective—namely explaining the past course of the subject alongside its imagined futures—has been the focus of this book, and equally a matter of reinvention and interpretation. In a subject riven by contrasting assessments and predictions, the only indisputable failures surrounding holography concerned the forecasts themselves.

These different holographic visions have influenced and even inspired thousands, and yet evoking that appeal may remain as elusive as the holographic image itself. As I have tried to argue, holography is not an aberrant case of modern science, technology, and art. Rather, it is an intriguing and important example of a technical subject that created and grew with its communities. Holography represents an important collective creation, combining thousands of disparate individuals with different visions, impelled by a consuming enthusiasm for an idea, a technique, and a product. Its complementary accounts show how a seductive new science and potent ideas can traverse and pervade culture.

Bibliography

INTERVIEWS

Conducted by Sean Johnston unless otherwise noted.

1. Carl Aleksoff, 9 Sep. 2003, Ann Arbor, MI, USA
2. Eric Ash (by Frederik Nebeker for IEEE), 25 Aug. 1994, London, UK
3. Brian Athey, 2 Aug. 2004, Glasgow, UK
4. Norm Barnett, 11 Sep. 2003, Ann Arbor, MI, USA
5. Eric Begleiter, 10 Jul. 2003, Cambridge, MA, USA
6. Stephen Benton, 11 Jul. 2003, Cambridge, MA, USA
7. Margaret Benyon, 21 Jan. 2003, Santa Clara, CA, USA and 30 Apr. 2005, London, UK
8. Hans Bjelkhagen, 18–19 Sep. 2002, Leicester, UK and 5–6 Jul. 2005, St Asaph, UK
9. Roger Brentnall, 20 Sep. 2002, Mobberley, UK
10. Nigel Briggs, 20 Sep. 2002, Mobberley, UK
11. Harriet Casdin-Silver, 3 Jul. 2003, Boston, MA, USA
12. H. John Caulfield, 21 Jan. 2003, Santa Clara, CA, USA
13. Clark Charnetski, 3 Sep. 2003, Ann Arbor, MI, USA
14. Gary Cochran, 6 and 8 Sep. 2003, Ann Arbor, MI, USA
15. Betsy Connors, 10 Jul. 2003, Cambridge, MA, USA
16. Melissa Crenshaw, 19 Nov. 2003, Vancouver, Canada
17. Duncan Croucher, 16 Aug. 2002, by telephone
18. Mary Dentschuk, 20 Sep. 2002, Mobberley, UK
19. Frank Denton, 1 May 2003, by telephone
20. Ed Dietrich, 19 Nov. 2003, Vancouver, Canada
21. Vincent DiBiase, 22–23 Jan. 2003, Santa Clara, CA, USA
22. William Fagan, 19 Sep. 2002, Leicester, UK
23. Don Gillespie, 29 Aug., 5, 6, and 11 Sep. 2003, Ann Arbor, MI, USA
24. Kenneth Haines, 23 Jan. 2003, Santa Clara, CA, USA
25. Parameswaran Hariharan, 9 Jul. 2003, Cambridge, MA, USA
26. Alan Hodgson, 20 Sep. 2002, Mobberley, UK
27. Randy James (by Jonathan Ross), Dec. 1980, CA, USA
28. Tung Jeong, 21 Jan. 2003, Santa Clara, CA, UK
29. Yasumasa Kamata, 19 Nov. 2003, Vancouver, Canada
30. Lewis Kontnik, 20 Nov. 2003, Vancouver, Canada
31. Linda Lane (by Jonathan Ross), Dec. 1980, Los Angeles, USA
32. Larry Leiberman (by Alan Rhody), Mar. 1998
33. Emmett Leith, 22–23 Jan. 2003, Santa Clara, CA, USA and 29 Aug.–12 Sep. 2003, Ann Arbor, MI, USA

34. Carl Leonard, 4 and 10 Sep. 2003, Ann Arbor, MI, USA
35. Roger Lessard, 24 May 2004, Kiev, Ukraine
36. Vladimir Markov, 26 May 2004, Kiev, Ukraine
37. Steve McGrew (by Jonathan Ross), Dec. 1980, Los Angeles, CA, USA
38. Lon Moore (by Jonathan Ross), Dec. 1980, Los Angeles, CA, USA
39. Rob Munday, 31 Mar. 2004, Richmond, Surrey, UK
40. Harry Owen, 9 Sep. 2003, Ann Arbor, MI, USA
41. Nick Phillips, 18 Sep. 2002, Leicester, UK
42. Steve Provence (by Jonathan Ross), Dec. 1980, Los Angeles, CA, USA
43. Amanda Ranalli, 4 Apr. 2003, London, UK
44. Ana Maria Richardson, 23 Jan. 2003, Santa Clara, CA, USA
45. Jonathan Ross, 4 Apr. 2003, London, UK
46. Graham Saxby, 16–17 Sep. 2002, Wolverhampton, UK
47. Chalmers Sherwin (by John Bryant for IEEE), 12 Jun. 1991, Cambridge, MA, USA
48. Lawrence Siebert, 4 Sep. 2003, Ann Arbor, MI, USA
49. Chris Slinger, 20 Nov. 2003, Vancouver, Canada.
50. Marat Soskin, 26 May 2004, Kiev, Ukraine.
51. Dmitry Staselko, 26 May 2004, Kiev, Ukraine.
52. George Stroke (by William Aspray for IEEE), 3 Jul. 1993, Munich, Germany
53. Charles Townes (by Frederik Nebeker for IEEE), 14–15 Sep. 1992, Berkeley, CA, USA
54. Jim Trolinger, 26 Apr. 2004, Kiev, Ukraine
55. Fred Unterseher, 22 Jan. 2003, Santa Clara, CA, USA
56. Charles Vest, 11 Jul. 2003, Cambridge, MA, USA
57. Gary Zellerbach (by Jonathan Ross), Dec. 1980, San Francisco, CA, USA

ARCHIVES CONSULTED

Publicly accessible collections

Bentley — Bentley Historical Library, Ann Arbor, Michigan, USA: archives of the University of Michigan Institute of Science and Technology, Willow Run Laboratory, and Optical Sciences Laboratory.

BC — Birkbeck College archives, London, UK: David Bohm papers.

CHF — Chemical Heritage Foundation: records of the Gordon Research Conferences.

IC — Imperial College archives, South Kensington, London, UK: Dennis Gabor papers.

MIT — MIT Museum, Cambridge, Massachusetts, USA: New York Museum of Holography collection.

Sci Mus — Science Museum Library, South Kensington, London, UK: Gordon L. Rogers papers.

Private collections

Bjelkhagen collection — Hans Bjelkhagen, Optik Technium, St. Asaph, UK.

Charnetski collection — Clark Charnetski, Ann Arbor, MI, USA.

Cochran collection — Gary Cochran, Ann Arbor, MI, USA.

Haines collection — Kenneth Haines, Santa Clara, CA, USA.

Leith collection — Emmett Leith, University of Michigan, Ann Arbor, MI, USA.

Leonard collection — Carl Leonard, Ann Arbor, MI, USA.

Ross collection — Jonathan Ross, London, UK.
Saxby collection — Graham Saxby, Wolverhampton, UK.
Upatnieks collection — Juris Upatnieks, Ann Arbor, MI, USA.
Numerous documents were supplied by other individuals, and are listed in individual references.

BOOKS

Abbott, Andrew Delano, *The System of Professions: An Essay on the Division of Expert Labor* (Chicago: University of Chicago Press, 1988)

Barger, M. Susan and William B. White, *The Daguerreotype: Nineteenth-Century Technology and Modern Science* (Washington: Smithsonian Institution Press, 1991)

Barnes, Barry, *Interests and the Growth of Knowledge* (London: Routledge and Kegan Paul, 1977)

Barrekette, Euval S., *Applications of Holography: Proceedings of the United States-Japan Seminar on Information Processing by Holography, held in Washington, D.C., October 13–18, 1969* (New York; London: Plenum Press, 1971)

Benthall, Jonathan, *Science and Technology in Art Today* (New York: Praeger Publishers, 1972)

Benton, Stephen A., *Holographic Imaging (draft)* http://splweb.bwh.harvard.edu:8000/courses/mas450/reading/chaptersPDF/, 2003)

Beran, M. J. and G. B. Parrent Jr, *Theory of Partial Coherence* (Eaglewood Cliffs, NJ: Prentice-Hall, 1964)

Berner, Jeff, *The Holography Book* (New York: Avon Books, 1980)

Bijker, Wiebe E. and John Law, *Shaping Technology/Building Society: Studies in Sociotechnical Change* (Cambridge, MA: MIT Press, 1992)

Bjelkhagen, Hans I., *Silver-Halide Recording Materials for Holography and Their Processing* (Berlin: Springer-Verlag, 1993)

Blackmore, Susan, *The Meme Machine* (Oxford: Oxford University Press, 1999)

Bloom Murray, Teigh, *The Brotherhood of Money: The Secret World of Bank Note Printers* (Port Clinton, OH: BNR Press, 1983)

Bloor, David, *Knowledge and Social Imagery* (London: Routledge and Kegan Paul, 1976)

Bohm, David, *Wholeness and the Implicate Order* (London: Routledge, 1980)

Bohm, David and F. David Peat, *Science, Order, and Creativity* (New York: Bantam Books, 1987)

Boyle, Sandy, *Light Fantastic 2: A New Exhibition of Holograms by Holoco* (London: Bergstrom-Boyle, 1978)

Brand, Stewart, *Whole Earth Catalog—Access to Tools* (Menlo Park, CA: Portola Institute Inc., 1970)

Brand, Stewart, *The Media Lab: Inventing the Future at MIT* (New York: Viking Penguin, 1987)

Breines, Wini, *Community and Organization in the New Left, 1962–1968: The Great Refusal* (New York: Praeger; J. F. Bergin, 1982)

Bromberg, Joan Lisa, *The Laser in America, 1950–1970* (Cambridge, MA: MIT Press, 1991)

Brook, Donald, *The Social Role of Art* (Adelaide: Experimental Art Foundation, 1981)

Bunzel, John H., *New Force on the Left: Tom Hayden and the Campaign Against Corporate America* (Stanford, CA: Hoover Institution Press Stanford University, 1983)

Burgmer, Brigitte, *Holographic Art: Perception, Evolution, Future* (La Coruña, Spain: Daniel Weiss, 1987)

Collins, H. M., *Changing Order: Replication and Induction in Scientific Practice* (London; Beverly Hills: Sage Publications, 1985)

Cosslett, V. E., *Practical Electron Microscopy* (London: Butterworths, 1951)

Crombie, A. C. (ed.), *Scientific Change: Historical studies in the intellectual, social and technical conditions for scientific discovery and technical invention, from antiquity to the present* (London: Heinemann, 1963)

Crowther, J. G., *Soviet Science* (London: Kegan Paul, Trench, Trubner & Co., 1936)

Darrah, William C., *The World of Stereographs* (Gettysburg, PA: Times & News Publishing Co., 1977)

Dawkins, Richard, *The Selfish Gene* (Oxford: Oxford University Press, 1976)

Day, Dwayne A., John M. Logsdon, and Brian Latell, *Eye in the Sky: The Story of the Corona Spy Satellites* (Washington, DC: Smithsonian Institution Press, 1998)

de Solla Price, Derek, *Little Science, Big Science* (Washington: Columbia University Press, 1963)

Denisyuk, Yu N., *Fundamentals of Holography* (Moscow: Mir, 1984)

DeVelis, John B. and George O. Reynolds, *Theory and Applications of Holography* (Reading, MA: Addison-Wesley Pub. Co., 1967)

Divall, Colin and Sean F. Johnston, *Scaling Up: The Institution of Chemical Engineers and the Rise of a New Profession* (Dordrecht: Kluwer Academic, 2000)

Douglas, Susan J., *Inventing American broadcasting: 1899–1922* (Baltimore, MD; London: Johns Hopkins University Press, 1987)

Dowbenko, George, *Homegrown Holography* (Garden City, NY: Amphoto, 1978)

Dudley, David D. and Computer Sciences Corporation, *Holography: A Survey* (Washington, DC: Technology Utilization Office National Aeronautics and Space Administration, 1973)

Duffieux, P. M., *L'intégrale de Fourier et ses Applications à l'Optique* (Besançon: Faculté des Sciences, Besançon; Societé Anonyme des Imprimeries Oberthur: Rennes, 1946)

Eastman Kodak Company, *Two new Kodak materials for holography: Kodak holographic plate, type 120-02: Kodak holographic film (Estar base) SO-173* (Rochester, NY: Eastman Kodak Co., 1973)

Eastman Kodak Company, *Reversal bleach process for producing phase holograms on Kodak spectroscopic plates, type 649-F* (Rochester, NY: Eastman Kodak Co., 1974)

Eastman Kodak Company, *Kodak materials for holography* (Rochester, NY: Eastman Kodak, 1976)

Eco, Umberto and William Weaver, *Travels in Hyper-Reality: Essays* (London: Picador, 1986)

Foucault, Michel and Alan Sheridan, *Discipline and Punish: The Birth of the Prison* (London: Allen Lane, 1977)

Françon, M., *Modern Applications of Physical Optics* (New York: John Wiley, 1963)

Gabor, Dennis, *The Electron Microscope: Its Development, Present Performance, and Future Possibilities* (London: Electronic Engineering, 1948)

Gabor, Dennis, *Inventing the Future* (London: Secker & Warburg, 1963)

Gabor, Dennis, *The Mature Society* (London: Secker and Warburg, 1972)

Galison, Peter, *Image and Logic: A Material Culture of Microphysics* (Chicago: University of Chicago Press, 1997)

Galison, Peter and Bruce Hevley (eds), *Big Science: The Growth of Large-Scale Research* (Stanford: Stanford University Press, 1992)

Gibson, William, *Burning Chrome* (New York: Ace Books, 1987)

Goodman, Joseph W., *Introduction to Fourier Optics* (New York: McGraw-Hill, 1968)

Halas, John, *Graphics in Motion: From the Special Effects Film to Holographics* (New York: Van Nostrand Reinhold, 1984)

Hariharan, Parameswaran, *Optical Holography: Principles, Techniques and Applications* (Cambridge: Cambridge University Press, 1996)

Hayward, Philip, *Culture, Technology & Creativity in the Late Twentieth Century* (London: J. Libbey, 1990)

Hecht, Gabrielle, *The Radiance of France: Nuclear Power and National Identity After World War II* (Cambridge, MA; London: The MIT Press, 1998)

Hecht, Jeff, *City of Light: The Story of Fiber Optics* (Oxford: Oxford University Press, 1999)

Hecht, Jeff, *Beam: The Race to Make the Laser* (Oxford: Oxford University Press, 2005)

Hentschel, Klaus, *Mapping the Spectrum: Techniques of Visual Representation in Research and Teaching* (Oxford: Oxford University Press, 2002)

Hercock, R. J. and George Alan Jones, *Silver by the Ton: The History of Ilford Limited, 1879–1979* (London; New York: McGraw-Hill Book Co., 1979)

Hughes, Thomas Parke, *Networks of Power: Electrification in Western Society, 1880–1930* (Baltimore: Johns Hopkins University Press, 1983)

Jay, Martin, *The Dialectical Imagination: A History of the Frankfurt School and the Institute of Social Research 1923–50* (London: Heinemann, 1973)

Jeong, Tung H., *Gaertner-Jeong Holography Manual* (Chicago: Gaertner Scientific Corp., 1968)

Jeong, Tung H., *A Study Guide on Holography (Draft)* (Lake Forest, Illinois: Lake Forest College, 1975)

Jeong, Tung H. and Francis E. Lodge, *Holography using a Helium–Neon Laser* (Metrologic Instruments, 1978)

Joerges, B. and T. Shinn (eds), *Instrumentation: Between Science, State and Industry* (Dordrecht: Kluwer Academic Press, 2001), pp. 17–28

Johnston, Sean F., *A History of Light and Colour Measurement: Science in the Shadows* (Bristol: Institute of Physics Publishing, 2001)

Jones, Caroline A. and Galison, Peter (eds), *Picturing Science, Producing Art* (New York: Routledge, 1998)

Jones, Robert and Oliver Marriott, *Anatomy of a Merger: A History of G.E.C, A.E.I. and English Electric* (London: Cape, 1970)

Kallard, Thomas, *Holography: State of the Art Review 1969* (New York: Optosonic Press, 1969)

Kallard, Thomas, *Holography: State of the Art Review 1970* (New York: Optosonic Press, 1970)

Kallard, Thomas, *Holography: State of the Art Review 1971–1972* (New York: Optosonic Press, 1972)

Kasper, Joseph Emil and Steven A. Feller, *The Hologram Book* (Englewood Cliffs, NJ: Prentice-Hall, 1985)

Kevles, Daniel J., *The Physicists: The History of a Scientific Community in Modern America* (Cambridge, MA; London: Harvard University Press, 1995)

Kock, Winston E., *Lasers and Holography; an Introduction to Coherent Optics* (Garden City, NY: Doubleday, 1969)

Kragh, Helge, *An Introduction to the Historiography of Science* (Cambridge: Cambridge University Press, 1987)

Kuhn, Thomas S., *The Structure of Scientific Revolutions* (Chicago: University of Chicago Press, 1962)

Laser Institute of America and Tung H. Jeong, *Holography: Principles and Applications* (Toledo, Ohio: The Institute, 1975)

Latour, Bruno, *Science in Action: How to Follow Scientists and Engineers Through Society* (Milton Keynes: Open University Press, 1987)

Latour, Bruno and Steve Woolgar, *Laboratory Life: The Construction of Scientific Facts* (Princeton, NJ; Chichester: Princeton University Press, 1986)

Law, J. and J. Hassard (eds), *Actor Network Theory and After* (Oxford: Blackwell, 1999)

Layton, Edwin, *Revolt of the Engineers: Social Responsibility and the American Engineering Profession* (Cleveland: Case Western Reserve University Press, 1971)

Lenoir, Timothy, *Instituting Science: The Cultural Production of Scientific Disciplines* (Stanford: Stanford University Press, 1997)

Leslie, Stuart W., *The Cold War and American Science: The Military–Industrial–Academic Complex at MIT and Stanford* (New York: Columbia University Press, 1993)

Lewis, Jonathan E., *Spy Capitalism: Itek and the CIA* (New Haven: Yale University Press, 2002)

Linfoot, E. H., *Fourier Methods in Optical Image Evaluation* (London: Focal Press, 1964)

Longhurst, Richard S., *Geometrical and Physical Optics* (London: Longmans, Green, 1957)

Lucie-Smith, Edward, *Alexander* (London: Art Books International, 1992)

MacDonald, Keith M., *The Sociology of the Professions* (London: Sage Publications, 1995).

MacDonald, Sharon, *Behind the Scenes at the Science Museum* (Oxford: Berg, 2002)

MacKenzie, Donald A. and Judy Wajcman (eds), *The Social Shaping of Technology: How the Refrigerator Got Its Hum* (Milton Keynes, England; Philadelphia: Open University Press, 1985)

Maiman, Theodore H., *The Laser Odyssey* (Blaine, WA: Laser Press, 2000)

McBurnett, Ted, *N Dimensional Space* (New York: Finch Museum of Art, 1970)

McNair, Don, *How To Make Holograms* (Blue Ridge Summit, PA: TAB, 1983)

Medvedev, Zhores A., *Soviet Science* (Oxford: Oxford University Press, 1979)

Merton, Robert K., *Social Theory and Social Structure* (Glencoe, IL: Free Press, 1957)

Merton, Robert K. and Norman William Storer, *The Sociology of Science: Theoretical and Empirical Investigations* (Chicago; London: University of Chicago Press, 1973)

Mertz, Lawrence, *Transformations in Optics* (New York: John Wiley, 1965)

Michelson, Albert A., *Studies in Optics* (Chicago: University of Chicago Press, 1927)

Miller, James, *'Democracy is in the Streets': From Port Huron to the Siege of Chicago* (New York; London: Simon & Schuster, 1987)

Miller, Larry, *Holographic Golf: Uniting the Mind and Body to Improve Your Game* (New York: HarperCollins, 1993)

Minnaert, M., *The Nature of Light and Colour in the Open Air* (New York: Dover, 1954)

Nelson, Elizabeth, *The British Counter-Culture, 1966–1973* (Basingstoke: MacMillan, 1989)

Nobel Prize Committee (ed.), *Les Prix Nobel En 1971* (Stockholm, 1971)

Noble, David F., *America by Design: Science, Technology, and the Rise of Corporate Capitalism* (New York: Knopf, 1979)

Nye, David E., *American Technological Sublime* (Cambridge, MA: MIT Press, 1994)

O'Neill, Edward L. (ed.), *Communication and Information Theory Aspects of Modern Optics* (Syracuse, NY: General Electric Company, 1962)

O'Neill, Edward L., *Introduction to Statistical Optics* (Reading, MA: Addison-Wesley Publishing Co., 1963)

Ostrovsky, Y. I., *Holography and its Application* (Moscow: Mir Publishers, 1977)

Outwater, Chris and Eric Van Hamersveld, *Guide to Practical Holography* (Beverly Hills, CA: Pentangle Press, 1974)

Pearce, Susan M., *Museums Objects and Collections: A Cultural Study* (Leicester: Leicester University Press, 1992)

Pepper, Andrew, 'Creative Holography Index,' (Nottingham: Monand Press, 1992–95)

Pethick, J., *On Holography and a Way to Make Holograms* (Burlington, Ontario: Belltower Enterprises, 1971)

Picardo, Robert, *Star Trek Voyager: The Hologram's Handbook* (New York: Pocket Books, 2002)

Polanyi, Michael, *Personal Knowledge: Towards a post-critical philosophy* (London: Routledge & Kegan Paul, 1958)

Popper, Frank, *Art of the Electronic Age* (London: Thames and Hudson, 1993)
Preston, Kendall, *Coherent Optical Computers* (New York: McGraw-Hill, 1972)
Rasmussen, Nicolas, *Picture Control: The Electron Microscope and the Transformation of Biology in America, 1940–1960* (Stanford, CA: Stanford University Press, 1997)
Richelson, Jeffrey, *The Wizards of Langley: Inside the CIA's Directorate of Science and Technology* (Boulder, CO: Westview Press, 2001)
Roberg, Jeffrey L., *Soviet Science Under Control: The Struggle for Influence* (Basingstoke: MacMillan, 1998)
Roszak, Theodore, *The Making of a Counter-Culture: Reflections on Technocratic Society and Its Youthful Opposition* (London: 1970)
Rudisill, Richard, *Mirror Image: The Influence of the Daguerreotype on American Society* (Albuquerque: University of New Mexico Press, 1971)
Saxby, Graham, *Holograms: How to Make and Display Them* (London: Focal Press, 1980)
Saxby, Graham, *Manual of Practical Holography* (Oxford: Focal Press, an imprint of Butterworth-Heinemann Ltd., 1991)
Saxby, Graham, *Practical Holography* (Bristol: Institute of Physics Press, 2004)
Smiles, Samuel, *Self-Help: With Illustrations of Character and Conduct* (London: John Murray, 1859)
Smith, Howard M., *Principles of Holography* (New York: Wiley-Interscience, 1969)
Snow, C. P., *The Two Cultures; and, A Second Look—An Expanded Version of the Two Cultures and the Scientific Revolution* (Cambridge: Cambridge University Press, 1964)
Staudenmaier, John M., *Technology's Storytellers: Reweaving the Human Fabric* (Cambridge, MA: MIT Press, 1985)
Taubman, Philip, *Secret Empire: Eisenhower, the CIA, and the Hidden Story of America's Space Espionage* (New York: Simon & Schuster, 2003)
Taylor, Nick, *Laser: The Inventor, the Nobel Laureate, and the Thirty-Year Patent War* (New York; London: Simon & Schuster, 2000)
Tchalakov, Ivan, *Da napravish holograma: kniga za svetlinata, uchenite if vsichko ostanalo (Making a Hologram: A Book About the Light, About the Scientists and Their World. (Development of opto-electronics in Bulgaria—1969–1998))* (Sofia: Marin Drinov Publishing House, 1998)
Thomas, J. R. and U. M. Kruse-Vaucienne (eds), *Soviet Science and Technology: Domestic and Foreign Perspectives* (Washington DC: George Washington University Press, 1977)
Tolanski, S., *Surface Microtopography* (London: Longmans, 1960)
Townes, Charles H., *How the Laser Happened: Adventures of a Scientist* (Oxford: Oxford University Press, 2002)
Stroke, George W., *An Introduction to Coherent Optics and Holography* (New York: Academic Press, 1966 and 2nd edition, 1969)
Unger, Irwin and Debi Unger, *The Movement: A History of the American New Left, 1959–1972* (New York: Dodd Mead, 1974)
Unterseher, Fred, Jeannene Hansen, and Bob Schlesinger, *Holography Handbook: Making Holograms the Easy Way* (Berkeley, CA: Ross Books, 1981)
Valenta, E., *Die Photographie in natüralichen Farben* (Germany: Halle, 1912)
Vest, Charles M., *Holographic Interferometry* (New York: Wiley, 1979)
Wang, Jessica, *American Science in an Age of Anxiety: Scientists, Anticommunism, and the Cold War* (Chapel Hill; London: University of North Carolina Press, 1999)
Weinberg, A. M., *Reflections on Big Science* (Boston: MIT Press, 1967)

Wilber, Ken, *The Holographic Paradigm and Other Paradoxes: Exploring the Leading Edge of Science* (Boulder: Shambhala, 1982)

Winston, Brian, *Technologies of Seeing: Photography, Cinematography and Television* (London: British Film Institute, 1996)

Wood, R. W., *Physical Optics* (New York: MacMillan, 1929)

Yaroslavsky, Leonid P. and N. S. Merzlyakov, *Methods of Digital Holography* (New York: Consultants Bureau, 1980)

Youngblood, Gene, *Expanded Cinema* (London: Studio Vista, 1970)

Zachary, G. Pascal, *Endless Frontier: Vannevar Bush, Engineer of the American Century* (New York: Free Press, 1997)

Zec, Peter, *Holographie: Geschichte, Technik, Kunst* (Köln: DuMont, 1987)

PAPERS, ARTICLES, CATALOGUES, DISSERTATIONS, AND BOOK CHAPTERS

'3-D lasography—the month old giant', *Laser Focus*, 1 Jan. 1965: 10

'3D photography coming to Europe, inventor pair says', *holosphere* 7 (1978): 2

'A book for all seasons: *Holography Handbook*', *L.A.S.E.R. News*, 4 (2) (1985): 6–7

'A holographic interface', *Laser Focus*, Jul. 1979: 12–22

'A laser-man's way to altered reality', *Daily Californian*, 17 Oct. 1977, 31

'A touch of the future', *Display World*, Jan. 1973: 3

'A. H. Prismatics', *Holography News*, 7 (1) (1993): 1

Abbe, Ernst, 'Beiträge zur Theorie des Mikroscops und der mikroscopischen Wahrnehmung', *Archiv für Mikroskopische Anatomic* 9 (1873): 413

Abramson, N., 'Moire patterns and hologram interferometry', *Nature (Physical Science)* 231 (1971): 65–7

Aebischer, N. and B. Carquille, 'White light holographic portraits (still or animated)', *Applied Optics* 17 (1978): 3698–700

Aebischer, N. and C. Bainier, 'Multicolor holography of animated scenes by motion synthesis using a multiplexing technique', *Proceedings of the SPIE* 402 (1983): 51–6

Albright, Thomas, 'Holography's accelerating impact', *S. F. Sunday Examiner and Chronicle*, 11 Jun. 1972: 30–1

Albright, Thomas, 'School of holography', *Radical Software* 2 (1972): 56–7

Alexander, Kent, 'Seller beware', *Holosphere* 13 (1985): 10

Allibone, T. E., 'Dennis Gabor 1900–1979', Memorial Service Address, Holy Trinity Church, 15 Mar. 1979

Allibone, T. E., 'Dennis Gabor', *Biographical Memoirs of the Fellows of the Royal Society* 26 (1980): 107–47

Allibone, T. E., 'Dennis Gabor: A biographical memorial lecture', *Holosphere* 10 (1981): 1, 4–6

Allibone, T. E., 'White and black elephants at Aldermaston', *Journal of Electronics and Control* 4 (1958): 179–92

Amateur Photographer, 167 (26) (25 Jun. 1983): cover

'An interview with Margaret Benyon', *Holosphere* 9 (1980): 1–4

'Ann Arbor holography firms', *Ann Arbor News*, 16 Feb. 1969: 4

Anait, 'My art in the domain of reflection holography', *Leonardo* 11 (1978): 306–7

Annulli, R. J. and J. T. Ziewacz, 'Single-beam 360 degrees holograms', *American Journal of Physics* 45 (1977): 493–4

Ansley, D. A. and L. D. Siebert, 'Portrait-holography by impulse lasers', *Laser* 1 (1969): 29–34

Aquarius Project, 'Revolutionary engineering: Towards a counter-technology', *Radical Software* 1 (1970): 7

'Art holography from Russia', *L.A.S.E.R. News*, 13 (1) 1993: 9

Artforum International 55 (1985): cover

Artner, Alan G., 'Rhetoric, not results, at holography show', *Chicago Tribume*, 12 Jun. 1977: 46

'Atari buys patent license for consumer holography', *holosphere* 9 (1980): 1–2

'Atari unveils Cosmos, a holographic game system', *holosphere* 10 (1981): 1

Ausherman, Dale A., 'Digital versus optical in synthetic aperture radar (SAR) data processing', *Optical Engineering* 19 (1980): 157–67

Baez, Albert V., 'A study in diffraction microscopy with special reference to x-rays', *Journal of the Optical Society of America* 42 (1952): 756–762

Baez, Albert V., 'Resolving power in diffraction microscopy', *Nature* 169 (1952): 963–4

Baez, Albert V., 'Anecdotes about the early days of x-ray optics', *Journal of X-Ray Science and Technology* 7 (1997): 90–7

Baez, Albert V., Tung H. Jeong, and Encyclopaedia Britannica Educational Corporation, 'Introduction to holography', videorecording, Encyclopaedia Britannica Educational Corp.

Bains, Sunny, 'Editorial', *Holographics International* 1 (1987): 4

Barachevsky, V. A., 'Holographic recording media: modern trends', *Proceedings of the SPIE—The International Society for Optical Engineering* 3011 (1997): 306–18

Barachevsky, V. A., 'Russian advances in holographic recording media', *Proceedings of the SPIE—The International Society for Optical Engineering* 3417 (1998): 142–53

Bartell, Lawrence S., 'A brief history of holographic hubris and hilarity', *The Chemical Intelligencer*, October 1998: 53–6

Bartolini, R., W. Hannan, D. Karlsons, and M. Lurie, 'Embossed hologram motion pictures for television playback', *Applied Optics* 9 (1970): 2283–90

'Bay Area holography: An historical view', *L.A.S.E.R. News* 2 (1985): 10–11

Becker, Howard S., 'Stereographs: local, national and international art worlds', in: E. Earle (ed.), *Points of View: The Stereograph in America—A Cultural History* (New York: Visual Studies Workshop Press, 1979), pp. 89–96

Becquerel, E., 'De l'image photographique colorée du spectre solaire', *Annales de Chimic et de Physique* 22 (1848): 451

Becsey, J. G., G. E. Maddux, N. R. Jackson, and J. A. Bierlein, 'Holography and holographic interferometry for thermal diffusion studies in solutions', *Journal of Physical Chemistry* 74 (1970): 1401–3

Beeching, Dave, 'Holopack-Holoprint Special', *Holography News* 15 (2001): 1

'Be-it-yourself works of art: laser exhibit at Cranbrook', *Detroit News*, 20 Nov. 1969: 21–2

Benton, S. A., S. M. Birner, and A. Shirakura, 'Edge-lit rainbow holograms', *Proceedings of the SPIE—The International Society for Optical Engineering* 1212 (1990): 149–57

Benton, Stephen A., 'Edwin Land, 3-D, and holography', *Optics & Photonics News*, 5 (10) (1994): 41

Benton, Stephen A., 'Hologram reconstructions with extended incoherent sources', presented at *1969 Annual Meeting of the Optical Society of America*, 1969

Benton, Stephen A., 'Rainbow holograms', *Journal of the Optical Society of America* 50 (1969): 1545–1546

Benton, Stephen A., Pat. No. 3,633,989 'Method for making reduced bandwidth holograms' (1972), assigned to Polaroid Corp.

Benton, Stephen A., 'From the inside looking out', *Applied Optics* 14 (1975): 2795

Benton, Stephen A., 'Holographic displays—a review', *Optical Engineering* 14 (1975): 402–7

Benton, Stephen A., 'Holographic displays. 1975–1980', *Optical Engineering* 19 (1980): 686–90

Benton, Stephen A., 'Ten years of white-light holography', presented at *Electro-optics/Laser International 1980*, London, UK, 1980

Benton, Stephen A., 'Holography: the second decade', *Optics News* (1977): 16–21

Benton, Stephen A. and Herbert S. Mingace, Jr, 'Silhouette holograms without vertical parallax', *Applied Optics* 9 (1970): 2812–13

Benyon, Margaret, 'Holography as an art medium', *Leonardo* 6 (1973): 1–9

Benyon, Margaret, 'Pulsed holographic art practice', presented at *Practical Holography*, Los Angeles, California, 1986

Benyon, Margaret, 'Do we need an aesthetics of holography?' *Leonardo* 25 (1992): 411–16

Benyon, Margaret, 'Holography as art: cornucopia', presented at *Fifth International Symposium on Display Holography*, Lake Forest, Illinois, 1994

Benyon, Margaret, *How is Holography Art?*, PhD thesis, Royal College of Art, London (1994)

Benyon, Margaret, 'DEFINING TRADITIONS 1969–1996. Living and working with holography', presented at *Art in Holography*, Nottingham, 1996

Benyon, Margaret and J. Webster, 'Pulsed Holography as Art', *Leonardo* 19 (1986): 185–91

Bilbro, James W., 'Letter from the President', *SPIE Member Guide*, (2004): 1

Billings, Loren, Vince DiBiase, Jerry Fox, Mark Holzbach, Chris Outwater, and Gary Zellerbach, 'Bay Area holography: An historical view', *L.A.S.E.R. News* 2 (1985): 10–1

Bjelkhagen, H. I., 'Holography and philately: postage stamps with embossed holograms', *Proceedings of the SPIE—The International Society for Optical Engineering* 4149 (2000): 12–31

Bjelkhagen, Hans I., 'Holographic portraits made by pulse lasers', *Leonardo* 25 (1992): 443–8

Blyth, Jeff, 'Holographic security', *holosphere* 13 (1985): 16

Bohm, David, 'Quantum theory as an indication of a new order in physics. II. Implicate and explicate order in physical law', *Foundations of Physics* 3 (1973): 139–68

Boone, Pierre M., 'Report of the nations: Belgium', presented at *International Symposium on Display Holography*, Lake Forest, Illinois, 1991

Born, Max and Emil Wolf, *Principles of Optics: Electromagnetic Theory of Propagation, Interference and Diffraction of Light* (London; New York: Pergamon Press, 1959)

Bosco, Mary C., *What Ever Happened to Holography?*, MA thesis, John F. Kennedy School of Government, Harvard (1981)

Bourdon, David, *Village Voice*, 21 Jul. 1975

Boyarchuk, A. A. and L. V. Keldysh, 'From a physics laboratory to the Division of General Physics and Astronomy', *Physics Uspekhi* 42 (1999): 1183–91

Bragg, W. L., 'A new type of X-ray microscope', *Nature* 143 (1939): 678

Bragg, W. L., 'The x-ray microscope', *Nature* 149 (1942): 470

Bragg, W. L., 'Microscopy by reconstructed wavefronts', *Nature* 166 (1950): 399–400

Bragg, W. L. and Gordon L. Rogers, 'Elimination of the unwanted image in diffraction microscopy', *Nature* 167 (1951): 190–3

Brandes, R. G., E. E. Francois, and T. A. Shankoff, 'Preparation of dichromated gelatin films for holography', *Applied Optics* 8 (1969): 2346–8

Brandt, G. B., 'Image plane holography', *Applied Optics* 8 (1969): 1421–9

Brandt, G. B. and A. K. Rigler, 'Reflection holograms of focused images', *Physics Letters* 25A (1967): 68–9

Bringolf, Peter H., 'Holography: a medium in the making', *Proceedings of the SPIE—The International Society for Optical Engineering* 2043 (1994): 319–21

Brown, G. M., 'Holographic nondestructive testing (HNDT) of rubber—and plastics—containing products', presented at *22nd annual conference of electrical engineering problems in the rubber and plastics industries*, Akron, Ohio, 1970

Brown, G. M., 'A review of holographic nondestructive testing of pneumatic tires', *Materials Evaluation* 31 (1973): 37A

Brown, B. R. and A. W. Lohmann, 'Computer-generated binary holograms', *IBM Journal of Research and Development* 13 (1969): 160–8

Brown, G. M., R. M. Grant, and G. W. Stroke, 'Theory of holographic interferometry', *Journal of the Acoustical Society of America* 45 (1969): 1166–79

Bryngdahl, O. and Adolf W. Lohmann, 'Interferograms are image holograms', *Journal of the Optical Society of America* 58 (1968): 141–2

Buckland, Michael K., 'Histories, heritages and the past: the case of Emanuel Goldberg', presented at *Second Conference on the History and Heritage of Scientific and Technical Information Systems*, Philadelphia, 2002

Burch, J. M., 'Laser speckle metrology', *Proceedings of the SPIE* 25 (1971): 149–56

Burns, Jody, 'Messages to the future http://www.holonet.khm.de/Holographers/Burns-Jody/text, consulted 11 Aug. 2005

Burns Jr, Joseph, 'Update on the New York School of Holography', *holosphere* 4 (1975): 3–4

Buschmann, H. T., 'The production of low noise, bright-phase holograms by bleaching', *Optik* 34 (1971): 242–55

Bush, Edward A., 'A conversation with Dan Schweitzer', *holosphere* 10 (1981): 3–6

Butters, J. N., 'Lasers in the visualization of surface strain and vibrations', *Proceedings of the Institution of Mechanical Engineers* 183 (1969): 67–74

Butters, J. N., 'Measurement techniques using laser holography', *Electronics and Power. Journal of the Institution of Electrical Engineers* 20 (1974): 585–8

Butters, J. N. and D. Denby, 'Some practical uses of laser beam photography in engineering', *Journal of Photographic Science* 18 (1970): 60–7

Casdin-Silver, Harriet, 'Of holography and art and artists', *Proceedings of the International Symposium on Display Holography* 2 (1985): 403–10

Casdin-Silver, Harriet, 'My first ten years as artist/holographer (1968–1977)', *Leonardo* 22 (1989): 317–26

Casdin-Silver, Harriet, 'Holographic installations: sculpting with light', *Sculpture* 10 (1991): 50–5

Caulfield, H. John, 'Holographic brain: a good analogy gone bad', *Proceedings of the SPIE—The International Society for Optical Engineering* 4737 (2002): 124–30

'Center for Experimental Holography completes first phase of research', *holosphere*, 7 (1) (1978): 4–5

Chang, M., 'Improved dichromated gelatin for holographic recording', presented at Optical Society of America Annual Conference, 1970

Chang, M., 'Dichromated gelatin of improved optical quality', *Applied Optics* 10 (1971): 2550–1

Charnetski, Clark, 'The impact of holography on the consumer', presented at *EASCON Convention*, Washington DC, 1970

Charnetski, Clark and Richard Wilt, 'Interviewed on holography and art by Ed Burroughs', audio recording, Jan. 1969, WUOM, 'Eleventh Hour'

Christakis, A., 'Musee de l'Holographie of Paris and its activities: 1980–1994', *Fifth International Symposium on Display Holography* 2333 (1994): 245–7

Close, D. H. and A. Graube, 'Holographic lens for pilots head-up display', report, Hughes Research Laboratories, Aug. 1974

Cochran, Gary D. and Robert D. Buzzard, 'The new art of holography', in: *Science Year: The World Book Science Annual 1967* (Chicago, Field Enterprises Educational Corp., 1967), pp. 200–11

Colgate, Gilbert, 'Patents—in the history of security embossed holography', *Holography News* 10 (1996): 3–5

Collier, R. J., E. T. Doherty and K. S. Pennington, 'Application of moire techniques to holography', *Applied Physics Letters* 7 (1965): 223

'Computer men see laser trick', *Times*, 7 Aug. 1968, 2

'Computermen hear five laser papers—a critique', *Laser Focus*, 1 May 1966: 11

'Conductron develops new techniques for hologram production', *Conductron-Missouri Antenna*, 1 (10), Aug. 1967: 1–2

Connes, Pierre, 'Silver salts and standing waves: the history of interference colour photography', *Journal of Optics* 18 (1987): 147–66

Connors, B. A., S. A. Benton, and W. Seamans, 'Report from the MIT Museum', *Fifth International Symposium on Display Holography* 2333 (1994): 146–51

Cort, David, *Focusing the Sun*, Electronic Arts Intermix, videotape (1977)

Couture, J. J. A. and R. A. Lessard, 'Diffraction efficiency of specular multiplexed holograms recorded on Kodak 649F plates', *Applied Optics* 21 (1979): 3652–60

'cover hologram', *Journal of the Society of Instrument and Control Engineers* 20 (1981)

Cowles, Susan, 'Museum of Holography: past, present and future', *Holographics International* (1989): 26–9

Cox, M. E. and R. G. Buckles, 'Evaluation of selected films for holography', presented at *Optical Society of America Spring Meeting*, Philadelphia, PA, 1970

Cranbrook Academy of Art, 'The Laser: Visual Applications,' Detroit, Michigan, 1969

Crombie, A. C. and M. A. Hoskin, 'A note on history of science as an academic discipline', in: A. C. Crombie (ed.), *Scientific Change: Historical studies in the intellectual, social and technical conditions for scientific discovery and technical invention, from antiquity to the present* (London, 1963), pp. 757–94

Cross, Lloyd G., 'The potential impact of the laser on the video medium', *Radical Software* 1 (1970): 6

Cross, Lloyd G., 'The Story of Multiplex', transcription from audio recording, Naeve collection, Spring 1976

Cross, Lloyd G. and Cecil Cross, 'HoloStories: Reminiscences and a prognostication on holography', *Leonardo* 25 (1992): 421–4

Croucher, Duncan, 'Agfa-Gevaert photographic materials for holography', *Proceedings of the International Symposium on Display Holography* 1 (1983): 71–8

Curran, R. K. and T. A. Shankoff, 'The mechanism of hologram formation in dichromated gelatin', *Applied Optics* 9 (1970): 1651–7

Cutrona, L. J., 'The relationship of off axis reference function in holography to the equivalent problem in radar processing', Conductron inter-office memo, Conductron Inc., 2 Aug. 1966

Cutrona, L. J., Emmett N. Leith, and L. J. Porcello, 'Data processing by optical techniques', presented at *National Convention on Military Electronics*, Washington DC, 1959

Cutrona, L. J., E. N. Leith, C. J. Palermo, and L. J. Porcello, 'Optical data processing and filtering systems', *IRE Transactions on Information Theory* IT-6 (1960): 386–400

Cutrona, L. J., W. E. Vivian, E. N. Leith, and G. O. Hall, 'A high-resolution radar combat-surveillance system', *IRE Transactions on Military Electronics* MIL 5 (1961): 127–31

Cutrona, L. J., E. N. Leith, L. J. Porcello, and W. E. Vivian, 'On the application of modern optical techniques to radar data processing', presented at *Proceedings of the 9th AGARD Symposium on Opto-Electronic Components and Devices*, Paris (1965)

Cutrona, L. J., Emmett N. Leith, L. J. Porcello, and W. E. Vivian, 'On the application of coherent optical processing techniques to synthetic-aperture radar', *Proceedings of the IEEE* 54 (1966): 1026

Dallas, W. J., 'Phase quantization in holograms-a few illustrations', *Applied Optics* 10 (1971): 674–6

Dawson, Paula, *The Concrete Holographic Image: An Examination of Spatial and Temporal Properties and their Application in a Religious Art Work*, PhD thesis, Fine Arts, University of New South Wales (2000)

Day, Dwayne A., 'The development and improvement of the CORONA satellite', in: Day, D. A., J. M. Logsdon, and B. Latell (eds), *Eye in the Sky: The Story of the Corona Spy Satellites* (Washington DC: Smithsonian Institution Press, 1998), pp. 48–85

de Marrais, Robert, 'Holography in the future tense', *holosphere* 12 (1984): 4, 6–7

de Solla Price, Derek J., 'Is technology historically independent of science? A study in statistical historiography', *Technology and Culture* 6 (1965): 553–68

DeBitetto, D. J., 'Bandwidth reduction of hologram transmission systems by elimination of vertical parallax', *Applied Physics Letters* 12 (1968): 176–8

Denby, D. and J. N. Butters, 'Holography as an engineering tool', *New Scientist* 45 (1970): 394–6

Denisyuk, Yu N., 'On the reflection of optical properties of an object in the wave field of light scattered by it', *Doklady Akademii Nauk SSSR* 144 (1962): 1275–8

Denisyuk, Yu N., 'On reflection of the optical properties of an object in wavefield of radiation scattered by it', *Optika i Spektroskopija* 15 (1963): 522–32

Denisyuk, Yu N., 'On reflection of the optical properties of an object in wavefield of radiation scattered by it. II', *Optika i Spektroskopija* 18 (1965): 276–83

Denisyuk, Yu N., 'On the problem of a photograph reproducing the full illusion of the reality of the object depicted (in Russian)', *Zhurnal Nauchnoi i Prikladnoi Fotografi i Kinematografi* 11 (1966): 46–56

Denisyuk, Yu N., 'The work of the State Optical Institute on holography', *Soviet Journal of Optical Technology* 34 (1967): 706–10

Denisyuk, Yu N., *Soviet Union* 9 (1970): 12–14

Denisyuk, Yu N., 'Holography motion pictures', *Soviet Physics—Technical Physics* 18 (1974): 1549–51

Denisyuk, Yu N., 'Holography and its prospects (review)', *Journal of Applied Spectroscopy* 33 (1980): 901–15

Denisyuk, Yu N., 'Holography at the State Optical Institute (GOI)', *Soviet Journal of Optical Technology* 56 (1989): 38–43

Denisyuk, Yu N., 'Denisyuk on holography in the USSR', *Holography News* 4 (1990): 2

Denisyuk, Yu N., 'Certain features of the development of display holography in the USSR', *Proceedings of the SPIE The International Society for Optical Engineering* 1600 (1991): 376–86

Denisyuk, Yu N., 'My way in holography', *Leonardo* 25 (1992): 425–30

Denisyuk, Yu N. and R. R. Protas, 'Improved Lippmann photographic plates for the record of standing light waves (in Russian)', *Optika i Spektroskopija* 14 (1963): 721–5

Denisyuk, Yu N. and V. Gurikov, 'Advancement of Holography, Investigations by Soviet Scientists', *History and Technology* 8 (1992): 127–32

Deschin, Jacob, 'No-lens pictures: photographic technique employs light alone', *New York Times*, 15 Dec. 1963: 143

Diamond, Mark, 'Holotalk interview', Internet radio interview, F. DeFreitas, 2003, www.holoworld.com.

'Dichromate and Integral holography in Japan', *holosphere* 7 (1978): 2

Dickson, Leroy D. and C. L. Stong, 'The Amateur Scientist: Stability of the apparatus: insuring a good hologram by controlling vibration and exposure', *Scientific American*, Jul. 1971: 110–12

Dietrich, Edward, 'The development of the holography program at the School of the Art Institute of Chicago', *Proceedings of the International Symposium on Display Holography* 2 (1985): 435–40

Dodd, Philip, 'A spy in the sky! Army unveils radar camera: it shoots enemy thru clouds of smoke', *Chicago Daily Tribune*, 20 Apr. 1960: 3

Dolgoff, E., 'Commercial holography', *Optical-Spectra* 9 (1975): 26–31

Domanski, G., 'Three-dimensional television', *Funkschau*, 25–26 (1981), 60–4

Duffy, Jonathan, 'Holograms: high art of just a gimmick?' *BBC News*, 23 Jun. 2004, webpage

Dulberger, Leon H. and Charles Wixom, 'Lensless optical system uses laser: opaque 3-D objects may be imaged without lenses using reflected light', *Electronics*, 27 Dec. 1963: 15

Dye, W. D., *Proceedings of the Royal Society* A 138 (1932): 1

Dyens, Georges, 'Art Holography—The Real Virtual 3D Images,' CD-ROM, Montreal, 2002

Eastman Kodak Company, 'Kodak High-speed Holographic Glass Plates 131CX and 131PX', Kodak, Jan. 2002

Edelstein, J. Y., 'Technology Forecast—Commercial Applications of Holography', *Proceedings of the Society of Photo-Optical Instrumentation Engineers* 523 (1985): 343–6

Edson, Lee, 'A Gabor named Dennis seeks Utopia', *Think*, Jan.–Feb. 1970: 23–7

'Edwin Land', *Physics Today*, Jan. 1982: 35

Elias, Peter, 'Optics and communication theory', *Journal of the Optical Society of America* 43 (1953): 229

El-Sum, Hussein M. A., *Reconstructed Wave-Front Microscopy*, PhD thesis, Physics, Stanford University (1952)

El-Sum, Hussein M. A., Pat. No. 3,083,6155 'Optical apparatus for making and reconstructing holograms' (1960), assigned to Lockheed Aircraft Corp.

El-Sum, Hussein M. A. and Paul Kirkpatrick, 'Microscopy by reconstructed wave-fronts', *Physical Review* 85 (1952): 763

Ennos, A. E., 'Holographic techniques in engineering metrology', *Proceedings of the Institution of Mechanical Engineers* 183 (1969): 5–12

Ennos, A. E. and E. Archbold, 'Techniques of hologram interferometry for engineering inspection and vibration analysis', presented at *The Engineering Uses of Holography*, University of Strathclyde, Glasgow, 1968

Ennos, A. E. and E. Archbold, 'Vibrating surface viewed in real time by interference holography', *Laser Focus* 4 (1968): 58–9

'Entrepreneurs urged to set up Holographic centers', *holosphere* (1976): 4 (1976): 8

Erickson, Ronald R., 'There is this 'attitude' in holography', *holosphere* 16 (1989): 4

Erickson, Ronald R., 'American Bank Note Holographics: 8 years of innovation', *Holosphere* 17 (1990): 14–5

Eskowitz, Henry, 'The making of an entrepreneurial university: the traffic among MIT, industry, and the military, 1860–1960', in: E. Mendelsohn, M. R. Smith, and P. Weingart (ed.), *Science, Technology, and the Military* (Dordrecht; London: Kluwer, 1988), pp. 515–40

Fairstein, John, 'The San Francisco School of Holography', http://www.jfairstein.com/SOH.html, accessed 28 Feb. 2003

Feldkamp, John C., 'Student life since 1945', in: W. B. Shaw (ed.), *The University of Michigan: An Encyclopedic Survey* (Ann Arbor: University of Michigan, 1974)

Feldman, Ruth Duskin, 'Trapping the light fantastic', *Dynamic Years* (1986): 66–70

Fink, W., P. A. Buger, and L. Schepens, 'Rock probe deformation measured by holographic interferometry', *Optics Technology* 2 (1970): 146–50

'Fire Diamond, hologram maker, moved to Richland, Washington in reorganization program', *holosphere* 7 (1978): 1–2

Forman, Paul, 'Inventing the maser in postwar America', *Osiris* 7 (1992): 105–34

Frateschi, G. and E. Gatier, 'Gigamemories based on recyclable holograms', *L'Antenna* 45 (1973): 446–52

Frecska, S. A., 'Characteristics of the Agfa-Gevaert type 10E70 holographic film', *Applied Optics* 7 (1968): 2315–17

Fujio, T., 'Optical communication and image technology, the new future', *Journal of the Institute of Television Engineers of Japan* 32 (1978): 349–54

Gabor, Dennis, 'Theory of communication', *Proceedings of the IEEE* 93 (1946): 429–57

Gabor, Dennis, Pat. No. 685,286 'Improvements in and relating to Microscopy' (1947), assigned to British Thomson-Houston

Gabor, Dennis, 'A new microscopic principle', *Nature* 161 (1948): 777–8

Gabor, Dennis, 'Optical synthetizer for electron microscope', Report, British Thomson-Houston, Sep. 1948

Gabor, Dennis, 'Microscopy by reconstructed wavefronts', *Proceedings of the Royal Society of London, Series A* 197 (1949): 454–87

Gabor, Dennis, 'Problems and prospects of electron diffraction microscopy', presented at *Conference on Electron Microscopy*, Delft, 1949

Gabor, Dennis, 'A new interference microscope', mimeographed report, Imperial College, 4 Dec. 1950

Gabor, Dennis, 'Diffraction microscopy. Full reconstruction by interpolation', report, Imperial College, 7 Jul. 1951

Gabor, Dennis, 'Microscopy by reconstructed wavefronts: II', *Proceedings of the Physical Society (London)* B64 (1951): 449–69

Gabor, Dennis, 'Progress in microscopy by reconstructed wavefronts', presented at *Conference on Electron Microscopy*, Washington, 1951

Gabor, Dennis, Pat. No. 2,770,166 'Improvements in and relating to optical apparatus for producing multiple interference patterns' (1956), assigned to National Research Development Corporation, UK

Gabor, Dennis, 'Gabor interviewed by Rex Keating', recording from BBC radio, *c.*1963, IC GABOR P/1

Gabor, Dennis, 'The outlook for holography', *Optik* 28 (1969): 437–41

Gabor, Dennis, 'Holography, 1948–1971', in: Nobel Prize Committee (ed.), *Les Prix Nobel En 1971* (Stockholm, Imprimerie Royale PA Norstedt & Soner, 1971), pp. 169–201

Gabor, Dennis, 'Holography, past, present and future', *Proceedings of the SPIE—The International Society for Optical Engineering* 25 (1971): 129–34

Gabor, Dennis, *Holographie 1973: Vortrag, gehalten an dem Mentorenabend der Carl Friedrich von Siemens Stiftung in München Nymphenburg am 15. Juni 1973* (München: Die Stiftung, 1973)

Gabor, Dennis, 'The history of holography', *Fizikai-Szemle* 24 (1974): 289–303

Gabor, Dennis and George W. Stroke, 'The theory of deep holograms', *Proceedings of the Royal Society of London, Series A (Mathematical and Physical Sciences)* 304 (1968): 275–89

Gabor, Dennis and George W. Stroke, 'Holography and its applications', *Endeavour* 28 (1969): 40–7

Gabor, Dennis and W. P. Goss, report, 1964, US Army Contract (European Research Office) DA-91–591-EUC-3886 OI 652–1251, at IC GABOR GD/2/5

Gabor, Dennis and W. P. Goss, 'Interference microscope with total wavefront reconstruction', *Journal of the Optical Society of America* 56 (1966): 849–58

Gabor, Dennis, W. E. Kock, and George W. Stroke, 'Holography', *Science* 173 (1971): 11–23

Gamble, Susan, *The Hologram and its Antecedents 1891–1965: The Illusory History of a Three-dimensional Illusion*, PhD thesis, History of Science, Cambridge University (2002)

Gambogi, W. J., A. M. Weber, and T. J. Trout, 'Advances and applications of DuPont holographic photopolymers', *Proceedings of the SPIE—The International Society for Optical Engineering* 2043 (1994): 2–13

Gates, J. W. C., R. G. N. Hall, and I. N. Ross, 'Holographic recording of rapid transient events and the problems of evaluation of the reconstructions', presented at *Eighth International Congress on High Speed Photography*, Teddington, UK, 1968

Gates, J. W. C., 'Holography, industry and the rebirth of optics', *Review of Physics in Technology* 2 (1971): 173–91

Gates, Max, 'Holography a 're-creation of reality'—defense debate', *Ann Arbor News*, 18 Mar. 1982: 2

'Generals tour WR Laboratories', *Ypsilanti Press*, 8 Jul. 1965

'Gennadi A. Sobolev', *Holography News*, 16 (7), Sep. 2002: 5

George, Nicholas and J. T. McCrickerd, 'Holographic stereogram from sequential component photographs', *Applied Physics Letters* 12 (1968): 10

George, Nicholas and J. T. McCrickerd, 'Holography and stereoscopy: the holographic stereogram', *Photographic Science and Engineering* 13 (1969): 342–50

Gibbons, G. W., 'Holography and the future tube', *Classical and quantum gravity* 17 (2000): 1071–80

Gigliotti, Davidson and Ira Schneider, 'Videocity: Summer 1973', http://www.radicalsoftware.org/e/volume2nr3.html, accessed 11 Nov. 2004

Glauber, R. J., 'Coherent and incoherent states of the radiation field', *Physical Review* 131 (1963): 2766

Goldberg, Shoshanah, 'Conductron and McDonnell Douglas', *holosphere* 15 (1987): 16–18

Goleman, Daniel and Karl Pribram, 'Holographic Memory', *Psychology Today*, 13 Feb. 1979: 71–84

Goodman, Joseph W., 'Film-grain noise in holography', presented at *Proceedings of the Symposium on Modern Optics*, Stanford University: CA, USA, 1967

Gorglione, Nancy, 'Lloyd Cross', *Holographics International* 1 (1987): 17, 29

Gorglione, Nancy, 'A partial view of a three-dimensional world', *Leonardo* 25 (1992): 407–9

Gorglione, Nancy, 'Forms of light: a personal history in holography', *Leonardo* 25 (1992): 473–80

Gorin, Peter A., 'ZENIT: The Soviet response to CORONA', in: D. A. Day, J. M. Logsdon, and B. Latell (eds), *Eye in the Sky: The Story of the Corona Spy Satellite* (Washington DC, 1998), pp. 157–70

Graham, Loren R., 'The place of the Academy of Sciences system in the overall organization of Soviet science', in: J. R. Thomas and U. M. Kruse-Vaucienne (eds), *Soviet Science and Technology: Domestic and Foreign Perspectives* (Washington DC: National Academy of Sciences, 1977), pp. 45–63

Graham, Loren R., 'Russian & Soviet Science and Technology', *History of Science Society Newsletter* 18 (1989): 1

Grant, R. M. and G. M. Brown, 'Holographic nondestructive testing (HNDT)', *Materials Evaluation* 27 (1969): 79–84

Grant, R. M. and G. M. Brown, 'Holographic nondestructive testing (HNDT) in the automobile industry', presented at *International Automotive Engineering Congress*, Detroit, Michigan, 1969

Green, Harvey, ' "Pasteboard masks": the stereograph in American culture 1865–1910', in: E. Earle (ed.), *Points of View: The Stereograph in America—A Cultural History* (New York: Visual Studies Workshop Press, 1979), pp. 109–15

Greguss, P., 'Thoughts on the future of holography in biology and medicine', *Optics and Laser Technology* 7 (1975): 253–7

Gurevich, S. B., 'Holographic television future', *Tekhnika Kino i Televideniya* 2 (1971): 67–9

Gurwitsch, A. G. and L. D. Gurwitsch, 'Twenty years of mitogenetic radiation', *Uspechi Biologii Nauk V.* 16 (1943): 305–34

Haig, N. D., 'Three-dimensional holograms by rotational multiplexing of two-dimensional films', *Applied Optics* 12 (1973): 419–20

Haine, M. E. and J. Dyson, 'A modification to Gabor's diffraction microscope', *Nature* 166 (1950): 315–16

Haine, M. E. and T. Mulvey, *Journal of the Optical Society of America* 42 (1952): 756

Haines, Kenneth, *The Analysis and Application of Hologram Interferometry*, PhD thesis, Electrical Engineering, University of Michigan (1966)

Haines, Kenneth and B. P. Hildebrand, 'Contour generation by wavefront reconstruction', *Physics Letters* 19 (1965): 10–11

Hakanson, Joy, 'They create art that isn't there', *The Sunday News Magazine*, 8 Feb. 1970: 18–20, 44

Hall, R. G. N., J. W. C. Gates, and I. N. Ross, 'Recording rapid sequences of holograms', *Journal of Physics E (Scientific Instruments)* 3 (1970): 789–91

Halle, M.W., S. A. Benton, M. A. Klug, and J. S. Underkoffler, 'Ultragram: a generalized holographic stereogram', *Practical Holography V* 1461 (1991): 132–41

Hammond, A. L., 'Holography: beginnings of a new art form or at least of an advertising bonanza', *Science* 180 (1973): 484–5

'Handwritten will expert to speak', *Jackson Citizen's Patriot*, 24 Mar. 1968: 6

Hariharan, Parameswaran, 'Photographic materials and coherent light', *Proceedings of the Indian National Science Academy* 37A (1971): 193–9

Hariharan, Parameswaran, 'Bleached reflection holograms', *Optics Communications* 6 (1972): 377–9

Hariharan, Parameswaran, 'Silver handle sensitized gelatin holograms: mechanism of hologram formation', *Applied Optics* 25 (1986): 2040–2

Harrison, G. R. and G. W. Stroke, 'Interferometric control of grating ruling with continuous carriage advance', *Journal of the Optical Society of America* 45 (1955): 112–21

Harvey, Thomas W., 'Holotron reports results of holography market survey', *holosphere* 4 (1975): 1–3

Harwood, Jonathan, 'Ludwig Fleck and the sociology of knowledge', *Social Studies of Science* 16 (1986): 173–87

Hauk, D. and A. W. Lohmann, 'Minimumstrahlkennzeichnung bei Gitterspektrographen', *Optik* 15 (1958): 275–7

Hay, W. C. and B. D. Guenther, 'Characterization of Polaroid's DMP-128 holographic recording photopolymer', *Proceedings of the SPIE* 883 (1988): 102–5

'Heat is on with new Du Pont photopolymers', *Holographics International* (1989): 26

Hecht, Jeff, 'Applications pioneer interview: Emmett Leith', *Lasers & Applications*, 5 Apr. 1986: 56–8

Heflinger, L. O., R. F. Wuerker, and R. E. Brooks, 'Holographic interferometry', *Journal of Applied Physics* 37 (1966): 642–9

Heflinger, L. O. and R. F. Wuerker, 'Holographic contouring via multifrequency lasers', *Applied Physics Letters* 15 (1969): 28–30

Herschel, J., *Philosophical Transactions* 130 (1840): 1

Heumann, Sylvain M. and C. L. Stong, 'The Amateur Scientist: How to make holograms and experiment with them or with ready-made holograms', *Scientific American*, Feb. 1967: 122–8

Hildebrand, B. P. and Kenneth Haines, 'Interferometric measurements using the wavefront reconstruction technique', *Applied Optics* 5 (1966): 172–3

Hildebrand, B. P., *A General Analysis of Contour Holography*, PhD thesis, Electrical Engineering, University of Michigan (1967)

Hildebrand, B. P. to R. Washburn and Woodcock Phelan and Washburn patent attorneys, 2 Feb. 1970, Philadelphia, PA, Leith collection

Hiley, B. J., 'David Joseph Bohm', *Biographical Memoirs of Fellows of the Royal Society* 43 (1997): 105–31

Hioki, R. and T. Suzuki, 'Reconstruction of wavefronts in all directions', *Japanese Journal of Applied Physics* 4 (1965): 816

Hockley, B. S. and J. N. Butters, 'Holography as a routine method of vibration analysis', *Journal of Mechanical Engineering Science* 12 (1970): 37–47

Holmes, Oliver Wendell, 'The stereoscope and the stereograph', *The Atlantic Monthly* III (1859): 738–48

'Holograms as giftware: the commercialization of artistic holography', *Holography Marketplace*, 8 (1999): 37–43

'Holograms don't work . . .' *Holography News* 16 (2002): 1–2

'Holographic model of brain function explicated by Pribram at conference', *holosphere* 7 (1978): 1–2

'Holography: an introduction and survey', presented at *Proceedings of the 7th Annual Biomedical Sciences Instrumentation Symposium on Imagery in Medicine*, University of Michigan, Ann Arbor, MI, US, 1969

'Holography: Ann Arbor shows how', *Detroit Free Press*, 6 Aug. 1967: 37

'Holography—Three Dimension Photos', *Red Bluff News*, 20 Aug. 1976: 6

'Holotron licensing outlined by Cravens', *holosphere* 7 (1978): 3, 6

Howard, John M., 'The early years of *Applied Optics*', *Optics & Photonics News*, 14 (9), Sep. 2003: 22–3

Hsu, D., J. Jiao, H. Tao, and P. Long, 'Recent developments on holography in China', *Three-Dimensional Holography: Science, Culture, Education* 1238 (1989): 13–7

Hsu, Dahsiung, 'Anti-counterfeiting holograms and government anti-piracy activities in China', *Proceedings of the SPIE The International Society for Optical Engineering* 3358 (1998): 318–21

Hsue, S. T., B. L. Parker, and M. Monahan, '360 degrees reflection holography', *American Journal of Physics* 44 (1976): 927–8

Huff, Lloyd, 'Holography in the Soviet Union', *holosphere* 12 (1983): 4–5

'IBM supermarket scanner uses holographic optics', *holosphere* 9 (1980): 1–2

Ilford Ltd, 'ILFORD: The leader in specialised holographic film, plates and chemistry', *Holographics International* 1 (1) (Autumn 1987): 9

Ilford Ltd, 'The integrated approach to holography giving you greater profitability with reduced labour and wastage. ILFORD', *Holographics International* 1 (4) (Summer 1988): 4

Ilford Ltd, 'ILFORD HOLOGRAPHY—In the last two years alone ILFORD research has created . . .', *Holographics International* 1 (7) (Winter 1989): 11

Ilford Ltd, 'ILFORD HOLOGRAPHY—HOTEC—A new family of holographic products to make holography simpler and safer—HOTEC designed with teaching in mind', *Holographics International* 1 (8) (Summer 1990): 30

Ingwall, R. T. and H. L. Fielding, 'Hologram recording with a new photopolymer system', *Optical Engineering* 24 (1985): 808–11

'IST establishes lab in new study field', *Ann Arbor News*, 2 Oct. 1963: 37

'It's Spring, and holo galleries are popping up from coast to coast', *holosphere* 8 (3) (Mar. 1979): 1

Ives, Harold E., 'An experimental study of the Lippmann color photograph', *Astrophysical Journal* 27 (1908): 323–32

Jackson, Myles, 'From theodolite to spectral apparatus: Joseph Von Fraunhofer and the invention of a German optical research-technology', in: B. Joerges and T. Shinn (eds), *Instrumentation: Between Science, State and Industry* (Dordrecht, Kluwer Academic Publishers, 2001), pp. 17–28

Jackson, Rosemary H., 'An accounting, 1976–1978', *holosphere* 8 (1979): 3, 5–6, 8–10

Jackson, Rosemary H., 'Workshops: Goldsmith's College', proposal, Jan. 1979

Jackson, Rosemary H., 'Off the wall', *holosphere* 9 (1980): 4–7

Jackson, Rosemary H., 'A thirty-five year account of the development of holography—Part I', *holosphere*, 12 (4), Summer 1983: 5–12

Jackson, Rosemary H., 'A thirty-five year account of the development of holography—Part II', *holosphere*, 12 (5), Fall 1983: 13–17

Jackson, Rosemary H., 'A thirty-five year account of the development of holography—Part III', *holosphere*, 12 (6), Winter 1984: 19–23

Jackson, Rosemary H., 'Posy writes . . .' *Holographics International* (1989): 4–5

Jacobson, A. D., V. Evtuhov and J. K. Neeland, 'Motion picture holography', *IEEE Journal of Quantum Electronics* QE-5 (1969): 334–5

James, Dorothy, 'Report from Rochester', *holosphere*, 13 (1), 1985, 13–5

James, Randy, 'Off the wall', *holosphere*, 8 (12), Dec. 1979: 3

Jansson, E., N. E. Molin, and H. Sundin, 'Resonances of a violin body studied by hologram interferometry and acoustical methods', *Physica Scripta* 2 (1970): 243–56

Jarrett, Steven M., 'Early ion laser development', *Optics and photonics news* 15 (2004): 24–7

Jeong, Tung H., P. Rudolf, and J. Luckett, '123 360 degree holography', *Journal of the Optical Society of America* 56 (1966): 1263

Jeong, Tung H., 'Cylindrical holography and some proposed applications', *Journal of the Optical Society of America* 57 (1967): 1396

Jeong, Tung H., 'Holography comes out of the cellar', *Optical Spectra*, 2 (6) (1968): 59–65

Jeong, Tung H., 'A demonstrated lecture on holography', sound recording, Instant Replay 1978

Jeong, Tung H., 'Future holography', *holosphere* 12 (1984): 18

Jeong, Tung H., *Directory of Holographic Companies, Institutes and Individuals—Worldwide*, report, Photics Corp, Sep. 2002

'John Perry: doyen of large format holography', *Holography News* 7 (8) (1993): 69–70

Johnson, C. and E. Briggs, 'Holography as applied to information storage and retrieval systems', *Journal of the American Society for Information Science* 22 (1971): 187–92

Johnston, Sean F. and C. L. Stong, 'Amateur Scientist: A high school student builds a recording spectrophotometer', *Scientific American*, January 1975: 118

Johnston, Sean F., 'Holographic animation apparatus', *American Journal of Physics* 47 (1979): 681–2

Johnston, Sean F., 'Making light work: practices and practitioners of light measurement', *History of Science* 34 (1996): 273–302

Johnston, Sean F., 'The construction of colorimetry by committee', *Science in Context* 9 (1996): 387–420

Johnston, Sean F., 'In search of space: Fourier spectroscopy 1950–1970', in: B. Joerges and T. Shinn (eds), *Instrumentation: Between Science, State and Industry* (Dordrecht, 2001), pp. 121–41

Johnston, Sean F., 'An unconvincing transformation? Michelson's interferential spectroscopy', *Nuncius* 18 (fasc. 2) (2003): 803–23

Johnston, Sean F., 'Stephen Benton on holography, Polaroid and MIT', *Optics & Photonics News* 15 (8) (Aug. 2004): 32–5

Kac, Eduardo, 'On Baudrillard's text "Hologrammes" ', *holosphere* 17 (1990): 25–6

Kakichashvili, S. D., 'Polarization holography: possibilities and the future', presented at *Holography '89: International Conference on Holography, Optical Recording, and Processing of Information*, Varna, Bulgaria, 1989

Kay, Ronald H., Pat. No. 2,982,176 'Information storage and retrieval system' (1961), assigned to IBM Corp.

King, M. C., 'Multiple exposure hologram recording of a 3-D image with a 360 degrees view', *Applied Optics* 7 (1968): 1641–2

Kirkpatrick, Paul, 'An approach to X-ray microscopy', *Nature* 166 (1950): 251

Kirkpatrick, Paul, 'History of holography', *Proceedings of the SPIE, Holography* 15 (1968): 9–12

Kirkpatrick, P. and H. M. A. El-Sum, 'Image formation by reconstructed wavefronts I. Physical principles and methods of refinement', *Journal of the Optical Society of America* 46 (1956): 825–31

Klimenko, I. S., E. G. Matinyan and G. I. Rukman, 'Double-exposure holographic interferometry with reconstruction in white light', *Optics and Spectroscopy* 29 (1970): 85–8

Kock, W. E., L. Rosen, and G. W. Stroke, 'Focused-image holography—a method for restoring the third dimension in the recording of conventionally-focused photographs', *Proceedings of the IEEE* 54 (1966): 80–1

Kogelnik, H. W., 'Optics at Bell Laboratories—lasers in technology', *Applied Optics* 11 (1972): 2426–34

Kojevnikov, Alexei, 'President of Stalin's Academy: the mask and responsibility of Sergei Vavilov', *Isis* 87 (1996): 18–50

Komar, V. G., 'Possibility of creating a theatre holographic cinematograph with 3-D colour image', *Tekhnika Kino i Televideniya* 4 (1975): 31–9

Komar, V. G. and O. B. Serov, 'Work on holographic cinematography in the USSR', presented at *Holography '89: International Conference on Holography, Optical Recording, and Processing of Information*, Varna, Bulgaria, 1989

Kompanets, I. N. and A. P. Yakimovich, 'Three-dimensional television (a review)', *Telecommunications and Radio Engineering, Part 2 (Radio Engineering)* 46 (1991): 40–7.

Konstantinov, B. P., S. B. Gurevich, G. A. Gavrilov, A. A. Kolesnikov, A. B. Konstantinov, V. B. Konstantinov, A. A. Rizkin, and D. F. Chernykh, 'The transmission of holograms on standard phototelegraphic channels with a limited member of sub-tones', *Zhurnal Tekhnicheskoi Fiziki* 39 (1969): 374–83

Kontnik, Lewis T., 'Commercial holography: a decade of emergence', *holosphere* 14 (1986): 12–6

Kontnik, Lewis T., 'A teacher of holography (Tung Jeong)', *holosphere* 15 (1987): 31

Kontnik, Lewis T., 'Governments underwrite holography industry', *Holography News* 1 (1987): 1

Kontnik, Lewis T., 'Personality profile: Ian M. Lancaster—Bend your knees!' *holosphere* 15 (1987): 27

Kontnik, Lewis T., 'Computers, pixels and holograms: the patents', *Holography News* 8 (1994): 3–5

Kozma, Adam, Emmett N. Leith and Norman G. Massey, 'Tilted-plane optical processor', *Applied Optics* 11 (1972): 1766–77

Kramer, H., 'Holography: a technical stunt', *New York Times*, 20 Jul. 1975, 1–2

Kubota, T., T. Ose, M. Sasaki and K. Honda, 'Hologram formation with red light in methylene blue sensitized dichromated gelatin', *Applied Optics* 15 (1976): 556–8

'Labs' fate linked to ROTC', *Ann Arbor News*, 10 Dec. 1969,

Lancaster, Ian M., 'An apologetic industry', *Holography News*, 8 (7) (Sep. 1994): 2

Lancaster, Ian M., 'Warning: rumours can seriously damage your health', *Holography News*, 9 (1) (Feb. 1995): 2

Lancaster, Ian M., 'Bad news is bad news for the holography industry', *Holography News*, 9 (9/10) (Dec. 1995): 2

Lancaster, I. M. and L. T. Kontnik, 'Market conditions for display holography in the USA and Europe', presented at *Practical Holography IV*, Los Angeles, CA, 1990

Lancaster, I. M. and L. T. Kontnik, 'Holography industry: market size and trends in commercial activity', presented at *Fifth International Symposium on Display Holography*, Lake Forest, Illinois, 1994

Lang, M. and H. Eschler, 'Gigabyte capacities for holographic memories', *Optics and Laser Technology* 6 (1974): 219–24

'Large format holography: a major producer of large format display holograms explains his trade', *Holography Marketplace*, 8 (1999): 30–6

Larkin, A. I., 'Holography and education in the USSR', *International Symposium on Display Holography* 1600 (1991): 412–7

'Laser photographic process uses no lenses, produces 3-D images', *Science Fortnightly*, 1 (9), 25 Dec. 1963: 1–2

Latour, Bruno, 'Where are the missing masses? The sociology of a few mundane artifacts', in: E. Bijker Wiebe and J. Law (eds), *Shaping Technology/Building Society* (Cambridge, 1992), pp. 225–58

Latour, Bruno, 'How to be iconophilic in Art, Science, and Religion?' in: C. A. Jones and P. Galison (eds), *Picturing Science, Producing Art* (New York, 1998), pp. 418–40

Latta, J. N., 'The bleaching of holographic diffraction gratings for maximum efficiency', *Applied Optics* 7 (1968): 2409–16

Law, Linda, 'David Katzive interview', *Wavefront*, Fall (1986): 2–3

Lawrence, Pamela J., 'Retailing holography in the U.S.' *holosphere*, 13 (3) (1985): 8–10

Lehmann, Matt, 'Holography's early days at Stanford University', *holosphere* 11 (1982): 1, 4–5

Leith, E. N., 'Overview of the development of holography', *The Journal of Imaging Science and Technology* 41 (1997): 201–4

Leith, Emmett N., 'A data processing system viewed as an optical model of a radar system', memo to W. A. Blikken of Willow Run Laboratories, University of Michigan, 22 May 1956

Leith, Emmett N. and Juris Upatnieks, 'Reconstructed wavefronts and communication theory', *Journal of the Optical Society of America* 52 (1962): 1123–30

Leith, Emmett N. and Juris Upatnieks,, 'Holograms: their properties and uses', *SPIE Technical Symposium* 10 (1965): 3–6

Leith, Emmett N., 'Applications of holography', presented at *Automotive Engineering Congress*, Detroit, Michigan, 1968

Leith, Emmett N., 'Modern holography', presented at *1968 IEEE international convention*, New York, 1968

Leith, Emmett N., 'Some recent results in holography', presented at *Northeast Electronics Research and Engineering Meeting*, Boston, MA, 1968

Leith, Emmett N., 'Electro-optics and how it grew in Ann Arbor', *Optical Spectra* (May 1971): 25–6

Leith, Emmett N., D. B. Brumm, and S. S. H. Hsiao, 'Holographic cinematography', *Applied Optics* 11 (1972): 2016–23

Leith, Emmett N., *The Origin and Development of the Carrier Frequency and Achromatic Concepts in Holography*, PhD thesis, Electrical Engineering, Wayne State University (1978)

Leith, Emmett N, 'The legacy of Dennis Gabor', *Optical Engineering* 19 (1980): 633–5

Leith, Emmett N., 'The legacy of Dennis Gabor', *Optical Engineering* 19 (1980): 633–5

Leith, Emmett N., 'Some highlights in the history of display holography', *Proceedings of the International Symposium on Display Holography* 1 (1983): 1–4

Leith, Emmett N., 'A short history of the Optics Group of the Willow Run Laboratories', in: A. Consortini (ed.), *Trends in Optics* (New York, 1996), pp. 1–26

Leith, Emmett N., 'Overview of the development of holography', *The Journal of Imaging Science and Technology* 41 (1997): 201–4

Leith, Emmett N., 'The evolution of information optics', *IEEE Journal of Selected Topics in Quantum Electronics* 6 (2000): 1297–304

Leith, Emmett N., 'Reflections on the origin and subsequent course of holography', *Proceedings of the SPIE* 5005 (2003): 431–8

Leith, Emmett N. and C. J. Palermo, 'Spatial filtering for ambiguity suppression, and bandwidth reduction (SECRET)', presented at *6th Annual Radar Symposium*, Ann Arbor News, MI, 1960

Leith, Emmett N. and Juris Upatnieks, 'New techniques in wavefront reconstruction', *Journal of the Optical Society of America* 51 (1961): 1469

Leith, Emmett N. and Juris Upatnieks, 'Wavefront reconstruction with continuous-tone transparencies', *Journal of the Optical Society of America* 53 (1963): 522

Leith, Emmett N. and Juris Upatnieks, 'Wavefront reconstruction with diffused illumination and three dimensional objects', *Journal of the Optical Society of America* 54 (1964): 1295–301

Leith, Emmett N. and Juris Upatnieks, 'Photography by laser', *Scientific American* 224 (1965): 24–36

Leith, Emmett N. and Juris Upatnieks, 'Holography at the crossroads', *Optical Spectra* 4 (1970): 21

Leith, Emmett N. and Juris Upatnieks, 'Progress in holography', *Physics Today* 25 (1972): 28–34

'Lensless photography uses laser beams to enlarge negatives, microscope slides', *Wall Street Journal*, 5 (Dec. 1963): 28

Leonard, Carl, Emmett N. Leith and Juris Upatnieks, 'Investigation of architectural applications of holography', University of Michigan, Apr. 1972, Leonard collection

Levenson, Edgar A., 'A holographic model of psychoanalytic change', *Contemporary Psychoanalysis* 12 (1976): 1–20

Lightfoot, D. Tulla, 'Contemporary art-world bias in regard to display holography: New York City', *Leonardo* 22 (1989): 419–24

Lin, L. H., 'Hologram formation in hardened dichromated gelatin films', *Applied Optics* 8 (1969): 963–6

Lin, L. H., 'Edge-illuminated hologram', *Journal of the Optical Society of America A* 60 (1970): 714A

Lin, R. H., K. S. Pennington, G. W. Stroke, and A. Labeyrie, 'Multi-color holographic image reconstruction with white-light illumination', *Bell System Technical Journal* 45 (1966): 659–61

Lindgren, Nilo, 'Search for a holography market', *Innovations* (1) (1969): 16–27

Lindvold, Lars, 'New uses for HOEs', *Holographics International* 1 (1988): 33

Lippmann, Gabriel, 'La photographie des couleurs', *Comptes Rendus de L'Academie des Sciences* 112 (1891): 274

Lippmann, Gabriel, 'Photographie des couleurs', *Journal de Physique* 3 (1894): 97–106

Lippmann, Gabriel, 'Sur la theorie de la photographie des couleurs simples et composés', *Journal de Physique* 3-e serie 3 (1894): 97–107

Lippmann, Gabriel, 'Colour photography', *1908 Nobel Prize Lecture*, 1908

Lippmann, Gabriel, 'Epreuves reversibles photographies integrales', *Comptes Rendus de l'Academie Francais* 146 (1908): 446–51

Lloyd, Scott, 'The Holography Unit at the Royal College of Art', *holosphere*, 14 (3) (Summer 1986): 15–16

Loewen, E. G. to Editor *Technology Review*, 1967, Bentley, Institute of Science and Technology Box 21: Electro-Optical Lab

Lohmann, Adolf W. and Horst Wegener, 'Theory of optical image formation using a plane waves expansion', *Zeitschrift für Physik* 143 (1955): 431–4

Lohmann, A. W. and D. P. Paris, 'Binary Fraunhofer holograms, generated by computer', *Applied Optics* 6 (1967): 1739–48

Lohmann, Adolf W., 'A new duality principle in optics, applied to interference microscopy, phase contrast etc.' *Optik* 11 (1954): 478–88

Lohmann, Adolf W., 'Optical single side band transmission applied to the Gabor microscope', *Optica Acta* 3 (1956): 97–9

Lohmann, Adolf W., 'The Abbe experiments as methods for measuring the absolute light phase', *Zeitschrift für Physik* 143 (1956): 533–7

Lohmann, Adolf W., 'The holographic principle', in: W.-M. Boerner (ed.), *Inverse Methods in Electromagnetic Imaging* (Boston: Reidel, 1985), pp. 1033–42

Lubrano, Linda L. and John K. Berg, 'Academy scientists in the USA and USSR: background characteristics, institutional and regional mobility', in: J. R. Thomas and U. M. Kruse-Vaucienne (ed.), *Soviet Science and Technology: Domestic and Foreign Perspectives* (Washington DC, 1977)

Lucas, George, 'Star Wars script (revised fourth draft)', 15 Jan. 1976

Lucie-Smith, Edward, 'A 3D triumph for the future', *The Spectator*, 277 (8777) (5 Oct. 1996): 57–8

'Lunch Box Review,' MIT Museum MoH 26/619, 1978

Lutz, William W., 'New discoveries at Michigan universities', *The Detroit News—The Passing Show*, Sunday, 23 Feb. 1964: 1

Mach, Ernst, 'Modifikation und Ausfuhrung des Jamin-Interferenz-Refraktometers', *Annalen der Akademic der Wissenschaften (Wien)* 28 (1891): 223–4

Maline, Sarah Radley, 'Eluding the aegis of science: art holography on its own', *International Symposium on Display Holography* 1600 (1991): 215–19

Maline, Sarah Radley, 'The aesthetic problem of figural holography', *Proceedings of the SPIE—The International Society for Optical Engineering* 1732 (1992): 438–42

Maline, Sarah Radley, *Art Holography, 1968–1993: A Theatre of the Absurd*, PhD thesis, Art History, University of Texas at Austin (1995)

Mallard, Alexandre, 'From the laboratory to the market: the metrological arenas of research-technology', in: B. Joerges and T. Shinn (eds), *Instrumentation: Between Science, State and Industry* (Dordrecht: Kluwer, 2001), pp. 219–40

Mandel, L. and E. Wolf, 'Coherence properties of optical fields', *Reviews of Modern Physics* 37 (1965): 231

'Many holographic roads lead to Ann Arbor', *Laser Focus*, Feb. 1969: 16–17

Maréchal, André and P. Croce, 'Amélioration de la perception des détails des images par filtrage optique des fréquences spatiales', in: (ed.), *Problems in Contemporary Optics* (Arcetri-Firenze: Istituto Nazionale di Ottica, 1956), pp. 76–82

Mariotti, D., '3D television', *Radio Industria* 36 (1969): 33–4

Markov, Vladimir B. and G. I. Mironyuk, 'Holography in museums of the Ukraine' *Three-Dimensional Holography: Science, Culture, Education*, Proceedings of the SPIE 1238 (1989): 340–7

Markov, Vladimir B., 'Display and applied holography in culture development', in: (ed.), *Holography: Commemorating the 90th Anniversary of the Birth of Dennis Gabor* (Bellingham, 1990), pp. 268–304

Martinez, Zulma Nelly, 'From a Mimetic to a Holographic Paradigm in Fiction: Toward a Definition of Feminist Writing', *Women's Studies: An Interdisciplinary Journal* 14 (1988): 225–45

Mayer, Anna-K., 'Setting up a discipline, II: British history of science and "the end of ideology", 1931–1948', *Studies in History and Philosophy of Science* 35 (2004): 41–72

McCrickerd, J. T., 'Comparison of stereograms: pinhole, fly's eye, and holographic types', *Journal of the Optical Society of America* 62 (1972): 64–70

McGrew, Steve, 'Custom embossed holograms', presented at *Proceedings of the International Symposium on Display Holography*, Lake Forest College, Illinois, 1983

McGrew, Steve, 'Mass produced holograms for the entertainment industry', *Proceedings of the SPIE* 391 (1983): 19–20

McGrew, Steve, 'Holography before the laser', presented at *Proceedings of the International Symposium on Display Holography*, Lake Forest College, Illinois, 1985

McGrew, Steve, 'My reflections on holography', unpublished manuscript, 5 Dec. 2004, SFJ collection

McLuskie, Caroline, 'Holography Community', *Wavefront* 2 (1987): 4

Mertz, Lawrence and N. Young, in: K. J. Habell (ed.), *Proceedings of the International Conference on Optical Instruments* (1961): 305–12

Metherell, A. F., 'Acoustical holography', *Scientific American* 221 (1969): 36–53

Metherell, A. F., 'The present status of acoustical holography', *Proceedings of the SPIE* 25 (1971): 137–46

Meyerhofer, D., 'Spatial resolution of relief holograms in dichromated gelatin', *Applied Optics* 10 (1971): 416–21

Meyerhofer, D., 'Phase holograms in dichromated gelatin', *RCA Review* 33 (1972): 110–30

Michalak, Michael W., 'Holography in the Test and Evaluation Division at Goddard Space Flight Center', presented at *Holographic Instrumentation Applications*, Ames Research Center, Moffett Field, CA, 1970

Mikaelyan, A. L., 'The present state and future prospects of holography', *Telecommunications and Radio Engineering, Part 2 (Radio Engineering)* 25 (1970): 66–74

Mikaelyan, A. L., 'The present state and future prospects for the development of optical memory systems', *Soviet Journal of Communications Technology & Electronics* 35 (1990): 21–39

Miller, Peter, 'An educator with some light at the end of the tunnel!' *L.A.S.E.R. News* 3 (3) (Fall/Winter 1986): 8–9

Misselbeck, Reinhold and Matthias Lauk, 'Holography in Germany', *L.A.S.E.R. News* 10 (1990): 8

Misselbeck, Reinhold, 'The Museum für Holographie und neue visuelle Medien and its influence on holography in Germany', *Leonardo* 25 (1992): 457–8

'Mobberley pioneers 3-D film', *Focus* 8, (June 1984): 1

Morgan, Mary, 'Research firm has colourful past: U-M faculty started today's ERIM International for government projects', *Ann Arbor News*, 12 June 1999: 19

Morgenstern, Arthur L., 'Review of *An Introduction to Coherent Optics and Holography*', *Electronics* (16 May 1966):

Murata, K. and K. Kunugi, 'Cone-shaped cover for 360 degrees holography', *Applied Optics* 16 (1977): 1798–800

Murray, R., 'Holography at the Royal College of Art, London', *International Symposium on Display Holography* 1600 (1991): 237–9

Museum für Holographie & neue visuelle Medien., *Museum für Holographie & neue visuelle Medien* (Pulheim: The Museum, 1989)

'National advertisers test display holograms throughout the country', *holosphere* 6 (1977): 1

'National Electronics Conference, Chicago', *Pontiac Press*, 28 Oct. 1965: 6

'N-Dimensional Space exhibition', *Village Voice*, 14 May 1970: 13

Nelson, Bryce, 'DNA double helix: photo sends controversy spiraling', *Science* 173 (1971): 800–1

'New camera operating without lens shows scientific promise', *New York Times International Edition*, 11 Dec. 1963: 14

'New microscope limns molecule: Britons impressed by paper combining optical principle with electron method', *New York Times*, 15 Sep. 1948: 35

New Scientist 117 (1988): cover

'New York company producing integral holos', *holosphere*, 7 (2) (February 1978): 4

'Nicholson to offer pulsed holograms to advertisers / sales promoters', *holosphere* 9 (1980): 1–2

Nicholson, J. P., A. F. Hogan and J. Irving, 'Electron-density profiles of a theta-pinched plasma by holographic interferometry', *Journal of Physics D (Applied Physics)* 3 (1970): 1387–91

Nisida, M. and H. Saito, *Scientific Papers of the Institute of Physical and Chemical Research (Tokyo)* 59 (1965): 5

'No more Ilford Holo Film', *Holography News*, 5 (5) (1991): 1, 7

Novotny, George V., 'The little train that wasn't', *Electronics*, 37 (30) (30 Nov. 1964): 86–9

Öhlmann, Dietmar and Niklas Möller, 'The new goals of the German Association for Holography, DGH', *Proceedings of the SPIE* 3358 (1989): 152–4

Oliva, J., P. G. Boj and M. Pardo, 'Dichromated gelatin holograms derived from Agfa 8E75 HD plates', *Applied Optics* 23 (1984): 196–7

Olson, Ron and Bernadette Olson, 'Playmates to primates', *Proceedings of the SPIE* 3358 (1989): 251–6

O'Neill, Edward L., 'Spatial filtering in optics', *IRE Transactions on Information Theory* IT-2 (1956): 56

'On Holosonics and H. Fondiller', *Business Week*, 28 Jan. 1980: 96–7

'Optics center established', *Radiation Ink*, Jun. 1969: 1

Osmundsen, John A., 'Scientists' camera has no lens', *New York Times*, 5 Dec. 1963: 55

Ostria, V., 'The Beginnings of Cinematic Holography', *Cahiers Du Cinema* (1982): R11

Ostrovskaya, G. V. and Y. I. Ostrovsky to E. N. Leith and J. Upatnieks, letter, 13 Sep. 1966, Leningrad, Leith collection

Palais, J. C. and M. E. Miller, 'Holographic movies', *Optical Engineering* 35 (1996): 2578–82

Pappu, R. and W. Plesniak, 'Haptic interaction with holographic video images', *Proceedings of the SPIE The International Society for Optical Engineering* 3293 (1998): 38–45

Parks, W. George, 'Doctor Gordon's serious thinkers', *Saturday Review*, 4 (Aug. 1956): 42–9

Pearson, Jean, 'The Army's New Eyes: U-M Scientists develop far-sighted radar unit', *Detroit Free Press*, 20 Apr. 1960

Pennington, K. S., 'Advances in holography', *Scientific American* 218 (1968): 40–8

Pennington, K. S., J. S. Harper, and G. Kappel, 'Forming high efficiency holograms', *IBM Technical Disclosure Bulletin* 13 (1971): 2282–3

Pennington, K. S. and J. S. Harper, 'Techniques for producing low-noise, improved efficiency holograms', *Applied Optics* 9 (1970): 1643–50

Pepper, Andrew, 'German museum celebrates 5th anniversary', *holosphere* 13 (1985): 9–10

Pepper, Andrew, 'Expansion of holographic film products', *holosphere* 15 (1987): 12–13

Pepper, Andrew, *Drawing in Space: A Holographic System to Simultaneously Display Drawn Images on a Flat Surface and in Three Dimensional Space*, PhD thesis, Fine Arts, University of Reading (1988)

Pepper, Andrew T. and E. P. Krantz, 'Art of collaboration: a conflict of disciplines or constructive relationship', presented at *Fifth International Symposium on Display Holography*, Lake Forest, Illinois, 1994

Perry, J. F., 'Design of large format commercial display holograms', *Proceedings of the SPIE* 1051 (1989): 2–5

Pescovitz, David, 'The future of holography', *Wired*, July 1995: 50

Pethick, Jerry, 'On sculpture and laser holography: a statement', *Arts Canada* 25 (1968): 70–1

Pethick, Jerry, 'Animals Dream', in: R. Amos (ed.), *Collection* (Victoria, BC: Open Space, 1999), pp. 2

Phillips, N. J. and D. Porter, 'An advance in the processing of holograms', *Journal of Physics E (Scientific-Instruments)* 9 (1976): 631–4

Phillips, N. J. and D. Porter, 'Organically accelerated bleaches: their role in holographic image formation', *Journal of Physics E (Scientific-Instruments)* 10 (1977): 96–8

Phillips, N. J., 'Modes of holographic image formation-major areas of progress in the various regimes', presented at *Colloquium on Holographic Displays*, Loughborough UK, 1978

Phillips, N. J., 'Bridging the gap between Soviet and Western holography', in: P. Greguss and T. H. Jeong (ed.), *Holography: Commemorating the 90th Anniversary of the Birth of Dennis Gabor* (Bellingham, 1990), pp. 206–14

'Photo shows work of new optic system: Viewers call exhibit startling, fantastic', *Chicago Tribune*, 27 Oct. 1965

'Physical Optics Notebook', *SPIE Journal* 3 (1964)

Pierattini, G., 'Real-time and double-exposure microholographic interferometry for observing the dynamics of phase variations in transparent specimens', *Optics Communications* 5 (1972): 41–4

Pinch, Trevor J. and Wiebe E. Bijker, 'The social construction of facts and artifacts: or how the sociology of science and the sociology of technology might benefit each other', in: W. E. Bijker, T. P. Hughes, and T. J. Pinch (ed.), *The Social Construction of Technological Systems* (Cambridge, MA: MIT Press, 1987), pp 17–50

Pizzanelli, David, 'Captain Cook in half-baked hologram', *Holographics International* 1 (1988): 7–8

Pizzanelli, David, 'Countering counterfeiting', *Holographics International* 2 (1988): 18–19

Pizzanelli, David, 'Public or patented', *Holographics International* 1 (1990): 33

Pizzanelli, David, 'Evolution of the mythical hologram', *Proceedings of the SPIE The International Society for Optical Engineering* 1732 (1992): 430–7

Plesniak, Wendy J. and Massachusetts Institute of Technology. Dept. of Architecture. Program in Media Arts and Sciences, *Haptic Holography: An Early Computational Plastic*, PhD thesis, (2001)

Powell, Robert L. and Karl A. Stetson, 'Interferometric vibration analysis by wavefront reconstruction', *Journal of the Optical Society of America* 55 (1965): 1593–8

Powell, Robert L., 'Discussion', in: E. R. Robertson and J. M. Harvey (eds), *The Engineering Uses of Holography* (Cambridge: Cambridge University Press, 1968), pp. 128–9

Pribram, Karl H., 'The neurophysiology of remembering', *Scientific American* 220 (1969): 73–86

Pribram, Karl, H., 'Rethinking Neural networks: Quantum Fields and Biological Data', presented at *First Appalachian Conference on Behavioral Neurodynamics.*, Hillsdale, New Jersey., 1993

Rabkin, Yakov M., 'Science studies as an area of scientific exchange', in: J. R. Thomas and U. M. Kruse-Vaucienne (eds), *Soviet Science and Technology: Domestic and Foreign Perspectives* (Washington DC: National Academy of Sciences, 1977)

'Radar Photos From Afar to Pierce Iron Curtain', *Detroit News*, 20 Apr. 1960: 1

Rallison, Richard D., 'Control of DCG and nonsilver holographic materials', presented at *International Symposium on Display Holography*, Lake Forest, Illinois, 1991

Rallison, Richard D., 'The history of dichromates', http://www.xmission.com/~ralcon/dichro-hist.html, accessed 31 Oct. 2002

'Rapid expansion in Chinese hologram production', *Holography News*, 7 (2) (1993): 13–14

Rayleigh, Lord, 'On the theory of optical images, with special reference to the microscope', *Philosophical Magazine* 42 (Series 5) (1896): 167

Razutis, Al, 'Detailed history of holography at Visual Alchemy', http://www.alchemists.com/visual_alchemy/holo_hist.html, accessed 6 Oct. 2004

Redman, J. D., 'Novel applications of holography', *Journal of Physics E (Scientific Instruments)* 1 (ser.2) (1968): 821–2

Redman, J. D., 'The three-dimensional reconstruction of people and outdoor scenes using holographic multiplexing', presented at *Holography Seminar Proceedings*, San Francisco, 1968

Redman, J. D., C. J. Norman, and W. P. Wolton, 'Holographic reconstruction of animate objects', *Nature* 222 (1969): 476–7

Reeves, Daniel Paul, *Holography with a sandbox optical bench system*, MA thesis, Loma Linda University (1970)

Reingand, Nadya, 'Holography in Russia and other FSU states', http://www.media-security.ru/rusholo/index.htm, accessed 27 Sep. 2004

Remis, X., 'Holography and the Cinema', *Revue Du Cinema* 383 (1983): 102–3

Reuterswärd, Carl Fredrik, 'Rubies and rubbish: an artist's notes on lasers and holography', *Leonardo* 22 (1989): 343–56

Roberts, Christopher, 'Preface', in: E. Ritscher, J. Reilly, J. Lambe, and R. MacArthur (eds), *Light Dimensions: The Exhibition of the Evolution of Holography* (Bath, UK, 1983), pp. ii

Robertson, Elliot R. and James M. Harvey (eds), *Symposium on Engineering Uses of Holography* (University of Strathclyde, 1968)

Rogero, S., B. J. Mathews and R. F. Wuerker, 'Pulsed laser holography-new instrumentation for use in the investigation of liquid rocket combustion', presented at *15th International ISA Aerospace Instrumentation Symposium*, Pittsburgh, 1969

Rogers, G. L., 'Gabor diffraction microscopy: the hologram as a generalized zone-plate', *Nature* 166 (1950): 237–8

Rogers, G. L., 'Experiments in diffraction microscopy', *Proceedings of the Royal Society (Edinburgh)* A63 (1952): 193

Rogers, G. L., 'Two hologram methods in diffraction microscopy', *Journal of the Optical Society of America* 56 (1956): 849–58

Rogers, G. L., 'Diffraction microscopy and the ionosophere', draft paper, Jan. 1957, Sci Mus ROGRS 4

Rogers, G. L. to D. Gabor, 'Holographed' letter, 27 Sep. 1949, Dundee, *Scientific Research* 4 Jan. 1968

Rogers, G. L., 'The word "holography" ', *Scientific Research* 8 Jan. 1968: 57–8

Rogers, G. L. and M. Benyon, 'Holographic recording of a complete closed surface', *Applied Optics* 12 (1973): 886–7

Rosen, L., 'Hologram of the aerial image of a lens', *Proceedings of the IEEE* 54 (1966): 79–80

Rosen, L., 'Focused-image holography with extended sources', *Applied Physics Letters* 9 (1966): 337–9

Ross, David A., 'Radical Software redux', http://www.radicalsoftware.org/e/ross.html, accessed 11 Nov. 2004

Rotz, F. B. and A. A. Friesem, 'Holograms with nonpseudoscopic real images', *Applied Physics Letters* 8 (1966): 146–8

Sadowski, Patrick, 'Holographic child's play', *holosphere* 15 (1987): 17

Salmon, Pippa, 'Commercial survey', *Holographics International* 1 (1987): 4

Salmon, Pippa and Kamala Sen, 'Holography in Hollywood?' *Holographics International* 1 (1988): 26–7

Salter, J. L. and M. F. Loeffler, 'Comparison of dichromated gelatin and Dupont HRF-700 photopolymer as media for holographic notch filters', *Proceedings of the SPIE The International Society for Optical Engineering* 1555 (1991): 268–78

Saxby, Graham, 'A visit to Applied Holographics plc', *British Journal of Photography* (1984): 667–8, 681

Saxby, Graham, 'British holography at Light Fantastic', *British Journal of Photography* 129 (1982): 476–7

Saxby, Graham, 'Ilford's new holographic emulsions', *British Journal of Photography* 132 (1985): 1286–7

Saxby, Graham, 'Display holography in Britain: 1991', presented at *International Symposium on Display Holography*, Lake Forest, Illinois, 1991

Sayce, L. A., 'Closing address', in: E. R. Robertson and J. M. Harvey (eds), *The Engineering Uses of Holography* (Cambridge: Cambridge University Press, 1968), p. 560

Scanlon, M. J. B., 'Holography: a simple physical account', *GEC Review* 8 (1992): 47–57

Schade, Otto. H., 'Electro-optical characteristics of television systems', *RCA Review* 9 (1948): 5 (Part I), 245 (Part II), 490 (Part III), 653 (Part IV)

'School of holography flourishes on west coast', *holosphere* 2 (1973): 1, 5–6

Schultz, Catherine, 'Holography: they do it all with lasers', *Berkeley Gazette*, 21 Oct. 1977

Seidel, Robert W., 'From glow to flow: A history of military laser research and development', *Studies in the History of the Physical and Biological Sciences* 17 (1986): 111–48

Senior, T. B. A., 'Radiation Lab History', http://www.eecs.umich.edu/RADLAB/labhistory.html, accessed 1 Sep. 2004

Shankoff, T. A., 'Phase holograms in dichromated gelatin', *Applied Optics* 7 (1968): 2101–5

Sherwin, Chalmers W., J. P. Ruina, and R. D. Rawcliffe, 'Some early developments in synthetic aperture radar systems', *IRE Transactions on Military Electronics* MIL-6 (1962): 111–15

Shinn, Terry, 'Failure or success? Interpretations of 20th century French physics', *Studies in the History of the Physical and Biological Sciences* 17 (1987): 361–3

Shinn, Terry, 'Formes de division du travail scientifique et convergence intellectuelle: La recherche technico-instrumentale', *Revue Française de Sociologie* 41 (2000): 447–73

Shinn, Terry, 'Strange cooperations: the U.S. Research-Technology perspective, 1900–1955', in: B. Joerges and T. Shinn (eds), *Instrumentation: Between Science, State and Industry* (Dordrecht: Kluwer Academic Press, 2001), pp. 69–95

'Shoppers get the picture: Hologram marketers test the water', *Union*, 20 Jan. 1977

Shushurin, S. S., 'On the history of holography', *Soviet-Physics—Uspekhi* 14 (1972): 655–7

Siebert, L. D., 'Large-scene front-lighted hologram of a human subject', *Proceedings of the IEEE* 56 (1968): 1242–3

Siebert, L. D., 'Coherence length curve for ruby oscillator-amplifier', *Applied Physics Letters* 16 (1970): 318–20

Siebert, L. D., 'Holographic coherence length of a pulse laser', *Applied Optics* 10 (1971): 632–7

Skande, Per, 'The development of commercial display holography in Sweden', *Proceedings of the International Symposium on Display Holography* 2 (1985): 47–50

Smith, C. Zoe, 'Fritz Goro: emigre photojournalist', *American Journalism* 3 (1986): 206–21

Smith, P. J., 'Between the Lines', *Leonardo* 22 (1989): 291–4

Smith, Howard and Brian Van der Horst, 'Scenes', *Village Voice*, 14 Feb. 1974

'Soviet holography conference shows high amount of research activity', *holosphere*, 7 (9), Sep. 1978: 1–2

Sowdon, Michael, 'The sixties revisited', *Wavefront* 1 (1986): 8

Speer, L., 'Before Holography—a Call for Visual Literacy', *Leonardo* 22 (1989): 299–306

SPIE, 'Seminar In-Depth on Holography', San Francisco, 1968

Staselko, D. I., V. G. Smirnov, and N. Denisyuk Yu, 'On obtaining holograms of an alive diffuse object with the help of single-mode ruby laser', *Zhurnal Nauchnoi i Prikladnoi Fotografii i Kinematografii* 13 (1968): 135–6

Staselko, D. I., A. G. Smirnov, and Yu. N. Denisyuk, 'Production of high-quality holograms of three-dimensional diffuse objects with the use of single-mode ruby lasers', *Optics and Spectroscopy* 25 (1969): 505–7

St.-Hilaire, P., S. A. Benton, M. E. Lucente, J. D. Sutter, and W. J. Plesniak, 'Advances in holographic video', presented at *Practical Holography VII: Imaging and Materials*, San Jose, California, 1993

Staselko, D. I., Yu N. Denisyuk and A. G. Smirnov, 'Holographic portrait of a man', *Zhurnal Nauchnoi i Prikladnoi Kinematografii* 15 (1970): 147

Stetson, K. A., 'A rigorous treatment of the fringes of hologram interferometry', *Optik* 29 (1969): 386–400

Stetson, Karl A., *A Study of Fringe Formation in Hologram Interferometry and Image Formation in Total Internal Reflection Holograms*, PhD thesis, Institut för Optisk Forskning (1969)

Stetson, Karl A., 'Some techniques for hologram interferometry of string instruments', presented at *79th meeting of the Acoustical Society of America*, Atlantic City, NJ, 1970

Stetson, Karl A., 'The origins of holographic interferometry', *Experimental Mechanics* 31 (1991): 15–18

Stetson, Karl A., 'The problems of holographic interferometry', *Experimental Mechanics* 39 (1999): 249–55

Stevenson, S. H., 'DuPont multicolor holographic recording films', *Proceedings of the SPIE The International Society for Optical Engineering* 3011 (1997): 231–41

Stirn, B. A., 'Recording 360 degrees holograms in the undergraduate laboratory', *American Journal of Physics* 43 (1975): 297–300

Storm, Carlyle B., Jimmie C. Oxley and Alexander M. Cruickshank, 'The Gordon Research Conferences: a brief history', presented at *GRC 50 Years in New Hampshire*, New Hampshire, 1997

'Strick will make laser film', *Los Angeles Times*, 21 Aug. 1970, 14

Stroke, George W., 'New optical principle of "lensless" x-ray microscopy', Press release for Symposium on Optical and Electro-Optical Information Processing, Boston, University of Michigan, 10 Nov. 1964

Stroke, George W., 'Theoretical and experimental foundations of electro-optical image formation, image modulation, and wave-front reconstruction imaging', presented at *Symposium on Optical and Electro-Optical Information Processing Technology*, Boston, MA, 1964

Stroke, George W., 'Lensless Fourier-transform method for optical holography', *Applied Physics Letters* 6 (1965): 201–3

Stroke, George W., 'Three advances in Fourier transform holography', *Journal of the Optical Society of America* 55 (1965): 1566 (conference abstract)

Stroke, George W., 'White-light reconstruction of holographic images using transmission holograms recorded with conventionally-focused images and 'in-line' background', *Physics Letters* 23 (1966): 325–7

Stroke, George W., 'Diffraction Gratings', in: S. W. Flugge (ed.), *Handbuch der Physik* (Berlin: Springer-Verlag, 1967), pp. 26–754

Stroke, George W., 'Recent advances in holography', *Technology Review* 69 (1967): 16–22

Stroke, George W., 'U.S.–Japan seminar on holography, Tokyo and Kyoto, 2–6 October 1967', *Applied Optics* 7 (1968): 622

Stroke, George W., 'Optical image deblurring methods', *Naval Research Reviews* (1971): 14–20

Stroke, George W., 'Ultrasonic imaging and holography: medical, sonar, and optical applications', presented at *United States-Japan Science Cooperation Seminar on Pattern Information Processing in Ultrasonic Imaging, 1973*, University of Hawaii, 1974

Stroke, George W., 'Optical engineering', in: *Encyclopaedia Britannica Yearbook of Science and the Future* (1980)

Stroke, George W. and A. Labeyrie, 'Two-beam interferometry by successive recording of intensities in a single hologram.' *Applied Physics Letters* 8 (1965): 42–4

Stroke, George W. and Antoine E. Labeyrie, 'White-light reconstruction of holographic images using the Lippmann-Bragg diffraction effect', *Physics Letters* 20 (1966): 368–70

Stroke, George W. and D. G. Falconer, 'Attainment of high resolutions in wavefront-reconstruction imaging', *Physics Letters* 13 (1964): 306–9

Stroke, George W., D. B. Brumm, A. T. Funkhouser, A. Labeyrie, and R. C. Restrick, 'On the absence of phase-recording or 'twin-image' separation problems in 'Gabor' (in-line) holography', *British Journal of Applied Physics* 17 (1966): 497–500

Supertizi, E. P. and A. K. Rigler, *Journal of the Optical Society of America* 56 (1966): 524

'Talk of the Town', *New Yorker*, 24 Aug. 1987:

Tanner, L. H., 'On the holography of phase objects', *Journal of Scientific Instruments* 43 (1966): 346

Tanner, P.G. and T. E. Allibone, 'The patent literature of Nobel laureate Dennis Gabor (1900–1979)', *Notes and Records of the Royal Society of London* 51 (1997): 105–20

Tchalakov, Ivan, 'Innovating in Bulgaria—two cases in the life of a laboratory before and after 1989', *Research Policy* 30 (2001): 391–402

Tchalakov, Ivan, 'The object and the other in holographic research—approaching passivity and responsibility of human actors', *Science, Technology and Human Values* 29 (2002): 64–87

Tchalakov, Ivan, 'The history of holographic optical storage at both sides of the Iron Curtain, 1969–1989', presented at *Society for the History of Technology Annual Meeting*, Amsterdam, Netherlands, 2004

'The incredible hologram', *Newsweek*, 29 Dec. 1969: 29

Theatre Design & Technology 14 (1978): cover

Thompson, E. P., 'The long revolution—II', *New Left Review*, (9) (May–Jun. 1961): 32

'Three colours from one Ilford emulsion', *Holographics International* 1 (1990): 14

Tipler, F. J., 'The ultimate future of the universe, black hole event horizon topologies, holography, and the value of the cosmological constant', *AIP Conference Proceedings* (2001): 769–74

Tomko, Martha, 'Letter to the editor—what's going on at the MoH?' *holosphere*, 17 (1) (Spring 1990): 4–5

Trion Instruments Inc / Lear-Siegler Inc, 'Trion Instruments Inc brochure', 1961

Tsujiuchi, J., 'Restitution des images aberrantes par le filtrage des frequencies spatiales I', *Optica Acta* 7 (1960): 243, 386

Tsujiuchi, J., 'Restitution des images aberrantes par le filtrage des frequencies spatiales II', *Optica Acta* 8 (1961): 161

Tsujiuchi, J., N. Takeya, and K. Matsuda, 'Measurement of the deformation of an object by holographic interferometry', *Optica Acta* 16 (1969): 707–20

Tyler, Douglas E., 'Holography: roots of a new vision', *Proceedings of the International Symposium on Display Holography* 2 (1985): 379–86

Upatnieks, Juris and A. M. Tai, 'Development of the holographic sight', *Proceedings of the SPIE The International Society for Optical Engineering* 2968 (1997): 272–81

Upatnieks, Juris and Emmett N. Leith, 'Wavefront reconstruction with continuous tone objects', *Journal of the Optical Society of America* 53 (1963): 1377–81

Upatnieks, Juris and Emmett N. Leith, 'Lensless, three-dimensional photography by wavefront reconstruction', *Journal of the Optical Society of America* 54 (1964): 579–80

'Update on the New York School of Holography and the Museum of Holography', *holosphere*, 4 (4), 3–4

van Heerden, Pieter J., 'Theory of optical information storage in solids', *Applied Optics* 2 (1963): 393–400

Vandervert, L. R., 'Chaos Theory and the Evolution of Consciousness and Mind—a Thermodynamic–Holographic Resolution to the Mind-Body Problem', *New Ideas in Psychology* 13 (1995): 107–27

'Variations of integral holos planned: Cross', *holosphere* 6 (1977): 1, 4–5

Vasilieva, N. V. and N. I. Kirillov, 'Requirements to high resolution photomaterials for holography', *Tekhnika Kino i Televideniya* 7 (1972): 3–9

Vivian, W. E., L. J. Cutrona, and E. N. Leith, 'Report of Project MICHIGAN: A Doppler Technique for Obtaining Very Fine Angular Resolution from a Side-Looking Airborne Radar', Report 2144-5-T, University of Michigan Willow Run Laboratory, July 1954

Vuotto, Joe, 'Holography—a new security device', *Holosphere* 12 (1983): 1, 3

Walker, Jearl, 'Amateur Scientist: Easy way to make holograms', *Scientific American*, Feb. 1980: 158

Walker, Jearl., 'Amateur Scientist: Rainbow holograms', *Scientific American*, Sep. 1986: 114

Walker, Jearl, 'Amateur Scientist: How to stop worrying about vibrations and make holograms viewable in white light', *Scientific American* (May 1989): 134

Wedlin, Randy, 'Scientific utopia: the Gordon Research Conferences', *Chemistry* (Aug. 2002): 25–9

Weitzman, C., 'Optical technologies for future computer system design', *Computer Design* 9 (1970): 169–75

Wesly, Ed, 'Silver anniversaries', *holosphere* 15 (1988): 12–13

Whittaker, Jeanne, 'On the beam . . . in color!' *Detroit Free Press*, 10 May 1970, 4–D

Wilfong, Joan, 'Hologram product display announced', *Conductron antenna*, 2 (Dec. 1968): 4

Wilfong, Joan, 'Holography use increases', *Conductron antenna*, 3 (Dec. 1969): 4

Wilfong, Joan, 'Conductron—Ann Arbor laser photography is "Belle of Ball" ', *Conductron antenna* 4 (1970): 7

Wilhelmsson, Hans, 'Holography: a new scientific technique of possible use to artists', *Leonardo* 1 (1968): 161–9

'Willow Run Research Paying Big Dividends', *Ann Arbor News*, 20 Apr. 1960, 4

'Willow Run future', *Detroit Daily News*, 4 Dec. 1969

Winner, Langdon, 'Do artifacts have politics?' in: (ed.), *The Whale and the Reactor: A Search for Limits in an Age of High Technology* (Chicago, 1986), pp. 19–39

Winthrop, John T., *The Formation of Diffraction Images: Fresnel Images, Compound Eye, and Holographic Microscopy*, PhD thesis, University of Michigan (1966)

Winthrop, John T. and C. Roy Worthington, *Physics Letters* 15 (1965): 124–6

Winthrop, John T. and C. Roy Worthington, 'Theory of Fresnel images I. Plane periodic objects in monochromatic light', *Journal of the Optical Society of America* 55 (1965): 373–81

Winzer, G. and N. Douklias, 'Improved holographic matched filter systems for pattern recognition using a correlation method', *Optics and Laser Technology* 4 (1972): 222–7

Wolf, Emil, review of *An Introduction to Coherent Optics and Holography* in: *Journal of the Optical Society of America* 57 (1967): 433

Wolf, Emil, 'Recollections of Max Born', *Optics News* 9 (1983): 10–14

Wolff, Michael, 'The birth of holography: a new process creates an industry', *Innovations* (1969): 4–15

Woltz, Robert L., 'Before the beginning of SPIE', *SPIE Journal* 5 (1967): 2–3

Wood, Glenn P., 'New silver halide holographic materials from Ilford Ltd', presented at *Proceedings of the International Symposium on Display Holography*, Lake Forest College, Illinois, 1985

Wood, Glenn P., 'Ilford holography—1986 progress report', *Proceedings of the SPIE* 747 (1986): 62–6

Wood, Glenn P., 'Ilford holographic consumables world strategy', presented at *Proceedings of the Third International Symposium on Display Holography*, 1988

Wood, Glenn P., 'Silver halide film supply: follow-up—Silver halide users let down the suppliers', *Holography News* 7 (5) (1991): 3

Wuerker, Ralph, 'Discussion', in: E. R. Robertson and J. M. Harvey (eds), *The Engineering Uses of Holography* (Cambridge: Cambridge University Press, 1968), p. 72

Wuerker, Ralph, 'Discussion', in: E. R. Robertson and J. M. Harvey (eds), *The Engineering Uses of Holography* (Cambridge: Cambridge University Press, 1968), p. 129

Wuerker, Ralph, 'Discussion', in: E. R. Robertson and J. M. Harvey (eds), *The Engineering Uses of Holography* (Cambridge: Cambridge University Press, 1968), p. 501

Wuerker, R. F., 'Holography and holographic interferometry: industrial applications', *Annals of the New York Academy of Sciences*. 168(3) (1969): 492–505

Wuerker, R. F., L. O. Heflinger, R. E. Brooks, and C. Knox, 'Action holography', presented at *Northeast Electronics Research and Engineering Meeting*, Boston, 1968

Yamaguchi, I. and H. Saito, 'Application of holographic interferometry to the measurement of Poisson's ratio', *Japanese Journal of Applied Physics* 8 (1969): 768–71

Yamaguchi, M., N. Ohyama, and T. Honda, 'Holographic three-dimensional printer: new method', *Applied Optics* 31 (1992): 217–22

Yaroslavsky, L. P., 'Digital holography: 30 years later', *SPIE Proceedings* 4659 (2002): 1–11

Yaroslavsky, Leonid P., 'Recollections on early works on digital holography in the former Soviet Union', unpublished report, 14 Jun. 2003, SFJ collection

Yefremov, Ivan Antonovich, 'Shadow of the past', in: I. A. Yefremov (ed.), *Stories* (Moscow: Foreign Languages Publishing House, 1954), pp. 34–5

Yefremov, Ivan Antonovich, 'Stellar Ships', in: I. A. Yefremov (ed.), *Stories* (Moscow: Foreign Languages Publishing House, 1954), pp. 258–9
Yinger, Milton, 'Contraculture and subculture', *American Sociological Review* 25 (1960): 629
Zagorskaya, Z. A., 'High-resolution photographic emulsion for recording three-dimensional holograms', *Soviet Journal of Optical Technology* 40 (1973): 134
Zec, P., 'The Aesthetic Message of Holography', *Leonardo* 22 (1989): 425–30
Zech, R. G. and L. D. Siebert, 'Pulsed laser reflection holograms', *Applied Physics Letters* 13 (1968): 417–18
Zech, R. G., L. D. Siebert, and H. C. Henze, 'A new holographic technique for medical and biomedical applications', presented at *7th annual biomedical sciences instrumentation symposium on imagery in medicine.*, Pittsburgh, 1969
Zehnder, Ludwig, 'Ein neuer Interfernz-Refraktor', *Zeitschrift für Instrumentenkunde* 11 (1891): 275–85.
Zellerbach, Gary A., 'Selling holographic art: an analysis of past, present and future markets', *Proceedings of the SPIE The International Society for Optical Engineering* 1600 (1991): 446–54
Zemtsova, E. G. and L. V. Lyakhovskaya, 'Recording of three-dimensional holograms on LO1-2-63 photographic plates', *Soviet Journal of Optical Technology* 41 (1974): 473–4
Zernike, Frits, 'How I discovered phase contrast', *Nobel Lecture, Nobel Prize for Physics* (1953)

UNPUBLISHED COMMUNICATIONS

Abraham, G. to MoH, letter, 13 Aug. 1983, MIT 2/27
Akopov, Edmund to SFJ, email, 24 Feb. 2004, SFJ collection
Aleksoff, Carl to SFJ, interview, 9 Sep. 2003, Ann Arbor, SFJ collection
Allibone, T. E. to D. Gabor, letter, 13 Sep. 1961, IC GABOR MS/6/3
American Institute of Physics, 'Press release: Lensless optics system makes clear photographs', 5 Dec. 1963, 2
American Institute of Physics, 'Press release: Objects behind others now visible in 3-D pictures made by new method', 25 Oct. 1964
Asimov, Isaac, 'erotic limerick', 26 Sep. 1978, MIT 6/619
Aspray, William, 'George Wilhelm Stroke, Electrical Engineer, an oral history conducted in 1993 by William Aspray, IEEE History Center, Rutgers University, New Brunswick, NJ, USA'. 1993
Athey, Brian to SFJ, interview, 2 Aug. 2004, Glasgow, SFJ collection
Baez, Albert V. to G. L. Rogers, letter, 16 Jan. 1951, Sci Mus ROGRS 6
Baez, Albert V. to SFJ, email, 13 Mar. 2003, SFJ collection
Barnett, Norm to SFJ, interview, 11 Sep. 2003, Ann Arbor, SFJ collection
Baxter, K. D. to R. H. Jackson, letter, 8 Oct. 1976, MIT 29/804
Begleiter, Eric to SFJ, interview, 10 Jul. 2003, Cambridge, MA, SFJ collection
Bender, David C. to E. Weitzen, letter, 12 Jun. 1986, Haines collection
Benton, Stephen A., 'Artist files,' MIT Museum 26/584, 1978
Benton, Stephen A. to T. Edwards, letter, 15 May 1978, MIT 26/584
Benton, Stephen A. to SFJ, interview, 11 Jul. 2003, Cambridge, MA, SFJ collection
Benyon, Margaret to P. Jackson, letter, 10 May 1983, MIT 26/587
Benyon, Margaret to SFJ, interview, 21 Jan. 2003, Santa Clara, CA, SFJ collection
Benyon, Margaret to SFJ, email, 4 May 2005, SFJ collection

Benyon, Margaret to SFJ, interview, 30 Apr. 2005, London, SFJ collection
Bjelkhagen, Hans I. to S. Johnston, interview, 18–19 Sep. 2002, SFJ collection
Blyth, Jeff to SFJ, letter, 7 May 2005, SFJ collection
'Board minutes', report, New York Museum of Holography, 1986
Bohm, David to D. Gabor, letter, 14 Mar. 1969, London, IC GABOR MB/9
Born, Max to D. Gabor, letter, 21 Feb. 1951, IC GABOR MB/10/3
Brady, Hugh to SFJ, letter, 23–24 Sep. 2003, SFJ collection
Bragg, W. L. to D. Gabor, letter, 5 Jul. 1948, IC GABOR EL/1
Bragg, W. L. to D. Gabor, letter, 27 Jul. 1948, IC GABOR EL/1
Bragg, W. L. to G. L. Rogers, letter, 25 Mar. 1952, Sci Mus ROGRS 6
Brentnall, Roger, Alan Hodgson, Nigel Briggs, and Mary B. Dentschuk to SFJ, interviews, 20 Sep. 2002, Mobberley, Cheshire, SFJ collection
Briggs, Nigel to SFJ, interview, 20 Sep. 2002, SFJ collection
Brown, William M., George W. Stroke and Emmett N. Leith to file, memo, 10 Dec. 1963, Ann Arbor, Bentley, Institute of Science and Technology Box 21: Electro-Optical Lab
Burns, Jody to C. Capa, by letter, New York, MIT 33/1022
Capa, Cornell to P. Jackson, letter, New York, MIT 33/1022
Casdin-Silver, Harriet to SFJ, interview, 3 Jul. 2003, Boston, MA, SFJ collection
Castelli, Leo to P. Cummings, interview, 14 May 1969, New York, Smithsonian Archives of American Art
Caulfield, H. John to A. M. Cruickshank, by letter, 15 May 1970, Sudbury, MA, Chemical Heritage Foundation
Caulfield, H. John to SFJ, email, 12 May 2003, SFJ collection
Caulfield, H. John to SFJ, interview, 21 Jan. 2003, Santa Clara, CA, SFJ collection
Charnetski, Clark to SFJ, interview, 3 Sep. 2003, Ann Arbor, MI, SFJ collection
Claudius, Peter, MIT 33/1028, 1976
Cochran, Gary D., 'New Ideas Notebook', Conductron Corporation, 1965
Cochran, Gary D. to J. Zorn, 18 May 1967, Ann Arbor, Cochran collection
Cochran, Gary D. to SFJ, interview, 6 and 8 Sep. 2003, Ann Arbor, MI, SFJ collection
Connors, Betsy to SFJ, interview, 10 Jul. 2003, Cambridge, MA, SFJ collection
Crane, Alison to P. Jackson, letter, 25 Apr. 1979, MIT 29/822
Crenshaw, Melissa to SFJ, interview, 19 Nov. 2003, Vancouver, SFJ collection
Cross, Lloyd G., MIT 30/843, 1977
Cross, Lloyd G. to SFJ, email, 25 Oct. 2003, SFJ collection
Cross, Lloyd G. to SFJ, email, 26 Oct. 2003, SFJ collection
Croucher, Duncan to S. Johnston, telephone interview, 16 Aug. 2002, SFJ collection
De, Manoranjan to SFJ, email, 31 May 2003, SFJ collection
Denisyuk, Yu N. to SFJ, email, 13 Apr. 2003, SFJ collection
Denisyuk, Yu N. to SFJ, email, 16 Apr. 2003, SFJ collection
Denisyuk, Yu N. to SFJ, email, 17 Apr. 2003, SFJ collection
Denisyuk, Yu N. to SFJ, email, 3 May 2003, SFJ collection
Denisyuk, Yu N. to SFJ, email, 8 Aug. 2003, SFJ collection
Denisyuk, Yu N. to SFJ, email, 23 Aug. 2003, SFJ collection
Denisyuk, Yu N. to SFJ, email, 8 Dec. 2004, SFJ collection
Denton, Frank to SFJ, telephone interview, 1 May 2003, SFJ collection

Dentschuk, Mary B., 'Visit report', Ilford Ltd, Aug. 1988
DiBiase, Vince to SFJ, interview, 21 Jan. 2003, Santa Clara, CA, SFJ collection
DiBiase, Vincent, unpublished report, 24 Jan. 2003, Santa Clara, California, SFJ collection
Dietrich, Ed to SFJ, interview, 19 Nov. 2003, Vancouver, SFJ collection
Dinsmore, Sydney, 'on MoH collection from curator', report, MIT 10/192, 18 Oct. 1990 and 22 Jan. 1991
Dinsmore, Sydney, 'strategic plan', report, MIT 10/192, Jul. 1990
Director, MoH, 'MoH attendance information', MIT Museum 10/192, 1983
Dow, W.G. to G. W. Stroke, letter, 30 Nov. 1962, Ann Arbor, Bentley Historical Library, Institute of Science and Technology Box 21: Personnel
Dow, W.G. to G. W. Stroke, memo, 26 Jan. 1967, Ann Arbor, Bentley Historical Library, Institute of Science and Technology Box 21: Electro-Optical Lab
Fagan, William and Hans Bjelkhagen to SFJ, interview, 19 Sep. 2002, SFJ collection
Fagan, William to SFJ, interview, 19 Sep. 2002, SFJ collection
Farris, H. W. to J. T. Wilson, 21 Jan. 1965, Ann Arbor, Bentley, Institute of Science and Technology Box 21: Electro-Optical Lab
Funkhouser, Arthur to G. H. McLaughlin, letter, 24 Nov. 1966, London, IC GABOR MM/3
Funkhouser, Arthur to SFJ, email, 9 Apr. 2003, SFJ collection
Gabor, Dennis to A. Lohmann, letter, 2 Jan. 1957, London, Lohmann collection
Gabor, Dennis to A. Lohmann, letter, 5 May 1957, London, Lohmann collection
Gabor, Dennis to Appointment Committee for Chair in Electron Physics, typewritten CV, 19 Feb. 1958, IC GABOR GB/1
Gabor, Dennis to C. R. Burch, letter, 11 Sep. 1950, IC GABOR EL/1
Gabor, Dennis to D. S. Strong, letter, 18 Jan. 1972, IC Gabor ML/16
Gabor, Dennis to E. N. Leith, letter, 3 Nov. 1966, IC Gabor ML/4
Gabor, Dennis to E. N. Leith, letter, 21 Nov. 1968, Leith collection
Gabor, Dennis to E. N. Leith, letter, 22 Apr. 1969, Leith collection
Gabor, Dennis to file, 1948, IC GABOR D/39
Gabor, Dennis to G. L. Rogers, letter, 18 Aug. 1950, Sci Mus ROGRS 6
Gabor, Dennis to G. L. Rogers, letter, 6 Nov. 1950, Sci Mus ROGRS 6
Gabor, Dennis to G. L. Rogers, letter, 25 Oct. 1954, Sci Mus ROGRS 6
Gabor, Dennis to G. L. Rogers, letter, 23 May 1958, Sci Mus ROGRS MS 1014/11
Gabor, Dennis to G. L. Rogers, letter, 5 Aug. 1966, Sci Mus MS 1014/15
Gabor, Dennis to I. C. Prof. Willis Jackson, letter, 28 Sep. 1948, Rugby, IC Gabor GB/1
Gabor, Dennis to I. Williams, letters, IC B/GABOR MA/13/1
Gabor, Dennis to J. B. Le Poole, letter, 5 Feb. 1948, IC GABOR EL/1
Gabor, Dennis to L. Cross, letter, 27 Apr. 1973, MIT 40 Correspondence XVIII
Gabor, Dennis to L. J. Davis, letter, 18 May 1951, IC GABOR B/36
Gabor, Dennis to M. Born, letter, 5 Dec. 1951, IC GABOR MB/10/3
Gabor, Dennis to M. E. Haine, letter, 18 Jun. 1949, Rugby, IC GABOR MA/6/1
Gabor, Dennis to M. E. Haine, letter, 14 Apr. 1950, Aldermaston, IC B/GABOR MH 2/2
Gabor, Dennis to M. E. Haine, letter, early 1951, IC B/GABOR MH 2/2
Gabor, Dennis to M. E. Haine, letter, 26 Nov. 1952, IC GABOR MH/2/2
Gabor, Dennis to M. E. Haine, letter, 28 Dec. 1955, IC GABOR MS/6/3
Gabor, Dennis to N. Calder, letters, IC GABOR EM/7

Gabor, Dennis to P. Goldmark, letter, 2 Mar. 1966, IC GABOR LA/9
Gabor, Dennis to P. Goldmark, letter, 30 May 1966, IC GABOR LA/9
Gabor, Dennis to P. Goldmark, letter, 27 Sep. 1966, IC GABOR LA/9
Gabor, Dennis to P. Goldmark, letter, 3 Jun. 1967, IC GABOR LA/9
Gabor, Dennis to P. Goldmark, letter, 26 Jul. 1967, IC GABOR LA/9
Gabor, Dennis to P. Goldmark, letters, 1966–67, IC GABOR LA/9
Gabor, Dennis to R. Peierls, letter, 15 Jun. 1948, IC GABOR MP/2
Gabor, Dennis to C. Darwin, letter, 29 Nov. 1948, Rugby, IC GABOR EL/2
Gabor, Dennis to W. L. Bragg, letter, IC GABOR EL/1, Mar.–Jul. 1948
Gabor, Dennis to W. L. Bragg, letter, 19 Jan. 1948, Rugby, IC GABOR EL/1
Gabor, Dennis to W. L. Bragg, letter, 7 Jul. 1948, IC GABOR EL/1
Gabor, Dennis to *Sunday Times*, letter, 14 Apr. 1968, IC B/GABOR/11/111
Gabor, Dennis to T. E. Allibone, letter, 9 Mar. 1933, IC GABOR MS/6/1
Gabor, Dennis to T. E. Allibone, letter, 4 May 1947, IC GABOR MS/6/1
Gabor, Dennis to T. E. Allibone, letters, 1932–1954, IC GABOR MA/6/1
Gabor, Dennis to T. E. Allibone, letter, 22 Mar. 1954, IC GABOR MA/6/1
Gabor, Dennis to T. Raison, letter, 15 Jul. 1961, IC GABOR MN/5-6
Gabor, Dennis to unspecified recipient, report, 7 Jul. 1951, IC GABOR EL/2
George, Nicholas to R. H. Jackson, 1978, MIT 42/1278
Gillespie, Donald to SFJ, interview, 29 Aug. and 4–6 Sep. 2003, Ann Arbor, MI, SFJ collection
Goodman, Joseph W. to A. M. Cruickshank, letter, 18 May 1970, CHF
Haine, Michael E. to D. Gabor, letter, 14 Mar. 1950, Aldermaston, B/GABOR MH 2/2
Haines, Kenneth to SFJ, interview, 21 Jan. 2003, Santa Clara, CA, SFJ collection
Harvey, Robin to P. Jackson, memo,1980, MIT 12/265
Howard, John M. to E. N. Leith, letter, 5 May 1967, Leith collection
Hlynsky, David to P. Jackson, letter, Nov. 1979, MIT 27/682
Jackson, Rosemary H., 'A proposal for the maintenance and development of the Museum of Holography under the Directorship of Rosemary H. Jackson', MIT 8/191, 1 Feb. 1976
Jackson, Rosemary H., 'History of the Museum of Holography Oct. 2, 1974–May 1, 1976', report, MIT 8/191
Jackson, Rosemary H. to T. E. Allibone, letter, 17 Dec. 1979, MIT 2/40
Jackson, Rosemary H. to S. Bains, letter, 14 Sep. 1989, MIT 28/743
Jackson, Rosemary H. to S. A. Benton, letter, 19 May 1978, New York, MIT 26/584
Jackson, Rosemary H. to H. Kramer, letter, Apr. 25 1977, MIT 2/25
Jackson, Rosemary H. to E. O'Neill, letter, 1990, MIT 12
Jackson, Rosemary H. to M. Tomko, letter, 1991, New York, MIT 12/265
Jackson, Rosemary H. to G. Zellerbach, letter, 16 Oct. 1978, New York, MIT 28/743
Jeong, Tung H. to P. Jackson, letter, 1975, New York, MIT 33/1022
Jeong, Tung H. to SFJ, email, 11 Jan. 2005, SFJ collection
Jeong, Tung H. to SFJ, interview, 21 Jan. 2003, Santa Clara, CA, SFJ collection
Kirkpatrick, Paul to G. L. Rogers, letter, 3 Dec. 1950, Sci Mus ROGRS 6
Kock, Winston E. to P. Jackson, letter, 1 Sep. 1977, MIT Museum 30/909
Kontnik, Lewis T. to SFJ, interview, 20 Nov. 2003, Vancouver, SFJ collection
Kozma, Adam to E. N. Leith, letter, 7 Oct. 1966, IC GABOR ML/4
Kozma, Adam to E. N. Leith, letter, 3 Jan. 1967, Leith collection

Kamata, Yasumasa to SFJ, email, 17 Dec. 2003, SFJ collection
Kamata, Yasumasa to SFJ, interview, 19 Nov. 2003, Vancouver, SFJ collection
Kan, Michael to SFJ, email, 29 Jan. 2004, SFJ collection
Kennard, A. H. W., The Patent Office to New Holographic Design Ltd, letter, 22 Feb. 1983, London, Waller-Bridge collection.
Labeyrie, Antoine E. to SFJ, email, 2 Jun. 2003, SFJ collection
Labeyrie, Antoine E. to SFJ, email, 22 Jul. 2003, SFJ collection
Lancaster, Ian M., Holopack-Holoprint 2003 presenters' documentation, Holopack
Lancaster, Ian M., 'The Museum of Holography: Its role and policies in a changing environment', report for Board of Trustees, MIT 8/191, 1986
Lane, Linda to J. Ross, interview, 1980, San Francisco, Ross collection
Le Poole, J. B. to D. Gabor, letter, 21 Jan. 1948, IC GABOR EL/1
Leen, Mary and Joan Whitlow to SFJ, discussions, 2–14 Jul. 2003, Cambridge, MA, SFJ collection
Lehmann, Matt, 'Holography—the early days', unpublished account to R. Jackson, MIT 29/790, 2 Jun. 1982
Leith, Emmett N. to A. M. Cruickshank, letter, 20 May 1970, CHF
Leith, Emmett N. to D. Gabor, letter, 26 Mar. 1965, Leith collection
Leith, Emmett N. to D. Gabor, letter, 6 Dec. 1966, London, IC Gabor ML/4
Leith, Emmett N. to D. Gabor, letter, 21 Feb. 1969, Ann Arbor, IC Gabor ML/4
Leith, Emmett N. to G. L. Rogers, letter, 6 Jul. 1965, Sci Mus ROGRS 4
Leith, Emmett N. to G. W. Stroke, letter, 20 Oct. 1964, Leith collection
Leith, Emmett N. to G. W. Stroke, letter, 30 Nov. 1964, Leith collection
Leith, Emmett N. to SFJ, email, 4 Mar. 2003, Ann Arbor, SFJ collection
Leith, Emmett N. to SFJ, email, 11 Mar. 2003, SFJ collection
Leith, Emmett N. to SFJ, email, 17 Mar. 2003, SFJ collection
Leith, Emmett N. to SFJ, email, 20 May 2003, SFJ collection
Leith, Emmett N. to SFJ, email, 5 Jun. 2003, SFJ collection
Leith, Emmett N. to SFJ, email, 6 Jun. 2003, SFJ collection
Leith, Emmett N. to SFJ, email, 20 Aug. 2003, SFJ collection
Leith, Emmett N. to SFJ, email, 8 Jan. 2004, SFJ collection
Leith, Emmett N. to SFJ, email, 20 May 2004, SFJ collection
Leith, Emmett N. to SFJ, interview, 22 Jan. 2003, Santa Clara, CA, SFJ collection
Leith, Emmett N. to SFJ, interviews, 29 Aug.–12 Sep. 2003, Ann Arbor, MI, SFJ collection
Leith, Emmett N. to SFJ, interview, 30 Aug. 2003, Ann Arbor, SFJ collection
Leith, Emmett N. to WRL administrators, memo, *c.* Summer 1965, Ann Arbor, Michigan, Leith collection
Leith, Emmett N., 'Proposal for Applications of the Wavefront Reconstruction Technique', Grant proposal ORA-63-1255-PB1, Institute of Science and Technology, University of Michigan to US Army Research Office, Durham, NC, 28 May 1963
Leith, Emmett N. and Juris Upatnieks, 'A communication theory of reconstructed wavefronts', memo to file, 10 Apr. 1961, Upatnieks collection
Leith, Emmett N. and Juris Upatnieks, 'Wavefront reconstruction disclosure', Patent disclosure report, University of Michigan Willow Run Laboratory, 22 Apr. 1964
Leith, Emmett N., 'Materials collected on 5/19/93 by S.A. Benton, at Ann Arbor', 19 May 1993, MIT 1993.061.004

Leith, Emmett N., Donald Gillespie and Brian Athey to SFJ, interview, 11 Sep. 2003, Ann Arbor, SFJ collection

Leonard, Carl to SFJ, email, 12 May 2003, SFJ collection

Leonard, Carl to SFJ, email, 15 May 2003, SFJ collection

Leonard, Carl to SFJ, interview, 4 Sep. 2003, Ann Arbor, MI, SFJ collection

Lohmann, Adolf W. to patent attorneys, patent declaration, Jun. 1987, Haines collection

Lohmann, Adolf W. to SFJ, fax, 13 May 2003, SFJ collection

Lohmann, Adolf W. to SFJ, fax, 20 May 2003, SFJ collection

Lohmann, Adolf W. to SFJ, fax, 18 Aug. 2003, SFJ collection

Lohmann, Adolf W. to SFJ, email, 4 Nov. 2003, SFJ collection

Lohmann, Adolf W. to SFJ, fax, 1 Dec. 2003, SFJ collection

Lohmann, Adolf W. to SFJ, fax, 19 Aug. 2003, SFJ collection

Lohmann, Adolf W. to SFJ, fax, 9 Oct. 2003, SFJ collection

Lohmann, Adolf W. to SFJ, fax, 19 Feb. 2004, SFJ collection

Lohmann, Adolf W. to SFJ, fax, 30 Jun. 2004, SFJ collection

Lohmann, Adolf W. to SFJ, fax, 9 Sep. 2004, SFJ collection

Lohmann, Adolf W. to SFJ, fax, 3 Apr. 2005, SFJ collection

Lohmann, Adolf W. to SFJ, fax, 11 Apr. 2005, SFJ collection

Lohmann, Adolf to SFJ, fax, 9 Aug. 2005, SFJ collection

LoVetri, Joan, IDS thesis, MIT 41/1238, 1982

Markov, Vladimir B. to SFJ, interview, 26 May 2004, Kiev, SFJ collection

McCann, B. and A. Denby to G. L. Rogers, Feb. 1967, Sci Mus ROGRS 4

MasterCard International, 'Artist files,' 1983, 26/802

McGrew, Steve to E. N. Leith, letter, 26 Mar. 1986

McGrew, Steve to J. Ross, interview, Dec. 1980, Los Angeles, Ross collection

McGrew, Steve to SFJ, email, 28 Oct. 2004, SFJ collection

McGrew, Steve, 'Steve McGrew comments', unpublished notes, 26 Oct. 2004, SFJ collection

McLaughlin, G. Harry to D. Gabor, letters, 22–24 Nov. 1966, London, IC GABOR MM/3

MIT Museum, 'DRAFT MIT Museum collecting guidelines for holography', MIT, Aug. 2001

MIT Museum, 'Strategic Plan, Nov. 2000–Jun. 2005', report, MIT

MoH, 'Strategic Plan', Museum of Holography, MIT 10/192, Jul. 1990

MoH, 'Exhibits: Holography '75,' MIT 33/1022, 1975

MoH, 'Large editions,' MIT 41/1253, 1985

MoH, 'MoH Board of Trustees,' 1976–1992, MIT 8/191

MoH, 'Workshops: Goldsmith's College,' MIT 42/1301, 1980

MoH, 'Board,' MIT 8/163, 1989–1992

MoH, 'Correspondence XVII–XXII,' MIT 2/39–44, 1988–1992

MoH, 'First International Congress for the Art of Holography—Program Proposal', Program Proposal, Museum of Holography, MIT 3/56, 1979

MoH, 'Holography: museum fur Holographie & Neue Visuelle Medien,' MIT 41/1264, 1981

MoH, 'Museums,' MIT 41/1263-1264, 1976–1992

Moore, Lon to J. Ross, interview, 1980, Los Angeles, Ross collection

Munday, Rob to SFJ, interview, 31 Mar. 2004, Richmond, UK, SFJ collection

Murray, Rod to SFJ, emails, 4 Jul. and 10 Sep. 2002, SFJ collection

Nebeker, Frederik, 'Eric Ash, Electrical Engineer, an oral history conducted in 1994 by Frederik Nebeker, IEEE History Center, Rutgers University, New Brunswick, NJ, USA.' www.ieee.org/organizations/history_center/oral_histories/transcripts/ash.html, accessed 10 Nov. 2004
Nicholson, Ana Maria to SFJ, interview, 21 Jan. 2003, Santa Clara, CA, SFJ collection
Owen, Harry to SFJ, interview, 9 Sep. 2003, Ann Arbor, MI, SFJ collection
Parsons, G. P. to S. F. D'Amato, letter, 21 May 1986, Haines collection
Peck, R. F. to G. P. Parsons, letter, 27 Jul. 1974, Cambridge, MA, Haines collection
Phillips, Nicholas J. to SFJ, interview, 18 Sep. 2002, Leicester, UK, SFJ collection
Pizzanelli, David to SFJ, email, 4 May 2005, SFJ collection
Powell, Robert L. to P. Jackson, letter, 14 Oct. 1976, MIT 30/850
Powell, Robert L. to S. Zussman, letter, 28 Mar. 1995, Leith collection
Provence, Steve to J. Ross, interview, 1980, Los Angeles, Ross collection
Rallison, Richard D. to SFJ, email, 22 Mar. 2004, SFJ collection
Reutersward, Carl Fredrik, 'Artist files,' MIT Museum 30/859-860
Rogers, G. L. to A. Boivin, letter, 1951, Sci Mus ROGRS 6
Rogers, G. L. to A. V. Baez, letter, 8 Mar. 1952, Sci Mus ROGRS 6
Rogers, G. L. to A. V. Baez, letter, 19 Jul. 1956, Sci Mus ROGRS 6
Rogers, G. L. to D. Gabor, letter, 12 Nov. 1949, Sci Mus ROGRS 6
Rogers, G. L. to D. Gabor, letter, 1 Nov. 1951, Sci Mus ROGRS 6
Rogers, G. L. to D. Gabor, letter, 8 Nov. 1954, Sci Mus ROGRS 6
Rogers, G. L. to D. Gabor, letter, 5 May 1964, Sci Mus MS 1014/13
Rogers, G. L. to E. N. Leith, letter, 15 Jul. 1965, Sci Mus ROGRS 4
Rogers, G. L. to D. Gabor, letter, 31 Oct. 1966, Sci Mus MS 1014/16
Rogers, G. L. to G. D. Preston, letter, 2 Sep. 1952, Sci Mus ROGRS 6
Rogers, G. L. to I. A. G. Le Bek, 15 Feb. 1966, Oliver and Boyd Ltd, Sci Mus ROGRS 4
Rogers, G. L. to K. E. Nicholds, 27 May 1966, Redditch, Worcs., Sci Mus ROGRS 4
Rogers, G. L. to M. E. Haine, letter, 25 Jun. 1951, Sci Mus ROGRS 6, Technical Correspondence
Rogers, G. L. to M. E. Haine, letter, 20 Oct. 1952, Sci Mus ROGRS 6
Rogers, G. L. to P. Kirkpatrick, letter, 13 Sep. 1950, Sci Mus ROGRS 6
Rogers, G. L. to S. L. Bragg, letter, 5 Sep. 1950, Sci Mus ROGRS 6
Rogers, G. L. to V. E. Cosslett, 23 Jun. 1966, Sci Mus ROGRS 4
Rogers, G. L., 'Comments by Dr. G. L. Rogers on Dr D. Gabor's Report B. T. H. No. L. 3696', report, Sci Mus ROGRS 6, 18 Mar. 1948
Rogers, G. L., lab notebook, 21 Jan. 1949–1950, Sci Mus ROGRS 2/10 and 2/11
Ross, Jonathan to SFJ, interview, 3 Apr. 2003, London, SFJ collection
Sapan, Jason to SFJ, email, 17 Mar. 2004, SFJ collection
Sapan, Jason to SFJ, email, 18 Mar. 2004, SFJ collection
Saxby, Graham to S. Johnston, interview, 16–17 Sep. 2002, SFJ collection
Schmidt, David to P. Jackson, 1978, MIT 40-1209: holography advertising
Schwider, Johannes to SFJ, email, 1 Jul. 2003, SFJ collection
Sherwin, Chalmers W. to J. Bryant, interview, 12 Jun. 1991, MIT Radiation lab, IEEE History Center
Siebert, L. D. to SFJ, interview, 4 Sep. 2003, Ann Arbor, MI, SFJ collection
Siegel, Keeve M., 'Speech to Conductron Missouri', audio recording, 14 Jun. 1966, C. Charnetski collection
Silberman, Rick, 'Artist files,' New York Museum of Holography Collection, MIT 30/889, 1980

Smith, Allen F. and Gordon Van Wylen to G. W. Stroke, Dec. 1966–Jan. 1967, Ann Arbor, Bentley, Library Institute of Science and Technology Box 21: Electro-Optical Science Laboratory

Smith, Howard M. to A. M. Cruickshank, by letter, 19 Jun. 1970, Rochester, NY, CHF

Snell, J. to R. Evaldson, letter, 4 Apr. 1966, Leith collection

Soskin, Marat, 'Optical holography in Chernivtsi University', unpublished manuscript, 20 Sep. 2004, SFJ collection

Stetson, Karl A. to M. A. Warga, 7 Mar. 1966, Ann Arbor, Bentley Historical Library, Institute of Science and Technology Box 21: Electro-Optical Lab

Stetson, Karl A. to SFJ, email, 30 Oct. 2004, SFJ collection

Stetson, Karl A. to SFJ, email, 1 Nov. 2004, SFJ collection

Stetson, Karl A. and J. M. Marks, 'Use of a Helium-Neon laser as a light source in processing side-looking radar data', report DDC No AD 336 356, Willow Run Laboratory, 1963

Stroke, George W. to C. W. Wixom, letter, 11 Apr. 1964, Leith collection

Stroke, George W. to C. G. Brown, letter, 16 Apr. 1964, Ann Arbor, Leith collection

Stroke, George W. to D. Gabor, letter, telegram and telephone conversation, 1 Nov. 1966, Ann Arbor, IC Gabor ML/4

Stroke, George W. to D. Gabor, letter, 23 Apr. 1967, IC GABOR MR7/5

Stroke, George W. to D. Gabor, telegram, Nov. 1971, Paris, IC Gabor MS/15

Stroke, George W. to D. L. MacAdam, letter, 12 Dec. 1964, Ann Arbor, Michigan, Leith collection

Stroke, George W. to H. W. Farris, memo, 24 Mar. 1966, Ann Arbor, Leith collection

Stroke, George W. to J. T. Wilson, memo, 12 Dec. 1963, Bentley Historical Library, Institute of Science and Technology Box 21: Electro-Optical Lab

Stroke, George W. to J. T. Wilson, memo, 30 Sep. 1963, Bentley Historical Library, Institute of Science and Technology Box 21: Electro-Optical Lab

Stroke, H. Henry to SFJ, email, 6 Dec. 2004, SFJ collection

Stroke, George W. to D. Gabor, telegram, Nov. 1971, Paris, IC Gabor MS/15

Stroke, George W. to J. T. Wilson, memo, 27 Aug. 1963, Ann Arbor, Bentley Historical Library, Institute of Science and Technology Box 21: Electro-Optical Lab

Stroke, George W., 'An Introduction to Optics of Coherent and Non-Coherent Electromagnetic Radiations', Course notes, Summer School, University of Michigan Ann Arbor, March 1965

Stroke, George W., 'Press Release: Breakthrough in "lensless photography" sets stage for 3-dimensional home color television and a possible multi-million industrial explosion in electro-optics', 12 Mar. 1966, Leith collection

Stroke, George W., 'To whom it may concern: lensless photography work at the University of Michigan', News release, Apr. 1965

Thompson, Brian J. to A. M. Cruickshank, by letter, 15 May 1970, Rochester, NY, CHF

Trolinger, James D., 'The history of the aerospace holography industry from my perspective', unpublished manuscript, May 2004, SFJ collection

Tsujiuchi, J. to SFJ, unpublished report, 20 Jun. 2003, SFJ collection

Tuchkevich, V. M. to E. N. Leith, letter, 12 Mar. 1979, Leningrad, Leith collection

Unterseher, Fred to SFJ, interview, 23 Jan. 2003, Santa Clara, CA, SFJ collection

Unterseher, Fred to SFJ, telephone interview, 11 Aug. 2005, SFJ collection

Upatnieks, Juris to SFJ, email, 15 May 2003, SFJ collection

Upatnieks, Juris to SFJ, email, 6 Jun. 2003, SFJ collection

Upatnieks, Juris to SFJ, email, 24 Jun. 2003, SFJ collection

Upatnieks, Juris to SFJ, email, 25 Jun. 2003, SFJ collection
Upatnieks, Juris to SFJ, email, 27 Jul. 2003, SFJ collection
Upatnieks, Juris to SFJ, email, 30 Sep. 2003, SFJ collection
Upatnieks, Juris to SFJ, email, 12 Jan. 2005, SFJ collection
Upatnieks, Juris to SFJ, email, 14 Jan. 2005, SFJ collection
Upatnieks, Juris to SFJ, email, 28 Jan. 2005, SFJ collection
Upatnieks, Juris to SFJ, email, 9 Jan. 2005, SFJ collection
Upatnieks, Juris, 'Artist files,' MIT 31/928, 1979
Upatnieks, Juris, 'Computation Notebook #768', lab notebook, beginning 16 Jan. 1963, Upatnieks collection
Upatnieks, Juris, 'Computation Notebook #768', lab notebook, 29 Jan. 1963, Upatnieks collection
Upatnieks, Juris, 'Computation Notebook #768', lab notebook, 1 Apr. 1963, Upatnieks collection
Upatnieks, Juris, 'Computation Notebook #768', lab notebook, 25 Apr. 1963, Upatnieks collection
Upatnieks, Juris, 'Computation Notebook #768', lab notebook, 20 Feb. 1964, Upatnieks collection
Upatnieks, Juris, 'Computation Notebook #768', lab notebook, 18 Jun. 1964, Upatnieks collection
Upatnieks, Juris, 'Development of holography from November 1962 to January 1965', Report for patent attorneys, Willow Run Laboratories, Feb. 1965
Upatnieks, Juris, 'Monthly progress report for February 1961', Institute of Science and Technology, University of Michigan, 27 Feb. 1961
Vest, Charles M. to SFJ, interview, 13 Jul. 2003, Cambridge, MA, SFJ collection
van Riper, Peter, MIT 31/930, 1976
Waller-Bridge, Michael to SFJ, internal reports and publicity documents, 9 Aug. 2005, SFJ collection
Williams, Ivor to D. Gabor, letter, 20 Dec. 1966, IC B/GABOR MA/13/1
Williams, Van Zandt to J. Shewell, letter, 23 Mar. 1966, Bentley Historical Library, Institute of Science and Technology Box 21: Electro-Optical Lab
Wilson, J. T. to A. G. Norman, *c*. Mar. 1964, Ann Arbor, Bentley Historical Library, Institute of Science and Technology Box 21: Electro-Optical Lab
Wilt, Richard, 'Holograms have a message', report, 1969, Charnetski collection
Wixom, Charles to L. H. Dulberger, letter, 11 Dec. 1963, Leith collection
Wood, Glenn P. to M. Tomko, by letter, 13 Jun. 1991, MIT 2/38
Worthington, C. Roy to SFJ, letter, 21 Sep. 2004, SFJ collection
Worthington, C. Roy to SFJ, email, 29 Sep. 2004, SFJ collection
Zellerbach, Gary to J. Ross, interview, 1980, Los Angeles, Ross collection
Zellerbach, Gary to SFJ, email, 2 Aug. 2003, SFJ collection
Zellerbach, Gary to SFJ, email, 9 Aug. 2003, SFJ collection

Appendix
Publication statistics

Over its first half century, holography has been influential at a wide range of scales. Numbers can put its impact into perspective. The subject has attracted thousands of researchers, designers, and other creators, and hundreds of discrete applications. Some 20,000 papers have been published, and over 10,000 conference presentations presented. Some 7000 patents have been granted, a thousand books published, nearly as many theses defended, and at least 500 exhibitions staged. These could be called 'public statistics'. But there are also private statistics. Perhaps market-tracking organizations like Reconnaissance International can suggest how much money has been generated or lost by holographic firms or provided by grant agencies, corporations, and other sponsors. This is one semi-public statistic that is still difficult to discern.

This scale is considerably greater than some other late twentieth-century technologies and nascent sciences. So large, in fact, that the scale itself allows certain trends to be tracked. One relevant approach is 'bibliometry'—the measuring of publications.

Databases of publications, maintained by academic libraries, institutions, and private indexing firms, suggest the rate of activity in the subject of holography. Figure A.1 plots the annual rate of publications of several kinds: books, dissertations, conference presentations, exhibitions, papers, and patents. No single source is complete, because papers on holography have been distributed among a wide variety of journals, magazines, publishers, organizers, and countries. And some forms of publication, such as art exhibitions, are poorly represented in such databases, being listed, if at all, in the form of exhibition catalogues. Even patents can be awkward to quantify, because they may be filed in more than one country.

This kind of analysis can be skewed inadvertently in other ways. A major contribution to inaccuracy is the fact that much research is unpublished. For example, the side-looking radar research that led to holography studies by Leith and Upatnieks was supported by the American military and is still confidential, in parts. The same is true of some of the expertise in Head Up Displays (HUDs) and Holographic Optical Elements (HOEs). Even more commonly, commercial research is often unpublished to ensure a business advantage.

Even with these potential drawbacks, the attempt to quantify publications in holography can still reveal useful information. It is apparent, for example, that the venue of publication shifted with time. Books (see Figure A.1(a)) appeared at a near-constant rate 1970–90, but declined thereafter. The completion of theses (see Figure A.1(b)) peaked in the early 1970s, during the first pulse of interest when research money was relatively plentiful, and again in the early 1990s, when a wider range of applications were identified and investigated. Perhaps surprisingly, the heyday of conference publications (see Figure A.1(c)) was between the mid-1970s and early 1980s, despite the proliferation of regular conferences under the auspices of the SPIE and IS&T from the 1980s. Holography exhibitions (see Figure A.1(d)), both artistic and for wider publics, were an important setting for information transfer during the 1980s, but have declined significantly since then. From the handful of yearly publications from 1947, papers in scientific journals (see Figure A.1(e)) exploded in 1965; indeed, the earlier publications by Gabor, El-Sum, and others are scarcely visible at this graphical scale. Papers peaked in 1971, about the time that some highly visible players such McDonnell Douglas were actively promoting holographic technology for advertising displays, pulsed-laser portraiture and non-destructive testing. McDonnell Douglas closed its production facility in 1973, reflecting a wider loss of confidence by American sponsors of holography. Scientific and engineering papers gradually rose to a higher maximum in the mid-1990s, but since that time they have fallen twofold. The granting of patents (see Figure A.1(f)) also evinced a peak in the early 1970s and then, from the early 1980s, has risen steadily.

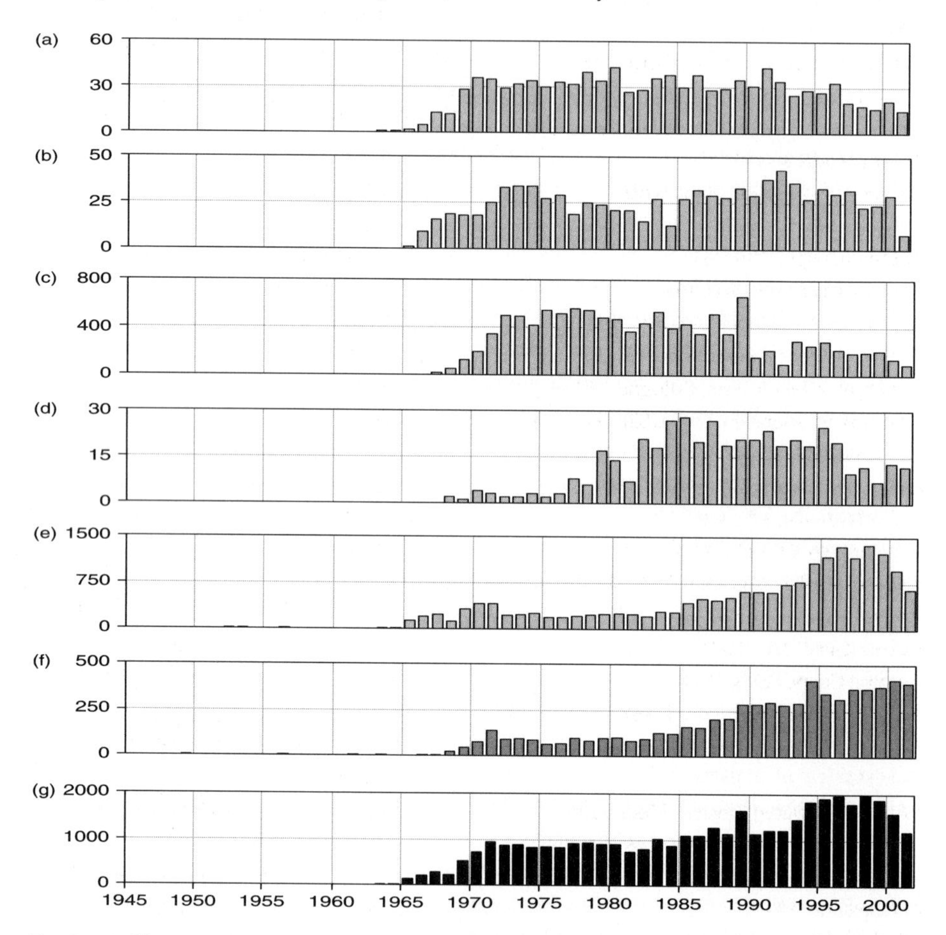

Fig. A.1 Publications per year, 1945–2001 inclusive. (a) Books; (b) Theses; (c) Conference presentations; (d) Exhibitions; (e) Papers; (f) Patents; (g) Total publications (c.40,000). Sources (not exhaustive): databases including ArticlesFirst, Arts & Humanities Citation Index, Dissertations Abstracts, INSPEC, MedLine, PsycINFO, Social Sciences Citation Index, and WorldCat; European and world patents; artists' CVs; and literature reviews.[1] These sources very likely under-represent publication activity, particularly during the early 1970s and for exhibitions. The data are also biased in that they omit publications that do not refer explicitly to holography or wavefront reconstruction in titles or abstracts.

[1] Kallard, Thomas, *Holography: State of the Art Review 1969* (New York: Optosonic Press, 1969); Kallard, Thomas, *Holography: State of the Art Review 1970* (New York: Optosonic Press, 1970); Kallard, Thomas, *Holography: State of the Art Review 1971–1972* (New York: Optosonic Press, 1972).

INDEX

2D-3D hologram 374, 386, 388–9
3D – *see* three-dimensional
A. F. Ioffe Physical-Technical Institute 74, 182–3, 327
Abbe, Ernst 22, 23, 46, 49, 50, 58, 62, 82, 85, 105
Abbott, Andrew D. 237
ABNH – *see* American Bank Note Holographics
Abraham, Nigel 339, 363
Abrams, Claudette 304, 309
Abramson, Nils Hugo 199, 278, 292, 442
Académie Française des Sciences 60
Academy of Media Arts, Cologne 305–6
Academy of Sciences of the USSR 60–3, 76–7, 181–4, 218–20
acoustical holography 45, 177, 188, 219, 243, 251–2, 374, 378, 380
acoustics 80, 95, 143, 157, 236, 251
actor-network theory (ANT) 8, 285, 423, 441
Adams, Ansel Easton 179
Adams, Gary 274–5, 278
Admiral Corporation 258
Advanced Holographics Ltd 361
Advanced Research Projects Agency (ARPA) 156
AEI: *see* Associated Electrical Industries
Aerodyne Research Inc 236, 246
aerospace industry 158–9, 174, 188, 239–41
Aerospatiale 199
aerosols 207
AGARD – *see* NATO Advisory Group for Aerospace Research and Development
aesthetic holographer – *see* holographers, aesthetic
aesthetics 278, 280, 287–8, 319, 324, 328, 330–2, 335, 360, 391, 396, 399, 424, 426, 428–9, 432, 440
Agfa emulsions – *see* emulsions, Agfa
Agfa Gavaert NV 166, 207, 221–3, 225, 227, 349, 369, 372, 391
A. H. Prismatics 361
Air Force, US 79, 81, 83, 151–4, 157–9, 238–41, 258
Alabama 240
Alcorn, Allan 381
Alda, Máximo 309
Aldermaston, UK 18, 22, 24, 33, 36–7, 52, 56, 142, 213
Aleksoff, Carl C. 97, 155, 256, 435–6
Alexander 309
Alice in the Light World exhibition 313, 317, back cover
Allibone, Thomas Edward 18, 19, 20, 24, 26, 34, 36, 53–4, 56, 58–9, 142, 396, 444
Altarum 83, 161
Amateur Photographer 375
Amateur Scientist 271
amateurs 226–7, 271, 283, 313, 326, 330, 369, 401
America 20, 37, 43–8, 62–3, 67, 70–1, 78, 80, 83, 101, 124–5, 131, 140, 143, 151, 154, 157–9, 168, 173, 175–6, 181, 184–6, 190, 199, 200–2, 209, 212, 232, 235–8, 241–4, 246, 249–50, 254–5, 257–8, 272, 274–5, 278, 309, 349, 351, 353, 355, 359–62, 364, 368, 376, 378, 380, 389, 393, 399, 400, 411–12, 422, 427. *See also* Ann Arbor; Arizona; California; Cambridge; Chicago; Los Angeles; New York; San Francisco
American Bank Note (ABN) Corporation 371, 374, 383–6, 391
American Bank Note Holographics (ABNH) Inc 154, 363, 374–7, 384, 386
American Institute of Electrical Engineers (AIEE) 232
American Institute of Physics (AIP) 101–2, 113, 118, 342, 435

American National Standards Institute (ANSI) 190
American Optical (AO) 131, 155, 298, 300
American Society of Photogrammetry (ASP) 237
Ames Research Center, NASA 159, 262
amplitude 23–5, 42, 67–9, 72, 86, 88, 96, 103, 107, 137, 141
Amsterdam, Netherlands 8, 305, 316, 358
anaglyph 294, 401
Anaït [Arutunoff Stephens] 275, 304
analogue electronics 84
analogy, holographic brain 277, 410–11
analogy, holographic universe 277, 317, 411–12
analogy, holograms as relics 434–5
analogy, photography 320–1, 403, 421–4, 432
Ann Arbor, MI 79, 96, 113, 116, 128, 140, 155–6, 160–3, 167–5, 178–9, 198, 201, 209, 243, 254–61, 266, 291, 303, 327, 350, 354, 356, 424, 435–6
Andrews, Matthew 309
Ant Farm 269, 272
anti-Semitism 74
Aoki, Y. 188, 242
Aphrodite hologram 302
Appleton, Sir Edward V. 33
Applied Optics 136, 236
Applied Physics Letters 236
Applied Holographics plc 222, 363–4, 368–71, 389
Applied Optical Technologies plc 369
Arago, Dominique-François-Jean 41
architecture applications 156
archives x, 26, 37, 111, 123, 139, 147–8, 329, 341, 412, 433, 435–7
Arecchi, F. T. 134
Argonne National Laboratory 116
Arizona 209, 213, 235, 272
Army Research Office (ARO) 128, 153, 157
Army Signal Corps 83
Arnold Engineering Development Center (AEDC) 159, 239
ARO – *see* Army Research Office
ARPA – *see* Advanced Research Projects Agency
art, abstract 294, 311–12, 319
Art and Technology movement 289, 306, 333
art critics 288, 312–13, 317–22, 336, 339, 365, 399
Artforum International 375
art holography 261, 285, 287, 290, 295, 297, 303, 306, 308, 310–13, 320–4, 328, 343–4, 360, 365–6, 377, 424, 427–8
Art Holography: The Real-Virtual 3D Images 310
art, Impressionist 294, 323
Art Institute of Chicago 306–7
Art Institute of San Francisco 272
art, mainstream 317, 366
art, mimetic 323
art, mixed media 317
art, Modern 289
art, op 294, 322
art, pop 294, 319
art, video 269–70, 289, 317, 340
artefacts, optical 100, 426
artefacts, technological 118, 149, 285, 393, 406, 434–6, 440
Arte Povera 322
artisans 6, 253, 259, 268–9, 271, 280, 283–4, 325, 332, 345, 393, 413, 424–5, 428, 431–2, 440
artisanal holographers 229, 259, 277, 280–1, 283–6
artists 6, 229, 246, 253, 259–1, 268–9, 270, 272, 276–8, 280, 285, 287–324, 325–6, 329–30, 332–4, 337, 344, 345, 423–9, 431–3, 440, 443
Artist in Residence (AIR) Programme 303, 338, 345
Artner, Alan G. 322
Artsimovich, Lev Andreevich 76
ARTTRANSITION 302
Asakura, T. 187
Ash, Eric 54
Asimov, Isaac 398
assimilative repetition 181
Associated Electrical Industries Inc (AEI) 18–20, 22, 24, 33–4, 37, 41, 48, 51–4, 56–8, 60, 100, 140
Athey, Brian 119
l'Association pour le Développement des Arts et des Techniques Holographiques (ADATH) 344

asymmetry, historical 3
Aston University 37, 56
Atari Inc 359, 377, 381–4
l'Atelier Holographique 279
atmospheric optics 394
Atomic Weapons Research Establishment (AWRE), UK 213–14
Aurorean studio 399
Australia 187, 275, 288, 293, 296, 306, 307, 309, 317, 360, 365, 376
Australian National University 296–7
Austria 243, 373
axicon 95

Bâ, Amadou Hampâté viii
Baez, Albert V. 40, 44–5, 47–8, 51, 57–8, 60, 71, 100, 141, 148, 187, 205, 250, 257, 332, 435
Baez, Joan 257
Bailey, Jean 306
Bains, Sunny 328
Baldwin, James T. (Jay) 269
bandwidth 172, 190, 210, 211
bankruptcy 359–61, 381, 383–4, 391
Barnett, Norman E. 156–7, 174
Barr, Pascal 394
Bartell, Lawrence S. xi, 133
Basingstoke, UK 386
Battelle Development Corporation 145–6, 152–3, 155, 178, 201, 255, 349, 377, 379–80, 438
Battelle Memorial Institute 152, 255, 379
Baudrillard, Jean 412
Baumber, Kevin 306
Bausch & Lomb 134, 235
Bay Area 257–9, 269, 270, 272–3, 281, 284–6, 289
Bay Area Sidewalk Astronomers 271
Bazargan, Kaveh 364
beam, laser 251–2, 261–2, 264–5, 270, 292, 295, 299, 318, 321, 354, 362, 373, 404
beam, object 67, 70, 73, 97, 104, 111, 147
beam, reference 67–8, 70, 73, 76, 96–8, 104, 147, 164, 200–1, 204–6, 210, 213, 216
beamsplitter 170, 172
Beame, Abraham 337
Becker, Howard 400
Beckman Instruments 258
Begleiter, Eric 391–2
Belgium 21, 134, 221–3, 240, 243
Békésy, Georg von 18
Bell Systems Journal 236
Bell Telephone Laboratories 91, 106, 133, 143, 197, 202, 212, 225, 236, 292, 349
Bendix Corporation 143, 160, 211, 256
Bennett, Margaret Pater 261
Benthall, Jonathan 287–9, 317
Bentley Historical Library, Ann Arbor MI xi, 435
Benton, Jeanne 179
Benton, Stephen A. 178–80, 209–12, 214–15, 250, 269, 278, 299–302, 306, 311, 317, 321, 333–4, 339, 342–3, 345, 372, 383, 413, 427, 430, 437, 442, 444
Benyon, Margaret xi, 8, 206, 288, 293–7, 301, 303, 306–10, 312–13, 318, 323, 365
Berkeley University 254, 257, 270
Berkhout, Rudie 275, 304–5, 309–10, 317, 343
Berlin, Germany 17–18, 21, 222, 243, 263, 304, 313
Berne, Switzerland 292
Besançon, France 183
Bessel functions 193
bibliometry 6, 425, 489
big science 61, 254
Billings, Loren 278
Bird in a Box hologram 296
Birmingham, College of Advanced Technology 37, 175
Bjelkhagen, Hans I. xi, 199, 209, 247, 278, 292, 304, 333, 343, 442, 448
Blackmore, Susan 403
bleach 72, 163, 175, 219, 224, 240, 358, 391
Blikken, Wendell 106, 154
Block Engineering Inc 242, 244
blockhouse, WRL 92–3, 156
Bloor, David 11
Blyth, Jeff xi, 307, 362–3, 387
Bohm, David Joseph 277, 317, 409, 411–12
Boissonet, Phillippe 309, 317
Boivin, Albéric 43

bombing, U-M 255
Bond, David 332
books 490
Booth, Bruce L. 226
Born, Max 32, 42–3, 52, 97, 134, 136
Boston 1, 116, 118–19, 121, 130, 156, 238, 242, 257–8, 299–300, 302, 359, 362, 387
Boston University 238, 242
Boston University Optical Research Laboratory (BUORL) 238
Boulton, Barrie 358
Bourdieu, Pierre 9
Boyd, Joseph Aubrey 173
Boyd, Patrick 306
Brady, Hugh 272, 275
Bragg, Sir (William) Lawrence 1, 15, 23, 26–7, 29, 31–5, 37–44, 55, 58–9, 62, 202
Brand, Stewart 269
Braunschweig, Germany 50, 232
Brazier, Pamela Eve 214, 274–5, 396
breakthroughs 417
BRH – *see* Bureau of Radiological Health
bribery, claims of 146
Bristol, UK 294
Britain 18, 20–1, 34, 50, 57, 59, 62–3, 176, 185–6, 196, 207, 221, 223, 232, 243, 259, 293–7, 307, 309–10, 319. 339, 341, 354, 361–2, 365, 367, 407, 411, 416, 442. *See also* Aldermaston; Basingstoke; Birmingham; Covent Garden; Edinburgh; Glasgow; London; Loughborough; Rugby; Teddington
British Aircraft Corporation 294, 349
British Association for the Advancement of Science (BAAS) 15, 33–4
British Thomson-Houston Co (BTH) 15, 18–20, 22, 24, 32–5, 37, 47, 56, 78
Broadbent, Donald xi, 226, 261, 276
Brook, Donald 288
Brookhaven National Laboratory, NY 209, 246
Brooks, Harvey 233–4
Brooks, Robert 196
Browers, Bill 233
Brown, Boveri & Co. 343
Brown, Byron R. 217
Brown, Gordon M. 174, 198
Brown, John 359, 362
Brown, William M. 87, 135, 152
Brown University 301
Brumm, Douglas B. 126, 129–30, 155, 179
Bryskin, V. 312
BTH – *see* British Thomson-Houston
Budapest, Hungary 17–18, 139
Buddha 205
Bulgaria 441
bulletin board, internet 320, 414
Bunker-Hunt, Nelson 367
Burch, C. R. 33, 37
Burch, Jim 196, 252, 295
Bureau des Poids et Mesures, France 62
Bureau of Radiological Health (BRH), USA 190, 318
Burgmer, Brigitte 309, 317
Burns, Joseph R. (Jody) 275, 278, 304, 310, 313, 320
Burridge, Mike 307
bus, hologram 313–14, 344
Business Week 381
Butters, John N. 176, 196, 295
Buzzard, Bob 163, 166

C Project 360
California 8, 37, 43–5, 51, 58–9, 187, 196, 205, 207, 213, 238, 241, 250, 254, 257, 266, 272, 273, 275, 278, 282, 289, 291, 305, 309, 329, 332, 352–3, 358, 366, 390–1, 404, 420, 426
California College of Arts and Crafts 272, 278
Cambridge University 1, 26–7, 32, 48, 76, 176
Cambridge, MA 178, 200, 203, 238, 257, 299–302, 357, 372, 412, 436
Cambridge Stereographic Society 299
Canada 21, 138, 186, 243, 245, 259, 260, 309, 406. *See also* Toronto; Quebec; Vancouver
Canada Council 259, 309, 432
Canadian Holographic Developments Ltd 380
Candidate of Science 63–4, 76, 99, 219
Canon Inc 242, 390
Canterbury Tales 435
Capa, Cornell 320

carrier wave 97, 213
Cartier necklace hologram 343, 352–3
Casdin-Silver, Harriet 293, 298–303, 309–10, 317, 320–1, 343, 396
Caulfield, H. John 244, 246, 249, 413, 442
CAVS – *see* Center for Advanced Visual Studies
CBS Research Laboratories, CT 138–9, 176–8, 236, 251, 290, 378, 424
Celestial Holograms 278–9
Cenco Inc 263
Center for Advanced Visual Studies (CAVS) 300, 302, 307, 437
Center for the Holographic Arts, New York 304, 345
Centerbeam installation 302
Central Electricity Generating Board (CEGB), UK 56, 297
Centre National de Recherche Scientifique (CNRS) 187
Cerenkov, Pavel A. 62
Cerenkov radiation 62
CGH – *see* Computer-Generated Hologram
Challinor, David 209
Chang, Milton 225, 263
Charnetski, Clark 160, 163, 166, 261, 350, 352, 448
Chernivtsi University, Ukraine 183
Cherry Optical Holography 279
Chicago, IL 168, 211, 263, 278, 306–7, 322, 396, 410, 416
Chicago, Art Institute of 306–7
Chicago, Fine Arts Research and Holographic Center 263, 278, 340, 355, 366
Chicago Holographic Cooperative 329
Chicago Tribune 322
China 44, 331, 390–1
China Lake, CA 241
chirped signals 93–4
Christakis, Anne-Marie 344
Christian Science Monitor 319
Chromagem Ltd 366
Cifelli, Dan 361
Cinema and Photographic Research Institute (NIKFI), Moscow 184, 201, 221–2, 248, 313, 327, 389
cinematography 20–1, 31, 35, 114, 274, 401, 403, 407
classical optics 46, 83, 86–7, 192–3, 233, 235
classified research 78–80, 83, 86, 90–2, 99, 107, 109, 113, 176, 186, 235, 238, 255–8, 266, 284, 291, 423, 435–6
Claudius, Peter W. 274–5, 396–8
Close, Don 225
CNRS – *see* Centre National de Recherche Scientifique
Cobweb Space hologram 300–301
Cochran, Gary D. 160, 162–6, 171–3, 205, 239, 264, 267, 404, 448
coherence, optical 1, 6, 10, 14, 24–5, 48, 51, 54, 58, 59, 73, 97, 100, 103–4, 106, 108, 114, 140, 194, 206–7, 234, 239, 410, 421
coherence, social 6, 10, 14, 242, 280
coherence length 25, 54, 58, 73, 100, 104, 107, 111, 141, 192, 207, 290, 298, 350
coherent glint – *see* speckle
coherent optical processing 86, 90, 92, 128, 146, 148, 241, 244, 264
coherent optics – *see* modern optics
Coherent Radiation Laboratories 178
cold fusion 408
cold war 78–9, 123, 126, 202, 241, 310, 327, 354, 431
collimated light 39, 70–3, 98, 147, 216, 385
Collins, Bob 170
Collins, Harry 11
Cologne, Germany 305–6, 340, 343–4, 355
Colorado 357
colorimetry 9
commercialization 16, 20, 34, 55–6, 120, 135, 146–7, 191, 194, 198, 202, 205, 207, 210, 212–13, 218, 220–1, 224–8, 236, 239, 242, 261–3, 266, 270, 273–8, 284, 349–92, 419–20, 422–4, 426–7, 429–2
Commission Internationale de l'Ėclairage (CIE) 234
communication theory 35, 50, 52, 85–7, 92, 97, 99, 107, 115, 119–20, 122, 235, 239, 264
Communications Research Centre (CRC), Ottawa 259

Commonwealth Scientific and Industrial Research Organisation (CSIRO), Australia 187
communities, segregation, of 285, 425–9
communities, technical 4, 229, 231, 257, 270, 280, 283–6, 320, 415, 418–19, 432, 440, 445
computers, digital 80, 84–5, 95, 217, 219–20
Computer-Aided Holometry Laboratory 174
computer-generated hologram, (CGH) 216–20, 388–9
Comte, Auguste 7, 415, 417
Conant, James B. 238
Conductron Corporation, MI 121, 146–7, 153, 159–69, 171–4, 185, 239, 261, 264, 350–2, 358, 370, 392, 396, 403, 422–3, 426, 436, 444
conferences 3–4, 122, 131, 134, 147–8, 242–50, 254, 280, 289, 291–3, 303, 325–7, 329–31, 333, 345, 365, 390, 411, 413, 425, 428, 430, 435, 490
conjugate image 40–2, 46, 59, 73–4, 96–7, 99, 187, 378
Congress, American 256–7
Conlon, Emerson W. 79
Connors, Betsy 309, 437
constructivism 288, 441
contextual channelling 253, 420
contextual screening 41, 59, 73–4, 89, 105, 196, 209, 211, 420, 421
contract, research and development 123–5, 128, 134, 152, 157, 164, 175, 236, 254, 368, 390
contract, exploratory 157–8
contract, industrial 18, 34, 53, 91, 153, 160, 162, 164, 366, 384
contract, military 18, 42, 58, 79–83, 83, 91–2, 123–6, 128, 134, 152–4, 156–8, 175–6, 236, 240, 246, 255–6, 353–4, 391, 422, 436
Convair 91
Cooper, Alice 214, 276, 289
Copp, James 369
CORONA surveillance satellite 5, 238–9, 258
Cossette, Marie-Andrée 306, 309
Cosslett, Vernon Ellis 27, 48, 176
counterculture 255, 257–9, 269, 271, 277, 280, 327–8, 409, 424
counterfeiting – *see* security applications
Covent Garden, UK 362
Cowan, Jack 410
Cowles, Susan 306, 309
Crafton, Joseph 174
Cranbrook School of Fine Art 261–2, 266, 273, 303, 313
Cravens, Bill 381
Creative Holography Index 309, 328
credit cards 356, 372, 374–5, 383, 386–7, 428, 438
Crenshaw, Melissa 280, 309, 366, 440
creoles 84
Crewe, Albert C. 133
Crimea, Ukraine 313
Cross, Lee 170–1
Cross, Lloyd G. 161, 170–2, 209, 213–16, 246, 259–62, 264–9, 289, 291, 303–4, 316, 318, 355, 359, 396
cross-correlation 84, 88
cryptography 121, 158, 183
Crystal Beginnings hologram 334
CSIRO – *see* Commonwealth Scientific and Industrial Research Organisation, Australia
Cullen, Gary xi, 380
curators 6, 110, 319, 322, 435, 437
Cutrona, Louis J. 83, 85, 87, 92, 138, 146

Daguerre, Louis 65
daguerreotype 65, 206, 399, 403, 420
Dai Nippon Printing Company 375, 386, 388, 390, 428
Dali, Salvador 167, 214, 278, 289, 335
Dalton, Michael 364
Daniels, Larry 368
DARPA – *see* Defense Research Projects Agency
Darwin, Sir Charles Galton 32, 36
Data Optics 170
Dausman, Gunther 376
Davies, L. J. 24, 36
Davis, Frank 220
Dawkins, Richard 402, 405
Dawson, Paula 8, 307, 309
DCG – *see* dichromated gelatin

d'Cruz, Wasy 386
De Bitetto, Dominick J. 209–10, 383
deblurring, holographic 133
De Broglie, Louis 32
DeFreitas, Frank 328–9
De La Rue Holographics Ltd 386
Defense Advanced Research Projects Agency (DARPA) 180, 246
Deem, Rebecca 309, 365
Denisyuk, Yuri Nicholaevitch xi, 3, 60–77, 78, 83, 95, 99, 118, 144, 200–2, 216, 221, 223–4, 231, 239, 310–13, 327, 379, 384–5, 387, 394, 403, 405, 420
Denius, Homer 173
Denmark 243, 309, 344
Dennison, David 132
Denton, Frank 125–6, 129, 135, 263
Department of Defense (DoD), USA 79, 151, 220, 236, 284, 353
Department of Scientific and Industrial Research (DSIR), UK 34, 48
De Solla Price, Derek J. 325
detour phase 217
Detroit, MI 83, 93, 256, 261–2, 303, 313, 316, 351
Deutsch, Oscar 20–1
Deutsche Gesellschaft für Holographie 328, 344
Deutsches Museum, Munich 8, 437
developer, photographic 182, 222, 224
DiBiase, Vincent 275–6
dialectical materialism 181
Diamond, Mark 304, 318
dichromated gelatin (DCG) 225–6, 332, 356–9, 362–3, 372–3, 380, 385, 390, 399
dichromates – *see* dichromated gelatin
Dietrich, Edward 307
diffraction 131, 232, 239, 264, 389
diffraction diagram 27, 32–3
diffraction grating 49, 73–4, 86, 96–7, 104, 124–5, 135, 169, 187, 210, 220, 263, 359, 374, 387, 389
diffraction microscopy 15, 24–5, 34–8, 42, 44, 46, 48, 51, 53, 56–9, 135, 140, 148
digital processing 90, 219
Di-Ho system 388
Dikrotek 357
dildos 301
Dimensional Foods Co. 391–2
Dinsmore, Sydney 342, 366
Diploma of Appreciation 184
Diploma of Discovery 184
Dirac, Paul A. M. 32
discipline 84, 87, 92, 119, 124, 229, 231–3, 235, 242, 244–5, 254, 271, 286, 291, 293, 438, 440–1, 416, 431
Discoverer satellite programme 238
Disney World 401
display holography 184–6, 163–4, 172, 191, 199, 223, 280, 287, 299, 306–7, 322, 332–4, 350, 353–5, 371, 384, 391, 428–32, 442
disputes, patent 121, 145–7, 378–86, 433
disputes, priority 121–3, 125, 131, 133, 143, 145, 147–8, 438
dissertations 8–9, 45, 64, 71–2, 125, 290, 323, 489–90
Dniepropetrovsk, Ukraine 61
Dobson, John 271
DoD – *see* Department of Defense
DOE – *see* diffractive optical element
Domebook 270
dot-matrix hologram 220, 388–9
DOVID (Diffractive Optically Variable Image Device) 389
Doppler effect 49, 81–2, 84, 88, 93, 119
Dortmund, Germany 292
Dow, William Gould 79, 123, 125
Dowbenko, George 274, 279, 281
DSIR – *see* Department of Scientific and Industrial Research 34, 48
Dubosq-Soleil, M. 399
Dundee, Scotland 33, 37, 41, 43–4, 51
Dungeons and Dragons game 370
Du Pont de Nemours Inc 226, 372–3, 377, 380
Dutch Holographic Laboratory B.V. 388
Dwyer, Craig 163
Dwyer, Wayne 203
Dyens, Georges 304, 309, 310, 317
Dyson, Jim 24, 37, 41, 48, 51, 53, 56–7
DZ Company 361

Eastman Kodak 110–11, 166, 202, 207, 221–2, 225–6, 235, 238–9, 246, 258, 264, 370
Eco, Umberto 412
École Supérieure d'Optique 124
ecology of the professions 126, 237, 249, 317
Edgerton, 'Doc' 179
Edinburgh, UK 4, 8, 42, 134
Editions Inc 260–1, 266, 273, 302–3, 313
Edmund Scientific Ltd 260, 274
Eidetic Images Inc 359, 374, 377
Eindhoven, Netherlands 292
Einstein, Albert 23, 411
Eldon Inc 161, 170, 442
Electric Umbrella 357
electrical engineers 17, 83–9, 119, 233, 236, 251, 284
electrical engineering 15, 24, 36, 55, 79, 83–4, 86–7, 89, 92, 96–7, 122–3, 125, 130, 135, 231–5, 239, 253, 256, 263, 288
Electro-Optical Sciences Laboratory – *see* Univ. of Michigan Electro-Optical Sciences Laboratory
Electro-Optics Center, Radiation Inc 174
electron microscope 15, 21–4, 30–1, 33–4, 37, 44, 52–4, 57
Electronics 112
Electronics Research Center (ERC), MA 203
Elias, Peter 87, 233–4
Elkin, Alexander E. 64
El-Sum, Hussein M. A. 40, 44–8, 51, 57–8, 71, 87, 89, 90, 95, 100, 105, 133, 141–2, 148, 257, 264, 417, 435, 490
embossing 358–9, 361–5, 370–8, 380–1, 383, 386–8, 390–1, 427–8, 431–2, 438
Emergency Medical Hologram 408
empiricism, logical – *see* positivism
emulsion, photosensitive 98, 103, 108, 115–16, 203–4, 207, 220–7, 308, 311
emulsions, Agfa 166, 207, 223, 226
emulsion, BB 363
emulsions, Kodak 111, 166, 176, 186–7, 202, 221–2, 225
649F emulsion (Kodak) 111, 166, 176, 186–7, 202, 221, 225
emulsion, Denisyuk 62, 67–70, 73, 221
emulsion, dichromated gelatine – *see* dichromated gelatine
emulsions, Ilford 222, 368–70
emulsion, LOI-2 (Vavilov) 221
emulsion, PE-2 (NIKFI) 184, 221
emulsions, Slavich 182, 224
Encyclopaedia Britannica 144, 332
Engine No. 9 hologram 302
engineer, electrical – *see* electrical engineer
engineer, electro-optical 123–4, 234–5
engineer, optical 92, 116, 235–9, 241, 250–1, 258, 260, 264, 277, 280–1, 284, 287–8, 291
engineer, opto-electronic 235
engineer, radio 232
engineers versus scientists 4, 233, 243, 325, 422, 426–8
Ennos, Tony 196, 295
entrepreneurs 275, 285, 315, 405, 408, 417–18, 431, 435, 443
Environmental Research Institute of Michigan (ERIM) 79, 82, 90, 96, 246, 248, 256–7, 435–6
EOSL – *see* Univ. of Michigan Electro-Optical Sciences Laboratory
epistemology 11, 287
Equivocal Forks hologram 302
Erickson, Ronald 325
ERIM – *see* Environmental Research Institute of Michigan
erotica 11, 396–9
Erskine College 332
ET holograms 374
ethics 136–7, 145, 287–8
Europe 133–4, 245–6, 292, 295, 322, 330, 337, 340, 343–4, 359, 361, 363–4, 389, 400, 427, 438
Europlex Holographics B. V. 358, 373
Evaldson, Rune L. 255
exhibitions 6, 266, 272, 275, 302, 306, 312–16, 325 347, 359–60, 361–2, 367, 391, 425, 431, 434–7, 444, 490
Exploratorium 272–3
Experiments in Art and Technology (E.A.T.) 288
Expo '70, Osaka, Japan 295

Fach, Heinrich 50
facsimile transmission 210
Fagan, William 295, 442
failure, attributing 3, 430–3
Fairchild Camera Inc 238
Fairstein, John, 267, 273, 275, 278–9
Falconer, David G. 121, 126, 130–1
Farris, H. W. 135, 138
Fedchuck, I. Yu 184
fellatio 396
Feynman diagrams 119
Field Enterprises 166
feminism 299, 300–1, 396
Festival of Erotic Art 398
filtering, spatial 50, 85, 87, 107, 127, 129, 133, 137, 143, 153, 186–7, 235, 264, 385
Finch Gallery 262, 266, 289, 303, 313
Fine Arts and Research and Holographic Center of Chicago 278
fine-art hologram 365–6, 377, 424, 428
fine-art holographers 290, 297, 307, 309–10, 320, 323–4, 424, 432
Finlay, Jim 315
Fire Diamond Inc 357–8, 380
First World War 62, 79, 400
First Cylindric Chromo-Hologram of Alice Cooper's Brain 289
Fischer, Michael 274–5
fixing of hologram image 224
Fizeau, Armand Hippolyte-Louis 41
FlashPrint, Ilford 370
Fleck, Ludwig 59
Fleischmann, Rudolf 49
fly's eye lens 66, 68, 259, 274
Fondiller, Harry 381
Ford Motor Company 79, 155, 174, 199
former Soviet Union (FSU) 224, 241, 389
Fornari, Arthur 304
Foster, Michael 373
Foucault, Jean Bernard Leon 41
Foucault, Michel 9
Fourier transform 41, 50, 69, 85, 87, 217, 234
Fourier transform hologram 131–2, 143, 158, 179, 253
Fourier transform spectroscopy 234, 242
Fournier, Jean-Marc 164
found-art aesthetic 289
'Fragments of a hologram rose' 406, 413
Framingham, MA 298
France 60, 62, 124, 134–5, 164, 183, 186, 199, 240, 243, 292, 309, 311, 329, 344–5, 362
Françon, Maurice 303
Franken, Peter 171
Frankfurt School 6
Free Speech Movement 254
Fries, Urs 310, 330
French-German Research Institute of Saint-Louis 240
frequency analysis 46
frequency compression 20
frequency doubling 182
Fresnel, Augustin-Jean 38, 96
Friedman, D. Rufus 274–6
Friesem, A. A. 126, 153, 155, 174, 211
fringes, interference 23–5, 29, 48, 67–8, 73, 97, 103–4, 111, 167, 192–9, 220, 224, 252, 263, 379
fringes, live 195
fringes, frozen 195
Fringe Research 309, 321, 366
Frumkin, Yuri 239
Fuji Photo Optical Company 390
Fujitsu 390
Fuller, Buckminster 269, 406
funding 78–9, 90, 125, 127, 151–2, 157, 170, 174, 188, 235–6, 238–9, 241–2, 246, 255, 257, 259, 267, 271, 296–7, 307, 321
Funkhouser, Arthur T. xi, 125–6, 129–30, 155, 169, 174, 201, 403
Furst, Anton 361

Gabor, André 396
Gabor, Dennis 1, 15–59, 60, 62, 64–5, 67–8, 70–4, 77, 78, 83, 100, 120–2, 124, 131, 136–48, 153, 158, 176–8, 180, 187, 203–4, 216–17, 221, 231–3, 242, 251, 257, 259, 264, 287, 289, 290, 309, 321, 339, 341, 349, 378, 384–5, 398–9, 404, 410–2, 417, 420–2, 424, 429, 432–3, 435, 444–5, 490
Gabor, Marjorie 339, 444–5

Gaboroscopy 16, 48, 57, 148
Gaertner Scientific 263, 332
Galison, Peter 6, 84, 88, 325, 330, 333
Gallagher, Terry 383
Gallery 1134 322
Gallery 286 343
Gamble, Susan 8, 307, 309
Gates, John W. C. 207
Gauchet, Pascal 309
GC Optronics Inc 157, 161, 174–5, 198, 263, 356, 424, 436
gender 296, 299, 310, 323
General Dynamics Inc 91, 161, 436
General Electric Company (GEC) 19
General Electric Inc 258
General Motors Inc 214, 351
General Tire Inc 353
GCO – *see* GC Optronics
Gentet, Yves 356
George, Nicholas 216, 235, 263
Georgia, Soviet 183
German Democratic Republic (GDR) 185–6
Germany 17–18, 20–1, 42–3, 48–51, 61–2, 70, 139, 165, 166, 199, 243, 272, 292, 306, 309, 323, 362, 391, 396, 432. *See also* Berlin; Braunschweig; Cologne; Göttingen; Hamburg; Munich; Pulheim
Gerritson, Henrik 301
Gestalt shift 417
Ghost Busters cereal package 370
ghost-image holography 183
Gibson, William 406, 413
Gillespie, Donald 135, 161, 163, 169–70, 442
Gillespie, John 170
Gilliam, Gene 261
Glasgow, UK 198, 243
Glass Balls hologram 302
Goddard Space Flight Center (GSFC), MD 159, 262
GOI – *see* Vavilov State Optical Institute
Goldberg, Bruce 209
Goldberg, Emanuel 134, 368
Goldberg, Jerry 358
Goldberg, Larry 358
Goldenberg, Carl 261
Goldmark, Peter Carl 139, 176–8, 251, 290
Goebbels, Pim 279
Goldsmith's College 209, 279–80, 306–7, 309, 339
Goodman, Joseph W. 158, 245, 258, 356
Goodyear Aircraft Corporation 96, 264
Gordon Research Conferences (GRC), USA xi, 244–9, 327, 432
Gorglione, Nancy 268, 275–6, 278–9, 283, 289, 304, 309, 323
Görlich, Paul 185
Goro, Fritz 263, 290–1, 439
Goss, W. P. 41, 51, 54–5, 58–9
Göttingen, Germany 50
Government Mechanical Laboratory, USSR 187, 242
Graham, Loren 61
granite optical table 157, 164, 262–5, 284, 321
Grant, Ralph M. 156–7, 174
grant, research 153, 256, 262
Great Exhibition 295, 399, 435
Green, Harvey 400
Groningen, Netherlands 313
Groupe de Recherche et d'Expérimentation en Photonique Appliquée (GREPA) 329
Grumman Aerospace Corporation 236, 246
Guide to Practical Holography 281
Guinness Inc 315
Gulbenkian Foundation 339, 341, 365
Gurevich, Simon B. 210

Haber, Alan 254
Haight-Ashbury Arts Workshop (HAAW) 329
Haine, Michael E. 22, 24, 34, 37, 41, 45, 47, 51–5, 57, 59, 140
Haines, Kenneth A. 87, 148, 153–5, 193–4, 197, 277, 359, 373–5, 377, 383, 388, 438, 448
Hallmark greeting cards 385
Hamburg, Germany 272
Hand and Jewels hologram – *see* Cartier necklace hologram
Hardy, Arthur C. 233–4
Harger, Robert 87
Hariharan, Parameswaran 186, 309
Harman, Mary 309

Harper, J. 225
Hammarskjold, Hans 292
Harris, Andy 339, 382
Harris Electro-Optics Center 174
Harris Inc Electronic Division 163, 236, 246
Harrison, George R. 124
Hart, Stephen 364
Hartman, Nile 153, 201
Harvard College Observatory 238
Harvard Optical Laboratory 238
Harvard University 179, 209, 233–4, 299
Hasselblad 239
Hawaii 209
Hayden, Tom 254
head-up display (HUD) 353, 371, 385, 391, 489
Health and Safety Executive 190
Hecht, Gabriel 6
Hector, Roger 381
Hedgecoe, John 307
Heflinger, Lee 196
Hefner, Hugh 168
Hildebrand, H. Percy 126, 153, 155, 193–5, 197
Hentschel, Klaus 9
Herman, Thomas 170
Hermitage Museum, Moscow 184
Hewlett-Packard 258
Hioki, Ryuichi 187, 199
historians 3–6, 8, 61, 74, 83, 110, 123, 177, 238, 257–8, 285, 287–8, 290, 323, 325, 393, 395, 400, 403, 415–19, 433–4, 437
historiography 2, 5, 122, 136, 140, 144–6, 148, 325, 384, 422, 430, 436
History and Philosophy of Science (HPS) 8
Hlynsky, David 309, 321
HNDT – *see* holographic non-destructive testing
hobby 270–1, 283, 285, 313, 315, 401
HOE – *see* holographic optical element
Hoefler, Donald C. 257
Hoffmann La Roche 351
Hokkaido University 188
Holex Corporation 357–9, 373–4, 391
holism 11, 269, 277, 285, 412, 414, 435
Hollusions Ltd 363
Holoblo 396–7
Holoco Ltd 315, 338, 343–4, 361–2
Holocom 272, 279, 426
Holoconcepts Corporation of America 343
HoloCopier, Ilford 369–70
Holocrafts Ltd 380, 399
holodeck 408
HoloFilm, Ilford 369
Holografi: Det 3-Dimensionel Mediet exhibition 313
Holografix 272, 278, 365
hologram 2D-3D 374, 386, 388–9
hologram, achromatic 334, 364
hologram, acoustic – *see* acoustic holography
hologram, adoption of term for 25, 29, 122
hologram, amplitude
hologram, animated 166–7, 179, 212, 214–15, 270, 352–3, 388, 398, 401, 406, 408
hologram, Benton – *see* hologram, rainbow
hologram, binary – *see* hologram, computer-generated
hologram, bleached 72, 163, 175, 219, 224, 240, 358, 391
hologram, cast 376, 384
hologram, colour 163–4, 178, 185, 202, 209, 221, 270, 278, 304, 370, 379, 422
hologram, computer-generated 216–20, 388–9
hologram, cylindrical 204–5, 215, 227
hologram, Denisyuk – *see* hologram, reflection
hologram, dichromate – *see* dichromated gelatin
hologram, digital – *see* hologram, computer-generated
hologram, dot-matrix – *see* dot-matrix hologram
hologram, edge-lit 334, 426
hologram, elemental – *see* hologram, dot-matrix
hologram, embossed – *see* embossing
hologram, erotic 395–9
hologram, fine-art – *see* fine-art hologram
hologram, first-generation – *see* hologram, master
hologram, focused – *see* hologram, image-plane
hologram, Fourier transform 131–2, 143, 253

hologram, Gabor – *see* hologram, in-line
hologram, greyscale 80, 100–3, 107–9, 115, 153
hologram, haptic 334
hologram, image-plane 184, 202–4, 227, 301, 351–2, 427
hologram, in-line 15–6, 26–9, 32–3, 37–42, 44, 48, 51–2, 56–7, 67, 71, 96–7, 101, 204–5
hologram, integral – *see* holographic stereogram
hologram, laser transmission 106–8, 169, 215, 296, 302, 366
hologram, Leith-Upatnieks 113, 118, 128, 136, 146, 178, 185, 200, 203–5, 211–12, 216, 290, 384, 386, 403, 409, 421–2. *See also* hologram, off-axis; hologram, greyscale; hologram, solid object; hologram, laser transmission
hologram, Lippmann – *see* hologram, reflection
hologram, Lippmann-Bragg – *see* hologram, reflection
hologram, master 166, 211, 359, 387, 428
hologram, multiplex 212–16, 267, 274–5, 277, 280, 289, 316, 352–3, 358–9, 363, 396, 398, 406–7
hologram, off-axis 67, 80, 96–102, 108, 111, 115–17, 145–6, 200, 241, 381, 384–5
hologram, phase 72, 175, 224–5, 372
hologram, rainbow 179, 209–12, 227, 283, 299–301, 310–11, 333–4, 339, 375–6, 381, 383, 394, 407, 413, 427
hologram, reflection 69, 72–4, 283, 304, 310–11, 313, 315, 339, 351–2, 358, 362–363, 368, 370–2, 376, 387, 394, 406, 427
hologram, solid object 80, 104, 109–13, 115–17
hologram, synthetic – *see* hologram, computer-generated
Hologram Tales 408
hologram, time-averaged 192–4
hologram, double-pulse 197
hologram, volume – *see* hologram, reflection
hologram, white-light 73, 179, 201–4, 210–12, 214, 227, 300, 310, 339, 352, 429
hologram, white-light transmission (WLT) – *see* hologram, rainbow
Holo-Gram, The 328
Holograma 389
holographers, aesthetic 11, 287, 289–90, 297–8, 305–6, 308–10, 312, 317, 323, 413, 426, 438
holographers, amateur enthusiast 270–1, 283, 285, 313, 315, 405
holographers, artisanal 11, 259–60, 268–9, 269, 270–1, 277, 280–1, 283–6
holographers, early 7, 11, 155, 165, 173, 177, 181, 183, 200, 236. *See also* Baez; Denisyuk; El-Sum; Gabor; Leith; Lohmann; Rogers; Upatnieks
holographers, female 296, 299, 310, 323
holographers, fine-art – *see* fine-art holographer
holographers, scientist-engineer 11, 229, 231, 243, 247, 249–52, 254, 258, 260–1, 270, 291
Holografi – Det 3-Dimensional Mediet exhibition 313
Holographic Arts Company 279, 355, 367
Holographic Communications Corporation of America 278
holographic cinema 167, 184–5, 190, 274
Holographic Developments Ltd 362
Holographic Display Artists and Engineers Club (HODIC) 329
Holographic Film Company Ltd 209, 216, 338, 367
Holographic Imaging Studios 385
holographic non-destructive testing (HNDT) 198–9, 355, 424
holographic optical element (HOE) 74, 353–5, 371, 390, 489
holographic paradigm 411–12, 443
holographic sight 158
holographic stereograms 209, 212–16, 275–6, 289
Holographic Sunrise hologram 374
holographic television 190, 202, 210, 270, 347, 354, 379, 401, 405, 409, 423, 429
Holographics International 328, 391, 399
Holographics North Inc 366, 414
Holographie Dreidimensionale Bilder exhibition 313
Holography 1975: The First Decade exhibition 313, 319, 429

holography, adoption of term for 13, 21, 121–2, 139
Holography Handbook 279, 281–2
Holography News 328, 365, 428
holography, polarization 183–4
holokitsch 319, 360, 398–401, 431
Hololight '79 exhibition 313
holo-litter 413
holometry – *see* holographic interferometry
Holonet website 310, 330
Holopack-Holoprint Conference 326, 365
HoloPak Inc 369
Holoprinter 369, 388
Holoptics 382
Holos! Holos! Valesquez! Gabor! hologram 289
Holos Corporation 184
Holos Gallery 355, 359–62, 371, 374, 398, 431
holoscope 27–9, 51, 58, 118
holoscopy 16, 22, 27, 29, 33–4, 38, 83, 122
Holosonics Inc 274, 277, 359, 377, 380–1, 384–5
holosphere 328, 341–2, 345, 391
holosuites 308
Holotape 373
Holotest, Agfa Gavaert 223
Holotron Inc 138, 153–5, 177–8, 349, 357–9, 373, 377, 380–5
Holovision AB 186, 209, 332, 367
Holoworld.com 329
Hollusions Ltd 363
Holtronics GmbH 376
Homegrown Holography 281
Homeland Security Special Interest Group, SPIE 241
Honda Inc 334
Hopkins, Harold H. 50, 137
HOTEC, Ilford 369
Howard, John 136, 237
How is Holography Art? 297
HPS – *see* history and philosophy of science
Hubel, Paul M. 370
Hughes Aircraft Company 106, 236, 246, 356, 385–6, 391
Hughes Power Products 391
Hughes Research Laboratories 106, 153, 170, 225
Hungary 17–18. *See also* Budapest
Huntley, Wright 157
Huygens, Christian 66, 72, 96
Huygens' Principle 22–3, 46, 66–8, 72–3
hybrid careers 84, 119, 234–5, 286, 420, 440

ICI Americas 353
Icon Holographics Ltd 364
Ideecentrum Foundation 279
IEEE – *see* Institute of Electrical and Electronics Engineers
Ilford Ltd 307, 367–2, 391
image, pseudoscopic 203
Image Plane 328
image, real 46, 114, 203, 211, 351
image, virtual 39, 46, 114, 334
Images in Time and Space 317
IMAX cinema 401
incommensurability 83–4
Indiana, Robert 303
Imperial College, London 34–7, 41, 43, 47, 50, 54, 60, 137, 155, 176–7, 361, 364, 378
impulse laser – *see* pulsed laser
Indebetouw, Guy 126, 134–5
Indiana, Robert 303
information storage and retrieval 58, 174, 354
information processing 122, 124
information theory 41, 46, 52, 234, 236, 271, 420
infrared 20, 76, 83, 106, 157, 160, 170, 235
Ingalls, Albert G. 162, 264
inner tube 164, 263, 266–7, 357
Institut D'Optique, France 201
Institute of Autometry, Siberia 220, 327
Institute of Electrical and Electronics Engineers (IEEE) 116, 144, 161, 165, 232
Institute of Industrial Science 187
Institute of Radio Engineers (IRE) 232
Institute of Science and Technology (IST, University of Michigan) 79, 83, 96, 125–6, 130, 135, 255
Institutet för Optisk Forskning, Sweden 155, 195
Instrument Society of America (ISA) 237
Integraf 330

integral photograph 66
intentionality 288, 290, 297
Interference Gallery 342, 366
interference, optical 10, 23, 46, 66–7, 197, 264
Interferenzen 328
interferometer 62, 186, 239
interferometer, Mach-Zehnder 54–5, 100
interferometry 62, 115, 155–7, 170, 174, 183, 185–6, 188, 234
interferometry, time-averaged 192–4, 199
interferometry, contour 194, 197
interferometry, double-pulse 197
interferometry, double-exposure 199
interferometry, real-time 195, 198–9
International Banknote 381
International Center for Photography (ICP) 313, 319–20, 336
International Commission for Optics (ICO) 27, 30, 187, 233
International Congress for the Art of Holography 339
International Hologram Manufacturers' Association (IHMA) 386
International Society for Optical Engineering – *see* Society of Photo-Optical Instrumentation Engineers (SPIE)
International Symposium on Display Holography – *see* Lake Forest
interstitial communities 439
interviews x, 6, 101, 109, 321, 447–8
Inventing the Future 444
investors 159, 163, 166–8, 171–4, 203, 205, 352, 355–6, 358, 362, 391, 422, 425
Ioffe Institute – *see* A. F. Ioffe Physical-Technical Institute
Ishii, Setsuko 303, 309
Isvestia 182
Italy 233, 243, 313, 376
Itek Corporation 238, 242, 298
Iudin, Eugenii 64
Ivanov, S. P. 184
Ives, Harold E. 70
Iwata, Fujio 387, 389

Jackson, Rosemary H. (Posy) 246, 304, 310, 313, 321–2, 335–42, 345, 365, 413, 444
Jackson, Willis 36
Jacobson, A. D. 153
James, Randy 275–6, 278, 279
Japan 139, 143, 187–8, 199, 242–3, 250, 275, 309, 329, 340, 360, 388, 390. *See also* Osaka; Sapporo; Tokyo; Tsukuba
Japanese Society of Applied Physics 243
Japanese Society for the Promotion of Sciences 243
jargon 83, 97, 114, 232
Jay, Martin 6
Javan, Ali 106
Jenkins, Harry 381
Jeong, Tung Hon (TJ) 250, 277–8, 281, 304, 307, 321, 330–3, 336–7, 385–6
Jet Propulsion Laboratory (JPL), CA 159, 366
Jodon Engineering Associates Ltd 135, 161, 163, 170, 174
Joerges, Bernward 439–40
Johansen, Frithioff 309
John, Pearl 306, 332, 440
Johns Hopkins University 152
Johns, Jasper 288–9, 323
JOSA – see *Journal of the Optical Society of America*
Journal of Applied Physics 236
Journal of the Optical Society of America (*JOSA*) 108, 133, 175, 236, 242
Judd, Deane 236
Jull, George Walter 138, 176, 259
Jung, Dieter 303–6, 309, 317, 323, 343

Kac, Eduardo 309
Kaiser Optical Systems 354
Karl Stetson Associates 155
Kakichashvili, Shermazan D. 183–4
Kan, Michael 214, 268, 274–5, 358
Kandidat – *see* Candidate of Science
Kansas City 357
Kapitsa, Petr L. 76, 181
Kaplan, Alice 209
Kármán, Theodor von 18
Kastler, Alfred 143
Katzive, David 340

Kaufman, John 303, 307, 309
Kay, Ronald 217
Keldysh, Mstislav V. 76–7, 181
Kepes, Gyorgy 302
Kharkov, Ukraine 61
Kiev, Ukraine 61, 184–5, 313, 327, 389
Kikuchi, Chihiro 135, 170–1
kinegram 389, 428
King Vitaman cereal 352, 399
King, M. C. 212
Kirkpatrick, Paul 2, 43–5, 47–8, 57, 60, 89, 95, 105, 140–1, 148, 251
Kirillov, N. I. 182, 221
Kiss II hologram 213, 274, 316
kitsch 319, 360, 398–401, 431
Klüver, Johan Wilhelm (Billy) 288
klystron tube 85
KMS Industries Inc 155, 160–2, 165, 171–4, 177, 259–60, 264, 267, 273, 350, 436
KMS Fusion Inc 161–2, 173–4
Knoedler Gallery 289, 292, 295
Knoll, Max 21
Kock, Winston E. 143–4
Koestler, Arthur 18
Komar, Viktor G. 184–5, 222
Kontnik, Lewis T. 364–5
Korad pulsed ruby laser 207
Korea 313
Koronkevich, W. 220
Kozma, Adam 87, 107, 126, 137–8, 154–5, 161, 174, 176
Kramer, Hilton 319–23
Krasnogorsk Optical-Mechanical Plant, Moscow 239
Kremlin Museum, Moscow 184
Krishnamurty, J. 411
Krystal Holographics Ltd 357
Kubota, Toshihiro 187, 199
Kuhn, Thomas S. 11, 83–4, 417
Kuybyshev Aviation Institute 220
Kwajalein launch facility, Marshall Islands 241

Labeyrie, Antoine E. 126, 134, 201
Laboratoire d'Optique de Besançon (LOBE) 164, 183
laboratories 125, 138, 164, 167, 177, 189, 191, 195, 226, 301–2, 306, 308, 344, 355–6, 379, 421–2, 428, 441
laboratories, artisanal 172, 264–271, 273, 278
laboratories, cottage industry 359, 363, 377, 382
laboratories, corporate –*see* American Optical; Bell Telephone; BTH; CBS; Eastman Kodak; Hughes Research; Polaroid
laboratories, scientific 134, 170, 180–7, 198–9, 201, 204, 206, 219–20, 236–8, 246, 250, 258, 262–4, 280–1, 295, 308, 313, 317, 327, 331. *See also* Electro-Optical Sciences Lab; MIT Media Lab; Willow Run Laboratories; S. I. Vavilov Institute
laboratory demonstrations 16, 32, 58, 82, 117, 156, 167, 171, 173, 205, 290, 311, 394, 422, 429
laboratory environment 254, 294, 358, 394, 399, 422, 428, 432
labour, division of 175, 239, 289, 441
Lake Forest Conferences 272, 278, 287, 304, 307, 326, 331–3, 336–7, 339, 345, 384–385
lamp, discharge 17–18
Lancaster, Ian M. 339–41, 363, 365
Land, Edwin H. 178–80, 211, 236, 238, 300, 334
Landis & Gyr 389
Landsat satellites 238
Lane, Linda 276, 315
Langford, Michael 307
Langley (NASA) Research Center, VA 159
Lasart Ltd 399
laser 4, 73, 97, 100, 104–8, 111, 113, 115–16, 202, 203, 206, 214, 218–19, 221–5, 251, 301, 307, 350, 356, 394, 409, 421–3, 425–6, 429, 442
Laser Arts Society for Education and Research (L.A.S.E.R.) 328–9
laser, argon 111, 117, 164, 167, 187, 190, 226, 291, 296
laser, continuous wave (CW) 106, 206, 208, 308
Laser Focus 169
laser, helium-neon (He-Ne) 100, 106–8, 162–4, 166, 170, 187, 190, 192, 206, 222, 225–6, 260, 291, 301, 357, 370

laser, krypton ion 178, 259–60, 308
Laser Light Concepts 355
L.A.S.E.R. News 328
Laser Point 363, 368
laser, pulsed 141, 163, 167–1, 177, 182, 184, 196–7, 206–9, 227, 261, 289, 292, 297, 301, 308, 315, 338, 350, 352–3, 368, 376, 396, 400, 423, 424, 426, 442
Laser Reflections 396, 426
laser, ruby 106, 170–2, 182, 196–7, 206–7, 234, 292, 308, 353, 423, 426
LaserFilm, Ilford 368
Lasergruppen Holovision AB 186, 209, 367
lasography 115
Latour, Bruno 8, 285, 288, 423, 441
Lau, Ernst 185
Lauk, Matthias 343–4
Laval University, Canada 43
Laza Holograms 308
League for Industrial Democracy 254
Lear-Siegler Corporation 171
Lebedev, Alexander Alexeyevich 62, 76
Lebedev, Dmitry S. 219
Lebedev Institute – see P. N. Lebedev Institute, Moscow
Lehmann, Matt 158, 195, 272, 295, 356
Leith, Emmett N. xi, 2, 52, 55, 78, 83–117, 120, 122–40, 142–4, 146–8, 151–7, 161–3, 168–9, 171–2, 174–80, 183, 186–7, 190, 191–3, 198, 201, 210, 231–5, 242, 244, 246, 251, 254, 256, 262–3, 277, 287, 290, 291, 301, 325, 327, 331, 349, 378–9, 384–5, 403, 417, 421, 434–5, 438–9, 442, 448, 489
Leith, Kim 100–2
Lenin Prize 181–2, 186
Lenin, Vladimir I. 181, 186
Leningrad, Russia 13, 60–4, 74, 76, 182–5, 202, 210, 312
Leningrad Institute of Fine Mechanics and Optics 64
Lenoir, Timothy 9
lens, aspheric 186
lens, axicon – *see* axicon
lens, camera 237–8
lens, cylindrical 95, 100, 267
lens, electron 21
lens, liquid 264, 267
lens, aberration of 24, 44, 74
lensless photography 100–2, 108–9, 112–13, 118–19, 127, 130–1, 148, 190, 204, 234, 349, 417, 421
lensless Fourier transform hologram – *see* hologram, Fourier transform
lenticular photograph 209, 212, 401
lenticular screen 184
Leonard, Carl D. xi, 126, 155–6, 449
Le Poole, Jan B. 27, 29
Leslie hologram 213
Leslie, Stuart 233
Leo Castelli Gallery 292, 323
Leonardo magazine 260, 290
Licht-Blicke exhibition 317
Lieberman, Larry 279, 360, 370
Life magazine 290–291
Light Dimensions exhibition 315, 361
Light Fantastic exhibitions 306, 313, 315, 338 361–3, 367
Light Fantastic Gallery 362, 398
Light Impressions Inc 279, 359, 363–4, 373, 381–3, 387, 390–1
light shows 172, 260–2, 298, 356, 362
light, white – *see* white light
Lilliana hologram 218
limerick 398
Lin, Lawrence H. 202
linear systems theory 85, 87, 119
Lin, Shu Min 309
Lines, Adrian 309
Linn, Douglas 171
Linnik, Vladimir P. 62, 76, 181
Lippmann, Gabriel 66–73
Lippmann photography – *see* photography, Lippmann
Lisson Gallery 295
Lite, Allyn 260–1
Lissack, Selwyn 276, 278
Liverpool Polytechnic 307
Llewellyn, Fred B. 108–9, 145, 152–3
Lockheed Corporation 187, 236, 246, 258
Lockheed Missiles and Space Company 238

Loewen, E. G. 134
Logan's Run 275, 406
logical empiricism 415
Lohmann, Adolf W. xi, 24, 48–51, 52, 54, 58–9, 85, 87, 100, 104, 120, 133, 142, 144–5, 147, 217–19, 232, 242, 247, 412, 417, 421
London UK 137, 242, 293, 295–6, 305–6, 309, 313, 355, 389–90, 396, 398, 410–11, 435, 437, 444
Loughborough University of Technology, UK 164, 176, 196, 198, 209, 224, 240, 295, 311, 315, 361
Los Angeles CA 275, 279, 292, 313, 315, 404, 443
Lowenthal, Serge 134, 186
Lucie-Smith, Edward 2, 309, 322
Luck, Thomas 309
Lunch Box Review exhibition 398
L'vov, Ukraine 313
Lysenko, Trofim 62

MacArthur, Ana xi, 304, 309, 317, 332, 399
Mach, Ernst 54
Mach-Zehnder interferometer 54–5
Maiman, Theodore 106, 170
Maksutov, Dmitri D. 62
Maline, Sarah 8, 320, 323
Manned Space Flight Center, TX 159
MARC – *see* Univ of Michigan Aeronautical Research Center 79, 160
Marconi 354
Maréchal, André 134
Markem Systems 361
Markov, Vladimir B. 184, 313, 389
Marks, R. 126, 153
Marshall (NASA) Space Flight Center, AL 159
Marx, Karl 186
maser 80, 105–6, 170, 196, 234
mass production 166, 203, 350, 352, 362, 373–4, 380, 425, 427–8
Massachusetts Institute of Technology (MIT) 124–6, 155, 168, 178–80, 233–4, 258–9, 269, 300, 302
Massey, Norman 91, 153, 155
MasterCard Inc 374–5, 383, 387
master hologram 166, 203, 211, 358–9, 361–3, 369, 384, 387
matched filtering 95, 107, 127, 174
material culture 6, 262, 435, 437
materialism 254
materials, found 267, 271, 289
Mathieu, Marie-Christiane 304, 309
Matsuda K. 187
Matsumoto K. 242
Mattel Inc 373
Mazzero, François 309, 394
McLuhan, Marshall 298
McCormack, Sharon 268, 274–5, 279, 309
McDonald's Inc 376
McDonnell Douglas Missouri 161, 164–9, 171, 185, 198, 206, 209, 214, 227, 236, 278, 338, 396, 424, 426, 490
McDonnell, Jim 161, 167, 171
McGrew, Steve 276–9, 303, 357–60, 362–3, 374, 381, 383–4, 386–8, 390–1, 443
McLaughlin, G. Harry 410
McLaughlin, Liz 296
McLellan, Rolf 129, 135
medium, holographic 25, 38–9, 47. *Dee also* emulsions and photopolymers
Medora, Mike 363
Medvedev, Zhores Aleksandrovich 61, 63, 181, 256
meetings 33–3, 36, 100, 113, 116–17, 119, 121–2, 130, 133, 176, 241, 242–50, 254, 256, 260, 296, 327, 329–30, 337, 353, 365, 379, 384
meme 402–403, 405, 409–10
memory, holographic 174, 183–4, 200, 368
Menning, Melinda 306
mercury arc lamp 17, 25, 46, 54, 69–73, 88, 164, 186, 190, 201, 214, 295, 311, 352
Merton, Robert K. 5
Mertz, Lawrence 242, 384
Merzlyakov, Nickolay 219
Metherell, Alexander 252
Metropolitan-Vickers Ltd 18–20, 22, 56
Metrovick – *see* Metropolitan-Vickers
Michalak, Michael 262–3
Michelson, Albert Abraham 41, 105, 263
microfilm 134

microhistories 5–6
microscope, x-ray 26–7, 58
microscopists 27, 30–1, 34, 46, 58, 101, 119, 133, 420
microscopy by wavefront reconstruction, *see* Diffraction Microscopy
microscopy, x-ray 26–7, 38, 43–5, 58, 153
microscopy, electron 22, 153
microscopy, optical 22, 120, 129
Microsoft Inc 387
microwaves 85, 90, 242
MIDAC 80
Milburn, Colin 430
militarism 254
Miller, Peter 307
Millman, Frank 360
Mingace Jr, Herbert S. 299
Ministry of Defence, UK 364
Ministry of Culture, Ukrainian 184, 313
Minneapolis, MN 357
Minsk, Russia 185
missile systems 79–80, 93, 162, 173
mirror, convex 69–72
Missouri 332; *see also* McDonnell Douglas Missouri
MIT – *see* Massachusetts Institute of Technology
MIT Media Lab 299, 306, 333, 391, 427, 437, 444
MIT Museum, Cambridge, MA x, 343, 412, 436–7
MIT Spatial Imaging Group 180, 220, 299
MIT Spectroscopy Laboratory 124
Mitamura, Shensuke 307, 309
Mitton, Jon 309, 398
mode, laser 106, 111, 192, 207
modern optics 59, 66, 124, 143, 155, 186, 217, 233–5, 251, 254, 331, 438
MoH – *see* Museum of Holography, New York
Molteni, Jr., William J. 209, 310, 336
monochromatic light 15, 23, 25, 29, 38, 40, 52, 62, 70, 73, 87, 99, 104, 113, 185, 202, 321, 426
moon glasses 169
Moore, Lon 273, 275, 278–9, 353
Moree, Sam 275, 303–5, 309, 310, 343
Moscow, Russia 61, 181, 184–5, 209, 220, 248–9, 327, 389
Mother Earth News 270
Motorola 91
movement, measurement of 192–200
Muehleisen, Roland 251
Multiplex Company 212–6, 260, 267–8, 274–7, 279–81, 326, 352–3, 355, 358–9, 363, 396–8, 406–7, 438
Multiplex printer 269, 275
Mulvey, Tom 24, 48, 56
Mumford, Ray 363
Munday, Rob 307–8, 388, 398, 414
Munday Spatial Imaging 308, 426
Munich, Germany 8, 257, 437
Murray, Rod 307
Musée de l'Holographie, Paris 340, 344–5, 390, 426
Museum für Holographie und neue visuelle Medien, Pulheim, Germany 340, 343–4, 355, 391, 432
Museum of Holography (MoH), New York 278–9, 335–343, 358, 361–3, 365, 391, 426, 432, 434, 436, 438, 443–5
museums 6, 60, 184, 311, 355–6, 358, 360–3, 365, 391, 432–7
Muth, August 332, 399
mythology 122, 141, 303, 404–5, 427

N-Dimensional Space exhibition 260, 262, 289, 303–4, 313, 318
Narodnyi Komissariat Vnutrennikh Del (NKVD) 76
NASA – *see* National Aeronautics and Space Administration
NASA Electronics Research Center 143, 203
National Aeronautics and Space Administration (NASA) 125, 151, 159, 175, 203, 214, 220, 236, 240, 251, 349, 366
National Bureau of Standards (NBS) 62
National Defense Research Committee (NDRC), USA 124
National Engineering Laboratory, UK 198
National Geographic magazine 166, 375–6

National Museum of Photography, UK 437
National Physical Laboratory (NPL), UK 26, 33–4, 36–7, 56, 62, 73, 155, 196, 198, 207, 252
National Research Council of Canada 186
National Science Foundation (NSF), USA 125–8, 134, 143, 151, 153, 156, 175, 220, 243, 332
NATO Advisory Group for Aerospace Research and Development (AGARD) 240
Nature 30–2, 37, 40–1, 236
Nauman, Bruce 167–8, 291–3, 295, 303, 313, 323, 352
Naval Ordnance Test Station (NOTS), CA 241
Naval Research Laboratory (NRL), Washington DC 91, 246
Navy, USSR 62, 64, 69, 74
Navy, US 83, 116, 151, 157, 174, 334, 349
Nelson, Elizabeth 258, 327
Neumann, John von 18
New Alchemy 270
New Engineer 235
Newman, Paul R. 309
Netherlands 279, 292, 298, 301, 313, 316
New Holographic Design Ltd 364
New Mexico 332, 399
New Scientist 56, 376, 388
New York Art Alliance Inc 278, 335–7, 358, 367
New York City 165, 262, 289, 292, 295, 301, 304, 309–10, 313, 315, 318–19, 323, 337, 343, 401
New York Holographic Laboratories 339
New York Institute of Technology 329
New York Museum of Holography – *see* Museum of Holography, New York
New York School of Holography 304–5, 335–6, 358
New York Times 15–16, 26, 101–2, 319, 321–2
Newport Research Corporation 199, 263, 381
Newport Holographic Camera 199
newspaper reports 15–16, 101–2, 104, 109, 113, 116, 118, 122, 157, 183, 256, 293, 319, 322–3, 352–3, 356, 360, 404–6, 409, 430
Newswanger, Craig 220, 353, 360
New Zealand 37, 154, 438
Nicholas Wilder Gallery 292
Nicholson, Ana Maria 209, 275, 304
Nicholson, Peter 209, 337
Niepce, Joseph 65
NIKFI – *see* Cinema and Photographic Research Institute
Nikon 242, 390
Nilsson, Nils Robert 292
Nimslo 3D camera 401
Nobel Prize 16, 21, 26, 43, 49, 59, 66, 78, 122, 141–4, 147, 177, 233–4, 378, 341, 420, 422, 429, 433, 435
noise, optical 46, 51–2, 96–7, 101, 103, 107–9, 224, 264
nomenclature 13, 21, 25, 29, 121–2, 139–40
Nottingham University 294–6, 313, 318
Novosibirsk, Russia 61, 327
NPL – *see* National Physical Laboratory
NSF – *see* National Science Foundation
Nuñez, Ruben 303, 309
Nye, David 393

Oak Ridge Laboratories 331
object beam – *see* beam, object
Obreimov, Ivan Vasilievich 76
occupation 249, 252, 280, 283–5, 309, 418, 421, 425, 431, 439–40
Ohnuma, Kazuhiko 387
Office of Advanced Research and Technology, NASA (OART), USA 159
Office of Naval Research (ONR), USA 116, 125, 157
Office of Scientific Research and Development (OSRD), USA 124
Offner, Abe 108
Olson, Ron and Bernadette 396
O'Neill, Edward N. 85, 87, 242
Office National d'Etudes et de Recherches Aérospatiales (ONERA), France 199, 240, 323, 309
Old Dominion Foils Ltd 374, 377
Olografia exhibition 313
onion radiation 50
ONR – *see* Office of Naval Research
Ontario College of Art 306

op art 294
OpSec Inc 369
Optec Design Ltd 375
Optica Acta 236
optical engineering 11, 124, 236–8, 241, 250, 252, 258, 260, 264, 284–5, 291, 431
Optical Engineering 236
optical filtering 187, 232
optical processing 83, 85–7, 90–2, 95, 99, 106–7, 115, 119, 128, 146, 148, 182, 232, 241, 244, 258, 263, 280, 413, 435
optical processor 85–6, 88, 90, 92, 95, 107–8, 162–3, 174, 264
Optical Society of America (OSA) 100–1, 108, 113, 116–17, 131, 133, 140, 145, 233, 237, 249
optical table 110, 157, 229, 263, 285, 295, 301
optical transform 85
optical transfer function (OTF) 50
optician 34, 36, 51
Optics & Spectroscopy 236
Optics Communications 236
Optik (book) 42–3
Optik (journal) 236
Optika i Spectroskopija (*Optics & Spectroscopy*) 73
oral history 3, 6, 435
Orazem, Vito 309
origin stories 13, 16, 64, 104, 109, 120–1, 136–41, 145–7, 422, 427, 434
orthochromatic emulsion 426
Orr, Edwina 304–5, 307–9, 370
Ortman, George 303
oscillograph 17–8, 21
Ose, T. 187, 242
OSRD – *see* Office of Scientific Research and Development
Ossipov, J. 62
Ostrovskaya, Galya V. 183
Ostrovsky, Yuri 183, 216
Outwater, Christopher 281, 406
Overhage, Carl F. J. 238
Oyo Buturi 236

P. N. Lebedev Institute, Moscow 327
Packaging 326, 365, 370, 374, 376, 387, 427–9
Page, Michael 309
Pacific Holographics 279
Pais, Abraham 416
Palermo, Carmen J. 87, 92, 153, 174
Palo Alto 187, 238
Pam and Helen hologram 396
papyrocentric culture 325
parallax 66, 110, 114, 118, 205, 210, 211, 300, 311, 364, 421, 428
Parallax Gallery 362, 368
Paris 124, 134, 135, 187, 201, 233, 303, 315, 340, 355, 367, 390
Parker, Rick 240
Parrent, George B. 158
patents 20–1, 27–31, 41, 55–6, 100, 110, 115, 120–1, 145, 148,153, 154, 158, 178, 349, 359, 364, 373–4, 377–86, 388–9, 391, 399, 422–3, 425, 433, 490
peace symbol 253
Peierls, Rudolph Ernest 32
Peine, Otto 302
Peking, China 344
Pennington, Keith S. 202, 225
Pentagon 152
Pepper, Andrew 8, 287, 303–4, 306–7, 309–10, 328, 345
periodization 418
peripheral science 9, 439–40
Perkin, Richard (Dick) 109, 233
Perkin Elmer Corporation, 1, 106, 108–9, 233, 236, 238
Perry Jr, Hart 216, 276, 310, 338
Perry Sr, Hart 216
Perry, John 309, 366–7, 414
pessimism 326
Peter the Great 60
Peterson, Russell 380
Pethick, Jerry T. B. 259–60, 276, 279–81, 283, 289, 295, 303–4, 317, 355
Petrograd 60–1
Phalli hologram 301, 396
phase 23–5, 27, 39, 42, 51, 67–8, 72, 86, 88, 96, 101, 103–4, 107, 113, 140–1, 186, 217, 267
phase contrast 49, 51, 59, 175
pheasant's eyes 38–40, 103
Philadelphia Society of Holographers 329

Philips Laboratories 210
Phillips, Nicholas J. 164, 209, 224, 240, 307, 311, 315, 325–7, 338–9, 361
Philosophical Magazine 32
philosophy 411–13, 415, 422
photograph, integral – *see* integral photograph
photograph, wave – *see* wave photograph
photographers 38, 112, 179, 207, 278, 319, 323, 399–400, 426
photographic medium – *see* medium, holographic
photography 1, 15, 23, 27, 31, 38, 40, 52, 64, 102, 112–14, 119, 178, 190, 205, 213, 238, 261, 307–8, 315, 320–1, 323, 396, 400–1, 421–3, 432, 443
photography, lensless 101–2, 108–9, 112–14, 119, 127, 130, 148, 190, 204, 349, 417
photography, Lippman 66, 68, 100, 120, 259, 420
photography, wave 60–77, 120, 180–1, 190, 200, 239, 417
photometry 9, 61, 63
photonics 440
photopolymers 226, 228, 278, 372–3
Photo-Secession movement 319
Photo-Technical Research (PTR) 126, 135, 161, 169–70, 174, 356
physical optics 10, 23, 62–3, 83–4, 87–9, 92, 97, 105, 119, 231–2, 235, 239, 241, 250. *See also* modern optics
physical shadow 29
physicists 16, 18, 46, 88, 97–8, 114–15, 143, 160, 199, 232–3, 236, 254, 283, 291, 300, 302–3, 319–21, 362, 404, 411, 422
physics 13, 15–16, 36–7, 42–5, 48–9, 59, 60, 63–4, 66, 68, 76, 84, 101, 119, 122, 124, 126, 130–1, 142–3, 154, 160, 184, 191, 203, 233–6, 278, 281, 298–9, 325, 331, 357, 411–12, 416, 420
Physics Today 122
Physikalisch-Technische Bendesanstalt (PTB), Gemany 50
Physikalisch-Technische Reichsanstalt (PTR), Gemany 62
Pictorialism 319
pidgins 84
Pilkington P.E., UK 307, 354, 385
Pizzanelli, David xi, 306, 309, 317, 363–4, 387
plagiarism 131, 134
Playboy 396
Polaroid Research Laboratories Inc 178–9, 200, 210–11, 226, 238, 333–4, 372–3, 383, 385, 422, 424
Poleyatev, Sergey 389
Pollard, Andy 339
Polyanskii, Vyacheslav K. 183
pop art 294, 319
popular culture 393–414
Porcelain Cat Group 330
Porcello, Leonard J. 87, 92
pornography 274, 276, 396–7, 401
portraits 169, 206–9, 278, 396, 400, 423, 426, 442, 445
positivism 7, 415–16, 433
post-war 18, 20–1, 32, 34, 49, 64, 78, 84, 124, 232–3, 237–8, 244, 254, 257, 283, 285
postage stamps 373, 376
Powell, Robert L. 111, 126, 131, 153, 155, 157, 174, 192–4, 197, 199, 438
Powell Associates 155
power, laser 106, 111, 190, 203, 252, 368
Practical Holography conferences 250
Pratt & Whitney 366
Pravda 183
predictions 164, 178, 350, 403, 405, 416, 422, 428–33, 444–5
Preston, George Dawson 33, 37
Preston, Kendall 236
Pribram, Karl H. 277, 409–12
Princess Leia 2, 406–7
priority disputes 121–3, 125, 129, 131, 133, 143, 145, 147–8, 193, 202, 378, 384, 389, 433, 438
Proceedings of the IEEE 236
Proceedings of the Physical Society 32
processing, chemical 30, 191, 198, 222, 224, 226
profession 4, 237, 241–2, 244, 249–50, 252–3, 258–60, 262, 270–1, 280, 283, 285, 287, 291, 299, 303, 308, 310, 315, 325, 345, 418, 428, 430–1, 433, 439–40

profit 121, 137, 145, 203, 223, 238, 329, 335–6, 349, 355–6, 368–9, 371–2, 428, 432, 436, 444
Project CORONA 5, 238–9, 258
Project MICHIGAN 83, 91, 174
Project WIZARD 79
Project WOLVERINE 81, 83
Prokhorov, Alexandr M. 143
Protas, Rebekka R. 70, 182, 202, 221, 223
Prostev, Aleksander 312
Provence, Steve 275, 358–9, 382
Providence, RI 301
Pruitt, Nancy 129, 135
pseudo-colour 212
public access holography 282
public understanding of science 8
publications 236, 281, 328, 341, 354, 391, 489
Pulheim, Germany 340, 343–4
pulse position modulation (PPM) 217
Pum III hologram 302

Qinetiq Ltd 220
Qingao Qimei Images Ltd 390
quadrature, optical 42, 54–5
Quaker Oats hologram 352
Quebec, Canada 43, 306, 309
Queen Elizabeth II hologram 413–14
Queen Victoria 399
Queen's University, Belfast 203

R2D2 2, 406–7
radar 19–20, 36, 49, 74, 81, 122, 206
radar, synthetic aperture – *see* synthetic aperture radar
Radar Research and Development Establishment (RRDE), UK 42
Radiation Inc 155, 161, 173–4, 186, 211
Radiation Laboratory (University of Michigan) 160–2, 170
Radical Caucus 256
Radical Software 270
Radioteknika i Elektronika 236
Rallison, Richard D. 226, 261, 356–8, 362, 380, 385, 399
Ramishvili, N. M. 183
Rauschenberg, Robert 288–9
Rayleigh, Lord 41
Raytheon Company 244, 386
Raven Holographics Ltd 357
Razutis, Al 255, 309, 319
RCA 225, 236, 349, 373
Reconnaissance International Ltd 326, 328, 341, 364–5, 428, 489
Red Bluff News 353
Red Dwarf 407
Redlands University 44–5, 47–8, 57, 250
Redondo Beach, CA 196
reference beam – *see* beam, reference
referenceless holography 183
relics, holograms as 434–5
Renaissance Center 316
Renault 199
research-technologists 243, 285, 440–1
research-technology 9–11, 59, 159, 286, 439–40
Restrick, Robert C. 126, 129
Reuterswärd, Carl Frederick 292–3, 303, 343
Revlon Cosmetics 214, 353
Revolution, Russian 60–1
Revolution, Scientific-Cultural 61
Revolution, Scientific-Technical 61
Rezny, Abe 310
Richardson, Martin 306–7, 309
Richland, WA 357
Richmond Holographic Studios Ltd 279, 370
Ritscher, Eve 315, 339, 343, 361
Ritschl, Rudolph 185
Robb, Jeffrey 306, 317, 414
Rochester, NY 42, 133, 134, 235, 242, 244, 246, 262
Rockwell Inc 236, 246
Rogers, Gordon L. 29, 37–44, 46–7, 51–2, 55–9, 60, 71, 87, 89, 94, 100, 121–2, 140, 141, 144, 148, 175–7, 180, 187, 196, 203, 221, 224–5, 294, 296, 417, 435
Rogers, Simon 368
Rolling Stone magazine 277
Rolls Royce Inc 240
Romanova, M. 62
Rome, Italy 313
Rongonen, Vera 70

Rosen, Lowell 203–4
Ross, David A. 270
Ross, Jonathan 339, 343, 364, 449
Roszak, Theodore 257
Rotz, Frederick B. 111, 126, 174
Royal Academy of Arts, UK 315
Royal College of Art (RCA), UK 259, 295, 297, 306–7, 332, 396, 398, 432
Royal Institute of Technology, Sweden 186, 247, 442
Royal Microscopical Society, UK 27
Royal Military College, Australia 296–7
Royal Photographic Society (RPS), UK 315, 319, 361
Royal Signals and Radar Establishment (RSRE), UK 220
Royal Society, UK 29, 32–3, 35–6, 41
royalties 178, 379, 383, 385
Ruchin, Cecile 278–9
Rugby, UK 13, 19, 32–3, 56
Ruiz, Julio 309
ruling engine, diffraction grating 126, 134
Ruska, Ernst August Friedrich 21, 143
Russia 60–2, 64–5, 184, 186, 202, 311, 314, 327, 330, 389, 426. *See also* Kuybyshev; Leningrad; Minsk; Moscow; Novosibirsk; St. Petersburg; Ulyanovsk

Saabor, Peter 339
safety, laser 252, 317–18, 350, 404, 426
safety, chemical 277, 369, 424, 438
Sagaut, Jean 303
St Petersburg 60, 65, 389
Sakane, Itsuo 315
San Diego Union 352
San Francisco, CA 3, 257, 260–2, 266, 269–73, 277, 281–2, 289, 291, 304–5, 396–8, 431, 438
San Francisco School of Holography 213, 278–1, 289, 291, 438
San Jose, CA 217
sand table 229, 266, 270, 273, 280–2, 284–5
sandbox – *see* sand table
Sandia National Laboratories, USA 246
Santa Clara County 257
Sapan, Jason 276, 353, 396
Sapporo, Japan 188
SAR – *see* synthetic aperture radar
Saxby, Graham 307–8, 449
scattering, optical 224–5
Schlieren photography 239
Schmidt, David 274–5, 279
Schweitzer, Daniel K. 303–5, 309–10, 343
Science and Technology Imaging Group (Ilford) 367
Science and Technology Studies (STS) 5, 8
science fiction 1, 64–5, 398, 405–8, 422, 429–31
Science Museum, London 361, 435, 7
science, speculative 430
Science Studies 7–8, 288
Science Year 166, 169, 350, 375, 423
Scientific Advances Ltd 377, 380
Scientific American magazine 271
Scientific Applications Inc 156
scientists, optical 115–16, 151, 233, 235, 252
SDS – *see* Students for a Democratic Society
Seamans, Warren 343
seamless web 441
Searl, Gil 363
Sears Robuck & Co. 352
SEASAT 90
Seattle, WA 357
Sebastopol, Ukraine 313
Second World War 10, 61–3, 76, 124–5, 185, 232, 234, 238, 240, 244, 257, 393, 419, 426
secrecy 139, 158, 186, 224, 228, 256, 266, 325–7, 364, 386, 436, 438
security applications 365, 370, 374–7, 381, 383–4, 386–9, 427–8
SEE3, 339, 343, 363–4, 368
'Shadow of the Past' 65, 68
shadowgram 206, 310
Shankoff, T. S. 225
Shapley, Harlow 238
Shaw, George 173
Shearer, Hamish 368
Shearwater Foundation 330, 343, 345
Sherwin, Chalmers 81–4
Shinn, Terry 9, 59, 159, 237, 243, 285, 439–40
shiny shit 428

shock and awe 393–4
side-looking radar – *see* synthetic aperture radar
Siebert, Lawrence D. (Larry) xi, 160, 163, 167, 172, 206–9, 292
Siegel, Keeve M. (Kip) 160–6, 171, 173, 205, 350, 392, 403, 405, 422, 430
Siemens & Halske 17–18, 21
signal processing 49, 84, 144, 217, 232
Silberman, Rick 303, 309–20, 343
Sigg, Eric 292
Silicon Valley 257
silver halide 70, 184, 222–8, 236, 357–8, 363–4, 367–8, 370, 372, 391
silver market 367
Sisakyan, I. N. 220
sit-ins 255
S. I. Vavilov State Optical Institute, USSR 3, 61–4, 70, 76, 78, 95, 142, 181–2, 207, 221, 239
Skande, Per 209
Skeletal Hand hologram 301
Slade School of Fine Art 293, 295
slander 134
Slavich 182
Smigielski, Paul 240
Smiles, Samuel 5, 416
Smith, Bailey 343
Smith, Howard M. 246
Smith, Thomas 34, 36–7
Smithsonian Museum, Washington DC 209, 214, 302, 312, 360, 365, 437
Snow, Charles Percy 8, 325
Snow, Michael 367
Sobolev, Gennadi A. 184, 221
Soboleva, Svetlana 184, 221, 249
social construction 288, 297, 420, 425
Social Construction of Technology (SCOT) 11, 419–20, 425
Society of Imaging Science and Technology (IS&T) 299, 489
Society of Motion Picture and Television Engineers (SMPTE) 237
Society of Photographic Engineers (SPE) 237
Society of Photographic Scientists and Engineers (SPSE) 249
Society of Photo-Optical Instrumentation Engineers *or* Society of Photographic Instrumentation Engineers (SPIE) 122, 237, 241, 249–50, 299, 306, 329, 333, 384, 413, 489
sociologists 4, 418, 441
sociology 7, 418–19, 421, 441
sociology of scientific knowledge (SSK) 11, 288, 419
sociology of the professions 11
sodium lamp 54, 192
Soho, NY 342
Soifer, Victor 220
Sonovision Inc 161, 172, 260–1, 266
Soskin, Marat S. 184
South Carolina 332
Soviet Academy of Sciences 60–3, 76–7, 181–2, 218–20
Soviet Union 59–62, 65, 76, 181–5, 210, 218, 220, 224, 228, 232, 239, 310–13, 327, 339–40, 356, 389. *See also* Georgia; Russia; Ukraine
Sovietization 60, 181
Space-Light exhibition 317
Spain 309
spatial filter 50, 87, 107, 153, 186–17, 235, 264, 385
spatial frequency 49, 58, 86, 97, 213, 264
speckle 107, 190, 201, 211, 254, 394–5, 425–6
Spectra-Physics Inc 106, 108, 115–7, 183, 298, 359
Spectratek 373
spectroscopists 83, 111, 185, 331
spectroscopy 27, 61–3, 73, 76, 83, 116, 182, 185, 221, 233–4, 239, 263, 390, 439
spectroscopy, Fourier transform – *see* Fourier transform spectroscopy
specular reflection 71, 104, 195, 221, 264, 385
Sperry Rand Research Center, MA 159, 244
SPIE – *see* Society of Photo-Optical Instrumentation Engineers
SPIE Journal 237–8, 241
Spicer, Peter 294
Spierings, Walter 388
SRI Research 258
Smith, Steve 385

Smithsonian Institution 209, 214, 353, 360, 365, 437
Sowdon, Michael 309, 366, 413
St John, Dan 380
St Mary's College, IN 306
stability, optical 52, 86, 98, 104, 111, 191, 206, 221, 310
stability, professional 229, 236–7, 283–5, 309, 425
standing wave 68, 70–1
Stanford Applied Electronics Laboratory, USA 258
Stanford Electronics Labs, USA 245
Stanford University, USA 2, 37, 43–5, 48, 158, 233, 238
'Stellar Ships' 65
Star Trek 408
Star Wars 1, 406–7
Staselko, Dimitri I. 182
State Optical Institute (GOI), USSR 60–3, 67, 76
State University of New York (SUNY) 135, 138, 243
Statistical Machine 134, 368
Steichen, Edward 319
Stephens, Anaït Arutunoff 304
stereograph 399–401
Stereo-Realist camera 118, 401
stereoscopy 118, 399, 420
Stetson, Karl 126, 131, 153, 155, 157, 174, 192–4, 197, 248
Stetson, Karl A. 87, 104, 106–7, 111, 126, 131, 153, 155, 157, 174, 192–5, 248
Stieglitz, Alfred 319
Stocker, Wolfgang 376
Stockholm, Sweden 71, 142, 155, 186–7, 209, 247, 292, 313
Stong, C. L. (Red) 271
Stony Brook, NY 135, 138, 243
Strachey, John 17
Strategic Defense Initiative (SDI) 241, 423
Strathclyde University 157
Strick, Joseph 404
Stroke, George Wilhelm 1, 3, 119, 121–47, 153, 156, 163, 168–70, 175–7, 179, 187, 187, 201–4, 233–5, 243, 263, 379
Stroke, Henry H. 124, 138
student protests 254–7
Students for a Democratic Society (SDS) 254–6
Study Guide on Holography 281
subcultures 5, 84, 116, 148, 199, 281, 284, 262, 304, 308–12, 330, 413, 432
Subversive Optical Society (SOS) 242
success, attributing 3, 423–33
Sunnyvale, CA 358–9, 381–2
Surrey, UK 307
surveillance satellite 5, 238–9, 242
Sushkin, P. P. 65
Suzuki, T. 187
Sverdlovsk, Urals 61
Sweden 71, 142, 155, 186–7, 190, 209, 243, 247, 292, 309, 313, 337, 362, 442. *See also* Stockholm
Switzerland 243, 292
Sydney College of the Arts 306
synthetic aperture radar (SAR) 80–96, 102, 109, 113, 115, 118, 123, 146, 148, 154, 162, 165, 170, 172, 175, 191, 199, 206, 235, 239, 241, 443, 489
synthetic hologram 186, 216–18,
synthetizer, optical 28, 34, 37
Systems Techniques Laboratory, CA 158
Szilard, Leo 18

table, granite 88, 110, 157
table, optical 110
table, sand 229, 266, 270, 273, 280–2, 284–5, 307, 332
tacit knowledge 264, 326
Tanaka, S. 187
Tblisi, Georgia 183
Tchalakov, Ivan 8, 441
Technische Hochschule Berlin-Charlottenburg 17, 21
Technology Review 134
technological determinism 7, 416, 419
technological progressivism 405, 416, 418, 422, 429–30, 433
Teller, Edward 18
Teddington, UK 295
Tennessee 166, 239–40, 331
Terman, Frederick 233

Texas Instruments Inc 90–1, 171, 236
textuality 6
The Hologram Place Ltd 339, 355
The Holography Image Studio (T.H.I.S.) 363
The Holography Studio 338, 366
The Kiss hologram 215, 316
The Laser: Visual Applications exhibition 313
The Man Who Fell to Earth 275
The Who 361
theft 327
theses – *see* dissertations
Third Dimension Ltd 341, 363
Third Eye Productions Inc 355, 367
Thomas, Carleton (Sandy) 163, 172
Thompson, Brian J. 158, 233, 244
Thompson, E. P. 280
Thornton, Donald K. 307
thought collective 59
three-dimensional cinema 20, 31, 74, 167, 184–5, 190, 219, 354, 379, 401, 420
three-dimensional imagery 31, 33, 66, 68, 73, 110, 112–15
three-dimensional television 31, 74, 165, 209, 354, 379, 429
Through the Looking Glass exhibition 313–14, 337
tilted-plane optical processor 94
Time magazine 263, 290–1
Titterington, Christopher 437
TJ – *see* Jeong, Tung Hon
Tokyo, Japan 187, 242, 260, 313, 375
Toppan Printing Company 375, 387–90
Tonka Toys Inc 370, 376
Toraldo di Francia, Guiliano 134, 233–4
Toronto, Canada 278, 309
torpedo 157
toy train 1, 110–11, 113–15, 117
Townes, Charles H. 233–4
toys 11, 352, 359, 370–1, 375–6, 381–4
trading zones 84, 88, 330, 333
Travels Bounty 365
Traynor, David 305, 308, 370
tribes, technical 84, 325
Tribillon, Gilbert 164
Trion Instruments Inc 161, 170–1, 260
Trocadero 362
Trolinger, James D. 158–9, 199, 240, 395
TRW Inc 159, 236, 254, 359
Tsujiuchi, Jumpei 187, 199, 242
Tsukuba University, Japan 307
Tsuruta T. 242
Tunnadine, Graham 306
twin image problem 33, 40–2, 51, 56–7, 59, 95–8, 115, 136, 140, 144, 187, 384–5, 420–1
two-dimensional 33, 67, 73, 94–7, 100–2, 107–9
Tyler, Douglas 309, 317
Tyrer, John 240

UK Atomic Energy Authority 349
Ukraine 61, 183–4, 313–14, 244, 344, 389. *See also* Crimea; Dniepropetrovsk; Kharkov; Kiev
ultragram 334, 426
Ulyanovsk, Russia 185, 310–11
Unbehaun, Klaus 328
'undisciplined' subjects 283, 285
United Aircraft Research Labs 155
United Technologies Research Center, CT 248
Université du Québec à Montréal 306
University of Michigan (U-M) 1, 78–9, 83, 87, 91–2, 100, 107, 116, 119, 151–3, 168, 170–1, 174, 201, 203, 233–4
Univ of Michigan Upper Atmospheric Physics Group 160
Univ of Michigan Electro-Optical Sciences Laboratory (EOSL) 123, 126–9, 133, 135, 138, 155, 201–2
Univ of Michigan Aeronautical Research Center (MARC) 79
Univ of Michigan North campus 79, 125–7, 160, 255–6
Univ of Michigan Radar and Optics Laboratory 83, 92, 152, 211
Univ of Michigan Students for a Democratic Society chapter (SDS) 254–5
Univ of Michigan Willow Run Laboratories (WRL) – *see* Willow Run Laboratories
University of New South Wales 306–7
Univ of Tennessee Space Institute (UTSI) 240
Univ of Vermont 366
Universag Technische AG 18
Unterseher, Fred 261, 268–9, 272–3, 275, 278–9, 303–5, 309, 359, 365, 375, 386, 408–9, 438, 440, 443

Upatnieks, Juris xi, 55, 96–120, 134, 136–7, 140–1, 144–8, 152–3, 155, 158, 162–3, 174–6, 179, 183, 187, 190–2, 194, 200–1, 210, 232, 240, 242, 248, 254, 256, 262, 287, 290–1, 349, 378–9, 384–5, 403, 417, 421, 449, 489
Upatnieks, Ojars 100
Uppsala University, Sweden 292
users' groups 231
Utah 357, 373
US Army Research Office (ARO) 128, 153, 157
U. S. Banknote Corporation (USBC) 383, 386
U. S. Holographics 361
U.S. –Japan Seminars on Holography 187, 243
U.S. Philips Corporation 349
Usanov, Uri 182
US Information Agency 337

Valenta, E. 70
Van Heerden, Pieter J. 200
Van Ligten, Raoul 298, 300–1
Van Riper, Peter 260
Vancouver, Canada 313, 328, 338, 366
Vandenberg Air Force Base, CA 241
VanderLugt, Anthony B. 87, 107, 126–7, 143, 154–5, 161, 174
Varian Associates 258
Varian, Elayne 262, 289
Varian, Russell 85
Vavilov Institute – *see* S. I. Vavilov State Optical Institute
Vavilov, Sergei Ivanovitch 62
velocimetry, holographic particle image 240
Verde Valley, UT 213, 272
Veret, Claude 240
Veridian 83, 161, 436
Vest, Charles M. 156–7, 343
vibration 52, 56, 104, 111, 157, 172, 193–4, 202, 206, 240, 263, 264, 266–7, 272–3, 335
Victoria and Albert Museum, London 312, 437
videotape player, holographic 373
Viénot, Jean Charles 164, 183
Vietnam war 255–6
View-Master 401
Vila, Doris 309
Village Voice 335
virtual reality 443
Visa Inc 375, 383
Visual Alchemy 309
Vivian, Weston E. 83, 85–6, 161–2
Volinski, Edward 170
Volkswagen 199
Voxel Inc 364

Waddell, Peter 295
Walker, Jearl 271
Wall Street Journal 101
Waller, Allen 371
Waller-Bridge, Michael xi, 364
Walter, William R. 209, 299
Warhol, Andy 288–9, 323
Warren, Sir Hugh 19–20
Watertown Arsenal 156
wave photographs 71–4, 77, 83, 99, 120, 148, 180–1, 190, 200, 239, 417, 420
Wavefront magazine 309, 328, 391
wavefront reconstruction 15–59, 137, 139–41, 149, 152, 162, 176, 180, 187, 236
Wayne State University 83
Weapons Systems Evaluation Group 152
Weathermen 255
Weber, Sally 307
websites 330
Webster, John 297
Wegener, Horst 49
Weitzen, Ed 374–5, 392
Wender, Dave 167
Wenyon, Michael 307, 309, 339
Wesley, Edward 307, 396
West Coast Artists in Light 309
Westech Development Laboratories 226
Westinghouse Corporation 152
Westwood Pharmaceuticals 353
Whig history 416
white light 23, 69, 70, 73, 166, 178–9, 201–4, 210–12, 214, 227, 300, 310, 339, 351, 352, 358, 364, 367, 379, 397, 406, 420, 427, 429, 444
white-light hologram – *see* hologram, rainbow and hologram, image-plane

white-light laser 178
White Light Works 279
White Sands, NM 241
Whole Earth Catalog 269–70, 277, 282
Whole Message 313
Wigner, Eugene 18
Wilber, Ken 411–12
Wiley, Carl 80, 84
Williams, Ivor 19–20, 24–6, 29–30, 34, 56
Willow Run Laboratory, MI 13, 78–96, 99–100, 106–18, 122, 125–8, 132, 135–7, 139, 140–3, 145, 147, 152–5, 157, 160–2, 168, 170, 174, 177, 232–3, 239, 251, 435
Wilhelmsson, Hans 290
Wilson, James 124
Wilt, Richard 168, 293, 303, 352
Winner, Langdon 285
Winthrop, John T. 131
Witchcraft 158
Wolf, Emil 24, 42–3, 134, 136, 242
Wolfe, Allan M. 364
Wolff, John 361
Wolfke, Miecislav 23
Wolverhampton Polytechnic, UK 308
Wood, Glenn P. 367, 369–70, 372, 392, 443
Woodd, Peter 362
Worthington, C. Roy 131–2, 138
Wright Patterson Air Force Base, OH 154
WRL – *see* Willow Run Laboratory
Wuerker, Ralph F. 196–7, 199, 252, 254, 312

x-ray microscope – *see* microscopy, x-ray
x-rays 23, 26–7, 43–5, 47–8, 83, 90, 127, 131
Xograph 401

Yalta, Ukraine 313
Yaroslavsky, Leonid 219–20
Yefremov, Yuri Antonovich 64–6, 68, 406
Yinger, Milton 258
Yorktown, NY 217
Young, Thomas 96
Yu, Francis T. S. 83
Yugoslavia 124, 313

Zagorskaya, Diana 182
Zap Comix 269
Zebra Books 376
Zec, Peter 323, 344
Zech, Richard G. 126, 163, 203, 206–7
Zehnder, Ludwig 54
Zeiss 54
Zeiss Ikon 134
Zeiss Jena 185–6
zeitgeist 412
Zellerbach, Gary A. 359–62, 371
Zemtsova, Ella G. 182
Zenit surveillance satellite 239
Zentrum für Kunst und Medientechologie, Germany 344
Zernike, Frits 49, 59
zone plate 38–40, 46, 71, 88, 93–4, 144, 186, 196, 384–5